Integrative Theory of Evolution - Aspects and Insights

Guenther Witzany
Editor

Integrative Theory of Evolution - Aspects and Insights

Springer

Editor
Guenther Witzany
Telos - Philosophische Praxis
Buermoos, Salzburg, Austria

ISBN 978-3-032-24294-5 ISBN 978-3-032-24295-2 (eBook)
https://doi.org/10.1007/978-3-032-24295-2

This Springer imprint is published by the registered company Springer Nature Switzerland AG
The registered company address is: Gewerbestrasse 11, 6330 Cham, Switzerland

Dedicated to Zeno Bucher (1907–1984), who, during my studies of philosophy, explained to me, with compelling arguments, why the diversity and beauty of the living world cannot be the result of random error replications.

Preface

This book assembles experts from diverse fields to highlight aspects of and insights into epigenetics, RNA networking, virology, genetics, and cell biology that must be part of the *Integrative Theory of Evolution.*

This theory illuminates how the abundance of epigenetic regulations, RNA group interactions, the role of viruses in cellular host organisms, and cell identity drive evolutionary processes. This book does not present a final *Integrative Theory of Evolution* but rather presents various aspects and insights that must be included to better integrate current empirical knowledge than previous attempts that still insist on the crucial role of error replication as the primary reason for variability.

Darwin's Theory of Evolution remains a scientific revolution; variations and selection are the reasons for new species and the organismal diversity on this planet. With the increasing empirical knowledge—especially in molecular biology and genetics—throughout the twentieth century, several updates have occurred to better integrate newly derived knowledge. The new synthesis, its extended form, and the integration of early epigenetics with the unified Theory of Evolution represent such attempts.

For nearly a century, the focus in biological disciplines such as molecular biology, biochemistry, genetics, and evolutionary theory has been on cellular life as a machine-like process in which mechanistic pathways regulate metabolism, genetic transcription, translation into proteins, and evolution through variations (random error replications) and selection. Modern biochemistry began with the cellular theory of life, and modern synthesis concentrated on cells as the starting point of life. The dominance of this paradigm lasted until a few decades ago, when the comeback of virology provided new empirical data and explanatory models illustrating how viruses influence cellular life through numerous parasite-host interactions that override cellular processes. The RNA world hypothesis demonstrated that RNA group interactions were fundamental to biological selection before cellular life emerged. Lastly, the central dogma of molecular biology collapsed when epigenetics revealed that the history and developmental experiences of the past can be epigenetically imprinted, serving as identity markers. In every replication process of any cell of any organism on this planet, timely and locally coordinated replication is regulated and orchestrated by these programs.

In the twenty-first century, empirical knowledge has increased dramatically. (i) Viruses, once viewed as disease-causing enemies of life, are now recognized as essential agents of life and of host genome formation, as documented in mobile genetic elements that act as natural genetic content operators. Viruses and related genetic parasites transport regulatory RNA networks to the host, and the genetic identities of cells in all cellular domains are strongly influenced by such genetic parasites. (ii) RNAs have transitioned from being an intermediate tool in cell replication to occupying center stage in the regulation of all steps and sub-steps of cellular life. RNA networks ensure both the transport of regulatory instructions and the reprogramming of these instructions within cells. (iii) Cellular identities are fundamental players in life on this planet, whether pro- and eukaryotic, with a wealth of competencies to actively adapt and modify various behavioral patterns (sensing, monitoring, interpreting, decision-making, memorizing, learning, and communication). (iv) Epigenetics has shown us that every cell and the entire organism's phenotype is epigenetically regulated in development, morphology, cell fate and identity, physiology, genetic instructions, immunity, memory/learning, and physical and mental health. Epigenetics demonstrates how the environment and social networks can directly impact phenotypic variation independent of genetic change.

We must acknowledge that biological variations have various causes that do not fit into the error-replication (mutation) concept. If learned skills of adaptive behavior can be inherited and lead to better-adapted organisms, the error-replication (mutation) narrative is insufficiently complex to account for that.

As Richard Hunter (Chap. 5 in this book) aptly put it:

> "No organism that requires the accumulation of random point mutations to evolve would survive in a world where its competitors can adapt in a few generations and thus change the environment in ways that random search would not even have had time to detect. If evolvability is the norm, then it must act on substrates that permits adaptation in canalized directions rather than entirely at random."
>
> As biologists have become more aware of the speed at which environments can change, a parallel awareness has emerged of the speed at which organisms can adapt to environmental shocks and stress: rapid evolution. The speed of these evolutionary changes, coupled with our deepening understanding of genomics, has driven significant changes in evolutionary theory and led us to a view of evolution that is less centered on random mutation of DNA sequence toward an emerging synthesis that gives greater prominence to regulatory RNA and the transposons that are the primary source of it.
>
> The ever-expanding number of examples of rapid evolution in both experimental and field contexts, coupled with the detail and volume of genomics, epigenetics and transcriptional data now available have made plain that evolution moves rapidly in response to changing environments. We are beginning to see how regularly evolutionary innovation works via the substrate of transposable elements, and how genomes have domesticated and exapted them via epigenetic mechanisms.

Buermoos, Austria
April 2026

Guenther Witzany

Contents

Chapter 1
A Brief Look at the Biological Theory of Evolution: What Determines How Life Changes Itself?

James A. Shapiro

Abstract Hereditary variation generally results from the action of biological genome reproduction and repair functions. These range from simple-nucleotide polymorphisms to genome-wide chromosome rearrangements. The molecular details of representative genome change operations reveal cellular informatic links between genome damage and repair functions that generate genome novelty. However, there is no evidence of whether there are adaptively useful informatic links at the whole-organism level to genome change systems. Genetic change by interspecific hybridization is proposed as an experimental proxy for evolutionary variation. Performed under systematically varied growth and mating conditions, hybrid speciation may indicate whether there are any external triggers for particular kinds of genome variation.

1 The Biological Nature of Genome Change Operations

The most important discovery about hereditary variation in evolution since 1953, when DNA structure was discovered, is that virtually all genome change results from the action of cellular biochemical systems. From single-nucleotide polymorphisms to massive multisite chromosome rearrangements or changes in chromosome number, new genome structures result from the action of cellular proteins. A list of examples includes the following biologically mediated genome change events:

- A role for membrane vesicles in the horizontal transfers of protein-coding sequences between different taxa (Biller et al. 2025).
- Integrase-mediated site-specific recombination in bacteria and archaea for viral insertion and excision, incorporation of ORF cassettes into integron arrays, and

---

J. A. Shapiro (✉)
Emeritus, Department of Biochemistry and Molecular Biology, University of Chicago, Chicago, IL, USA
e-mail: jsha@uchicago.edu

G. Witzany (ed.), *Integrative Theory of Evolution - Aspects and Insights*,
https://doi.org/10.1007/978-3-032-24295-2_1

phenotype switching (Vit et al. 2021; Hackl et al. 2023; Trzilova and Tamayo 2021; Badel et al. 2021).
- Transposition and retrotransposition of mobile elements, including multiple events that "wire" (format) coordinated transcription networks (Cowley and Oakey 2013; Glinsky 2018; Guffanti et al. 2018; Qiu and Köhler 2020).
- Intron transposition (Gozashti and Corbett-Detig 2025; Ares et al. 2024; Hiller et al. 2000; Muñoz et al. 2001; Torriani et al. 2011; Zhao and Pyle 2017).
- Exon shuffling to generate novel protein domain combinations (Saccone et al. 2025; Kariyayama et al. 2025; Patthy 2021).
- Elaborate routines for the repair of double-strand DNA breakage involving DNA synthesis (Bantele et al. 2025; Burssed et al. 2022; Carvajal-Garcia et al. 2020; Marnef et al. 2017; Saad et al. 2014).
- Reverse transcription of cellular RNAs and pseudogene genome integration (Yan et al. 2023; Hallast et al. 2023; Troskie et al. 2021; Sisu et al. 2020).
- Untemplated and composite "template-switching tandem" array synthesis of linker DNA joints connecting chromosome double-strand (DS) break ends joined by alternative end joining (altEJ) with template-switching DNA polymerases (Umbreit et al. 2020; Begnis et al. 2018; Northam et al. 2014; Lowden et al. 2011; Schmidt et al. 2010).
- Chromoanagenesis ("chromosome rebirth"), a diverse series of apparently simultaneous multisite events in germlines that produce complex genome rearrangements, occurs repeatedly (Pellestor 2025; Castellano-Pozo and Comaills 2025; Guo et al. 2023).
- Whole-genome duplications (WGDs) that involve the suppression of mitotic divisions following DNA replication (Au et al. 2025; Holland et al. 2017).

2 Implications of the Biological Nature of Genome Change

Given these and numerous additional cellular capacities for genome reorganization, it should be clear that the multiple known processes of genome variation occur in the biological realm. That statement carries several important implications: (1) DNA change is subject to biological controls, (2) DNA changes cannot be treated as random stochastic events, and (3) biocognitive processes render DNA changes sensitive to life history events. We know that implication 3 is correct because there is experiential control for certain DNA change operations:

- Extracellular quorum-sensing pheromones control horizontal DNA transfers in bacterial biofilms (van Gestel et al. 2021).
- DNA-damaging agents and certain gut chemicals induce site-specific recombination in bacteriophage excision from bacterial genomes (Henrot and Petit 2022; Bruce et al. 2021).
- Conjugally transferred single-strand DNA induces site-specific cassette recombination into integrons (Fonseca É and Vicente 2022; Cambray et al. 2011).

- Micronucleus localization activates chromothripsis (single chromosome or fragment "shattering" into multiple fragments) (Krupina et al. 2024; Degrassi and Russo 2025; Umbreit et al. 2020).
- Interspecific hybridization activates transposable element mobility and chromosome rearrangement functions (Madlung et al. 2005; Smukowski Heil et al. 2021; Romero-Soriano et al. 2016).

3 Roles of Biocognition in Genome Change

There is also evidence that biological perceptions and biocognition play roles in the execution of DNA changes. Central to chromosome restructuring events are the identification of the broken ends of damaged chromosomes and their juxtaposition for joining them in new chimeric combinations. Barbara McClintock spoke of this process in her Nobel Prize address in explicitly "aware" terms:

> The conclusion seems inescapable that cells are able to sense the presence in their nuclei of ruptured ends of chromosomes, and then to activate a mechanism that will bring together and then unite these ends, one with another. And this will occur regardless of the initial distance in a telophase nucleus that separated the ruptured ends. The ability of a cell to sense these broken ends, to direct them toward each other, and then to unite them so that the union of the two DNA strands is correctly oriented, is a particularly revealing example of the sensitivity of cells to all that is going on within them. They make wise decisions and act upon them (McClintock 1984).

Today, we can identify molecular and cellular features needed for these genome repair processes that indicate that they have the potential to generate and utilize spatial and functional guidance. For example, H2AX histones around DNA breakage sites undergo chemical–physical marking by phosphorylation to identify the site of genome damage within the nucleus (Panigrahi et al. 2025). Distinct broken chromosomes are frequently mobilized to the same subnuclear DNA repair factory, enabling nonhomologous end joining (NHEJ) and the formation of novel chimeric chromosome structures (Meister et al. 2003; Helt et al. 2005; Marnef and Legube 2017; Rothkamm et al. 2015; Rulten and Grundy 2017).

4 Outstanding Problems in the Nature of Organismal Biocognition in Genome Change and a Method to Study Them

Despite our growing knowledge of the molecular machinery of DNA damage signaling and repair at the cellular level, we remain largely ignorant of how damage signals are interpreted at the organismal level. We do know that organismal stress activates natural genetic engineering and mobile DNA elements (Milyaeva et al. 2023; Shin et al. 2022; Dubin et al. 2018). But we do not know if there is any biocognitive input into the DNA repair process that might bias it toward outcomes that have functional benefits. Fortunately, there is a reliable method of inducing evolutionary change, which makes it possible to test empirically whether organismal perception or experience can affect the results of genome restructuring. This method is interspecific hybridization, where individuals from similar but distinct species mate and produce one or more hybrid offspring. These hybrids show elevated levels of mutagenesis, transposable element activity, chromosome restructuring, and changes in ploidy (especially in plants) and often produce novel species with new traits and reproductive isolation from their parents (Stebbins 1951; Steensels et al. 2021; Gámez-Visairas et al. 2020; Selz and Seehausen 2019; Gallone et al. 2019; Alix et al. 2017).

5 Proposal for How to Test the Role of Organismal Sensing in Biological Genome Change

Interspecific hybridization provides a broadly applicable experimental tool to test for the effects of organismal experience on genome diversification. Interspecific hybrids can be generated in yeast, butterflies, other arthropods, and plants under diverse conditions of illumination, nourishment, temperature, etc. and the resulting genomic and phenotypic changes cataloged. If different conditions produce different arrays of phenotypic or genotypic outcomes, such as different colored wing patterns in butterflies depending upon illumination, then we will know that our theories of how genome change may be guided require expansion to integrate the impacts of organismal experience.

6 Summary

Biologically mediated genome change accounts for most of evolutionary variability, often by processes that were inconceivable 73 years ago. The biological engines of genome change operate biocognitively in at least two ways: (1) activation in response to genomic or organismal distress and (2) proper localization of the genomic sites to be changed. But nothing is known about the possibility of biocognitive functions linking evolutionary change to functional outcomes for the organism as a whole. The use of interspecific hybridization as a trigger for variation under different conditions can determine whether whole-organism responses can influence the nature of genomic changes.

Conflicts of Interest The author has no conflicts of interest.

References

Alix K, Gérard PR, Schwarzacher T, Heslop-Harrison JSP (2017) Polyploidy and interspecific hybridization: partners for adaptation, speciation and evolution in plants. Ann Bot 120(2): 183–194. https://doi.org/10.1093/aob/mcx079

Ares M Jr, Igel H, Katzman S, Donohue JP (2024) Intron lariat spliceosomes convert lariats to true circles: implications for intron transposition. Genes Dev 38(7–8):322–335. https://doi.org/10.1101/gad.351764.124

Au HM, Nong W, Hui JHL (2025) Whole genome duplication in the genomics era: the hidden gems in invertebrates? Genome Biol Evol 17(5). https://doi.org/10.1093/gbe/evaf073

Badel C, Da Cunha V, Oberto J (2021) Archaeal tyrosine recombinases. FEMS Microbiol Rev 45(4). https://doi.org/10.1093/femsre/fuab004

Bantele S, Mordini I, Biran A, Alcaraz N, Zonderland G, Wenger A, Krietenstein N, Groth A, Lukas J (2025) Repair of DNA double-strand breaks leaves heritable impairment to genome function. Science 390(6773):eadk6662. https://doi.org/10.1126/science.adk6662

Begnis M, Apte MS, Masuda H, Jain D, Wheeler DL, Cooper JP (2018) RNAi drives nonreciprocal translocations at eroding chromosome ends to establish telomere-free linear chromosomes. Genes Dev 32(7–8):537–554. https://doi.org/10.1101/gad.311712.118

Biller SJ, Ryan MG, Li J, Burger A, Eppley JM, Hackl T, DeLong EF (2025) Distinct horizontal gene transfer potential of extracellular vesicles versus viral-like particles in marine habitats. Nat Commun 16(1):2126. https://doi.org/10.1038/s41467-025-57276-w

Bruce JB, Lion S, Buckling A, Westra ER, Gandon S (2021) Regulation of prophage induction and lysogenization by phage communication systems. Curr Biol 31(22):5046–5051.e5047. https://doi.org/10.1016/j.cub.2021.08.073

Burssed B, Zamariolli M, Bellucco FT, Melaragno MI (2022) Mechanisms of structural chromosomal rearrangement formation. Mol Cytogenet 15(1):23. https://doi.org/10.1186/s13039-022-00600-6

Cambray G, Sanchez-Alberola N, Campoy S, Guerin É, Da Re S, González-Zorn B, Ploy MC, Barbé J, Mazel D, Erill I (2011) Prevalence of SOS-mediated control of integron integrase expression as an adaptive trait of chromosomal and mobile integrons. Mob DNA 2(1):6. https://doi.org/10.1186/1759-8753-2-6

Carvajal-Garcia J, Cho JE, Carvajal-Garcia P, Feng W, Wood RD, Sekelsky J, Gupta GP, Roberts SA, Ramsden DA (2020) Mechanistic basis for microhomology identification and genome scarring by polymerase theta. Proc Natl Acad Sci USA 117(15):8476–8485. https://doi.org/10.1073/pnas.1921791117

Castellano-Pozo M, Comaills V (2025) Chromosomal instability and chromoanagenesis as forces for genomic evolution. Methods Mol Biol 2968:591–613. https://doi.org/10.1007/978-1-0716-4750-9_36

Cowley M, Oakey RJ (2013) Transposable elements re-wire and fine-tune the transcriptome. PLoS Genet 9(1):e1003234. https://doi.org/10.1371/journal.pgen.1003234

Degrassi F, Russo A (2025) Origin and fate of micronuclei on the road to chromoanagenesis. Methods Mol Biol 2968:361–372. https://doi.org/10.1007/978-1-0716-4750-9_21

Dubin MJ, Mittelsten Scheid O, Becker C (2018) Transposons: a blessing curse. Curr Opin Plant Biol 42:23–29. https://doi.org/10.1016/j.pbi.2018.01.003

Fonseca ÉL, Vicente AC (2022) Integron functionality and genome innovation: an update on the subtle and smart strategy of integrase and gene cassette expression regulation. Microorganisms 10(2). https://doi.org/10.3390/microorganisms10020224

Gallone B, Steensels J, Mertens S, Dzialo MC, Gordon JL, Wauters R, Theßeling FA, Bellinazzo F, Saels V, Herrera-Malaver B, Prahl T, White C, Hutzler M, Meußdoerffer F, Malcorps P, Souffriau B, Daenen L, Baele G, Maere S, Verstrepen KJ (2019) Interspecific hybridization facilitates niche adaptation in beer yeast. Nat Ecol Evol 3(11):1562–1575. https://doi.org/10.1038/s41559-019-0997-9

Gámez-Visairas V, Romero-Soriano V, Martí-Carreras J, Segarra-Carrillo E, García Guerreiro MP (2020) Drosophila interspecific hybridization causes a deregulation of the piRNA pathway genes. Genes (Basel) 11(2). https://doi.org/10.3390/genes11020215
van Gestel J, Bareia T, Tenennbaum B, Dal Co A, Guler P, Aframian N, Puyesky S, Grinberg I, D'Souza GG, Erez Z, Ackermann M, Eldar A (2021) Short-range quorum sensing controls horizontal gene transfer at micron scale in bacterial communities. Nat Commun 12(1):2324. https://doi.org/10.1038/s41467-021-22649-4
Glinsky GV (2018) Contribution of transposable elements and distal enhancers to evolution of human-specific features of interphase chromatin architecture in embryonic stem cells. Chromosom Res 26(1–2):61–84. https://doi.org/10.1007/s10577-018-9571-6
Gozashti L, Corbett-Detig R (2025) Double-stranded DNA viruses may serve as vectors for horizontal transfer of intron-generating transposons. Mob DNA 16(1):25. https://doi.org/10.1186/s13100-025-00363-y
Guffanti G, Bartlett A, Klengel T, Klengel C, Hunter R, Glinsky G, Macciardi F (2018) Novel bioinformatics approach identifies transcriptional profiles of lineage-specific transposable elements at distinct loci in the human dorsolateral prefrontal cortex. Mol Biol Evol 35(10): 2435–2453. https://doi.org/10.1093/molbev/msy143
Guo W, Comai L, Henry IM (2023) Chromoanagenesis in plants: triggers, mechanisms, and potential impact. Trends Genet 39(1):34–45. https://doi.org/10.1016/j.tig.2022.08.003
Hackl T, Laurenceau R, Ankenbrand MJ, Bliem C, Cariani Z, Thomas E, Dooley KD, Arellano AA, Hogle SL, Berube P, Leventhal GE, Luo E, Eppley JM, Zayed AA, Beaulaurier J, Stepanauskas R, Sullivan MB, DeLong EF, Biller SJ, Chisholm SW (2023) Novel integrative elements and genomic plasticity in ocean ecosystems. Cell 186(1):47–62.e16. https://doi.org/10.1016/j.cell.2022.12.006
Hallast P, Ebert P, Loftus M, Yilmaz F, Audano PA, Logsdon GA, Bonder MJ, Zhou W, Höps W, Kim K, Li C, Hoyt SJ, Dishuck PC, Porubsky D, Tsetsos F, Kwon JY, Zhu Q, Munson KM, Hasenfeld P, Harvey WT, Lewis AP, Kordosky J, Hoekzema K, O'Neill RJ, Korbel JO, Tyler-Smith C, Eichler EE, Shi X, Beck CR, Marschall T, Konkel MK, Lee C (2023) Assembly of 43 human Y chromosomes reveals extensive complexity and variation. Nature 621(7978): 355–364. https://doi.org/10.1038/s41586-023-06425-6
Helt CE, Wang W, Keng PC, Bambara RA (2005) Evidence that DNA damage detection machinery participates in DNA repair. Cell Cycle 4(4):529–532. https://doi.org/10.4161/cc.4.4.1598
Henrot C, Petit MA (2022) Signals triggering prophage induction in the gut microbiota. Mol Microbiol 118(5):494–502. https://doi.org/10.1111/mmi.14983
Hiller R, Hetzer M, Schweyen RJ, Mueller MW (2000) Transposition and exon shuffling by group II intron RNA molecules in pieces. J Mol Biol 297(2):301–308. https://doi.org/10.1006/jmbi.2000.3582
Holland PW, Marlétaz F, Maeso I, Dunwell TL, Paps J (2017) New genes from old: asymmetric divergence of gene duplicates and the evolution of development. Philos Trans R Soc Lond Ser B Biol Sci 372(1713). https://doi.org/10.1098/rstb.2015.0480
Kariyayama H, Kawashima T, Wada H, Ozaki H (2025) Domain-shuffling in the evolution of cyclostomes and gnathostomes. J Exp Zool B Mol Dev Evol 344(2):59–79. https://doi.org/10.1002/jez.b.23282
Krupina K, Goginashvili A, Cleveland DW (2024) Scrambling the genome in cancer: causes and consequences of complex chromosome rearrangements. Nat Rev Genet 25(3):196–210. https://doi.org/10.1038/s41576-023-00663-0
Lowden MR, Flibotte S, Moerman DG, Ahmed S (2011) DNA synthesis generates terminal duplications that seal end-to-end chromosome fusions. Science 332(6028):468–471. https://doi.org/10.1126/science.1199022
Madlung A, Tyagi AP, Watson B, Jiang H, Kagochi T, Doerge RW, Martienssen R, Comai L (2005) Genomic changes in synthetic Arabidopsis polyploids. Plant J 41(2):221–230. https://doi.org/10.1111/j.1365-313X.2004.02297.x

Marnef A, Legube G (2017) Organizing DNA repair in the nucleus: DSBs hit the road. Curr Opin Cell Biol 46:1–8. https://doi.org/10.1016/j.ceb.2016.12.003

Marnef A, Cohen S, Legube G (2017) Transcription-coupled DNA double-strand break repair: active genes need special care. J Mol Biol 429(9):1277–1288. https://doi.org/10.1016/j.jmb.2017.03.024

McClintock B (1984) The significance of responses of the genome to challenge. Science 226(4676): 792–801

Meister P, Poidevin M, Francesconi S, Tratner I, Zarzov P, Baldacci G (2003) Nuclear factories for signalling and repairing DNA double strand breaks in living fission yeast. Nucleic Acids Res 31(17):5064–5073. https://doi.org/10.1093/nar/gkg719

Milyaeva PA, Kukushkina IV, Kim AI, Nefedova LN (2023) Stress induced activation of LTR retrotransposons in the Drosophila melanogaster genome. Life (Basel) 13(12). https://doi.org/10.3390/life13122272

Muñoz E, Villadas PJ, Toro N (2001) Ectopic transposition of a group II intron in natural bacterial populations. Mol Microbiol 41(3):645–652. https://doi.org/10.1046/j.1365-2958.2001.02540.x

Northam MR, Moore EA, Mertz TM, Binz SK, Stith CM, Stepchenkova EI, Wendt KL, Burgers PM, Shcherbakova PV (2014) DNA polymerases zeta and Rev1 mediate error-prone bypass of non-B DNA structures. Nucleic Acids Res 42(1):290–306. https://doi.org/10.1093/nar/gkt830

Panigrahi R, Edwards R, Islam MT, Lu J, Kulepa A, Kim TH, Mark Glover JN (2025) Structural insights into γH2Ax containing nucleosomes. Nucleic Acids Res 53(19). https://doi.org/10.1093/nar/gkaf1028

Patthy L (2021) Exon shuffling played a decisive role in the evolution of the genetic toolkit for the multicellular body plan of Metazoa. Genes (Basel) 12(3). https://doi.org/10.3390/genes12030382

Pellestor F (2025) Chromoanagenesis and beyond: catastrophic events shaping the genome. Methods Mol Biol 2968:65–73. https://doi.org/10.1007/978-1-0716-4750-9_4

Qiu Y, Köhler C (2020) Mobility connects: transposable elements wire new transcriptional networks by transferring transcription factor binding motifs. Biochem Soc Trans 48(3):1005–1017. https://doi.org/10.1042/bst20190937

Romero-Soriano V, Burlet N, Vela D, Fontdevila A, Vieira C, García Guerreiro MP (2016) Drosophila females undergo genome expansion after interspecific hybridization. Genome Biol Evol 8(3):556–561. https://doi.org/10.1093/gbe/evw024

Rothkamm K, Barnard S, Moquet J, Ellender M, Rana Z, Burdak-Rothkamm S (2015) DNA damage foci: meaning and significance. Environ Mol Mutagen 56(6):491–504. https://doi.org/10.1002/em.21944

Rulten SL, Grundy GJ (2017) Non-homologous end joining: common interaction sites and exchange of multiple factors in the DNA repair process. BioEssays 39:3. https://doi.org/10.1002/bies.201600209

Saad H, Gallardo F, Dalvai M, Tanguy-le-Gac N, Lane D, Bystricky K (2014) DNA dynamics during early double-strand break processing revealed by non-intrusive imaging of living cells. PLoS Genet 10(3):e1004187. https://doi.org/10.1371/journal.pgen.1004187

Saccone S, Brancato D, Bruno F, Coniglio E, Sturiale V, Federico C (2025) Origin and evolution of genes in eukaryotes: mechanisms, dynamics, and functional implications. Genes (Basel) 16(6). https://doi.org/10.3390/genes16060702

Schmidt KH, Viebranz E, Doerfler L, Lester C, Rubenstein A (2010) Formation of complex and unstable chromosomal translocations in yeast. PLoS One 5(8):e12007. https://doi.org/10.1371/journal.pone.0012007

Selz OM, Seehausen O (2019) Interspecific hybridization can generate functional novelty in cichlid fish. Proc Biol Sci 286(1913):20191621. https://doi.org/10.1098/rspb.2019.1621

Shin E, Ye Q, Lee SJ (2022) Active transposition of insertion sequences in prokaryotes: insights from the response of Deinococcus geothermalis to oxidative stress. Antioxidants (Basel) 11(3). https://doi.org/10.3390/antiox11030481

Sisu C, Muir P, Frankish A, Fiddes I, Diekhans M, Thybert D, Odom DT, Flicek P, Keane TM, Hubbard T, Harrow J, Gerstein M (2020) Transcriptional activity and strain-specific history of mouse pseudogenes. Nat Commun 11(1):3695. https://doi.org/10.1038/s41467-020-17157-w
Smukowski Heil C, Patterson K, Hickey AS, Alcantara E, Dunham MJ (2021) Transposable element mobilization in interspecific yeast hybrids. Genome Biol Evol 13:3. https://doi.org/10.1093/gbe/evab033
Stebbins JGL (1951) Cataclysmic evolution. Sci Am 184(4):54–59
Steensels J, Gallone B, Verstrepen KJ (2021) Interspecific hybridization as a driver of fungal evolution and adaptation. Nat Rev Microbiol 19(8):485–500. https://doi.org/10.1038/s41579-021-00537-4
Torriani SF, Stukenbrock EH, Brunner PC, McDonald BA, Croll D (2011) Evidence for extensive recent intron transposition in closely related fungi. Curr Biol 21(23):2017–2022. https://doi.org/10.1016/j.cub.2011.10.041
Troskie RL, Faulkner GJ, Cheetham SW (2021) Processed pseudogenes: a substrate for evolutionary innovation: retrotransposition contributes to genome evolution by propagating pseudogene sequences with rich regulatory potential throughout the genome. BioEssays 43(11):e2100186. https://doi.org/10.1002/bies.202100186
Trzilova D, Tamayo R (2021) Site-specific recombination - how simple DNA inversions produce complex phenotypic heterogeneity in bacterial populations. Trends Genet 37(1):59–72. https://doi.org/10.1016/j.tig.2020.09.004
Umbreit NT, Zhang CZ, Lynch LD, Blaine LJ, Cheng AM, Tourdot R, Sun L, Almubarak HF, Judge K, Mitchell TJ, Spektor A, Pellman D (2020) Mechanisms generating cancer genome complexity from a single cell division error. Science 368(6488). https://doi.org/10.1126/science.aba0712
Vit C, Richard E, Fournes F, Whiteway C, Eyer X, Lapaillerie D, Parissi V, Mazel D, Loot C (2021) Cassette recruitment in the chromosomal integron of vibrio cholerae. Nucleic Acids Res 49(10):5654–5670. https://doi.org/10.1093/nar/gkab412
Yan Y, Tian Y, Wu Z, Zhang K, Yang R (2023) Interchromosomal colocalization with parental genes is linked to the function and evolution of mammalian retrocopies. Mol Biol Evol 40(12). https://doi.org/10.1093/molbev/msad265
Zhao C, Pyle AM (2017) The group II intron maturase: a reverse transcriptase and splicing factor go hand in hand. Curr Opin Struct Biol 47:30–39. https://doi.org/10.1016/j.sbi.2017.05.002

Chapter 2
DNA Methylation and Evolution

Moshe Szyf

Abstract This chapter discusses the evolutionary implications of DNA methylation. DNA methylation is a covalent modification that alters the chemical identity of DNA and regulates gene expression, providing a mechanism for cell type–specific gene expression programs in complex multicellular organisms, as well as increased genomic plasticity and adaptability. This adaptability has been fundamental to evolutionary processes, enhancing organismal fitness through improved responses to environmental changes. DNA methylation is established post-replication by DNA methyltransferases (DNMTs), which transfer methyl groups to cytosine or adenine bases. In bacteria, these modifications occur at specific sequences, enabling faithful restoration of methylation patterns after cell division. In vertebrates, methylation predominantly targets CG dinucleotides, with maintenance enzymes ensuring semi-conservative inheritance of these patterns during replication. The persistence of DNA methylation patterns through cell divisions raises questions about their potential role as an independent mechanism in evolution and speciation, beyond genetic variation alone. Furthermore, environmental and social cues can influence DNA methylation, echoing Lamarckian ideas of environmentally directed evolution, though factors such as the erasure of methylation during gametogenesis and the association of DNA methylation variation with genetic sequence variation remain challenges. Synthetic theories suggest that DNA methylation may drive phenotypic diversity that is later genetically fixed. While it remains uncertain whether DNA methylation directly causes permanent evolutionary change, its emergence has undeniably expanded the range of phenotypes from identical genetic sequences, significantly enhancing adaptability, fitness, and the evolutionary potential of species.

M. Szyf (✉)
EpiMedTech Global Ltd., Singapore, Singapore

Zhejiang University-Lishui Joint Innovation Center for Life and Health & Lishui Lvgu Institute for Life and Health Zhiezhang University, Hangzhou, China

G. Witzany (ed.), *Integrative Theory of Evolution - Aspects and Insights*,
https://doi.org/10.1007/978-3-032-24295-2_2

1 Introduction: Natural History of DNA Methylation

1.1 Methylated Cytosine and Adenine in DNA

Methylated cytosine and adenine are classified as "minor" bases, present at significantly lower concentrations compared to the four principal bases of DNA. These minor bases are distinct from the canonical bases that constitute the genetic code. The earliest identified minor base was discovered in calf thymus DNA by Hotchkiss, who referred to it as "epicytosine" (Hotchkiss 1948). Subsequently, Gerard Wyatt characterized this base as 5-methylcytosine (5mC) (Wyatt 1950). Notably, 5-methylcytosine was chemically identified in a biomolecule initially extracted from tuberculosis two decades earlier, originally termed tuberculinic acid; this substance is now recognized as DNA from tuberculosis (Johnson and Coghill 1925). Additionally, N6-methyladenine (6 mA) was identified in bacterial DNA and lower eukaryotes. Dunn and Smith used two-dimensional paper chromatography to separate the hydrolyzed bases of DNA from *Escherichia coli* and *Aerobacter aerogenes* and identified 6 mA, noting its significant abundance (up to 2.5% of all adenine in *Aerobacter aerogenes*). This established 6 mA as the most prevalent DNA modification in prokaryotes.

1.2 Natural History of 5mC DNA Methyltransferases: The "Writers"

The biosynthesis of methylated bases proceeds through an enzymatic transfer of a methyl group from the universal donor S-adenosylmethionine (SAM) to either the 5′ position of cytosine or the N6 position of adenine following DNA replication. The enzymes responsible for these modifications are known as DNA methyltransferases (DNMTs). The evolutionary emergence of both S-adenosylmethionine and DNMTs was fundamental to increasing the chemical diversity of the DNA molecule. Evidence indicates that both entities evolved early and are present throughout the entire evolutionary phylogeny, spanning from Archaea to primates. This widespread occurrence supports their ubiquitous critical role in enhancing biological fitness and survivability.

DNA methylation systems emerged early in evolution, with DNA methyltransferases found in Archaea (Anton and Roberts 2023). The role of restriction methylation systems was first described in bacteria (Luria and Human 1952). Restriction–modification (R–M) is a defense mechanism that evolved in bacteria to protect against bacteriophage invasion and maintain genomic integrity. These systems involve a sequence-specific DNA methyltransferase and a restriction enzyme that recognizes the same sequence but is inhibited by DNA methylation at the recognition site. Since the host DNA is methylated and the invading DNA is unmethylated, the invading DNA is destroyed by the restriction enzyme at its recognition sequence. The same sequence is methylated by the DNA methyltransferase in the host DNA and is protected from restriction (Arber and Dussoix 1962).

DNA methyltransferases are also found in Archaea. Although, as in bacteria, lateral gene transfer (LGT) of restriction–modification systems is common, the presence of DNA methyltransferases in Archaea cannot be explained solely by recent bacterial LGT because "persistent" DNA methyltransferases are conserved within specific archaeal lineages. Persistent archaeal methyltransferases—for example, the C(m5C)WGG system in Crenarchaeota—support a model of in-domain radiation rather than recent acquisition from bacteria (Anton and Roberts 2023). This raises the possibility that DNA methyltransferases were already present in a common ancestor predating the divergence of Archaea and bacteria. It is reasonable to assume that such restriction–modification systems were selected early in evolution to protect nascent cellular life from lethal invasion by early viruses. As we will see below, SAM-dependent enzymes appeared very early in evolution, and it is plausible that the first DNA methyltransferases arose within this ancient SAM-dependent protein family.

DNA methyltransferases are present in the three domains of life: Archaea, Bacteria, and Eukarya, suggesting an ancient origin (Jurkowski and Jeltsch 2011). Eukaryotic maintenance DNA methyltransferase, DNMT1, which copies CG methylation during DNA replication, and DNMT3, de novo methyltransferase families, originated independently from different restriction–modification DNA methyltransferases in prokaryotic ancestors (Fig. 2.1). The wide distribution of these DNA methyltransferases in eukaryotes implies that they were already present in the last common eukaryotic ancestor (LCEA). Since DNMT1 is separated from the DNMT2/DNMT3 group, it was proposed that DNMT1 and DNMT2/3 descended from distinct bacterial DNA methyltransferases (Jurkowski and Jeltsch 2011). DNMT3 later underwent duplication, leading to the development of DNMT3A and DNMT3B. DNMT3A is a de novo methyltransferase capable of methylating non-CpG sites, which plays a crucial role in gene silencing in vertebrate brains (Liu et al. 2020; De Mendoza et al. 2021). This expansion of the methylome to the abundant non-CG sites may have contributed to increased cognitive capacity (De Mendoza et al. 2021) by vastly increasing the number of sites in the genome that could be methylated during learning and memory acquisition.

1.3 Natural History of the Methyl Donor S-Adenosylmethionine

The donor of methyl groups in DNA methylation reactions is S-adenosylmethionine (SAM). SAM can form spontaneously and is believed to have existed even in the prebiotic times (Laurino and Tawfik 2017). It is thought to have formed from methionine and adenosine, which were present in the prebiotic mix. The enzyme

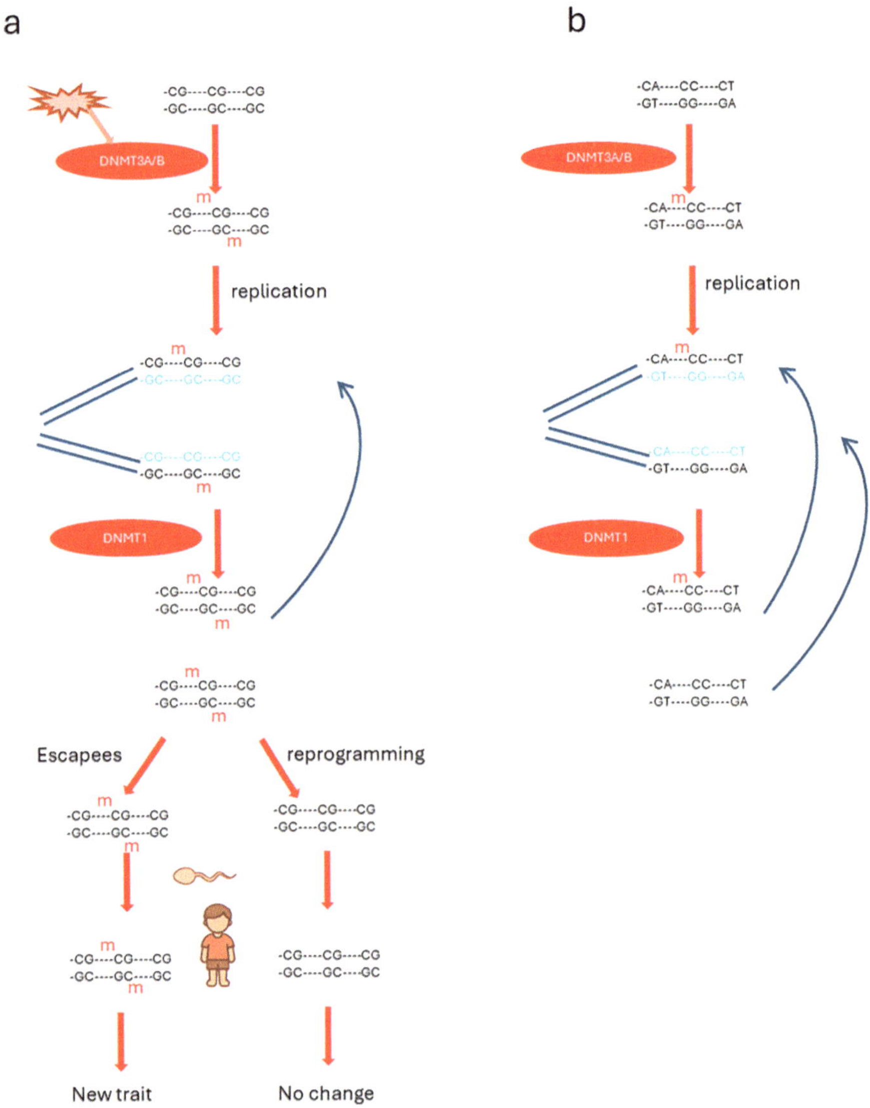

Fig. 2.1 Heritability of DNA methylation in vertebrates. (**a**) *CpG methylation and somatic maintenance*: De novo CpG methylation (m) induced by physiological or environmental signals is preserved through cell division because the CpG dyad is palindromic (5′-CG-3′ on one strand and 3′-GC-5′ on the complement). After replication, the duplex is hemimethylated—the parental strand is methylated, while the daughter strand (blue) is not—allowing DNMT1 (with its cofactors) to copy the mark onto the daughter strand using the parental methylation as a template. This mechanism maintains the pattern across somatic divisions. During gametogenesis and early embryogenesis, however, genomes undergo extensive methylation reprogramming, so most somatic marks are not transmitted to offspring. A subset of loci—especially certain retroelements (e.g., SVA)—can partially escape reprogramming and are plausible carriers of methylation states and potentially associated traits across generations. (**b**) *Non-CpG methylation and limited copying*: DNMT3A-mediated non-CpG methylation (mCH) is asymmetric and lacks a complementary CpG on the opposite strand; consequently, DNMT1 cannot use a template to copy these marks after replication. Persistence of mCH, therefore, requires ongoing de novo activity, and it is not reliably maintained through mitosis or across generations

for SAM synthesis, S-adenosylmethionine synthetase (SAMS), alternatively named methionine adenosyltransferase (MAT), evolved early to enhance the chemical reaction, and the selection of this gene may have been due to the increased fitness it provided, with SAM emerging as an essential metabolite. The overall structure and active-site residues of SAMS are highly conserved across vast evolutionary distances. For instance, the *Plasmodium falciparum* SAMS protein shares high homology and similar active-site amino acids with the human hepatic SAMS and the *Escherichia coli* SAMS, suggesting strong selective pressure to maintain its core function (Chiang et al. 1999).

SAM is an essential metabolic intermediate, and phylogenetic analysis suggests that the last universal common ancestor (LUCA) of all cellular life (which lived approximately 3.5–3.8 billion years ago) already possessed several SAM-binding proteins and the mechanisms for SAM formation (Kozbial and Mushegian 2005).

SAMS enzymes originated early, maintaining a conserved catalytic mechanism but diversifying to serve specialized roles in various organisms and tissues. DNMTs, which bind SAM, evolved from ancient SAM-binding proteins (Kozbial and Mushegian 2005).

1.4 Natural History of Proteins that Distinguish Methylated from Unmethylated DNA ("Readers")

The methylated bases in DNA (5mC and 6 mA) are chemically distinct from their unmodified counterparts, yet the DNA replication machinery—particularly DNA polymerases—does not discriminate between them. This allows faithful replication of the genetic information encoded in the four canonical bases without interference from methylation. For methylation to have biological meaning, its effects must instead be exerted through proteins that interact with DNA.

The earliest known evolutionary role of DNA methylation was to modulate the binding of sequence-specific nucleases in restriction–modification (R–M) systems (Arber and Dussoix 1962). Methylation of host DNA prevents cleavage by restriction enzymes that degrade invading bacteriophage DNA, thereby protecting "self" DNA while leaving foreign DNA vulnerable. In this context, methylation acts by altering protein–DNA recognition, either blocking or permitting binding. This principle—changing the chemical identity of a recognition sequence so that proteins either prefer or avoid it—has persisted throughout evolution. Proteins may evolve to bind the *methylated* version of a sequence, or the *unmethylated* version, as seen for restriction enzymes and, later, transcription factors.

As eukaryotes evolved, transcription factors became central to both establishing and interpreting DNA methylation patterns. Some factors protect their binding sites from methylation during replication and differentiation, while others are themselves sensitive to methylation of their recognition motifs. In both cases, they act as methylation "readers" or "anti-readers," thereby greatly expanding the regulatory information encoded by the genome.

Over time, three major classes of methylated DNA–binding proteins emerged: (Su(var)3-9, Enhancer-of-zeste, Trithorax) domain–containing methyltransferases (SET)- and (Really Interesting New Gene) domain (RING)-associated (SRA) domain proteins, methyl CpG–binding domain (MBD) proteins, and zinc finger proteins with methylation sensitivity.

DNA methylation "writers" (DNMTs) are ancient and conserved, whereas reader proteins are a mosaic of independently evolved domains that became progressively specialized to interpret increasingly complex methylation landscapes.

SRA Domain Proteins: Ancient Methylation Readers

The earliest proteins known to recognize methylated DNA are those containing SET and RING-associated (SRA) domains. These occur in eukaryotes—including plants, animals, and fungi—and in some prokaryotes, often fused to nuclease domains. In bacteria, SRA proteins help discriminate 5mC-modified DNA from unmodified self-DNA and protect against invading methylated DNA (Han et al. 2015).

In eukaryotes, SRA domains act as molecular bridges between DNA methylation and chromatin states, recruiting histone-modifying complexes to methylated DNA (Hashimoto et al. 2009; Johnson et al. 2007). A key evolutionary innovation was the emergence of SRA-containing proteins with a preference for hemimethylated DNA, the intermediate state generated during DNA replication. In mammals, the SRA domain of ubiquitin-like with Plant Homeodomain (PHD) and RING finger domains 1 (UHRF1) targets hemimethylated sites and recruits DNMT1, enabling accurate copying of methylation patterns onto the daughter strand (Arita et al. 2008). This mechanism underlies the heritable maintenance of DNA methylation profiles through cell division and allows stable epigenetic states to be fixed and selected during evolution.

Different SRA domains later evolved distinct binding preferences for CG, CHG, or CHH contexts, a diversification particularly apparent in plants, where multiple methylation contexts coexist and play specialized regulatory roles (Ebbs and Bender 2006; Johnson et al. 2007).

Zinc Finger Readers: Expanding Regulatory Complexity

Zinc finger proteins are among the most abundant and diverse protein families in eukaryotes. The ability to use zinc ions to stabilize small DNA-binding domains is ancient: Even Archaea, from which eukaryotes evolved, possess zinc-binding proteins reminiscent of eukaryotic zinc fingers (Guilliere et al. 2013). In early eukaryotes, this scaffold diversified into the "classical" C2H2-type zinc finger, present in the last eukaryotic common ancestor (LECA) and subsequently used as a building block for a massive expansion of transcription factors across animals, plants, and fungi (Krishna et al. 2003; Bohm et al. 1997).

The evolution of zinc finger proteins capable of discriminating between methylated and unmethylated DNA added a new dimension to the methylome. Methylation-sensitive zinc finger proteins—including Kaiso (Prokhortchouk et al. 2001), ZBTB4, and ZBTB38 (Filion et al. 2006), and the insulator protein CCCTC-binding factor (CTCF) (Bell and Felsenfeld 2000; Hark et al. 2000)—can either bind specifically to methylated CpGs or be repelled by methylation within their motifs. This created a layered reader system in which DNA sequence and methylation state jointly determine transcription factor binding, dramatically enriching the regulatory code.

The rapid expansion of zinc finger genes, in some lineages driven in part by transposable element (TE) activity (e.g., proliferation of CTCF sites in rodents), generated a rich substrate for evolving novel protein–DNA interactions within a diversifying methylation landscape.

MBD Proteins: Eukaryote-Specific CpG Readers

Methyl CpG–binding domain (MBD) proteins appear to have arisen as a novel eukaryotic domain (Hendrich and Bird 1998). They are found in lineages separated by over a billion years of evolution, including plants (Springer and Kaeppler 2005), vertebrates (Bilinovich et al. 2025), and invertebrates (Roder et al. 2000), but not in fungi, suggesting that MBDs either evolved after the divergence of fungi or were secondarily lost in that lineage. To date, no true MBD-domain proteins have been documented in bacteria.

MBD proteins evolved to recognize and bind methylated CpG dinucleotides. The MBD domain itself is a conserved motif that docks on methylated CpGs, while additional domains (e.g., CXXC, transcriptional repression domains, and glycosylase domains) were acquired to couple methylation recognition to chromatin remodeling, transcriptional repression, or DNA repair (Du et al. 2015a, b). Gene duplications further expanded the family, for example, the duplication that produced MBD2 and MBD3. MBD3 subsequently lost specific methyl CpG binding but retained the ability to recruit the NuRD chromatin–*remodeling complex, thereby modulating chromatin independently of direct methylation recognition (Hendrich et al. 2001).

Together, SRA, MBD, and zinc finger proteins form a diverse reader network that translates DNA methylation patterns into functional outcomes, enhancing phenotypic plasticity and providing a strong selective advantage.

1.5 Diversity, Convergence, and Loss of DNA Methylation Systems

Despite the ancient origin of DNA methyltransferases and DNA methylation machinery (Sadler 2023), eukaryotes show striking diversity in the abundance of methylated cytosines, the sequence context of methylation (CG, CHG, and CHH), the genomic distribution of methylation, and, in some lineages, complete loss of 5mC- and DNMT1-family enzymes (De Mendoza et al. 2020; Provataris et al. 2018; Rosic et al. 2018).

These differences can occur even within the same broad taxonomic group, whereas convergent patterns sometimes appear in distantly related lineages. For example, *Drosophila melanogaster* has essentially undetectable 5mC (Urieli-Shoval et al. 1982), whereas other insects, such as the honeybee, retain substantial CpG methylation, largely in gene bodies (Bewick et al. 2017; Provataris et al. 2018). The nematode *Caenorhabditis elegans* similarly lacks canonical DNA methylation and DNMT1/3 homologs (Simpson et al. 1986). By contrast, vertebrates exhibit high global CpG methylation, and a remarkably "vertebrate-like" highly methylated CpG methylome has independently evolved in the sponge *Amphimedon queenslandica* (De Mendoza et al. 2019). Such patterns marked variation within clades, and convergence across distant clades likely reflects a combination of convergent evolution under similar selective pressures and multiple secondary losses of an ancestral methylation system (De Mendoza et al. 2019; Sadler 2023).

The existence of organisms that have entirely dispensed with DNA methylation, such as *Saccharomyces*, suggests that this mark confers advantages in some contexts but is dispensable or even disadvantageous in others (Sadler 2023). Conversely, the extensive diversification and specialization of 5mC "writers" (DNMTs) and "readers" (methyl CpG–binding proteins) in vertebrates and mammals point to a significant adaptive benefit in these lineages (Liu et al. 2020; Molaro et al. 2020). DNA methylation increases the informational complexity of the genome, enabling regulatory nuance beyond the primary DNA sequence. Unlike other epigenetic mechanisms (histone modifications, noncoding RNAs, and chromatin remodelers), DNA methylation is covalently embedded in the DNA itself, likely contributing to its stability across cell divisions and developmental time.

In mammals, the duplication and functional diversification of DNMT3 enzymes (Liu et al. 2020; De Mendoza et al. 2021), including the specialized role of DNMT3A in the brain (Wu et al. 2010; Zocher et al. 2021; Lister et al. 2013), together with the clear association of DNMT3A mutations with neurodevelopmental and cognitive defects in humans (Tatton-Brown et al. 2014), suggest that 5mC methylation was crucial for the evolution of increasingly complex body plans and nervous systems. This is particularly relevant to the emergence of advanced neural functions and the exceptional cognitive capacity of humans.

1.6 Natural History of N6-Methyladenine (6 mA)

Both 5-methylcytosine (5mC) and N6-methyladenine (6 mA) are abundant in prokaryotic DNA, where they participate in restriction–modification systems. Bacteria, for example, have evolved 6 mA methyltransferases (Dam enzymes) that use S-adenosylmethionine (SAM) as a methyl donor, analogous to cytosine methyltransferases. In eukaryotes, 5mC is widespread across major clades, including animals, plants, and fungi, whereas 6 mA is absent in animals, and low-level 6 mA signals in vertebrates and other eukaryotes have been controversial.

Genome-wide mapping of 6 mA at single-base resolution was greatly improved by the advent of nanopore sequencing, which can distinguish methylated from unmethylated cytosines and adenines (Romero Charria et al. 2025). A recent study demonstrated that 6 mA is abundant in certain eukaryotic groups, including ciliates, early-branching fungi, and *Chlamydomonas*, but is essentially absent in plants, animals, and the so-called Solaris fungi clade. These data support the idea that both 6 mA and 5mC were present in the last eukaryotic common ancestor (LECA) and were subsequently lost in animals and plants. Phylogenetic analysis further suggests that a single bacterial 6 mA methyltransferase, MT-A70, gave rise to five adenine methyltransferases in eukaryotes (Romero Charria et al. 2025). Notably, AMT1 functions as a maintenance DNA methyltransferase: Analogous to DNMT1, which copies 5mC across replication at CG/GC palindromes, AMT1 targets AT/TA palindromes and propagates adenine methylation from the parental to the daughter strand (Romero Charria et al. 2025). High levels of 6 mA are found only in species that possess AMT1, such as ciliates, and can, therefore, maintain stable methylation patterns through cell division. In these organisms, 6 mA marks actively transcribed genes, is enriched downstream of transcription start sites, correlates strongly with H3K4me1, and is positioned between nucleosomes (Romero Charria et al. 2025).

These observations suggest that adenine methylation may have provided early eukaryotes with a dual advantage: distinguishing actively transcribed genes from silent regions and enabling faithful inheritance of methylation patterns during mitosis. In contrast, 5mC carries the drawback of enhanced mutability because spontaneous deamination converts 5mC to thymine, generating frequent C $\rightarrow$ T transition mutations, whereas 6 mA is not similarly prone to mutational decay. Thus, although 6 mA likely arose in prokaryotes as part of restriction–modification systems that protected host genomes from viral invasion, it also became a useful signal for differentiating identical DNA sequences according to their transcriptional status. Over evolutionary time, however, multicellular eukaryotes appear to have favored 5mC-based regulation, likely due to its versatility and compatibility with increasingly complex chromatin architectures. Plants, animals, and fungi (Dikarya) retained only 5mC; taxa such as prasinophytes, *Chlorella*, and *Amoebidium* preserved both 5mC and 6 mA; ciliates, Ichthyosporea, and *Trichomonas* maintained only 6 mA; and organisms such as *Dictyostelium* and *Saccharomyces* seem to have dispensed with DNA methylation altogether. These patterns underscore that DNA methylation provides clear advantages in many evolutionary contexts but is not universally required and that different forms of DNA methylation have been independently retained or lost at multiple evolutionary stages.

2 DNA Methylation, Gene Expression, and Cellular Differentiation

2.1 Waddington and the Concept of Epigenetics

While gene function is encoded in DNA sequence, the sequence alone is insufficient to determine when, where, and under what conditions a gene should be expressed. This limitation becomes particularly evident as organisms encounter varying environmental and developmental contexts. With the evolution of multicellular life, tissues and organs required cell type–specific gene expression programs, and the process by which cells acquire stable, lineage-specific phenotypes became known as terminal differentiation. Because the classical genetic paradigm linked the phenotype directly to the DNA sequence, an early question was whether differentiation mirrors evolution or whether cells progressively accumulate DNA changes that "fix" the gene expression programs characteristic of their specialized state. This analogy, often discussed within the framework of "evo–devo," highlighted superficial similarities between evolution (where unicellular organisms diversify into complex life forms) and development (where a single fertilized egg gives rise to a complex organism). However, evolution involves genetic variation and natural selection, whereas development is a highly predictable, orderly, and conserved process. Crucially, John Gurdon's landmark nuclear transfer experiments in *Xenopus* demonstrated that nuclei from so-called terminally differentiated cells retain the full genomic complement necessary to direct embryogenesis: when transferred into an enucleated egg, a differentiated intestinal epithelial nucleus could produce a feeding tadpole (Gurdon 1962). This result showed unambiguously that differentiation does not entail permanent loss or alteration of genetic information. Instead, it implies that additional, reversible layers govern the establishment and maintenance of cell identity.

If the DNA sequence remains unchanged during development, how can the same genome generate such a wide array of cellular phenotypes? Waddington, who grappled with this question in the early 1940s, introduced the term "epigenetics" to describe the regulatory processes through which genes interact with each other and with their cellular environment to produce differentiated states (Waddington 1942). His influential epigenetic landscape metaphor depicted a pluripotent cell rolling down a series of branching valleys, each representing a developmental pathway that culminates in a stable and restricted cell fate. At the time, no biochemical mechanisms were known that could account for these stable changes in gene activity. Today, we recognize that differentiation is orchestrated by multiple layers of epigenetic regulation, including DNA methylation, histone modifications, chromatin remodeling, higher-order chromatin structure, and noncoding RNAs that modulate genome accessibility and gene expression without altering the underlying DNA sequence. These molecular mechanisms provide the structural and regulatory framework that stabilizes lineage-specific transcriptional programs throughout development (Cedar and Bergman 2011; Lister et al. 2013).

2.2 *Restriction–Modification Systems and the Silencing of Repetitive Elements in Eukaryotes: The Evolutionary Cradle of DNA Methylation*

DNA methylation most likely originated in restriction–modification systems, as discussed above, where it enabled cells to distinguish "self-" from foreign ("ectopic") DNA. This represents one of the earliest examples in evolution in which identical DNA sequences acquire distinct functional identities through a covalent epigenetic mark. The need to maintain genomic integrity in the face of invading, self-replicating DNA created strong selective pressure for the evolution of methylation-based discrimination, thereby introducing DNA methylation into evolution.

Once established, this modification endowed cells with a new layer of information encoded in the genome by adding two minor bases to DNA. Because these modified bases did not interfere with faithful replication of the underlying genetic sequence, they allowed diversification of genomic function without compromising genetic integrity. This innovation had the potential to dramatically expand the phenotypic repertoire that a given genome can generate, extending the role of DNA methylation from genome protection per se to the regulation of gene expression and the diversification and expansion of phenotypes expressed by DNA as cells responded to dynamically changing environmental conditions and as complex multicellular organisms emerged.

In this discussion, we focus specifically on DNA methylation, but the same conceptual framework applies to the wider array of epigenetic mechanisms that arose during evolution to diversify genomic functions.

Restriction–modification systems introduced the first fundamental principles of how DNA methylation operates. First, minor bases create a chemical mark on DNA that allows otherwise identical sequences to be distinguished based on their methylation status. Second, this difference becomes biologically informative by altering protein–DNA interactions: methylation can either repel a protein by disrupting its recognition sequence or attract it by creating a new binding site for proteins that specifically recognize methylated DNA. In the earliest systems, these proteins were restriction enzymes, whose binding and activity were either inhibited or facilitated by the presence of DNA methylation.

A selective pressure favoring the conservation of DNA methylation in eukaryotes was likely the protection against invasion of the genome by foreign, self-replicating

DNA elements that accumulate as repetitive sequences (Bestor 1990; Yoder et al. 1997). Canonical bacterial-style restriction–modification systems are largely absent from eukaryotic nuclear genomes, but their DNA methyltransferase components have been repeatedly captured and integrated into eukaryotic regulatory pathways (de Mendoza et al. 2018; Schmitz et al. 2019).

As transposable and viral elements became stably integrated into host chromosomes, endonuclease-based restriction alone was no longer sufficient to eliminate them without collateral damage to host DNA. Instead, horizontally acquired and ancestrally inherited DNA methyltransferases were co-opted to repress these elements epigenetically, building on molecular principles that first evolved in prokaryotic defense systems (De Mendoza et al. 2018). Consistent with this view, 5-methylcytosine-based DNA methylation is widely used to silence transposable and other repetitive elements across diverse eukaryotic lineages, including animals, plants, diatoms such as *Phaeodactylum*, dikaryan fungi, prasinophyte and chlorophyte algae such as *Chlorella*, and holozoan protists such as *Amoebidium* (Zemach et al. 2010; Romero Charria et al. 2025).

Comparative evolutionary analyses suggest that land plants and vertebrates, which exhibit extensive genomic DNA methylation, are under strong selective pressure to maintain methylation-based repression of transposable elements (TEs) because sexually outcrossing reproduction allows TEs to spread efficiently through populations (Zemach and Zilberman 2010). In contrast, predominantly asexual unicellular eukaryotes, including many green algae and fungi, can more readily lose DNA methylation–mediated TE silencing with relatively modest fitness costs and, correspondingly, often show a strong reduction or complete absence of TE methylation (Zemach and Zilberman 2010). In plants and animals, DNA methylation is deployed both to silence TEs (including those inserted within or near genes) and mark the bodies of many constitutively expressed genes. Genome-wide evolutionary analyses indicate that gene body methylation is an ancient and conserved feature of eukaryotic genomes (Zemach et al. 2010). However, because cytosine methylation is intrinsically mutagenic, 5-methylcytosine spontaneously deaminates to thymine and gene body methylation appears to be comparatively dispensable and has been independently lost early in fungal evolution and later in several plant and animal lineages (Zemach and Zilberman 2010).

Once DNA methylation was repurposed as a mechanism of gene silencing of repetitive and retroviral elements, it became a new tool for expanding phenotypic landscapes beyond what is dictated solely by the genetic sequence, thereby opening additional avenues for evolution.

2.3 DNA Methylation and Gene Expression

Three main lines of work converged to implicate DNA methylation in the control of gene expression. First, in vitro methylation experiments showed that methylation can directly silence promoters. Vectors carrying viral (Vardimon et al. 1982) or cellular promoters, for example, the hamster *adenine phosphoribosyltransferase* (*aprt*) gene (Stein et al. 1982), were methylated at CpG sites in vitro and then transfected into mouse cells. Clones harboring methylated constructs were largely transcriptionally silent, whereas the same sequences in an unmethylated state were actively expressed despite comparable levels of genomic integration.

Second, correlational studies used methylation-sensitive restriction enzymes such as HpaII, together with its methylation-insensitive isoschizomer MspI (Danna and Nathans 1971), and Southern blot analysis to assess the methylation state of tissue-specific genes in different mouse and human tissues. These studies showed that specific CpG sites in many genes are methylated in tissues where the gene is silent but unmethylated in tissues where the gene is expressed. A classic example is the human β-globin locus, which is heavily methylated in non-erythroid cells but relatively unmethylated in erythroid precursors that express β-globin (Van Der Ploeg and Flavell 1980). Because most early work relied on bulk tissue DNA, interpretation was sometimes confounded by cellular heterogeneity: a minority of highly expressing cells mixed with a majority of non-expressing, methylated cells could give the misleading impression of a "methylated but highly expressed" gene. This problem was particularly acute in brain tissue, which contains many neuronal and glial subtypes. In addition, restriction enzymes such as HpaII interrogate only a small subset of CpG sites and can miss key regulatory regions.

The observation that genome-wide correlations between DNA methylation and gene expression were only partial challenged the hypothesis that DNA methylation universally inhibits transcription. Analyses of bulk tissues revealed many genes whose promoters appeared both methylated and actively expressed. This seemingly paradoxical pattern is now understood to arise from cellular heterogeneity within complex organs, particularly the brain, where multiple cell types coexist and exhibit distinct epigenetic signatures (Massart et al. 2017).

To resolve the methylation status of actively transcribed promoters at single-molecule resolution, Massart and colleagues used bisulfite sequencing, next-generation sequencing, and chromatin immunoprecipitation with an antibody recognizing RNA polymerase II phosphorylated on serine-5 of its C-terminal domain. This serine-5-phosphorylated form of RNA polymerase II marks transcription initiation and early elongation at the 5′ end of mRNAs. Their analysis demonstrated that promoters bound by serine-5-phosphorylated RNA polymerase II were uniformly unmethylated, whereas the same promoters showed methylation only in the non-immunoprecipitated fraction of bulk genomic DNA that was not actively transcribed (Massart et al. 2017). These findings support the conclusion that promoter DNA methylation is incompatible with active transcription and functions as a silencing signal.

The study also uncovered an additional principle: Although promoter methylation blocks transcription, demethylation alone does not guarantee gene activation. Promoter demethylation is necessary but not sufficient for expression (Massart et al. 2017). Transcription requires the availability of appropriate transcription factors, which may be absent under basal conditions. It has, therefore, been proposed that demethylation "primes" genes for future activation, enabling rapid and robust transcription only when the relevant transcription factor is expressed. This mechanism is particularly relevant for inducible gene systems, such as stress-responsive pathways, immune activation, and neuronal plasticity underlying learning and memory. An unmethylated promoter can be induced, whereas a methylated promoter cannot. Thus, two individuals with identical basal mRNA levels may exhibit markedly different transcriptional and phenotypic responses to environmental or physiological challenges. If such methylation differences arise and persist over evolutionary timescales, they could contribute to variation in fitness and adaptive potential.

The third line of evidence came from studies using the DNA methylation inhibitor 5-azacytidine (and later 5-aza-2′-deoxycytidine). These nucleoside analogs inhibit DNA methyltransferases, cause genome-wide DNA hypomethylation, and can induce differentiation and changes in gene expression in cultured cells such as Friend erythroleukemia and HL-60 cells (Jones and Taylor 1980; Creusot et al. 1982; Chiu and Blau 1985). However, these experiments are also difficult to interpret mechanistically because 5-azacytidine has broad effects on DNA replication and cell viability (Juttermann et al. 1994), and it is challenging to link demethylation of a specific locus causally to the activation of a specific gene.

A critical set of experiments from the laboratories of Razin and Cedar provided the first direct biochemical demonstration that DNA methyltransferase 1 (DNMT1) can faithfully copy a preexisting CpG methylation pattern from the parental DNA strand onto the newly synthesized daughter strand during replication. Using hemimethylated DNA substrates and purified mammalian DNMT activity, they showed that DNMT1 has a strong preference for hemimethylated CpG sites over unmethylated sites, thereby ensuring that methylation patterns are heritably propagated through cell division (Gruenbaum et al. 1982; Bestor and Ingram 1983). This discovery provided definitive mechanistic evidence that DNA methylation constitutes a stable epigenetic mark of cellular identity, capable of maintaining lineage-specific gene expression states across generations of somatic cells.

Taken together, these complementary approaches supported the idea that DNA methylation, particularly 5-methylcytosine at CpG dinucleotides in regulatory regions, can silence gene expression and that stable, cell type–specific methylation patterns accumulate during development. On the basis of such evidence, Razin and Riggs hypothesized that DNA methylation functions as a heritable epigenetic signal that helps lock in gene expression programs during cellular differentiation (Razin and Riggs 1980) and, thereby, contributes to the evolution and stability of multicellular organisms.

DNA methylation has therefore been proposed to facilitate the expansion of the evolutionary ladder toward increasingly complex multicellular organisms, a role that is particularly evident in the remarkable diversification and sophistication of DNA methylation patterns in mammals and especially the human brain.

2.4 *Interpretation of the DNA Methylation Signal*

A methyl group added to cytosine or adenine alters the chemical identity of a DNA sequence and can modify how proteins recognize that sequence. In bacterial restriction–modification systems, such methylation determines whether DNA-binding proteins are recruited or excluded, enabling distinction between methylated "self-"DNA and unmethylated foreign DNA. A similar principle applies in eukaryotic transcriptional regulation: DNA methylation can directly modulate transcription factor (TF) binding by reducing or abolishing their affinity for methylated recognition motifs.

Classic studies showed that the methylation of CpG sites within TF-binding elements inhibits binding of AP-family transcription factors; for example, Comb and Goodman (1990) demonstrated that CpG methylation strongly diminishes AP-2 binding to its recognition sequence in the human proenkephalin (PENK) gene promoter (~ −72 bp relative to the transcription start site). These experiments showed that DNA methylation can directly influence gene expression by altering the binding landscape of sequence-specific transcription factors.

Genome-wide analyses have made it possible to systematically screen transcription factor (TF)–binding sites in relation to their local DNA methylation state, revealing numerous TFs whose binding is either inhibited or enhanced by CpG methylation of their recognition sequences (Yin et al. 2017; Domcke et al. 2015). Some factors are repelled by methylation, whereas others preferentially recognize the methylated motif. Across the genome, there is a strong correlation between the methylation status of TF recognition elements and TF occupancy in vivo, as shown by integrating DNA methylation maps with ChIP–Seq profiles for specific factors such as NRF1 and others (Domcke et al. 2015).

Some transcription factors, such as EGR1 (NGFI-A), act as "pioneer factors": they can bind their recognition elements in relatively closed chromatin and either shield these sites from maintenance methylation or recruit active DNA demethylation machinery (Weaver et al. 2007). In the developing and activity-stimulated brain, EGR1 recruits TET1 to a subset of its binding sites, promoting iterative oxidation of 5-methylcytosine, followed by repair, thereby driving local DNA demethylation and long-lasting changes in gene expression programs (Sun et al. 2019). This site-specific demethylation increases the affinity of these elements for TF binding and helps establish new transcriptional landscapes during development and neuronal plasticity.

Methylation-sensitive TFs, therefore, participate both in shaping the DNA methylation landscape and in executing DNA methylation–dependent transcriptional programs. Genome-wide studies indicate that the pattern of methyl-sensitive TF occupancy is highly tissue-specific, providing a mechanism by which differential DNA methylation at regulatory elements contributes to tissue-specific gene regulation (Domcke et al. 2015; Schubeler 2015).

In parallel, a distinct but complementary mechanism evolved, in which proteins bind directly to methylated cytosines, particularly 5-methylcytosine at CpG dinucleotides, and recruit repressive chromatin-modifying complexes. Members of the methyl CpG–binding domain (MBD) family, including MeCP2, MBD1, MBD2, and MBD4, were identified as methylation "readers" that recognize methylated CpG sites and often co-localize with heavily methylated genomic regions (Hendrich and Bird 1998; Nan et al. 1998). For example, MeCP2 and MBD2 can target methylated DNA and recruit multiprotein repressors such as the NuRD complex, which couples histone deacetylation and nucleosome remodeling to DNA methylation (Nan et al. 1998; Zhang et al. 1999; Wood and Zhou 2016). Through these interactions, methyl CpG–binding proteins provide a physical link between DNA methylation and other epigenetic systems such as histone deacetylation and histone methylation, creating partially redundant but mutually reinforcing layers of repression. DNA methylation is closely linked to other chromatin modifications. For example, histone deacetylation can trigger DNA demethylation (D'alessio et al. 2007). Furthermore, various histone modifications can recruit DNA methyltransferases and facilitate DNA methylation (Esteve et al. 2006; Vire et al. 2006; Du et al. 2015a). This redundancy is thought to stabilize silencing and restore local epigenetic states when individual marks are stochastically perturbed, thereby contributing to robust epigenetic memory over cell divisions and potentially across generations (Bird 2002).

2.5 DNA Methylation and Cellular Identity

Genome-scale DNA methylation analyses have generated high-resolution maps of methylation patterns across distinct cell types and provided compelling evidence that cellular identity can be inferred from its characteristic DNA methylation landscape. Lister et al. (2009) compared human embryonic stem cells and differentiated fibroblasts, revealing widespread differences in DNA methylation, consistent with the hypothesis that differentiation involves extensive remodeling of methylation patterns. Importantly, enhancers that regulate cell type–specific transcription are also differentially methylated, and differentially methylated regions frequently overlap transcription factor–binding motifs. In general, enhancers are demethylated in the cell types in which they are active. Promoters associated with CpG islands are predominantly unmethylated in both embryonic stem cells and differentiated cells, although a minority shows differential methylation between these states (Lister et al. 2009). In contrast, promoters with low CpG density are far more likely to be methylated and associated with transcriptional repression, indicating a fundamental distinction between CpG-island and low-CpG promoter classes (Weber et al. 2007). During brain development, non-CG (mCH) methylation accumulates within gene bodies during late prenatal and postnatal stages and is strongly associated with transcriptional repression, particularly in long neuronal genes, whereas the relationship between CG gene body methylation and expression is weaker and context-dependent (Lister et al. 2013).

Marked differences between embryonic and differentiated cells were also observed at the genome-wide level. Gene body methylation correlates positively with transcription in differentiated fibroblasts, whereas this relationship is weak or absent in pluripotent stem cells. In addition, non-CG (mCH) methylation is abundant in embryonic stem cells but largely absent in fibroblasts. The enrichment of non-CG methylation in stem cells corresponds to elevated expression of the de novo methyltransferases DNMT3A and DNMT3B. De novo methylation is not restricted by hemimethylated template recognition, unlike maintenance methylation mediated by DNMT1, and therefore, can modify cytosines in non-CG contexts (Lister et al. 2009).

DNMT3A and DNMT3B originated from a gene duplication event in early vertebrates, and DNMT3A acquired specialized functions in postnatal brain development. Although non-CG methylation is not efficiently propagated through cell division and is therefore lost from most proliferating somatic tissues, it accumulates in neurons, largely postmitotic cells, during early life. The emergence of DNMT3A-dependent non-CG methylation in neurons greatly expanded the number of methylatable positions in the brain, increasing the potential plasticity of the neuronal methylome. Accumulating evidence indicates that DNMT3A-dependent non-CG methylation in neurons plays a role in neuronal maturation, repression of activity-dependent genes, synaptic function, and cognitive processes, with genetic and experimental perturbations impairing learning, memory, and neurodevelopment (Lister et al. 2013; Tatton-Brown et al. 2014; Stroud et al. 2017).

Using bisulfite next-generation sequencing, investigators have identified distinct DNA methylation profiles for different human cell types, showing that DNA methylation is a key determinant of cellular identity (Loyfer et al. 2023). Perhaps the strongest support for the Razin and Riggs model (Razin and Riggs 1980), which proposed that changes in DNA methylation drive cellular differentiation, establish cell type–specific transcriptional programs, and are guided by transcription factors, comes from studies examining the evolution of DNA methylation patterns in the developing brain. As neurons and glia diverge, they acquire distinct methylation landscapes characterized by numerous cell-type- and site-specific differentially methylated positions that preferentially localize to DNase I–hypersensitive regions, enhancers, and transcription factor–binding motifs (Lister et al. 2013). Although most promoters remain unmethylated regardless of transcriptional status, a defined subset exhibits differential promoter methylation between neurons and glia that correlates with expression patterns.

During late gestation and in the early postnatal life, early neurodevelopmental genes become transcriptionally silenced through gene body methylation, particularly via DNMT3A-dependent non-CG (mCH) methylation, which is a hallmark of neuronal maturation (Lister et al. 2013). Machine learning analyses that infer sequence determinants and variants of DNA methylation demonstrate that predictive motifs map to transcription factor–binding positions and overlap with neurodevelopmental and psychiatric risk variants, including those associated with schizophrenia and autism (Zhou et al. 2022; Zhou et al. 2025). Consistent with the idea that DNA methylation encodes cellular identity, single-cell methylome profiling has resolved neuronal subtypes that cluster exclusively on the basis of their methylation signatures (Luo et al. 2017), and more recent work has shown dynamic and cell type–specific methylation changes across prenatal and postnatal cortical development (Franklin et al. 2025). Moreover, transcription factor EGR1, together with Tet dioxygenases, is required for activity-dependent and developmental DNA demethylation in neurons, supporting the view that transcription factors direct targeted demethylation during neurodevelopment (Sun et al. 2019).

DNA methylation provided multicellular organisms with an additional toolkit for establishing diverse, cell type–specific gene expression programs, thereby facilitating the evolution of complex tissues and organ systems in vertebrates and mammals. It is proposed that DNA methylation could accelerate evolutionary change by amplifying the genetic signal through downstream epigenetic consequences, thereby expanding the substrate on which natural selection can act during the evolution of complex brains, including the human brain. Although some invertebrates, such as *Drosophila melanogaster*, have lost canonical 5-methylcytosine DNA methyltransferases (Urieli-Shoval et al. 1982; Raddatz et al. 2013) and rely predominantly on other chromatin-based mechanisms (for example, histone modifications) for cell type–specific gene regulation (Vidaurre and Chen 2021), the widespread conservation and diversification of DNA methylation machinery in vertebrates and other complex multicellular lineages suggests that DNA methylation has conferred a substantial selective advantage in vertebrates and mammals.

2.6 Parental Imprinting

Using DNA methylation to mark the parental origin of specific genes (Sapienza et al. 1987; Stoger et al. 1993) is an excellent illustration of how DNA methylation defines an exquisite identity of genes beyond their DNA sequence and how these marks in the germline can pass and persist into the next generation. However, they are erased again during gametogenesis and reestablished to mark the new parental origin for the following generation. The reestablishment of the imprint involves DNMT3A and DNMT3L being targeted to the imprinting control region (ICR) (Bourc'his et al. 2001; Hata et al. 2002) by specific protein complexes, such as those containing Tripartite Motif–Containing 28 (TRIM28), that are recruited by sequence-specific factors recognizing motifs in the ICR (Alexander et al. 2015). To serve as a marker of parental origin in the next generation, the imprint needs to be protected from erasure following fertilization, when the majority of both paternal and maternal methylomes is erased. Dedicated proteins such as PGC7/Stella protect the imprints from demethylation in early embryogenesis (Nakamura et al. 2007).

This function of DNA methylation in parental imprinting emerged relatively late in evolution in live-bearing therian mammals, placental organisms, and marsupials (Killian et al. 2000). Parentally imprinted genes such as Insulin-like Growth Factor 2 Receptor (Igf2r) and the Insulin-like Growth Factor 2 (Igf2)/H19 pair are usually involved in controlling growth (Reik et al. 1995), placental function, and the allocation of resources to the developing embryo (Tabano et al. 2010). It is hypothesized that the methylation and silencing of parent-of-origin-specific genes evolved to protect the mother from increasing demands by the offspring that could deplete maternal resources. While the paternal genome is selected to maximize the resources allotted to its offspring, the maternal genome is selected to ensure that maternal resources are not exhausted and remain available to the mother and her other offspring, including those sired by different fathers. This competition is especially pronounced in placental animals, where the embryo develops in the mother's womb (Constancia et al. 2004).

Thus, DNA methylation confers parent-of-origin identity on genes, in addition to their cellular identity, thereby expanding the information encoded in the genome for selective advantage. The fact that DNA methylation in certain regions can survive reprogramming and be transmitted across generations raises the possibility that a similar mechanism could underlie stable transgenerational inheritance of DNA methylation and epigenetic inheritance of traits.

3 Environmental Programming of the Genome by DNA Methylation and Epigenetic Mechanisms

3.1 The Challenge of Stable Environmental Adaptation of Traits

Natural selection has been the principal mechanism driving adaptation to environmental conditions in evolution (Darwin 1859; Mayr 1982; Fisher 1930). However, the timescale of such selection is typically long, often spanning many generations. Darwinian selection offers a solution at the level of the population or species by favoring the fittest individuals, but it does not readily explain how individual organisms can acquire adaptive traits over just a few generations. Nor does it, by itself, account for stable differences among individuals who experience distinct environmental conditions within the same overall environment, for example, interindividual differences in social experiences or physical exposures.

This raises a key question: Does the genome possess mechanisms that allow it to respond and adapt to these diverse and rapidly changing conditions affecting different individuals? Because environments can change on timescales much faster than those of genetic selection, it is essential to consider what mechanisms might enable the genome to adjust to such changes. Are there directional mechanisms through which environmental factors can alter genome function in a stable, generational, or even multigenerational manner? This question is closely related to the broader issue of whether stable interindividual phenotypic differences can emerge in the absence of underlying genetic variation.

In contrast to Darwinian "blind" natural selection, Lamarck proposed that the environment can directly induce changes in organisms, which are then inherited, leading to adaptive modifications at the species level (Lamarck 1809/1963). As molecular genetics developed, Lamarck's concept of environmentally induced, directed adaptations fell out of favor because the discovery of spontaneous mutations and understanding DNA sequence variation provided a compelling framework for environmental adaptation via selection on genetic variation (Fisher 1930; Mayr 1982). Nonetheless, Lamarckian ideas of directional environmental influences on traits have repeatedly resurfaced, in part because the notion of direct environmental guidance of phenotypic evolution is conceptually appealing, even if it appears improbable within the classical framework of genetics (Burkhardt Jr. 2013; Jablonka and Lamb 2005).

3.2 The Impact of Early Environments on Lifelong Health Trajectories

Numerous sociological, economic, and population-health studies in humans have shown that early-life socioeconomic position has long-term effects on health (Hertzman and Wiens 1996; Hertzman 1999; Power et al. 2012). In humans, adverse childhood experiences (ACEs) have a lasting impact on life-course health trajectories, influencing adult morbidity and mortality in a clear dose–response manner (Felitti et al. 1998). Importantly, these patterns can extend across generations: Parents' exposure to ACEs is associated with poorer physical and mental health outcomes in their children, consistent with intergenerational transmission of risk (Le-Scherban et al. 2018). Similarly, in nonhuman primates, early-life social adversity in mothers and infants leads to enduring alterations in offspring stress physiology and behavior, and recent work in rhesus macaques has identified cross-generational methylomic signatures of infant maltreatment in newborns, indicating an epigenetic trace of early adversity (Lardenoije et al. 2025). Comparable findings have been reported in rodents, where early-life maltreatment produces long-lasting changes in DNA methylation and stress-related behaviors (Roth et al. 2009). Together, these observations suggest that interindividual variation in exposure to social environments can direct stable phenotypic changes that may even extend to the next generation. They pose a challenge to a strictly gene-centric view, in which the genome responds passively to environmental challenges solely through random sequence variation and selection, and instead, point to the existence of mechanisms for the adaptation of traits to environmental change and for the transmission of such influences across generations, which are not determined exclusively by the DNA sequence.

3.3 Epigenetic Programming by Maternal Care

Maternal care in rats—quantified primarily as the frequency of licking and grooming (LG) and arched-back nursing (ABN)—shows stable interindividual variation, with dams distributed along a continuum from high to low LG-ABN. Offspring of high-

LG-ABN mothers differ markedly from those of low-LG-ABN mothers: As adults, they show increased hippocampal glucocorticoid receptor (GR) expression, enhanced glucocorticoid negative-feedback sensitivity, and reduced adrenocorticotropic hormone (ACTH) and corticosterone *responses to acute stress*, while basal hormone levels are unchanged (Liu et al. 1997). Cross-fostering experiments, in which pups born to high-LG-ABN dams are reared by low-LG-ABN dams and vice versa, demonstrate that these differences in stress responsivity are determined by the fostering mother rather than the biological mother, indicating a nongenomic, gamete-independent mode of behavioral inheritance (Francis et al. 1999). This behavioral transmission is mediated, at least in part, by epigenetic programming in the offspring. In the hippocampus, variations in maternal care are associated with stable differences in DNA methylation and histone acetylation at the exon 17 ($GR1_7$) promoter of the glucocorticoid receptor gene (Nr3c1), altering NGFI-A binding, GR expression, and hypothalamic–pituitary–adrenal (HPA) stress responses (Weaver et al. 2004). These effects extend beyond a single locus: maternal care is linked to broad, coordinated changes in DNA methylation, histone acetylation, and gene expression across large chromosomal regions, including protocadherin clusters and other genes involved in neuroplasticity (Weaver et al. 2006; Mcgowan et al. 2011). The maternal behavior phenotype itself is also transmitted from mother to daughter. High-LG-ABN mothers produce female offspring that, as adults, are themselves high-LG-ABN dams, a process associated with differential methylation of the estrogen receptor-α1b (ERα1b) promoter and ERα expression in the medial preoptic area, a key brain region for maternal behavior (Champagne et al. 2006) (Fig. 2.2).

Comparable epigenetic correlates of early caregiving have been reported in other species. In rhesus macaques, differences in early maternal rearing are associated with distinct DNA methylation signatures in the prefrontal cortex and T cells (Provencal et al. 2012). In humans, childhood abuse is associated with increased methylation of a neuron-specific NR3C1 promoter and reduced GR expression in the hippocampus, paralleling the rodent findings (Mcgowan et al. 2009). There are conserved patterns of early-experience-sensitive methylation across the NR3C1 locus in both rats and humans (Suderman et al. 2012).

This mechanism of behavioral inheritance, which does not involve the germline, differs from classical genetically mediated inheritance that drives evolutionary processes. Although this form of inheritance is independent of germline DNA, it proceeds through pathways that translate the sensing of behavior into epigenetic modifications of genes in the brain via activation of the serotonergic pathways that induce changes in DNA methylation and chromatin structure in the brain and possibly other tissues, such as the immune system. In summary, this alternative mechanism of inheritance is driven by maternal behavior, which induces DNA methylation changes that, in turn, produce phenotypic effects, including alterations in the maternal behavior of female offspring, who then program their own offspring. It remains unclear for how many generations such effects can persist and whether DNA methylation–based behavioral inheritance can become a stable phenotype on an evolutionary timescale, thereby adding a mechanism for environmentally directed acquisition of inherited traits. An interesting aspect of these DNA methylation–

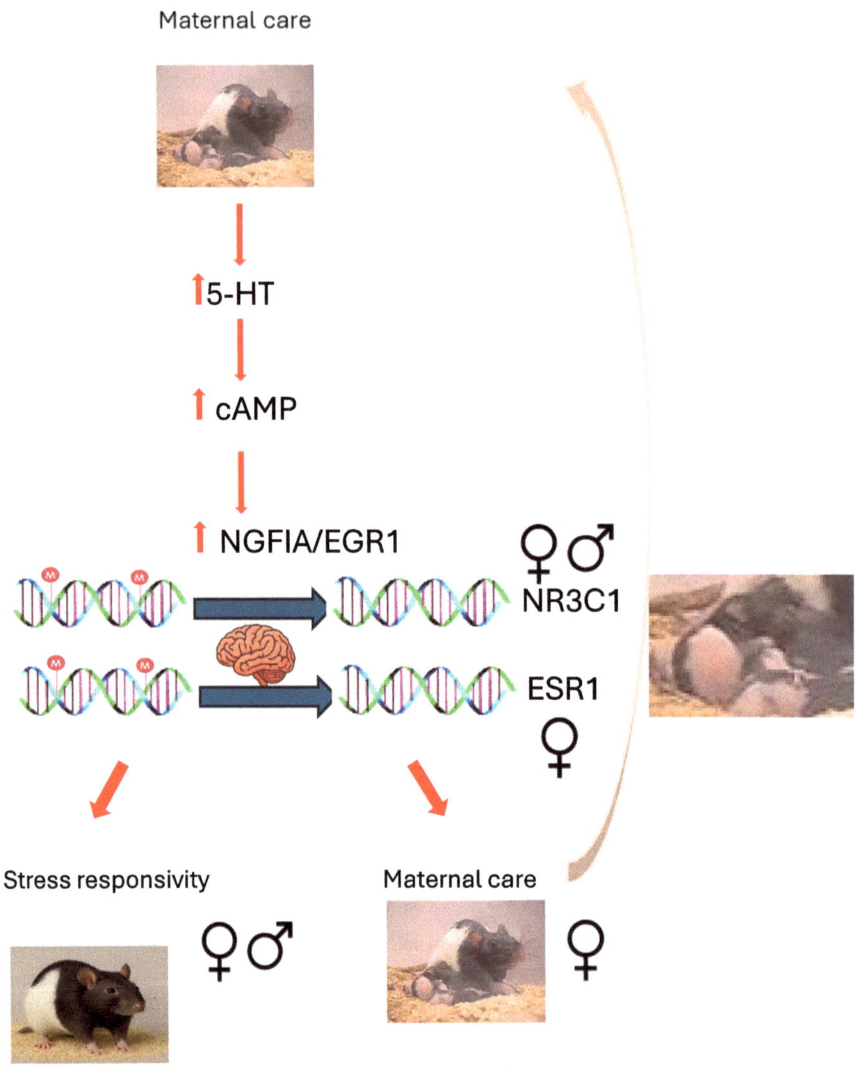

Fig. 2.2 Behavioral transmission of DNA methylation states and stress responsivity across generations. Variation in maternal care (e.g., licking/grooming and arched-back nursing) mediates nongenetic inheritance of stress-related phenotypes. Enhanced maternal care activates hippocampal serotonergic signaling, elevating cAMP–PKA activity and inducing the transcription factor NGFI-A (EGR1). NGFI-A binds to the exon 17 promoter of NR3C1 (glucocorticoid receptor), promoting site-specific DNA demethylation, increased receptor expression, and attenuated HPA-axis stress responses in the offspring. In parallel, maternal care modifies DNA methylation and expression of ESR1 in the medial preoptic area (MPOA), shaping the daughter's maternal behavior. As adults, these daughters tend to reproduce similar caregiving, thereby perpetuating the associated methylation patterns and stress responsivity traits across generations via behavior rather than germline transmission

mediated, behaviorally inherited traits is that they are reversible, both through behavioral interventions such as cross-fostering and through the use of epigenetic agents. This reversibility makes them fundamentally different from classical inherited traits fixed by genetic variation and natural selection.

3.4 Cross-Generational Embedding of Adverse Childhood Experiences in DNA Methylation and the Potential for Germline Transmission

As discussed above, adverse childhood experiences (ACEs) are known to have long-term and intergenerational phenotypic effects (Sect. 3.2). Animal models show that these effects involve changes in DNA methylation that persist in the offspring. Roth et al. (2009) reported that abusive maternal behavior in rats induces enduring alterations in DNA methylation of the Bdnf gene in the offspring's prefrontal cortex, accompanied by reduced Bdnf expression. Importantly for our discussion, this study showed that, in contrast to the maternal care studies described above—where maternal effects could be reversed by cross-fostering—the Bdnf methylation changes were only partially rescued by cross-fostering. This suggests that the effect is at least partly mediated by factors associated with the biological mother (e.g., prenatal environment or germline epigenetic marks) and could potentially involve maternal germline contributions, although this was not directly demonstrated (Roth et al. 2009). Similarly, in rhesus macaques, Klengel and colleagues showed that maternal exposure to abuse during early life was associated with hypomethylation of the *FK506-binding protein 5* (*Fkbp5*) gene in the blood of offspring, and this effect persisted despite randomized cross-fostering to competent foster mothers (Klengel et al. 2019). FKBP5 is a key regulator of glucocorticoid receptor sensitivity, and allele-specific differences in *Fkbp5* DNA methylation have been linked to an increased risk of stress-related psychiatric disorders, including PTSD, in individuals exposed to childhood abuse (Klengel et al. 2013). Recently, these findings were extended by genome-wide profiling of DNA methylation in newborn offspring of rhesus macaque mothers who had been abused early in life, using the same randomized cross-fostering design to distinguish germline- or prenatal-mediated epigenetic effects from postnatal behavioral transmission. Lardenoije et al. (2025) reported cross-generational methylomic signatures enriched in pathways related to glucocorticoid receptor sensitivity, steroid biosynthesis, and inflammatory signaling, consistent with an environmentally targeted effect in which epigenetic changes map onto functional pathways relevant to the original stress exposure. Moreover, mediation analyses suggested nominal mediation by DNA methylation of offspring cortisol dynamics (Lardenoije et al. 2025).

These data are consistent with but do not conclusively prove a model in which acquired stress response traits can be transmitted via the germline for at least one generation. However, this study did not assess DNA methylation in oocytes, so it remains unclear whether the methylation changes observed in the offspring's peripheral blood cells were driven by germline differences in DNA methylation. An alternative hypothesis is that these DNA methylation changes are mediated by an altered intrauterine environment in mothers who were exposed to ACEs early in life, with this environment secondarily shaping the embryo's DNA methylation profiles. The cross-fostering design cannot rule out this interpretation. Only transferring in vitro–fertilized embryos from such mothers into the uterus of an unexposed surrogate could exclude intrauterine effects as an explanation and provide stronger evidence for epigenetic germline transmission (Szyf 2025).

The idea that paternal traumatic experiences could be transmitted epigenetically through the germline and affect future generations has profound implications for humans, raising the question of how abusive or traumatic environments might shape human evolution. Human data are consistent with cross-generational DNA methylation changes associated with parental exposure to extreme trauma. For example, the children of survivors of the Holocaust (Yehuda et al. 2016; Yehuda et al. 2014) and of the Rwandan genocide (Musanabaganwa et al. 2022) show altered DNA methylation profiles. In the offspring of Holocaust survivors, genes involved in stress responsivity, such as *NR3C1* (the glucocorticoid receptor) (Yehuda et al. 2014) and *FKBP5* (Yehuda et al. 2016), are differentially methylated, again consistent with a targeted effect on the stress response pathway that matches the nature of the parental environmental insult. However, human studies face important limitations, including small effect sizes, a lack of access to brain tissue, relatively small sample sizes, and numerous potential confounders (such as genetic stratification and unmeasured environmental factors), which complicate causal interpretation.

3.5 Sperm DNA Methylation in Cross-Generational Responses

If DNA methylation is to serve as a force in the stable inheritance of environmentally directed traits over evolutionary time, there should be evidence for persistent DNA methylation alterations in gametes that are transmitted across multiple generations. First, there must be a conduit that translates environmental and experiential exposures into alterations in DNA methylation in sperm or oocytes. Second, these methylation marks in gametes must evade the wave of postfertilization DNA methylation erasure. Third, they must either persist through the extensive remodeling of DNA methylation that accompanies cellular differentiation or, alternatively, bias the developmental trajectory of cell type–specific methylation patterns, producing a predictable, exposure-specific methylation change in particular tissues. Fourth, for these marks to be inherited by subsequent generations, the environmentally induced methylation pattern must itself survive the reprogramming events of gametogenesis. Fifth, it remains unclear whether inheritance of such DNA methylation profiles would follow Mendelian patterns or whether environmentally driven methylation marks could behave in a dominant, "in trans" fashion, imposing the same methylation profile on chromosomes inherited from the unaffected parent.

Comparative data suggest that large-scale erasure and reestablishment of DNA methylation during gametogenesis and early embryogenesis are not ancestral features of all animals but emerged during vertebrate evolution and reach their most extreme form in mammals (Xu et al. 2019). In many invertebrates, DNA methylation is more limited, and there is little evidence for the kind of global reprogramming seen in mammalian development (Xu et al. 2019). This raises the possibility that the inheritance of acquired DNA methylation changes—and the associated traits—may have been more common before the evolution of robust methylation reprogramming mechanisms in vertebrates and could have played a greater role in the acquisition of environmentally guided traits earlier in evolution. The subsequent evolution of extensive epigenetic reprogramming in mammals may have been selected, at least in part, to restrict long-term inheritance of environmentally induced methylation changes and their phenotypic consequences. Nevertheless, reprogramming may be selectively relaxed at specific genomic domains that retain the capacity to mediate environmentally responsive, epigenetically based inheritance in mammals.

Cross-Generational Transmission of Behavioral Response to Early-Life Maternal Separation and Maternal Unpredictable Stress (MSUS)

There is evidence in rodents that sperm DNA methylation can be altered by environmental exposure and that methylation changes detected in sperm can also be observed in relevant tissues in the offspring. Franklin et al. (2010) showed that chronic, unpredictable maternal separation, combined with maternal stress (MSUS) during early life, induces depressive-like behavior in mice that persists into adulthood and is transmitted via the paternal line to offspring that are themselves reared under normal conditions by unstressed dams. Depressive-like behaviors were expressed in F1 MSUS males, in F2 females (but not males), and again in F3 males, indicating a complex, sex-dependent pattern of transgenerational transmission. At the molecular level, MSUS altered DNA methylation in the promoters of several candidate genes, *methyl CpG–binding protein 2 (Mecp2), cannabinoid receptor 1 (Cb1)*, and *corticotropin-releasing factor receptor 2 (Crfr2),* in the sperm of F1 males, and similar directionally consistent methylation changes were found in the cortex of F2 female offspring, accompanied by corresponding changes in gene expression, suggesting functional consequences. In addition, altered methylation of some of these loci (*Mecp2* and *Crfr2*) was also detected in the sperm of F2 males. This study, therefore, provides evidence that early-life stress can modify DNA methylation in the male germline and that related methylation changes can appear in the brains of offspring derived from that germline, consistent with sperm-mediated, cross-generational transmission of stress-induced epigenetic alterations (Franklin et al. 2010).

However, several questions remain. First, even if the sperm and brain show similar methylation changes at these loci, the study did not follow the developmental trajectory of these marks from fertilization through embryogenesis and postnatal differentiation. Because DNA methylation is extensively reprogrammed after fertilization and during development, it is unclear whether these alterations persist unchanged through all stages and tissues or whether they are erased and then reestablished in a lineage- or tissue-specific manner. One possibility is that the same methylation pattern is faithfully maintained throughout development and across tissues; another is that the pattern is preserved only along the brain lineage but lost elsewhere; a third is that the original sperm methylation changes are erased, but other epigenetic signals (such as alternative DNA methylation changes at different loci, chromatin modifications, or noncoding RNAs) bias the subsequent establishment of methylation at *Mecp2, Cb1*, and *Crfr2,* specifically in the brain.

Moreover, the study did not provide genome-wide methylation data, did not examine methylation in other somatic tissues in a systematic way, and did not report DNA methylation profiles in the F3 generation or beyond. Thus, we do not yet know whether the observed methylation changes and their associated behavioral phenotypes persist across additional generations or how they are sculpted during development. Neither was there proof that sperm or sperm DNA methylation played a causal role in the offspring phenotype. In vitro fertilization with sperm from MSUS animals could have addressed the causal role of sperm in the transmission of the phenotype. At this stage, mechanistic interpretations about exactly how these sperm methylation changes bypass postfertilization reprogramming and shape brain methylation remain speculative.

Subsequent work from the same group shifted the focus from DNA methylation to noncoding RNAs in sperm as candidate carriers of the MSUS-induced phenotype. In follow-up studies, purified sperm RNAs from MSUS males were injected into fertilized oocytes of unexposed parents and were sufficient to reproduce key behavioral and endocrine alterations in the offspring, closely resembling those observed in the natural MSUS lineage (Gapp et al. 2014). Further dissection of RNA fractions revealed distinct phenotypic effects of different RNA species: Injection of the long RNA fraction from the sperm of MSUS-exposed males affected food intake, glucose/insulin responses, and risk-taking behavior, whereas the small RNA fraction influenced other traits such as body weight and behavioral despair (Gapp et al. 2014). These findings indicate that sperm RNAs act as a functional vehicle of transgenerational inheritance in this model and suggest that the DNA methylation changes observed in the sperm and brain in the original Franklin et al. (2010) study likely represent one layer of a broader epigenetic cascade rather than the sole information-bearing mark. If RNA is indeed involved in transgenerational inheritance, it is unclear how it is inherited across generations in the male germline. However, unlike DNA, RNA does not have a dedicated mechanism for faithful, long-term replication across cell generations. It is, therefore, tempting to speculate that the "memory" for specific RNA expression patterns resides in a heritable epigenetic layer, such as DNA methylation and associated chromatin modifications, that regulates the transcription of these RNAs in each generation (Fig. 2.3).

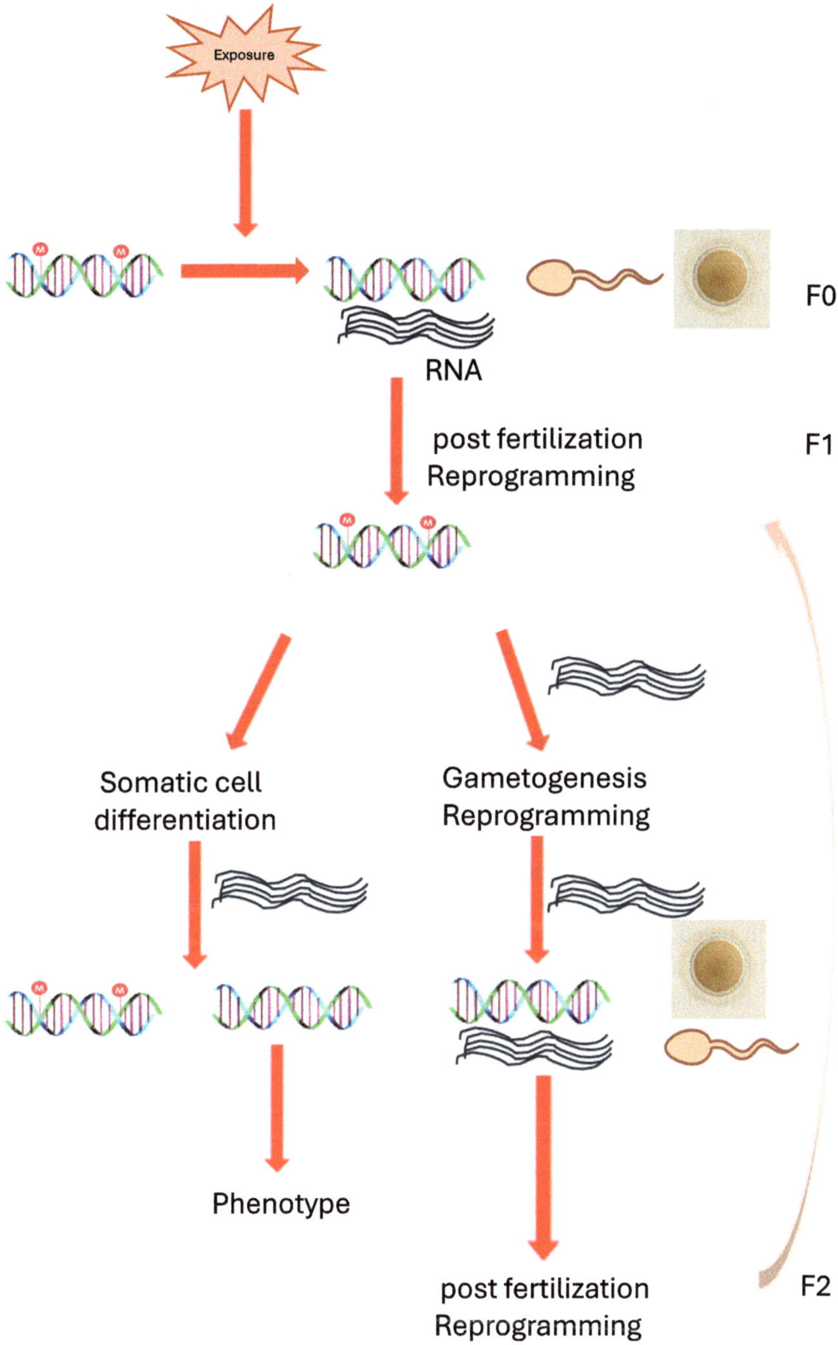

Fig. 2.3 Cross-generational epigenetic programming. Environmental exposures can alter the epigenome of gametes—sperm, oocytes, or both—by changing DNA methylation and, plausibly, the cargo of small noncoding RNAs and mRNAs. After fertilization, most DNA methylation is globally reprogrammed; however, gamete-derived RNAs are hypothesized to influence early

Cross-Generational Transmission of Acquired Odor Sensitivity

Another example of transgenerational, nongenomic, sperm-mediated inheritance of an acquired trait was provided by Dresler's group studying olfactory sensitivity (Dias and Ressler 2014). Animal freezing in response to specific odors, like a mouse freezing in response to fox urine, is a prime example of genetic natural selection in action. A strong selective pressure by a predator results in a genetic selection of individuals that recognize the odor. Could recognition of an odor be acquired and mediated by a directed epigenetic process that translates exposure to an odor to a specific methylation/demethylation in sperm and transmit it to the next generation? The odor–fear conditioning work of Dias and Ressler provides a striking experimental model of nongenomic inheritance of an acquired trait (Dias and Ressler 2014). Adult mice (F_0) were conditioned to fear the odor of acetophenone by pairing it with mild foot shocks. This odor is detected by the olfactory receptor Olfr151. The conditioned animals developed a robust, odor-specific startle response to acetophenone but not to a control odor (propanol, detected by a different receptor, Olfr6). Remarkably, their F_1 and F_2 descendants—never exposed to the shocks—also showed an enhanced startle response specifically to the conditioned odor but not to the control odor, whether the exposed parent was the mother or the father. This demonstrates cross-generational transmission of an acquired, odor-specific fear response via the germline, without changes in DNA sequence.

A key line of evidence for molecular specificity is the targeted DNA methylation change in the germline (Dias and Ressler 2014). In sperm from acetophenone-conditioned males, Dias and Ressler found selective demethylation of Olfr151 but not of *Olfr6*; *Olfr151* receptor–specific methylation signature was also detectable in the sperm of F_1 and F_2, indicating that it had escaped both postfertilization and primordial germ cell waves of methylation erasure. Consistent with a functional consequence, the offspring showed increased numbers and enlarged glomeruli of Olfr151-expressing neurons in the olfactory bulb, i.e., an anatomical amplification of the circuit tuned to the conditioned odor. Crucially, when the sperm from conditioned males was used for in vitro fertilization of naïve females, the resulting progeny still displayed enlarged odor-specific glomeruli, strongly supporting the idea that the sperm carries sufficient epigenetic information to transmit the odor-specific neuroanatomical phenotype.

Fig. 2.3 (continued) embryonic pathways and guide the reestablishment of DNA methylation states during lineage specification. This can imprint exposure-linked methylation changes in specific somatic tissues, contributing to the offspring phenotype. In the paternal line, a similar process is proposed during spermatogenesis in the next generation: Following the normal demethylation–remethylation cycle, sperm RNAs (e.g., tRNA fragments, miRNAs, and piRNA-like species) may help restore exposure-dependent methylation profiles. Fertilization by these sperm can, therefore, reproduce the molecular and phenotypic effects in F1 and beyond, enabling cross-generational transmission without stable germline DNA methylation inheritance. Because most methylation marks are reset each generation and RNA-mediated effects may decay, these transmitted influences are expected to attenuate over successive generations unless reinforced by continued exposure

This model thus fulfills several criteria for nongenomic inheritance of an acquired trait: a learned association (odor + fear) in parents, a corresponding behavioral and anatomical phenotype in unexposed descendants, and a receptor-specific epigenetic mark in sperm that tracks with the trait. At the same time, it raises exactly the mechanistic questions about transmission and developmental unfolding. First, epigenetic marks are highly tissue-specific, yet the initial change is detected in sperm, while the ultimate phenotype is expressed in a subset of olfactory receptor neurons and in higher-order fear circuits. How does a site-specific methylation change in the germline survive global reprogramming and then guide differentiation so that only the appropriate olfactory and fear-related neurons are altered? Second, the methylation difference is not evident in the mature olfactory receptor neurons themselves, suggesting that the critical "memory" may operate at earlier developmental stages and then be overwritten once the circuitry is established. Third, the inherited phenotype involves not just odor sensitivity but odor–fear association, implying that additional, as-yet-unidentified epigenetic marks in sperm must encode information about the emotional valence of the experience, not only receptor identity.

One plausible scenario that was proposed is that exposure to odor plus shock triggers systemic signals—such as circulating hormones or small RNAs—that reach the germline, where they direct locus-specific DNA methylation changes at olfactory receptor genes and other, unknown loci involved in fear circuitry (Szyf 2014). These marks must then be protected during early embryogenesis and repurposed during differentiation to bias the assembly of odor-specific and fear-related neural networks. Exactly how this information is written into sperm, preserved through reprogramming, and read out in a cell type–specific manner during brain development remains unresolved.

It is also unknown whether this altered odor recognition remains a stable trait across many generations or gradually fades. Does the methylation signal extend beyond the F2 generation? What is the penetrance of the trait? Do all sperm in the F1 and F2 generations carry the methylation mark, and if not, is there any indication that the probability of inheritance corresponds to the fraction of sperm bearing this mark?

Moreover, it remains unclear how a sperm methylation mark could be maintained through all stages of somatic differentiation and then through gametogenesis in the next generation. Addressing these questions is crucial for evaluating the hypothesis that DNA methylation can drive the stable acquisition of traits in the absence of genetic changes. The available data are still limited, and many key aspects of this mechanism remain unresolved.

Transgenerational Response to Maternal Prenatal Exposure to Pesticides Vinclozolin and Methoxychlor

Skinner and colleagues (Anway et al. 2005) developed one of the most influential models of toxicant-induced transgenerational epigenetic inheritance using the agricultural fungicide vinclozolin (and, in some experiments, methoxychlor). In their landmark study, pregnant rats (F0) were transiently exposed during gonadal sex determination (E8–E14). Male offspring from the F1 generation showed reduced spermatogenic capacity, decreased sperm counts and motility, and increased spermatogenic cell apoptosis. Strikingly, similar reproductive defects were observed in F2, F3, and F4 males that were never directly exposed, with high penetrance (>90%) in the male line. Methylation-specific PCR analysis was performed on F1 DNA, detecting changes in methylation in 25 regions, and then a selected region that showed altered methylation in F1 was shown to show altered methylation in F2 and F3 sperm. The persistence of sperm DNA methylation alterations at this stage suggests that the methylation changes escaped the reprogramming during spermatogenesis and after fertilization. Outcross experiments indicated that transmission occurred through the paternal germline (Anway et al. 2005).

Follow-up work extended these observations to a broad spectrum of adult-onset diseases. F1–F3 descendants of vinclozolin-exposed females exhibited increased incidence of testis and prostate abnormalities, kidney disease, immune defects, tumor development, and, in females, pregnancy abnormalities and mammary tumors, again without further exposure beyond the F0 generation (Nilsson et al. 2008).

To identify a mechanistic substrate, Guerrero-Bosagna et al. (2010) analyzed promoter DNA methylation in the sperm from F3 vinclozolin-line males—the first generation whose germ cells were never directly exposed in utero. Using MeDIP-chip and bisulfite validation, they found a set of differential DNA methylation regions (DMRs) in F3 sperm, including 16 confirmed promoter DMRs, which were consistent within the vinclozolin lineage and distinct from controls. This established that stable "sperm epimutations" can be detected three generations after ancestral exposure, in parallel with the disease phenotypes. Similar F3 sperm DMRs have subsequently been reported for other toxicants (plastics, dioxin, jet fuel, and hydrocarbons), again with exposure-specific DMR sets and associated reproductive and metabolic abnormalities (Manikkam et al. 2012).

More recent studies have refined this picture. King et al. (2019) analyzed DDT-line rats and showed that F3 descendants exhibit heightened susceptibility to multiple pathologies, each associated with a partially distinct panel of F3 sperm DMRs that can be viewed as disease-specific sperm "epigenetic biomarkers" of environmental toxicant exposures. Unique sets of DMRs were associated with late-onset puberty, prostate disease, kidney disease, testis disease, obesity, and multiple disease pathologies. Ben Maamar et al. (Ben Maamar et al. 2018) further showed that the transgenerational inheritance of environmental toxicant exposure involves not only DNA methylation but also an altered retention of histones and changes in the sperm noncoding RNA landscape, including microRNAs and other small RNAs. These studies support a model in which ancestral environmental toxin exposure induces a coordinated, multilevel reprogramming of the male germline that is stable to at least the F3 generation.

Independent work in mice and other systems provides partial replication of specific aspects of this model. Stouder and Paoloni-Giacobino (2010) reported that vinclozolin exposure in mice alters the methylation of several imprinted genes in sperm across generations, consistent with the transgenerational effects on germline methylation (Stouder and Paoloni-Giacobino 2010).

At the same time, the vinclozolin model remains controversial. Schneider et al. (2008), using different exposure routes and doses, did not detect robust transgenerational antiandrogenic effects or reproductive phenotypes in F2–F4 males following maternal vinclozolin exposure and concluded that they found no evidence for the transgenerational inheritance of antiandrogenic effects under their conditions. It is unclear whether the differences in results are caused by different routes and timing of exposure. Iqbal et al. (2015), in a study of endocrine disruptor exposure in mice, observed clear epigenetic perturbations (including DNA methylation changes) in the germline of directly exposed embryos (G1), but these changes were not maintained in the F2 germline or F3 soma, leading them to argue that germline reprogramming normally corrects most exposure-induced epimutations rather than transmitting them (Iqbal et al. 2015). However, it is unclear whether a subset of sites do remain differentially methylated and escape germline epigenetic reprogramming, as reported by Skinner's group.

Broader critical reviews have raised concerns about small sample sizes, multiple-testing issues in genome-wide DMR analyses, strain and exposure differences between laboratories, and the difficulty of ruling out subtle genetic or non-epigenetic confounders (Horsthemke 2018; Nadeau 2015). Thus, while the Skinner vinclozolin data are extensive and internally consistent, the field does not yet view them as unambiguous proof that transgenerational DNA methylation epimutations are common or easily induced in mammals.

With respect to the mechanistic questions raised above (see section "Cross-Generational Transmission of Acquired Odor Sensitivity"), the vinclozolin studies provide partial but not complete answers. First, they clearly demonstrate that both phenotypes and sperm DMRs can persist for at least three or, sometimes, four generations in the absence of continued exposure, strengthening the case that environmentally induced germline epimutations can be multigenerationally stable under some conditions (Anway et al. 2005; Guerrero-Bosagna et al. 2010; King et al. 2019). Second, work in DDT and vinclozolin models suggests that at least a subset of sperm DMRs retain differential methylation in early embryos and differ between directly exposed and transgenerational generations, implying that at least a subset of germline marks can escape the global demethylation waves (Ben Maamar et al. 2018). DNA methylation in these studies was profiled in sperm (and, for DDT, selected early embryonic stages) but not systematically in adult somatic tissue. This directly addresses the question of whether sperm DNA methylation marks can survive early reprogramming in principle.

However, several key questions remain. The existing sperm methylation data are almost entirely from bulk samples; we have no single-cell measurements of what fraction of sperm carry a given DMR or how this relates quantitatively to trait penetrance in offspring. We also still lack a complete developmental trajectory that traces specific sperm DMRs through the zygote, blastocyst, primordial germ cells, and somatic lineages into defined adult tissues (e.g., specific brain nuclei or kidney cell types). Although there are data showing altered transcriptomes and epigenomes in F3 germ cells, gonads, and some somatic tissues (e.g., prostate, testis, and kidney) after ancestral vinclozolin exposure, these remain largely correlational and do not map one-to-one from individual germline epimutations to particular somatic phenotypes.

Moreover, there is currently no demonstration, to my knowledge, that transferring purified F3 sperm or isolated epigenetic fractions (e.g., DMR-enriched DNA, specific small RNAs, or chromatin components) from vinclozolin-line males into naïve embryos is sufficient to recreate the full disease phenotype. Mechanistic issues regarding penetrance, developmental unfolding, and the detailed "readout" of germline epimutations into highly tissue-specific phenotypes remain unanswered.

In summary, the vinclozolin model provides some of the strongest current evidence for toxicant-induced, sperm-mediated transgenerational phenotypes in mammals and demonstrates that stable sperm DNA methylation differences can persist to at least the F3 generation. At the same time, replication attempts have yielded mixed results (Schneider et al. 2008; Iqbal et al. 2015), mechanistic mapping from sperm to specific tissues remains incomplete, and important questions about per-sperm penetrance, developmental dynamics, and causal sufficiency of DNA methylation versus other epigenetic layers are still unanswered.

Transgenerational Inheritance of DNA Methylation Acquired by CRISPR Editing

In order to directly test whether a DNA methylation change acquired in one generation can be inherited by subsequent generations, something that is difficult to resolve in naturally occurring, exposure-induced methylation changes, Takahashi et al. (2023) engineered stable methylation of two promoter-associated CpG islands (CGIs) of metabolism-related genes, *ankyrin repeat domain 26 (Ankrd26*) and *low-density lipoprotein receptor* (*Ldlr*), in mouse embryonic stem cells (ESCs). Using CRISPR/Cas9-mediated homologous recombination, they inserted a CpG-free ("CG-less") cassette into each CGI; such inserts were previously shown to trigger de novo methylation of the surrounding CGI. Once the CGI became methylated and silenced, the cassette was excised, leaving behind an otherwise intact sequence that remained stably methylated.

These DNA methylation–edited ESCs were injected into blastocysts to generate chimeric animals and, subsequently, DNA methylation–edited mouse lines. Takahashi et al. reported that promoter methylation and transcriptional silencing of *Ankrd26* and *Ldlr*, as well as associated repressive chromatin features and metabolic abnormalities, were maintained and transmitted across multiple generations (at least up to F4) in somatic tissues. At the same time, they observed that the targeted CGI methylation was erased in primordial germ cells (PGCs) and in germ cells during the normal waves of epigenetic reprogramming and then reestablished in the next generation at postimplantation stages. Thus, the methylation mark itself is not continuously maintained through the germline, yet the same CGI ultimately reacquires methylation in descendants, together with the metabolic phenotype.

These data suggest that a methylation state introduced at a defined locus in one generation can be maintained as a heritable "memory" across several generations and can drive phenotypic effects. At the same time, the fact that targeted CGI methylation is erased in PGCs implies that DNA methylation per se is unlikely to be the sole heritable signal. Some additional, as-yet-unknown mechanism must encode the information that instructs remethylation of the same CGI during embryonic development. Conceptually, this is important: It shows that the "memory" of an acquired methylation state can be transmitted even when the methylation mark itself is periodically removed. This could be a way for transgenerational methylation signals to be protected against reprogramming. However, the mechanism mediating such "memory" is unknown.

A major strength of this study is that the starting event is a well-defined, experimentally imposed methylation edit, in contrast to exposure studies, where DNA methylation changes are correlated with an environmental insult but may be secondary to other processes. Here, the direction of causality is clear at the outset: methylation and silencing are engineered first, and metabolic phenotypes follow. The identity of the affected CpG islands is known, the effect sizes are large, and their methylation state can be tracked across generations. On this basis, the authors argue that their work provides strong evidence that CGI methylation can be inherited across generations and can transmit a novel phenotype transgenerationally.

However, despite the elegance of the design, an alternative interpretation has been proposed. Sapozhnikov and Szyf (2024) reexamined the data and argued that the apparent heritability of the acquired methylation state may be driven by genetic rather than purely epigenetic changes introduced by the editing procedure itself. They point out that DNA methylation is highly dependent on local sequence context and highlight several potential genetic confounders: mutations at the cassette excision site, the use of a homologous recombination arm derived from a different mouse strain, and possible off-target Cas9 activity. Their reanalysis of genome-wide sequencing data identifies substantial sequence differences that could, in principle, explain the stable remethylation of the targeted CGI across generations (Sapozhnikov and Szyf 2024).

If the "memory" of the methylation state is in fact encoded primarily by these DNA sequence alterations, this would explain how methylation can be erased during germline reprogramming yet be reproducibly reestablished later in development and why methylation is "inherited" in somatic tissues but not maintained in sperm. In that scenario, DNA methylation would be a readout of an underlying genetic change rather than the primary heritable signal.

In summary, Takahashi et al.'s (2023) study offers the most controlled demonstration so far that a defined CGI methylation state, imposed in ESCs, can recur across multiple generations and be associated with a stable phenotype. At the same time, the plausible genetic confounds raised by Sapozhnikov and Szyf must be resolved before this model can be regarded as definitive proof that DNA methylation itself, independent of underlying sequence changes, is the transgenerationally heritable mark.

Although these findings suggest that detecting DNA methylation marks in sperm may not, by itself, be a decisive test for DNA methylation–driven transgenerational inheritance of traits, the possibility that the observed effects in this model are genetically mediated argues that we should *continue* to examine, with particular care, whether epigenetic inheritance can be mediated by stable DNA methylation changes in sperm.

4 The Role of DNA Methylation in Human Evolution and Speciation

4.1 *Is DNA Methylation a Mechanism of Evolution?*

DNA methylation at CG dinucleotides is heritable and can be faithfully replicated, and it can alter gene function without any underlying genetic change. In addition, the data discussed in the previous section suggest that environmental signals—whether physical or behavioral—can modify DNA methylation and that some of these effects may be transmitted across generations. The key question is whether DNA methylation can drive the evolution of new traits and new species on an evolutionary time scale or whether the effects described above mainly represent adaptations to changing environments over one or a few generations. In the latter case, DNA methylation would contribute to evolution indirectly, by increasing fitness and facilitating short-term adaptation, but would not itself be a primary driver of new, stable traits over deep evolutionary time. The appeal of DNA methylation as an evolutionary mechanism lies in the idea that it can be directed by the environment, allowing organisms to acquire traits without relying solely on natural selection—potentially a more targeted, efficient, and rapid process, much in the spirit of what Lamarck originally envisioned.

As discussed above, the main barrier to transgenerational inheritance of a newly acquired DNA methylation profile in mammals is the extensive reprogramming of DNA methylation that occurs twice: shortly after fertilization and again during germ cell development (in PGCs and during spermatogenesis/oogenesis) (Gkountela et al. 2015; Seisenberger et al. 2012). The canonical, near-global "two-wave" reprogramming appears to be a relatively late evolutionary innovation that is most fully developed in mammals (and, in modified forms, other vertebrates). This mechanism could have evolved to protect the germline from stable incorporation of changes induced by transient environmental conditions.

By contrast, many plants and several invertebrates (such as nematodes and insects) either lack global DNA methylation reprogramming or show more limited resetting, and there are clear examples in these taxa where epigenetic effects can persist for multiple generations (Feng et al. 2010; Xu et al. 2019). It is, therefore, plausible that epigenetic mechanisms may play a more prominent and direct role in evolutionary change in such organisms, whereas in mammals, the potential for DNA methylation to drive long-term evolutionary innovation may have been constrained as complex organisms evolved mechanisms to shield the germline from short-lived, environmentally induced changes. From this perspective, it may be advantageous not to fix, over many generations, a hyperresponsive stress phenotype via stable germline DNA methylation if the environmental trigger is transient.

At the same time, as discussed above, there is evidence that environmentally induced DNA methylation changes in sperm do occur in mammals in response to diverse signals, and cross-generational DNA methylation differences in peripheral tissues have been reported in humans as well. However, these effects tend to be modest in magnitude, often confounded by genetic and environmental variability, and the penetrance of the associated phenotypes, as well as their precise relationship to specific sperm methylation changes, remains poorly characterized.

One would anticipate that any DNA methylation change contributing to speciation would be robust, present in sperm, and either be widespread across the body or have a very predictable and stable effect on the downstream evolution of cell type–specific methylation profiles. Although demethylation in primordial germ cells (PGCs) is almost global, certain sequences remain persistently methylated. For example, specific families of repeat elements and transposons retain methylation in mouse PGCs (Seisenberger et al. 2012) and in human PGCs (Gkountela et al. 2015). In humans, the HERVK9-Int and HERVK11-Int subfamilies, as well as the transposon bodies of young, human-specific LINE-1 elements (L1HS) and their immediate ancestors, L1PA2 and L1PA3, remain persistently methylated (Gkountela et al. 2015). These sequences might provide a substrate for stable, locus-specific methylation states across generations (Fig. 2.1a).

In addition to retroelements such as SINE–VNTR–Alu (SVA), Tang et al. identified non-repetitive, single-copy CGI promoters in human and mouse PGCs that remain $\geq$30% methylated when most CGIs are nearly fully demethylated—these are termed "escapee loci" (Tang et al. 2015). They are enriched for genes linked to metabolic and neurological functions. Notably, these functions are the functions that would be anticipated to be the most important adaptive response to social and physical environmental signals.

4.2 DNA Methylation Differences Between Humans and Other Hominoids

The evolution of DNA methylation enzymes and machinery in humans is consistent with DNA methylation playing an important role in human evolution, particularly in relation to the complexity of the human brain. Early evidence that DNA methylation might contribute to primate evolution and especially to the emergence of Homo sapiens came from studies examining ancient DNA methylation profiles in bones from *Homo sapiens* and archaic humans, including *H. neanderthalensis* and Denisovans (Gokhman et al. 2014). These studies identified close to 2000 differentially methylated regions (DMRs) between modern humans and archaic humans. Notably, three DMRs were located within the HOXD locus and two others in putative enhancer regions 20 and 25 kb upstream of the HOXD cluster, which includes five genes (HOXD9–HOXD13). The HOXD cluster is involved in limb development, and interestingly, some of the distinctions between modern humans and archaic humans involve differences in limb size and structure. Importantly, DMRs were also enriched in genes implicated in neurological disease, suggesting that many of the genes differentially methylated in modern humans are involved in neurodevelopment (Gokhman et al. 2014). This is consistent with the idea that changes in DNA methylation during human evolution are closely related to brain evolution and increased cognitive capacity in humans.

Further support for the link between human-specific DNA methylation patterns and brain function comes from the observation that recently evolved human DMRs are enriched for genetic risk variants for schizophrenia (Banerjee et al. 2018). More generally, DNA methylation profiles appear tightly related to species-specific traits. For example, aspects of Denisovan anatomy have been reconstructed from DNA methylation maps (Gokhman et al. 2019). Moreover, human-specific DNA methylation profiles can shed light on anatomical differences between modern humans and archaic humans, such as differences in methylation in genes involved in vocal and facial anatomy in modern humans (Gokhman et al. 2020). Together, these findings provide strong evidence that changes in DNA methylation during human evolution are involved in establishing uniquely human traits and, thus, have been involved in speciation.

Since DNA methylation is known to be responsive to the environment, it is tempting to speculate that some of these evolutionary changes in DNA methylation were environmentally driven. It has even been proposed that aspects of past environmental conditions might be inferred from ancient DNA methylation maps (Gokhman et al. 2017). This raises the possibility that at least some of the DNA methylation differences between modern humans and archaic humans evolved in response to environmental pressures.

However, the central question is whether heritable DNA methylation differences were the primary driving force behind these evolutionary changes or whether the methylation differences were themselves largely downstream of genetic variation, amplifying the functional impact of DNA sequence changes and, thereby, increasing the selective advantage of particular alleles during natural selection. Indeed, many of the DMRs that distinguish modern and archaic humans overlap genomic regions that also differ in sequence between these groups (Gokhman et al. 2014).

4.3 Sperm Methylation in Humans and Other Primates and the DNA Methylation–Based Inheritance of Acquired Traits

If DNA methylation alone is responsible for maintaining specific methylation profiles across generations, we would expect to see evidence that such patterns are present in gametes (e.g., sperm) and are maintained independently of underlying genetic variation. When comparing archaic humans to modern humans, this is impossible to test directly because only bone and tooth material are available.

However, it *is* possible to address this question by comparing sperm methylation in humans and other primates (Molaro et al. 2011).

Molaro and colleagues showed that over ~6 million years of evolution, most methylation patterns in male germ cells are highly conserved between humans and chimpanzees. Nevertheless, they identified numerous differentially methylated regions (DMRs) between the two species despite the fact that they share >95% of DNA sequence identity. The most striking differences were found in retrotransposon subfamilies, particularly SINE–VNTR–Alu (SVA) elements, which showed strong species-specific differences in DNA methylation. The youngest SVA families (D–F) displayed the largest divergence and, importantly, were consistently hypomethylated in humans relative to chimpanzees. Across orthologous SVA loci, ~35% of human SVAs had methylation levels below 50%, compared with only about 6% in chimpanzee. Many SVA insertions are located near promoters; thus, SVA hypomethylation in humans could, in principle, lead to human-specific activation of nearby promoters. It is unclear whether hypomethylation arose along the human lineage or whether increased methylation was acquired along the chimpanzee lineage (Molaro et al. 2011).

There is very little primary sequence divergence between humans and chimpanzees within many of these SVA elements, yet their methylation levels differ markedly. More broadly, Molaro et al. (2011) reported multiple genomic regions in which the underlying DNA sequence is conserved but DNA methylation diverges, as well as regions where the most pronounced methylation differences coincide with the greatest sequence divergence. Long-term high methylation is known to promote CpG depletion (through deamination of 5mC), so shifts in methylation can feed back on local CpG density, while sequence changes can, in turn, alter the propensity for methylation. Indeed, regions that are strongly methylated in one species tend to show a marked loss of CpG dinucleotides, indicating that such changes can occur on relatively rapid evolutionary timescales. Taken together, these observations suggest that the genome and the methylome do not evolve independently: There is substantial cross talk, with sequence divergence shaping methylation patterns and, conversely, methylation history influencing CpG content and local mutation rates.

SVA elements that appear to have lost methylation in humans without major sequence alteration may represent some of the most compelling candidates for transgenerationally stable DNA methylation changes. Because these hypomethylated states are present in sperm and many SVA insertions lie near promoters, they could provide a route by which heritable DNA methylation differences alter gene regulation and contribute to evolutionary changes in genome function during speciation.

Beyond their striking species-specific hypomethylation in the human germline documented by Molaro et al. (2011), SVA elements also show functional relevance in somatic tissues and human disease. The clearest example comes from X-linked dystonia–parkinsonism (XDP), where a disease-causing SVA-F insertion in *TAF1* disrupts normal splicing and expression (Makino et al. 2007; Aneichyk et al. 2018). This demonstrates that an SVA insertion can act as a stable, allele-specific regulatory element, whose presence alters gene expression and contributes directly to human disease. Additional human disorders involve SVA insertions that alter gene regulation through promoter interference or splice disruption, as documented across a range of Mendelian diseases (Hancks and Kazazian Jr. 2016). Although detailed methylation maps of these disease-associated SVAs are limited, these cases collectively show that SVA elements can act as regulatory "hot spots," whose methylation state modulates gene expression. Taken together with their germline hypomethylation in humans, these findings support the view that SVA elements are a class of sequences capable of carrying heritable, species-specific, and functionally relevant DNA methylation signatures that may contribute to both disease susceptibility and evolutionary divergence.

It would be of particular interest to determine whether these unmethylated SVAs remain unmethylated throughout somatic differentiation or whether they influence the emergence of cell type–specific methylation patterns and gene expression programs. However, Molaro et al. (2011) did not systematically follow these methylation profiles across development beyond analyzing sperm and pluripotent cells, so their potential impact on tissue-specific methylomes remains to be explored.

Beyond germ cells, comparative methylome studies in somatic tissues of humans and great apes support the same overall picture. Genome-wide analyses of blood and multiple other tissues in humans, chimpanzees, gorillas, and orangutans show that most DNA methylation patterns are deeply conserved, but they also reveal sets of human-specific DMRs enriched near genes involved in development, immunity, and brain function (360 hypomethylated and 210 hypermethylated DMRs) (Hernando-Herraez et al. 2015; Hernando-Herraez et al. 2013). There is a significant positive relationship between the rate of coding sequence variation and alterations in promoter methylation. Additional evidence for interplay between genetics and DNA methylation comes from the enrichment of sequence differences within human DMRs at transcription factor–binding sites, suggesting that nucleotide changes that create or abolish such sites may precipitate the observed methylation differences. Moreover, acquisition of DNA hypermethylation on the human lineage is frequently coupled with rapid nucleotide evolution in the vicinity of these CpG sites, although it remains unclear which change precedes the other.

Importantly, 184 genes are perfectly conserved at the protein level between humans and chimpanzees yet show significant epigenetic differences between the two species, suggesting that divergence in epigenetic regulation may be an important determinant of species differences. This highlights the role of DNA methylation in evolution, irrespective of whether methylation changes are the primary trigger or secondary to sequence divergence. In either case, altered DNA methylation can provide functional consequences for underlying sequence changes and, thereby, influence how those variants are filtered by natural selection.

As shown by Molaro et al. (2011), in sperm, hominoid-specific retroelements, including SVA, contribute to species-specific methylation differences in somatic tissues (Hernando-Herraez et al. 2015, Hernando-Herraez et al. 2013).

These tissue-based data (Hernando-Herraez et al. 2013, 2015) complement the Molaro et al. sperm study (Molaro et al. 2020) by indicating that, against a largely conserved background, hundreds of loci have undergone coordinated, human-specific changes in DNA methylation, suggesting that DNA methylation is a force in speciation that may contribute to lineage-specific traits.

5 Conclusions: What Is the Role of DNA Methylation in Evolution?

5.1 Origins of DNA Methylation and Its Early Defensive Role

The DNA methylation machinery has been present since the last universal common ancestor (LUCA), and its methyl donor, S-adenosylmethionine (SAM), likely predates the emergence of the first living organisms (Laurino and Tawfik 2017). DNA methylation enzymes initially evolved to defend against invading DNA and preserve genomic integrity (Arber and Dussoix 1962). The early appearance of DNA methylation in evolution expanded the chemical information encoded in DNA by adding minor, modified bases to the four canonical nucleotides (Hotchkiss 1948). By decoupling DNA replication, which follows the Watson–Crick rules and is largely oblivious to DNA modifications, from the copying of DNA methylation patterns, which is governed by distinct enzymes and rules, evolution created a mechanism for conferring different identities on otherwise identical DNA sequences. Two DNA molecules can share the same sequence of the four bases yet carry distinct patterns of methylated bases. While DNA methylation was first used to protect "self-"DNA, it was later repurposed to regulate gene expression. This added an additional layer of information to the genome, enabling diverse gene expression programs in multicellular organisms and, thereby, amplifying the impact of genetic evolution. DNA methylation and other epigenetic mechanisms, not discussed in detail in this chapter, facilitated the emergence of multicellular organisms with differentiated cell types and functions. It is plausible that the transition from a purely "restriction–modification" function to gene silencing evolved initially to repress ectopic retroviral sequences and repetitive elements as an alternative means of maintaining genomic integrity (Bestor 1990). Methylation of repetitive elements is widespread across evolution and relies on the same basic principle as restriction–modification systems: DNA methylation alters the chemical properties of recognition sequences for DNA-binding proteins. Initially, these proteins were restriction enzymes; later, they included transcription factors. In both cases, bacteriophages in bacteria and retroviral elements in eukaryotes, restriction enzymes or transcription factors distinguish self from nonself through differences in DNA methylation profiles.

5.2 *Repurposing for Gene Regulation, Multicellularity, and Brain Evolution*

This mechanism, originally used to silence foreign elements via DNA methylation, was repurposed to silence endogenous genes in a cell type–specific manner, enabling the development of multicellular organisms with highly differentiated tissues and organs. In turn, this regulatory versatility facilitated the evolution of complex organisms, such as humans, and, in particular, supported the extraordinary diversification and complexity of the human brain. By adding a heritable epigenetic layer on top of the nucleotide sequence, DNA methylation amplified the informational capacity of the genome and, thereby, expanded its evolutionary potential.

The evolution and diversification of DNA methylation enzymes, including gene duplications that gave rise to DNMT3A, have been especially important for the emergence of the rich landscape of neuron subtype–specific and glial methylation patterns in the mammalian brain (De Mendoza et al. 2021). DNA methylation has thus been central to the evolution of the human brain, and consistent with this, the disruption of DNA methylation machinery—such as mutations in DNMT3A or in the methyl CpG–binding protein MeCP2—is associated with severe neurodevelopmental and neuropsychiatric disorders (Amir et al. 1999; Tatton-Brown et al. 2014).

5.3 *Environmentally Responsive Methylation and Epigenetic Memory*

The ability of DNA methylation to distinguish between otherwise identical DNA sequences was later repurposed again to create an "experiential identity," providing the genome with a mechanism for relatively rapid adjustment to changing social, biological, and physical environments. Environmental signals can act through hormonal and neural pathways, activating intracellular signaling cascades and transcription factors that recruit or modulate the DNA methylation machinery and associated chromatin regulators (Weaver et al. 2004). In this way, gene expression programs can be durably adjusted in response to environmental conditions.

This kind of epigenetic programming should be distinguished from short-term, reversible physiological adaptations mediated by hormones such as insulin or glucocorticoids, which typically trigger transient gene expression changes that fade once the signal is removed. In contrast, environmentally induced changes in DNA methylation can be long-lasting, persisting across the lifespan of a cell lineage and, in some cases, across one or even several generations. The discovery that environmental exposures can induce stable epigenetic changes, with evidence for cross-generational effects, has revived interest in ideas reminiscent of Lamarck's inheritance of acquired characteristics and the environmentally influenced component of evolution that complements, rather than replaces, classical Darwinian selection on genetic variation (Szyf 2014). DNA methylation is a particularly attractive candidate mechanism for such processes because it is heritable through cell division, can modulate gene function without altering the underlying DNA sequence, and is responsive to environmental cues.

5.4 Reprogramming Barriers and Escapees

This raises the question of whether DNA methylation can act as a primary driver in the emergence of new traits and participate in speciation. If so, it would have major implications for how we understand the timescales of trait evolution, the likelihood of "directed evolution," and the nature of adaptive responses to the environment.

The main obstacle to accepting this hypothesis is the extensive reprogramming of DNA methylation that occurs during early embryogenesis and gametogenesis in mammals. Notably, such global reprogramming appears to have arisen relatively late in evolution and is not a universal feature across taxa; for example, plants do not undergo the same kind of genome-wide erasure and resetting seen in mammals (Xu et al. 2019). It is tempting to speculate that this reprogramming evolved, at least in part, to prevent sporadic environmental events from permanently altering DNA methylation in the germline and, thereby, becoming fixed in future generations. Indeed, environmentally driven epigenetic changes could become maladaptive if environmental conditions later change.

However, it is also clear that some genomic sequences escape this reprogramming, most notably certain retrotransposons and endogenous retroviral elements, including relatively "young" or species-specific repetitive elements in humans and other mammals. Evidence for cross-generational transmission of DNA methylation changes in response to social stress, toxins, or diet is intriguing but not yet robust enough to conclude that the specific differentially methylated sites observed across generations are simply "escapees" of the reprogramming process (Tang et al. 2015). Moreover, it remains unclear whether these methylation differences are primary causal drivers of new traits or whether the apparent transmission of epigenetic effects to the second generation reflects direct copying of DNA methylation patterns by DNA methyltransferases. Experimental studies in several model systems suggest that various forms of RNA, including small noncoding RNAs, may act as primary signals in transgenerational inheritance, with DNA methylation differences emerging secondarily as part of a broader epigenetic memory system.

5.5 Behavioral Transmission as an Alternative Intergenerational Mechanism

I would propose that the DNA methylation responses to environmental exposures reviewed in this chapter, including those observed across generations, evolved to increase the preparedness of offspring for anticipated social and environmental stressors. These responses appear to be limited in temporal scope, typically affecting one or only a few generations. Because many environmental challenges are transient and heterogeneous across individuals, it is plausible that mechanisms evolved to restrict the duration of such environmentally induced epigenetic effects, preventing them from becoming permanently fixed in the lineage.

A different paradigm of environmental influence on DNA methylation—one that is inherited behaviorally rather than through the germline—is the effect of maternal care described above. In this case, maternal behavior programs epigenetic changes in the offspring's brain (and, in some cases, other tissues). These epigenetic and phenotypic outcomes can be maintained across generations because maternal behavior itself is transmitted behaviorally: offspring that experience a particular style of care tend to exhibit similar maternal behaviors toward their own young. As long as the behavioral pattern is maintained, the associated DNA methylation alterations are reestablished in each generation. Here, behavior serves as the "vector of inheritance," with DNA methylation acting as a mediator of phenotypic change in the brain and other tissues, rather than germline methylation directly transmitting the trait (Fig. 2.2).

Such a mechanism helps ensure that methylation patterns and resulting phenotypes remain relevant to prevailing environmental conditions. It confers a selective advantage by allowing phenotypes to be tuned to the environment while avoiding the long-term maladaptive consequences of fixing an outdated response once conditions change. It is, therefore, tempting to speculate that both germline methylation reprogramming and behaviorally mediated transmission of epigenetic states evolved as "short-term" adaptive strategies on an evolutionary timescale. These forms of inheritance do not represent classic Lamarckian evolution but rather time-limited environmental adaptations suited to conditions that fluctuate across individuals and are not pervasive or permanent.

5.6 Disentangling Genetic Variation and Epigenetic Change in Human Evolution

The most robust data, from an evolutionary perspective, come from species-specific DNA methylation differences (DMRs) between modern and archaic humans (Gokhman et al. 2014). Several of these DMRs are linked to clear phenotypic differences. Because sperm from archaic humans is obviously unavailable, these studies could not directly assess germline methylation, and it remains uncertain whether the observed DMRs are inherited through the germline or are established later in development.

An alternative and, often, favored hypothesis is that many of these DMRs arise secondarily from underlying genetic variation. Genetic differences are well-known to drive DNA methylation differences—for example, sequence variants that alter transcription factor–binding sites can change factor occupancy and, thereby, affect local methylation states (Martin-Trujillo et al. 2020; Villicana and Bell 2021). Indeed, many, though not all, of the identified DMRs are associated with nearby sequence variants. In such cases, the emergence of these DMRs reflects a conventional evolutionary process in which genetic variants are subject to selection, and the resulting changes in DNA methylation are downstream functional consequences.

However, DNA methylation changes may markedly amplify the phenotypic impact of such variants by reshaping gene regulation, thereby enhancing the selective advantage (or disadvantage) associated with the underlying genetic change. Thus, a likely major role of DNA methylation in evolution is to magnify the functional consequences of genetic variants and their contribution to fitness. It is conceivable that a relatively small number of sequence variants can propagate their effects across a much broader methylation landscape, extending their influence over multiple genes and pathways.

Not all DMRs between modern and archaic humans are associated with identifiable genetic differences. However, in the absence of methylation data from sperm or early germ cells, it is difficult to determine whether these DMRs are directly inherited via semiconservative copying of DNA methylation patterns across generations; whether they arise secondarily from other, as yet undetected, genetic variants; or whether they are maintained across generations by additional epigenetic mechanisms.

5.7 *Bidirectional Interaction Between DNA Methylation and Genetic Mutation*

In evolution, DNA methylation variation and sequence variation are tightly intertwined. It is well-established that genetic variation can precipitate changes in DNA methylation, for example, by altering transcription factor–binding sites or chromatin context. Conversely, DNA methylation can influence the rate and pattern of genetic mutation. The best-documented example is the spontaneous deamination of methylated cytosines to thymidines such that hypermethylation accelerates the loss of CpG dinucleotides and their conversion to TpG/CpA, which in turn reshapes the local methylation landscape (Bird 1980).

The most intriguing open question is whether environmentally induced DMRs can sometimes be stably transmitted across generations in a manner that directly contributes to the emergence of new traits or even, in rare cases, to speciation.

5.8 Retroelements as Epigenetic Substrates for Evolution

As discussed earlier, one of the earliest roles of DNA methylation in eukaryotes was likely the silencing of repetitive and retroviral elements. Importantly, some retrotransposons—particularly more recent elements that emerged in primates, such as SINE–VNTR–Alu (SVA) elements—are at least partially protected from genome-wide demethylation during gametogenesis and early embryogenesis. Subsets of L1, Alu, SVA, and related elements can retain methylation ("escape" reprogramming), providing a potential route for methylation patterns at these loci to be transmitted through the germline. Moreover, SVAs frequently insert within or near genes, where they can influence gene expression. In this sense, SVAs and related retroelements are strong candidates for DNA methylation–based contributions to the acquisition of new traits.

Consistent with this idea, several studies have shown that some of the most marked and persistent species-specific DMRs between humans and chimpanzees occur at SVA and other retrotransposon loci, including in sperm (Molaro et al. 2011). Many SVA insertions lie close to genes, including genes implicated in brain function and neurological disease, supporting the potential role of these elements in shaping primate brain phenotypes. Although these findings do not demonstrate causality, they highlight SVAs as plausible substrates through which DNA methylation and retrotransposition could interact to influence evolutionary change.

Can the environment direct the acquisition of a new trait by inducing heritable, DNA methylation–based changes? The data reviewed so far do not provide conclusive evidence, but they are consistent with several models that merit further investigation. One model involves environmentally contingent DNA methylation changes that depend on preexisting genetic variants. The *Fkbp5* gene provides a well-studied example: A specific risk allele interacts with childhood adversity to produce allele-specific DNA demethylation and altered stress responsiveness, with phenotypic consequences for PTSD risk (Klengel et al. 2013). Analogously, genetic variants that arose during human evolution might predispose certain loci to undergo environmentally dependent methylation changes. If such methylation changes have functional effects that confer a selective advantage under particular environmental conditions, the underlying genetic variants could be favored by natural selection. In this scenario, DNA methylation is a key functional mediator of selection, but the initial trigger remains genetic variation rather than a purely Lamarckian, environment-driven acquisition of traits.

A second, more speculative model is that heritable changes in DNA methylation leading to new traits in response to environmental cues might involve SVAs and other repetitive elements that escape reprogramming during gametogenesis. In principle, environmental signals could alter methylation at SVA loci, modulating their regulatory activity; SVA insertions located near promoters or enhancers could then change nearby gene expression in a way that is transmissible through the germline if the relevant methylation marks persist or are reliably reconstructed in each generation. The enrichment of persistent interspecies DMRs at SVA loci between humans and chimpanzees is consistent with the idea that these elements can carry stable, lineage-specific methylation signatures (Molaro et al. 2011), but it does not yet demonstrate that such signatures are originally established by environmental exposures.

It is therefore intriguing but still highly speculative to propose that environmentally guided DNA methylation changes in retroelements might, in rare cases, feed back onto the genome via new insertions or stable regulatory changes, thereby approximating a "Lamarckian-like" mechanism in which experience shapes heritable variation. Any such process would operate against a background of strong constraints imposed by germline reprogramming and classical Darwinian selection. Given that humans and chimpanzees share over 95% of sequence identity at the nucleotide level, with only ~1–3% of the genome differing by single-nucleotide changes and indels, it is tempting to speculate that differences in the regulation and activity of retroelements, together with their associated methylation landscapes, could have contributed disproportionately to the qualitative divergence between the species. This hypothesis, however, remains to be tested. It will require systematic mapping of DNA methylation and retroelement activity across multiple primate species, developmental stages, tissue types, and, critically, germ cells, as well as functional studies to establish causality.

5.9 Summary

In summary, DNA methylation is tightly intertwined with genetic change during evolution and speciation. An intriguing possibility is that DNA methylation amplifies the functional impact of genetic variants, thereby accelerating the emergence of new traits. In addition, repetitive elements, particularly retrotransposons, may have provided a substrate for environmentally responsive, heritable epigenetic modifications, allowing DNA methylation to mediate a limited form of environmentally influenced trait acquisition.

References

Alexander KA, Wang X, Shibata M, Clark AG, Garcia-Garcia MJ (2015) TRIM28 controls genomic imprinting through distinct mechanisms during and after early genome-wide reprogramming. Cell Rep 13:1194–1205

Amir RE, Van Den Veyver IB, Wan M, Tran CQ, Francke U, Zoghbi HY (1999) Rett syndrome is caused by mutations in X-linked MECP2, encoding methyl-CpG-binding protein 2. Nat Genet 23:185–188

Aneichyk T, Hendriks WT, Yadav R, Shin D, Gao D, Vaine CA, Collins RL, Domingo A, Currall B, Stortchevoi A, Multhaupt-Buell T, Penney EB, Cruz L, Dhakal J, Brand H, Hanscom C, Antolik C, Dy M, Ragavendran A, Underwood J, Cantsilieris S, Munson KM, Eichler EE, Acuna P, Go C, Jamora RDG, Rosales RL, Church DM, Williams SR, Garcia S,

Klein C, Muller U, Wilhelmsen KC, Timmers HTM, Sapir Y, Wainger BJ, Henderson D, Ito N, Weisenfeld N, Jaffe D, Sharma N, Breakefield XO, Ozelius LJ, Bragg DC, Talkowski ME (2018) Dissecting the causal mechanism of X-linked dystonia-parkinsonism by integrating genome and transcriptome assembly. Cell 172:897–909 e21

Anton BP, Roberts RJ (2023) A survey of archaeal restriction-modification systems. Microorganisms 11(10):2424

Anway MD, Cupp AS, Uzumcu M, Skinner MK (2005) Epigenetic transgenerational actions of endocrine disruptors and male fertility. Science 308:1466–1469

Arber W, Dussoix D (1962) Host specificity of DNA produced by Escherichia coli: I. Host controlled modification of bacteriophage λ. J Mol Biol 5:18–36

Arita K, Ariyoshi M, Tochio H, Nakamura Y, Shirakawa M (2008) Recognition of hemi-methylated DNA by the SRA protein UHRF1 by a base-flipping mechanism. Nature 455:818–821

Banerjee N, Polushina T, Bettella F, Giddaluru S, Steen VM, Andreassen OA, Le Hellard S (2018) Recently evolved human-specific methylated regions are enriched in schizophrenia signals. BMC Evol Biol 18:63

Bell AC, Felsenfeld G (2000) Methylation of a CTCF-dependent boundary controls imprinted expression of the Igf2 gene. Nature 405:482–485

Ben Maamar M, Sadler-Riggleman I, Beck D, Mcbirney M, Nilsson E, Klukovich R, Xie Y, Tang C, Yan W, Skinner MK (2018) Alterations in sperm DNA methylation, non-coding RNA expression, and histone retention mediate vinclozolin-induced epigenetic transgenerational inheritance of disease. Environ Epigenet 4:dvy010

Bestor TH (1990) Dna methylation: evolution of a bacterial immune function into a regulator of gene expression and genome structure in higher eukaryotes. Philos Trans R Soc Lond Ser B Biol Sci 326:179–187

Bestor TH, Ingram VM (1983) Two DNA methyltransferases from murine erythroleukemia cells: purification, sequence specificity, and mode of interaction with DNA. Proc Natl Acad Sci USA 80:5559–5563

Bewick AJ, Vogel KJ, Moore AJ, Schmitz RJ (2017) Evolution of DNA methylation across insects. Mol Biol Evol 34:654–665

Bilinovich SM, Chhetri SB, Mitchell JT, Leighton GO, Jdanov V, Zieba JK, Cook TW, Engel KL, Vaughan RM, Bhatti H, Heeringa SL, Wilstermann AM, Vanderweele S, Das AS, Sherry EC, Duncan SA, Uhl KL, Hinds DC, Mackiewicz M, Meadows SK, Newberry KM, Partridge EC, Cline SG, Shepard CJ, Smith DC, Bradley A, Vogt D, Campbell DB, Pozzo-Miller L, Bupp CP, Goodyke AJ, Rajasekaran S, Knickmeyer RC, Blair B, Krawczyk CM, Triche TJ Jr, Myers RM, Mendenhall EM, Williams DC Jr, Prokop JW (2025) Methyl-CpG-binding domain as a protein interaction partner in promoter regulation and neurodevelopment through evolutionary expanded entanglement. Am J Physiol Cell Physiol 329:C848–C867

Bird AP (1980) DNA methylation and the frequency of CpG in animal DNA. Nucleic Acids Res 8: 1499–1504

Bird A (2002) DNA methylation patterns and epigenetic memory. Genes Dev 16:6–21

Bohm S, Frishman D, Mewes HW (1997) Variations of the C2H2 zinc finger motif in the yeast genome and classification of yeast zinc finger proteins. Nucleic Acids Res 25:2464–2469

Bourc'his D, Xu GL, Lin CS, Bollman B, Bestor TH (2001) Dnmt3L and the establishment of maternal genomic imprints. Science 294:2536–2539

Burkhardt RW Jr (2013) Lamarck, evolution, and the inheritance of acquired characters. Genetics 194:793–805

Cedar H, Bergman Y (2011) Epigenetics of haematopoietic cell development. Nat Rev Immunol 11: 478–488

Champagne FA, Weaver IC, Diorio J, Dymov S, Szyf M, Meaney MJ (2006) Maternal care associated with methylation of the estrogen receptor-alpha1b promoter and estrogen receptor-alpha expression in the medial preoptic area of female offspring. Endocrinology 147:2909–2915

Chiang PK, Chamberlin ME, Nicholson D, Soubes S, Su X, Subramanian G, Lanar DE, Prigge ST, Scovill JP, Miller LH, Chou JY (1999) Molecular characterization of plasmodium falciparum S-adenosylmethionine synthetase. Biochem J 344(Pt 2):571–576
Chiu CP, Blau HM (1985) 5-Azacytidine permits gene activation in a previously noninducible cell type. Cell 40:417–424
Comb M, Goodman HM (1990) CpG methylation inhibits proenkephalin gene expression and binding of the transcription factor Ap-2. Nucleic Acids Res 18:3975–3982
Constancia M, Kelsey G, Reik W (2004) Resourceful imprinting. Nature 432:53–57
Creusot F, Acs G, Christman JK (1982) Inhibition of DNA methyltransferase and induction of Friend erythroleukemia cell differentiation by 5-azacytidine and 5-aza-2′-deoxycytidine. J Biol Chem 257:2041–2048
D'alessio AC, Weaver IC, Szyf M (2007) Acetylation-induced transcription is required for active DNA demethylation in methylation-silenced genes. Mol Cell Biol 27:7462–7474
Danna K, Nathans D (1971) Specific cleavage of simian virus 40 DNA by restriction endonuclease of hemophilus influenzae. Proc Natl Acad Sci USA 68:2913–2917
Darwin C (1859) On the origin of species by means of natural selection or the preservation of favoured races in the struggle for life. John Murray, London
De Mendoza A, Bonnet A, Vargas-Landin DB, Ji N, Li H, Yang F, Li L, Hori K, Pflueger J, Buckberry S, Ohta H, Rosic N, Lesage P, Lin S, Lister R (2018) Author correction: recurrent acquisition of cytosine methyltransferases into eukaryotic retrotransposons. Nat Commun 9: 1811
De Mendoza A, Hatleberg WL, Pang K, Leininger S, Bogdanovic O, Pflueger J, Buckberry S, Technau U, Hejnol A, Adamska M, Degnan BM, Degnan SM, Lister R (2019) Convergent evolution of a vertebrate-like methylome in a marine sponge. Nat Ecol Evol 3:1464–1473
De Mendoza A, Lister R, Bogdanovic O (2020) Evolution of DNA methylome diversity in eukaryotes. J Mol Biol 432:1687–1705
De Mendoza A, Poppe D, Buckberry S, Pflueger J, Albertin CB, Daish T, Bertrand S, De La Calle-Mustienes E, Gomez-Skarmeta JL, Nery JR, Ecker JR, Baer B, Ragsdale CW, Grutzner F, Escriva H, Venkatesh B, Bogdanovic O, Lister R (2021) The emergence of the brain non-CpG methylation system in vertebrates. Nat Ecol Evol 5:369–378
Dias BG, Ressler KJ (2014) Parental olfactory experience influences behavior and neural structure in subsequent generations. Nat Neurosci 17:89–96
Domcke S, Bardet AF, Adrian Ginno P, Hartl D, Burger L, Schubeler D (2015) Competition between DNA methylation and transcription factors determines binding of NRF1. Nature 528: 575–579
Du J, Johnson LM, Jacobsen SE, Patel DJ (2015a) DNA methylation pathways and their crosstalk with histone methylation. Nat Rev Mol Cell Biol 16:519–532
Du Q, Luu PL, Stirzaker C, Clark SJ (2015b) Methyl-CpG-binding domain proteins: readers of the epigenome. Epigenomics 7:1051–1073
Ebbs ML, Bender J (2006) Locus-specific control of DNA methylation by the Arabidopsis SUVH5 histone methyltransferase. Plant Cell 18:1166–1176
Esteve PO, Chin HG, Smallwood A, Feehery GR, Gangisetty O, Karpf AR, Carey MF, Pradhan S (2006) Direct interaction between DNMT1 and G9a coordinates DNA and histone methylation during replication. Genes Dev 20:3089–3103
Felitti VJ, Anda RF, Nordenberg D, Williamson DF, Spitz AM, Edwards V, Koss MP, Marks JS (1998) Relationship of childhood abuse and household dysfunction to many of the leading causes of death in adults. The Adverse Childhood Experiences (ACE) study. Am J Prev Med 14: 245–258
Feng S, Jacobsen SE, Reik W (2010) Epigenetic reprogramming in plant and animal development. Science 330:622–627
Filion GJ, Zhenilo S, Salozhin S, Yamada D, Prokhortchouk E, Defossez PA (2006) A family of human zinc finger proteins that bind methylated DNA and repress transcription. Mol Cell Biol 26:169–181

Fisher R (1930) The genetical theory of natural selection. Clarendon Press, Oxford, UK
Francis D, Diorio J, Liu D, Meaney MJ (1999) Nongenomic transmission across generations of maternal behavior and stress responses in the rat. Science 286:1155–1158
Franklin TB, Russig H, Weiss IC, Graff J, Linder N, Michalon A, Vizi S, Mansuy IM (2010) Epigenetic transmission of the impact of early stress across generations. Biol Psychiatry 68: 408–415
Franklin A, Davies JP, Clifton NE, Blake GET, Bamford R, Walker EM, Chioza B, Frith M, APEX Consortium, Youth-GEMs Consortium, Burrage J, Owens N, Prabhakar S, Dempster E, Hannon E, Mill J (2025) Cell-type-specific DNA methylation dynamics in the prenatal and postnatal human cortex. Cell Genom 5(12):101010
Gapp K, Jawaid A, Sarkies P, Bohacek J, Pelczar P, Prados J, Farinelli L, Miska E, Mansuy IM (2014) Implication of sperm RNAS in transgenerational inheritance of the effects of early trauma in mice. Nat Neurosci 17:667–669
Gkountela S, Zhang KX, Shafiq TA, Liao WW, Hargan-Calvopina J, Chen PY, Clark AT (2015) DNA demethylation dynamics in the Human prenatal germline. Cell 161:1425–1436
Gokhman D, Lavi E, Prufer K, Fraga MF, Riancho JA, Kelso J, Paabo S, Meshorer E, Carmel L (2014) Reconstructing the DNA methylation maps of the Neandertal and the Denisovan. Science 344:523–527
Gokhman D, Malul A, Carmel L (2017) Inferring past environments from ancient Epigenomes. Mol Biol Evol 34:2429–2438
Gokhman D, Mishol N, De Manuel M, De Juan D, Shuqrun J, Meshorer E, Marques-Bonet T, Rak Y, Carmel L (2019) Reconstructing Denisovan anatomy using DNA methylation maps. Cell 179:180–192 e10
Gokhman D, Nissim-Rafinia M, Agranat-Tamir L, Housman G, Garcia-Perez R, Lizano E, Cheronet O, Mallick S, Nieves-Colon MA, Li H, Alpaslan-Roodenberg S, Novak M, Gu H, Osinski JM, Ferrando-Bernal M, Gelabert P, Lipende I, Mjungu D, Kondova I, Bontrop R, Kullmer O, Weber G, Shahar T, Dvir-Ginzberg M, Faerman M, Quillen EE, Meissner A, Lahav Y, Kandel L, Liebergall M, Prada ME, Vidal JM, Gronostajski RM, Stone AC, Yakir B, Lalueza-Fox C, Pinhasi R, Reich D, Marques-Bonet T, Meshorer E, Carmel L (2020) Differential DNA methylation of vocal and facial anatomy genes in modern humans. Nat Commun 11:1189
Gruenbaum Y, Cedar H, Razin A (1982) Substrate and sequence specificity of a eukaryotic DNA methylase. Nature 295:620–622
Guerrero-Bosagna C, Settles M, Lucker B, Skinner MK (2010) Epigenetic transgenerational actions of vinclozolin on promoter regions of the sperm epigenome. PLoS One 5:e13100
Guilliere F, Danioux C, Jaubert C, Desnoues N, Delepierre M, Prangishvili D, Sezonov G, Guijarro JI (2013) Solution structure of an archaeal DNA binding protein with an eukaryotic zinc finger fold. PLoS One 8:e52908
Gurdon JB (1962) The developmental capacity of nuclei taken from intestinal epithelium cells of feeding tadpoles. Development 10:622–640
Han T, Yamada-Mabuchi M, Zhao G, Li L, Liu G, Ou HY, Deng Z, Zheng Y, He X (2015) Recognition and cleavage of 5-methylcytosine DNA by bacterial SRA-HNH proteins. Nucleic Acids Res 43:1147–1159
Hancks DC, Kazazian HH Jr (2016) Roles for retrotransposon insertions in human disease. Mob DNA 7:9
Hark AT, Schoenherr CJ, Katz DJ, Ingram RS, Levorse JM, Tilghman SM (2000) CTCF mediates methylation-sensitive enhancer-blocking activity at the H19/Igf2 locus. Nature 405:486–489
Hashimoto H, Horton JR, Zhang X, Cheng X (2009) UHRF1, a modular multi-domain protein, regulates replication-coupled crosstalk between DNA methylation and histone modifications. Epigenetics 4:8–14
Hata K, Okano M, Lei H, Li E (2002) Dnmt3l cooperates with the Dnmt3 family of de novo DNA methyltransferases to establish maternal imprints in mice. Development 129:1983–1993

Hendrich B, Bird A (1998) Identification and characterization of a family of mammalian methyl-CpG binding proteins. Mol Cell Biol 18:6538–6547

Hendrich B, Guy J, Ramsahoye B, Wilson VA, Bird A (2001) Closely related proteins MBD2 and MBD3 play distinctive but interacting roles in mouse development. Genes Dev 15:710–723

Hernando-Herraez I, Prado-Martinez J, Garg P, Fernandez-Callejo M, Heyn H, Hvilsom C, Navarro A, Esteller M, Sharp AJ, Marques-Bonet T (2013) Dynamics of DNA methylation in recent human and great ape evolution. PLoS Genet 9:e1003763

Hernando-Herraez I, Heyn H, Fernandez-Callejo M, Vidal E, Fernandez-Bellon H, Prado-Martinez-J, Sharp AJ, Esteller M, Marques-Bonet T (2015) The interplay between DNA methylation and sequence divergence in recent human evolution. Nucleic Acids Res 43:8204–8214

Hertzman C (1999) The biological embedding of early experience and its effects on health in adulthood. Ann N Y Acad Sci 896:85–95

Hertzman C, Wiens M (1996) Child development and long-term outcomes: a population health perspective and summary of successful interventions. Soc Sci Med 43:1083–1095

Horsthemke B (2018) A critical view on transgenerational epigenetic inheritance in humans. Nat Commun 9:2973

Hotchkiss RD (1948) The quantitative separation of purines, pyrimidines, and nucleosides by paper chromatography. J Biol Chem 175:315–332

Iqbal K, Tran DA, Li AX, Warden C, Bai AY, Singh P, Wu X, Pfeifer GP, Szabo PE (2015) Deleterious effects of endocrine disruptors are corrected in the mammalian germline by epigenome reprogramming. Genome Biol 16:59

Jablonka E, Lamb MJ (2005) Evolution in four dimensions: genetic, epigenetic, behavioral, and symbolic variation in the history of life. MIT Press, Cambridge, MA

Johnson TB, Coghill RD (1925) Researches on pyrimidines. C111. The discovery of 5-methyl-cytosine in tuberculinic acid, the nucleic acid of the tubercle bacillus1. J Am Chem Soc 47: 2838–2844

Johnson LM, Bostick M, Zhang X, Kraft E, Henderson I, Callis J, Jacobsen SE (2007) The SRA methyl-cytosine-binding domain links DNA and histone methylation. Curr Biol 17:379–384

Jones PA, Taylor SM (1980) Cellular differentiation, cytidine analogs and DNA methylation. Cell 20:85–93

Jurkowski TP, Jeltsch A (2011) On the evolutionary origin of eukaryotic DNA methyltransferases and Dnmt2. PLoS One 6:e28104

Juttermann R, Li E, Jaenisch R (1994) Toxicity of 5-aza-2′-deoxycytidine to mammalian cells is mediated primarily by covalent trapping of DNA methyltransferase rather than DNA demethylation. Proc Natl Acad Sci USA 91:11797–11801

Killian JK, Byrd JC, Jirtle JV, Munday BL, Stoskopf MK, Macdonald RG, Jirtle RL (2000) M6P/IGF2R imprinting evolution in mammals. Mol Cell 5:707–716

King SE, Mcbirney M, Beck D, Sadler-Riggleman I, Nilsson E, Skinner MK (2019) Sperm epimutation biomarkers of obesity and pathologies following DDT induced epigenetic transgenerational inheritance of disease. Environ Epigenet 5:dvz008

Klengel T, Mehta D, Anacker C, Rex-Haffner M, Pruessner JC, Pariante CM, Pace TW, Mercer KB, Mayberg HS, Bradley B, Nemeroff CB, Holsboer F, Heim CM, Ressler KJ, Rein T, Binder EB (2013) Allele-specific FKBP5 DNA demethylation mediates gene-childhood trauma interactions. Nat Neurosci 16:33–41

Klengel T, Morin EL, Howell BR, Bramlett S, Guzman D, Meyer JS, Ressler KJ, Sanchez MM (2019) Molecular and behavioral effects of infant maltreatment across generations in rhesus monkeys. bioRxiv:791152

Kozbial PZ, Mushegian AR (2005) Natural history of S-adenosylmethionine-binding proteins. BMC Struct Biol 5:19

Krishna SS, Majumdar I, Grishin NV (2003) Structural classification of zinc fingers: survey and summary. Nucleic Acids Res 31:532–550

Lamarck J (1809/1963) Zoological philosophy: an exposition with regard to the natural history of animals. Hafner, New York, NY

Lardenoije R, Smulders M, Morin EL, Howell BR, Guzman D, Meyer JS, Ressler KJ, Sanchez M, Klengel T (2025) A cross-generational methylomic signature of infant maltreatment in newborn rhesus macaques. Biol Psychiatry 98:767–778

Laurino P, Tawfik DS (2017) Spontaneous emergence of S-Adenosylmethionine and the evolution of methylation. Angew Chem Int Ed Engl 56:343–345

Le-Scherban F, Wang X, Boyle-Steed KH, Pachter LM (2018) Intergenerational associations of parent adverse childhood experiences and child health outcomes. Pediatrics 141(6):e20174274

Lister R, Pelizzola M, Dowen RH, Hawkins RD, Hon G, Tonti-Filippini J, Nery JR, Lee L, Ye Z, Ngo QM, Edsall L, Antosiewicz-Bourget J, Stewart R, Ruotti V, Millar AH, Thomson JA, Ren B, Ecker JR (2009) Human DNA methylomes at base resolution show widespread epigenomic differences. Nature 462:315–322

Lister R, Mukamel EA, Nery JR, Urich M, Puddifoot CA, Johnson ND, Lucero J, Huang Y, Dwork AJ, Schultz MD, Yu M, Tonti-Filippini J, Heyn H, Hu S, Wu JC, Rao A, Esteller M, He C, Haghighi FG, Sejnowski TJ, Behrens MM, Ecker JR (2013) Global epigenomic reconfiguration during mammalian brain development. Science 341:1237905

Liu D, Diorio J, Tannenbaum B, Caldji C, Francis D, Freedman A, Sharma S, Pearson D, Plotsky PM, Meaney MJ (1997) Maternal care, hippocampal glucocorticoid receptors, and hypothalamic-pituitary-adrenal responses to stress. Science 277:1659–1662

Liu J, Hu H, Panserat S, Marandel L (2020) Evolutionary history of DNA methylation related genes in chordates: new insights from multiple whole genome duplications. Sci Rep 10:970

Loyfer N, Magenheim J, Peretz A, Cann G, Bredno J, Klochendler A, Fox-Fisher I, Shabi-Porat S, Hecht M, Pelet T, Moss J, Drawshy Z, Amini H, Moradi P, Nagaraju S, Bauman D, Shveiky D, Porat S, Dior U, Rivkin G, Or O, Hirshoren N, Carmon E, Pikarsky A, Khalaileh A, Zamir G, Grinbaum R, Abu Gazala M, Mizrahi I, Shussman N, Korach A, Wald O, Izhar U, Erez E, Yutkin V, Samet Y, Rotnemer Golinkin D, Spalding KL, Druid H, Arner P, Shapiro AMJ, Grompe M, Aravanis A, Venn O, Jamshidi A, Shemer R, Dor Y, Glaser B, Kaplan T (2023) A DNA methylation atlas of normal human cell types. Nature 613:355–364

Luo C, Keown CL, Kurihara L, Zhou J, He Y, Li J, Castanon R, Lucero J, Nery JR, Sandoval JP, Bui B, Sejnowski TJ, Harkins TT, Mukamel EA, Behrens MM, Ecker JR (2017) Single-cell methylomes identify neuronal subtypes and regulatory elements in mammalian cortex. Science 357:600–604

Luria SE, Human ML (1952) A nonhereditary, host-induced variation of bacterial viruses. J Bacteriol 64:557–569

Makino S, Kaji R, Ando S, Tomizawa M, Yasuno K, Goto S, Matsumoto S, Tabuena MD, Maranon E, Dantes M, Lee LV, Ogasawara K, Tooyama I, Akatsu H, Nishimura M, Tamiya G (2007) Reduced neuron-specific expression of the TAF1 gene is associated with X-linked dystonia-parkinsonism. Am J Hum Genet 80:393–406

Manikkam M, Tracey R, Guerrero-Bosagna C, Skinner MK (2012) Dioxin (TCDD) induces epigenetic transgenerational inheritance of adult onset disease and sperm epimutations. PLoS ONE 7(9):e46249.https://doi.org/10.1371/journal.pone.0046249

Martin-Trujillo A, Patel N, Richter F, Jadhav B, Garg P, Morton SU, Mckean DM, Depalma SR, Goldmuntz E, Gruber D, Kim R, Newburger JW, Porter GA Jr, Giardini A, Bernstein D, Tristani-Firouzi M, Seidman JG, Seidman CE, Chung WK, Gelb BD, Sharp AJ (2020) Rare genetic variation at transcription factor binding sites modulates local DNA methylation profiles. PLoS Genet 16:e1009189

Massart R, Suderman M, Mongrain V, Szyf M (2017) DNA methylation and transcription onset in the brain. Epigenomics 9:797–809

Mayr E (1982) The growth of biological thought: diversity, evolution, and inheritance. Harvard University Press, Cambridge, MA

Mcgowan PO, Sasaki A, D'alessio AC, Dymov S, Labonte B, Szyf M, Turecki G, Meaney MJ (2009) Epigenetic regulation of the glucocorticoid receptor in human brain associates with childhood abuse. Nat Neurosci 12:342–348

Mcgowan PO, Suderman M, Sasaki A, Huang TC, Hallett M, Meaney MJ, Szyf M (2011) Broad epigenetic signature of maternal care in the brain of adult rats. PLoS One 6:e14739

Molaro A, Hodges E, Fang F, Song Q, Mccombie WR, Hannon GJ, Smith AD (2011) Sperm methylation profiles reveal features of epigenetic inheritance and evolution in primates. Cell 146:1029–1041

Molaro A, Malik HS, Bourc'his D (2020) Dynamic evolution of de novo DNA methyltransferases in rodent and primate genomes. Mol Biol Evol 37:1882–1892

Musanabaganwa C, Wani AH, Donglasan J, Fatumo S, Jansen S, Mutabaruka J, Rutembesa E, Uwineza A, Hermans EJ, Roozendaal B, Wildman DE, Mutesa L, Uddin M (2022) Leukocyte methylomic imprints of exposure to the genocide against the Tutsi in Rwanda: a pilot epigenome-wide analysis. Epigenomics 14:11–25

Nadeau JH (2015) The nature of evidence for and against epigenetic inheritance. Genome Biol 16: 137

Nakamura T, Arai Y, Umehara H, Masuhara M, Kimura T, Taniguchi H, Sekimoto T, Ikawa M, Yoneda Y, Okabe M, Tanaka S, Shiota K, Nakano T (2007) PGC7/Stella protects against DNA demethylation in early embryogenesis. Nat Cell Biol 9:64–71

Nan X, Ng HH, Johnson CA, Laherty CD, Turner BM, Eisenman RN, Bird A (1998) Transcriptional repression by the methyl-CpG-binding protein MeCP2 involves a histone deacetylase complex. Nature 393:386–389

Nilsson EE, Anway MD, Stanfield J, Skinner MK (2008) Transgenerational epigenetic effects of the endocrine disruptor vinclozolin on pregnancies and female adult onset disease. Reproduction 135:713–721

Power C, Thomas C, Li L, Hertzman C (2012) Childhood psychosocial adversity and adult cortisol patterns. Br J Psychiatry 201:199–206

Prokhortchouk A, Hendrich B, Jorgensen H, Ruzov A, Wilm M, Georgiev G, Bird A, Prokhortchouk E (2001) The p120 catenin partner Kaiso is a DNA methylation-dependent transcriptional repressor. Genes Dev 15:1613–1618

Provataris P, Meusemann K, Niehuis O, Grath S, Misof B (2018) Signatures of DNA methylation across insects suggest reduced DNA methylation levels in holometabola. Genome Biol Evol 10: 1185–1197

Provencal N, Suderman MJ, Guillemin C, Massart R, Ruggiero A, Wang D, Bennett AJ, Pierre PJ, Friedman DP, Cote SM, Hallett M, Tremblay RE, Suomi SJ, Szyf M (2012) The signature of maternal rearing in the methylome in rhesus macaque prefrontal cortex and T cells. J Neurosci 32:15626–15642

Raddatz G, Guzzardo PM, Olova N, Fantappie MR, Rampp M, Schaefer M, Reik W, Hannon GJ, Lyko F (2013) Dnmt2-dependent methylomes lack defined DNA methylation patterns. Proc Natl Acad Sci USA 110:8627–8631

Razin A, Riggs AD (1980) DNA methylation and gene function. Science 210:604–610

Reik W, Brown KW, Schneid H, Le Bouc Y, Bickmore W, Maher ER (1995) Imprinting mutations in the Beckwith-Wiedemann syndrome suggested by altered imprinting pattern in the IGF2-H19 domain. Hum Mol Genet 4:2379–2385

Roder K, Hung MS, Lee TL, Lin TY, Xiao H, Isobe KI, Juang JL, Shen CJ (2000) Transcriptional repression by drosophila methyl-CpG-binding proteins. Mol Cell Biol 20:7401–7409

Romero Charria P, Navarrete C, Ovchinnikov V, Xu L, Sarre LA, Shabardina V, Ksiezopolska E, Casacuberta E, Lara-Astiaso D, Sebe-Pedros A, De Mendoza A (2025) Adenine DNA methylation associated with transcriptionally permissive chromatin is widespread across eukaryotes. Nat Genet 57:3126

Rosic S, Amouroux R, Requena CE, Gomes A, Emperle M, Beltran T, Rane JK, Linnett S, Selkirk ME, Schiffer PH, Bancroft AJ, Grencis RK, Jeltsch A, Hajkova P, Sarkies P (2018) Evolutionary analysis indicates that DNA alkylation damage is a byproduct of cytosine DNA methyltransferase activity. Nat Genet 50:452–459

Roth TL, Lubin FD, Funk AJ, Sweatt JD (2009) Lasting epigenetic influence of early-life adversity on the BDNF gene. Biol Psychiatry 65:760–769

Sadler KC (2023) Epigenetics across the evolutionary tree: new paradigms from non-model animals. BioEssays 45:e2200036

Sapienza C, Peterson AC, Rossant J, Balling R (1987) Degree of methylation of transgenes is dependent on gamete of origin. Nature 328:251–254

Sapozhnikov DM, Szyf M (2024) Genetic confounds of transgenerational epigenetic inheritance in mice. Epigenetics 19:2318519

Schmitz RJ, Lewis ZA, Goll MG (2019) DNA methylation: shared and divergent features across eukaryotes. Trends Genet 35:818–827

Schneider S, Kaufmann W, Buesen R, Van Ravenzwaay B (2008) Vinclozolin--the lack of a transgenerational effect after oral maternal exposure during organogenesis. Reprod Toxicol 25:352–360

Schubeler D (2015) Function and information content of DNA methylation. Nature 517:321–326

Seisenberger S, Andrews S, Krueger F, Arand J, Walter J, Santos F, Popp C, Thienpont B, Dean W, Reik W (2012) The dynamics of genome-wide DNA methylation reprogramming in mouse primordial germ cells. Mol Cell 48:849–862

Simpson VJ, Johnson TE, Hammen RF (1986) Caenorhabditis elegans DNA does not contain 5-methylcytosine at any time during development or aging. Nucleic Acids Res 14:6711–6719

Springer NM, Kaeppler SM (2005) Evolutionary divergence of monocot and dicot methyl-CpG-binding domain proteins. Plant Physiol 138:92–104

Stein R, Razin A, Cedar H (1982) In vitro methylation of the hamster adenine phosphoribosyltransferase gene inhibits its expression in mouse L cells. Proc Natl Acad Sci USA 79:3418–3422

Stoger R, Kubicka P, Liu CG, Kafri T, Razin A, Cedar H, Barlow DP (1993) Maternal-specific methylation of the imprinted mouse Igf2r locus identifies the expressed locus as carrying the imprinting signal. Cell 73:61–71

Stouder C, Paoloni-Giacobino A (2010) Transgenerational effects of the endocrine disruptor vinclozolin on the methylation pattern of imprinted genes in the mouse sperm. Reproduction 139:373–379

Stroud H, Su SC, Hrvatin S, Greben AW, Renthal W, Boxer LD, Nagy MA, Hochbaum DR, Kinde B, Gabel HW, Greenberg ME (2017) Early-life gene expression in neurons modulates lasting epigenetic states. Cell 171:1151–1164 e16

Suderman M, Mcgowan PO, Sasaki A, Huang TC, Hallett MT, Meaney MJ, Turecki G, Szyf M (2012) Conserved epigenetic sensitivity to early life experience in the rat and human hippocampus. Proc Natl Acad Sci USA 109(Suppl 2):17266–17272

Sun Z, Xu X, He J, Murray A, Sun MA, Wei X, Wang X, Mccoig E, Xie E, Jiang X, Li L, Zhu J, Chen J, Morozov A, Pickrell AM, Theus MH, Xie H (2019) Egr1 recruits Tet1 to shape the brain methylome during development and upon neuronal activity. Nat Commun 10:3892

Szyf M (2014) Lamarck revisited: epigenetic inheritance of ancestral odor fear conditioning. Nat Neurosci 17:2–4

Szyf M (2025) Cross-generational embedding of adverse Childhood experiences. Biol Psychiatry 98:728–730

Tabano S, Colapietro P, Cetin I, Grati FR, Zanutto S, Mando C, Antonazzo P, Pileri P, Rossella F, Larizza L, Sirchia SM, Miozzo M (2010) Epigenetic modulation of the IGF2/H19 imprinted domain in human embryonic and extra-embryonic compartments and its possible role in fetal growth restriction. Epigenetics 5:313–324

Takahashi Y, Morales Valencia M, Yu Y, Ouchi Y, Takahashi K, Shokhirev MN, Lande K, Williams AE, Fresia C, Kurita M, Hishida T, Shojima K, Hatanaka F, Nunez-Delicado E, Esteban CR, Izpisua Belmonte JC (2023) Transgenerational inheritance of acquired epigenetic signatures at CpG islands in mice. Cell 186:715–731 e19

Tang WW, Dietmann S, Irie N, Leitch HG, Floros VI, Bradshaw CR, Hackett JA, Chinnery PF, Surani MA (2015) A unique gene regulatory network resets the human germline epigenome for development. Cell 161:1453–1467

Tatton-Brown K, Seal S, Ruark E, Harmer J, Ramsay E, Del Vecchio Duarte S, Zachariou A, Hanks S, O'brien E, Aksglaede L, Baralle D, Dabir T, Gener B, Goudie D, Homfray T, Kumar A, Pilz DT, Selicorni A, Temple IK, Van Maldergem L, Yachelevich N, Childhood Overgrowth Consortium, Van Montfort R, Rahman N (2014) Mutations in the DNA methyltransferase gene DNMT3A cause an overgrowth syndrome with intellectual disability. Nat Genet 46:385–388

Urieli-Shoval S, Gruenbaum Y, Sedat J, Razin A (1982) The absence of detectable methylated bases in Drosophila melanogaster DNA. FEBS Lett 146:148–152

Van Der Ploeg LH, Flavell RA (1980) DNA methylation in the human gamma delta beta-globin locus in erythroid and nonerythroid tissues. Cell 19:947–958

Vardimon L, Kressmann A, Cedar H, Maechler M, Doerfler W (1982) Expression of a cloned adenovirus gene is inhibited by in vitro methylation. Proc Natl Acad Sci USA 79:1073–1077

Vidaurre V, Chen X (2021) Epigenetic regulation of drosophila germline stem cell maintenance and differentiation. Dev Biol 473:105–118

Villicana S, Bell JT (2021) Genetic impacts on DNA methylation: research findings and future perspectives. Genome Biol 22:127

Vire E, Brenner C, Deplus R, Blanchon L, Fraga M, Didelot C, Morey L, Van Eynde A, Bernard D, Vanderwinden JM, Bollen M, Esteller M, Di Croce L, De Launoit Y, Fuks F (2006) The Polycomb group protein EZH2 directly controls DNA methylation. Nature 439:871–874

Waddington CH (1942) Canalization of development and the inheritance of acquired characters. Nature 150:563–565

Weaver IC, Cervoni N, Champagne FA, D'alessio AC, Sharma S, Seckl JR, Dymov S, Szyf M, Meaney MJ (2004) Epigenetic programming by maternal behavior. Nat Neurosci 7:847–854

Weaver IC, Meaney MJ, Szyf M (2006) Maternal care effects on the hippocampal transcriptome and anxiety-mediated behaviors in the offspring that are reversible in adulthood. Proc Natl Acad Sci USA 103:3480–3485

Weaver IC, D'alessio AC, Brown SE, Hellstrom IC, Dymov S, Sharma S, Szyf M, Meaney MJ (2007) The transcription factor nerve growth factor-inducible protein a mediates epigenetic programming: altering epigenetic marks by immediate-early genes. J Neurosci 27:1756–1768

Weber M, Hellmann I, Stadler MB, Ramos L, Paabo S, Rebhan M, Schubeler D (2007) Distribution, silencing potential and evolutionary impact of promoter DNA methylation in the human genome. Nat Genet 39:457–466

Wood KH, Zhou Z (2016) Emerging molecular and biological functions of MBD2, a reader of DNA methylation. Front Genet 7:93

Wu H, Coskun V, Tao J, Xie W, Ge W, Yoshikawa K, Li E, Zhang Y, Sun YE (2010) Dnmt3a-dependent nonpromoter DNA methylation facilitates transcription of neurogenic genes. Science 329:444–448

Wyatt GR (1950) Occurrence of 5-methylcytosine in nucleic acids. Nature 166:237–238

Xu X, Li G, Li C, Zhang J, Wang Q, Simmons DK, Chen X, Wijesena N, Zhu W, Wang Z, Wang Z, Ju B, Ci W, Lu X, Yu D, Wang QF, Aluru N, Oliveri P, Zhang YE, Martindale MQ, Liu J (2019) Evolutionary transition between invertebrates and vertebrates via methylation reprogramming in embryogenesis. Natl Sci Rev 6:993–1003

Yehuda R, Daskalakis NP, Lehrner A, Desarnaud F, Bader HN, Makotkine I, Flory JD, Bierer LM, Meaney MJ (2014) Influences of maternal and paternal PTSD on epigenetic regulation of the glucocorticoid receptor gene in holocaust survivor offspring. Am J Psychiatry 171:872–880

Yehuda R, Daskalakis NP, Bierer LM, Bader HN, Klengel T, Holsboer F, Binder EB (2016) Holocaust exposure induced intergenerational effects on FKBP5 methylation. Biol Psychiatry 80:372–380

Yin Y, Morgunova E, Jolma A, Kaasinen E, Sahu B, Khund-Sayeed S, Das PK, Kivioja T, Dave K, Zhong F, Nitta KR, Taipale M, Popov A, Ginno PA, Domcke S, Yan J, Schubeler D, Vinson C, Taipale J (2017) Impact of cytosine methylation on DNA binding specificities of human transcription factors. Science 356(6337):eaaj2239

Yoder JA, Walsh CP, Bestor TH (1997) Cytosine methylation and the ecology of intragenomic parasites. Trends Genet 13:335–340

Zemach A, Zilberman D (2010) Evolution of eukaryotic DNA methylation and the pursuit of safer sex. Curr Biol 20:R780–R785

Zemach A, Mcdaniel IE, Silva P, Zilberman D (2010) Genome-wide evolutionary analysis of eukaryotic DNA methylation. Science 328:916–919

Zhang Y, Ng HH, Erdjument-Bromage H, Tempst P, Bird A, Reinberg D (1999) Analysis of the Nurd subunits reveals a histone deacetylase core complex and a connection with DNA methylation. Genes Dev 13:1924–1935

Zhou J, Chen Q, Braun PR, Perzel Mandell KA, Jaffe AE, Tan HY, Hyde TM, Kleinman JE, Potash JB, Shinozaki G, Weinberger DR, Han S (2022) Deep learning predicts DNA methylation regulatory variants in the human brain and elucidates the genetics of psychiatric disorders. Proc Natl Acad Sci USA 119:e2206069119

Zhou J, Weinberger DR, Han S (2025) Deep learning predicts DNA methylation regulatory variants in specific brain cell types and enhances fine mapping for brain disorders. Sci Adv 11:eadn1870

Zocher S, Overall RW, Berdugo-Vega G, Rund N, Karasinsky A, Adusumilli VS, Steinhauer C, Scheibenstock S, Handler K, Schultze JL, Calegari F, Kempermann G (2021) De novo DNA methylation controls neuronal maturation during adult hippocampal neurogenesis. EMBO J 40: e107100

Chapter 3
Developmental Dynamics: Shaping the Pathways of Evolution

Gianluca Ursini and Giovanna Punzi

Abstract Development influences evolution not merely by translating genes into phenotypes. It actively channels evolutionary trajectories, mediates environmental interactions, and generates phenotypic variation in structured ways. This chapter explores how developmental processes bias variant production, how developmental plasticity enables organisms to construct and modify their selective environments, and how perturbations during critical windows can have cascading effects across generations. Drawing on evidence from experimental evolution, placental genomics, and mechanical morphogenesis, we discuss how development is a dynamic participant in evolutionary change, whose mechanisms transcend simple error replication narratives. We examine how recently evolved structures like the human brain may be particularly vulnerable to developmental decanalization and how understanding developmental dynamics provides new insights into the origins of complex disorders. This perspective reveals evolution as emerging from the interplay between developmental systems and environments, where genetic information is not merely transmitted but actively interpreted through developmental processes.

1 Introduction: Development as an Evolutionary Process

Over the past two million years, the human neocortex has tripled in size, enabling language, social cognition, and symbolic thought. Yet this expansion has come at a cost: longer development, extended vulnerability, and increased risk of disorders. The reason for this resides in the fact that development is not *infinitely plastic*. Newly evolved structures cannot instantly acquire the buffering mechanisms that stabilize ancient, conserved development. The human brain thus embodies a fundamental

G. Ursini (✉) · G. Punzi
Lieber Institute for Brain Development, Johns Hopkins Medical Campus, Baltimore, MD, USA

Department of Psychiatry and Behavioral Sciences, Johns Hopkins University School of Medicine, Baltimore, MD, USA
e-mail: gianluca.ursini@libd.org

G. Witzany (ed.), *Integrative Theory of Evolution - Aspects and Insights*,
https://doi.org/10.1007/978-3-032-24295-2_3

truth about evolution: It proceeds not just through changes in genes but through changes in how genes are deployed during development.

For much of the twentieth century, evolutionary biology treated development as a "black box"—a process that faithfully translates genetic information into the adult form. In this view, evolution was fundamentally about genes: Random mutations generated variation, natural selection sorted variants, and development merely executed the genetic program. This gene-centric narrative obscures critical aspects of how evolution actually works. Development does not passively transmit genetic information; it dynamically interprets it, modulates it, and generates structured variation that biases evolutionary trajectories.

This chapter explores how developmental dynamics shape evolution, placing emphasis on three interrelated mechanisms at play. First, developmental architecture biases which variants are possible and probable, channeling evolution along certain trajectories while constraining others. Second, developmental plasticity enables organisms to respond to environmental variation, and these responses can themselves become targets of selection or sources of heritable variation. Third, perturbations during critical windows can destabilize and decanalize developmental trajectories, increasing phenotypic variance and disease risk in ways that persist across generations. Together, these three mechanisms see development as a contributor to evolutionary change, not merely a passive vehicle for genetic information. By reframing evolution as a process that unfolds through development, modern science has gained a more dynamic perspective on both adaptation and the origins of vulnerability.

2 Why Development Was Excluded: A Historical Detour

In the nineteenth century, comparative embryology was central to evolutionary thought. Darwin himself drew on embryological evidence (Darwin and Kebler 1859), and Haeckel's biogenetic law (Haeckel 1866)—ontogeny recapitulates phylogeny—made development the key to understanding phylogenetic relationships. By the mid-twentieth century, however, development had been pushed to the margins of evolutionary theory.

Why? Gerson identifies two key factors (Gerson 2009). First, technical limitations made it impossible to resolve the fine details needed to answer fundamental questions about early development. Microscopy could not adequately visualize fertilization or cleavage stages; fossils rarely preserved developmental sequences directly. As research focused on ever-smaller differences—from organ systems to organs to cells to molecules—the tools available could not keep pace with the questions being asked.

Second, conceptual shifts enabled development to be put aside. Weismann's germ–soma distinction (Weismann et al. 1893), the idea that only germ cells contribute hereditary information to the next generation, made somatic development practically irrelevant to transmission. Fisher's population genetics showed that evolution could be modeled statistically without reference to individual lineages or life histories (Fisher 1930). Organisms were seen as a temporary vessel for enduring genes and development as the process by which genetic instructions were executed, that is, a process important for understanding how organisms work, not how they evolve.

What was lost was the recognition that organisms are life cycles, that selection operates at every developmental stage, and that development has its own dynamics that can bias and channel evolutionary change. The evo–devo revolution (Carroll 2005; Gerson 2009) of the late twentieth century began recovering these insights, revealing development as central to understanding both the constraints on and the directions of evolutionary change.

3 Developmental Bias: How Ontogeny Channels Evolution

Not all evolutionary trajectories are equally accessible. Developmental architecture—the genetic networks, regulatory interactions, and physical constraints that shape ontogeny—biases which variants are readily produced and which are effectively impossible. Wallace Arthur captured this insight in his concept of developmental bias (Arthur 2002): The direction of evolutionary change is determined by the interaction between the structure of variation (biased by development) and the direction of selection (imposed by ecology).

The evidence is strong at multiple scales. At the morphological level, Brakefield's experiments on butterfly eyespots reveal that artificial selection can push traits in almost any direction, but some directions require far less genetic change than others (Brakefield et al. 1996). Eyespot size shows high evolvability: strong responses to selection, ability to evolve along or against genetic correlations, and rapid divergence. However, the eyespot shape shows much lower evolvability. The reason is that size is specified by the strength of signals from a central organizer, while shape requires coordinating changes across multiple developmental modules. The developmental architecture makes size easy to evolve and shape hard.

At the molecular level, the pattern is even clearer. Complex morphological evolution proceeds primarily through changes in gene regulatory regions, not protein-coding sequences. For example, the *yellow* gene in *Drosophila melanogaster* serves as a regulatory "hot spot", where different cis-regulatory elements mediate independent evolutionary changes in wing spot patterns (Prud'homme et al. 2007). This is not because mutations preferentially occur in regulatory regions but because developmental architecture makes regulatory mutations more likely to produce viable, functionally integrated phenotypes. Mutations in highly pleiotropic genes are more often lethal; mutations in modular regulatory elements can tweak one trait without disrupting others.

Critically, developmental bias itself evolves. Traits that share developmental pathways will be coupled in their evolution; those that have evolved separate regulatory modules can evolve independently. This means evolution is not just exploring a landscape of phenotypic possibilities; it is dynamically reshaping the landscape, changing the rules by which variation is generated.

4 Physical Forces as Developmental Constraints

Developmental bias operates through physical as well as molecular mechanisms. Recent studies on crocodile head scales provide a striking example: Rather than emerging from genetically determined placodes through chemical Turing patterning (as do body scales, feathers, and hair), these irregular polygonal domains arise purely from compressive folding (Milinkovitch et al. 2013; Santos-Duran et al. 2025). Between embryonic days 48–63, the epidermis progressively keratinizes, becoming substantially stiffer than the underlying dermis. When both layers undergo near-homogeneous growth with differential stiffness, in-plane compressive stress generates surface buckling—the skin folds inward, creating scale boundaries.

The experimental validation of this research is elegant: Treating embryos with epidermal growth factor (EGF) increases epidermal keratinization and growth, producing extensively convoluted “brainy” folding rather than normal polygons (Santos-Duran et al. 2025). This confirms compression-driven patterning: Exacerbating epidermal stiffness intensifies folding. When EGF treatment is arrested, the pattern partially relaxes into smaller polygonal scales resembling those of spectacled caimans, revealing a simple evolutionary mechanism. By varying just two parameters (relative growth rates and elastic moduli of epidermis versus dermis), a model-generated range of forms spans the full range of head-scale patterns across crocodilian species.

Thus, developmental constraints emerge from multiple independent channels: chemical (reaction–diffusion), molecular (regulatory architecture), and physical (mechanical properties). Evolution has produced orthogonal mechanisms to achieve similar outcomes, patterned integument, by operating through different biases. Mechanical parameters like stiffness can vary continuously, potentially enabling smoother phenotypic transitions than discrete regulatory changes. The same genome can produce radically different patterns through purely mechanical perturbation, suggesting that evolutionary divergence may reflect changes in developmental timing rather than patterning genes per se.

The crocodile example also explains why evolution often produces multiple solutions to similar problems. Body scales and head scales in the same organism emerge through completely different mechanisms—chemical patterning for the former and mechanical instabilities for the latter—suggesting that different regions of the body are subject to different developmental constraints. The head scales’ purely mechanical origin may reflect the rapid jaw elongation during embryogenesis: Chemical patterning systems require time to establish morphogen gradients, but mechanical instabilities emerge immediately from growth and material properties (Santos-Duran et al. 2025).

More broadly, the recognition that self-organized mechanical patterning operates alongside genetic patterning systems reveals development as integrating information from multiple physical scales. Genes specify protein sequences, proteins determine tissue material properties, material properties and growth rates generate mechanical forces, forces produce deformations and instabilities, and geometric patterns emerge. At each level, the system interprets information from the previous level through physical laws. Developmental bias thus reflects not just which genetic variants are viable but which combinations of material properties, growth dynamics, and geometric constraints are physically stable.

This perspective suggests that understanding evolutionary change requires understanding not just regulatory evolution but material evolution, that is, how tissue properties, extracellular matrix architecture, and mechanical forces evolve. For crocodilian head scales, evolutionary diversification operates through the parameter space of differential growth and stiffness. For other structures, different mechanical principles may dominate: buckling, folding, branching, or adhesion. The developmental bias imposed by physics is as real as that imposed by genetics, and evolution navigates both simultaneously.

5 Developmental Plasticity: From Environmental Response to Evolutionary Change

Development is not a rigid program but a responsive process. Developmental systems extract information from internal and external environments and use this information to shape phenotypes. This developmental plasticity—the capacity to produce different phenotypes from the same genotype depending on environmental conditions—creates a second route through which development shapes evolution.

The human brain exemplifies experience-expectant development: Optimal brain maturation requires specific environmental inputs during critical windows. Visual cortex development requires patterned visual input, language acquisition requires exposure to speech during early childhood, and social cognition requires interaction with other minds (West-Eberhard 2003). When the environment fails to provide expected inputs, development decanalizes, buffering capacity is overwhelmed, and trajectories diverge from optimal paths.

This creates an intimate connection between developmental plasticity and niche construction. Organisms do not merely respond to environments; they actively construct the developmental environments their offspring will experience. Human cultural evolution provides the most dramatic example: Over 50,000 years, humans created novel environments—agricultural, urban, and technologically complex—that exposed developing individuals to radically new inputs, which then shaped human genetic evolution.

Lactose tolerance provides a classic example. The capacity to digest milk sugars normally disappears after weaning, a conserved mammalian developmental program. But in populations that adopted dairy farming, cultural practices created a new developmental environment where milk remained available throughout life. This occurrence imposed strong selection (coefficients of 0.09–0.19) on genetic variants that maintained lactase expression into adulthood (Bersaglieri et al. 2004). Within 9000 years, the lactose tolerance allele swept to high frequency in dairy-farming populations.

Critically, gene–culture coevolution proceeds through developmental plasticity. Infants exposed to milk learn that it is nutritious (behavioral plasticity), populations that rely on dairy develop physiological tolerance through developmental accommodation, and over generations, genetic variants that enhance this capacity spread. At each stage, developmental processes mediate the interaction between genes, environment, and evolution. The trajectory is not gene → phenotype → selection but gene × development × environment → phenotype → selection → modified environment → modified development.

6 Developmental Trade-Offs: The Evolutionary Legacy of Life History

Development proceeds through a series of trade-offs. Resources invested in one structure or function cannot be invested elsewhere: developmental programs that optimize early survival may compromise late-life function, and variants that enhance reproduction may accelerate aging. These trade-offs are central to evolution because selection operates at every developmental stage.

Antagonistic pleiotropy (Williams 1957), where a genetic variant has opposite effects at different stages, emerges naturally from these developmental constraints. The classic example is sickle cell anemia: Heterozygotes gain malaria resistance during childhood but carry the risk of the disease (Allison 1954). The phenomenon is far broader. Variants at cardiovascular disease loci show signatures of positive selection because alleles that increase heart disease risk also increase reproductive success through effects on fertility, age at menarche, and offspring survival (Byars et al. 2010). The *APOE* ε4 allele offers another clear case: It is associated with higher fecundity, and it seems to enhance childhood neurodevelopment and cognitive function but triples late-life Alzheimer's risk (and cardiovascular disease), with signatures of ancient positive selection likely tied to immune or reproductive benefits (Lu et al. 2021; Trumble et al. 2023; Lockwood et al. 2024).

This concept generalizes to environmentally induced trade-offs in the developmental origins of health and disease (DOHaD) framework (Barker 2007). Fetuses facing nutritional stress make developmental "trade-offs": prioritizing brain development over body size, shifting metabolic set points to match anticipated environments, and altering stress response systems (Gluckman et al. 2008). These adaptive responses reflect developmental plasticity shaped by selection; however, when the developmental environment mismatches the adult environment, disease risk may increase.

Remarkably, these developmental effects can span three generations: The egg that eventually became us was formed while our mother was a fetus in our maternal grandmother's womb. Environmental exposures during our grandmother's pregnancy may have affected our mother's developing oocytes, which later affected our own development. Such transgenerational reach means that evolutionary pressures on one generation can shape developmental trajectories of descendants through purely developmental mechanisms, independent of genetic change.

7 The Placenta: Developmental Interface for Gene × Environment Interactions

If development mediates interactions between genes and environments, then the placenta exemplifies this mediation. The placenta integrates maternal environment, fetal genome, and evolutionary history to shape development in ways that persist throughout life. In this regard, our previous studies reveal how developmental dynamics at the placental interface influence psychiatric disease risk decades later (Ursini et al. 2018, 2021, 2023).

The initial discovery demonstrated a striking gene-by-environment interaction: genomic risk of schizophrenia (measured as polygenic risk scores) explained 11.2% of variance in individuals with early life complications (ELCs, that is, severe pregnancy, labor/delivery, and early neonatal complications) but only 0.8% in those without ELCs, a more than fivefold difference (Ursini et al. 2018). This interaction replicated across five international cohorts and was specifically driven by genes expressed in the placenta. Schizophrenia risk genes were highly expressed across all placental compartments and differentially upregulated in placentae from complicated pregnancies. "Placental genomic risk scores" (PlacGRS), i.e., generated with only these placenta-expressed genes, strongly interacted with ELCs on schizophrenia risk, while nonplacental genomic risk scores showed no interaction.

Follow-up studies traced these effects developmentally (Ursini et al. 2021). In 242 neonates with ELC histories, PlacGRS negatively predicted intracranial volume at birth and cognitive development at the age of 1 year, effects consistently stronger in males. The effects weakened by the age of 2 when environmental factors dominated, suggesting that most infants "recanalize" toward normal development. Indeed, in adult patients who developed schizophrenia, PlacGRS remained associated with reduced brain volume, specifically in males, while healthy controls showed no association, consistent with successful developmental rescue in most individuals.

Molecular analyses suggest the possible mechanisms: Placenta-specific schizophrenia risk genes converge on nutrient-sensing pathways (Ursini et al. 2023). Risk genes were enriched for the inhibition of mTOR (mechanistic target of rapamycin) signaling—the central hub integrating amino acids, glucose, oxygen, folate, and growth factors—to regulate trophoblast function. The emerging picture is that genetic risk impairs placental nutrient sensing, making placental functioning vulnerable to environmental perturbations like infection or malnutrition, with possible effects on altered trophoblast invasiveness. Sex-specific pathway analyses reveal differential mechanisms of risk: In females, altered insulin signaling may provide compensatory metabolic flexibility; in males, inflammatory signaling may exacerbate trophoblast dysfunction. Specificity of these findings to schizophrenia (associations weak for autism, ADHD, or other conditions (Ursini et al. 2023)) suggests that the placental–genomic–environmental nexus may be unique to this disorder's etiology.

The placenta's evolutionary history sheds light on these vulnerabilities. Endogenous retroviruses—genomic fossils of ancient infections—have been extensively co-opted for placental function, requiring elevated tolerance for genomic instability (Dupressoir et al. 2012). This created an organ whose developmental programs are simultaneously ancient and evolutionarily labile, integrating genetic variants, environmental perturbations, and stochastic retroviral element expression. When these align—high genetic risk, adverse environment, and dysregulated development—brain development may be decanalized, increasing disease risk.

The placenta may also exemplify complex antagonistic pleiotropy. Our findings are compatible with genomic risk conferring fetal resilience under severe ELCs and stress (Ursini et al. 2018). At the same time, schizophrenia risk genes show the inhibition of mTOR signaling (Ursini et al. 2023), a pathway activated in cancer, fitting with the placenta's cancer-like invasiveness during trophoblast invasion (Costello and Fisher 2021). Lacking further studies in support, such as ELC-stratified research on cancer, it remains tempting to speculate the following evolutionary trade-offs: in the absence of ELCs, placenta- and brain-beneficial variants would spread, despite increased cancer risk; in the presence of ELCs, embryo-survival variants would spread, conferring increased schizophrenia risk but decreased cancer risk. The placenta thus exemplifies how development mediates evolution by integrating genetic variation and environmental perturbation, generating both adaptive diversity and disease vulnerability.

8 Decanalization: When Developmental Buffering Fails

Waddington's epigenetic landscape provides a powerful metaphor for developmental dynamics (Waddington 1942). Ontogeny unfolds as a ball rolling down a contoured surface, with valleys channeling development toward optimal end points. The depth of valleys reflects the buffering capacity, the ability to maintain developmental fidelity despite perturbation. Decanalization occurs when this buffering is overwhelmed: Development diverges from the optimal trajectory.

McGrath and colleagues argue that the brain is particularly vulnerable to decanalization (McGrath et al. 2011). First, brain development is extraordinarily complex, requiring orchestrated expression of over half the genome. Second, relatively recently evolved structures like the expanded human neocortex have had insufficient evolutionary time to acquire robust buffering mechanisms. Third, brain development is intensely experience-expectant, requiring specific environmental inputs during critical windows. Fourth, these critical windows are brief and tightly regulated, leaving few opportunities for compensatory development.

The epidemiology of schizophrenia may reveal decanalization in action. ELCs are the most important environmental risk factor for schizophrenia and psychosis, and they may also contribute to destabilize early neurodevelopmental trajectories (Cannon et al. 2002; Davies et al. 2020) also in interaction with genomic factors (Ursini et al. 2018, 2021). Another example is provided by the early life deficiency of vitamin D. Dark-skinned migrants to northern latitudes show a six- to ninefold increased risk of schizophrenia, urban birth confers a twofold increased risk, and winter–spring births carry an elevated risk (McGrath et al. 2004). These patterns suggest developmental vitamin D deficiency as a unifying mechanism (Horsdal et al. 2025); vitamin D is a potent regulator of brain development, and prenatal deficiency produces persistent alterations in brain structure and behavior.

Critically, vitamin D deficiency represents the absence of an expected developmental signal. Human ancestors evolved in equatorial Africa, where vitamin D was abundant year-round. Migration to higher latitudes and adoption of urban lifestyles created environments where this expected signal is lacking (McGrath et al. 2010). From a developmental perspective, this is not an unexpected toxin but a missing instruction, and developmental systems expecting certain cues may fail catastrophically when those are missing. The result is increased developmental variance: more individuals develop atypically, and more cross phenotypic thresholds into disorder.

When genetic risk, placental and early life complications, and signal deprivation converge, developmental bias, plasticity, and buffering may collapse together, revealing development as evolution's active interpreter at work.

9 Synthesis: Development as an Evolutionary Mediator

These perspectives converge on a unified view: Development mediates evolution by structuring how genetic and environmental variation maps onto phenotypic variation. This mediation operates through multiple mechanisms that interact across timescales (Fig. 3.1).

At the shortest timescale—within individual development—developmental processes interpret genetic and environmental information to produce phenotypes. This interpretation is context-dependent: The same genes produce different phenotypes in different developmental environments (plasticity), and the same environment produces different phenotypes depending on the developmental stage (critical windows) and genetic background (gene–environment interaction). The phenotypic outcome reflects integration across several sources, mediated by developmental dynamics.

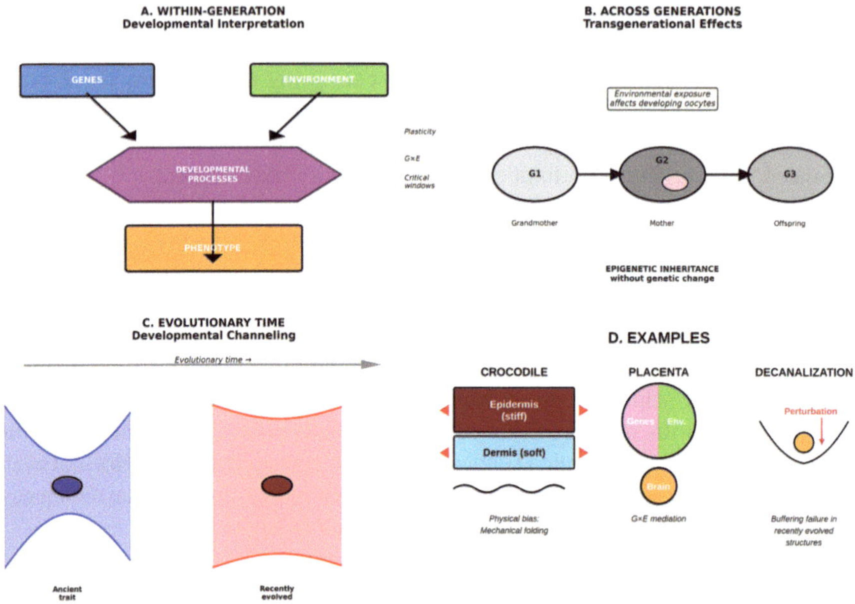

Fig. 3.1 Development mediates evolution across multiple timescales. (**a**) Within individual development, developmental processes interpret genetic and environmental information to produce phenotypes, with outcomes depending on the genetic background, environmental context, and developmental stage. (**b**) Across two to three generations, environmental effects can be transmitted through epigenetic modifications without genetic change. (**c**) Over evolutionary time, developmental architecture channels evolutionary change by biasing which variants are produced and viable. Deep valleys represent ancient, canalized traits resistant to perturbation; shallow valleys represent recently evolved, evolvable traits vulnerable to decanalization. (**d**) Examples: Crocodile head scales emerge from physical forces (compressive folding); the placenta mediates gene–environment interactions affecting brain development; decanalization occurs when recently evolved structures lack robust buffering

At intermediate timescales—across generations—developmental processes may transmit environmental effects through epigenetic inheritance. Maternal stress alters offspring stress reactivity, paternal diet affects offspring metabolism, and grandmaternal nutrition influences grandoffspring disease risk. These transgenerational effects operate through developmental mechanisms: altered methylation patterns, modified histone marks, and changed small RNA content in gametes.

At evolutionary timescales—across countless generations—developmental architecture shapes which evolutionary trajectories are accessible. Mutations in modular regulatory elements can produce viable phenotypic variants; mutations in pleiotropic genes are more often lethal. Developmental programs that evolved under stable selection become canalized, buffered against perturbation; newly evolved programs lack this buffering and remain evolvable.

What unifies these timescales is the concept of developmental interpretation. Genes do not directly specify phenotypes; they specify developmental processes that interpret environmental contexts to produce phenotypes. Evolution does not just change genes; it changes how developmental systems interpret genetic and environmental information. Variation in phenotypes reflects not just genetic variation but variation in developmental interpretation, i.e., how robustly developmental systems buffer perturbation, how sensitively they respond to environmental cues, and how completely they execute genetic instructions.

10 Implications for Evolutionary Theory and Medicine

If development mediates evolution through these multiple mechanisms, what follows for how we study evolution and disease?

For evolutionary theory, the implication is that models treating development as a black box are incomplete. The genotype–phenotype map is not a simple translational process but a complex developmental interpretation that varies with environmental context, developmental stage, and evolutionary history. Incorporating development means modeling how developmental architecture biases variation, how plasticity enables environmental responsiveness, how critical windows constrain when environmental effects act, and how buffering mechanisms shape phenotypic robustness.

For medical genetics, the implication is that gene–disease associations depend critically on the developmental context. The same genetic variant may increase disease risk in one developmental environment but not another (gene–environment interaction). Risk may be highest when genetic vulnerability coincides with environmental perturbation during critical developmental windows (decanalization). Effects may span generations through developmental programming. This implies that genetic studies should stratify by developmental exposures, that prevention should target developmental windows, and that etiology should be framed in terms of developmental trajectories rather than static end points.

For schizophrenia, this framework suggests tangible avenues. Ensuring adequate maternal vitamin D status in pregnancy, especially in high-risk populations, may prevent a subset of cases by restoring expected developmental signals, as suggested by convergent cohort data (Horsdal et al. 2025). Monitoring and protecting placental health—through prenatal care, nutrition support, and infection prevention—may be as important as refining genetic risk scores, given the evidence that placental pathways mediate a part of schizophrenia heritability. The detection of placental biomarkers in maternal blood may trigger opportunities for early screening and prevention (Ursini et al. 2018, 2021, 2023).

11 Conclusion

Development shapes evolution not by passively transmitting genetic information but by dynamically mediating how genes and environments interact to produce phenotypes. This mediation operates through developmental bias (both molecular and physical), developmental plasticity, and developmental buffering that can fail under duress. Examples span from molecular regulatory hot spots to physical forces generating crocodile scale patterns and organ-level mediation of gene–environment interactions in the placenta.

Understanding these developmental dynamics transforms how we think about evolution and disease. Evolution emerges not from genes alone but from developmental systems interpreting genetic and environmental information across contexts and timescales. Disease risk reflects not just genetic variants but how developmental processes handle those variants in specific environments during critical windows. Prevention opportunities arise not just from correcting genetic defects but from supporting optimal developmental environments.

The integration of development into evolutionary theory reveals that the modern synthesis was incomplete, not because it was wrong but because it bracketed questions that turned out to be central. Organisms are life cycles, not just adult forms; selection operates on developmental trajectories, not mere final phenotypes; and the genotype–phenotype map is a complex developmental interpretation, not a simple translation. Toward an integrative theory of evolution means toward a theory that takes development seriously, not as genetic execution but as evolutionary mediation.

References

Allison AC (1954) Protection afforded by sickle-cell trait against subtertian malareal infection. Br Med J 1(4857):290–294

Arthur W (2002) The emerging conceptual framework of evolutionary developmental biology. Nature 415(6873):757–764

Barker DJ (2007) The origins of the developmental origins theory. J Intern Med 261(5):412–417

Bersaglieri T, Sabeti PC, Patterson N, Vanderploeg T, Schaffner SF, Drake JA, Rhodes M, Reich DE, Hirschhorn JN (2004) Genetic signatures of strong recent positive selection at the lactase gene. Am J Hum Genet 74(6):1111–1120

Brakefield PM, Gates J, Keys D, Kesbeke F, Wijngaarden PJ, Monteiro A, French V, Carroll SB (1996) Development, plasticity and evolution of butterfly eyespot patterns. Nature 384(6606): 236–242

Byars SG, Ewbank D, Govindaraju DR, Stearns SC (2010) Colloquium papers: natural selection in a contemporary human population. Proc Natl Acad Sci U S A 107 Suppl 1(Suppl 1):1787–1792

Cannon M, Jones PB, Murray RM (2002) Obstetric complications and schizophrenia: historical and meta-analytic review. Am J Psychiatry 159(7):1080–1092

Carroll SB (2005) Endless forms most beautiful : the new science of evo devo and the making of the animal kingdom. Norton, New York

Costello JF, Fisher SJ (2021) The placenta - fast, loose, and in control. N Engl J Med 385(1):87–89

Darwin C, Kebler L (1859) On the origin of species by means of natural selection, or, the preservation of favoured races in the struggle for life. J. Murray, London
Davies C, Segre G, Estrade A, Radua J, De Micheli A, Provenzani U, Oliver D, Salazar de Pablo G, Ramella-Cravaro V, Besozzi M, Dazzan P, Miele M, Caputo G, Spallarossa C, Crossland G, Ilyas A, Spada G, Politi P, Murray RM, McGuire P, Fusar-Poli P (2020) Prenatal and perinatal risk and protective factors for psychosis: a systematic review and meta-analysis. Lancet Psychiatry 7(5):399–410
Dupressoir A, Lavialle C, Heidmann T (2012) From ancestral infectious retroviruses to bona fide cellular genes: role of the captured syncytins in placentation. Placenta 33(9):663–671
Fisher RA (1930) The genetical theory of natural selection. The Clarendon press, Oxford
Gerson EM (2009) Why development does not matter in evolution and development. In: Barberousse AM, Pradeu T (eds) Mapping the future of biology. Springer, Dordrecht, pp 83–98
Gluckman PD, Hanson MA, Cooper C, Thornburg KL (2008) Effect of in utero and early-life conditions on adult health and disease. N Engl J Med 359(1):61–73
Haeckel E (1866) Generelle Morphologie der Organismen : allgemeine Grundzüge der organischen Formen-Wissenschaft, mechanisch begründet durch die von Charles Darwin reformirte Descendenz-Theorie. G. Reimer, Berlin
Horsdal HT, Albinana C, Zhu Z, Boelt SG, Borbye-Lorenzen N, Cohen AS, Skogstrand K, Melgaard L, MacSween NJ, Thorbek MJ, Plana-Ripoll O, Petersen LV, Bulik CM, Rglum ADB, Mors O, Nordentoft M, Werge T, Moen GH, D'Urso S, Wray NR, Vilhjalmsson BJ, Agerbo E, Pedersen CBC, Mortensen PB, McGrath JJ (2025) Convergent evidence linking neonatal vitamin D status and risk of neurodevelopmental disorders: a Danish case-cohort study. Lancet Psychiatry 12(6):410–420
Lockwood C, Vo AS, Bellafard H, Carter AJR (2024) More evidence for widespread antagonistic pleiotropy in polymorphic disease alleles. Front Genet 15:1404516
Lu K, Nicholas JM, Pertzov Y, Grogan J, Husain M, Pavisic IM, James SN, Parker TD, Lane CA, Keshavan A, Keuss SE, Buchanan SM, Murray-Smith H, Cash DM, Malone IB, Sudre CH, Coath W, Wong A, Henley SMD, Fox NC, Richards M, Schott JM, Crutch SJ (2021) Dissociable effects of APOE-epsilon4 and beta-amyloid pathology on visual working memory. Nat Aging 1(11):1002–1009
McGrath J, Saari K, Hakko H, Jokelainen J, Jones P, Jarvelin MR, Chant D, Isohanni M (2004) Vitamin D supplementation during the first year of life and risk of schizophrenia: a Finnish birth cohort study. Schizophr Res 67(2–3):237–245
McGrath JJ, Burne TH, Feron F, Mackay-Sim A, Eyles DW (2010) Developmental vitamin D deficiency and risk of schizophrenia: a 10-year update. Schizophr Bull 36(6):1073–1078
McGrath JJ, Hannan AJ, Gibson G (2011) Decanalization, brain development and risk of schizophrenia. Transl Psychiatry 1(6):e14
Milinkovitch MC, Manukyan L, Debry A, Di-Poi N, Martin S, Singh D, Lambert D, Zwicker M (2013) Crocodile head scales are not developmental units but emerge from physical cracking. Science 339(6115):78–81
Prud'homme B, Gompel N, Carroll SB (2007) Emerging principles of regulatory evolution. Proc Natl Acad Sci U S A 104 Suppl 1(Suppl 1):8605–8612
Santos-Duran GN, Cooper RL, Jahanbakhsh E, Timin G, Milinkovitch MC (2025) Self-organized patterning of crocodile head scales by compressive folding. Nature 637(8045):375–383
Trumble BC, Charifson M, Kraft T, Garcia AR, Cummings DK, Hooper P, Lea AJ, Eid Rodriguez D, Koebele SV, Buetow K, Beheim B, Minocher R, Gutierrez M, Thomas GS, Gatz M, Stieglitz J, Finch CE, Kaplan H, Gurven M (2023) Apolipoprotein-epsilon4 is associated with higher fecundity in a natural fertility population. Sci Adv 9(32):eade9797
Ursini G, Di Carlo P, Mukherjee S, Chen Q, Han S, Kim J, Deyssenroth M, Marsit CJ, Chen J, Hao K, Punzi G, Weinberger DR (2023) Prioritization of potential causative genes for schizophrenia in placenta. Nat Commun 14(1):2613
Ursini G, Punzi G, Chen Q, Marenco S, Robinson JF, Porcelli A, Hamilton EG, Mitjans M, Maddalena G, Begemann M, Seidel J, Yanamori H, Jaffe AE, Berman KF, Egan MF, Straub

RE, Colantuoni C, Blasi G, Hashimoto R, Rujescu D, Ehrenreich H, Bertolino A, Weinberger DR (2018) Convergence of placenta biology and genetic risk for schizophrenia. Nat Med 24(6): 792–801

Ursini G, Punzi G, Langworthy BW, Chen Q, Xia K, Cornea EA, Goldman BD, Styner MA, Knickmeyer RC, Gilmore JH, Weinberger DR (2021) Placental genomic risk scores and early neurodevelopmental outcomes. Proc Natl Acad Sci USA 118(7)

Waddington CH (1942) Canalization of development and the inheritance of acquired characters. Nature 150(3811):563–565

Weismann A, Parker WN, Rōnníeldt H (1893) The germ-plasm: a theory of heredity. C. Scribner's sons, New York

West-Eberhard MJ (2003) Developmental plasticity and evolution. Oxford University Press, Oxford\New York

Williams GC (1957) Pleiotropy, natural selection, and the evolution of senescence. Evolution 11(4): 398–411

Chapter 4
Modular Complementarity as an Evolutionary Organizer: Beyond Base Pairing

Robert Root-Bernstein

Abstract The possible permutations of small molecules, gene and protein sequences that could have participated in the origins of living systems, and the further permutations of their possible interactions are astronomical. Living systems utilize only a tiny fraction of these to form modern interactomes. How, in the face of extraordinary permutational possibilities, did these interactomes evolve, and how do they retain their integrated functions against persistent and ongoing mutational variation? This chapter focuses on a particular aspect of this larger problem, which is how the receptor and transporter systems essential to cellular life evolved. Did receptors and transporters evolve independently of their ligands, in which case, how did they become integrated and functional? If receptor– or transporter–ligand pairs coevolved by selection for specific functions, by what process? This chapter argues the process that limits permutational possibilities, creates interactomes, and endows them with functions involves a combination of two mechanisms: 1) modularity, using small, repeating units of structure to build more complex molecules; and 2) selection for molecular complementarity, non-covalent interactions among molecules mediated by stereochemical fit. The molecular paleontology inherent in modern interactomes demonstrates selection for molecularly complementary modules (MCMs). Two case studies explore how MCM was used to bootstrap complex systems from small-molecule complexes: the evolution of (1) glucose regulation and (2) adrenergic and opioid receptor–ligand pairs and their cross talk.

1 Introduction

One of the most puzzling problems bedeviling modern evolutionary theory is the combinatorial explosion problem. Combinatorial explosion refers to the fact that as the number of components of a system increases linearly, the possible permutations of these components increase exponentially (Schuster 2000). Thus, the more

R. Root-Bernstein (✉)
Michigan State University, East Lansing, MI, USA
e-mail: rootbern@msu.edu

G. Witzany (ed.), *Integrative Theory of Evolution - Aspects and Insights*,
https://doi.org/10.1007/978-3-032-24295-2_4

complex a system, the more difficult it should become to maintain its organization and the less likely that organization is to arise by chance. From a purely probabilistic point of view, it is difficult to imagine how cellular life, which is characterized by highly integrated interactomes, could have arisen, not only by chance mechanisms but also under the influence of mutational processes that would seem to work against the maintenance of organization (Moorhead and Kaplan 1967). This chapter will demonstrate that the probabilistic approach is based on incorrect assumptions. The resulting conclusion that organized systems are highly unlikely to have evolved spontaneously is not, therefore, warranted. In fact, a combination of several natural mechanisms working together, including, most notably, molecular complementarity and modularity, constrains the combinatorial possibilities so drastically as to make self-organizing chemical systems subjected to a range of environmental selection processes highly likely. Moreover, the molecularly complementary module (MCM) approach to understanding the origins and evolution of biological order also provides direct insights into the reasons that particular molecules were selected to perform specific functions. Evidence for MCM can be found not only in the molecular paleontology that has shaped modern molecular systems but also in the fundamental chemical properties of the sets of molecules making up these systems. This approach goes beyond both the Modern Synthesis, which focused on the effects of natural selection on genetic variations, and the New (or Extended Evolutionary) Synthesis, which has integrated development, plasticity, and niche construction, by focusing on how selection for interactivity between and among the components of living systems has constrained and shaped their evolution. Evolution, according to this approach, selects for increasingly integrated networks, not ever-diverging trees. Those components that cannot integrate are selected out, while those that do become the material from which novel structures and functions emerge (Root-Bernstein and Root-Bernstein 2016b; Root-Bernstein and Bernstein 2024).

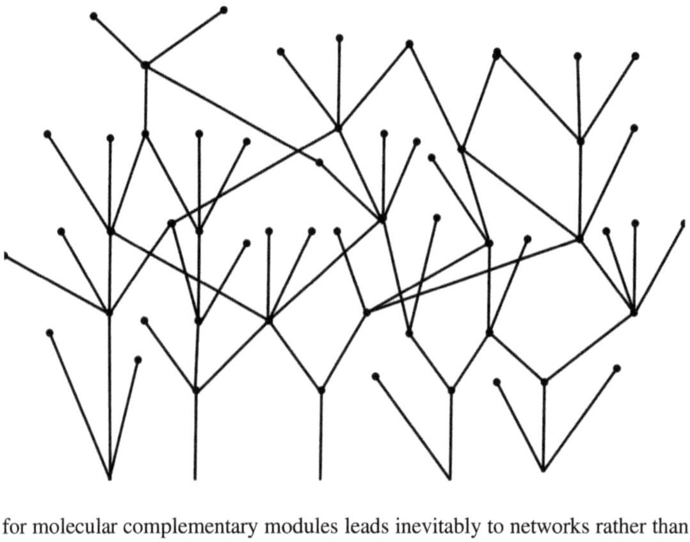

Selection for molecular complementary modules leads inevitably to networks rather than trees and convergence of structure and function rather than divergence and, thus, explains the origins of interactomes. The result is a complex tree diagram with multiple branching lines and nodes, illustrating hierarchical relationships. The diagram features interconnected branches that diverge into smaller sub-branches that, in turn converge and diverge, to create a networked system structure.

2 Combinatorial Explosions: The Problems of Accounting for Selection for Organization

The combinatorial explosion problem comes in many forms. The most basic involves calculations of the number of permutations possible from even a limited number of elements, nucleotides, or amino acids. Over 100 million molecules can be made up of 11 atoms of C, N, O, and halogens (Fink and Reymond 2007) and 166 billion from 17 atoms if sulfur or phosphorus are added to the mix (Ruddigkeit et al. 2012). The probability of generating just a single molecule of any given protein is on the order of $1/10^{600}$ (Jacobson 1955, 125; Cairns-Smith 1971; Crick 1981; Hoyle 1984), and Hoyle and Wickramasinghe (1981) calculated that the probability of obtaining a cell's approximately 2000 different enzymes in a random trial is about one in $10^{40,000}$. In contrast, the number of such small molecules (amino acids, nucleic acids, sugars, and their metabolites) in a modern cell has been calculated to be on the order of a few hundreds to perhaps a thousand (Morowitz et al. 2000), and only 20,000–30,000 unique protein sequences are actually transcribed in mammalian cells and only a few thousand in bacteria, yeasts, and insects (Oliver and Leblanc 2003; Engel et al. 2014; Forde et al. 2014). So one aspect of the combinatorial explosion problem is how evolution has limited its use of myriad possible small molecules to such a small number and then retained just 300 highly conserved gene families across all cellular life (Zhang et al. 2018).

The combinatorial explosion is exacerbated by the problem of evolving interactomes. Interactomes may be defined as the set of nonrandom, non-covalent (and therefore reversible), chemical interactions among the molecules and/or macromolecular assemblages, organelles, or cells making up the components of a living system (Sanchez et al. 1999; Root-Bernstein and Root-Bernstein 2016b; Yehorova et al. 2024). Every type of molecule in a living system interacts specifically with one or more (often many more) to create highly conserved networks, and this process leads to another type of combinatorial explosion called "Levinthal paradox of the interactome" (Tompa and Rose 2011). The Levinthal paradox of the interactome results from calculating the possible permutations of molecular interactions or complexes that could be formed by any set of molecules or molecular aggregates. Tompa and Rose (2011) have calculated that the theoretical number of possible specific interactions for n proteins is $n!/(2^{n/2} \times (n/2)!)$, which, for an interactome of 4500 proteins, gives 10^{7200} different possibilities. Given that each of these proteins requires a gene and each gene a promoter region, all of which need to be properly linked and controlled through additional interactomes, the probability of evolving a complex interactome would appear to be nearly zero.

However, the validity of the Levinthal paradox, like the original combinatorial explosion problem, depends, as all such calculations depend, on the assumptions made in setting up the statement of the problem. In this case, Tompa and Rose (2011) have assumed, as do all combinatorial explosion calculations, that every possible variant is generated *before any selection begins*. Thus, according to these calculations, evolution has to find the proverbial needle in a haystack. But that assumption is clearly incorrect. Evolution by natural selection always starts with a fit system (i.e., one that has already been selected for adaptation to a local environment) and then makes variations upon it. This is why evolutionary systems are shaped by historical contingency (Gould 2002; Blount et al. 2018). By starting with a fit system, the number of possible variants on it is not every possible permutation of all its components and the components of every other possible similar system (which is how the calculations cited above were made). Rather, the possible permutations are limited to variations based on both the already limited set of selected components and the fitness of their organization within the existing system (Root-Bernstein and Bernstein 2024). I shall return to this crucial point near the end of this essay.

Since evolution bootstraps modifications off of already fit systems, severely limiting the possible permutations to modifications of those fit systems, the problem to be addressed in the origin of the first living systems is therefore reduced to the problem of identifying the selection criteria present during the initial emergence of the first chemical systems and determining how these served to prune the set of possible permutations during system evolution. In addition, these selection processes should have the ability to maintain functional organization against mutational variations without also limiting evolvability. My argument in a nutshell, then, is that only when we have a good understanding of prebiotic selection can we begin to calculate the actual probabilities of living systems arising de novo. Blindly calculating odds in the way that those who have raised the specters of combinatorial explosions do is both pointless and misleading.

3 Prebiotic Selection

Some progress has been made toward identifying some specific selection factors involved in pruning back the tree of combinatorial possibilities. Molecules are made in different amounts at different rates. The amount of amino acids, lipids, and other molecules made in Miller-like prebiotic chemical experiments varies more than a thousandfold, and not every possible small molecule is generated in measurable quantities (Miller 1955; Bada 2013; Root-Bernstein et al. 2023). Solubility, reactivity, stability, ionizability, and other chemical properties further limit which molecules can actually be useful to a living system (Green and Goldberger 1967; Morowitz et al. 2000; Weber 2004). Moreover, selection among proteinogenic amino acids and natural nucleotides from complex mixtures has been reported under prebiotic conditions so that not every molecule survives (Fairchild et al. 2024; Szabla 2021; Islam et al. 2017). Thus, the number of chemical permutations is significantly less than the calculations described above make out. Put simply, the fact that a compound could theoretically be made does not mean that it was made or survived long enough to participate in further (bio)chemistry.

However, cutting down the remaining chemical possibilities to the tens of thousands of molecules actually used by living systems from the astronomical number of possible permutations of their synthesizable, stable components still requires pruning those permutations by hundreds of orders of magnitude, which current criteria fail to do. That networks can self-organize in mathematical models of systems (Kauffman 1993; Gardner 1970) and that hypercycles in metabolic cycles clearly act as seeds for crystallizing efficient, constrained organized biochemical systems (Eigen and Schuster 1979; Morowitz 1999) is beyond doubt, but these approaches fail to provide (bio)chemical mechanisms for the origins of such networks. In particular, living systems deploy these metabolic cycles by means of highly efficient sets of enzymes that are required to increase energy efficiency while also preventing decay of the cycles through the bleeding off of metabolites to non-useful or possibly poisonous side reactions. These enzymes do not operate independently from each other but rather self-organize to form enzyme complexes that "channel" substrates from one enzyme to another, allowing them to "tunnel" from one to the other without free diffusion (Srivastava and Bernhard 1986; Ovádi et al. 1989; Beeckmans et al. 1990; Pareek et al. 2021). How do the reaction cycles involving metabolites become linked into integrated sets of self-organizing proteins capable of catalyzing them? Do the uncatalyzed cycles evolve first (as assumed by Eigen, Schuster, and Morowitz), the enzymes independently evolving afterward to optimize each separate reaction? Are these enzymes selected for optimizing cycle functions from a permuted pool of every possible gene for every possible protein sequence by randomly interacting with the substrates of the cycles? If so, is selection for enzyme–enzyme interactions (channeling) an independent and later third step, and how was this third step genetically integrated? Or do the enzymes (or the genes encoding them) for each metabolic reaction evolve first, with the channeling cycle "accidentally" forming as a result of selection for enzymatic cycles? Or, alternatively, do the enzymes, their genes, and their metabolites coevolve and, if so, by what mechanism? How does a substrate "select" a peptide or protein sequence to catalyze its conversion to a product useful in a cycle? And does a coevolution process involve concurrent selection for the enzyme–enzyme interactions required for channeling as well? These remain important, unresolved questions.

The problems just raised with metabolites and their enzymes have myriad analogs throughout the evolution of living systems. Cells must interact with their environments, regulating the uptake of nutrients, excluding or excreting harmful or useless compounds, and secreting "messages." These processes required the evolution of specific receptors and transporters and their ligands. Were the small-molecule ligands for these complex receptors and transporters selected first and the larger proteins and/or their genes afterward? If so, what properties permitted the selection of the ligands, appropriate proteins, and their respective genes? Or did the receptors and transporters evolve first and, afterward, select their ligands? If receptors and transporters or their genes evolved independently of ligands, what possible factors would have led to their integration as a functional system? Or did the ligands, their proteins, and genes coevolve and, if so, by what mechanism? Was every possible combination of ligand, receptor, or transporter and their respective genes explored (returning us to the combinatorial explosion problem), or were there other factors at work that limited the possibilities to some manageable set? Moreover, regardless of which of these three mechanisms was at work, how did the receptor or transporter and its ligands acquire the specific functions that they perform? Is the function of a molecule "assigned" randomly by the evolutionary process, or is there some physicochemical "reason" that molecules perform their particular functions with a living system? More generally, how did feedback inhibition, allosteric regulation of enzyme and receptor activity, and other metabolic control mechanisms that involve specific interactions between unlike molecules evolve?

Astoundingly, very few of these questions have yet been addressed. I have been unable to find any explanations for the origins of enzymatic channeling or even how enzymes became paired with specific substrates. The literature concerning how receptors first came into existence and were paired with an appropriate ligand is extremely sparse (e.g., Bost et al. (1985a, b), Topiol (1987), Dwyer (1989), and Root-Bernstein (2005)). Instead, virtually the entire literature on the evolution of enzymes, receptors, and transporters consists of studies of how these types of molecules have varied through evolutionary time *after* they had already become functional sets. Their origins remain obscure. And yet that would seem to be the key question that needs to be addressed if we are ever to make systems biology a real science and trace evolution back to its pre-cellular origins.

I cannot claim to have all the answers to the questions I have just posed above, but I have spent my career looking deeply into a few of them, and this paper will summarize some of the little progress I have made. The most important aspect of my approach is that it lies somewhere between the mechanistic reductionist approach of modern molecular biology and the wholistic approach that dominates cellular or organismal biology. Thus, the types of questions I have posed above can only be answered, it seems to me, by studying and understanding the principles guiding the interactions among molecules and selection for those interactions—a philosophical position that stands somewhere between molecular biology and cellular or organismal biology. Investigating interactome evolution requires a third way (Noble 2006; Henney et al. 2015; Witzany 2016; Witzany 2021).

4 Selection for Functionality Through Selection for Molecularly Complementary Modules (MCMs): A Way Out of Infinities

I propose that the solution to the combinatorial explosion and Levinthal paradox lies in the integration of two principles that are abundantly evident in modern living systems: simultaneous selection for modularity and complementarity. Both of these principles severely constrain molecular variations and, in combination, can explain selection for interactome evolution. Moreover, the properties that result from modular complexes result in new molecular properties that cannot be predicted from their individual components. Thus, modular complementarity provides a means of limiting permutations, forming interactive networks, and bootstrapping the emergence of novel molecular functions. I will treat modularity first and then complementarity, before demonstrating through case studies the implications of their combination for understanding the evolution of functional interactomes.

4.1 Modularity

Modularity can be defined as a property of a system's components characterized by the ability of the components to be separated and recombined in a limited number of permutations. The modular assembly of many biological macromolecules has been well-established, with recognition of duplication of sequences within proteins having been established as long ago as the 1970s (Fitch 1966; Waterson and Konigsberg 1974; Jue et al. 1980; Doolittle et al. 1982; Doolittle 1992; Burton 2018). Modularity may assist proteins to fold correctly through preprogrammed stages (Burton 2018; Im et al. 2023). Many proteins have also been found to be genetically encoded in sequentially independent modules, which are then spliced, sometimes in alternative orderings, during transcription (Braun et al. 2018; Roy et al. 2013). Ribosomal RNA appears to have been assembled from multiple tRNA-like modules (Root-Bernstein and Root-Bernstein 2015; Prosdocimi et al. 2021; Opron and Burton 2018), while tRNAs themselves appear to have been assembled from proto-tRNA mini-helices (Root-Bernstein et al. 2016b; Burton 2020). All polysaccharides display obvious modular construction. And modularity has been educed as a selective agent in the evolution of carbon fixation cycles (Braakman and Smith 2013). These, and many other examples, have established modularity as a fundamental source for generating diversity and increasing complexity in evolutionary systems generally (Schiller 2016). Additionally, such modularity has been strongly associated with the evolvability of proteins by providing stable, robust kernels around which modifications leading to new functions can be made rapidly (Rorick and Wagner 2011).

These implications of modularity for streamlining adaptation have gotten much less attention than they warrant. The Nobel Prize–winning economist, artificial intelligence pioneer, and systems expert Herb Simon (1962, 1981) provided the most general analysis of how modularity can drastically increase the efficiency by which any system can be assembled and provided clear applications to evolutionary systems.

Alluding to the time-honored use of watchmakers in philosophical discussions of evolution, Simon explained the utility of modularity by means of an analogy to the way two watchmakers, Hora and Tempus, might assemble watches. Each watchmaker must assemble a watch composed of 1000 pieces. Tempus has designed his watch so that he must assemble the whole thing in one sitting. If he is interrupted at any point, he loses all of his work because the partially assembled watch is unstable and falls apart. Hora, however, designs his watch in a series of stable subassemblies of about ten components each. Groups of ten of these subassemblies can then be organized into more complex assemblies that are also stable, and then these sub-assemblies can be put together to form the completed watch. If Hora is interrupted at any point, only the subassembly he is then working on falls apart.

Simon provided a general equation for comparing the two processes: "The time required for the evolution of a complex form from simple elements depends critically on the number and distribution of potential intermediate stable forms. In particular, if there exists a hierarchy of potential stable 'sub-assemblies', with about the same span, S [the number of components making up the sub-assembly], at each level of the hierarchy, then the time required for a sub-assembly can be expected to be about the same at each level—that is, proportional to $1/(1 - p) \exp S$" where p is the probability of being interrupted" (Simon 1981, p. 203). If one assumes that it takes 10 min to add each part, Tempus will need 170 h to assemble all 1000 parts—realistically, an impossibility for a human being who needs to sleep, eat, etc. Hora, however, can put together a stable subassembly every 100 min, completing a watch, even accounting for interruptions, on average every 25 days. Even if each part can be added in just a minute, Hora will complete 20,000 times as many watches as Tempus (Simon 1962). So the efficiency provided by modularity vastly improves the odds of achieving any assembly process. By analogy, Tempus represents the difficulty of assembling the exact sequence of any protein or gene from its subunits, while Hora represents the benefits of assembling proteins and genes from previously selected modules.

Simon developed his work on modularity as a direct response to Jacobson's (1955) argument that the amount of information in even the simplest protein made evolution impossible (see above). His analysis of the benefits of modularity provided a way out of the combinatorial explosion problem, and it would be easy to stop at this point and conclude that the problem is, therefore, solved. However, Simon's analysis is, itself, flawed by two questionable assumptions. First, he embraces what might be called "the engineering fallacy." Second, he provides no mechanism for providing the stability that Hora's stable subunits require.

The engineering fallacy first. Unlike Simon's Hora and Tempus, evolution is a *blind* watchmaker that does not have a blueprint to guide the assembly of its watches. Evolution does not know at the outset what order to assemble the parts, how many parts a watch contains, or even that it is making a watch. Therefore, such a blind watchmaker would have to explore every possible combination of Simon's 1000-part watches—the equivalent of assembling the functional interactome of a 1000-component living cell. The possibilities are so staggering that it is worth looking at a much simpler problem of assembling a watch comprising a mere 25 parts. A watchmaker like Tempus, who does not know what order to put the pieces together, how many there are, or even what he is making—and, moreover, who needs to assemble the whole working mechanism in one go!—would have to explore 25! or 1.55×10^{25} possibilities. A 100-piece watch would require exploring 9.33×10^{155}

permutations. One thousand pieces could be arranged in 4×10^{2567} permutations. This is the Levinthal paradox in a different form and just as seemingly intractable.

Surprisingly, modularity actually becomes more effective in such a totally "blind" case than when assuming the engineering fallacy. Sticking with our 25-piece watch for the moment, let us assume that Hora assembles it in stable modules of five parts. Each five-part subunit needs to explore only 5! or 120 combinations. Hora would then need to explore 5! possible combinations of the five subunit modules—another 120 possibilities. The total is six sets of 120 combinations, or 720 possibilities. Compare that with the 25! or 1.55×10^{25} possibilities that Tempus has to explore. If stable five-piece subunits were used to assemble a 100-piece watch, only 2904 possibilities would need to be explored instead of 9.33×10^{155} permutations. The smaller the modules, the more effective modularity appears to be: If stable subunits composed of ten pieces were used for the 100-piece watch instead of five-piece subunits, a total of 3.99×10^{7} possibilities would need to be explored—still a savings of 148 orders of magnitude compared with Tempus. So, for a truly blind watchmaker who does not even know he is assembling a watch, modularity is far more effective than even Simon thought.

Does modularity, therefore, solve the combinatorial explosion problem and the Levinthal paradox? Not quite! The second and perhaps more serious flaw in Simon's analysis of modularity (as with Kauffman's (1993) analyses of self-organizing systems) is that he offered no mechanism by which Hora could produce the necessary stable subassemblies. There is no chemistry or biochemistry to enable self-organization. In Simon's scenario, modules just miraculously appear. So, modularity could literally be a lifesaver for evolutionary processes but only if we know how it is implemented.

Enter molecular complementarity.

4.2 *Molecular Complementarity and the Origins of Information and Functionality*

The key to implementing modularity is complementarity and, in the case of bioevolution, more specifically, molecular complementarity (Root-Bernstein and Dillon 1997). Stable modules require their parts to fit stereochemically and stick together non-covalently. Molecular complementarity is present universally in every biochemical system (Csermely 2006). It is present in polynucleotides, enabling replication and transcription; in the translation system; in enzyme–substrate interactions; the formation of polyenzymatic channeling cycles; receptor–ligand specificity; transporter–metabolite specificity; antigen–antibody and antigen–T cell receptor recognition; lipid bilayer self-assembly; the formation of lipid rafts; the formation of receptor homo- and heterodimers; and so on. Indeed, complementarity in molecular and cellular interactions is ubiquitous and extends to small-molecule interactions, such as aromatic compounds binding to each other (Root-Bernstein and Westall 1984a; Root-Bernstein 1987; Root-Bernstein and Dillon 1997; Dillon et al. 2006; Root-Bernstein and Dillon 2008; Root-Bernstein 2012), aromatic compounds binding to peptides (Root-Bernstein 1984; Root-Bernstein and Westall 1984b; Root-Bernstein 1987; Root-Bernstein and Dillon 1997; Root-Bernstein and Dillon 2008; Root-Bernstein 2012; Root-Bernstein and Dillon 2014), and glucose binding to insulin (Anzenbacher and Kalous 1975; Yu and Caspar 1998; Zoete et al. 2004). Peptides bind to peptides (Root-Bernstein and Westall 1984c; Root-Bernstein and Westall 1986; Loo et al. 1994; Root-Bernstein and Holsworth 1998; Siemion et al. 2004; Štambuk et al. 2014; Štambuk and Konjevoda 2020; Dayhoff 2nd et al. 2020; Biruš et al. 2025), and peptides binding to RNA aptamers (Sugaya et al. 2008). Such small-molecular complementarity was, I will argue below, essential to the origins of most, if not all, biomolecular systems and provided the preselected, fit modules from which more complex, interactive macromolecules evolved (Root-Bernstein and Dillon 1997; Hunding et al. 2006; Root-Bernstein 2012).

One of the most important functions of molecular complementarity is to mediate self-organization and information capture within molecular systems by creating stable networks (Raine et al. 2003). Well-known examples include virus and ribosome self-assembly. Additionally, through self-organization, molecular complementarity provides a mechanism for creating and storing information. It is often assumed that only polynucleotides contain all necessary biological information. This is not the case. Information can be defined as how unlikely (surprising) a particular event or outcome is. Thus, anything that makes a system less random increases the information capacity of the system (Brillouin 1953; Segre et al. 2000; Norris and Root-Bernstein 2009). Every self-assembling system or subsystem—every interactome, regardless of the nature of its components—therefore, encodes information. Molecular complementarity, no matter how or where it is observed—not just in DNA or RNA—results in information storage and has the potential for information transmission by replication of its organization.

In this regard, it is important to remember that many cellular structures replicate, and some essential peptides are synthesized independently of genetic information. For example, cell membranes and their components are simply divided between daughter cells (Lopez et al. 2020; Warren 1993), while mitochondria (Chen et al. 2020; Christodoulou 2000), chloroplasts (Hennis and Birky Jr 1984), ribosomes (Tsai et al. 2012), and the nuclear envelope (Tsai et al. 2012; Chaudhary and Courvalin 1993) are stochastically assorted during replication. Modular assemblies that are present in small copy numbers, such as the Golgi apparatus (Misteli 1996), vacuoles (Weisman 2003), actin filaments (Sheahan et al. 2004), and enzyme assemblies that carry out channeling, are broken down into their modular components just prior to cell division; the components are randomly or stochastically distributed to the daughter cells; and the organelles and molecular aggregates are spontaneously reassembled (Rabouille et al. 1995; Chaudhary and Courvalin 1993) within the daughter cells. De novo synthesis of the components during cell growth provides additional components for the modules for all of these assemblages, which are incorporated by means of their molecular complementarity. Thus, molecular complementarity can mediate the accumulation of information and its transfer via self-organization and compositional replication. Evolution by random variation and selection does not, therefore, require an RNA or DNA world (Huh and Paulsson 2011; Mayer et al. 2024).

Molecular complementarity also buffers systems not only in the classical sense of moderating pH but also structurally by permitting aggregates to form, decompose, or rearrange depending on environmental conditions and stresses. pH buffering has been demonstrated for many small-molecule complexes, which respond to titration just like classical buffers and can serve as a measure of complementarity (Root-Bernstein and Westall 1986; Root-Bernstein 1987; Root-Bernstein and Dillon 1997; Root-Bernstein and Dobbelstein 2001). Supramolecular systems also buffer themselves, not just against changes in pH but also against temperature and concentration alterations, the number of bonds between the molecules generally determining the strength of the buffering (Paffen et al. 2015, 2017; Desrosiers et al. 2022). In aggregate, the ubiquity of molecular complementary complexes within living systems results in the maintenance of homeostasis (Dillon and Root-Bernstein 1997; Schmitt 2004; Root-Bernstein 2010; Hancock and Ang 2019; Mochizuki 2022). Thus, Shapiro (2006) has concluded that small-molecule interactions were essential to the origins of life, while Singh (2024) has summarized the importance of molecular complementarity this way: "The postulates of biological complementarity are that (1) it was an essential condition in the origin of life; (2) it provided physiological flexibility that allowed organisms to mount self-protection response and complexity to evolve in the face of deleterious mutations before the evolution of bi-parental sex; (3) it laid the foundation for the evolution of a choice of response when confronted with threat; and (4) it applies to all levels of biological organizations and, thus, can serve as a basis for the unification of mechanistic and holistic biology" (Singh 2024).

5 Case Studies in MCM

The key question remains: How do modularity and molecular complementarity synergize to enable selection for interactomes? The remainder of this chapter will examine, by means of two case studies, some of the ways in which integrating modularity with molecular complementarity provides mechanisms for bootstrapping complex interactomes from naturally occurring small-molecule interactions. I will demonstrate that the specific functions these interactomes perform follow necessarily from the nature of the components involved in these interactions and the ways in which the resulting complementary modules are then mixed and matched to provide novel variants upon which natural selection can work. Two case studies are provided here, and two additional ones have recently been published (Root-Bernstein 2026).

5.1 Interactome Evolution Case 1: Glucose and its Regulatory Molecules

One of my long-term projects has been to try to reconstruct how glucose regulation may have evolved. An assumption has been that there is a molecular paleontology that left its imprints on modern living systems and from which clues can be derived about the origins of those systems. In the case of glucose regulation, one question is why insulin, rather than some other protein, regulates glucose. Molecular complementarity provides a mechanism for this selection. Glucose and insulin exhibit a little-known example of molecular complementarity: Insulin contains six glucose-binding sites, two high-affinity and four low-affinity (Anzenbacher and Kalous 1975; Yu and Caspar 1998; Zoete et al. 2004). These sites share sequence similarities, suggesting that they derive from one or two common glucose-binding peptide modules. Thus, it appears that insulin evolved as a regulator of glucose metabolism from the ligation of smaller glucose-binding peptides. The binding of glucose to these modules and to insulin itself would have served initially as a storage mechanism to protect the glucose from being degraded or oxidized. Since the binding of glucose to insulin is reversible, insulin could also have regulated the local glucose concentration by means of taking up and releasing glucose. In addition, insulin is amphoteric, with affinity for both water and lipids, permitting it potentially to have acted as a simple glucose transporter for delivering glucose through lipid membranes prior to the evolution of specific glucose transport proteins.

3-D representation of X-ray crystallographic figure showing five of the six locations where glucose (circled) binds to insulin (ribbon structure). The images hows a complex arrangement of molecular bonds, peptide side chains and rings representing amino acids, some of them labelled with the amino acid names. Some of the ring structures represent glucose molecules that have bound to various places on the peptide structure. (Adapted from Zoete et al. 2004)

Another principle upon which I have been operating is that molecularly complementary molecules exhibit complementary physiological functions. Stated conversely, if two molecules have complementary functions, they will often be found to be molecularly complementary (Root-Bernstein and Dillon 2008). This principle appears to apply to glucose regulation. While insulin controls the passage of glucose into cells, glucagon controls the release of glucose from cells. Insulin and glucagon are molecularly complementary, forming complexes that are stable enough to persist even through multiple rounds of crystallization (Root-Bernstein and Dobbelstein 2001; Dillon et al. 2006). In fact, glucagon was discovered as a contaminant of early preparations of insulin, and its presence in insulin preparations used therapeutically lasted until insulin began to be made by recombinant means (Bloch 1994). The binding of glucagon to insulin blocks many of the glucose-binding sites and would, therefore, have acted as a means of promoting glucose release and preventing glucose storage in a primitive metabolic system. Indeed, nonenzymatic glycation of insulin resulting in covalent linkage has been documented, which is reversible over the short term but results in the inactivation of insulin due to the eventual formation of Maillard reagents (Rhinesmith et al. 2017; O'Harte et al. 2000; Walke et al. 2021). Glucagon and glucagon-like peptide also glycate, impairing their activity (O'Harte et al. 1998). Thus, insulin concentrations exert direct chemical feedback regulation on both insulin and glucagon activity. Moreover, our unpublished data show that glucagon binds to glycogen and shares significant homologies with glycogen phosphorylase (the enzyme that breaks down glycogen) but not with glycogen synthase (which makes glycogen). Thus, glucose, insulin, and glucagon were, from the outset, inextricably chemically linked through a series of complementary interactions that mediated their eventual physiological functions.

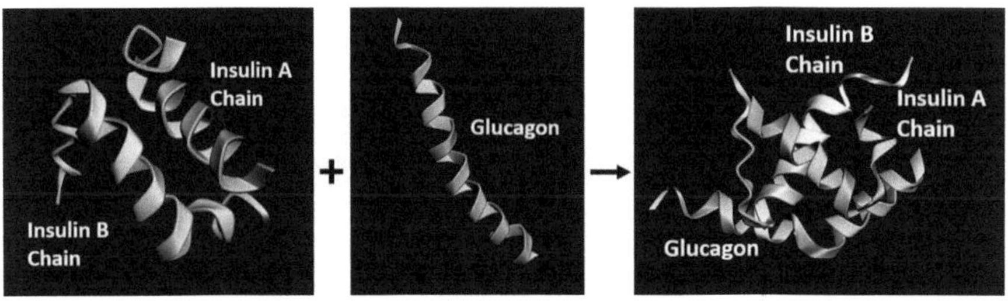

3-D AlphaFold2 structural models of insulin, glucagon, and the complex that forms when the two bind to each other. The insulin A and B chains are shown in the left-hand panel; the glucagon alpha helical structure in the middle panel; and the complex that forms when the insulin A and B chains bind to the glucagon helix is show in the right-hand panel.

Clearly, the kind of primitive metabolic system just outlined would not have been particularly efficient compared to modern glucose regulation, so the next question is how that primitive system might have been bootstrapped to yield the modern system. The fact that glycogen phosphorylase integrates a glucagon module that mediates glycogen binding provides a model for how such systems can be bootstrapped. Two additional types of proteins were also needed: glucose transport proteins and insulin receptors to regulate them. Modularity played crucial roles in the origins of both.

Glucose transporters come in many varieties, but all share one common feature, which is that they are composed of either six or 12 transmembrane regions that form a pore through which glucose is transported. The 12-transmembrane structures are essentially a pair of connected six-transmembrane versions. The six transmembrane regions may provide the key to understanding the origins of these receptors, for insulin itself exists in three forms: as single molecules, dimers, and, notably, hexamers. Insulin hexamers also produce a structure with a pore at the center, with several glucose-binding sites facing into the pore. Ligating six insulin molecules would, therefore, form a simple glucose transporter, and in fact, homology searching reveals that all insulin transporters have multiple insulin homologies located precisely so that they align to form the glucose transport core (Root-Bernstein 2007). A simple explanation for the origins of glucose transporters is, therefore, that at some point in evolution, copying of the insulin gene resulted in a string of six insulin-encoding regions that, when properly folded, produced a glucose transport protein. Duplicating this new protein gene then gave rise to the 12-transmembrane versions. Such copying and ligating of peptide modules—the genetic equivalent of musical ostinato—has frequently been documented in the evolution of novel peptides and proteins (Delihas 2007; Dwyer 1998; Doolittle 1992).

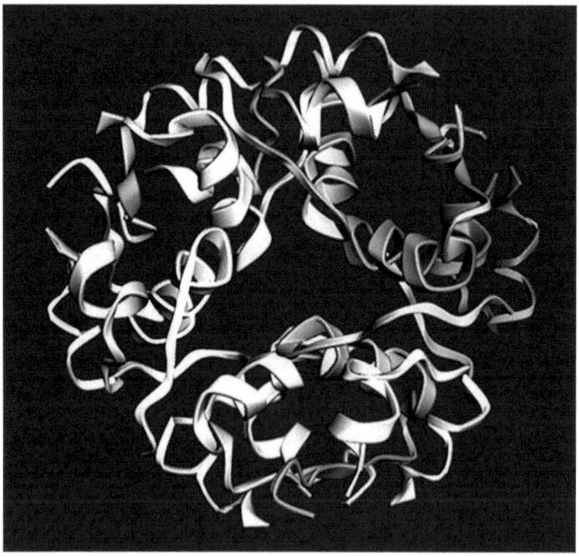

3-D AlphaFold2 ribbon model of insulin hexamer showing the complex folding and binding structure that forms from the interactions between the individual insulin helices and loops. The result is an open core through which glucose molecules might be able to pass. Such hexamers, if ligated, could have formed the basis for glucose transporter (GLUT) transport cores: GLUTs consist of a pair of hexamer-based transport cores consisting of multiple, high-homology sequences with insulin. (Root-Bernstein 2007)

Insulin receptors may have evolved from similar variations on modular complementarity. Dwyer (1989, 1998, 2021) has suggested that peptide receptors evolved from self-aggregating peptides such that the ligand became the basis for its own receptor. Thus, he predicted that peptide receptors should incorporate regions of high homology with their peptide ligand, specifically in the receptor-binding region. Dwyer demonstrated the principle with regard to acetylcholine receptor binding to the peptide bungarotoxin (an acetylcholine-binding peptide). Indeed, a sequence mimicking bungarotoxin appears at the acetylcholine receptor–binding site. Dwyer has since extended the application of his theory to other peptide receptors as well (Dwyer 1989, 1998, 2021), and I have expanded Dwyer's idea from homo-complementary to hetero-complementary ligands of any molecular type (e.g., monoamines and alkaloids) that can bind to specific peptide sequences (Root-Bernstein 2002; Root-Bernstein 2005; Root-Bernstein and Vonck 2009).

Insulin and glucagon receptors are two (among many (Root-Bernstein 2002, 2005)) of the receptor types to which Dwyer's principle appears to apply. Insulin, as noted above, self-aggregates into both dimers and hexamers. According to Dwyer's principle, the insulin receptor should, therefore, contain one or more insulin-like sequences within its binding site. This is, in fact, the case. The insulin receptor–binding site is actually quite complex, consisting of several insulin-like modules of high homology within the insulin-binding site. Additionally, two glucagon-like homologies also appear in the insulin receptor–binding site. As noted above, glucagon is complementary to insulin and binds to it with high affinity. Thus, both homo- and hetero-complementary modules appear to have played roles in the construction of the insulin receptor (Root-Bernstein 2002; Root-Bernstein 2005).

A number of testable implications follow from the modular complementary origins just described for the insulin receptor. One is that the insulin-like regions of the insulin receptor should bind to glucose and, therefore, that glucose should modulate insulin binding. Both of these implications have been tested and shown to be correct (Root-Bernstein and Vonck 2009, 2010). Second, because exposure of proteins to glucose can result in nonenzymatic glycation of the protein and the insulin receptor has insulin-like regions, it should be particularly susceptible to such glycation. It is (Rhinesmith et al. 2017). A third implication is that glucagon, being complementary to insulin and also to itself, should bind to the insulin receptor modifying insulin activity. In fact, combinations of insulin with glucagon have been found to significantly increase the uptake of glucose into muscle cells, strongly suggesting that the hormones enhance each other's activity (Elrick et al. 1958). Indeed, insulin–glucagon cross talk has been observed in many types of experiments, and glucagon does alter insulin binding to its receptor (Yamauchi and Hashizume 1986). Moreover, some experiments suggest that insulin binds directly to the glucagon receptor (Blecher et al. 1974; Giorgio et al. 1974; Blecher and

Goldstein 1977), though this possibility is not part of the standard explanation of insulin–glucagon cross talk (Habegger 2022).

The same mechanism just described for the origins of the insulin receptor also appears to contribute to the origins of the glucagon receptor and the classes of insulin-like and glucagon-like ligand–receptor pairs as well. Glucagon self-aggregates, and the glucagon receptor contains several glucagon-like homologies at its binding site (Root-Bernstein 2002; Root-Bernstein 2005). It also contains two insulin-like regions that may contribute to glucagon binding through hetero-complementarity (Root-Bernstein 2002). As noted above, insulin–glucagon cross talk has been observed in many types of experiments, but the specific possibility of insulin binding to the glucagon receptor has never been tested directly (Habegger 2022). It is also possible that the hetero-complementarity expressed in the two receptors may contribute to the formation of insulin receptor–glucagon receptor dimerization, which could be an (as yet untested) alternative explanation for the observed cross talk.

Regarding insulin–glucagon receptor cross talk, several observations are relevant. First, insulin is just one of three related peptides with overlapping functions and binding specificities. The other two peptides are insulin-like peptides I and II, which are highly homologous to insulin. Their respective receptors are also highly homologous to the insulin receptor. It is thought that the insulin-like growth factors (IGFs) and their receptors evolved by gene duplications, followed by variations on the insulin–insulin receptor pair (Navarro et al. 1999). Second, through a similar process of gene duplications, glucagon and its receptor became the evolutionary precursors to a series of glucagon-like peptides and their receptors, which have since taken on other functions (Irwin and Wong 2005; Irwin 2014). Third, the receptor of one of these glucagon-like peptides, glucagon-like peptide-1 (GLP-1), is known to heterodimerize with the insulin receptor, integrating insulin and GLP-1 regulation (Wang et al. 2022). Such heterodimerization probably occurs between other insulin-like and glucagon-like receptors as well. Thus, selection for insulin–glucagon binding during prebiotic or early biotic evolution has provided the basis for broader networking of hormones derived by duplication from the relevant genes that probably involves many, if not all, of the insulin-like and glucagon-like receptors. Notably, during this gene duplication process, a very high degree of homology is maintained between the ligands, while the receptors are much more likely to undergo greater diversification, suggesting that novel functions arise mainly in the nonbinding regions of the receptors (such as their second-messenger contact regions) rather than in the production of novel ligands per se (Hoyle 1999; Root-Bernstein and Churchill 2021). A more in-depth discussion of receptor diversification and heterodimerization will be made in the second case study below.

In sum, some of the many important characteristics of the modern glucose regulatory system can be explained in terms of an evolutionary process in which modular complementarity between glucose, insulin, and glucagon molecules was bootstrapped to create increasingly complex chimers that, because of MCMs, retained their ability to interact in a highly networked fashion.

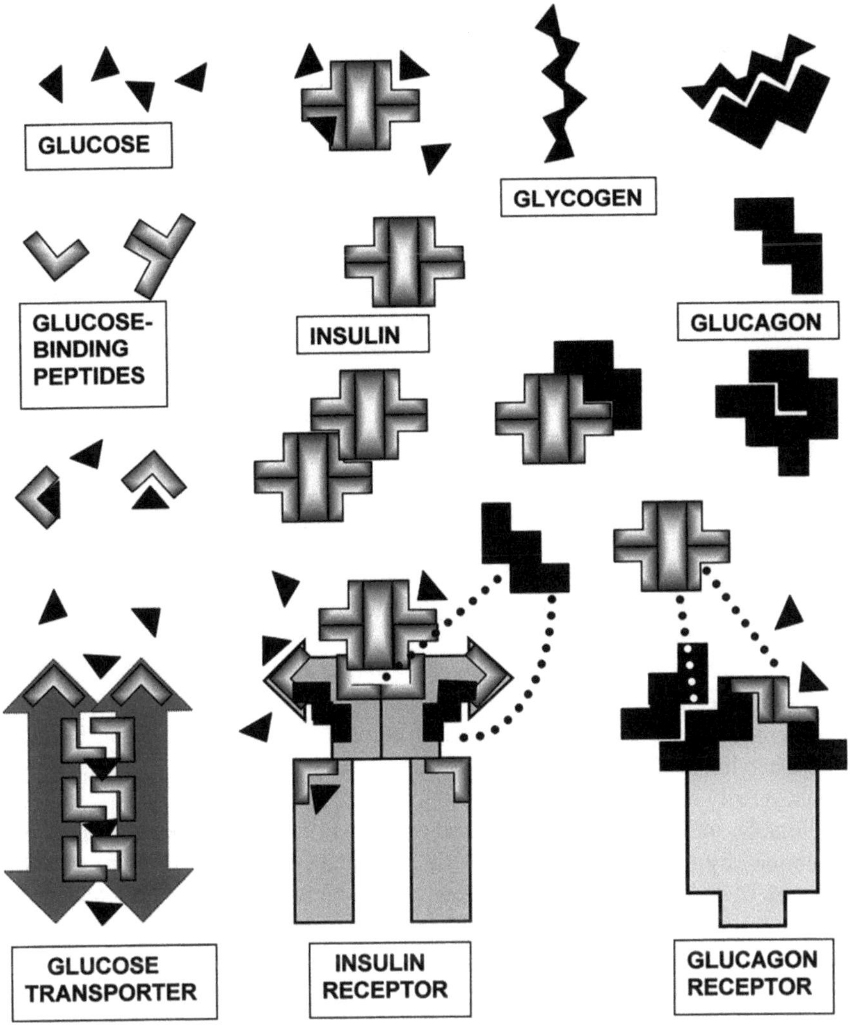

Schematic model of how molecularly complementary modules may have led to the evolution of the core of the glucose regulatory system. Glucose binds to multiple sites on insulin, which consist of six sequentially similar regions. Multiple copies of these insulin-like regions appear in all glucose transporters (GLUTs) and insulin; notably, they are amphoteric and, therefore, able to insert into lipid bilayers. Insulin binds to itself, and insulin-like regions appear in the insulin receptor. Glucagon binds to itself, and multiple glucagon-like regions appear in the glucagon receptor. Insulin and glucagon bind to each other, and copies of each appear in the other's receptor. Since glucagon and glucose both bind to insulin, they compete with each other, possibly creating a primitive glucose release mechanism. Additionally, glucagon binds to glycogen, and a glucagon-like region is found at the glycogen-binding region of the glycogen phosphorylase enzyme that breaks down glycogen into glucose. Insulin and glucagon, therefore, exhibit not only complementarity of structure but also complementarity of function with regard to glucose. Cross talk between the insulin and glucagon receptors may involve each ligand binding weakly to the other (dotted lines) and/or dimerization of the receptors, altering each other's affinity for their own ligands

5.2 Interactome Evolution Case 2: Opioids, Adrenergic Compounds, and Their Receptors

The second system that I have explored from a modular complementarity perspective involves the coevolution of opioid and adrenergic receptor systems and the mechanism by which they interact. As with the glucose–insulin system, a case can be made that ligands and receptors evolved in tandem. Three principles once again apply: first, that existing cellular systems reflect a molecular paleontology; second, that complementary molecules may serve as the basis for the evolution of ligand–receptor pairs (Dwyer's principle); and third, that molecules that alter each other's physiological functions will likely be molecularly complementary and vice versa (Root-Bernstein's principle).

Let us begin with the observation that enkephalins self-aggregate into homodimers by forming an antiparallel beta-ribbon, that pro-enkephalins self-aggregate to form amyloid fibrils, and that beta-endorphins also aggregate into amyloid fibrils. Additionally, morphine can bind to enkephalins and endorphins (reviewed in Root-Bernstein and Churchill (2021)). From Dwyer's principle, opioid peptides might, therefore, have been the evolutionary ancestors of opioid receptors, in which case, there should be opioid–peptide-like sequences in the binding regions of opioid receptors. In fact, the beta-endorphin and pro-enkephalin peptides are highly homologous not only to each other but also to the mu- and delta-opioid receptors, specifically within the peptide-binding regions of those receptors, and these homologies are conserved from fish through human beings (Root-Bernstein and Churchill 2021). Thus, homo-complementary modules link opioid ligands evolutionarily to their receptors.

Notably, the evolution of opioid and adrenergic receptors appears to have been evolutionarily linked, and although aminergic and peptidergic receptors are often assigned to different evolutionary clades, when one focuses on modular, rather than global, sequence similarities, the evidence suggests that the two classes of G-protein coupled receptors (GPCRs) may have coevolved (Wolf and Grünewald 2015; Root-Bernstein and Churchill 2021). One unusual type of evidence in this instance is that opioid and adrenergic compounds synergize and that synergism is mediated by receptor cross talk involving heterodimerization between the two types of receptors. If ligand–receptor evolution occurs randomly, then the emergence of an interactome involving heterodimers composed of two different receptors that permit cross talk between two unrelated ligands would be highly unlikely. Coevolution of opioid and adrenergic receptors as a result of shared molecularly complementary modules, however, could provide a simple resolution.

In this case, there is a well-known phenomenon in pharmacology that opioid and adrenergic drugs synergize (Root-Bernstein et al. 2019; Root-Bernstein 2022). The principle that compounds that affect each other's physiological functions are likely to be molecularly complementary (Root-Bernstein and Dillon 2008) suggests that opioids and catecholamines might, therefore, be complementary and that, as in the glucose–insulin case, this complementarity may have jump-started the selection for opioid–adrenergic interactions. Indeed, opioids bind to catecholamines such as epinephrine, norepinephrine (NorEpi), dopamine, and albuterol and produce very stable complexes that protect the participating molecules against oxidation. Opioids that can form such stable complexes include both morphine-like compounds as well as peptides such as enkephalins and endorphins (Root-Bernstein 1987; Dillon et al. 2006; Root-Bernstein and Churchill 2021).

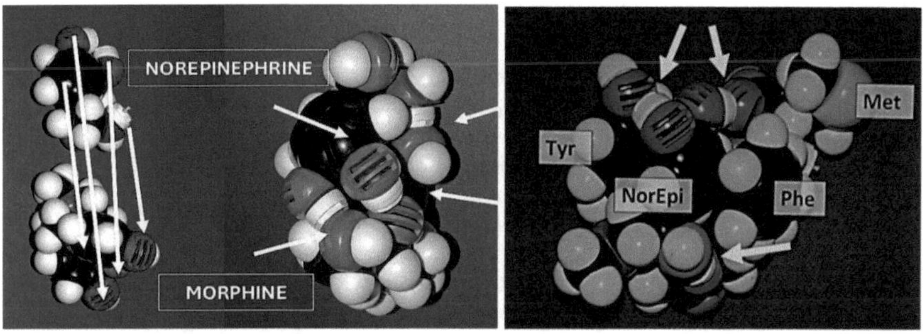

Space-filling 3-D CPK models of norepinephrine (NorEpi) binding to morphine (left) and into a pocket formed by the tyrosine (Tyr) and phenylalanine (Phe) of the opioid–peptide methionine (Met) enkephalin (right). Every available hydroxyl and amine of NorEpi is engaged in a hydrogen or ionic bond with both morphine and MetEnk, and pi–pi bonds form between the rings of the NorEpi with morphine or with the Tyr and Phe of the MetEnk. The result is an ultrastable complex that prevents oxidation of NorEpi and protects the opioids from degradation.

Yet another aspect of adrenergic–opioid binding that evolution has taken advantage of is the co-storage and co-release of the two compounds in neurons (Charnay et al. 1982; De Potter et al. 1987; Ceccatelli et al. 1989). Opioids are synthesized not only in neurons, often along with adrenergics, but also in the adrenal medulla, which is the main source of epinephrine production for the body. Opioids are co-stored and co-released with adrenergic compounds in both adrenal chromaffin cells and paraneurons (Lemaire et al. 1984; Lang et al. 1982; Livett et al. 1982). In fact, opioid–adrenergic co-storage and co-release is just one of the many examples of molecularly complementary small molecules being co-stored and released together in various tissues and cell types (Hökfelt et al. 1987). Some of these molecular pairs are known to be molecularly complementary, e.g., dopamine and neurotensin (Schenk et al. 1991) and thyrotropin-releasing hormone and adrenocorticotropic hormone with serotonin (Sharif et al. 1983; Root-Bernstein and Westall 1984b). As noted above for both insulin–glucose and opioid–adrenergic binding, such co-storage may protect such molecular complexes against degradation and may also increase the storage capacity of the granules by effectively "precipitating" the complex so that higher concentrations of the molecules can be achieved. Thus, co-storage and co-release may be further paleontological clues as to the molecularly complementary origins of metabolic and regulatory systems.

One prediction that follows from opioid–adrenergic complexing, in combination with my extension of Dwyer's principle, is that enkephalins or endorphins might, therefore, have served as proto-receptors for catecholamines. These proto-receptors may have served as foundational modules for the evolution of modern adrenergic receptors, from which it follows that opioid–peptide-like sequences should occur in the binding site of adrenergic receptors. This is the case. Adrenergic receptors exhibit significant homologies to both endorphins and pro-enkephalins within the extracellular loops and transmembrane regions that form the pocket for adrenergic binding (Root-Bernstein and Churchill 2021). Moreover, in vitro studies demonstrate that adrenergic compounds bind to these receptor-derived peptide sequences (Root-Bernstein and Churchill 2021).

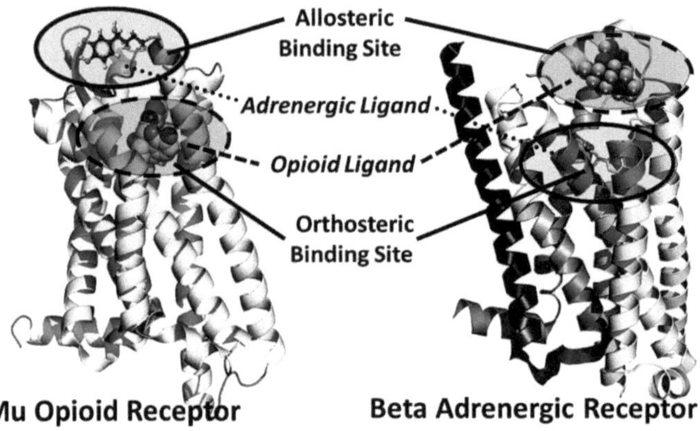

3-D AlphaFold2 structural models of the two proteins, the mu-opioid receptor and the beta-adrenergic receptors modified to show where the allosteric and orthosteric binding sites of epinephrine and morphine are located. The image highlights binding sites on each receptor. The morphine binds to the external allosteric site of the adrenergic receptor and the internal orthosteric site of the opioid receptor. Epineprhine binds to the external allosteric site of the opioid receptor and the internal orthosteric binding site of the adrenergic receptor. Thus, the opioid and adrenergic receptors each bind to the other molecule but at different places on the receptor. Modularity has permitted the allosteric and orthosteric binding sites to be rearranged, permitting each receptor to be enhanced by combinations of opioids and catecholamines. The receptors additionally share modular transmembrane sequences that permit them to both homo- and heterodimerize. The structures are depicted with ribbon diagrams and the ligands are shown as space-filling models.

Notably, opioid compounds such as morphine, as well as opioid antagonists naloxone and naltrexone, also bind to extracellular loops of adrenergic, dopaminergic, and histaminergic receptors (Root-Bernstein et al. 2019; Root-Bernstein et al. 2020; Root-Bernstein 2022). This binding does not activate the receptors but results, through allosteric modulation, in enhanced activity of the aminergic compound, increasing its effectiveness and duration of action (Gu et al. 1990; Caffrey et al. 1990a; Caffrey et al. 1990b; Gu et al. 1993; Root-Bernstein and Dillon 2002). Conversely, adrenergic compounds bind to the opioid receptors at modules that mimic the adrenergic receptor–binding sequences, and this binding acts as an enhancer of opioid activity (Goyagi and Nishikawa 1995; Wu et al. 1999; Root-Bernstein et al. 2018; Root-Bernstein et al. 2019; Root-Bernstein et al. 2020; Root-Bernstein 2022). We have therefore concluded that the opiate receptors, which arise evolutionarily at a later time than catecholamine receptors, were derived from adrenergic receptors through a process of gene duplication and rearrangement of the constituent modules (Root-Bernstein and Churchill 2021). Wolf and Grünewald (2015) have reached a similar conclusion.

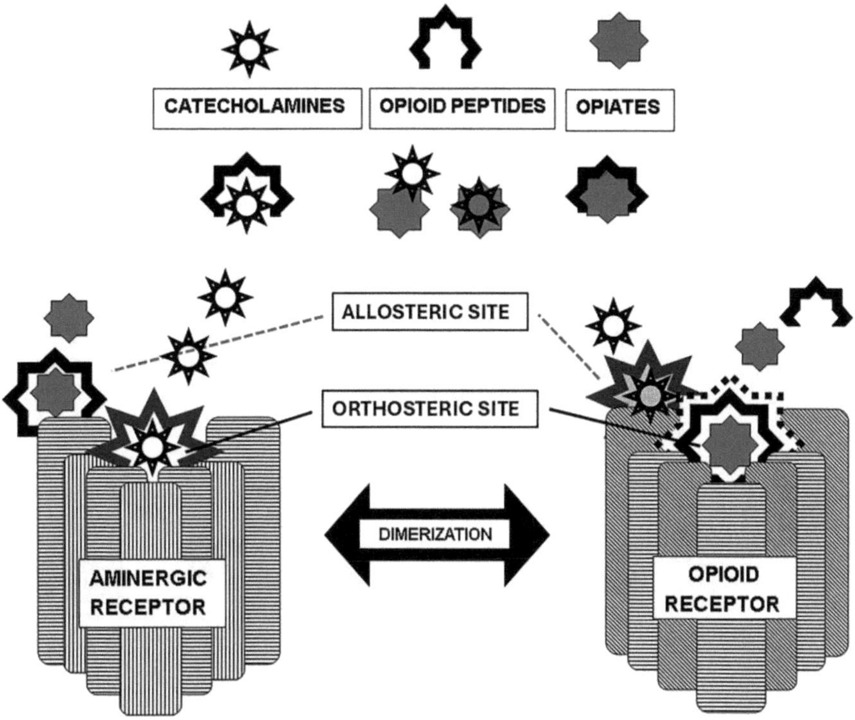

Schematic diagram illustrating the interactions between catecholamines such as epinephrine, norepinephrine, and dopamine and opioid peptides such as the enkephalins and endorphins or opiates such as morphine and fentanyl. Catecholamines bind to opiates and opioids. Opiates also bind to opioids, and each class of drugs is able to self-aggregate. Opioid peptides thus became conserved molecularly complementary modules for binding opioids, opiates, and catecholamines. Thus, endorphin- and pro-enkephalin-like sequences are found at the orthosteric binding sites of the opioid and aminergic receptors, as well as in extracellular loops of these receptors, that act as allosteric enhancers upon binding of the complementary ligand. Furthermore, because the opioid and aminergic receptors share conserved transmembrane regions involved in homodimerization, they are also able to heterodimerize. Thus, molecular complementarity expressed through conserved modules that have been rearranged in the two receptor classes mediates their cross talk through multiple mechanisms

The phenomenon of receptor homo- and heterodimer formation is yet another aspect of adrenergic–opioid co–evolution, which is also interesting to consider in the broader context of receptor–ligand evolution in general. Receptors often aggregate within lipid rafts on cell surfaces, and the aggregation is controlled mainly by interactions of the transmembrane-spanning sequences of the receptors. These transmembrane sequences are highly conserved across all types of receptors and not only self-associate to form the compact receptor structures seen in seven-transmembrane receptors but also mediate homo- and heterodimerization between such receptors (Zviling et al. 2007; Ramesh and Soliman 2015). Notably, these transmembrane regions are formed from modules about 20–22 amino acids in length that appear to have been duplicated within receptor genes. Even more intriguing is the fact that some single-span (bitopic) receptors are known to exist, and these single-span sequences share the highly conserved 20–22-amino-acid modularity seen in multi-span receptors such as seven-transmembrane G-protein coupled receptors as well (Arkin and Brunger 1998). Additionally, many single-span bitopic receptors are capable of oligomerization by means of specific interactions between their transmembrane alpha-helices (Arkin 2002; del Sol and Carbonell 2007). Thus, once again, it appears that selection occurred very early in membrane evolution for peptide modules capable of spanning a lipid bilayer, and these modules were then conserved (with some variation) in the construction of more complex receptors, mediating dimerization. It is not, therefore, surprising that adrenergic and opioid receptors are capable of both homo- and heterodimer receptor complexing or that this dimerization mediates their functional cross talk (Fujita et al. 2014; Vilardaga et al. 2008; Lavoie and Hébert 2003; Parmar et al. 2017).

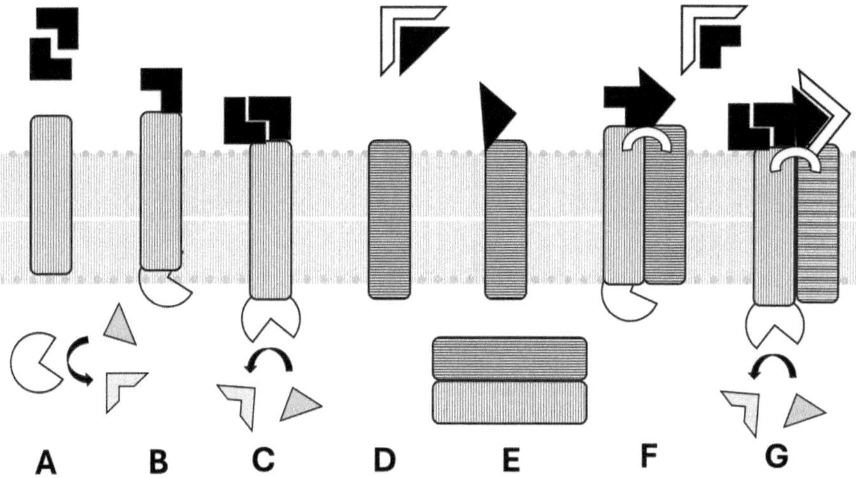

Schematic model based on Dwyer's theory of how the first receptors may have evolved. In (**a**), selection for molecular complementarity would have formed stable modules involving (from left to right) homo- or hetero-complementarity between peptides or between peptides and other small molecules, selection for peptides that could span lipid bilayers, and selection for peptides that could catalyze reactions. In (**b**), these modules become ligated (presumably, most permutations of the modules are nonfunctional and are selected out), creating a linear polypeptide that, in (**c**), is capable of binding a ligand on one side and altering its conformation so as to activate the catalytic site on the other side of the membrane. This is, in essence, the kind of bitopic receptor described in the text. Variants of transmembrane-spanning peptides coevolved (**d**) and were also ligated to various ligands (**e**). Additionally, some transmembrane-spanning peptides would have been selected for their propensity to form dimers or higher-order complexes that would have further stabilized their fitness. Ligating several of these polypeptides (**f**) would then lead to the multi-span transmembrane receptors that characterize GPCR and other receptor classes. Some of these multi-span receptors would have incorporated complementary binding regions optimized for binding complementary ligands (**g**), resulting in allosteric control, enhancement, and other forms of cross talk. The complementarity between membrane-spanning modules (**e**) would also have promoted receptor oligomerization and, thus, additional mechanisms of receptor cross talk (not shown). (Adapted and expanded from (Root-Bernstein 2002))

6 Bootstrapping as the Ultimate Limit on Randomness

In light of these case studies, let us return to the question of how MCM pares down combinatorial explosions to manageable levels. Critical readers will have realized that there is an apparent flaw in both Simon's and my analyses of the benefits of modularity. Modularity is beneficial in constraining permutations only if the modules that are initially created are functional. Simon assumed this feature when he said that Hora "designed" his watches to be built in stable modules—another hidden form in which the engineering fallacy intruded into his calculations. I made a different, also hidden, assumption, which is that there is always an evolutionarily selected, functional, stable module that can be formed from any set of five or ten components. Thus, it followed that our blind watchmaker simply has to test all of the permutations of the provided set to find a useful module. The apparent fallacy in both Simon's and my case is that any random set of five or ten of the 100 or 1000 pieces of the watch may not yield any stable modules or, perhaps worse, might yield stable modules that are useless or even detrimental to building the watch. One needs the right preselected sets of components in order to generate the right sets of stable modules to obtain a watch at the end of the process. If one cannot preselect the component sets, then one theoretically has to explore *every possible component set* in order to find the stable modules one needs to build the watch, which sets us back in a combinatorial explosion scenario.

However, the argument just provided and its conclusion rest upon another false assumption, one that, in fact, flaws many evolutionary arguments. That assumption is that our blind watchmaker knows that he is building a watch. This is the equivalent of saying that evolution always had as its "goal" to build a cell or plant or an insect, starfish, or human being. Evolution does not have a goal. Having no goal, evolution has no predetermined set of components and stable modules that it needs to generate. Evolution is opportunistic. So natural selection acts upon whatever stable modules are most fit for their environment at any given moment in evolutionary time and selects upon variants of these. Thus, the assumption that one must be able to select the "correct" set of components to produce the set of stable modules required to produce a specific watch is wrong. Evolution tinkers with whatever stable set of stable modules is available in any given mixture of compounds and builds with those. Our blind watchmaker is really just a tinkerer that improvises, varying, mixing, and matching the surviving complementary modules. That a glucose regulatory system or an adrenergic receptor or a banana, a human being, or a "watch", is an end result is not something that evolution foresaw or planned but an opportunistic result of exploring what could be made with the complexes available within the geochemical ecology.

So, in fact, Simon and I are actually correct in our calculations of how beneficial modularity is because, as this chapter has demonstrated with its two case studies, chemistry naturally provides the stable, functional modules from which evolution can build. Evolution is therefore a "MacGyver," a resourceful agent that makes use of whatever materials are at hand to create whatever functional assemblages can be created, retaining whatever is most fit. What is most fit are molecularly complementary modules, and what results are functional interactomes. Those that are fit and evolvable become the basis for the next round of variations.

In conclusion, I want to emphasize that the MCM approach to understanding the evolution of interactomes is generalizable to cases beyond the two discussed here. I have explored the evolution of two other systems using the MCM approach, one involving the evolution of the ascorbate–glutathione system (Root-Bernstein and Dillon 2014; Root-Bernstein et al. 2016a) and the other involving the evolution of the ribosome (Root-Bernstein and Root-Bernstein 2015; Root-Bernstein and Root-Bernstein 2016a; Root-Bernstein et al. 2016a), as well as describing a general model for the evolution of epistatic interactomes (Root-Bernstein and Bernstein 2024). Additionally, two other groups have been working along parallel lines and have reached similar conclusions based on very different sets of examples. One is led by Peter Csermely, whose book, *Weak Links: Stabilizers of Complex Systems from Proteins to Social Networks,* lays out the importance of molecular (and other forms of) complementarity as essential mechanisms of interactome formation: "Weak links came as a great surprise to me. It was not so much their existence, which is obvious, but their importance that was the new message from my reading and which constitutes the main message of this book. These links form most of the links inside and around us … Weak links stabilize all complex systems. Weak links give us a universal key to understanding network diversity and stability, and they are the major actors…." (Csermely 2006, xi & 275). More broadly, Caetano-Anollés has recently written: "Networks describe how parts associate with each other to form integrated systems which often have modular and hierarchical structure. In biology, network growth involves two processes, one that unifies and the other that diversifies… In the first phase, parts are at first weakly linked and associate variously. As they diversify, they compete with each other and are often selected for performance. The emerging interactions constrain their structure and associations. This causes parts to self-organize into modules with tight linkage. In the second phase, variants of the modules diversify and become new parts for a new generative cycle of higher level organization. The paradigm predicts the rise of hierarchical modularity in evolving networks at different timescales and complexity levels" (Caetano-Anollés et al. 2019).

References

Anzenbacher P, Kalous V (1975) Binding of D-glucose to insulin. Biochim Biophys Acta 386:603–607

Arkin IT (2002) Structural aspects of oligomerization taking place between the transmembrane alpha-helices of bitopic membrane proteins. Biochim Biophys Acta 1565(2):347–363. https://doi.org/10.1016/s0005-2736(02)00580-1

Arkin IT, Brunger AT (1998) Statistical analysis of predicted transmembrane alpha-helices. Biochim Biophys Acta 1429(1):113–128. https://doi.org/10.1016/s0167-4838(98)00225-8

Bada JL (2013) New insights into prebiotic chemistry from Stanley Miller's spark discharge experiments. Chem Soc Rev 42(5):2186–2196. https://doi.org/10.1039/c3cs35433d

Beeckmans S, Van Driessche E, Kanarek L (1990) Clustering of sequential enzymes in the glycolytic pathway and the citric acid cycle. J Cell Biochem 43(4):297–306. https://doi.org/10.1002/jcb240430402

Biruš I, Šeba T, Marić M, Gabričević M, Weitner T (2025) Design and binding affinity of antisense peptides for snake venom neutralization. Molecules 30(4):903. https://doi.org/10.3390/molecules30040903

Blecher M, Giorgio NA, Johnson CB (1974) Hormone receptors. IV. Solubilization and purification of a liver plasma membrane glucagon-binding protein. Adv Enz Reg 12:289–309. https://doi.org/10.1016/0065-2571(74)90018-1

Blecher M, Goldstein S (1977) Hormone receptors: VI. On the nature of the binding of glucagon and insulin to human circulating mononuclear leukocytes. Mol Cell L'Endocrinologo 8(4):301–315. https://doi.org/10.1016/0303-7207(77)90005-3

Bloch K (1994) Blondes in venetian paintings, the nine-banded armadillo, and other essays in biochemistry. Yale University Press, New Haven, pp 78–83

Blount ZD, Lenski RE, Losos JB (2018) Contingency and determinism in evolution: replaying life's tape. Science 362(6415):eaam5979. https://doi.org/10.1126/scienceaam5979

Bost KL, Smith EM, Blalock JE (1985a) Similarity between the corticotropin (ACTH) receptor and a peptide encoded by an RNA that is complementary to ACTH mRNA. Proc Natl Acad Sci U S A 82:1372–1375

Bost KL, Smith EM, Blalock JE (1985b) Regions of complementarity between the messenger RNAs for epidermal growth factor, transferring, interleukin-2 and their respective receptors. Biochem Biophys Res Commun 128:1373–1380

Braakman R, Smith E (2013) The compositional and evolutionary logic of metabolism. Phys Biol 10(1):011001. https://doi.org/10.1088/1478-3975/10/1/011001

Braun JE, Friedman LJ, Gelles J, Moore MJ (2018) Synergistic assembly of human pre-spliceosomes across introns and exons. eLife 7:e37751. https://doi.org/10.7554/eLife37751

Brillouin L (1953) Negentropy principle of information. J Appl Phys 24(9):1152–1163

Burton ZF (2018) Evolution since coding: cradles, halos, barrels, and wings. Academic Press, New York

Burton ZF (2020) The 3-Minihelix tRNA evolution theorem. J Mol Evol 88(3):234–242. https://doi.org/10.1007/s00239-020-09928-2

Caetano-Anollés G, Aziz MF, Mughal F, Gräter F, Koç I, Caetano-Anollés K, Caetano-Anollés D (2019) Emergence of hierarchical modularity in evolving networks uncovered by phylogenomic analysis. Evol Bioinformatics Online 15:1176934319872980. https://doi.org/10.1177/1176934319872980

Caffrey JL, Hathorne LF, Carter GC, Sinclair RJ (1990a) Naloxone potentiates contractile responses to epinephrine in isolated canine arteries. Circ Shock 31(3):317–332

Caffrey JL, Stol ST, Sinclair RJ, Barron BA (1990b) (+) naloxone enhances vascular contractile responses to added epinephrine. Prog Clin Biol Res 328:375–378

Cairns-Smith AG (1971) The life puzzle on crystals and organisms and on the possibility of a crystal as an ancestor. Oliver and Boyd, Edinburgh

Ceccatelli S, Millhorn DE, Hökfelt T, Goldstein M (1989) Evidence for the occurrence of an enkephalin-like peptide in adrenaline and noradrenaline neurons of the rat medulla oblongata. Exp Brain Res 74(3):631–640. https://doi.org/10.1007/BF00247366

Charnay Y, Léger L, Dray F, Bérod A, Jouvet M, Pujol JF, Dubois PM (1982) Evidence for the presence of enkephalin in catecholaminergic neurones of cat locus coeruleus. Neurosci Lett 30(2):147–151. https://doi.org/10.1016/0304-3940(82)90287-7

Chaudhary N, Courvalin JC (1993) Stepwise reassembly of the nuclear envelope at the end of mitosis. J Cell Biol 122(2):295–306. https://doi.org/10.1083/jcb1222295

Chen Z, Wang ZH, Zhang G, Bleck CKE, Chung DJ, Madison GP, Lindberg E, Combs C, Balaban R, Xu H (2020) Mitochondrial DNA segregation and replication restrict the transmission of detrimental mutation. J Cell Biol 219(7):e201905160. https://doi.org/10.1083/jcb201905160

Christodoulou J (2000) Genetic defects causing mitochondrial respiratory chain disorders and disease. Human Reprod 15(Suppl 2):28–43. https://doi.org/10.1093/humrep/15suppl_228
Crick FHC (1981) Life itself. Simon and Schuster, New York
Csermely P (2006) Weak links stabilizers of complex systems from proteins to social networks. Springer, Berlin
Dayhoff GW 2nd, van Regenmortel MHV, Uversky VN (2020) Intrinsic disorder in protein sense-antisense recognition journal of molecular recognition. JMR 33(10):e2868. https://doi.org/10.1002/jmr2868
De Potter WP, Coen EP, De Potter RW (1987) Evidence for the coexistence and co-release of [Met] enkephalin and noradrenaline from sympathetic nerves of the bovine vas deferens. Neuroscience 20(3):855–866. https://doi.org/10.1016/0306-4522(87)90246-6
del Sol A, Carbonell P (2007) The modular organization of domain structures: insights into protein-protein binding. PLoS Comput Biol 3(12):e239. https://doi.org/10.1371/journalpcbi0030239
Delihas N (2007) Enterobacterial small mobile sequences carry open reading frames and are found intragenically--evolutionary implications for formation of new peptides. Gene Regul Syst Biol 1:191–205
Desrosiers A, Derbali RM, Hassine S, Berdugo J, Long V, Lauzon D, De Guire V, Fiset C, DesGroseillers L, Leblond Chain J, Vallée-Bélisle A (2022) Programmable self-regulated molecular buffers for precise sustained drug delivery. Nat Commun 13(1):6504. https://doi.org/10.1038/s41467-022-33491-7
Dillon PF, Root-Bernstein R, Lieder CM (2006) Molecular shielding of electric field complex dissociation. Biophys J 90(4):1432–1438. https://doi.org/10.1529/biophysj105071969
Doolittle RF (1992) Stein and Moore Award address reconstructing history with amino acid sequences. Protein Sci 1(2):191–200. https://doi.org/10.1002/pro5560010201
Doolittle RF, Woodbury NW, Jue RA (1982) Ribosomal protein S1 is the product of a series of contiguous duplications. Biosci Rep 2(6):405–412. https://doi.org/10.1007/BF01119303
Dwyer DS (1989) Amino acid sequence homology between ligands and their receptors: potential identification of binding sites. Life Sci 45(5):421–429. https://doi.org/10.1016/0024-3205(89)90628-0
Dwyer DS (1998) Assembly of exons from unitary transposable genetic elements: implications for the evolution of protein-protein interactions. J Theor Biol 194(1):11–27. https://doi.org/10.1006/jtbi19980676
Dwyer DS (2021) Protein receptors evolved from homologous cohesion modules that self-associated and are encoded by interactive networked genes. Life (Basel, Switzerland) 11(12):1335. https://doi.org/10.3390/life11121335
Eigen M, Schuster P (1979) The Hypercycle: a principle of natural self-organization. Springer Verlag, Berlin, Heidelberg, New York
Elrick H, Arai Y, Hlad CJ Jr (1958) The action of glucagon-insulin mixtures in diabetic patients. J Clin Endocrin Metab 18(8):825–833. https://doi.org/10.1210/jcem-18-8-825
Engel SR, Dietrich FS, Fisk DG, Binkley G, Balakrishnan R, Costanzo MC, Dwight SS, Hitz BC, Karra K, Nash R, Weng S, Wong ED, Lloyd P, Skrzypek MS, Miyasato SR, Simison M, Cherry JM (2014) The reference genome sequence of Saccharomyces cerevisiae: then and now. G3 (Bethesda, Md) 4(3):389–398. https://doi.org/10.1534/g3113008995
Fairchild J, Islam S, Singh J, Bučar DK, Powner MW (2024) Prebiotically plausible chemoselective pantetheine synthesis in water. Science 383(6685):911–918. https://doi.org/10.1126/scienceadk4432
Fink T, Reymond JL (2007) Virtual exploration of the chemical universe up to 11 atoms of C, N, O, F: assembly of 264 million structures (1109 million stereoisomers) and analysis for new ring systems, stereochemistry, physicochemical properties, compound classes, and drug discovery. J Chem Inform Model 47(2):342–353. https://doi.org/10.1021/ci600423u
Fitch WM (1966) Evidence suggesting a partial, internal duplication in the ancestral gene for heme-containing globins. J Mol Biol 16(1):17–27. https://doi.org/10.1016/s0022-2836(66)80259-0

Forde BM, Ben Zakour NL, Stanton-Cook M, Phan MD, Totsika M, Peters KM, Chan KG, Schembri MA, Upton M, Beatson SA (2014) The complete genome sequence of Escherichia coli EC958: a high quality reference sequence for the globally disseminated multidrug resistant E coli O25b:H4-ST131 clone. PLoS One 9(8):e104400. https://doi.org/10.1371/journalpone0104400

Fujita W, Gomes I, Devi LA (2014) Revolution in GPCR signalling: opioid receptor heteromers as novel therapeutic targets: IUPHAR review 10. Br J Pharmacol 171(18):4155–4176. https://doi.org/10.1111/bph12798

Gardner M (1970) The fantastic combinations of John Conway's new solitaire game 'life. Sci Am 223(4):120–123. https://doi.org/10.1038/scientificamerican1070-120

Giorgio NA, Johnson CB, Blecher M (1974) Hormone receptors. 3. Properties of glucagon-binding proteins isolated from liver plasma membranes. JBC 249(2):428–437

Gould SJ (2002) The structure of evolutionary theory. Harvard University Press, Cambridge, MA

Goyagi T, Nishikawa T (1995) The addition of epinephrine enhances postoperative analgesia by intrathecal morphine. Anesth Analg 81(3):508–513. https://doi.org/10.1097/00000539-199509000-00014

Green DE, Goldberger RF (1967) Molecular insights into the living process. Academic Press, New York

Gu H, Barron BA, Gaugl JF, Caffrey JL (1993) (+)Naloxone potentiates the inotropic effect of epinephrine in the isolated dog heart. Circ Shock 40(3):206–211

Gu H, Gaugl JF, Barron BA, Caffrey JL (1990) Naloxone enhances cardiac contractile responses to epinephrine without altering epinephrine uptake from plasma. Circ Shock 32(4):257–271

Habegger KM (2022) Cross talk between insulin and glucagon receptor signaling in the hepatocyte. Diabetes 71(9):1842–1851. https://doi.org/10.2337/dbi22-0002

Hancock EJ, Ang J (2019) Frequency domain properties and fundamental limits of buffer–feedback regulation in biochemical systems. Automatica 103:330–336. https://doi.org/10.1016/jautomatica201902017

Henney A, Hunter P, McCulloch A, Noble D (2015) Multi-bio and multi-scale systems biology. Prog Biophys Mol Biol 117(1, 1):–3. https://doi.org/10.1016/jpbiomolbio201503001

Hennis AS, Birky CW Jr (1984) Stochastic partitioning of chloroplasts at cell division in the alga Olisthodiscus, and compensating control of chloroplast replication. J Cell Sci 70:1–15. https://doi.org/10.1242/jcs7011

Hökfelt T, Millhorn D, Seroogy K, Tsuruo Y, Ceccatelli S, Lindh B, Meister B, Melander T, Schalling M, Bartfai T (1987) Coexistence of peptides with classical neurotransmitters. Experientia 43(7):768–780. https://doi.org/10.1007/BF01945354

Hoyle CH (1999) Neuropeptide families and their receptors: evolutionary perspectives. Brain Res 848(1–2):1–25. https://doi.org/10.1016/s0006-8993(99)01975-7

Hoyle F (1984) The intelligent universe. Holt, Rinehart, and Winston, New York

Hoyle F, Wickramasinghe NC (1981) Evolution from space. JM Dent Sons, London

Huh D, Paulsson J (2011) Random partitioning of molecules at cell division. Proc Natl Acad Sci USA 108(36):15004–15009. https://doi.org/10.1073/pnas1013171108

Hunding A, Kepes F, Lancet D, Minsky A, Norris V, Raine D, Sriram K, Root-Bernstein R (2006) Compositional complementarity and prebiotic ecology in the origin of life. BioEssays 28(4): 399–412. https://doi.org/10.1002/bies20389

Im J, Hillenaar T, Yeoh HY, Sahasrabudhe P, Mijnders M, van Willigen M, Hagos A, de Mattos E, van der Sluijs P, Braakman I (2023) ABC-transporter CFTR folds with high fidelity through a modular, stepwise pathway. Cell Mol Life Sci 80(1):33. https://doi.org/10.1007/s00018-022-04671-x

Irwin DM (2014) Evolution of receptors for peptides similar to glucagon. Gen Comp Endocrinol 209:50–60. https://doi.org/10.1016/jygcen201403002

Irwin DM, Wong K (2005) Evolution of new hormone function: loss and gain of a receptor. J Hered 96(3):205–211. https://doi.org/10.1093/jhered/esi024

Islam S, Bučar DK, Powner M (2017) Prebiotic selection and assembly of proteinogenic amino acids and natural nucleotides from complex mixtures. Nat Chem 9:584–589. https://doi.org/10.1038/nchem2703

Jacobson H (1955) Information, reproduction and the origin of life. Am Sci 43:119–127

Jue RA, Woodbury NW, Doolittle RF (1980) Sequence homologies among E coli ribosomal proteins: evidence for evolutionarily related groupings and internal duplications. J Mol Evol 15(2):129–148. https://doi.org/10.1007/BF01732666

Kauffman SA (1993) The origins of order: self-organization and selection in evolution. Oxford University Press, Oxford

Lang RE, Taugner G, Gaida W, Ganten D, Kraft K, Unger T, Wunderlich I (1982) Evidence against co-storage of enkephalins with noradrenaline in bovine adrenal medullary granules. Eur J Pharmacol 86(1):117–120. https://doi.org/10.1016/0014-2999(82)90408-3

Lavoie C, Hébert TE (2003) Pharmacological characterization of putative beta1-beta2-adrenergic receptor heterodimers. Can J Physiol Pharmacol 81(2):186–195. https://doi.org/10.1139/y02-167

Lemaire S, Day R, Dumont M, Chouinard L, Calvert R (1984) Dynorphin and enkephalins in adrenal paraneurones Opiates in the adrenal medulla. Can J Physiol Pharmacol 62(4):484–492. https://doi.org/10.1139/y84-078

Livett BG, Day R, Elde RP, Howe PR (1982) Co-storage of enkephalins and adrenaline in the bovine adrenal medulla. Neuroscience 7(5):1323–1332. https://doi.org/10.1016/0306-4522(82)91138-1

Loo JA, Holsworth DD, Root-Bernstein R (1994) Use of electrospray ionization mass spectrometry to probe antisense peptide interactions. Biol Mass Spectrom 23(1):6–12. https://doi.org/10.1002/bms1200230103

Lopez A, Fayolle D, Fiore M, Strazewski P (2020) Chemical analysis of lipid boundaries after consecutive growth and division of supported giant vesicles. iScience *23*(11):101677. https://doi.org/10.1016/jisci2020101677

Mayer C, Lancet D, Markovitch O (2024) The GARD prebiotic reproduction model described in order and complexity. Life 14(3):288. https://doi.org/10.3390/life14030288

Miller SL (1955) Production of some organic compounds under possible primitive earth conditions. J Am Chem Soc 77:2351–2361

Misteli T (1996) Molecular mechanisms in the disassembly and reassembly of the mammalian Golgi apparatus during M-phase. FEBS Lett 389(1):66–69. https://doi.org/10.1016/0014-5793(96)00518-2

Mochizuki A (2022) A structural approach to understanding enzymatic regulation of chemical reaction networks. Biochem J 479(11):1265–1283. https://doi.org/10.1042/BCJ20210545

Morowitz HJ, Kostelnik JD, Yang J, Cody GD (2000) The origin of intermediary metabolism. Proc Natl Acad Sci USA 97(14):7704–7708. https://doi.org/10.1073/pnas110153997

Moorhead PS, Kaplan MM (1967) Mathematical challanges to the neo-darwinian interpretation of evolution. Wistar Institute Press, Philadelphia

Morowitz HJ (1999) A theory of biochemical organization, metabolic pathways, and evolution. Complexity 4(6):39–53. https://doi.org/10.1002/(SICI)1099-0526(199907/08)4:6<39::AID-CPLX8>30CO;2-2

Navarro I, Leibush B, Moon TW, Plisetskaya EM, Baños N, Méndez E, Planas JV, Gutiérrez J (1999) Insulin, insulin-like growth factor-I (IGF-I) and glucagon: the evolution of their receptors. Comp Biochem Physiol B Biochem Mol Biol 122(2):137–153. https://doi.org/10.1016/s0305-0491(98)10163-3

Noble D (2006) The music of life. Oxford University Press, Oxford

Norris V, Root-Bernstein R (2009) The eukaryotic cell originated in the integration and redistribution of hyperstructures from communities of prokaryotic cells based on molecular complementarity. Int J Mol Sci 10:2611–2632. https://doi.org/10.3390/ijms10062611

O'Harte FP, Abdel-Wahab YH, Conlon JM, Flatt PR (1998) Glycation of glucagon-like peptide-1 (7-36)amide: characterization and impaired action on rat insulin secreting cells. Diabetologia 41(10):1187–1193. https://doi.org/10.1007/s001250051050
O'Harte FP, Boyd AC, McKillop AM, Abdel-Wahab YH, McNulty H, Barnett CR, Conlon JM, Højrup P, Flatt PR (2000) Structure, antihyperglycemic activity and cellular actions of a novel diglycated human insulin. Peptides 21(10):1519–1526. https://doi.org/10.1016/s0196-9781(00)00306-5
Oliver B, Leblanc B (2003) How many genes in a genome? Genome Biol 5(1):204. https://doi.org/10.1186/gb-2003-5-1-204
Opron K, Burton ZF (2018) Ribosome structure, function, and early evolution. Int J Mol Sci 20(1):40. https://doi.org/10.3390/ijms20010040
Ovádi J, Tompa P, Vértessy B, Orosz F, Keleti T, Welch GR (1989) Transient-time analysis of substrate-channelling in interacting enzyme systems. Biochem J 257(1):187–190. https://doi.org/10.1042/bj2570187
Paffen TFE, Teunissen AP, de Greef TFA, Meijer EW (2017) Model-driven engineering of supramolecular buffering by multivalency. Proc Natl Acad Sci USA 114(49):12882–12887. https://doi.org/10.1073/pnas1710993114
Paffen TF, Ercolani G, de Gree TF, Meijer EW (2015) Supramolecular buffering by ring-chain competition. J Am Chem Soc 137(4):1501–1509. https://doi.org/10.1021/ja5110377
Pareek V, Sha Z, He J, Wingreen NS, Benkovic SJ (2021) Metabolic channeling: predictions, deductions, and evidence. Mol Cell 81(18):3775–3785. https://doi.org/10.1016/jmolcel202108030
Parmar VK, Grinde E, Mazurkiewicz JE, Herrick-Davis K (2017) Beta2-adrenergic receptor homodimers: role of transmembrane domain 1 and helix 8 in dimerization and cell surface expression. Biochim Biophys Acta Biomembr 1859(9 Pt A):1445–1455. https://doi.org/10.1016/jbbamem201612007
Prosdocimi F, José MV, de Farias ST (2021) The theory of chemical symbiosis: a margulian view for the emergence of biological systems (Origin of Life). Acta Biotheor 69(1):67–78. https://doi.org/10.1007/s10441-020-09388-7
Rabouille C, Misteli T, Watson R, Warren G (1995) Reassembly of Golgi stacks from mitotic Golgi fragments in a cell-free system. J Cell Biol 129(3):605–618. https://doi.org/10.1083/jcb1293605
Raine DJ, Grondin Y, Thellier M, Norris V (2003) Networks as constrained thermodynamic systems. C R Biol 326(1):65–74. https://doi.org/10.1016/s1631-0691(03)00009-x
Ramesh M, Soliman ME (2015) G-protein coupled receptors (GPCRs): a comprehensive computational perspective. Comb Chem High Throughput Screen 18(4):346–364. https://doi.org/10.2174/1386207318666150305155545
Rhinesmith T, Turkette T, Root-Bernstein R (2017) Rapid non-enzymatic glycation of the insulin receptor under hyperglycemic conditions inhibits insulin binding in vitro: implications for insulin resistance. IJMS 18(12):2602. https://doi.org/10.3390/ijms18122602
Root-Bernstein R (1984) Molecular sandwiches' as a basis for structural and functional similarities for interferons, MSH, ACTH, LHRH, myelin basic protein, and albumins. FEBS Lett 168(2):208–212. https://doi.org/10.1016/0014-5793(84)80247-1
Root-Bernstein R (1987) Catecholamines bind to enkephalins, morphiceptin, and morphine. Brain Res Bull 18(4):509–532. https://doi.org/10.1016/0361-9230(87)90116-x
Root-Bernstein RS, Dillon PF (1997) Molecular complementarity I: the complementarity theory of the origin and evolution of life. J Theor Biol 188(4):447–79. https://doi.org/10.1006/jtbi.1997.0476
Root-Bernstein R (2002) Molecular complementarity III. Peptide complementarity as a basis for peptide receptor evolution: a bioinformatic case study of insulin, glucagon and gastrin. J Theor Biol 218(1):71–84. https://doi.org/10.1006/jtbi20023056
Root-Bernstein R, Dillon PF (2002) Fostering adventure research. A case study of the discovery that ascorbic acid enhances adrenergic drug activity. Drug Dev Res 57:58–74

Root-Bernstein R (2005) Peptide self-aggregation and peptide complementarity as bases for the evolution of peptide receptors: a review. JMR 18(1):40–49. https://doi.org/10.1002/jmr690
Root-Bernstein R (2007) An insulin-like modular basis for the evolution of glucose transporters (GLUT) with implications for diabetes. Evol Bioinformatics Online 3:317–331
Root-Bernstein R (2010) Molecular complementarity in an ecosystems-first theory of the origin of life. In: Busiak V, Navorro-Gonzalez R (eds) Astrobiology: from simple molecules to primitive life. American Scientific Publishers, Valencia, CA
Root-Bernstein R (2012) A modular hierarchy-based theory of the chemical origins of life based on molecular complementarity. Acc Chem Res 45(12):2169–2177. https://doi.org/10.1021/ar200209k
Root-Bernstein R, Root-Bernstein M (2016a) The ribosome as a missing link in prebiotic evolution II: ribosomes encode ribosomal proteins that bind to common regions of their own mRNAs and rRNAs. J Theor Biol 397:115–127. https://doi.org/10.1016/j.jtbi.2016.02.030
Root-Bernstein R (2022) Biased, bitopic, opioid-adrenergic tethered compounds may improve specificity, lower dosage and enhance agonist or antagonist function with reduced risk of tolerance and addiction. Pharmaceuticals (Basel) 15(2):214. https://doi.org/10.3390/ph15020214
Root-Bernstein R (2026) Selection for molecularly complementary modules (MCMs) drives the origins and evolution of pleiofunctional, epistatic interactomes (PEIs). Life (Basel) 16(1):170. https://doi.org/10.3390/life16010170
Root-Bernstein R, Baker AG, Rhinesmith T, Turke M, Huber J, Brown AW (2023) "Sea water" supplemented with calcium phosphate and magnesium sulfate in a long-term Miller-type experiment yields sugars, nucleic acids bases, nucleosides, lipids, amino acids, and oligopeptides. Life 13(2):265. https://doi.org/10.3390/life13020265
Root-Bernstein R, Bernstein MI (2024) Evolutionary poker': an agent-based model of interactome emergence and epistasis tested against Lenski's long-term E coli experiments. J Physiol 602(11): 2511–2535. https://doi.org/10.1113/JP284421
Root-Bernstein R, Brown AW (2022) Novel apparatuses for incorporating natural selection processes into origins-of-life experiments to produce adaptively evolving chemical ecosystems. Life 12(10):1508. https://doi.org/10.3390/life12101508
Root-Bernstein R, Churchill B (2021) Co-evolution of opioid and adrenergic ligands and receptors: shared, complementary modules explain evolution of functional interactions and suggest novel engineering possibilities. Life 11(11):1217. https://doi.org/10.3390/life11111217
Root-Bernstein R, Churchill B, Turke M (2020) Glutathione and glutathione-like sequences of opioid and aminergic receptors bind ascorbic acid, adrenergic and opioid drugs mediating antioxidant function: relevance for anesthesia and abuse. IJMS 21(17):6230. https://doi.org/10.3390/ijms21176230
Root-Bernstein R, Churchill B, Turke M, Subhramanyam UKT, Labahn J (2019) Mutual enhancement of opioid and adrenergic receptors by combinations of opioids and adrenergic ligands is reflected in molecular complementarity of ligands: drug development possibilities. IJMS 20(17):4137. https://doi.org/10.3390/ijms20174137
Root-Bernstein R, Dillon PF (1997) Molecular complementarity I: the complementarity theory of the origin and evolution of life. J Theor Biol 188(4):447–479. https://doi.org/10.1006/jtbi19970476
Root-Bernstein R, Dillon PF (2008) Small molecule complementarity as a source of novel pharmaceutical agents and combination therapies. Curr Pharm Des 14(1):55–62. https://doi.org/10.2174/138161208783330727
Root-Bernstein R, Dillon PF (2014) A common molecular motif characterizes extracellular allosteric enhancers of GPCR aminergic receptors and suggests enhancer mechanism of action. Curr Med Chem 21(32):3673–3686. https://doi.org/10.2174/0929867321666140826120604
Root-Bernstein R, Dobbelstein C (2001) Insulin binds to glucagon forming a complex that is hyper-antigenic and inducing complementary antibodies having an idiotype-antiidiotype relationship. Autoimmunity 33(3):153–169. https://doi.org/10.3109/08916930109008044

Root-Bernstein R, Fewins J, Rhinesmith T, Koch A, Dillon PF (2016a) Enzymatic recycling of ascorbic acid from dehydroascorbic acid by glutathione-like peptides in the extracellular loops of aminergic G-protein coupled receptors. JMR 29(7):296–302. https://doi.org/10.1002/jmr.2530

Root-Bernstein R, Holsworth DD (1998) Antisense peptides: a critical mini-review. J Theor Biol 190(2):107–119. https://doi.org/10.1006/jtbi19970544

Root-Bernstein R, Kim Y, Sanjay A, Burton ZF (2016b) tRNA evolution from the proto-tRNA minihelix world. Transcription 7(5):153–163. https://doi.org/10.1080/21541264201612355527

Root-Bernstein M, Root-Bernstein R (2015) The ribosome as a missing link in the evolution of life. J Theor Biol 367:130–158. https://doi.org/10.1016/j.jtbi.2014.11.025

Root-Bernstein R, Root-Bernstein M (2016a) The ribosome as a missing link in prebiotic evolution II: ribosomes encode ribosomal proteins that bind to common regions of their own mRNAs and rRNAs. J Theor Biol 397:115–127. https://doi.org/10.1016/j.jtbi.2016.02.030

Root-Bernstein R, Root-Bernstein M (2016b) From compositional chemical ecologies to self-replicating ribosomes and on to functional trait ecological networks. In: Panterotti P (ed) Evolutionary biology - 19th meeting 2015. Springer, New York, London, pp 327–346

Root-Bernstein R, Turke M, Subhramanyam UKT, Churchill B, Labahn J (2018) Adrenergic agonists bind to adrenergic-receptor-like regions of the mu opioid receptor, enhancing morphine and methionine-enkephalin binding: a new approach to "biased opioids"? IJMS 19(1):272. https://doi.org/10.3390/ijms1901027

Root-Bernstein R, Vonck J (2009) Glucose binds to the insulin receptor affecting the mutual affinity of insulin and its receptor. CMLS 66(16):2721–2732. https://doi.org/10.1007/s00018-009-0065-8

Root-Bernstein R, Vonck J (2010) Modularity in receptor evolution: insulin- and glucagon-like peptide modules as binding sites for insulin and glucose in the insulin receptor. J Recept Ligand Channel Res 3:87–96. https://doi.org/10.2147/JRLCRS6737

Root-Bernstein R, Westall FC (1984a) Fenfluramine binds 5-hydroxytryptophan. Brain Res Bull 12(1):17–22. https://doi.org/10.1016/0361-9230(84)90209-0

Root-Bernstein R, Westall FC (1984b) Serotonin binding sites I. Structures of sites on myelin basic protein, LHRH, MSH, ACTH, interferon, serum albumin, ovalbumin and red pigment concentrating hormone. Brain Res Bull 12(4):425–436. https://doi.org/10.1016/0361-9230(84)90115-1

Root-Bernstein R, Westall FC (1984c) Fibrinopeptide a binds Gly-pro-Arg-pro. Proc Natl Acad Sci USA 81(14):4339–4342. https://doi.org/10.1073/pnas81144339

Root-Bernstein R, Westall FC (1986) Bovine pineal antireproductive tripeptide binds to luteinizing hormone-releasing hormone: a model for peptide modulation by sequence specific peptide interactions? Brain Res Bull 17(4):519–528. https://doi.org/10.1016/0361-9230(86)90219-4

Rorick MM, Wagner GP (2011) Protein structural modularity and robustness are associated with evolvability. Genome Biol Evol 3:456–475. https://doi.org/10.1093/gbe/evr046

Roy B, Haupt LM, Griffiths LR (2013) Review: alternative splicing (AS) of genes as an approach for generating protein complexity. Curr Genomics 14(3):182–194. https://doi.org/10.2174/1389202911314030004

Ruddigkeit L, van Deursen R, Blum LC, Reymond JL (2012) Enumeration of 166 billion organic small molecules in the chemical universe database GDB-17. J Chem Inform Modeling 52(11): 2864–2875. https://doi.org/10.1021/ci300415d

Sanchez C, Lachaize C, Janody F, Bellon B, Röder L, Euzenat J, Rechenmann F, Jacq B (1999) Grasping at molecular interactions and genetic networks in Drosophila melanogaster using FlyNets, an Internet database. Nucl Acids Res 27(1):89–94. https://doi.org/10.1093/nar/27189

Schenk JO, Morocco MT, Ziemba VA (1991) Interactions between the argininyl moieties of neurotensin and the catechol protons of dopamine. J Neurochem 57(5):1787–1795. https://doi.org/10.1111/j.1471-4159.1991.tb06382.x

Schiller MR (2016) The minimotif synthesis hypothesis for the origin of life. J Transl 2(5):289–296. https://doi.org/10.15761/JTS1000154

Schmitt BM (2004) The concept of "buffering" in systems and control theory: from metaphor to math. Chembiochem 5(10):1384–1392. https://doi.org/10.1002/cbic200400126
Schuster P (2000) Taming combinatorial explosion. Proc Natl Acad Sci USA 97(14):7678–7680. https://doi.org/10.1073/pnas150237097
Segre D, Ben-Eli D, Lancet D (2000) Compositional genomes: prebiotic information transfer in mutually catalytic noncovalent assemblies. Proc Natl Acad Sci USA 97:4112–4117
Shapiro R (2006) Small molecule interactions were central to the origin of life. Q Rev Biol 81(2): 105–125
Sharif NA, Burt DR, Towle AC, Mueller RA, Breese GR (1983) Codepletion of serotonin and TRH induces apparent supersensitivity of spinal TRH receptors. Eur J Pharm 95(3–4):301–304. https://doi.org/10.1016/0014-2999(83)90651-9
Sheahan MB, Rose RJ, DW MC (2004) Organelle inheritance in plant cell division: the actin cytoskeleton is required for unbiased inheritance of chloroplasts, mitochondria and endoplasmic reticulum in dividing protoplasts. The Plant J 37(3):379–390. https://doi.org/10.1046/j1365-313x200301967x
Siemion IZ, Cebrat M, Kluczyk A (2004) The problem of amino acid complementarity and antisense peptides. Curr Prot Pept Sci 5(6):507–527. https://doi.org/10.2174/1389203043379413
Simon HA (1962) The architecture of complexity. Proc Am Phil Soc 106(6):467–482
Simon HA (1981) The sciences of the artificial. MIT Press, Cambridge, MA
Singh R (2024) A concept of complementarity between complexity and redundancy can account for Kant's biological teleology and unify mechanistic and finalistic biology. J Mol Evol 92(3): 258–265. https://doi.org/10.1007/s00239-024-10169-w
Srivastava DK, Bernhard SA (1986) Enzyme-enzyme interactions and the regulation of metabolic reaction pathways. Curr Top Cell Reg 28:1–68. https://doi.org/10.1016/b978-0-12-152828-750003-2
Štambuk N, Konjevoda P (2020) The temperature dependence of amino acid hydrophobicity data is related to the genetic coding algorithm for complementary (sense and antisense) peptide interactions. Data Brief 30:105392. https://doi.org/10.1016/jdib2020105392
Štambuk N, Manojlović Z, Turčić P, Martinić R, Konjevoda P, Weitner T, Wardega P, Gabričević M (2014) A simple three-step method for design and affinity testing of new antisense peptides: an example of erythropoietin. IJMS 15(6):9209–9223. https://doi.org/10.3390/ijms15069209
Sugaya M, Nishimura F, Katoh A, Harada K (2008) Tailoring the peptide-binding specificity of an RNA by combinations of specificity-altering mutations. Nucleosides Nucleotides Nucleic Acids 27(5):534–545. https://doi.org/10.1080/15257770801944493
Szabla R (2021) Rethinking UV-induced prebiotic selection of biomolecules. In: saija, franz cassone, giuseppe (eds) prebiotic photochemistry : from Urey–Miller-like experiments to recent findings. Royal Society of Chemistry, Chapter 5, pp 79–106
Tompa P, Rose GD (2011) The Levinthal paradox of the interactome. Protein Sci 20(12): 2074–2079. https://doi.org/10.1002/pro.747
Topiol S (1987) The deletion model for the origin of receptors. Trends Biochem Sci 12:419–421
Tsai YJ, Lee HI, Lin A (2012) Ribosome distribution in HeLa cells during the cell cycle. PLoS One 7(3):e32820. https://doi.org/10.1371/journalpone0032820
Vilardaga JP, Nikolaev VO, Lorenz K, Ferrandon S, Zhuang Z, Lohse MJ (2008) Conformational cross-talk between alpha2A-adrenergic and mu-opioid receptors controls cell signaling. Nat Chem Biol 4(2):126–131. https://doi.org/10.1038/nchembio64
Walke PB, Bansode SB, More NP, Chaurasiya AH, Joshi R, Kulkarni MJ (2021) Molecular investigation of glycated insulin-induced insulin resistance via insulin signaling and AGE-RAGE axis. Biochim Biophys Acta 1867(2):166029. https://doi.org/10.1016/jbbadis2020166029
Wang Y, Song X, Wang Y, Wang N (2022) Specific interaction of insulin receptor and GLP-1 receptor mediates crosstalk between their signaling. Biochem Biophys Res Commun 636(Pt 2): 31–39. https://doi.org/10.1016/jbbrc202210094

Warren G (1993) Membrane partitioning during cell division. Annu Rev Biochem 62:323–348. https://doi.org/10.1146/annurevbi62070193001543

Waterson RM, Konigsberg WH (1974) Peptide mapping of aminoacyl-tRNA synthetases: evidence for internal sequence homology in Escherichia coli leucyl-tRNA synthetase. Proc Natl Acad Sci U S A 71(2):376–380. https://doi.org/10.1073/pnas712376

Weber AL (2004) Kinetics of organic transformations under mild aqueous conditions: implications for the origin of life and its metabolism. Orig Life Evol Biosph 34(5):473–495. https://doi.org/10.1023/b:orig000004312830559fe

Weisman LS (2003) Yeast vacuole inheritance and dynamics. Annu Rev Genet 37:435–460. https://doi.org/10.1146/annurevgenet37050203103207

Witzany G (2016) Crucial steps to life: from chemical reactions to code using agents. Bio Syst 140: 49–57. https://doi.org/10.1016/jbiosystems201512007

Witzany G (2021) Preliminary remarks to an integrative theory of evolution. Theor Biol Forum 114(2):41–59. https://doi.org/10.19272/202111402004

Wolf S, Grünewald S (2015) Sequence, structure and ligand binding evolution of rhodopsin-like G protein-coupled receptors: a crystal structure-based phylogenetic analysis. PLoS One 10(4): e0123533. https://doi.org/10.1371/journalpone0123533

Wu YW, Seah YS, Chung KT, Liu MD (1999) Postoperative pain relief in primigravida caesarean section patients--combination of intrathecal morphine and epinephrine. Acta Anaesthesiol Sin 37(3):111–114

Yamauchi K, Hashizume K (1986) Glucagon alters insulin binding to isolated rat epididymal adipocytes: possible role of adenosine 3′,5′-monophosphate in modification of insulin action. Endocrinology 119(1):218–223. https://doi.org/10.1210/endo-119-1-218

Yehorova D, Crean RM, Kasson PM, Kamerlin SCL (2024) Key interaction networks: identifying evolutionarily conserved non-covalent interaction networks across protein families. Protein Sci 33(3):e4911. https://doi.org/10.1002/pro4911

Yu GB, Caspar DL (1998) Structure of cubic insulin crystals in glucose solutions. Biophys J 74: 616–622

Zhang C, Phillips APR, Wipfler RL, Olsen GJ, Whitaker RJ (2018) The essential genome of the crenarchaeal model Sulfolobus islandicus. Nat Commun 9(1):4908. https://doi.org/10.1038/s41467-018-07379-4

Zoete V, Meuwly M, Karplus M (2004) Investigation of glucose binding sites on insulin proteins. 55:568–581

Zviling M, Kochva U, Arkin IT (2007) How important are transmembrane helices of bitopic membrane proteins? Biochim Biophys Acta 1768(3):387–392. https://doi.org/10.1016/jbbamem200611019

Chapter 5
Transposons, Stress, and the Evolution of Adaptability

Richard Hunter

Abstract Lines of evidence from divergent fields of biology have begun to demonstrate that evolutionary adaptation can happen with a speed that is inexplicable via random mutation of protein-coding genes. Evolution appears to happen most rapidly in regulatory RNA networks—networks that are primarily derived from transposons. Many transposons appear to be intimately involved in the evolution of both cellular and organismal stress response systems and, in turn, are regulated by stress. Further, in the organs most involved in flexible adaptation to changing environmental demands, for example, the mammalian brain, exapted transposons play a central role in the emergence of plasticity in response to experience.

1 Introduction

As biologists have become more aware of the speed at which environments can change, a parallel awareness has emerged of the speed at which organisms can adapt to environmental shocks and stress: rapid evolution has now been observed in numerous species, including vertebrates, such as the iconic Darwin's finches (Stroud and Ratcliff 2025). The speed of these evolutionary changes, coupled with our deepening understanding of genomics, has driven significant changes in evolutionary theory (Pigliucci et al. 2010) and led us to a view of evolution that is less centered on the random mutation of DNA sequence and toward an emerging synthesis that gives greater prominence to regulatory RNA and the transposons that are the primary source of it (Hunter et al. 2015; Monaco 2022; Mattick and Amaral 2023). Stress is a challenge to organisms, even in relatively stable environments; physiological stress responses have evolved to help organisms adapt to environmental change and threat. These stress responses are examples of evolved adaptability. Another example of evolved adaptability is the vertebrate nervous system, which is notable for its

R. Hunter (✉)
University of Massachusetts Boston, Department of Psychology, Developmental and Brain Sciences Program, Boston, MA, USA
e-mail: richard.hunter@umb.edu

G. Witzany (ed.), *Integrative Theory of Evolution - Aspects and Insights*,
https://doi.org/10.1007/978-3-032-24295-2_5

capacity for lasting plasticity and memory in response to the environment. In the past quarter century or so, it has become clear that much of the adaptive plasticity of the nervous system is under epigenetic control. Epigenetic mechanisms are thought to have evolved for the control of transposable elements in the genome (Fedoroff 2012). Stress and stress hormones have a significant role in regulating epigenetic landscapes in the brain and other tissues, and the stress response has evolved in interaction with transposons in a variety of different organisms (Griffiths and Hunter 2014; Hunter et al. 2015; Gray et al. 2017; Bartlett et al. 2019; Rahman and McGowan 2022).

Since their discovery in the late 1940s, controlling elements, or as we now call them, transposons (TE), have been linked to environmental stressors and shocks (McClintock 1950, 1984). Transposons, or mobile genetic elements, are segments of the genome that are able to move around the genome. They come in a variety of flavors and sizes, reflecting their diverse and, sometimes, mysterious evolutionary origins. For technical and historical reasons, it is only in the past decade or so that we have begun to understand their functional significance at either the organismal or evolutionary level. The technical change has been the emergence of modern sequencing technologies in the twenty-first century, which has allowed us to examine transcriptomes and genomic transpositions in single-base detail. The publication of the first full transcriptomes by the ENCODE consortium more than a decade ago demonstrated that more than 80% of our genome is transcribed, suggesting a much higher level of functionality than had been previously thought possible (The ENCODE Project Consortium 2012). Sequencing has not only allowed us to observe transposition in something close to real time but has also revealed a vast world of potentially functional noncoding RNAs (ncRNAs) derived from transposons (Kelley and Rinn 2012; Carlevaro-Fita et al. 2019). The data derived from these new technologies has driven a change from the historical view that transposons are "junk" to the emerging view that they often have function and may contribute much more to biological function than was appreciated during the past century (Ohno 1972; Joly-Lopez and Bureau 2014; Hunter 2020). Contrary to the expectation that genomic junk would be taken to the evolutionary curb, most transposons seem to be under nearly neutral selective pressure (Arkhipova 2018), though many genomes show evidence of "accordion-like" acquisition and deletion of transposons, suggesting that they may require periodic spring-cleanings (Gilbert and Feschotte 2018) and that transposons do have a cost. This change in the view of transposons has corresponded with a larger change in evolutionary theory, the emergence of the "Extended Synthesis," which derives from discoveries in diverse fields of biology that have highlighted the deficiencies of the Modern Synthesis of the twentieth century (Pigliucci et al. 2010).

Transposons constitute a large fraction of most eukaryotic genomes, up to 85% in some cases, though they are closer to 50% of most genomes described (Jurka et al. 2007). This, in contrast to the roughly 2% constituted by protein-coding genes, is one clue to their potential importance in genome evolution and evolution more generally. Transposon taxonomists divide transposable elements into two major classes: class I retrotransposons and class II DNA transposons. Members of class I tend to replicate via "copy and paste" mechanisms, thus tending to multiply their numbers once resident in a host genome, while class II transposons tend to move via "cut and paste" and are typically less common, at least in mammalian genomes. Class I retroelements are further subdivided into long terminal repeats/endogenous retroviruses (LTR/ERV) and non-LTR retrotransposons, which include long interspersed nuclear elements (LINEs) and short interspersed nuclear elements (SINEs), among other families. Class II elements comprise four major families, of which two, DDE (named for the conserved aspartate-aspartate-glutamate sequence in their transposase) and Helitron transposons, are most common in eukaryotes (Jurka et al. 2007; Wells and Feschotte 2020). Beyond these two major classes, a distinction is typically made between autonomous and nonautonomous transposons, that is to say those that encode their own transposition machinery and those that do not. The nonautonomous transposons therefore require the presence of an autonomous transposon in the host genome if they are to transpose. While transposition is an important part of their contribution to genome evolution, their contribution to a host of novel RNA elements, as well as a few important protein-coding genes, is arguably a more significant contribution to host adaptability (Hunter et al. 2015; Hunter 2024).

In this chapter, I will outline how transposons have played a vital role in the evolution of adaptability to environmental stress, with attention to what is likely the most adaptable of all of evolution's creations, the mammalian brain. Further, I will examine the potential molecular mechanisms by which the vertebrate stress response system might work to capture and direct the activity of transposons toward adaptive ends.

2 Transposons, Epigenetics, Development, and Evolution

Perhaps the most eloquent argument for transposons' role in evolution is the discovery, a decade ago, that the melanistic mutation in British peppered moths during the industrial revolution was due to a transposon (Hof et al. 2016). This transposition event was estimated to have occurred in 1819, when Charles Darwin was a boy of ten. Given that this is quite literally a textbook example of adaptive evolution, it becomes difficult to see why transposons would still be ignored from an evolutionary perspective. Even in the era when transposons were regarded as junk, they were regarded as important to genome evolution, if only as factors that genomes needed protection against. Nina Fedoroff, a mentee of McClintock's, has argued that they were the main drivers in the evolution of the molecular epigenetic mechanisms that are common to all eukaryotic genomes but are largely absent in prokaryotes (Fedoroff 2012). In prokaryotes, the need to maintain relatively small genome sizes for replication efficiency and the widespread availability of horizontal gene transfer have meant that transposons are largely selected against. In eukaryotes, these constraints do not hold. Some of the oldest retrotransposon families, the LINE elements, are thought to have emerged at the same time as eukaryotic life. Further, in mammals at least, LINE RNA appears necessary for the development of embryos beyond the two-cell stage (Percharde et al. 2018), suggesting that these elements may play a vital role in the evolution of multicellularity. Since Waddington coined the term epigenetics, these mechanisms have been posited to play a role in development, specifically in the interplay between the genome and the environment (Waddington 1942; Tronick and Hunter 2016). The role of transposons in development continues to be delineated, with more evidence emerging in recent years, with one study identifying tens of thousands of LTR-driven genomic regulatory modules in the human genome, many with developmentally specific activity (Babarinde et al. 2025; Glinsky 2025a). Rather than being obligate parasites, some transposons may be fundamental to the emergence of cellular cooperativity needed for the evolution of multicellular life. Another function that transposons may have is to compensate for the lack of horizontal gene transfer in eukaryotes by providing an intra-genomic "adaptive toolkit" (Hunter et al. 2015). While this may be the case, it has become evident in recent years that not only is horizontal gene transfer much more common in eukaryotic genomes than previously thought, but it is also largely driven by transposons and plays a substantial role in their evolution and in the evolution of host genomes (Gilbert and Feschotte 2018). This transfer has occurred between widely divergent lineages, such as the tree shrew (*Tupaia belangeri)* and the harvester ant (*Pogonomyrmex barbatus),* which share a mariner transposon with 87% sequence identity (Oliveira et al. 2012). More recent work suggests that these events may be fairly common (Shen et al. 2020). Such observations complicate the assumptions of any phylogenetics based on the conservation of coding sequences.

Epigenetic control of transposons might then be more than bagging up the garbage or suppressing genomic parasites; it might be directed toward exaptation and exploitation of these genomic hangers-on, perhaps analogously to our domestication of the dogs and cats that devoured the scraps around our ancestors' fires and chased off the rats competing for them, though the rats never really left us or our genomes. If suppression of TE transcription was the purpose of epigenetic control of TEs, the ENCODE data suggests that this suppression has been at best incomplete (The ENCODE Project Consortium 2012). Our own data has shown that epigenetic control of certain LINEs, SINEs, and the rat ERV IAP (intracisternal A-particle, so named for the virus like particles these TEs are still capable of forming) RNA expression via histone methylation changes rapidly over the course of a few hours in response to stress signaling. Further, one of these elements, the rodent B2 SINE, interacts directly with the glucocorticoid stress hormone receptor (GR), binding to it and changing its transcriptional activity (Hunter et al. 2009, 2012; Bartlett et al. 2021, 2023). Such precise control does not fit with the parasitism model.

In the germline, transposons have long been regarded as particularly dangerous. Mechanisms for controlling them include PIWI (P-element Induced WImpy testis)-interacting RNA (piRNA) elements, which have traditionally been regarded as "transposon traps," protecting genomes from transposon invasion. However, recent findings in *Drosophila* fail to confirm the trap model (Scarpa and Kofler 2023; Wierzbicki and Kofler 2023; Selvaraju et al. 2024), opening the door to the possibility that these may be corrals built for domestication as much as traps oriented toward suppression. As piRNAs are highly active in the testes, where they had been thought to suppress transposon activity, it is surprising that in *Drosophila,* they appear to be associated with the production of new noncoding RNA genes, typically with effects on male reproductive fitness and spermatogenesis (Mohammed et al. 2014). The testes are an interesting test bed for new genes, as the cost of a few defective sperm is relatively low compared to an oocyte or a somatic cell. Further, the number of potential evolutionary experiments is high, given the large volume of sperm production. This is striking evidence that in these animals at least, piRNAs and related transposon control mechanisms may have evolved to evolve new genes out of the transposome (Werner et al. 2015). This constitutes more evidence that genomes may be evolved to evolve in less random and more constrained ways than simple random mutation would suggest and supports the idea that genomes are evolved to be evolvable. This idea, that evolution has selected for what Waddington called "trait adaptability", has emerged in recent years to explain observations that demonstrate the capacity of organisms to constrain their own evolution in more adaptive rather than deleterious directions. As John Mattick has suggested, if this capacity has evolved anywhere it would likely predominate across the biosphere, therefore such capacities are likely more norm than exception in biology (Wagner 2015, 2023; Mattick and Amaral 2023; Hunter 2024). Interestingly, there is at least one example of a sex-determining gene arising via transposon exaptation (Hayashi et al. 2022). Indeed, the number of different epigenetic control mechanisms and their differential effectiveness against different TE families allow for the possibility of selective control or cultivation of useful TEs and suppression or deletion of more harmful ones. On the other hand, TEs can selectively create new transcription factor–binding sites and chromatin domains, altering local gene expression, perhaps most powerfully via the creation of enhancers and topologically associated domains. In mammals, it appears that the expansion of CCCTC-binding factor (CTCF)–binding sites and, therefore, enhancers and large genomic structures has been driven largely by transposable elements (Schmidt et al. 2012; Lanciano and Mirouze 2018; Ohtani and Iwasaki 2021; Glinsky 2025a, b).

3 Stress and Rapid Adaptability

Previously, I have defined stress as "any environmental force an organism perceives it must adapt to" (Hunter 2020). Part of the utility of this definition is that it allows us to think about stress in both organismal and evolutionary terms. It also calls attention to both the diversity of potential stressors and the importance of perceiving or detecting threats in order to mount an adaptive response to them. This connects stress with Conrad Waddington's "trait adaptability," the idea that phenotypic adaptability itself might be selected in certain circumstances. Trait adaptability bears a resemblance to the more recent and broader concept of evolvability (Wagner 2023), for which evidence continues to emerge. Indeed, John Mattick and Paolo Amaral have argued that if any mechanism of evolvability arose at any point in evolution, it would come to dominate the "evolutionary landscape and, . . . be part of the toolkit of most, if not all, extant lineages in the biosphere" (Mattick and Amaral 2023). While some stressors are relatively predictable and allow for strongly determined physiological and behavioral solutions, such as hunger and thirst, the total list of possible stressors is an open set. Infection is an example of this, as the list of potential pathogens for any species is continually changing, and known pathogens constantly evolve to evade host defense mechanisms. The primate social environment is another sphere of constant change and potential distress; as an individual in any social species depends on its social standing for its survival, effective social stress responses are as vital as immune responses. Indeed, these two potential threats can interact with unfortunate consequences for those at the bottom of the social hierarchy (Marmot and Sapolsky 2014). As we shall discuss below, in mammals, the two organ systems responsible for dealing with infection and the social world, the immune and nervous systems, show signs of extensive exaptation of transposons in their evolution. The stress response in mammals, coordinated by the hypothalamic–pituitary–adrenal (HPA) axis and the sympathetic nervous system, is another example of an evolved form of adaptability that permits transition into what has been called an "emergency life history stage" to adapt to more extreme threats like predatory attack or aggression from members of one's own species (Wingfield et al. 1998). This response ensures that blood flows to the skeletal muscles and that abundant glucose is available to fuel them, but it also suppresses the immune system and constrains cognition—thus focusing on and coordinating the immune and nervous systems toward the immediate goal of survival.

Beyond the immediate survival of a stressed organism, it is possible to propose that stress is a signal of environmental change that might require an adaptive response from the genome in the broader evolutionary sense. Thus, stress signaling systems within the organism might activate transposon-derived trait adaptation or evolvability modules. At the geological timescales abstracted from the fossil record, such rapid, constrained change would not seem necessary. However, since the idea of punctuated equilibria was broached in 1972, it has been apparent that the gradualist view of evolution had serious gaps (Eldredge and Gould 1972). This has only become more obvious in recent decades as experimental evolution has shown rapid phenotypic change in the course of a few generations in multiple species, including vertebrates. While rapid evolution in prokaryotes has long been accepted due to their rapid generation time and near-ubiquitous horizontal gene transfer (Lenski 2023), evidence for rapid evolution and speciation in multicellular organisms has taken longer to accumulate, but it has accumulated. Reproductive isolation of a new population of "big bird" Darwin's finches in the Galapagos occurred in three generations (Lamichhaney et al. 2018), and multiple examples of rapid evolution have emerged from the iconic genus (Grant and Grant 2014). Rapid evolution has also been observed in Trinidadian guppies (Reznick and Travis 2019) and Caribbean anole lizards (Schoener et al. 2017; Losos 2018) and numerous other clades (Stroud and Ratcliff 2025). In *Anolis,* phenotypic changes have been observed in the overall body plan, size, and limb morphology, often emerging in ways that promote adaptation to local environmental changes (Losos 2018). This observation would require a genome capable of producing modular changes in the developmental plan as opposed to random, largely deleterious mutations. In long-term studies like those just referenced, it is evident that evolution is not just rapid but fluctuating such that phenotypes might emerge on short time scales, alter the local environment via changing predation or niche exploitation, and then change again in response to altered local conditions (Stroud and Ratcliff 2025). If the findings in these cases are general, it would argue that evolution is happening on short scales all the time, and the evolution of evolvability is a necessary correlate of this. No organism that requires the accumulation of random point mutations to evolve would survive in a world where its competitors can adapt in a few generations and thus change the environment in ways that random search would not even have had time to detect. If evolvability is the norm, then it must act on substrates that permit adaptation in canalized directions rather than entirely at random. Given that most evolution appears to occur in regulatory networks, the ready-made banks of regulatory RNA tools that the transposome provides seem the most likely place to look for evolvability modules (Wagner 1996; Bolker 2000; Coyne et al. 2004; Mallarino and Abzhanov 2012; Mattick and Amaral 2023; Hunter 2024). In *Anolis* lizards, accumulation of transposons has been observed during adaptive radiation, and they have landed in precisely the places our hypothesis here would predict—in the Hox genes important for the development of the body plan and limb patterning (Feiner 2016).

4 Environmental Stress, Transposons, and the Origin of Adaptive Organ Systems

Since their discovery, the responsiveness of McClintock's controlling elements to environmental stress has been a central feature of their behavior (McClintock 1984). This sensitivity to stress is shared with many lysogenic viruses, many of which contribute to the pool of LTR transposons in vertebrate and other genomes. This sensitivity to host stress is dangerous when the mobile element remains undomesticated, as is the case with HIV or Herpes viruses, but if brought under host control, it offers the possibility to create new stress-responsive regulatory networks and enhance host adaptability. This has happened in a number of ways across evolution, including the creation of transcription factor–binding site enhancers, telomere growth, regulatory RNA networks, and larger-scale genomic rearrangements (Long et al. 2016; Horváth et al. 2017; Lanciano and Mirouze 2018; Hunter 2024; Wang et al. 2025).

Immediate early genes (IEGs) are another group of stress-responsive genes, typically transcription factors, which appear to often have viral or transposon origins (Hawker et al. 1993; Dudley et al. 2002; Hunter 2024). Many of these genes, for example, the Fos family, are evolutionarily ancient and ubiquitous mediators of eukaryotic stress responses (Alfonso-Gonzalez and Riesgo-Escovar 2018). Many, if not most, immediate responses to cellular activation in complex eukaryotes are guided by IEG-mediated transcriptional programs. IEGs are also vital to the function of the vertebrate nervous system and are intimately involved in synaptic plasticity and memory formation in the brain (Minatohara et al. 2016). Expression of the c-Fos gene is a commonly used means for detecting activated neurons in modern neuroscience (Chung 2015), for example. Fos is also intimately involved in memory formation in the mammalian brain (Miyashita et al. 2018; Pettit et al. 2022). The *Arc* gene is another example of a retrotransposon-derived gene with a significant role in memory formation. *Arc* is derived from a Ty3 retrotransposon *Gag* gene and retains the capacity to form virus-like capsid particles; the *Drosophila Arc* and the vertebrate version appear to have arisen independently, suggesting that convergent forces have discovered the utility of capsid formation in the formation of synapses in nervous systems. Indeed, *Arc's* ability to form capsids may be vital to its function, which appears to involve transporting its mRNA between neurons in a very virus-like fashion (Sullivan et al. 2025). Mice with *Arc* gene knockouts show pronounced deficits in memory consolidation and foraging behavior in spite of apparently normal gross neuroanatomy (Plath et al. 2006; Managò et al. 2016; Wall et al. 2018; Ravens et al. 2023). These findings highlight one of the challenges of establishing genotype–phenotype relationships for transposable element–derived genes, even those like *Arc*, that code for proteins: The phenotypes often relate to complex cognitive or developmental outcomes that are often missed in assays that only examine viability or gross anatomy. Only about 25% of protein-coding genes result in failures of viability (White et al. 2013; The International Mouse Phenotyping Consortium et al. 2016) and phenotypic outputs of those that are often missed. For example, mice knockouts reported to have spatial memory deficits, which turned out to be merely blind. If, as has been argued by John Mattick and others, organismal complexity goes hand in hand with a large noncoding genome, then we should look for phenotypes at higher levels of organization, development, and cognition (Mattick and Amaral 2023).

As the examples of *Arc* and other IEGs show, transposons have clearly had a major impact on the evolution and function of neurons and brains. The activity of transposons in the vertebrate brain has been a topic of interest for a decade or more (Reilly et al. 2013; Ahmadi et al. 2020). This interest began with discoveries by Fred Gage and colleagues that showed that most of the transcriptional differences between neural progenitor cells and fully differentiated neurons in the rodent hippocampus were derived from transposable elements (Muotri et al. 2005; Erwin et al. 2014). Work since has validated and expanded the important contributions of retrotransposons to both pluripotent stem cell lineages and neurons (Wang et al. 2020; Whiteley and Shepherd 2025), and transposons seem particularly active in neurodevelopment and in reproductive tissues like testes and placenta (Whiteley and Shepherd 2025). *Arc* is not the only neuronal gene that seems to leverage its viral origins or the only one to form virus-like particles; more than a dozen other Ty3-derived genes from the *paraneoplastic Ma antigen (PNMA)*, PEG10/Marts/RTL, and the Bik gene families are all known to produce virus-like particles, and some (Bik-1 and Bik-2) are required for normal neurodevelopment, while others appear to play roles in both neurodevelopment and the placenta (Chang et al. 2025; Whiteley and Shepherd 2025).

The placenta is a hot spot for transposon activity and, of course, the principal site for maternal–fetal communication during development (Hunter et al. 2015; Keighley et al. 2023; Whiteley and Shepherd 2025). Indeed, the placenta itself seems to have been created by exaptation of the retrotransposon-derived syncytin and PEG genes (Mi et al. 2000; Kitazawa et al. 2017). Transposons also serve as species-specific enhancer elements in the placenta (Chuong et al. 2013). The recent discovery of RNA-containing extracellular trafficking in both directions across the placenta (Adamova et al. 2023) suggests that it may also be a means by which both mother and fetus might communicate transposon-derived information with one another, allowing for environmental and epigenetic programming of both during gestation. Milk also seems to be a means of informational exchange, also co-opting retrotransposon-derived tools in the process (Irmak et al. 2012). Transposons certainly play a role in maternal adaptation during pregnancy, for instance, in promoting increased hematopoiesis (Phan et al. 2024). It is also the case that fetal DNA can take up residence in the mother, likely via transpositional mechanisms. It remains an open question if the content of these vesicles is controlled by maternal stress or other environmental stimuli but a question that will hopefully be answered in the not-too-distant future. The accumulation of observations that transposon activity is concentrated in tissues like the placenta and testes is significant. The placenta is a location where maternal information about the external environment can help program the fetus's phenotype for the expected demands it will face (and the fetus can counterprogram its mother to enhance its own prospects). The testes, on the other hand, via the piRNA system, may represent a laboratory for adaptive variation, which allows males to create variation without cost to them as individuals (though, of course, it can and does alter the fitness of their offspring). Transposons and their hosts would seem to have evolved a cooperative system that enhances the survival of both via reproductive programming and experimentation. The other areas of high transposon activity or exaptation, the nervous and immune systems, have greater influence on individual survival than transgenerational communication, but the two are responsible for dealing with the most variable aspects of the external environment, where rigid programs derived from a limited number of protein-coding genes are unlikely to outcompete more flexible systems derived from the regulatory RNA modules provided by transposons (Kelley and Rinn 2012; Hunter 2020).

Retrotransposon activity has clearly played a role in a variety of evolutionary and developmental processes across diverse species, and this appears to be particularly true in human evolution. Transposon expansions and invasions have occurred multiple times across primate evolution, and many seem to be human-specific. Indeed, it appears that the majority of primate- and human-specific regulatory sequences in the genome are transposon-derived (Jacques et al. 2013; Pontis et al. 2022). These regulatory sequences are heavily represented in modules governing the nervous and immune systems (Guffanti et al. 2018; Glinsky 2025a, b; Ye et al. 2025), the two adaptive systems we noted earlier in this chapter. In excess of 85% of primate-specific transcription factor–binding sites are transposon-derived (Andrews et al. 2023), an astounding figure if we hold the old view that these are junk or parasites, less so if we take the emerging view that they are drivers of evolutionary innovation. Human transposon–derived regulatory networks show a strong bias toward neural development and synapse formation and function (Glinsky 2025a, b), and many transposon transcripts are expressed in regionally specific patterns in the brain and elsewhere (Faulkner and Carninci 2009). In the human dorsolateral prefrontal cortex, more than 1500 human-specific transposon RNA transcripts are expressed (Guffanti et al. 2018). The expansion of the dorsolateral prefrontal cortex is arguably the most substantial change in human evolution since our divergence from the great apes. The human hippocampus, the major organ of long-term explicit memory formation, is also about 50% larger than that of our close relatives, and it too shows significant evidence of transposon activity in its evolution and function (Erwin et al. 2014; Linker et al. 2017; Trizzino et al. 2018; Patoori et al. 2022; Cakir et al. 2025). Long noncoding RNAs derived from transposons have been shown to have a role in a variety of aspects of brain function and development (Alammari et al. 2024).

A few actual transposon RNAs (as opposed to ncRNAs exapted from transposons) have also been implicated in brain function, notably the B2 SINE in rodents, which plays a role in transcriptional regulation via interactions with RNA polymerase II (Walters et al. 2009) and has more recently been shown to be under precise epigenetic control by glucocorticoid stress hormones in the brain (Hunter et al. 2012; Bartlett et al. 2021) and, in turn, has been shown to alter glucocorticoid-mediated transcription via a direct interaction with the glucocorticoid receptor (GR), a ligand-activated transcription factor (Bartlett et al. 2023). As many steroid receptor–binding sites in the genome are transposon-derived and many steroid response elements are specifically SINE-derived, the fact that the homology between the RNA- and the DNA-binding sites exists is perhaps unsurprising, but the fact that it has functional consequences, in at least one case, converting GR from an activator to a repressor, is surprising. This finding is yet another example of the high degree of reciprocal tuning between transposable elements and cell and organismal stress responses. Interestingly, B2 SINE RNA has also recently been shown to have a necessary role in the maintenance of neuronal structure. Zahavi and colleagues have shown that the RNA is necessary for the regeneration of peripheral motor neuron axons (Zahavi et al. 2025). In parallel, our own lab has shown that B2 SINE RNA is necessary for full dendritic arborization in male but not female hippocampal neurons (Richter et al. 2025). The primate analog of B2, Alu SINE, comprises around 11% of the human genome and is thought to have played a substantial role in the evolution of the human brain (Larsen et al. 2018). Alu has also been implicated in neurodevelopmental disorders like autism, which implies that they have a functional role in normal development as well (Saeliw et al. 2018; Saeliw et al. 2023).

Together, these are evidence that these transposons are a part of the machinery of adaptive structural plasticity in mammalian neurons and form a part of a broader picture where transposons are central to the evolution of adaptive systems like the mammalian brain; however, many mysteries remain on this front.

Perhaps the most important gap has been in our understanding of the molecular chain of communication between stressful environmental changes and changes in transposon activity in the genome. However, in recent years, these mechanisms have begun to reveal themselves, at least at the level of cellular stress response (Horváth et al. 2017; Hunter 2020, 2024). Mojica and Kültz (2022) have identified increased mutation rate, histone modification, DNA modification, chromoanagenesis, and transposon activity as mechanisms for stress-induced evolution. Of course, all of these mechanisms can and do involve transposons (Fedoroff 2012; Hunter et al. 2012, 2015; Werner et al. 2015; Pellestor 2019; Monaco 2022). The testes as a site of potentially transposon-driven variation emerge here too in the literature around transgenerational epigenetic inheritance (Perez and Lehner 2019). These mechanisms also interact with steroid-driven stress signaling, especially via glucocorticoids, but it is likely that sex steroids and other systems are involved as well (Bartlett et al. 2019; Chiang et al. 2022; Mojica and Kültz 2022). The glucocorticoid receptor is involved in many interactions with both the epigenome and with transposable elements and is itself regulated by transposon-derived ncRNAs like B2 and Gas5 (Daskalakis et al. 2018; Bartlett et al. 2019, 2021; Casciaro et al. 2023). However, the glucocorticoid receptor is part of a larger system of cellular stress response that includes its chaperones, the heat shock proteins HSP90 and HSP70. Of course, heat is an environmental stress, and heat and cold are among the most common stressors studied with regard to transposon-driven adaptation. HSP90 and HSP70 have emerged as major factors both in the control of transposon-driven genomic instability and the modulation of piRNA pathways in the germline, as well as their stress-induced activation (Cappucci et al. 2019; Specchia and Bozzetti 2021). Both of these chaperone proteins are directly involved in the maturation and function of not only glucocorticoid receptors but of the androgen, estrogen, and progesterone receptors as well (Echeverria and Picard 2010), meaning they provide a nexus for potential cross talk between these systems and the transposome. Indeed, HSP90 has been referred to as an "epigenetic capacitor," enabling phenotypic change in response to environmental challenges (Condelli et al. 2019). Another major driver of phenotypic plasticity, as well as an important factor in genomic structure, is CCCTC-binding factor (CTCF), which also participates in interactions with GR to create topologically associated domains in the genome that coordinate gene expression and cell differentiation (Clarisse et al. 2024). As noted above, many CTCF in mammalian genomes are transposon-derived, suggesting a multilayered interaction between transposable elements and the stress system to create new gene expression modules and tweak the activity of existing ones. The idea of epigenetic potential has been invoked in the past to explain the capacity of populations to adapt to new environments (Kilvitis et al. 2018). The examples above, GR transposon cross-talk, HSP90 driven epigenetic capacitance etc., offer a sketch of the mechanisms by which that adaptation in response to environmental challenge might occur and what some of the players may be.

5 Conclusions

Our understanding of the pace and mechanisms of evolution is changing at a pace not seen since the mid-twentieth century. The ever-expanding number of examples of rapid evolution in both experimental and field contexts, coupled with the detail and volume of genomics, epigenetics, and transcriptional data now available, has made it plain that evolution moves rapidly in response to changing environments. We are beginning to see how regularly evolutionary innovation works via the substrate of transposable elements and how genomes have domesticated and exapted them via epigenetic mechanisms. I have argued here that the mechanisms by which environmental signals can be converted into adaptive phenotypic and genomic change are already partly understood and that these mechanisms interact directly with the corticosteroid-driven stress axis via the glucocorticoid receptor and its partners. Of course, the GR system is not the only stress or environmental sensing system, and it is likely that as more work is done, linkages to other systems will emerge; indeed, much remains unknown about GR's molecular partners, particularly when it comes to ncRNA.

Nonetheless, it seems possible to outline a system where environmental stress, acting via the GR and other stress detectors, activates genomic innovation in the testes and alters developmental programming via the placenta (in placental mammals at least) and other parent–offspring interactions to create new, adaptive phenotypes in a few generations. These changes are likely epigenetic in nature in early generations, before becoming more firmly established by selection once the relation between a population and its environment becomes stable enough. Of course, this is not to argue that all transposon-driven change will be adaptive; the list of transposition-driven diseases is long, as is the list of stress-related disorders (Korte et al. 2005; McEwen 2012; Kazazian and Moran 2017; Bhat et al. 2026). Undoubtedly, we are in the process of a transformative shift in the way we think about evolution and the organism's relation to its environment and its own genomic passengers.

References

Adamova P, Lotto RR, Powell AK, Dykes IM (2023) Are there foetal extracellular vesicles in maternal blood? Prospects for diagnostic biomarker discovery. J Mol Med 101(1–2):65–81. https://doi.org/10.1007/s00109-022-02278-0

Ahmadi A et al (2020) Transposable elements in brain health and disease. Ageing Res Rev 64: 101153. https://doi.org/10.1016/j.arr.2020.101153

Alammari F et al (2024) Hidden regulators: the emerging roles of lncRNAs in brain development and disease. Front Neurosci 18:1392688. https://doi.org/10.3389/fnins.2024.1392688

Alfonso-Gonzalez C, Riesgo-Escovar JR (2018) Fos metamorphoses: lessons from mutants in model organisms. Mech Dev 154:73–81. https://doi.org/10.1016/j.mod.2018.05.006

Andrews G et al (2023) Mammalian evolution of human cis-regulatory elements and transcription factor binding sites. Science 380(6643):eabn7930. https://doi.org/10.1126/science.abn7930

Arkhipova IR (2018) Neutral theory, transposable elements, and eukaryotic genome evolution Kumar S, editor. Mol Biol Evol 35(6):1332–1337. https://doi.org/10.1093/molbev/msy083

Babarinde IA et al (2025) Transposable element expression and sub-cellular dynamics during hPSC differentiation to endoderm, mesoderm, and ectoderm lineages. Nat Commun 16(1):7670. https://doi.org/10.1038/s41467-025-63080-3

Bartlett AA, Lapp HE, Hunter RG (2019) Epigenetic mechanisms of the glucocorticoid receptor. Trends Endocrinol Metab 30(11):807–818. https://doi.org/10.1016/j.tem.2019.07.003

Bartlett AA et al (2021) Corticosterone dynamically regulates retrotransposable element expression in the rat hippocampus and C6 cells. Neurobiol Stress 15:100397. https://doi.org/10.1016/j.ynstr.2021.100397

Bartlett AA, Guffanti G, Hunter RG (2023) B2 SINE RNA as a novel regulator of glucocorticoid receptor transcriptional activity. Neurobiol Stress 23:100522. https://doi.org/10.1016/j.ynstr.2023.100522

Bhat MI et al (2026) Impact of transposable elements on DNA double-strand break repair and genomic stability. Mol Cell Biol:1–22. https://doi.org/10.1080/10985549.2025.2594182

Bolker JA (2000) Modularity in development and why it matters to Evo-Devo. Am Zool 40(5): 770–776. https://doi.org/10.1093/icb/40.5.770

Cakir U et al (2025) Erythropoietin reshapes the adult hippocampal chromatin landscape to promote neurogenesis. [accessed 2026 Jan 11]. http://biorxiv.org/lookup/doi/10.1101/2025.10.13.682070

Cappucci U et al (2019) The Hsp70 chaperone is a major player in stress-induced transposable element activation. Proc Natl Acad Sci USA 116(36):17943–17950. https://doi.org/10.1073/pnas.1903936116

Carlevaro-Fita J et al (2019) Ancient exapted transposable elements promote nuclear enrichment of human long noncoding RNAs. Genome Res 29(2):208–222. https://doi.org/10.1101/gr.229922.117

Casciaro C, Hamada H, Kostaki A, Matthews SG (2023) Glucocorticoid exposure modifies the miRNA profile of sperm in the Guinea pig: implications for intergenerational transmission. FASEB J 37(4):e22879. https://doi.org/10.1096/fj.202201784R

Chang N-C et al (2025) Gag proteins encoded by endogenous retroviruses are required for zebrafish development. Proc Natl Acad Sci USA 122(18):e2411446122. https://doi.org/10.1073/pnas.2411446122

Chiang VS-C, DeRosa H, Park JH, Hunter RG (2022) The role of transposable elements in sexual development. Front Behav Neurosci 16:923732. https://doi.org/10.3389/fnbeh.2022.923732

Chung L (2015) A brief introduction to the transduction of neural activity into Fos signal. Dev Reprod 19(2):61–67. https://doi.org/10.12717/DR.2015.19.2.061

Chuong EB, Rumi MAK, Soares MJ, Baker JC (2013) Endogenous retroviruses function as species-specific enhancer elements in the placenta. Nat Genet 45(3):325–329. https://doi.org/10.1038/ng.2553

Clarisse D et al (2024) Glucocorticoid receptor signaling: intricacies and therapeutic opportunities. Trends Biochem Sci 49(5):431–444. https://doi.org/10.1016/j.tibs.2024.01.012

Condelli V et al (2019) HSP90 molecular chaperones, metabolic rewiring, and epigenetics: impact on tumor progression and perspective for anticancer therapy. Cells 8(6):532. https://doi.org/10.3390/cells8060532

Coyne JA, Orr HA, Orr AH (2004) Speciation. Sinauer Associates, Inc. Publ
Daskalakis NP, Provost AC, Hunter RG, Guffanti G (2018) Noncoding RNAs: stress, glucocorticoids, and posttraumatic stress disorder. Biol Psychiatry 83(10):849–865. https://doi.org/10.1016/j.biopsych.2018.01.009
Dudley J et al (2002) What retroviruses teach us about the involvement of c-Myc in leukemias and lymphomas. Leukemia 16(6):1086–1098. https://doi.org/10.1038/sj.leu.2402451
Echeverria PC, Picard D (2010) Molecular chaperones, essential partners of steroid hormone receptors for activity and mobility. Biochim Biophys Acta Mol Cell Res 1803(6):641–649. https://doi.org/10.1016/j.bbamcr.2009.11.012
Eldredge N, Gould SJ (1972) Speciation and Punctuated Equilibria: An Alternative to Phyletic Gradualism [3rd draft]. [accessed 2025 Dec 31]. http://hdl.handle.net/2246/6567. https://doi.org/10.5531/sd.paleo.7
Erwin JA, Marchetto MC, Gage FH (2014) Mobile DNA elements in the generation of diversity and complexity in the brain. Nat Rev Neurosci 15(8):497–506. https://doi.org/10.1038/nrn3730
Faulkner GJ, Carninci P (2009) Altruistic functions for selfish DNA. Cell Cycle 8(18):2895–2900. https://doi.org/10.4161/cc.8.18.9536
Fedoroff NV (2012) Presidential address. Transposable elements, epigenetics, and genome evolution. Science 338(6108):758–767. https://doi.org/10.1126/science.338.6108.758
Feiner N (2016) Accumulation of transposable elements in *Hox* gene clusters during adaptive radiation of *Anolis* lizards. Proc R Soc B 283(1840):20161555. https://doi.org/10.1098/rspb.2016.1555
Gilbert C, Feschotte C (2018) Horizontal acquisition of transposable elements and viral sequences: patterns and consequences. Curr Opin Genet Dev 49:15–24. https://doi.org/10.1016/j.gde.2018.02.007
Glinsky GV (2025a) Genome-wide, chromosome-naïve compartmentalization of major developmental networks tailored by modular placement of human embryo-regulatory Long Terminal Repeats. Acad Mol Biol Genom 2(2) [accessed 2025 Dec 27]. https://www.academia.edu/129922697/Genome_wide_chromosome_na%C3%AFve_compartmentalization_of_major_developmental_networks_tailored_by_modular_placement_of_human_embryo_regulatory_Long_Terminal_Repeats. https://doi.org/10.20935/AcadMolBioGen7754
Glinsky GV (2025b) Transposable elements seed transcription factor binding sites to sequence-specific double-stranded DNA binding TF networks contributing to governance of primate brain evolution. J Mol Evol [published online ahead of print] [accessed 2025 Dec 30]. https://link.springer.com/10.1007/s00239-025-10285-1
Grant PR, Grant BR (2014) 40 years of evolution: Darwin's finches on Daphne Major Island. Princeton University Press
Gray JD, Kogan JF, Marrocco J, McEwen BS (2017) Genomic and epigenomic mechanisms of glucocorticoids in the brain. Nat Rev Endocrinol 13(11):661–673. https://doi.org/10.1038/nrendo.2017.97
Griffiths BB, Hunter RG (2014) Neuroepigenetics of stress. Neuroscience 275:420–435. https://doi.org/10.1016/j.neuroscience.2014.06.041
Guffanti G et al (2018) Novel bioinformatics approach identifies transcriptional profiles of lineage-specific transposable elements at distinct loci in the human dorsolateral prefrontal cortex Arkhipova I, editor. Mol Biol Evol 35(10):2435–2453. https://doi.org/10.1093/molbev/msy143
Hawker KL et al (1993) Transformation by the fos or jun oncogene does not increase AP-1 DNA-binding activity. J Virol 67(9):5487–5495. https://doi.org/10.1128/jvi.67.9.5487-5495.1993
Hayashi S et al (2022) Neofunctionalization of a noncoding portion of a DNA transposon in the coding region of the chimerical sex-determining gene *dm-W* in *Xenopus* frogs. Takahashi A, editor. Mol Biol Evol 39(7):msac138. https://doi.org/10.1093/molbev/msac138
Hof AEV et al (2016) The industrial melanism mutation in British peppered moths is a transposable element. Nature 534(7605):102–105. https://doi.org/10.1038/nature17951

Horváth V, Merenciano M, González J (2017) Revisiting the relationship between transposable elements and the eukaryotic stress response. Trends Genet 33(11):832–841. https://doi.org/10.1016/j.tig.2017.08.007
Hunter RG (2020) Stress, adaptation, and the deep genome: why transposons matter. Integr Comp Biol:icaa050. https://doi.org/10.1093/icb/icaa050
Hunter RG (2024) Transposons as environmental stress detectors, are eukaryotic genomes evolved to evolve? In: Witzany G (ed) Epigenetics in biological communication. Springer Nature Switzerland, pp 289–298. [accessed 2024 Sept 25]. https://link.springer.com/10.1007/978-3-031-59286-7_13
Hunter RG et al (2009) Regulation of hippocampal H3 histone methylation by acute and chronic stress. Proc Natl Acad Sci 106(49):20912–20917. https://doi.org/10.1073/pnas.0911143106
Hunter RG et al (2012) Acute stress and hippocampal histone H3 lysine 9 trimethylation, a retrotransposon silencing response. Proc Natl Acad Sci 109(43):17657–17662. https://doi.org/10.1073/pnas.1215810109
Hunter RG, Gagnidze K, McEwen BS, Pfaff DW (2015) Stress and the dynamic genome: steroids, epigenetics, and the transposome. Proc Natl Acad Sci USA 112(22):6828–6833. https://doi.org/10.1073/pnas.1411260111
Irmak MK, Oztas Y, Oztas E (2012) Integration of maternal genome into the neonate genome through breast milk mRNA transcripts and reverse transcriptase. Theor Biol Med Model 9(1):20. https://doi.org/10.1186/1742-4682-9-20
Jacques P-É, Jeyakani J, Bourque G (2013) The majority of primate-specific regulatory sequences are derived from transposable elements. Feschotte C, editor. PLoS Genet 9(5):e1003504. https://doi.org/10.1371/journal.pgen.1003504
Joly-Lopez Z, Bureau TE (2014) Diversity and evolution of transposable elements in Arabidopsis. Chromosom Res 22(2):203–216. https://doi.org/10.1007/s10577-014-9418-8
Jurka J, Kapitonov VV, Kohany O, Jurka MV (2007) Repetitive sequences in complex genomes: structure and evolution. Annu Rev Genom Hum Genet 8(1):241–259. https://doi.org/10.1146/annurev.genom.8.080706.092416
Kazazian HH, Moran JV (2017) Mobile DNA in health and disease. N Engl J Med 377(4):361–370. https://doi.org/10.1056/NEJMra1510092
Keighley LM et al (2023) Unveiling the hidden players: the crucial role of transposable elements in the placenta and their potential contribution to pre-eclampsia. Placenta 141:57–64. https://doi.org/10.1016/j.placenta.2023.05.017
Kelley D, Rinn J (2012) Transposable elements reveal a stem cell-specific class of long noncoding RNAs. Genome Biol 13(11):R107. https://doi.org/10.1186/gb-2012-13-11-r107
Kilvitis HJ, Ardia DR, Thiam M, Martin LB (2018) Corticosterone is correlated to mediators of neural plasticity and epigenetic potential in the hippocampus of Senegalese house sparrows (Passer domesticus). Gen Comp Endocrinol 269:177–183. https://doi.org/10.1016/j.ygcen.2018.09.014
Kitazawa M, Tamura M, Kaneko-Ishino T, Ishino F (2017) Severe damage to the placental fetal capillary network causes mid- to late fetal lethality and reduction in placental size in *Peg11/Rtl1* KO mice. Genes Cells 22(2):174–188. https://doi.org/10.1111/gtc.12465
Korte SM, Koolhaas JM, Wingfield JC, McEwen BS (2005) The Darwinian concept of stress: benefits of allostasis and costs of allostatic load and the trade-offs in health and disease. Neurosci Biobehav Rev 29(1):3–38. https://doi.org/10.1016/j.neubiorev.2004.08.009
Lamichhaney S et al (2018) Rapid hybrid speciation in Darwin's finches. Science 359(6372):224–228. https://doi.org/10.1126/science.aao4593
Lanciano S, Mirouze M (2018) Transposable elements: all mobile, all different, some stress responsive, some adaptive? Curr Opin Genet Dev 49:106–114. https://doi.org/10.1016/j.gde.2018.04.002

Larsen PA et al (2018) Warning SINEs: Alu elements, evolution of the human brain, and the spectrum of neurological disease. Chromosom Res 26(1–2):93–111. https://doi.org/10.1007/s10577-018-9573-4
Lenski RE (2023) Revisiting the design of the long-term evolution experiment with Escherichia coli. J Mol Evol 91(3):241–253. https://doi.org/10.1007/s00239-023-10095-3
Linker SB et al (2017) Examining non-LTR retrotransposons in the context of the evolving primate brain. BMC Biol 15(1):68. https://doi.org/10.1186/s12915-017-0409-z
Long HK, Prescott SL, Wysocka J (2016) Ever-changing landscapes: transcriptional enhancers in development and evolution. Cell 167(5):1170–1187. https://doi.org/10.1016/j.cell.2016.09.018
Losos JB (2018) Improbable destinies: fate, chance, and the future of evolution. First Riverhead trade paperback edition. Riverhead Books
Mallarino R, Abzhanov A (2012) Paths less traveled: Evo-Devo approaches to investigating animal morphological evolution. Annu Rev Cell Dev Biol 28(1):743–763. https://doi.org/10.1146/annurev-cellbio-101011-155732
Managò F et al (2016) Genetic disruption of Arc/Arg3.1 in mice causes alterations in dopamine and neurobehavioral phenotypes related to schizophrenia. Cell Rep 16(8):2116–2128. https://doi.org/10.1016/j.celrep.2016.07.044
Marmot MG, Sapolsky R (2014) Of baboons and men: social circumstances, biology, and the social gradient in health. In: Weinstein M, Lane MA (eds). Committee on Population; Division of Behavioral and Social Sciences and Education; National Research Council. National Academies Press (Comparative Biodemography: A Collection of Papers.). https://www.ncbi.nlm.nih.gov/sites/books/NBK242456/
Mattick J, Amaral P (2023) RNA, the epicenter of genetic information. Taylor & Francis
McClintock B (1950) The origin and behavior of mutable loci in maize. Proc Natl Acad Sci USA 36(6):344–355. https://doi.org/10.1073/pnas.36.6.344
McClintock B (1984) The significance of responses of the genome to challenge. Science 226(4676):792–801. https://doi.org/10.1126/science.15739260
McEwen BS (2012) Brain on stress: how the social environment gets under the skin. Proc Natl Acad Sci USA 109(supplement_2):17180–17185. https://doi.org/10.1073/pnas.1121254109
Mi S et al (2000) Syncytin is a captive retroviral envelope protein involved in human placental morphogenesis. Nature 403(6771):785–789. https://doi.org/10.1038/35001608
Minatohara K, Akiyoshi M, Okuno H (2016) Role of immediate-early genes in synaptic plasticity and neuronal ensembles underlying the memory trace. Front Mol Neurosci 8. [accessed 2025 Dec 28]. http://journal.frontiersin.org/Article/10.3389/fnmol.2015.00078/abstract
Miyashita T, Kikuchi E, Horiuchi J, Saitoe M (2018) Long-term memory engram cells are established by c-Fos/CREB transcriptional cycling. Cell Rep 25(10):2716–2728.e3. https://doi.org/10.1016/j.celrep.2018.11.022
Mohammed J et al (2014) Adaptive evolution of testis-specific, recently evolved, clustered miRNAs in *Drosophila*. RNA 20(8):1195–1209. https://doi.org/10.1261/rna.044644.114
Mojica EA, Kültz D (2022) Physiological mechanisms of stress-induced evolution. J Exp Biol 225 (Suppl_1):jeb243264. https://doi.org/10.1242/jeb.243264
Monaco AP (2022) The selfish environment meets the selfish gene: coevolution and inheritance of RNA and DNA pools: a model for organismal life incorporating coevolution, horizontal transfer, and inheritance of internal and external RNA and DNA pools. BioEssays 44(2):2100239. https://doi.org/10.1002/bies.202100239
Muotri AR et al (2005) Somatic mosaicism in neuronal precursor cells mediated by L1 retrotransposition. Nature 435(7044):903–910. https://doi.org/10.1038/nature03663
Ohno S (1972) So much "junk" DNA in our genome. Brookhaven Symp Biol 23:366–370
Ohtani H, Iwasaki YW (2021) Rewiring of chromatin state and gene expression by transposable elements. Develop Growth Differ 63(4–5):262–273. https://doi.org/10.1111/dgd.12735
Oliveira SG, Bao W, Martins C, Jurka J (2012) Horizontal transfers of Mariner transposons between mammals and insects. Mob DNA 3(1):14. https://doi.org/10.1186/1759-8753-3-14

Patoori S et al (2022) Young transposable elements rewired gene regulatory networks in human and chimpanzee hippocampal intermediate progenitors. Development 149(19):dev200413. https://doi.org/10.1242/dev.200413

Pellestor F (2019) Chromoanagenesis: cataclysms behind complex chromosomal rearrangements. Mol Cytogenet 12(1):6. https://doi.org/10.1186/s13039-019-0415-7

Percharde M et al (2018) A LINE1-nucleolin partnership regulates early development and ESC identity. Cell 174(2):391–405.e19. https://doi.org/10.1016/j.cell.2018.05.043

Perez MF, Lehner B (2019) Intergenerational and transgenerational epigenetic inheritance in animals. Nat Cell Biol 21(2):143–151. https://doi.org/10.1038/s41556-018-0242-9

Pettit NL, Yap E-L, Greenberg ME, Harvey CD (2022) Fos ensembles encode and shape stable spatial maps in the hippocampus. Nature 609(7926):327–334. https://doi.org/10.1038/s41586-022-05113-1

Phan J et al (2024) Retrotransposons are co-opted to activate hematopoietic stem cells and erythropoiesis. Science 386(6722):eado6836. https://doi.org/10.1126/science.ado6836

Pigliucci M, Müller G, Konrad Lorenz Institute for Evolution and Cognition Research, editors (2010) Evolution, the extended synthesis. MIT Press

Plath N et al (2006) Arc/Arg3.1 is essential for the consolidation of synaptic plasticity and memories. Neuron 52(3):437–444. https://doi.org/10.1016/j.neuron.2006.08.024

Pontis J et al (2022) Primate-specific transposable elements shape transcriptional networks during human development. Nat Commun 13(1):7178. https://doi.org/10.1038/s41467-022-34800-w

Rahman MF, McGowan PO (2022) Cell-type-specific epigenetic effects of early life stress on the brain. Transl Psychiatry 12(1):326. https://doi.org/10.1038/s41398-022-02076-9

Ravens A et al (2023) Arc regulates a second-guessing cognitive bias during naturalistic foraging through effects on discrete behavior modules. iScience 26(5):106761. https://doi.org/10.1016/j.isci.2023.106761

Reilly MT et al (2013) The role of transposable elements in health and diseases of the central nervous system. J Neurosci 33(45):17577–17586. https://doi.org/10.1523/JNEUROSCI.3369-13.2013

Reznick DN, Travis J (2019) Experimental studies of evolution and Eco-Evo dynamics in guppies (*Poecilia reticulata*). Annu Rev Ecol Evol Syst 50(1):335–354. https://doi.org/10.1146/annurev-ecolsys-110218-024926

Richter TA et al (2025) Sex differences in B2 SINE RNA expression and its role in hippocampal development. [accessed 2025 Dec 30]. http://biorxiv.org/lookup/doi/10.1101/2025.11.18.689111

Saeliw T et al (2018) Integrated genome-wide Alu methylation and transcriptome profiling analyses reveal novel epigenetic regulatory networks associated with autism spectrum disorder. Mol Autism 9(1):27. https://doi.org/10.1186/s13229-018-0213-9

Saeliw T et al (2023) Epigenetic gene-regulatory loci in Alu elements associated with autism susceptibility in the prefrontal cortex of ASD. IJMS 24(8):7518. https://doi.org/10.3390/ijms24087518

Scarpa A, Kofler R (2023) The impact of paramutations on the invasion dynamics of transposable elements. [accessed 2025 Dec 28]. http://biorxiv.org/lookup/doi/10.1101/2023.03.14.532580

Schmidt D et al (2012) Waves of retrotransposon expansion remodel genome organization and CTCF binding in multiple mammalian lineages. Cell 148(1–2):335–348. https://doi.org/10.1016/j.cell.2011.11.058

Schoener TW et al (2017) A multigenerational field experiment on eco-evolutionary dynamics of the influential lizard *Anolis sagrei*: a mid-term report. Copeia 105(3):543–549. https://doi.org/10.1643/CE-16-549

Selvaraju D, Wierzbicki F, Kofler R (2024) Experimentally evolving *Drosophila erecta* populations may fail to establish an effective piRNA-based host defense against invading *P* -elements. Genome Res genome;gr.278706.123v2. https://doi.org/10.1101/gr.278706.123

Shen D et al (2020) Multiple invasions of visitor, a DD41D family of Tc1/mariner transposons, throughout the evolution of vertebrates Betran E, editor. Genome Biol Evol 12(7):1060–1073. https://doi.org/10.1093/gbe/evaa135

Specchia V, Bozzetti MP (2021) The role of HSP90 in preserving the integrity of genomes against transposons is evolutionarily conserved. Cells 10(5):1096. https://doi.org/10.3390/cells10051096

Stroud JT, Ratcliff WC (2025) Long-term studies provide unique insights into evolution. Nature 639(8055):589–601. https://doi.org/10.1038/s41586-025-08597-9

Sullivan KR, Ravens A, Walker AC, Shepherd JD (2025) Arc—a viral vector of memory and synaptic plasticity. Curr Opin Neurobiol 91:102979. https://doi.org/10.1016/j.conb.2025.102979

The ENCODE Project Consortium (2012) An integrated encyclopedia of DNA elements in the human genome. Nature 489(7414):57–74. https://doi.org/10.1038/nature11247

The International Mouse Phenotyping Consortium et al (2016) High-throughput discovery of novel developmental phenotypes. Nature 537(7621):508–514. https://doi.org/10.1038/nature19356

Trizzino M, Kapusta A, Brown CD (2018) Transposable elements generate regulatory novelty in a tissue-specific fashion. BMC Genomics 19(1):468. https://doi.org/10.1186/s12864-018-4850-3

Tronick E, Hunter RG (2016) Waddington, dynamic systems, and epigenetics. Front Behav Neurosci 10 [accessed 2020 Sept 15]. http://journal.frontiersin.org/Article/10.3389/fnbeh.2016.00107/abstract.

Waddington CH (1942) The epigenotype. Endeavour 1:18–20

Wagner GP (1996) Homologues, natural kinds and the evolution of modularity. Am Zool 36(1):36–43. https://doi.org/10.1093/icb/36.1.36

Wagner A (2015) Arrival of the fittest: how nature innovates. Current

Wagner A (2023) Evolvability-enhancing mutations in the fitness landscapes of an RNA and a protein. Nat Commun 14(1):3624. https://doi.org/10.1038/s41467-023-39321-8

Wall MJ et al (2018) The temporal dynamics of arc expression regulate cognitive flexibility. Neuron 98(6):1124–1132.e7. https://doi.org/10.1016/j.neuron.2018.05.012

Walters RD, Kugel JF, Goodrich JA (2009) Inv *Alu* able junk: the cellular impact and function of Alu and B2 RNAs. IUBMB Life 61(8):831–837. https://doi.org/10.1002/iub.227

Wang J, Huang J, Shi G (2020) Retrotransposons in pluripotent stem cells. Cell Regen 9(1):4. https://doi.org/10.1186/s13619-020-00046-4

Wang C et al (2025) Multi-omics analysis reveals critical cis-regulatory roles of transposable elements in livestock genomes. iScience 28(3):112049. https://doi.org/10.1016/j.isci.2025.112049

Wells JN, Feschotte C (2020) A field guide to eukaryotic transposable elements. Annu Rev Genet 54(1):539–561. https://doi.org/10.1146/annurev-genet-040620-022145

Werner A, Piatek MJ, Mattick JS (2015) Transpositional shuffling and quality control in male germ cells to enhance evolution of complex organisms. Ann N Y Acad Sci 1341(1):156–163. https://doi.org/10.1111/nyas.12608

White JK et al (2013) Genome-wide generation and systematic phenotyping of knockout mice reveals new roles for many genes. Cell 154(2):452–464. https://doi.org/10.1016/j.cell.2013.06.022

Whiteley AM, Shepherd JD (2025) Retrotransposons unplugged: rewiring the nervous system and wreaking havoc. Neuron:S0896627325007470. https://doi.org/10.1016/j.neuron.2025.09.034

Wierzbicki F, Kofler R (2023) The composition of piRNA clusters in Drosophila melanogaster deviates from expectations under the trap model. BMC Biol 21(1):224. https://doi.org/10.1186/s12915-023-01727-7

Wingfield JC et al (1998) Ecological Bases of Hormone—Behavior Interactions: The "Emergency Life History Stage.". Am Zool 38(1):191–206. https://doi.org/10.1093/icb/38.1.191

Ye M, Rotival M, Amigorena S, Zueva E (2025) Transposon invasion of primate genomes shaped human inflammatory enhancers and susceptibility to inflammatory diseases. Nat Commun 16(1):9674. https://doi.org/10.1038/s41467-025-64690-7
Zahavi EE et al (2025) Repeat-element RNAs integrate a neuronal growth circuit. Cell 188(16): 4350–4365.e22. https://doi.org/10.1016/j.cell.2025.04.030

Chapter 6
Transgenerational Programming of Brain Development and Behaviour in Response to Early-Life Adversity

Jörg Bock

Abstract The functional development and maturation of the brain and behaviour are based on a complex, coordinated interaction between genetic predispositions and environmental influences. Stressful experiences or trauma in early life can negatively interfere with these developmental processes and, as a result, impair brain function, especially in prefrontal-limbic areas that are implicated in the regulation of stress response and socio-emotional behaviour. Perinatal stress, therefore, represents a significant risk factor for the development of mental disorders, such as depression, anxiety disorders or ADHD. It is important to emphasize that the described effects often occur in a sex-specific manner. However, successful coping with early negative experiences can also lead to positive adaptive changes in neural structures, which may improve stress coping and, thus, promote resilience in later life. There is considerable evidence that the interaction between genetic predispositions and the programming influences of early-life stress (ELS) on brain development is mediated by epigenetic mechanisms. A growing number of studies also show that the stress-induced, epigenetically regulated changes in brain function and behaviour can be passed on to subsequent generations in the sense of an inter- or transgenerational epigenetic inheritance. Thus, early-life stress not only affects the directly exposed individual but also represents a transgenerational programming factor that determines an epigenetic predisposition for the development of an individual in subsequent generations. This chapter summarizes major findings with respect to the environmentally driven development of functional brain circuits and their underlying epigenetic mechanisms and also discusses the potential, epigenetically mediated, inter- and transgenerational transmission of stress-induced neuronal and behavioural adaptations.

J. Bock (✉)
PG Epigenetics and Structural Plasticity, Institute of Biology, Otto von Guericke University Magdeburg, Magdeburg, Germany

Center for Behavioral Brain Sciences (CBBS) Magdeburg, Magdeburg, Germany
e-mail: joerg.bock@ovgu.de

G. Witzany (ed.), *Integrative Theory of Evolution - Aspects and Insights*,
https://doi.org/10.1007/978-3-032-24295-2_6

1 Introduction

The classic question, "What is innate, and what is acquired?" is based on the idea that some characteristics of an individual are explained by its genetically pre-programmed intrinsic nature, while other characteristics are the result of adaptive processes in response to environmental influences. However, this ancient thinking has been shifted into the concept that the development of individual phenotypes is the result of a complex interaction of genetically determined developmental processes and predispositions with environmental influences that may mediate and programme development throughout the whole lifespan.

The postulated influences of the environment are a core component of the "Developmental Origins of Health and Disease" (DOHaD) concept, claiming that specific environmental influences during distinct critical phases of development "shape" or "programme" the formation of an individual's functional phenotype and, thus, influence health and, also, susceptibility to disease in later phases of life (Gluckman and Hanson 2006). This postulate is based on the fact that various physiological systems, including the brain, show a particularly high degree of plasticity and adaptability in prenatal and early postnatal phases of life. Originally, this concept was primarily related to nutritional factors or influences during prenatal developmental phases, as summarized in the so-called foetal programming hypothesis (Barker 1998). This original hypothesis was later expanded to a broader perinatal programming concept that includes a wide spectrum of environmental factors, such as nutrition, various chemical influences, medication, noise, infections during pregnancy and, specifically, the influence of "toxic stress" during distinct phases of pre- and postnatal development (Heindel et al. 2015).

The perinatal programming concept is also relevant for the functional development of the brain and behaviour since environmental influences are of fundamental importance for the development and maintenance of functional synaptic networks in the developing brain (Miguel et al. 2019; Kolb et al. 2017; Bock et al. 2014). While genetic programmes determine the basic properties of nerve cells as well as the basic development of the general circuitry of the brain, fine-tuning and functional optimization of these neuronal and synaptic networks are critically dependent on environmental influences. However, the balance of this interplay between endogenous and exogenous factors shifts during the course of brain development. While genetically determined factors predominate at early developmental phases, environmental influences become gradually more relevant at later development, specifically at later gestational phases and early postnatal life periods. Newly acquired sensory, motor and emotional experiences now interact with the genetically pre-programmed processes and influence further development of the brain and behaviour to guarantee optimal adaptation to the given and potentially changing environmental and living conditions of the growing individual. In the case of early adverse experiences, such as early-life stress (ELS), maladaptations may lead to the development of dysfunctional neuronal networks, resulting in an increased risk for mental and behavioural disorders later in life (Agorastos et al. 2019; Bock et al. 2014, 2015; Maccari et al.

2014). When defining early-life stress, it is important to note that it is not a uniform phenomenon but a rather complex one. There are different dimensions of early-life stress, such as the timing of stress exposure (pregestational, prenatal, neonatal, adolescence), the type or modality of stress (traumatic experiences, childhood maltreatment (CM)), stress intensity and duration (short, chronic, repeated) and the number of stress episodes (Birnie and Baram 2025). Consequently, the effects of early-life stress on the brain and behaviour are very heterogeneous and strongly dependent on the respective dimensions and the interactions between these (see also Fig. 6.1). Additionally, the outcome of experienced early-life adversity is strongly influenced by the sex of the exposed individual since males and females show an innate difference in stress sensitivity and responsiveness (Bath 2020; Braun et al. 2020; Hodes and Epperson 2019; Bock et al. 2014).

It is important to emphasize at this point that early adverse experiences not only exert negative effects on brain development, leading to dysfunctional neuronal networks and behavioural disorders later in life. In principle, the experience of early stressful events provides an essential programming factor and is required for the biologically essential functional development of stress response systems. This adaptation to given environmental conditions has to be understood as an anticipation of future environmental conditions. Indeed, an increasing number of studies reveal that depending on the type, time point, duration or chronicity, early adverse experiences may also lead to positive functional adaptations and, in the sense of "stress inoculation," promote resilience or better stress coping later in life (Torres-Berrío et al. 2026; O'Brien et al. 2026; Santarelli et al. 2017; Daskalakis et al. 2013).

Increasing evidence reveals that the gene × environment interactions, which are required for the functional or dysfunctional development of an individual phenotype, are critically mediated by epigenetic mechanisms; Thus, epigenetic mechanisms provide the essential interface between genetic predispositions and environmental influences. Such epigenetically mediated programming processes, particularly those that are induced by early adverse experiences, are also a hallmark of the experience-driven development of functional brain circuits and, thus, define behavioural phenotypes in later life periods (Peña 2026; Arzate-Mejia et al. 2024; Bock 2024; Alyamani and Murgatroyd 2018; Babenko et al. 2015; Provençal and Binder 2015).

Similar to the concepts for the development of an individual (ontogenesis), the classical theories of evolution (phylogenesis) are also based on a distinction between genetically determined and environmentally driven processes. The neo-Darwinian theory, in extension to Darwin's original concept, is based on spontaneous, random genetic mutations in the DNA sequence, leading to phenotypic variations that can be inherited to the descendants, which may then fit better into the given environment and drive evolution in the sense of natural selection. In contrast, Lamarck's theory of evolution proposed that individual phenotypes are the result of adaptations to the environment during the lifetime that can be directly transmitted to the descendants. A unified theory of evolution now combines the two classical concepts, proposing an interaction of genetically and environmentally driven adaptations that includes gene mutations but also epigenetic changes and epimutations in response to a given environmental challenge (Korolenko and Skinner 2024; Skinner 2015).

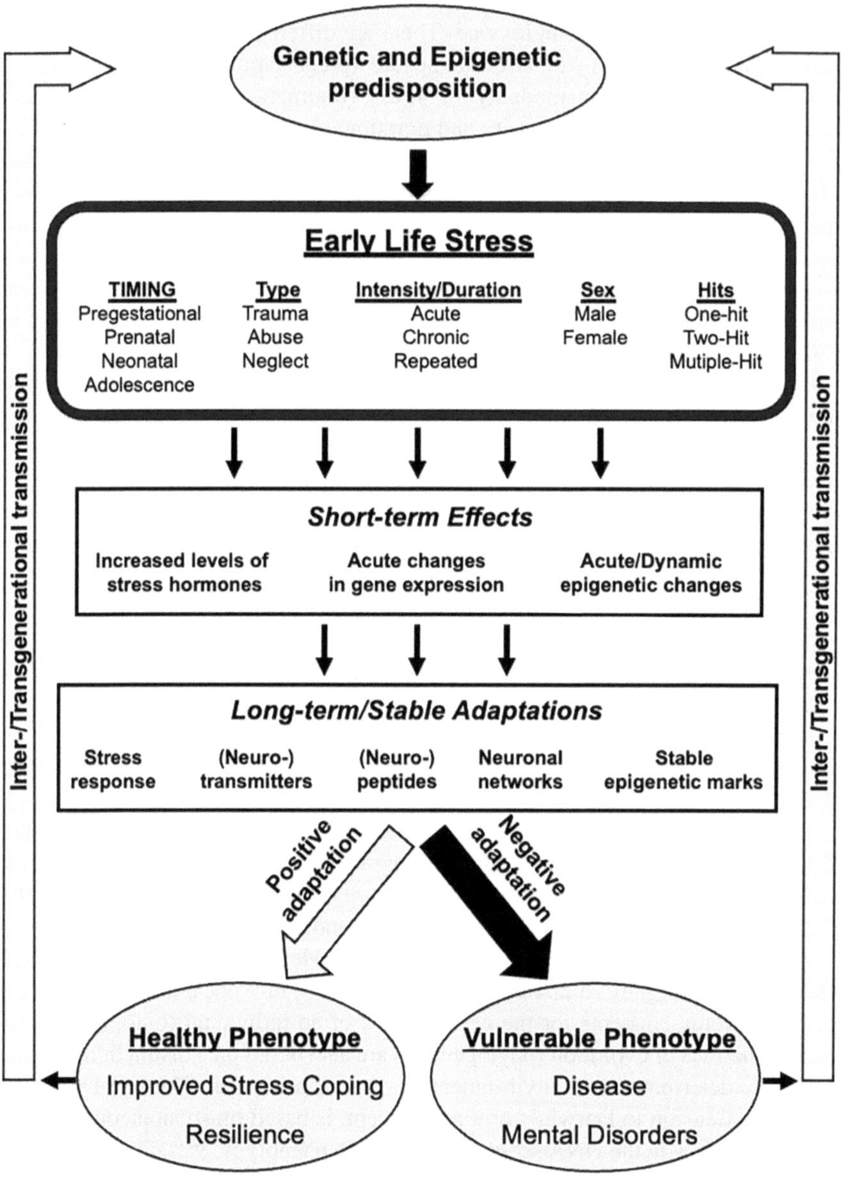

Fig. 6.1 Transgenerational programming of the brain and behaviour by early-life stress. Early-life stress influences the functional development of neuronal networks in a multi-step process by inducing a number of short-term and long-term changes that are mediated by epigenetic mechanisms. Depending on individual predispositions and the various dimensions of the experienced stress, this determines the development of healthy or diseased phenotypes later in life. The stress-induced adaptations can influence stress response and development in future generations by inter- and transgenerational epigenetic inheritance (for details, see text)

Interestingly, there is also increasing evidence for an inter- and transgenerational transmission of stress-induced adaptations of the brain and behaviour, mediated by stable, inheritable epigenetic changes that are passed down to several offspring generations, potentially providing a crucial mechanism in the evolutionary adaptation of brain function and mental health and disease (Liao et al. 2025; Abou Merhi et al. 2025; Braun et al. 2020; Jawaid et al. 2018).

2 Critical Time Windows: The Developing Brain as a Shifting Target for Early-Life Stress

An important aspect with regard to the experience-induced functional development and maturation of synaptic and neuronal networks is the existence of sensitive and critical time windows. They are characterized by a pronounced synaptic plasticity potential in early life and, thus, represent "windows of opportunity," during which positive and stimulating environmental challenges but also controllable adverse experiences induce a "positive programming" effect on developing neuronal networks. These processes lead to an adaptation to the respective environmental conditions and, thus, promote resilience and mental health in later life periods (Andersen 2003). In parallel, these phases also provide "windows of vulnerability," during which synaptic networks can be "negatively programmed" by adverse environmental influences, resulting in psychopathological behaviour later in life. The life cycle model of stress hypothesizes that the windows of opportunity and vulnerability are brain region–specific since stress experienced during different stages of life may affect the developmental profiles of brain areas, such as the amygdala, hippocampus and frontal cortex, in a different manner depending on the time point of stress exposure and the maturational status of a given brain region (Joseph 1999; Lupien et al. 2009). Indeed, results from clinical studies in humans and from experimental approaches in animal models provide evidence for this hypothesis and show that the neuro-structural changes and behavioural alterations in later life phases are strongly dependent on the developmental time point of ELS-exposure (Teicher et al. 2016; Pechtel et al. 2014; Andersen and Teicher 2008; Gos et al. 2008; Bock et al. 2005). The general principle of the environmentally driven functional brain development is classically discussed within the concept of experience-dependent development. However, in addition, there is also the concept of experience-expectant development, claiming the existence of distinct developmental time periods, during which the brain is experience-expectant and, thus, "waits" for environmental influences (Joseph 1999; Greenough et al. 1987). According to this concept, the brain can be adapted and functionally optimized only if it is exposed to a certain amount of experience. Without the required environmental influences during experience-expectant time windows, functional brain development will not occur sufficiently, with negative consequences for mental and cognitive capabilities in later life. Although the cellular and molecular mechanisms defining sensitive and critical time windows for experience-dependent and experience-expectant brain development are still poorly understood, there is some evidence that the emergence and maturation of inhibitory parvalbumin (PV) cells provide a "timing hub" and define critical periods of brain development (Reh et al. 2020; Marín 2016). Interestingly, the timeline of PV cell maturation is brain region–specific and, thus, may contribute to the time window–dependent sensitivity towards early environmental challenges. In addition, it is tempting to speculate that pre-programmed, region-specific epigenetic mechanisms are potentially involved in the transcriptional regulation of these processes and, thus, may also play an important role in the regulation of critical and sensitive time windows.

3 Pre- and Postnatal Epigenetic Programming of the Brain and Behaviour

The crucial question arising from the concepts described above concerns the underlying mechanisms. How are the environmentally driven development of an individual phenotype and, specifically, the experience-induced functional maturation of the brain and behaviour mediated? With regard to adverse early experiences, including prenatal and postnatal early-life stress, there is considerable evidence for the involvement of epigenetic mechanisms that translate these early-life experiences into partly long-lasting alterations of gene expression, neuronal (mal-)adaptations and consequently altered socio-emotional and cognitive behaviour in later life (Sulyok et al. 2023; Zhou and Ryan 2023; Rahman and McGowan 2022; Braun et al. 2020; Szyf 2019; Alyamani and Murgatroyd 2018; Provençal and Binder 2015; Bock et al. 2014). However, the analysis of stress-induced epigenetic changes in the brain is challenging, specifically with regard to studies in humans, since they are often limited by analysis in peripheral tissues such as blood and saliva that provide only limited predictiveness for brain-specific mechanisms. Post-mortem studies allow for brain-specific analyses but are often retrospective, specifically with regard to stressful or traumatic experiences in early childhood. Experimental animal models of prenatal stress and postnatal early-life stress provide an opportunity to gain a more detailed view on epigenetic programming processes in neuronal tissue.

3.1 *Prenatal Stress (PS)*

PS-Induced Disturbances of Neuronal Development

Prenatal stress (PS) experiences during different stages of pregnancy critically influence the development of neuroendocrine stress response systems in offspring, as well as the functional development of specific brain structures and neural networks involved in the control of socio-emotional and parental behaviour, anxiety, reward and addiction (Braun et al. 2020; van den Bergh et al. 2020; O'Donnell and Meaney 2017; Bock et al. 2015; Weinstock 2016; Maccari et al. 2014; Buss et al. 2012). Recent systematic reviews, summarizing prenatal stress effects on human foetuses, infants, children and adolescents, reveal that, particularly, areas of the limbic system, such as the hippocampus and amygdala, and also the prefrontal and insular cortex develop PS-induced structural and functional aberrations (Baroutis et al. 2025; Mandl et al. 2024). This is reflected by volumetric changes or hypo- or hyperactivity and also by disturbed functional connectivity within a higher-order network of the affected brain areas. It can be speculated that the high density of glucocorticoid receptors (GRs) in these brain areas makes them particularly sensitive towards the increased levels of stress hormones during PS experience. Thus, there is a clear link between PS-induced alterations of the brain and behavioural development, with the consequence of an increased risk of developing psychopathologies, including depression, anxiety disorders and ADHD later in life. In addition, PS-induced effects on the development of the hypothalamic–pituitary–adrenal (HPA) axis in the offspring have been described, leading to a dysfunctional HPA axis regulation and, thus, disturbed stress coping.

Since the analysis of PS-induced neurodevelopmental and cellular alterations in humans is limited, animal models of prenatal stress in rodents and non-human primates have been established to allow more detailed analyses of the underlying mechanisms (Braun et al. 2020; Maccari et al. 2017; Weinstock 2016; Bock et al. 2015). In these animal experiments, it has been consistently shown that PS-exposed offspring develop behavioural dysfunctions, specifically symptoms of enhanced depressive-like behaviour and anxiety that are accompanied by disturbed neuronal and synaptic development, particularly in the prefrontal and limbic brain areas, such as the hippocampus and amygdala (for review, Braun et al. 2020; Bock et al. 2015). An important outcome of these animal studies was the finding of sex-specific effects, which is in line with findings in humans (Bock et al. 2011, 2014; Murmu et al. 2006).

PS-Induced Epigenetic Changes

As already mentioned above, epigenetic mechanisms are proposed to play an important role at the interface between genes and environment and, thus, may also mediate the effects of prenatal stress. Indeed, a number of studies in humans revealed that different types of prenatal stress exposure can induce widespread changes in DNA methylation in the PS-exposed offspring. For instance, in blood cells from offspring of mothers who were pregnant during the "Dutch Hunger Winter," suffering from undernutrition and stressful experiences, a number of DNA methylation changes in different genomic regions were described (Tobi et al. 2009; Heijmans et al. 2008). Also, a series of studies within "Project Ice Storm" revealed that prenatal maternal stress, experienced during this natural disaster, resulted in long-lasting DNA methylation changes in a large number of genes in the peripheral tissues of the offspring, which mediate the association between PS exposure and the occurrence of a number of disease symptoms (Cao-Lei et al. 2014, 2025). Support for the mediating role of epigenetic mechanisms in the effects of prenatal stress on brain development comes from a series of human clinical studies showing that different types of prenatal stress exposure induce changes in DNA methylation in genes, which encode for factors involved in the regulation of neuronal and synaptic development. Also, alterations of DNA methylation in the promoter region of the glucocorticoid receptor (GR) gene (NR3C1) have been described in offspring of mothers exposed to prenatal stress, indicating a disturbed neuroendocrine regulation of the stress response (Braun et al. 2020; Kertes et al. 2016; Bock et al. 2015; Palma-Gudiel et al. 2015; Mulligan et al. 2012; Oberlander et al. 2008). In addition to the findings in humans, which are mostly restricted to analyses in peripheral tissues, there is also consistent evidence from animal studies, which allow for analyses in brain tissue, that PS-induced epigenetic modifications are crucial for the neuronal and neuroendocrine changes in PS offspring (Braun et al. 2020; Bock et al. 2015). For instance, a series of studies in rats revealed that PS-induced changes of DNA methylation in the brain are sex- and region-specific, depend on the intensity of the stressor and are associated with altered behaviour (Mychasiuk et al. 2011a, b, 2013). In the brain of PS-exposed mouse offspring changes of DNA methylation at the promoter regions of genes encoding for stress modulation factors could be detected, which were associated with disturbed hypothalamic–pituitary–adrenal (HPA) stress axis function (Mueller and Bale 2008). A direct effect of stress hormone exposure on the brain was described in guinea pigs, where glucocorticoid treatment during pregnancy affected the transcriptional and epigenetic machinery of the developing foetal hippocampus (Constantinof et al. 2019; Crudo et al. 2013).

In sum, studies in human and animal studies provide clear evidence for epigenetically mediated effects of prenatal stress on functional brain maturation, which, at least in part, may underlie the development of the described behavioural and mental disorders.

3.2 *Postnatal Early-Life Stress (ELS)*

ELS-Induced Disturbances of Neuronal Development

It is becoming increasingly apparent that environmental challenges during early postnatal developmental phases critically influence the functional maturation of the brain and behaviour and, thereby, determine the development of individual resilient (healthy) or vulnerable (disease) phenotypes. The early postnatal life represents a critical and sensitive time window towards the effects of early-life stress (ELS) or early traumatic experiences. For a newborn, the first and most significant formative life experience is provided by the contact and relationship with its parents or caregivers, the mother–child or parent–child interaction. Disturbed or adverse socio-emotional experiences in these early life phases represent a critical risk factor for the development of behavioural or mental disorders, such as depression, anxiety disorders, ADHD or dysfunctional social behaviour later in life (Baldwin et al. 2023;

Cay et al. 2022; Ehlert 2013; Bale et al. 2010; Lupien et al. 2009; Beers and De Bellis 2002). Similar to the findings for prenatal stress, the behavioural dysfunctions in response to ELS are associated with neuro-structural and brain functional changes (Miguel et al. 2019; Agorastos et al. 2019; Teicher et al. 2016; Baker et al. 2013). A summary of findings from clinical studies revealed that early postnatal stress and trauma experiences, such as childhood maltreatment or social isolation, particularly affect stress-sensitive brain areas, including the limbic areas hippocampus and amygdala, as well as the prefrontal and anterior cingulate cortex (Vaidya et al. 2024). In this context, it has to be emphasized that changes of brain structure and functionality are dependent on the type of childhood maltreatment (CM) (Tomoda et al. 2024).

In animal models, postnatal ELS is typically simulated by an experimentally induced disturbance of mother/parent–child interactions and relationships; also, natural variations of maternal care in rodents are used for experimental approaches (Gröger et al. 2016; Bock et al. 2014; Schmidt 2010; Oitzl et al. 2010). Since the predominant number of studies is focused on mother–child interactions without taking into account the role of the father, a number of animal models in rodents have also been established to mimic biparental care and investigate additional paternal care behaviour (Pohl et al. 2019; Feldman et al. 2019; Bales and Saltzman 2016; Braun and Champagne 2014). Similar to the findings in humans, animal studies revealed that ELS critically interferes with the functional maturation of the brain and, consequently, also with the development of behaviour in later life. Since there is a high variability in the applied models of postnatal ELS, it is not surprising that there is no overall conclusive outcome. Nevertheless, offspring of mothers who display less intensive, unstructured or even aversive maternal care generally tended to develop dysfunctional behaviour, specifically increased fear and anxiety, behavioural defensiveness and disturbed cognitive development (Short and Baram 2019; Gröger et al. 2016; Szyf et al. 2007; Roth and Sullivan 2005).

Similarly, offspring exposed to maternal separation (MS) as an ELS model develop increased depressive-like and anxiety behaviour, as well as changes in addictive and reward behaviour (Tractenberg et al. 2016; Nishi et al. 2014; Schmidt 2011). A study on a biparental animal model, the degu (*Octodon degus*), showed that repeated parental separation during the early postnatal life period induced ADHD-type behavioural symptoms, particularly in male offspring (Bock et al. 2017).

Overall, the existing animal models of postnatal ELS have gained consistent evidence for an ELS-induced disturbance of functional development and maturation of brain areas and neuronal networks, again specifically in limbic and prefrontal cortical brain areas. The structural and functional aberrations in these areas are usually reflected by ELS-induced changes of neuronal structure, specifically altered dendritic length and complexity and increased or decreased numbers of excitatory and inhibitory synapses (Short and Baram 2019; Gröger et al. 2016; Matas et al. 2016; Bock et al. 2014). For instance, a series of experiments applying parental separation as ELS exposure in the biparental rodent degu revealed that ELS-exposed degu pups show altered levels of excitatory dendritic spine synapses in a number of areas in the prefrontal cortex but also in the amygdala and hippocampus, major structures of the limbic system (Poeggel et al. 2003; Helmeke et al. 2001). Besides the described changes of excitatory spine synapses, there is also additional evidence that inhibitory synapses and transmitter systems are changed in response to ELS, overall inducing an excitation–inhibition (E/I) imbalance (Bock et al. 2014). The ELS-induced synaptic changes and E/I imbalance were accompanied by dramatically reduced metabolic brain activity and an aberrant functional interaction among the affected brain areas and were directly associated with the aforementioned development of ADHD-like symptoms in ELS-exposed individuals (Bock et al. 2012, 2017). In general, an E/I imbalance in the prefrontal-limbic brain systems is considered to be a hallmark of neurodevelopmental disorders, including autism, schizophrenia, depression and ADHD. The ELS-induced synaptic changes and E/I imbalance were accompanied by dramatically reduced metabolic brain activity and an aberrant functional interaction among the affected brain areas and were directly associated with the aforementioned development of ADHD-like symptoms in ELS-exposed individuals (Bock et al. 2012, 2017).

ELS-Induced Epigenetic Changes

Similar to the findings, after prenatal stress, there is increasing and considerable evidence that childhood adversity, including ELS, induces epigenetic changes, specifically alterations in DNA methylation patterns, histone modifications and non-coding RNA expression. However, it has to be emphasized that epigenetic adaptations in response to early-life experiences are a multi-step process, which starts with immediate but mostly dynamic and transient epigenetic changes. However, this process may also end up in permanent, stable epigenetic marks that mediate gene expression in later life periods (Dudley et al. 2011; Meaney and Ferguson-Smith 2010). For instance, dynamic epigenetic changes have been described for the expression of histone deacetylases (HDACs) and specific histone H4 acetylation in the mouse brain after repeated 3-h maternal separation during the first 2 weeks of life (Levine et al. 2012). Another study in mice showed that a short period of repeated maternal separation (3 h per day from postnatal day 14–16) induces a rapid increase of histone H3 and H4 acetylation in the hippocampus, which was associated with an increased expression of synaptic plasticity genes (Xie et al. 2013). As a long-term consequence, this particular ELS exposure induced increased dendritic length and dendritic spine numbers on neurons in the hippocampus. The results of this study provide direct evidence that ELS-induced rapid epigenetic changes in the brain mediate the expression of synaptic plasticity genes and, thus, are the initial step in the functional adaptation of neuronal networks in response to early-life experiences.

With regard to long-term, stable epigenetic marks, the analysis of large-scale or whole-genome DNA methylation patterns in the peripheral tissues of humans revealed that childhood trauma and early-life adversity induce altered methylation patterns at CpG sites within a wide range of gene loci, including genes encoding for factors of stress axis regulation and cell development (Provençal et al. 2014;

Suderman et al. 2014; Mehta et al. 2013; Essex et al. 2013; Bick et al. 2012). More specific analyses describe altered DNA methylation patterns as a result of childhood adversity experience, specifically in genes that are relevant for HPA axis regulation, including *FKBP5*, *NR3C1* and *KITLG* (Turecki and Meaney 2016; Houtepen et al. 2016; Palma-Gudiel et al. 2015; Klengel et al. 2013), but also in genes that are essential for neuronal development and synaptic plasticity, such as the brain-derived neurotrophic factor (BDNF) (Kundakovic et al. 2015). A pioneering post-mortem study revealed that suicide victims with a history of childhood abuse displayed increased DNA methylation in the GR receptor (*NR3C1*) gene associated with decreased expression levels of this receptor in neurones of the hippocampus, indicating disturbed stress regulation (McGowan et al. 2009). In an animal model focusing on naturally occurring variations of maternal care in rats, it has been shown that high licking/grooming or supportive maternal care positively stimulates the development of neuroendocrine and behavioural responses in the offspring (Liu et al. 1997). These alterations were associated with decreased DNA methylation and increased histone acetylation at the *NR3C1* gene in the hippocampus, resulting in an increased expression of glucocorticoid receptors (Meaney and Szyf 2005; Weaver et al. 2004). A subsequent study revealed that the epigenetic changes were not limited to the *NR3C1* but could also be detected at different gene sites, mostly associated with higher gene expression (McGowan et al. 2011). An alternative animal model showed that adult offspring of abusive rat mothers displayed persisting changes in DNA methylation at circumscribed loci at the BDNF promoter, resulting in altered *BDNF* gene expression in a sex-specific manner (Blaze and Roth 2017; Roth et al. 2009). Increased DNA methylation at the NPY-Y2 receptor gene, a regulator of the NPY system that is discussed to be a mediator of vulnerable vs. resilient adaptations, was found in the hippocampus of male mice exposed to maternal separation as an ELS model (Kocamaz et al. 2022). Different types of maternal separation were also shown to induce changes in hippocampal H3 acetylation that were associated with expression changes of DARPP-32, a modulator of dopaminergic neurotransmission (Köhler et al. 2019).

4 ELS-Induced Epigenetic Programming and Priming of Stress Sensitivity

In "real life," an individual's life is exposed not only to a single stressor at a given time point, but usually, it experiences a number of consecutive adverse experiences throughout life. Thus, the "two (or multiple)-hit" concept claims that early adverse experiences such as ELS may have a "(re-)programming" effect on stress responses and sensitivity to adverse experiences later in life (Jaric et al. 2019; Santarelli et al. 2017; Gruber et al. 2015; Nederhof and Schmidt 2012; Taylor 2010). In an extension of this hypothesis, the three-hit concept includes gene–environment interactions during critical phases of perinatal and juvenile brain development and defines the three hits as follows: hit-1: genetic predisposition, hit-2: early-life stress, and hit-3: stress in later life (Daskalakis et al. 2013). Support for the two-hit concepts comes from experiments in rodents showing that exposure to ELS as a first hit influences the sensitivity and, thus, the response to a second-hit stressor in adolescence or adulthood. For instance, in mice, it was shown that ELS exerts a programming effect towards changes in gene expression of the hippocampal oxytocin receptor (OxtR) in response to adult stress (Lesse et al. 2017). In this study, ELS or adult stress, applied as single stressors, had no effect on gene expression in adulthood. However, combining ELS (first hit) with consecutive stress exposure in adulthood (second hit) resulted in elevated hippocampal OxtR gene expression, indicating a programming effect of ELS. Similarly, it was shown that ELS exposure induces altered transcriptomic patterns in the reward circuit of male and female mice, resulting in an altered susceptibility to subsequent adult stress, as reflected by increased behavioural sensitivity (Pena et al. 2019).

A more detailed analysis of the mechanisms that may underlie the programming effects of ELS comes from a series of experiments in mice and rats. In male mice, increased NPY-Y2 receptor gene expression in the hippocampus was detected after a single exposure to adult swim stress (AS) (Kocamaz et al. 2022). Interestingly, a pre-exposure to ELS "buffered" the stress-induced effects in adulthood since animals exposed to both stressors (ELS + AS) displayed expression levels similar to those in unstressed controls. In rats, a more complex interaction of ELS with a consecutive stressor in adolescence was found. Behavioural profiling, applying tests for anxiety symptoms, depressive-like behaviour and social interaction, showed that animals exposed only to ELS or only to adolescent stress as single stressors were significantly affected when compared to unstressed controls. Interestingly, behavioural profiles of animals exposed to both stressors consecutively were unchanged since they were comparable to those of unstressed controls (Demaili et al. 2023). Thus, in this particular experiment, ELS appears to buffer or protect against the negative consequences of adolescent stress exposure on stress-related behaviour in later life. The findings on the behavioural level were accompanied by ELS-induced gene expression changes of factors of the endocannabinoid system, the endocannabinoid receptor 1 (CB1R) and the fatty acid amide hydrolase (FAAH) in the medial prefrontal cortex (mPFC). Analysis of CpG-site-specific DNA methylation revealed that both, the detected stress-induced gene expression changes of NPY-Y2 in the hippocampus of mice as well as of CB1R and FAAH in the mPFC of rats are mediated by CpG-site-specific changes of DNA methylation. Importantly, altered DNA methylation could only be detected after a single exposure to one of the stressors at different life periods. When animals were exposed to consecutive exposure to two of the respective stressors, DNA methylation was normalized back to the levels of the control animals. Thus, the specific ELS-induced changes in DNA methylation may represent a possible mechanism that protects or buffers ELS-exposed animals from the consequences of additional stress experiences later in life. This interpretation is consistent with a model of epigenetic programming claiming that early-life experiences trigger specific epigenetic changes that anticipate responses to future environmental signals (Szyf 2019). However, the concept of epigenetic programming in response to early-life stress is not restricted to changes in DNA methylation but has to be extended to histone modifications and changes in the activity of histone-modifying enzymes. This was shown in a recent study, again revealing that early-life stress influences stress sensitivity to subsequent stress exposure in later life periods (Geiger et al. 2025). The authors found that early-life stress alters histone dynamics in the ventral tegmental area (VTA), a dopaminergic brain region in the brainstem. Specifically, they describe stress-induced enhanced H3K4 monomethylation associated with an open chromatin state and enhanced levels of Setd7, a monomethylase. Experimental overexpression of H3K4 monomethylation and Setd7 was shown to sensitize transcriptional, physiological and behavioural responses to adult stress exposure. In addition, ELS induced a variety of other histone modifications, specifically altered histone methylation and histone acetylation. In sum, ELS in this particular experiment primed (epigenetic priming) or programmed a chromatin state that facilitated transcriptional response and, thereby, also the sensitivity to subsequent stress exposures later in life. However, it is tempting to speculate that early-life stress may increase the sensitivity not only to subsequent stressful events but also to non-stressful, positive environmental influences, providing an opportunity for therapeutic interventions to overcome the negative influences of early-life adversity and promote resilience.

5 Stress-Induced Inter- and Transgenerational Epigenetic Inheritance

The hypothesis that the development of an individual phenotype is not only defined by inherited genotypes but that phenotypic adaptations acquired in response to the given environmental conditions are also transmitted into the next generations influenced and changed the thinking about the drivers of evolutionary processes over the past centuries. Classically, Charles Darwin's concept of evolution postulated that new evolving phenotypes are the result of a natural selection, an increased adaptation of an organism to an existing environment over time. The individuals with the best-fitting characteristics will survive, reproduce at a higher rate and pass on the corresponding phenotype to subsequent generations. The neo-Darwinian theory extends this concept onto the molecular level, proposing that evolution involves genetic alterations and spontaneous mutations; in particular, it states that random inheritable alterations in the DNA sequence define phenotypic variations and the development and evolution of new phenotypes (Korolenko and Skinner 2024; Skinner 2015). However, this more or less random natural selection appears rather inefficient and slow, especially with regard to adaptations that are required and essential for survival in response to a rapidly changing environment, since the frequency of spontaneous mutations with a potentially beneficial effect in a given environment is rather low (Jablonka and Raz 2009). Therefore, the concept of transgenerational epigenetic inheritance of acquired phenotypic characteristics has come into focus. This concept proposes that phenotypic adaptations are driven by environmental influences that, on the molecular level, act via epigenetic modifications and that can be transmitted to the next generations (Jablonka and Raz 2009; Skinner 2015). This concept provides a molecular explanation for Lamarck's theory of evolution, which was established at the beginning of the nineteenth century, proposing that the environment can directly alter phenotypes during the lifetime and that these phenotypes can be directly transmitted to the descendants. Along with this concept, there is increasing evidence from human and animal studies that perinatal stress exposures (pregestational, prenatal and postnatal) not only affect the development of the brain and behaviour in the directly exposed individuals but that the stress-induced changes can be passed down and are transmitted across multiple generations. This process may provide a programming factor for health and disease in subsequent generations, potentially mediated by inter- and transgenerational epigenetic inheritance (Abou Merhi et al. 2025; Švorcová 2023; Braun et al. 2020; Franklin et al. 2010; Richards 2006).

In this context, it appears necessary to define the terms inter- and transgenerational transmission or inheritance in more detail (see Gemmati et al. 2025; Szyf 2015; Heard and Martienssen 2014). With regard to prenatal stress, an epigenetic transmission across two generations has to be termed intergenerational or multigenerational. In this case, a gestating mother (F0) is exposed to an environmental challenge (e.g. stress) during pregnancy and, consequently, also the developing fetus (F1 generation) as well as its developing germ cells (future F2). Affected germ cells of F1 may transmit the acquired epigenetic marks and, thereby, affect the phenotype of the F2 generation. Thus, not only the F0 mother but also the F1 and F2 generations are directly affected by the stress exposure. A true transgenerational inheritance after prenatal stress can be observed only after F3 since these animals are the first that were not exposed to the initial environmental challenge. In the case of postnatal stress exposure, such as early-life stress, where both non-pregnant F0 female as well as F0 male individuals are directly exposed to an environmental challenge, the respective germ cells may also be directly affected, thus influencing the F1 generation (intergenerational transmission). Transgenerational epigenetic inheritance in this case can be observed in the F2 and following generations since the F2 generation would be the first that is not directly exposed to stress.

With regard to the concept of transgenerational epigenetic inheritance, it has also to be emphasized that a transgenerational (or intergenerational) transmission of adaptive epigenetic marks, as well as an adapted phenotype, can occur via one of two possible routes of transmission (for reviews, see Steg and Mansuy 2025; Blaze and Roth 2015; Kundakovic and Champagne 2015; Bohacek et al. 2013). At first, there may be a pure behavioural or social transmission, which exists when a particular parental behaviour affects the phenotype of the offspring that again transmits this behaviour into the next generation and so on. Such a behavioural/social transmission occurs independent of the germline. On the other hand, there is evidence for germline transmission, the transmission of epigenetic marks that are induced by environmental influences such as early-life stress, through either the male or female germline over up to five generations. Considering transgenerational epigenetic inheritance in mammals and, in particular, human populations, an interaction of both routes has to be proposed, specifically with regard to transmission via the maternal germline.

6 Transgenerational Transmission of Early-Life Trauma and Stress

6.1 Transgenerational Transmission of Early-Life Trauma and Stress in Humans

Evidence for a real transgenerational transmission of stress and trauma experiences via epigenetic inheritance in humans is still limited and, thus, a subject of active debate and discussion (Banushi et al. 2025; Klibaner-Schiff et al. 2024; Horsthemke 2018). As already described above, the exposure to a variety of prenatal stress experiences, including antenatal depression or exposure to natural disasters, dramatically influences the development of the offspring and induces long-term, stable epigenetic marks in the descendants, specifically, altered DNA methylation (Cao-Lei et al. 2014, 2025; Tobi et al. 2009; Heijmans et al. 2008). Exposure to a variety of adverse factors during pregnancy interferes with the development of the brain and behaviour in the offspring, which increases the risk of neurodevelopmental and mental disorders in later life. In addition to the classic studies mentioned above, there is clear evidence in humans that the effects of prenatal stress on neurodevelopment are mediated by epigenetic mechanisms, which are not limited to DNA methylation but also include histone modifications and non-coding RNAs (ncRNAs) (Álvarez-Mejía et al. 2025). However, as outlined above, these studies cannot be considered to provide evidence for transgenerational transmission or transgenerational epigenetic inheritance since the developing offspring are also potentially directly exposed to the stress that is experienced by their mother during pregnancy. Thus, in the following chapters, I will focus on potential transgenerational effects in response to early postnatal stress experiences.

However, because of limitations in the experimental set-up and ethical issues, most of the human studies so far are focused on intergenerational transmission (Zhou and Ryan 2023) and of course are based on retrospective and correlational analyses and thus lack a proof of causality. Evidence for the intergenerational transmission of trauma exposure in humans comes from a classic study on a cohort of Holocaust survivors and their adult offspring, revealing that exposure to Holocaust experience was associated with altered blood DNA methylation within the FKBP5 gene, encoding for the FK506-binding protein 5, a key modulator of the HPA axis (Yehuda et al. 2016). Dysregulations of FKBP5 are implicated in the aetiology of stress-related disorders such as PTSD. Interestingly, altered DNA methylation at specific DNA methylation sites within the FKBP5 gene was not only found in the Holocaust-exposed parents but also in their adult offspring. Moreover, the DNA methylation levels for Holocaust-exposed parents and their offspring were significantly correlated. A recent study now reveals that the experienced Holocaust trauma not only affects the direct adult descendants in the first generation but is also likely to be passed on to subsequent generations. In this study, it was shown that third-generation offspring of Holocaust survivors had altered saliva DNA methylation levels in different genes associated with the regulation of the oxytocinergic system and genes implicated in the regulation of the HPA axis, such as FKBP5, CRH and NR3C1 (Oren et al. 2025). However, the outcome of such studies is influenced by the type of early-life adversity (stress, trauma) the parent is exposed to and potentially also by the time of assessment of biological and behavioural parameters of the descendants (newborn, childhood, adolescence, adulthood). For instance, in a study on mother–infant dyads, specific patterns of altered blood DNA methylation at a number of genes involved in stress regulation, including FKBP5, CRHR1 and NR3C1, were found in mothers with their own history of childhood maltreatment (sexual abuse, neglect). In contrast to the findings in the mothers, no changes of DNA methylation at the selected genes could be detected in the newborns, indicating that stress- or trauma-induced epigenetic changes are not always directly transmitted to the descendants or basically are just not detectable at this early-life periods (Ramo-Fernández et al. 2019).

With regard to the concept of potential inter- and transgenerational epigenetic inheritance of stress and trauma experiences in humans, there is still a lack of knowledge about the exact mechanisms of such a transmission. It is unclear why and how some of the acquired epigenetic and behavioural adaptations are transmitted and affect the following generations, while other experiments point to buffering effects that protect the offspring from the negative consequences in their stress-exposed parents. Interpreting these human findings is further complicated since inter- and transgenerational effects in response to early-life stress may be influenced by additional factors such as a rapidly changing cultural background and environment but also a collective family experience of stress and trauma that may programme an individual's stress-coping behaviour. Also, most of the human studies are limited to the analysis of epigenetic changes in peripheral tissues such as blood, and there is still a debate about the predictiveness of these findings for changes in the brain and, thus, for their relevance with regard to neuronal development. Thus, a number of experimental animal models have been developed and established to gain more detailed insights into the mechanisms of the proposed transgenerational transmission of early-life stress and trauma and the concept of transgenerational epigenetic inheritance.

6.2 *Transgenerational Transmission of Early-Life Trauma and Stress in Experimental Animal Studies*

A classic example with regard to a potential behavioural/social transmission is provided by the seminal experiments by Meaney and colleagues, focusing on naturally occurring variations of maternal care (high licking/grooming vs. low licking/grooming) in rats. In a series of experiments, it has been shown that the intensity of licking and grooming, as a measurement of maternal care behaviour, is transmitted from one generation to the next. Female offspring of high licking/grooming mothers showed the same intense care behaviour of their mothers, whereas female offspring from low licking/grooming mothers showed the reduced maternal care behaviour of their mothers (Champagne 2008; Champagne and Meaney 2007; Francis et al. 1999). Thereby, high licking/grooming behaviour induced beneficial adaptations of neuronal and endocrine outcomes that were paralleled by genome-wide changes of DNA methylation and gene-target-specific changes of DNA methylation and histone acetylation in the hippocampal GR promoter (McGowan et al. 2011; Weaver et al. 2004). These changes were passed on to the descendants, and cross-fostering studies revealed that this was clearly dependent on maternal care behaviour, reinstating the epigenetic marks in the respective offspring generations. Additional evidence for a behavioural intergenerational transmission of maternal care comes from a series of experiments in rats, applying the "scarcity–adversity model" that induces maltreatment behaviour by the female caregiver (Walker et al. 2017; Roth et al. 2009). Interestingly, female rats that experienced severe maltreatment by their mothers in early childhood displayed a similar aversive maternal care behaviour towards their own offspring. Additionally, it was shown that maltreatment experience induced changes in gene expression, specifically the BDNF gene, associated with altered DNA methylation in brain areas implicated in the control of maternal behaviour. In a more recent study, it could be shown that the changes of maternal behaviour, gene expression and DNA methylation in dams with a history of maltreatment could be normalized by the application of a DNA methyltransferase inhibitor (Keller et al. 2019), which, in turn, prevents the intergenerational transmission of the acquired behavioural changes and demonstrates the relevance of epigenetic mechanisms in this process. Overall, the studies cited reveal that the quality of maternal care that a female and future mother experiences during her own childhood influences the quality and intensity of maternal care she provides to her own offspring.

Evidence for the second route, a germline-dependent transmission of epigenetic marks and their effects on brain function and behaviour, is derived from approaches applying unpredictable maternal separation combined with unpredictable additional maternal stress (MSUS) in mice to mimic traumatic events in early postnatal life (Jawaid et al. 2018; Bohacek and Mansuy 2015; Bohacek et al. 2013). These seminal studies revealed that stress-induced changes of depressive-like behaviour are transmitted to offspring until the third generation and changes of risk-taking behaviour and metabolic function, such as glucose regulation, until the fifth generation

(Boscardin et al. 2022; van Steenwyk et al. 2018). Although a transmission up to the second generation has also been described for the maternal line, it is important to point out here that the transgenerational transmission of these stress-induced phenotypes up to the fifth generation was observed for exposed male individuals and in the patriline, indicating a potential transmission of epigenetic marks via the male germline. This is supported by findings that promoter-associated CpG islands within genes implicated in stress response and emotionality displayed altered DNA methylation patterns in the "stress-exposed" male germline (F1). In addition, it was shown that some of the alterations can be maintained and passed to the F2 offspring, and alterations can also be detected in the brain, where they correlate with altered expression of specific genes (Bohacek et al. 2013; Franklin et al. 2010). Besides alterations in DNA methylation, non-coding RNAs (ncRNAs), specifically microRNAs (miRNAs), also appear to be implicated in epigenetic modifications underlying germ cell reprogramming, providing an alternative mechanism for transgenerational epigenetic inheritance. In an elegant experimental study, it was demonstrated that early-life stress affects miRNA expression in the sperm, serum and brain in male mice, which was paralleled by behavioural and metabolic abnormalities that are passed down to subsequent generations. Interestingly, the described transgenerational transmission of stress-induced behavioural alterations in this model can also be induced by injecting miRNA from traumatized males into fertilized oocytes, revealing a causal relation between germline miRNA changes and transgenerational transfer of trauma (Gapp et al. 2014).

Additional evidence for a pure germline transmission via the patriline, independent of behavioural influences, comes from a classical but widely discussed study using olfactory fear conditioning to assess epigenetic transgenerational inheritance. In this study, the F1 and F2 generation descendants of mice, trained to identify an odour associated with mild foot shocks, showed increased responsiveness to the same odour during fear-potentiated startle, indicating transgenerational inheritance of an acquired negative association (Dias and Ressler 2014). It was also shown that the gene encoding for the olfactory receptor that is activated by the olfactory signal during the experiment was hypomethylated in the sperm of trained male animals and that this state of altered DNA methylation could also be detected in the sperm of F1 and F2 offspring. This suggests that the transgenerational transmission observed in this particular paradigm appears to be mediated by germline-dependent reprogramming of DNA methylation and provides evidence for epigenetic inheritance in the sense of Lamarck's original concept (Szyf 2014).

Evidence for epigenetic inheritance via the maternal line comes, for instance, from studies on preconception stress (PCS), in which future female mothers were exposed to stress shortly (up to 2 weeks) before conception. It was shown that PCS induced altered CRF1mRNA expression in the brain of the offspring at birth and in adulthood (Zaidan et al. 2013). Moreover, in the same study, it was found that PCS affects CRF1 expression not only in the brain but also in ova. Also, the levels of corticosterone were altered not only in first- but also in second-generation offspring, alongside reduced CRF1 mRNA expression at birth and alterations in anxiogenic behaviour in adulthood (Zaidan and Gaisler-Salomon 2015). In a slightly modified paradigm of pre-reproductive stress, where adolescent female rats were exposed to stress and mated roughly 3 weeks after stress exposure, changes of global DNA methylation were found in the brain, specifically the amygdala, in the stress-exposed dams and also in their F1 and F2 offspring in a sex-dependent manner (Zaidan et al. 2023). Furthermore, an altered expression of DNA-methylating enzymes in the brain as well as in blood and oocytes was observed. The findings in oocytes indicate a potential transgenerational transmission of PCS experience via the female germline.

6.3 The Transgenerational Switch of Epigenetic Inheritance

The classic concept of transgenerational epigenetic inheritance of early stress and trauma is based on the idea that the transmitted epigenetic marks and the associated changes in gene expression appear in more or less the same manner in the descendants. Thus, a specific epigenetic mark in germline or somatic cells, including neurones, should be found in the subsequent generations. However, findings from our own lab indicate a potential switch of the effects between the generations. For instance, we found that histone 3 acetylation (H3ac) and the expression of oxytocin receptors (OxtR) in the hippocampus were increased in females exposed to maternal separation as early-life stressor (F0 generation) when compared to stress-naive controls. Interestingly, their direct non-stressed female offspring (F1 generation) showed a decrease of OxtR expression and H3ac histone acetylation in the hippocampus, thus in an opposite direction than in their mothers. In F2 females, the OxtR expression switched back to an increase when compared to controls, thus having the same effect as in their grandmothers but opposite to their mothers. H3ac was unaltered in the F2 females and, thus, also in contrast to the alterations in F0 and F1. The idea of a switch between the generations is also supported by the above-mentioned MSUS model, where it was observed that some of the MSUS-induced effects are detectable in one generation but not in the next and reappear in later generations (Franklin et al. 2010). Similarly, some of the consequences of pre-reproductive stress on HPA axis function, brain region–specific CRF1 expression and anxiety behaviour were not stable across generations, and some of the effects switched between generations (Zaidan and Gaisler-Salomon 2015).

In line with findings in animal studies and further supporting the transgenerational switch concept, the classic study in Holocaust survivors and their descendants showed that DNA methylation at the same CpG sites of the FKBP5 gene changed in opposite directions. While parents directly exposed to the Holocaust showed increased DNA methylation, their unexposed offspring displayed decreased DNA methylation (Yehuda et al. 2016).

7 Summary and Conclusion

The environmentally driven functional development of an individual phenotype represents an essential aspect of ontogenesis, as it guarantees that the developing individual can optimally adapt to the given environmental conditions, promoting a healthy and resilient phenotype in later life. However, adverse environmental conditions at critical prenatal or early postnatal time windows can negatively interfere with developmental processes and, as a consequence, increase the risk for a number of diseases.

This concept can also be applied to the functional development of the brain and behaviour, specifically for the influence of early-life stress (ELS) as a critical and modulating factor (Fig. 6.1). The individual effects of ELS on the developing brain are pre-programmed by genetic and epigenetic predispositions and are also strongly dependent on the dimensions of the stress experience, such as timing during distinct developmental time windows and the type, intensity and duration of the adverse experience. In addition, multiple consecutive stressful experiences at different developmental periods may interact with each other, and finally, the effects of ELS are highly sex-specific. ELS, at first, induces short-term effects, reflected by increased levels of stress hormones that induce rapid, acute and partly transient epigenetic changes that result in altered gene expression profiles in the brain in a region-specific manner. The short-term effects are then transferred into long-term adaptations of the neuronal networks and partly into stable epigenetic marks. These adaptations finally determine the stress response and socio-emotional behaviour in later life and, consequently, the establishment of adapted, healthy or vulnerable, diseased phenotypes.

In line with the concept of inter- and transgenerational inheritance, increasing evidence suggests that the ELS-induced adaptations can be transmitted to the next generations and thereby determine an environmentally induced predisposition of the developing individual. This process results in "transgenerational epigenetic programming," which reflects the sum of adverse experiences across generations (from parents, grandparents etc.) and defines a transgenerational familial predisposition for susceptibility to adverse environmental challenges.

So far, the detailed mechanisms underlying the concepts described above are still far from being fully understood. Therefore, future studies will require refined experimental approaches to investigate these open questions in more detail, particularly with regard to the question of acquired epigenetic marks that potentially drive evolutionary adaptations.

Acknowledgement Financial support for JB's work on stress-induced epigenetic changes and transgenerational epigenetic inheritance was provided by grants from the Deutsche Forschungsgemeinschaft (DFG) to JB (BO 1674/7-1 and BO 1674/8-1) and Prof. Katharina Braun (BR 1692/11-1 and BR 1692/12-1).

References

Abou Merhi A, Sivan V, Koueik J et al (2025) Transgenerational non-genomic transmission of acquired nervous system phenotypes: narrative review and epigenetic insights. Neurochem Int 191:106088

Agorastos A, Pervanidou P, Chrousos GP, Baker DG (2019) Developmental trajectories of early life stress and trauma: a narrative review on neurobiological aspects beyond stress system dysregulation. Front Psych 10:118

Álvarez-Mejía D, Rodas JA, Leon-Rojas JE (2025) From womb to mind: prenatal epigenetic influences on mental health disorders. Int J Mol Sci 26(13):6096

Alyamani RAS, Murgatroyd C (2018) Epigenetic programming by early-life stress. Prog Mol Biol Transl Sci 157:133–150

Andersen SL (2003) Trajectories of brain development: point of vulnerability or window of opportunity? Neurosci Biobehav Rev 27:3–18

Andersen SL, Teicher MH (2008) Stress, sensitive periods and maturational events in adolescent depression. Trends Neurosci 31:183–191

Arzate-Mejia RG, Carullo NVN, Mansuy IM (2024) The epigenome under pressure: on regulatory adaptation to chronic stress in the brain. Curr Opin Neurobiol 84:102832

Babenko O, Kovalchuk I, Metz GA (2015) Stress-induced perinatal and transgenerational epigenetic programming of brain development and mental health. Neurosci Biobehav Rev 48:70–91

Baker LM, Williams LM, Korgaonkar MS et al (2013) Impact of early vs. late childhood early life stress on brain morphometrics. Brain Imaging Behav 7(2):196203

Baldwin JR, Wang B, Karwatowska L et al (2023) Childhood maltreatment and mental health problems: a systematic review and meta-analysis of quasiexperimental studies. Am J Psychiatry 180(2):117–126

Bale TL, Baram TZ, Brown AS et al (2010) Early life programming and neurodevelopmental disorders. Biol Psychiatry 68(4):314–319

Bales KL, Saltzman W (2016) Fathering in rodents: neurobiological substrates and consequences for offspring. Horm Behav 77:249–259

Banushi B, Collova J, Milroy H (2025) Epigenetic echoes: bridging nature, nurture, and healing across generations. Int J Mol Sci 26(7):3075

Barker DJ (1998) In utero programming of chronic disease. Clin Sci (Lond) 95(2):115–128

Baroutis D, Sotiropoulou IM, Mantzioros R et al (2025) Prenatal maternal stress and long-term neurodevelopmental outcomes: a narrative review. J Perinat Med 53(9):1159–1171

Bath KG (2020) Synthesizing views to understand sex differences in response to early life adversity. Trends Neurosci 43(5):300–310

Beers SR, De Bellis MD (2002) Outcomes of child abuse. Neurosurg Clin N Am 13:235–241

Bick J, Naumova O, Hunter S et al (2012) Childhood adversity and DNA methylation of genes involved in the hypothalamus-pituitary-adrenal axis and immune system: whole-genome and candidate-gene associations. Dev Psychopathol 24(4):14171425

Birnie MT, Baram TZ (2025) The evolving neurobiology of early-life stress. Neuron 113(10): 1474–1490

Blaze J, Roth TL (2015) Evidence from clinical and animal model studies of the longterm and transgenerational impact of stress on DNA methylation. Semin Cell Dev Biol 43:76–84

Blaze J, Roth TL (2017) Caregiver maltreatment causes altered neuronal DNA methylation in female rodents. Dev Psychopathol 2:477–489

Bock J (2024) Perinatal epigenetic programming of functional brain circuits. In: Witzany G (ed) Epigenetics in Biological Communication. Springer, Cham, pp 197–218

Bock J, Gruss M, Becker S, Braun K (2005) Experience-induced changes of dendritic spine densities in the prefrontal and sensory cortex: correlation with developmental time windows. Cereb Cortex 15:802–808

Bock J, Murmu MS, Biala Y et al (2011) Prenatal stress and neonatal handling induce sex-specific changes in dendritic complexity and dendritic spine density in hippocampal subregions of prepubertal rats. Neuroscience 193:34–43

Bock J, Riedel A, Braun K (2012) Differential changes of metabolic brain activity and interregional functional coupling in prefronto-limbic pathways during different stress conditions: functional imaging in freely behaving rodent pups. Front Cell Neurosci 6:19

Bock J, Rether K, Gröger N et al (2014) Perinatal programming of emotional brain circuits: an integrative view from systems to molecules. Front Neurosci 8:11

Bock J, Wainstock T, Braun K, Segal M (2015) Stress in utero: prenatal programming of brain plasticity and cognition. Biol Psychiatry 78(5):315–326

Bock J, Breuer S, Poeggel G, Braun K (2017) Early life stress induces attentiondeficit hyperactivity disorder (ADHD)-like behavioral and brain metabolic dysfunctions: functional imaging of methylphenidate treatment in a novel rodent model. Brain Struct Funct 222(2):765–780

Bohacek J, Mansuy IM (2015) Molecular insights into transgenerational non-genetic inheritance of acquired behaviours. Nat Rev Genet 16(11):641–652

Bohacek J, Gapp K, Saab BJ, Mansuy IM (2013) Transgenerational epigenetic effects on brain functions. Biol Psychiatry 73(4):313–320

Boscardin C, Manuella F, Mansuy IM (2022) Paternal transmission of behavioural and metabolic traits induced by postnatal stress to the 5th generation in mice. Environ Epigenet 8(1):dvac024

Braun K, Champagne FA (2014) Paternal influences on offspring development: behavioural and epigenetic pathways. J Neuroendocrinol 26(10):697–706

Braun K, Bock J, Wainstock T et al (2020) Experience-induced transgenerational (re)programming of neuronal structure and functions: impact of stress prior and during pregnancy. Neurosci Biobehav Rev 117:281–296

Buss C, Entringer S, Wadhwa PD (2012) Fetal programming of brain development: intrauterine stress and susceptibility to psychopathology. Sci Signal 5(245):pt7

Cao-Lei L, Massart R, Suderman MJ et al (2014) DNA methylation signatures triggered by prenatal maternal stress exposure to a natural disaster: Project Ice Storm. PLoS One 9(9):e107653

Cao-Lei L, Elgbeili G, Laplante DP et al (2025) DNA Methylation mediates the mssociation between prenatal maternal stress and the broad autism phenotype in human adolescents: Project Ice Storm. Int J Mol Sci 26(19):9468

Cay M, Gonzalez-Heydrich J, Teicher MH et al (2022) Childhood maltreatment and its role in the development of pain and psychopathology. Lancet Child Adolesc Health 6(3):195–206

Champagne FA (2008) Epigenetic mechanisms and the transgenerational effects of maternal care. Front Neuroendocrinol 29(3):386–397

Champagne FA, Meaney MJ (2007) Transgenerational effects of social environment on variations in maternal care and behavioral response to novelty. Behav Neurosci 121(6):1353–1363

Constantinof A, Boureau L, Moisiadis VG et al (2019) Prenatal glucocorticoid exposure results in changes in gene transcription and DNA methylation in the female juvenile guinea pig hippocampus across three generations. Sci Rep 9(1):18211

Crudo A, Suderman M, Moisiadis VG et al (2013) Glucocorticoid programming of the fetal male hippocampal epigenome. Endocrinology 154(3):1168–1680

Daskalakis NP, Bagot RC, Parker KJ et al (2013) The three-hit concept of vulnerability and resilience: toward understanding adaptation to early-life adversity outcome. Psychoneuroendocrinology 38:1858–1873

Demaili A, Portugalov A, Dudai M et al (2023) Epigenetic (re)programming of gene expression changes of CB1R and FAAH in the medial prefrontal cortex in response to early life and adolescence stress exposure. Front Cell Neurosci 17:1129946

Dias BG, Ressler KJ (2014) Parental olfactory experience influences behavior and neural structure in subsequent generations. Nat Neurosci 17(1):89–96

Dudley KJ, Li X, Kobor MS et al (2011) Epigenetic mechanisms mediating vulnerability and resilience to psychiatric disorders. Neurosci Biobehav Rev 35:1544–1551

Ehlert U (2013) Enduring psychobiological effects of childhood adversity. Psychoneuroendocrinology 38(9):1850–1857
Essex MJ, Boyce WT, Hertzman C et al (2013) Epigenetic vestiges of early developmental adversity: childhood stress exposure and DNA methylation in adolescence. Child Dev 84(1): 58–75
Feldman R, Braun K, Champagne FA (2019) The neural mechanisms and consequences of paternal caregiving. Nat Rev Neurosci 20(4):205–224
Francis D, Diorio J, Liu D, Meaney MJ (1999) Nongenomic transmission across generations of maternal behavior and stress responses in the rat. Science 286(5442):1155–1158
Franklin TB, Russig H, Weiss IC et al (2010) Epigenetic transmission of the impact of early stress across generations. Biol Psychiatry 68(5):408–415
Gapp K, Jawaid A, Sarkies P et al (2014) Implication of sperm RNAs in transgenerational inheritance of the effects of early trauma in mice. Nat Neurosci 17(5):667–669
Geiger LT, Balouek JR, Barrett MR et al (2025) Early-life stress alters chromatin modifications in VTA to prime stress sensitivity. bioRxiv 1:2024.03.14.584631
Gemmati D, Villanova M, Scarpellini F et al (2025) Epigenetic modifications and transgenerational inheritance in women victims of violence (EWVV). Environ Epigenet 11(1):dvaf025
Gluckman P, Hanson M (2006) Developmental origins of health and disease. Cambridge University Press. https://doi.org/10.1017/CBO9780511544699
Gos T, Bock J, Poeggel G, Braun K (2008) Stress-induced synaptic changes in the rat anterior cingulate cortex are dependent on endocrine developmental time windows. Synapse 62:229–232
Greenough WT, Black JE, Wallace CS (1987) Experience and brain development. Child Dev 58(3): 539–559
Gröger N, Matas E, Gos T et al (2016) The transgenerational transmission of childhood adversity: behavioral, cellular, and epigenetic correlates. J Neural Transm (Vienna) 123(9):1037–1052
Gruber D, Gilling KE, Albrecht A et al (2015) 5-HT receptor-mediated modulation of granule cell inhibition after juvenile stress recovers after a second exposure to adult stress. Neuroscience 293:67–79
Heard E, Martienssen RA (2014) Transgenerational epigenetic inheritance: myths and mechanisms. Cell 157(1):95–109
Heijmans BT, Tobi EW, Stein AD et al (2008) Persistent epigenetic differences associated with prenatal exposure to famine in humans. Proc Natl Acad Sci USA 105(44):17046–17049
Heindel JJ, Balbus J, Birnbaum L et al (2015) Developmental origins of health and disease: integrating environmental influences. Endocrinology 156(10):3416–3421
Helmeke C, Poeggel G, Braun K (2001) Differential emotional experience induces elevated spine densities on basal dendrites of pyramidal neurons in the anterior cingulate cortex of Octodon degus. Neuroscience 104:927–931
Hodes GE, Epperson CN (2019) Sex differences in vulnerability and resilience to stress across the life span. Biol Psychiatry 86(6):421–432
Horsthemke B (2018) A critical view on transgenerational epigenetic inheritance in humans. Nat Commun 9(1):2973
Houtepen LC, Vinkers CH, Carrillo-Roa T et al (2016) Genome-wide DNA methylation levels and altered cortisol stress reactivity following childhood trauma in humans. Nat Commun 7:10967
Jablonka E, Raz G (2009) Transgenerational epigenetic inheritance: prevalence, mechanisms, and implications for the study of heredity and evolution. Q Rev Biol 84(2):131–176
Jaric I, Rocks D, Cham H et al (2019) Sex and estrous cycle effects on anxiety- and depression-related phenotypes in a two-hit developmental stress model. Front Mol Neurosci 12:74
Jawaid A, Roszkowski M, Mansuy IM (2018) Transgenerational Epigenetics of traumatic stress. Prog Mol Biol Transl Sci 158:273–298
Joseph R (1999) Environmental influences on neural plasticity, the limbic system, emotional development and attachment: a review. Child Psychiatry Hum Dev 29:189–208

Keller SM, Doherty TS, Roth TL (2019) Pharmacological manipulation of DNA methylation normalizes maternal behavior, DNA methylation, and gene expression in dams with a history of maltreatment. Sci Rep 9(1):10253
Kertes DA, Kamin HS, Hughes DA et al (2016) Prenatal maternal stress predicts methylation of genes regulating the hypothalamic-pituitary-adrenocortical system in mothers and newborns in the Democratic Republic of Congo. Child Dev 87(1):61–72
Klengel T, Mehta D, Anacker C et al (2013) Allele-specific FKBP5 DNA demethylation mediates gene-childhood trauma interactions. Nat Neurosci 16(1):33–41
Klibaner-Schiff E, Simonin EM, Akdis CA et al (2024) Environmental exposures influence multigenerational epigenetic transmission. Clin Epigenetics 16(1):145
Kocamaz D, Franzke C, Gröger N et al (2022) Early life stress-induced epigenetic programming of hippocampal NPY-Y2 receptor gene expression changes in response to adult stress. Front Cell Neurosci 16:936979
Köhler JC, Gröger N, Lesse A et al (2019) Early-life adversity induces epigenetically regulated changes in hippocampal dopaminergic molecular pathways. Mol Neurobiol 56(5):3616–3625
Kolb B, Harker A, Gibb R (2017) Principles of plasticity in the developing brain. Dev Med Child Neurol 59(12):1218–1223
Korolenko A, Skinner MK (2024) Generational stability of epigenetic transgenerational inheritance facilitates adaptation and evolution. Epigenetics 19(1):2380929
Kundakovic M, Champagne FA (2015) Early-life experience, epigenetics, and the developing brain. Neuropsychopharmacology 40(1):141–153
Kundakovic M, Gudsnuk K, Herbstman JB et al (2015) DNA methylation of BDNF as a biomarker of early-life adversity. Proc Natl Acad Sci USA 112(22):6807–6813
Lesse A, Rether K, Gröger N et al (2017) Chronic postnatal stress induces depressive-like behavior in male mice and programs second-hit stress-induced gene expression patterns of OxtR and AvpR1a in adulthood. Mol Neurobiol 54(6):4813–4819
Levine A, Worrell TR, Zimnisky R, Schmauss C (2012) Early life stress triggers sustained changes in histone deacetylase expression and histone H4 modifications that alter responsiveness to adolescent antidepressant treatment. Neurobiol Dis 45:488–498
Liao H, Lu D, Reisinger SN, Mehrabadi MR, Gubert C, Hannan AJ (2025) Epigenetic effects of paternal environmental exposures and experiences on offspring phenotypes. Trends Genet 41(9):735–761
Liu D, Diorio J, Tannenbaum B et al (1997) Maternal care, hippocampal glucocorticoid receptors, and hypothalamic-pituitary-adrenal responses to stress. Science 277:1659–1662
Lupien SJ, McEwen BS, Gunnar MR, Heim C (2009) Effects of stress throughout the lifespan on the brain, behaviour and cognition. Nat Rev Neurosci 10:434
Maccari S, Krugers HJ, Morley-Fletcher S et al (2014) The consequences of early-life adversity: neurobiological, behavioural and epigenetic adaptations. J Neuroendocrinol 26(10):707–723
Maccari S, Polese D, Reynaert ML et al (2017) Early-life experiences and the development of adult diseases with a focus on mental illness: the Human Birth Theory. Neuroscience 342:232–251
Mandl S, Alexopoulos J, Doering S et al (2024) The effect of prenatal maternal distress on offspring brain development: a systematic review. Early Hum Dev 192:106009
Marín O (2016) Developmental timing and critical windows for the treatment of psychiatric disorders. Nat Med 22(11):1229–1238
Matas E, Bock J, Braun K (2016) The impact of parent-infant interaction on epigenetic plasticity mediating synaptic adaptations in the infant brain. Psychopathology 49(4):201–210
McGowan PO, Sasaki A, D'Alessio AC et al (2009) Epigenetic regulation of the glucocorticoid receptor in human brain associates with childhood abuse. Nat Neurosci 12(3):342–348
McGowan PO, Suderman M, Sasaki A et al (2011) Broad epigenetic signature of maternal care in the brain of adult rats. PLoS One 6(2):e14739
Meaney MJ, Ferguson-Smith AC (2010) Epigenetic regulation of the neural transcriptome: the meaning of the marks. Nat Neurosci 13:1313–1318

Meaney MJ, Szyf M (2005) Maternal care as a model for experience-dependent chromatin plasticity? Trends Neurosci 9:456–463
Mehta D, Klengel T, Conneely KN et al (2013) Childhood maltreatment is associated with distinct genomic and epigenetic profiles in posttraumatic stress disorder. Proc Natl Acad Sci USA 110(20):8302–8307
Miguel PM, Pereira LO, Silveira PP, Meaney MJ (2019) Early environmental influences on the development of children's brain structure and function. Dev Med Child Neurol 61(10): 1127–1133
Mueller BR, Bale TL (2008) Sex-specific programming of offspring emotionality after stress early in pregnancy. J Neurosci 28:9055–9065
Mulligan CJ, D'Errico NC, Stees J, Hughes DA (2012) Methylation changes at NR3C1 in newborns associate with maternal prenatal stress exposure and newborn birth weight. Epigenetics 7(8): 853–857
Murmu MS, Salomon S, Biala Y et al (2006) Changes of spine density and dendritic complexity in the prefrontal cortex in offspring of mothers exposed to stress during pregnancy. Eur J Neurosci 24(5):1477–1487
Mychasiuk R, Gibb R, Kolb B (2011a) Prenatal stress produces sexually dimorphic and regionally specific changes in gene expression in hippocampus and frontal cortex of developing rat offspring. Dev Neurosci 33:531–538
Mychasiuk R, Ilnytskyy S, Kovalchuk O et al (2011b) Intensity matters: brain, behaviour and the epigenome of prenatally stressed rats. Neuroscience 180:105–110
Mychasiuk R, Harker A, Ilnytskyy S, Gibb R (2013) Paternal stress prior to conception alters DNA methylation and behaviour of developing rat offspring. Neuroscience 241:100–105
Nederhof E, Schmidt MV (2012) Mismatch or cumulative stress: toward an integrated hypothesis of programming effects. Physiol Behav 106:691–700
Nishi M, Horii-Hayashi N, Sasagawa T (2014) Effects of early life adverse experiences on the brain: implications from maternal separation models in rodents. Front Neurosci 8:166
O'Brien KJ, Huque ZM, Pike MR et al (2026) Synergistic pathways to psychosis: understanding developmental risk and resilience factors. Neuropsychopharmacology 51(1):273–292
O'Donnell KJ, Meaney MJ (2017) Fetal origins of mental health: the developmental origins of health and disease hypothesis. Am J Psychiatry 174(4):319–328
Oberlander TF, Weinberg J, Papsdorf M et al (2008) Prenatal exposure to maternal depression, neonatal methylation of human glucocorticoid receptor gene (NR3C1) and infant cortisol stress responses. Epigenetics 3(2):97–106
Oitzl MS, Champagne DL, van der Veen R, de Kloet ER (2010) Brain development under stress: hypotheses of glucocorticoid actions revisited. Neurosci Biobehav 34:853–866
Oren G, Shoshani A, Samra NN et al (2025) From trauma to resilience: psychological and epigenetic adaptations in the third generation of holocaust survivors. Sci Rep 15(1):26193
Palma-Gudiel H, Córdova-Palomera A, Eixarch E et al (2015) Maternal psychosocial stress during pregnancy alters the epigenetic signature of the glucocorticoid receptor gene promoter in their offspring: a meta-analysis. Epigenetics 10(10):893–902
Pechtel P, Lyons-Ruth K, Anderson CM, Teicher MH (2014) Sensitive periods of amygdala development: the role of maltreatment in preadolescence. NeuroImage 97:236–244
Peña CJ (2026) Epigenetic regulation of brain development, plasticity, and response to early-life stress. Neuropsychopharmacology 51(1):5–15
Pena CJ, Nestler EJ, Bagot RC (2019) Environmental programming of susceptibility and resilience to stress in adulthood in male mice. Front Behav Neurosci 13:40
Poeggel G, Helmeke C, Abraham A et al (2003) Juvenile emotional experience alters synaptic composition in the rodent cortex, hippocampus, and lateral amygdala. Proc Natl Acad Sci USA 100:16137–16142
Pohl TT, Young LJ, Bosch OJ (2019) Lost connections: oxytocin and the neural, physiological, and behavioral consequences of disrupted relationships. Int J Psychophysiol 136:54–63

Provençal N, Binder EB (2015) The effects of early life stress on the epigenome: from the womb to adulthood and even before. Exp Neurol 268:10–20
Provençal N, Suderman MJ, Guillemin C et al (2014) Association of childhood chronic physical aggression with a DNA methylation signature in adult human T cells. PLoS One 9(4):e89839
Rahman MF, McGowan PO (2022) Cell-type-specific epigenetic effects of early life stress on the brain. Transl Psychiatry 12(1):326
Ramo-Fernández L, Boeck C, Koenig AM et al (2019) The effects of childhood maltreatment on epigenetic regulation of stress-response associated genes: an intergenerational approach. Sci Rep 9(1):983
Reh RK, Dias BG, Nelson CA 3rd et al (2020) Critical period regulation across multiple timescales. Proc Natl Acad Sci USA 117(38):23242–23251
Richards EJ (2006) Inherited epigenetic variation–revisiting soft inheritance. Nat Rev Genet 7(5): 395–401
Roth TL, Sullivan RM (2005) Memory of early maltreatment: neonatal behavioral and neural correlates of maternal maltreatment within the context of classical conditioning. Biol Psychiatry 57(8):823–831
Roth TL, Lubin FD, Funk AJ, Sweatt JD (2009) Lasting epigenetic influence of earlylife adversity on the BDNF gene. Biol Psychiatry 65:760–769
Santarelli S, Zimmermann C, Kalideris G et al (2017) An adverse early life environment can enhance stress resilience in adulthood. Psychoneuroendocrinology 78:213–221
Schmidt MV (2010) Molecular mechanisms of early life stress–lessons from mouse models. Neurosci Biobehav Rev 34(6):845–852
Schmidt MV (2011) Animal models for depression and the mismatch hypothesis of disease. Psychoneuroendocrinology 36:330–338
Short AK, Baram TZ (2019) Early-life adversity and neurological disease: age-old questions and novel answers. Nat Rev Neurol 15(11):657–669
Skinner MK (2015) Environmental epigenetics and a unified theory of the molecular aspects of evolution: a neo-Lamarckian concept that facilitates neo-Darwinian evolution. Genome Biol Evol 7(5):1296–1302
Steg LC, Mansuy IM (2025) A father's legacy: the sperm epigenome, preimplantation development, and paternal environment. Epigenomics 17(17):1267–1280
Suderman M, Borghol N, Pappas JJ et al (2014) Childhood abuse is associated with methylation of multiple loci in adult DNA. BMC Med Genet 7:13
Sulyok E, Farkas B, Bodis J (2023) Pathomechanisms of prenatally programmed adult diseases. Antioxidants (Basel, Switzerland) 12(7):1354
Švorcová J (2023) Transgenerational epigenetic inheritance of traumatic experience in mammals. Genes (Basel) 14(1):120
Szyf M (2014) Lamarck revisited: epigenetic inheritance of ancestral odor fear conditioning. Nat Neurosci 17(1):2–4
Szyf M (2015) Nongenetic inheritance and transgenerational epigenetics. Trends Mol Med 21(2): 134–144
Szyf M (2019) The epigenetics of perinatal stress. Dialogues Clin Neurosci 21(4):369–378
Szyf M, Weaver I, Meaney M (2007) Maternal care, the epigenome and phenotypic differences in behavior. Reprod Toxicol 24(1):9–19
Taylor SE (2010) Mechanisms linking early life stress to adult health outcomes. Proc Natl Acad Sci USA 107:8507–8512
Teicher MH, Samson JA, Anderson CM, Ohashi K (2016) The effects of childhood maltreatment on brain structure, function and connectivity. Nat Rev Neurosci 17(10):652–666
Tobi EW, Lumey LH, Talens RP et al (2009) DNA methylation differences after exposure to prenatal famine are common and timing- and sex-specific. Hum Mol Genet 18(21):4046–4053
Tomoda A, Nishitani S, Takiguchi S et al (2024) The neurobiological effects of childhood maltreatment on brain structure, function, and attachment. Eur Arch Psychiatry Clin Neurosci 275(7):1939–1958

Torres-Berrío A, Bortolami A, Peña CJ, Nestler EJ (2026) Neurobiology of resilience to early life stress. Neuropsychopharmacology 51(1):29–45
Tractenberg SG, Levandowski ML, de Azeredo LA et al (2016) An overview of maternal separation effects on behavioural outcomes in mice: evidence from a four-stage methodological systematic review. Neurosci Biobehav Rev 68:489–503
Turecki G, Meaney MJ (2016) Effects of the social environment and stress on glucocorticoid receptor gene methylation: a systematic review. Biol Psychiatry 79(2):87–89
Vaidya N, Marquand AF, Nees F et al (2024) The impact of psychosocial adversity on brain and behaviour: an overview of existing knowledge and directions for future research. Mol Psychiatry 29(10):3245–3267
Van den Bergh BRH, van den Heuvel MI, Lahti M et al (2020) Prenatal developmental origins of behavior and mental health: the influence of maternal stress in pregnancy. Neurosci Biobehav Rev 117:26–64
van Steenwyk G, Roszkowski M, Manuella F et al (2018) Transgenerational inheritance of behavioral and metabolic effects of paternal exposure to traumatic stress in early postnatal life: evidence in the 4th generation. Environ Epigenet 4(2):dvy023
Walker CD, Bath KG, Joels M et al (2017) Chronic early life stress induced by limited bedding and nesting (LBN) material in rodents: critical considerations of methodology, outcomes and translational potential. Stress 20(5):421–448
Weaver IC, Cervoni N, Champagne FA et al (2004) Epigenetic programming by maternal behaviour. Nat Neurosci 7:847–854
Weinstock M (2016) Prenatal stressors in rodents: effects on behavior. Neurobiol Stress 6:3–13
Xie L, Korkmaz KS, Braun K, Bock J (2013) Early life stress-induced histone acetylations correlate with activation of the synaptic plasticity genes Arc and Egr1 in the mouse hippocampus. J Neurochem 125:457–464
Yehuda R, Daskalakis NP, Bierer LM et al (2016) Holocaust exposure induced intergenerational effects on FKBP5 methylation. Biol Psychiatry 80(5):372–380
Zaidan H, Gaisler-Salomon I (2015) Prereproductive stress in adolescent female rats affects behavior and corticosterone levels in second-generation offspring. Psychoneuroendocrinology 58:120–129
Zaidan H, Leshem M, Gaisler-Salomon I (2013) Prereproductive stress to female rats alters corticotropin releasing factor type 1 expression in ova and behavior and brain corticotropin releasing factor type 1 expression in offspring. Biol Psychiatry 74(9):680–687
Zaidan H, Wnuk A, Aderka IM et al (2023) Pre-reproductive stress in adolescent female rats alters maternal care and DNA methylation patterns across generations. Stress 26(1):2201325
Zhou A, Ryan J (2023) Biological embedding of early-life adversity and a scoping review of the evidence for intergenerational epigenetic transmission of Stress and trauma in humans. Genes (Basel) 14(8):1639

Chapter 7
Maternal Transmission of Atopic Disease Susceptibility

Loekie Engels and Colin Logie

Abstract Maternal stress during pregnancy is a critical factor influencing foetal development and long-term health outcomes in offspring. This chapter reviews the role of maternal stress in the development of the foetal immune system, with a focus on foetal mast cells and atopic diseases such as eczema and asthma. It explores the physiological stress response, glucocorticoid regulation and how the placenta protects the foetus from overexposure to glucocorticoids. Further, it investigates how chronic maternal stress can disrupt these processes and cause immune system dysfunction, increasing the risk of developing childhood atopic diseases. An evolutionary perspective is included to examine the origins of this vulnerability. By revealing the underlying epigenetic mechanisms and target cell types, the importance of stress reduction during pregnancy can be highlighted, and an effort can be made to improve prenatal mental healthcare.

1 Introduction

Many physical and psychological stimuli can elicit a stress response in humans, activating different pathways for the adaptation and maintenance of homeostasis. Stressors can be either acute or chronic and stem from various sources. In modern times, a large portion of stressors are psychosocial such as social exclusion, conflict or traumatic events (Kogler et al. 2015), although stress can also be caused by physiological sources, such as illness, starvation, heat or exercise (Haas et al. 2022). Temporary stress can cause a short-term adaptive response, but prolonged or repeated stress exposure can cause long-term detrimental effects to health, especially during sensitive windows in development. Pregnancy represents one such critical window in development, in which exposure to stress is often increased and the consequences of stress are heightened. In addition to common life stressors, there are

L. Engels · C. Logie (✉)
Department of Molecular Biology, Faculty of Science, Radboud Institute for Molecular Life Sciences, Radboud University, Nijmegen, The Netherlands
e-mail: Colin.Logie@ru.nl

G. Witzany (ed.), *Integrative Theory of Evolution - Aspects and Insights*,
https://doi.org/10.1007/978-3-032-24295-2_7

unique challenges to pregnancy, including concerns for foetal health, hormonal fluctuations and anticipation of caregiving responsibilities (Saur and Dos Santos 2021). It has been found that up to 70% of women experience symptoms of stress and anxiety during pregnancy (Pascal et al. 2023), and over 10% of women experience prenatal depression (Woody et al. 2017). Studies have also found that women in countries with a lower economic level are impacted the most (Aziz et al. 2025).

Stress elicits a complicated physiological response that builds up to the release of hormones such as adrenaline and cortisol. These hormones have widespread effects throughout the body, including the regulation of energy availability, alteration of behaviour and an increase in the heart rate and blood pressure (Joëls and Baram 2009; White and Porterfield 2013). These effects enable a person to adapt to environmental challenges. Although this is important for survival, chronic stress can lead to adverse effects on mental and physical health. For example, chronic stress has been found to impair the immune system and promote inflammation, which is associated with heart disease and mental health disorders (Raza 2025).

Beyond their role in acute stress responses, stress hormones also play a critical role in the regulation of developmental processes during pregnancy. In particular, glucocorticoids, such as cortisol, have an important role in the regulation of foetal organ maturation and immune system programming. This review focusses on glucocorticoids rather than catecholamines, such as adrenaline and noradrenaline, due to the ability of glucocorticoids to cross the placenta and influence foetal development directly (Zhao et al. 2024). Catecholamines play a crucial role in the maternal stress response but are unable to cross the placental barrier, limiting their direct impact on the foetus (Rakers et al. 2020). Since glucocorticoids are only partially regulated by the placenta, these hormones are of great interest in the understanding of how maternal stress can shape the health of the offspring.

Given the central role of glucocorticoids in mediating the effects of maternal stress on foetal development, this review aims to explore the mechanisms further. Specifically, this review will focus on how maternal stress alters foetal glucocorticoid exposure and how these changes influence the foetal immune system, with particular attention to mast cells and atopic diseases such as eczema and asthma. To explore this topic, the general stress response and the function of glucocorticoids will be discussed. Then, a deeper dive will be taken into the effect of maternal stress on glucocorticoid levels and how this may affect the foetus. Later sections focus on the influence altered glucocorticoid levels have on foetal mast cell development and how this may be implicated in atopic disorders such as eczema and asthma. An evolutionary perspective is then introduced that may explain why developing foetal mast cells may be particularly vulnerable to glucocorticoid overexposure. Finally, the review will discuss the clinical implications of the findings and provide future perspectives for research and intervention.

1.1 General Stress Response

When a stressful stimulus is experienced, it is initially registered by the amygdala, the part of the limbic system responsible for the processing of emotions and the regulation of anxiety, aggression and emotional memory (Kaiser and Jaillardon 2023). From the amygdala, there are two main responses activated, a quick response with short-term effects mediated by the sympathetic–adreno–medullar (SAM) system and a slow response with long-term effects mediated by the hypothalamic–pituitary–adrenal (HPA) axis.

The quick SAM system response starts when the signals from the amygdala reach the hypothalamus, which sends further signals to the adrenal medulla to activate it. The activated adrenal medulla secretes the catecholamines adrenaline and noradrenaline into the circulation (Joëls and Baram 2009). These hormones bind to adrenergic receptors, a class of G protein–coupled receptors (GPCRs) on the surface of the target cells. Upon binding of adrenaline or noradrenaline to adrenergic receptors, vascular smooth muscle cells contract, leading to vaso-constriction and elevated blood pressure and heart rate. Activation of adrenergic receptors also stimulates gluconeogenesis in the liver and kidneys and glycogenolysis in the liver and muscle cells to increase glucose levels and expand energy availability (Dufour et al. 2009). Activation of the SAM system may also lead to behavioural changes like increased alertness, cognition, attention and focus (Joëls and Baram 2009).

The first step of the slower HPA axis response is the release of corticotropin-releasing hormone (CRH) from the paraventricular nucleus of the hypothalamus upon receiving stress signals from the amygdala (Lovejoy 2024). CRH is a neuropeptide that signals to the pituitary gland to release the adrenocorticotropic hormone (ACTH). This hormone stimulates the adrenal cortex to secrete glucocorticoid hormones, such as cortisol (Sheng et al. 2020; Kaiser and Jaillardon 2023). Similar to adrenaline, cortisol increases glucose availability by stimulating gluconeogenesis and glycogenolysis in the liver, kidneys and muscle cells, ensuring sufficient glucose supply to critical organs during stress (Chourpiliadis and Aeddula 2023). The hormone is also known to increase blood pressure and cardiac output; it decreases the function of reproductive organs, increases bone resorption, inhibits fibroblast proliferation and collagen formation, and it is known to inhibit antidiuretic hormone secretion and action (White and Porterfield 2013; McNamara et al. 2018). Glucocorticoid effects on the immune system have been investigated for many years, with evidence being found for both an immunosuppressive function and an enhancing function (Cain and Cidlowski 2017; Wang et al. 2019). A model proposed for these contradictory findings is a concentration-dependent response from immune cells, with low concentrations stimulating the immune system and higher concentrations having a suppressive effect (Munck and Náray-Fejes-Tóth 1992; Sapolsky et al. 2000). This model also states that short exposure is thought to be stimulating, while prolonged exposure to glucocorticoids has suppressive effects on the immune system. Of note, the HPA axis is regulated by a negative feedback loop, where elevated cortisol levels inhibit the release of CRH in the hypothalamus and ACTH in the pituitary gland. This prevents the overproduction of glucocorticoids (Gjerstad et al. 2018) to maintain homeostasis.

Though catecholamines play a large role in the maternal stress response, research has found that these hormones are unable to cross the placenta into the foetal circulation, limiting the direct effect on foetal development (Rakers et al. 2020). Glucocorticoids, however, have been found to cross the placenta to enter the foetal circulation throughout pregnancy (Zhao et al. 2024). This makes it more likely that these stress hormones may affect the foetus during development. As a result, glucocorticoids are the central focus in studies of prenatal stress and foetal programming.

1.2 Glucocorticoids

Beyond the role in mediating the physiological stress response, glucocorticoids are steroid hormones produced by the adrenal cortex, with an essential role in maintaining homeostasis of metabolic, immune and developmental processes (Timmermans et al. 2019). In humans, cortisol is the key glucocorticoid, with widespread effects on glucose metabolism, inflammation and immune function (Chourpiliadis and Aeddula 2023). Under non-stress conditions, glucocorticoid regulation by the HPA axis follows a circadian rhythm (Lehmann et al. 2023). Cortisol levels are at a maximum early in the morning and slowly decline during the day (Oster et al. 2016). Importantly, this rhythm is altered during pregnancy, when precise regulation of glucocorticoids is essential for foetal maturation and organ development (Solano and Arck 2019).

In pregnancy, glucocorticoids support both maternal adaptation and foetal development. Basal cortisol levels contribute to foetal organ maturation, including the skin, lungs and immune system (Zhao et al. 2024). Cortisol is also critical to prepare the foetus for postnatal life, and levels increase vastly at the end of pregnancy (Rog-Zielinska et al. 2015). However, foetal tissues are especially sensitive to glucocorticoid levels, so poor regulation could affect normal developmental processes (Tegethoff et al. 2009). This sensitivity is also relevant for the developing immune system, where glucocorticoids can affect immune cell differentiation and programming (Moisiadis and Matthews 2014; Solano and Arck 2019). This means that a tight regulation of foetal glucocorticoid exposure is necessary during gestation to ensure normal immune development and function.

2 Maternal Stress, Glucocorticoids and Foetal Mast Cell Development in the Context of Atopic Diseases

2.1 *Maternal Stress and Glucocorticoid Alterations*

Although precise regulation of glucocorticoid levels is critical for maternal health and foetal development, maternal stress can impact this balance significantly by the overactivation of the HPA axis. Normally, the placenta can partially protect the foetus against increased cortisol levels through a buffering mechanism. The placenta contains 11β-hydroxysteroid dehydrogenase type 2 (11β-HSD2), an enzyme that metabolizes bioactive cortisol into inactive cortisone. This partially prevents maternal cortisol from crossing the placenta into the foetal circulation (Salvante et al. 2017; Zhao et al. 2024). However, studies have shown that maternal stress can decrease the expression and activity of placental 11β-HSD2, increasing foetal exposure to maternal cortisol (O'Donnell et al. 2012; Jensen Peña et al. 2012).

The effects of maternal stress depend strongly on duration and intensity. Acute stress causes a rapid surge of cortisol, which is quickly resolved by the negative feedback loop of the HPA axis itself. The placental 11β-HSD2 is generally able to buffer such short surges of cortisol well. However, during times of chronic stress, the HPA axis is persistently activated, causing a prolonged increase in cortisol levels. This prolonged increase can overwhelm the placental buffering mechanism and even downregulate placental 11β-HSD2, causing more maternal cortisol to enter the foetal circulation during critical periods of development.

Excessive exposure to glucocorticoids can cause long-term effects in the foetal HPA axis, which is often referred to as foetal 'programming'. As discussed before, the cortisol produced by the HPA axis has many functions throughout the body and during foetal development, including immune development. This suggests that dysregulation of the foetal HPA axis can have lifelong effects in immune cells.

2.2 *Foetal Mast Cell Development*

Mast cells are innate immune cells that play a critical role in immune surveillance, inflammation and tissue homeostasis. In contrast to many other immune cells, mast cells leave the bone marrow as immature progenitors and complete their differentiation within a specific tissue under the influence of cues from their local environment (Kitamura et al. 2007; Kalesnikoff and Galli 2008). This tissue-specific maturation leads to heterogeneity in mast cell phenotypes, allowing the cells to function in a tissue-specific manner. There are two main types of mast cells, mucosal mast cells and connective tissue mast cells. The mucosal mast cells reside in the intestinal and airway mucosa, while the connective tissue mast cells are found in connective tissues like the skin (Fong 2023).

In mice, it has been found that mast cells begin entering foetal tissues during mid-gestation (Hallgren and Gurish 2007). In humans, immature mast cells appear in many tissues at the end of the first trimester, where they will mature and increase in number throughout gestation (Vaculik et al. 2012; Msallam and Redegeld 2023). The cells are present prenatally to prepare the foetus for defence against exposure to pathogens after birth (Chia et al. 2023).

Upon activation by a possible pathogen, mast cells release inflammatory mediators like histamines, proteases and cytokines to instigate inflammation and recruit other immune cells (Krystel-Whittemore et al. 2016; Fong 2023). Beyond their role in immune responses, these mediators are also important for the maintenance of the epithelial barrier. Proteases are able to remodel the extracellular matrix, and the cytokines increase vascular permeability to support immune cell infiltration (Bilgic et al. 2026). Cytokines are also crucial for barrier repair by controlling the proliferation, migration and differentiation of epithelial cells (Barada et al. 2023; Andrews et al. 2018). Dysregulation of the release of these mediators could thus lead to tissue damage and chronic inflammation.

Beyond their main function in immune surveillance and inflammation, mast cells also secrete cytokines and growth factors to promote tissue morphogenesis and angiogenesis, aiding in the maintenance of epithelial integrity and vascular development by recruiting endothelial cells (Chia et al. 2023; Mukai et al. 2018). This function highlights the importance of mast cells during foetal development.

Importantly, mast cells contain many receptors, including glucocorticoid receptors. These receptors make the mast cells sensitive to altered glucocorticoid exposure due to maternal stress, potentially dysregulating their function in inflammation, the maintenance of barrier integrity and tissue development. The next section will explore how glucocorticoids affect mast cells and how altered exposure during foetal development may lead to long-term adverse effects.

2.3 Glucocorticoid's Effects on Mast Cells and the Pathologies

For many years, synthetic glucocorticoids have been used as anti-allergic and anti-inflammatory compounds to treat inflammatory conditions, including asthma and eczema (Santoro et al. 2019; Oppong et al. 2013; Eklund et al. 1997). These effects are achieved by binding to glucocorticoid receptors expressed in immune cells, like mast cells. When glucocorticoids bind to receptors in a mast cell, the receptor undergoes conformational changes to allow translocation to the nucleus, where it interacts with glucocorticoid response elements on the DNA (Oppong et al. 2013). Through this interaction, genes involved in mast cell maturation, degranulation and cytokine production are regulated, and transcription factors associated with inflammation are suppressed (Santoro et al. 2019). This limits mast cell activation and the release of mediators like histamines, proteases and cytokines.

During pregnancy, maternal stress can lead to the overactivation of the maternal HPA axis, leading to increased cortisol levels. If the maternal stress is chronic, the sustained increase in cortisol can overwhelm the placental buffering system, leading to increased levels of cortisol in the foetal circulation. Importantly, excessive glucocorticoid exposure during sensitive developmental windows can interfere with the maturation of mast cells. For example, glucocorticoids reduce the production of stem cell factor, the ligand for the c-Kit receptor, which stimulates the differentiation and migration of mast cells (Oppong et al. 2013). Though there has been no specific research on stem cell factor inhibition in foetal mast cells, it has been found that treatment with dexamethasone, a widely used synthetic glucocorticoid, inhibits mast cell development from human foetal liver cells and human cord blood by up to 85% (Irani et al. 1995; Smith et al. 2002). Given these results, it would be of interest to see if glucocorticoid exposure also limits mast cell maturation during foetal development.

Beyond direct effects on gene expression, glucocorticoids can also induce epigenetic changes in mast cells, including DNA methylation and histone acetylation (Wang et al. 2019; Paes et al. 2024). When occurring in foetal cells, such epigenetic programming may persist after birth, causing long-term consequences. A recent study by Serhan et al. (2025) demonstrated that prenatal maternal stress in mice led to elevated glucocorticoid levels in the amniotic fluid. This led to changes in both gene expression and epigenetic marks in the skin mast cells of the offspring. The mast cells showed premature degranulation and hyperactivity. It also led to heightened sensitivity of the offspring's skin to harmless stimuli like friction, creating eczema-like lesions. After several months, the offspring had renewed mast cells, and the eczema-like symptoms faded.

These findings appear contradictory to the evidence showing that glucocorticoids suppress mast cell maturation and activity. However, several factors may explain this inconsistency. Firstly, the effects of glucocorticoids could be dependent on developmental timing. The studies by Irani et al. (1995) and Smith et al. (2002) investigated glucocorticoid exposure in mast cells that were not yet fully developed, whereas Serhan et al. (2025) focussed on more mature skin mast cells. As discussed previously, there is also a model suggesting that the effects of glucocorticoids on the immune system depend on the dose and duration of exposure, which could explain the differing outcomes found in mast cells. Although this model has not been demonstrated in foetal mast cells specifically, it suggests that there is a need for further research into how varying doses and durations of glucocorticoid exposure alter mast cell development and function.

Returning to the clinical relevance of the study by Serhan et al. (2025), eczema, also known as atopic dermatitis, is a common atopic disease with a lifetime prevalence of up to 20%, but it most commonly occurs in children (Kellogg and Smogorzewski 2023). It is an atopic disease that arises from both environmental factors and genetic predisposition (Andersson et al. 2016). Because this predisposition is present at birth, it is possible for prenatal factors, such as maternal stress and glucocorticoid exposure, to alter the susceptibility to atopic dermatitis and potentially other atopic conditions.

2.4 Similarities Between Asthma and Eczema

Similar to eczema, asthma is also an atopic disease prevalent in children (Chen and Chen 2021). The disease is characterized by inflammation of the airways, which causes difficulty in breathing. In recent reviews by Chen and Chen (2021) and Al-Hussainy and Mohammed (2021), a link was made between prenatal maternal stress and an increased risk of developing childhood asthma. Since eczema and asthma are both atopic diseases involving mast cells, the mechanisms described in the research by Serhan et al. (2025) may also explain the link between maternal stress and childhood asthma.

Although eczema primarily affects the skin and asthma affects the airways, these two diseases share some fundamental features that classify them as atopic diseases. A large similarity is their origin in the epithelial tissue. Both conditions involve dysfunction of the epithelial barrier, which increases the sensitivity to environmental allergens. In eczema, the impaired skin barrier enhances allergen penetration, which activates the immune system and causes inflammation (Kim and Leung 2018). In asthma, the airway epithelium is disrupted, and the epithelial cells are often damaged, making the tissue more permeable to allergens (Heijink et al. 2020).

Mast cells are also involved in both diseases, though in different ways. In asthma, mast cells have been implicated for decades (Méndez-Enríquez and Hallgren 2019). The cells have been found to accumulate in the airway smooth muscle tissue of asthma patients, where the release of histamines causes chronic inflammation. It has also been found that tryptase produced by mast cells contributes to airway remodelling, a process where the cellular structure and extracellular matrix in the airways are permanently altered (Berlin et al. 2023; Hough et al. 2020). In eczema, the role of mast cells is still less well-understood. However, research from Sehra et al. (2016) found that mast cells are important for the homeostatic expression of epidermal differentiation genes, which are essential to maintain barrier integrity. Dysfunction of the mast cells could, therefore, disrupt the skin homeostasis and weaken the epidermal barrier.

Importantly, maternal stress during pregnancy could predispose mast cells to overactivity in both tissue types. As described before, overexposure of developing mast cells to glucocorticoids can lead to genetic and epigenetic changes, causing sensitization of the cells. These changes could lead to hyperactivity or activation of the cells by harmless allergens, causing the release of excessive histamines, proteases and cytokines. The inflammation caused by these mediators could damage the epithelial barrier, increasing permeability for allergens. This mechanism could explain why maternal stress increases the chance of developing either eczema or asthma.

The shared mechanisms between eczema and asthma underscore that developing mast cells have a distinctive sensitivity to early-life disruptions, such as stress signals, a property that may be rooted in the evolutionary history of mast cells and glucocorticoid signalling.

3 Discussion

3.1 Evolutionary Perspective

From an evolutionary perspective, glucocorticoids are ancient hormones found in jawless vertebrates like lampreys, which diverged over 500 million years ago (Close et al. 2010). Mast cells also have an ancient origin but appear later, with clear evidence of their presence in bony fish, which evolved over 400 million years ago (Dobson et al. 2008; Hu et al. 2023). This mismatch in evolutionary timing suggests that the effects glucocorticoids have on mast cells are not shaped by direct evolutionary selection. Instead, these effects are likely a by-product of the expression of glucocorticoid receptors in nearly all cell types in the context of circadian gene regulation (Shimba and Ikuta 2020), which may explain why excessive levels of glucocorticoids during gestation derail foetal mast cell epigenetic programming and thus contribute to postnatal immune dysfunction in mice.

For the immune system to develop in a healthy manner, epigenetic regulation must remain both flexible and stable. Epigenetic marks need to be reversible to respond to environmental cues yet remain steady in processes like cell differentiation. We speculate that this delicate balance makes mast cells particularly vulnerable to disturbance, leading to long-lasting effects in their development. Because these changes persist throughout gestation and cannot be reversed before birth, the affected mast cells become overactive. This leads to inflammation, causing barrier dysfunction and early-life atopic disease (Serhan et al. 2025).

The pleiotropic actions of glucocorticoids may have made it evolutionarily impossible to fully optimize their influence across every cell type. Hence, as described in this chapter, prenatal maternal stress can lead to a maladaptive result in mice, namely sensitivity to atopic disease in progeny that is mediated by aberrant foetal epigenetic programming of mast cells. However, as shown by Serhan et al. (2025), the dysfunctional mouse foetal mast cells are postnatally replaced by bone marrow–derived mast cells within several months. This replacement serves as a 'reset' on the mast cell activity by removing the cells bearing the pathogenic epigenetic modifications acquired during gestation. Hence, even when mothers experience high stress during pregnancy, the offspring can still develop a healthy mast cell population after birth, although early-life symptoms may still occur. Whether, for this reason, very long-lived foetal mast cell populations are under negative evolutionary pressure remains to be established.

3.2 Clinical Implications and Future Perspectives

Understanding how the proposed maternal stress–related mast cell dysfunction and overactivity disrupts epithelial homeostasis and barrier integrity could support the earlier detection of children at increased risk of developing atopic disease. Recently, the importance and benefit of early intervention in eczema were highlighted by Vakirlis et al. (2024). This study stated that early systemic intervention could minimize inflammation and the burden this has on lymphoid organs. Early intervention could also stop the progression of the 'atopic march' often seen in young eczema patients. The atopic march is a well-known phenomenon in patients with childhood eczema, where they progress into developing food allergies, asthma or allergic rhinitis (Hill and Spergel 2018). Similarly, early diagnosis and treatment of asthma can provide control over symptoms and prevent the worsening of the condition (Batard et al. 2025). Reducing the inflammation caused by mast cell overactivity could also help reduce airway remodelling (Tiotiu et al. 2025).

Though early detection and treatment of atopic diseases can improve long-term health, prevention is much more effective when the cause is addressed before birth. By highlighting the consequences of maternal stress during pregnancy, it should become clear that reducing maternal stress can be essential for the well-being of the mother but also for preventing developmental issues and long-term health consequences in the child. There has been some research into methods for the reduction of maternal anxiety, for instance, by providing mindfulness training (Guardino et al. 2014), which showed positive effects. However, mental health is a complex issue with varying causes in each patient, making prevention and reduction a complicated topic of research.

While this review has focussed on the consequences of maternal stress on mast cells and atopic diseases, the impact of maternal stress extends far beyond this scope. Prenatal maternal stress has been implicated in many different pathologies, including an increased risk of preterm birth (Dole et al. 2003), altering foetal neurodevelopment that may predispose to behavioural disorders (Maccari et al. 2014) and causing adverse cardiovascular effects in offspring (Zhao et al. 2024). These findings highlight the importance of preventing and reducing maternal stress for both immediate and lifelong outcomes in the mother and child. The attention this issue has received in recent years is encouraging, and future research should prioritize developing methods for effective stress reduction and prevention during pregnancy.

4 Conclusion

Despite more research being focussed on the relationship between maternal stress and foetal development, significant gaps in knowledge remain. When focussing on the mast cells, direct research on the influence of glucocorticoids on mast cell development still requires further investigation. Additionally, most of the findings within this review stem from animal models, highlighting the fact that human model studies are still required to verify the mechanisms. Beyond research into the mechanisms, epidemiological studies are essential to further investigate how maternal stress is associated with atopic disease risk at the population level.

Understanding how maternal stress can negatively impact foetal immune system development could inform strategies for early detection and treatment of atopic diseases to lessen the progression of these conditions and the suffering caused by them. On the other hand, raising awareness among healthcare providers and mothers about the risks of maternal stress could elicit more effort into preventative measures. Offering stress reduction counselling or programmes may offer a promising way to improve maternal mental health and reduce the negative impact of maternal stress on the offspring.

References

Al-Hussainy A, Mohammed R (2021) Consequences of maternal psychological stress during pregnancy for the risk of asthma in the offspring. Scand J Immunol 93(1):e12919. https://doi.org/10.1111/sji.12919

Andersson NW, Hansen MV, Larsen AD, Hougaard KS, Kolstad HA, Schlünssen V (2016) Prenatal maternal stress and atopic diseases in the child: a systematic review of observational human studies. Allergy 71(1):15–26. https://doi.org/10.1111/all.12762

Andrews C, McLean MH, Durum SK (2018) Cytokine tuning of intestinal epithelial function. Front Immunol 9:1270. https://doi.org/10.3389/fimmu.2018.01270

Aziz HA, Yahya HDB, Ang WW, Lau Y (2025) Global prevalence of depression, anxiety, and stress symptoms in different trimesters of pregnancy: a meta-analysis and meta-regression. J Psychiatr Res 181:528–546. https://doi.org/10.1016/j.jpsychires.2024.11.070

Barada O, Salomé-Desnoulez S, Madouri F, Deslée G, Coraux C, Gosset P, Pichavant M (2023) IL-20 cytokines are involved in the repair of airway epithelial barrier: implication in exposure to cigarette smoke and in COPD pathology. Cells 12(20):2464

Batard T, Taillé C, Guilleminault L, Bozek A, Floch VB, Pfaar O, Canonica WG, Akdis C, Shamji MH, Mascarell L (2025) Allergen immunotherapy for the prevention and treatment of asthma. Clin Exp Allergy 55(2):111–141. https://doi.org/10.1111/cea.14575

Berlin F, Mogren S, Ly C, Ramu S, Hvidtfeldt M, Uller L, Porsbjerg C, Andersson CK (2023) Mast cell Tryptase promotes airway remodeling by inducing anti-apoptotic and cell growth properties in human alveolar and bronchial epithelial cells. Cells 12(10):1439

Bilgic HA, Bek M, Kleuskens M, Redegeld F (2026) Mast cells in digestive diseases: new insights to keep them under control. Pharmacol Res 223:108069. https://doi.org/10.1016/j.phrs.2025.108069

Cain DW, Cidlowski JA (2017) Immune regulation by glucocorticoids. Nat Rev Immunol 17(4):233–247. https://doi.org/10.1038/nri.2017.1

Chen S, Chen S (2021) Are prenatal anxiety or depression symptoms associated with asthma or atopic diseases throughout the offspring's childhood? An updated systematic review and meta-analysis. BMC Pregnancy Childbirth 21(1):435. https://doi.org/10.1186/s12884-021-03909-z

Chia SL, Kapoor S, Carvalho C, Bajénoff M, Gentek R (2023) Mast cell ontogeny: from fetal development to life-long health and disease. Immunol Rev 315(1):31–53. https://doi.org/10.1111/imr.13191

Chourpiliadis CA, Aeddula NR (2023) Physiology, Glucocorticoids. NCBI Bookshelf

Close DA, Yun S-S, McCormick SD, Wildbill AJ, Li W (2010) 11-Deoxycortisol is a corticosteroid hormone in the lamprey. Proc Natl Acad Sci 107(31):13942–13947. https://doi.org/10.1073/pnas.0914026107

Dobson JT, Seibert J, Teh EM, Da'as S, Fraser RB, Paw BH, Lin T-J, Berman JN (2008) Carboxypeptidase A5 identifies a novel mast cell lineage in the zebrafish providing new insight into mast cell fate determination. Blood 112(7):2969–2972. https://doi.org/10.1182/blood-2008-03-145011

Dole N, Savitz DA, Hertz-Picciotto I, Siega-Riz AM, McMahon MJ, Buekens P (2003) Maternal stress and preterm birth. Am J Epidemiol 157(1):14–24. https://doi.org/10.1093/aje/kwf176

Dufour S, Lebon V, Shulman G, Petersen K (2009) Regulation of net hepatic glycogenolysis and gluconeogenesis by epinephrine in humans. Am J Phys Endocrinol Metab 297:E231–E235. https://doi.org/10.1152/ajpendo.00222.2009

Eklund KK, Humphries DE, Xia Z, Ghildyal N, Friend DS, Gross V, Stevens RL (1997) Glucocorticoids inhibit the cytokine-induced proliferation of mast cells, the high affinity IgE receptor-mediated expression of TNF-alpha, and the IL-10-induced expression of chymases. J Immunol 158(9):4373–4380. https://doi.org/10.4049/jimmunol.158.9.4373

Fong C (2023) Histology: mast cells. StatPearls

Gjerstad JK, Lightman SL, Spiga F (2018) Role of glucocorticoid negative feedback in the regulation of HPA axis pulsatility. Stress 21(5):403–416. https://doi.org/10.1080/10253890.2018.1470238

Guardino CM, Dunkel Schetter C, Bower JE, Lu MC, Smalley SL (2014) Randomised controlled pilot trial of mindfulness training for stress reduction during pregnancy. Psychol Health 29(3): 334–349. https://doi.org/10.1080/08870446.2013.852670

Haas A, Borsook D, Adler G, Freeman R (2022) Stress, hypoglycemia, and the autonomic nervous system. Auton Neurosci 240:102983. https://doi.org/10.1016/j.autneu.2022.102983

Hallgren J, Gurish MF (2007) Pathways of murine mast cell development and trafficking: tracking the roots and routes of the mast cell. Immunol Rev 217:8–18. https://doi.org/10.1111/j.1600-065X.2007.00502.x

Heijink IH, Kuchibhotla VNS, Roffel MP, Maes T, Knight DA, Sayers I, Nawijn MC (2020) Epithelial cell dysfunction, a major driver of asthma development. Allergy 75(8):1902–1917. https://doi.org/10.1111/all.14421

Hill DA, Spergel JM (2018) The atopic march: critical evidence and clinical relevance. Ann Allergy Asthma Immunol 120(2):131–137. https://doi.org/10.1016/j.anai.2017.10.037

Hough KP, Curtiss ML, Blain TJ, Liu RM, Trevor J, Deshane JS, Thannickal VJ (2020) Airway Remodeling in Asthma. Front Med (Lausanne) 7:191. https://doi.org/10.3389/fmed.2020.00191

Hu C-B, Wang J, Hong Y, Li H, Fan D-D, Lin A-F, Xiang L-X, Shao J-Z (2023) Single-cell transcriptome profiling reveals diverse immune cell populations and their responses to viral infection in the spleen of zebrafish. FASEB J 37(6):e22951. https://doi.org/10.1096/fj.202201505RRRR

Irani AA, Nilsson G, Ashman LK, Schwartz LB (1995) Dexamethasone inhibits the development of mast cells from dispersed human fetal liver cells cultured in the presence of recombinant human stem cell factor. Immunology 84(1):72–78

Jensen Peña C, Monk C, Champagne FA (2012) Epigenetic effects of prenatal stress on 11-β-hydroxysteroid dehydrogenase-2 in the placenta and fetal brain. PLoS One 7(6):e39791. https://doi.org/10.1371/journal.pone.0039791

Joëls M, Baram TZ (2009) The neuro-symphony of stress. Nat Rev Neurosci 10(6):459–466. https://doi.org/10.1038/nrn2632

Kaiser M, Jaillardon L (2023) Pathogenesis of the crosstalk between reproductive function and stress in animals—part 1: Hypothalamo–pituitary–adrenal axis, sympatho-adrenomedullary system and kisspeptin. Reprod Domest Anim 58(S2):176–183. https://doi.org/10.1111/rda.14444

Kalesnikoff J, Galli SJ (2008) New developments in mast cell biology. Nat Immunol 9(11): 1215–1223. https://doi.org/10.1038/ni.f.216

Kellogg C, Smogorzewski J (2023) Update on atopic dermatitis. Adv Pediatr Infect Dis 70(1): 157–170. https://doi.org/10.1016/j.yapd.2023.03.006

Kim BE, Leung DYM (2018) Significance of skin barrier dysfunction in atopic dermatitis. Allergy Asthma Immunol Res 10(3):207–215. https://doi.org/10.4168/aair.2018.10.3.207

Kitamura Y, Oboki K, Ito A (2007) Development of mast cells. Proc Jpn Acad Ser B Phys Biol Sci 83(6):164–174. https://doi.org/10.2183/pjab.83.164

Kogler L, Müller VI, Chang A, Eickhoff SB, Fox PT, Gur RC, Derntl B (2015) Psychosocial versus physiological stress – meta-analyses on deactivations and activations of the neural correlates of stress reactions. NeuroImage 119:235–251. https://doi.org/10.1016/j.neuroimage.2015.06.059

Krystel-Whittemore M, Dileepan KN, Wood JG (2016) Mast cell: a multi-functional master cell. Front Immunol 6. https://doi.org/10.3389/fimmu.2015.00620
Lehmann M, Haury K, Oster H, Astiz M (2023) Circadian glucocorticoids throughout development. Front Neurosci 17. https://doi.org/10.3389/fnins.2023.1165230
Lovejoy DA (2024) Corticotropin-releasing hormone peptide family. In: Reference module in biomedical sciences. Elsevier. https://doi.org/10.1016/B978-0-443-13825-6.00136-9
Maccari S, Krugers HJ, Morley-Fletcher S, Szyf M, Brunton PJ (2014) The consequences of early-life adversity: neurobiological, behavioural and epigenetic adaptations. J Neuroendocrinol 26(10):707–723. https://doi.org/10.1111/jne.12175
McNamara KM, Kannai A, Sasano H (2018) Possible roles for glucocorticoid signalling in breast cancer. Mol Cell Endocrinol 466:38–50. https://doi.org/10.1016/j.mce.2017.07.004
Méndez-Enríquez E, Hallgren J (2019) Mast cells and their progenitors in allergic asthma. Front Immunol 10. https://doi.org/10.3389/fimmu.2019.00821
Moisiadis VG, Matthews SG (2014) Glucocorticoids and fetal programming part 2: mechanisms. Nat Rev Endocrinol 10(7):403–411. https://doi.org/10.1038/nrendo.2014.74
Msallam R, Redegeld FA (2023) Mast cells-fetal mast cells crosstalk with maternal interfaces during pregnancy: Friend or foe? Pediatr Allergy Immunol 34(4):e13943. https://doi.org/10.1111/pai.13943
Mukai K, Tsai M, Saito H, Galli SJ (2018) Mast cells as sources of cytokines, chemokines, and growth factors. Immunol Rev 282(1):121–150. https://doi.org/10.1111/imr.12634
Munck A, Náray-Fejes-Tóth A (1992) The ups and downs of glucocorticoid physiology permissive and suppressive effects revisited. Mol Cell Endocrinol 90(1):C1–C4. https://doi.org/10.1016/0303-7207(92)90091-J
O'Donnell KJ, Bugge Jensen A, Freeman L, Khalife N, O'Connor TG, Glover V (2012) Maternal prenatal anxiety and downregulation of placental 11β-HSD2. Psychoneuroendocrinology 37(6):818–826. https://doi.org/10.1016/j.psyneuen.2011.09.014
Oppong E, Flink N, Cato ACB (2013) Molecular mechanisms of glucocorticoid action in mast cells. Mol Cell Endocrinol 380(1):119–126. https://doi.org/10.1016/j.mce.2013.05.014
Oster H, Challet E, Ott V, Arvat E, de Kloet ER, Dijk D-J, Lightman S, Vgontzas A, Van Cauter E (2016) The functional and clinical significance of the 24-hour rhythm of circulating glucocorticoids. Endocr Rev 38(1):3–45. https://doi.org/10.1210/er.2015-1080
Paes T, Feelders RA, Hofland LJ (2024) Epigenetic mechanisms modulated by glucocorticoids with a focus on Cushing syndrome. J Clin Endocrinol Metabol 109(6):e1424–e1433. https://doi.org/10.1210/clinem/dgae151
Pascal R, Casas I, Genero M, Nakaki A, Youssef L, Larroya M, Benitez L, Gomez Y, Martinez-Aran A, Morilla I, Oller-Guzmán TM, Martín-Asuero A, Vieta E, Crispi F, Gratacos E, Gomez-Roig MD, Crovetto F (2023) Maternal stress, anxiety, well-being, and sleep quality in pregnant women throughout gestation. J Clin Med 12(23):7333. https://doi.org/10.3390/jcm12237333
Rakers F, Rupprecht S, Dreiling M, Bergmeier C, Witte OW, Schwab M (2020) Transfer of maternal psychosocial stress to the fetus. Neurosci Biobehav Rev 117:185–197. https://doi.org/10.1016/j.neubiorev.2017.02.019
Raza ML (2025) The stress-immune system axis: exploring the interplay between stress and immunity. Prog Brain Res 291:289–317. https://doi.org/10.1016/bs.pbr.2025.01.003
Rog-Zielinska EA, Craig MA, Manning JR, Richardson RV, Gowans GJ, Dunbar DR, Gharbi K, Kenyon CJ, Holmes MC, Hardie DG, Smith GL, Chapman KE (2015) Glucocorticoids promote structural and functional maturation of foetal cardiomyocytes: a role for PGC-1α. Cell Death Differ 22(7):1106–1116. https://doi.org/10.1038/cdd.2014.181
Salvante KG, Milano K, Kliman HJ, Nepomnaschy PA (2017) Placental 11 β-hydroxysteroid dehydrogenase type 2 (11β-HSD2) expression very early during human pregnancy. J Dev Orig Health Dis 8(2):149–154. https://doi.org/10.1017/s2040174416000611
Santoro T, Azevedo CT, Silva PMRE, Martins MA, Carvalho VF (2019) Glucocorticoids decrease the numbers and activation of mast cells by inducing the transactivation receptors of AGEs. J Leukoc Biol 105(1):131–142. https://doi.org/10.1002/jlb.3a0917-364rr

Sapolsky RM, Romero LM, Munck AU (2000) How do glucocorticoids influence stress responses? Integrating permissive, suppressive, stimulatory, and preparative actions. Endocr Rev 21(1): 55–89. https://doi.org/10.1210/edrv.21.1.0389

Saur AM, Dos Santos MA (2021) Risk factors associated with stress symptoms during pregnancy and postpartum: integrative literature review. Women Health 61(7):651–667. https://doi.org/10.1080/03630242.2021.1954132

Sehra S, Serezani APM, Ocaña JA, Travers JB, Kaplan MH (2016) Mast cells regulate epidermal barrier function and the development of allergic skin inflammation. J Invest Dermatol 136(7): 1429–1437. https://doi.org/10.1016/j.jid.2016.03.019

Serhan N, Abdullah NS, Gheziel N, Loste A, Ekren R, Labit E, Gonzalez A-A, Oliva G, Tarot P, Petitfils C, Payros G, D'Avino P, Voisin A, Tinsley HFG, Gentek R, Brosseau C, Bodinier M, Reber L, Val P, Akdis CA, Mitamura Y, Andiappan AK, Chan JKY, Ginhoux F, François A, Cénac N, Basso L, Gaudenzio N (2025) Maternal stress triggers early-life eczema through fetal mast cell programming. Nature 646(8083):161–170. https://doi.org/10.1038/s41586-025-09419-8

Sheng JA, Bales NJ, Myers SA, Bautista AI, Roueinfar M, Hale TM, Handa RJ (2020) The hypothalamic-pituitary-adrenal Axis: development, programming actions of hormones, and maternal-fetal interactions. Front Behav Neurosci 14:601939. https://doi.org/10.3389/fnbeh.2020.601939

Shimba A, Ikuta K (2020) Glucocorticoids regulate circadian rhythm of innate and adaptive immunity. Front Immunol 11:2143. https://doi.org/10.3389/fimmu.2020.02143

Smith SJ, Piliponsky AM, Rosenhead F, Elchalal U, Nagler A, Levi-Schaffer F (2002) Dexamethasone inhibits maturation, cytokine production and Fc epsilon RI expression of human cord blood-derived mast cells. Clin Exp Allergy 32(6):906–913. https://doi.org/10.1046/j.1365-2745.2002.01418.x

Solano ME, Arck PC (2019) Steroids, pregnancy and fetal development. Front Immunol 10:3017. https://doi.org/10.3389/fimmu.2019.03017

Tegethoff M, Pryce C, Meinlschmidt G (2009) Effects of intrauterine exposure to synthetic glucocorticoids on fetal, newborn, and infant hypothalamic-pituitary-adrenal Axis function in humans: a systematic review. Endocr Rev 30(7):753–789. https://doi.org/10.1210/er.2008-0014

Timmermans S, Souffriau J, Libert C (2019) A general introduction to glucocorticoid biology. Front Immunol 10. https://doi.org/10.3389/fimmu.2019.01545

Tiotiu A, Steiropoulos P, Novakova S, Nedeva D, Novakova P, Chong-Neto H, Fogelbach GG, Kowal K (2025) Airway remodeling in asthma: mechanisms, diagnosis, treatment, and future directions. Arch Bronconeumol 61(1):31–40. https://doi.org/10.1016/j.arbres.2024.09.007

Vaculik C, Prior M, Fiala C, Mildner M, Eppel W, Stingl G, Elbe-Bürger A (2012) Phenotypic characterization of leukocytes in prenatal human dermis. J Invest Dermatol 132:2581–2592. https://doi.org/10.1038/jid.2012.187

Vakirlis E, Gregoriou S, Bakirtzi K, Paschou E, Tsiogka A, Papadimitriou I, Sotiriou E (2024) Insights into early systemic treatment in atopic dermatitis: scientific facts and practical considerations. Dermatol Ther (Heidelb) 14(3):563–568. https://doi.org/10.1007/s13555-024-01110-6

Wang C, Nanni L, Novakovic B, Megchelenbrink W, Kuznetsova T, Stunnenberg HG, Ceri S, Logie C (2019) Extensive epigenomic integration of the glucocorticoid response in primary human monocytes and in vitro derived macrophages. Sci Rep 9(1):2772. https://doi.org/10.1038/s41598-019-39395-9

White AP, Porterfield SP (2013) The adrenal gland. In: Endocrine and reproductive physiology, pp 147–176. https://doi.org/10.1016/B978-0-323-08704-9.00007-5

Woody CA, Ferrari AJ, Siskind DJ, Whiteford HA, Harris MG (2017) A systematic review and meta-regression of the prevalence and incidence of perinatal depression. J Affect Disord 219: 86–92. https://doi.org/10.1016/j.jad.2017.05.003

Zhao C, He L, Li L, Deng F, Zhang M, Wang C, Qiu J, Gao Q (2024) Prenatal glucocorticoids exposure and adverse cardiovascular effects in offspring. Front Endocrinol (Lausanne) 15: 1430334. https://doi.org/10.3389/fendo.2024.1430334

Chapter 8
Small RNA Inheritance in *Caenorhabditis elegans*: Phenomenology, Constraints, and Evolutionary Logic

Germano Cecere

Abstract Small RNA–mediated inheritance has emerged as a prominent mechanism by which regulatory states can be transmitted across generations without changes in DNA sequence. Studies in *Caenorhabditis elegans* show that such inheritance is tightly regulated in its duration, specificity, and reversibility. In this chapter, I examine the molecular logic underlying small RNA inheritance, emphasizing how RNA amplification, RNA modification, subcellular organization, and chromatin engagement interact to shape the lifetime of heritable silencing. Rather than viewing epigenetic inheritance as a parallel mode of heredity, the chapter highlights how RNA- and chromatin-based pathways actively constrain the persistence of regulatory states, enabling transient, conditional, or long-lasting memory depending on the context. Placing these mechanisms within an evolutionary framework, I propose that small RNA inheritance functions as a regulated buffering system that modulates the temporal gap between environmental or genomic perturbations and long-term fixation. This perspective underscores how selection can act on regulatory architectures that control the stabilization and erasure of epigenetic states, thereby shaping phenotypic variation across generations while preserving the central role of genetic context.

1 Introduction

The discovery that small RNAs can mediate heritable gene regulation across generations has profoundly influenced contemporary thinking about inheritance. In *Caenorhabditis elegans*, small RNA–induced gene silencing can persist beyond the generation directly exposed to an environmental or experimental trigger, sometimes lasting for multiple generations (Fire et al. 1998; Grishok et al. 2000; Vastenhouw et al. 2006; Alcazar et al. 2008). These observations have provided

G. Cecere (✉)
Mechanisms of Epigenetic Inheritance, Department of Developmental and Stem Cell Biology, Institut Pasteur, CNRS UMR3738, Paris, France
e-mail: germano.cecere@pasteur.fr

G. Witzany (ed.), *Integrative Theory of Evolution - Aspects and Insights*,
https://doi.org/10.1007/978-3-032-24295-2_8

some of the clearest experimental evidence that heritable information can be transmitted without changes in DNA sequence, challenging strictly mutation-centered views of heredity.

At the same time, studies of small RNA inheritance in *C. elegans* have revealed that such inheritance operates within narrow and tightly regulated boundaries. In most cases, small RNA–induced effects are transient, variable, and strongly dependent on the regulatory pathway engaged and the identity of the targeted gene. Stable, indefinite inheritance is rare and appears to be confined to specific biological contexts, particularly those related to genome defense. These constraints reflect fundamental properties inherent to the organization and regulation of small RNA pathways (for a mechanistic overview, see (Cecere 2021)).

This combination of clear evidence for epigenetic inheritance and equally clear constraints has important implications for evolutionary theory. On the one hand, it challenges the notion that heredity operates exclusively through changes in DNA sequence. On the other hand, it resists simplistic interpretations that environmentally induced traits are freely and permanently transmitted across generations.

In this chapter, I focus on the phenomenology of small RNA inheritance in *C. elegans*, emphasizing not only what is inherited but also what is not and why. Rather than constituting an alternative to genetic inheritance, small RNA pathways provide a complementary mode of biological memory, whose limits are as informative as its capabilities.

The observations presented in this chapter pose a challenge to strictly mutation-centered views of heredity but not by replacing genetic inheritance with a fundamentally different mechanism. Small RNA inheritance does not provide a simple alternative mechanism for generating heritable variation that bypasses genetic change. Rather, it reveals that heritable regulatory states can be established, maintained, and erased through mechanisms that operate alongside and in concert with genetic variation. The key insight is not that acquired characteristics can be inherited—a claim that oversimplifies both the phenomenology and the constraints—but that biological systems possess layered mechanisms for transmitting regulatory information across generations, each with distinct temporal dynamics and functional logic. These mechanisms do not replace genetic inheritance but complement it, providing organisms with the capacity to respond to environmental and genomic perturbations at timescales where genetic adaptation would be too slow or where permanent genetic change would be inappropriate.

2 Why *C. elegans* Is Uniquely Suited to Study Small RNA Inheritance

The prominence of *C. elegans* in the study of small RNA inheritance is not accidental. Three features of its biology make it uniquely permissive for uncovering multigenerational RNA-based regulatory phenomena.

First, *C. elegans* encodes RNA-dependent RNA polymerases (RdRPs), which amplify small RNA signals by using target transcripts as templates (Sijen et al. 2001; Pak and Fire 2007). This amplification allows silencing signals to persist beyond the initiating trigger and provides a mechanistic basis for inheritance across generations. In most animals, including flies and mammals, the absence of RdRPs severely limits the duration of small RNA–mediated effects, making inheritance difficult to detect or sustain.

Second, *C. elegans* has a short generation time, approximately three days under laboratory conditions. Its fast development and rapid life cycle allow inheritance dynamics to be followed directly over many generations within weeks. Crucially, it enables the distinction between transient, multigenerational effects and more stable transgenerational phenomena. In organisms with longer generation times, reversible inheritance might easily be mistaken for permanent change.

Third, *C. elegans* reproduces primarily through self-fertilizing hermaphrodites, allowing the propagation of genetically isogenic lineages over many generations. This minimizes genetic variation and segregation, making it possible to attribute heritable phenotypic changes with unusual confidence to epigenetic mechanisms. Equally important, when inheritance fails to persist, this failure cannot be dismissed as genetic dilution, rendering negative results biologically meaningful.

These features together create a system in which the dynamics of epigenetic inheritance are unusually visible. At the same time, they caution against overgeneralization, underscoring the importance of comparative approaches for understanding how broadly these principles apply across species.

3 Functional Architecture of Small RNA Pathways in *C. elegans*

Small RNA–mediated gene silencing in *C. elegans* can give rise to heritable regulatory states that vary widely in their stability across generations. Rather than representing a single mode of inheritance, these states span a continuum ranging from transient, reversible repression to long-term, self-maintaining silencing. Understanding this diversity requires distinguishing between the regulatory layers that govern how silencing is initiated, amplified, and, ultimately, maintained or erased. Within this framework, distinct small RNA pathways provide access to different regimes of heritable regulation.

Experimental and endogenous triggers reveal two major entry points into this architecture: silencing initiated by double-stranded RNAs (dsRNAs) and silencing initiated by PIWI-interacting RNAs (piRNAs). Although both pathways engage small RNA amplification and can transmit silencing across generations, they differ profoundly in the duration and stability of the inherited state.

3.1 *Initiation, Amplification, and Maintenance: Regulatory Layers of Small RNA Silencing*

Small RNA–mediated gene silencing in *C. elegans* can be conceptually decomposed into distinct regulatory layers that include initiation, amplification, and maintenance. Although these layers are mechanistically interconnected, they are functionally separable and subject to independent regulation. Distinguishing among them provides a useful framework for understanding how silencing is established and why its persistence varies across targets, pathways, and contexts.

Initiation refers to the initial engagement of a transcript by small RNA pathways. This step can be triggered experimentally by exposure to dsRNAs or endogenously through piRNAs that recognize germline transcripts. In both cases, initiation involves the loading of small RNAs into Argonaute proteins—dsRNA-derived siRNAs into primary Argonautes associated with the RNA interference (RNAi) pathway and piRNAs into PIWI proteins (Yigit et al. 2006). Importantly, Argonaute loading at this stage is not sufficient to enforce durable repression of target transcripts. Rather than acting as direct effectors, these Argonaute–small RNA complexes primarily function to mark transcripts as eligible for downstream regulatory processing.

A key role of Argonaute-bound small RNAs during initiation is to recruit additional silencing machinery rather than to execute silencing directly (Sijen et al. 2001; Yigit et al. 2006; Pak and Fire 2007). Both dsRNA-derived siRNAs and piRNAs act as molecular guides that connect target transcripts to amplification pathways, thereby routing them into processes that generate secondary small RNAs. Through this coupling, initial targeting events are converted into substrates for RdRPs and associated RNA-processing reactions. The resulting secondary small RNAs are subsequently loaded into downstream Argonaute proteins that function as the principal effectors of silencing (Yigit et al. 2006).

Amplification enables silencing signals to be sustained beyond the initiating interaction. In *C. elegans*, RdRPs use targeted transcripts as templates to produce secondary small RNAs, reinforcing and extending the silencing response (Sijen et al. 2001). This step provides the molecular basis for the spread of silencing across cell divisions and, in some cases, across generations. However, amplification alone does not dictate the long-term behavior of silencing, as amplified signals can still be lost when permissive conditions or substrates are no longer available.

Maintenance refers to the regulatory processes that determine whether a silenced state is reestablished and sustained across developmental and generational transitions. It governs whether silencing is reinforced, allowed to decay, or actively reset over time. In *C. elegans*, maintenance can involve renewed rounds of small RNA amplification after inheritance, the establishment of chromatin-based repression, or both. Access to these mechanisms varies between targets and pathways, shaping the long-term outcome of silencing.

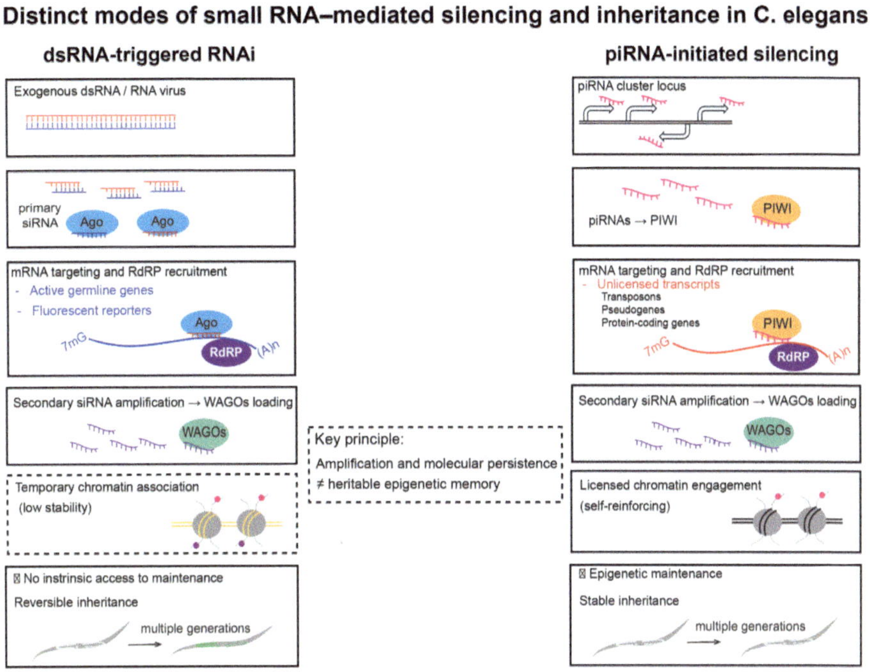

Fig. 8.1 Distinct modes of small RNA–mediated silencing and inheritance in *C. elegans*. Schematic representation of small RNA–mediated silencing pathways in *C. elegans*, highlighting key differences between dsRNA-triggered RNAi and piRNA-initiated silencing. Both pathways involve Argonaute loading, RNA-dependent RNA polymerase (RdRP)–mediated amplification, and convergence on worm-specific secondary Argonautes (WAGOs). Despite these shared features, access to long-term epigenetic maintenance is restricted and context-dependent. dsRNA-triggered RNAi typically leads to transient chromatin association and reversible inheritance, whereas piRNA-initiated silencing can, in specific regulatory contexts, engage self-reinforcing chromatin states that support stable inheritance across generations. Unlicensed transcripts refer to RNAs lacking protective germline identity and, therefore, are permissive to piRNA-mediated surveillance. The figure emphasizes that signal amplification and molecular persistence alone are insufficient to establish heritable epigenetic memory

By framing small RNA–mediated silencing as a progression through initiation, amplification, and maintenance, it becomes possible to separate transcript recognition from regulatory outcome. The following sections apply this framework to dsRNA-triggered RNA interference and piRNA-initiated silencing, illustrating how differential engagement of these layers gives rise to distinct modes of heritable regulation. The conceptual distinctions between initiation, amplification, and maintenance and the differential outcomes of dsRNA- and piRNA-triggered silencing despite shared amplification mechanisms are summarized schematically in Fig. 8.1.

3.2 dsRNA-Triggered RNAi: Transient But Heritable Repression

RNAi triggered by dsRNA represents the most experimentally accessible mode of small RNA–mediated inheritance in *C. elegans*. Exposure to dsRNA can induce robust silencing of target genes in the treated generation and in subsequent generations that have not encountered the original trigger (Grishok et al. 2000; Vastenhouw et al. 2006). This capacity to transmit repression across generations provided some of the earliest evidence that RNA-based regulatory states can be inherited independently of DNA sequence change.

A defining characteristic of dsRNA-triggered RNAi is that inheritance is multigenerational but typically reversible (Vastenhouw et al. 2006; Buckley et al. 2012). Silencing can persist for several generations after removal of the initiating dsRNA, yet it progressively weakens and, eventually, disappears. This behavior has been observed across a broad range of targets, including endogenous genes and reporter constructs, and is reproducible across laboratories. Thus, dsRNA-induced RNAi clearly supports heritable regulation, but it does not generally establish stable or permanent changes in gene expression.

The reversibility of dsRNA-induced silencing does not reflect weak engagement of the silencing machinery. Exposure to dsRNA leads to the efficient production of secondary small RNAs through RdRP activity and can access nuclear silencing pathways. In several cases, dsRNA-triggered RNAi engages nuclear Argonaute–dependent silencing pathways, leading to the deposition of repressive chromatin marks at targeted loci, such as H3K9me3, a hallmark of transcriptionally silent chromatin (Buckley et al. 2012; Gu et al. 2012a). These observations demonstrate that dsRNA-induced silencing can extend beyond cytoplasmic posttranscriptional repression and, in specific contexts, give rise to transcriptional gene silencing.

Despite this engagement, chromatin modifications induced by dsRNA are usually transient. Both repressive histone marks and transcriptional repression tend to decay within one or two generations once the initiating trigger is removed (Gu et al. 2012a). The loss of chromatin repression parallels the decline of small RNA signals, indicating that chromatin engagement alone is insufficient to maintain silencing across extended generational timescales in this context.

An additional feature of dsRNA-triggered RNAi that bears directly on its evolutionary interpretation is the pronounced variability in silencing efficiency across both genes and genetic backgrounds. Systematic RNAi experiments targeting essential embryonic genes have revealed substantial gene-to-gene differences in phenotypic severity, even under standardized experimental conditions (Paaby et al. 2015). These differences indicate that susceptibility to RNAi is not an intrinsic property of the pathway itself but depends on features of the targeted gene, such as expression level, transcript structure, or regulatory context.

This variability becomes even more apparent when RNAi responses are examined across natural *C. elegans* isolates. Studies comparing diverse wild strains have shown that the degree of RNAi-induced embryonic lethality varies widely for the same targeted genes, reflecting pervasive polygenic modifier effects (Paaby et al. 2015). Thus, the outcome of dsRNA-induced silencing is strongly shaped by standing genetic variation present in natural populations. These findings underscore that RNAi responsiveness is highly context-dependent and genetically conditioned, reinforcing the view that dsRNA-triggered silencing represents a flexible, contingent regulatory response rather than a robust, hardwired mode of heritable regulation.

From a biological perspective, the transient nature of dsRNA-induced inheritance is consistent with the proposed role of this pathway as a response to RNA-based perturbations, such as viral infection (Ashe et al. 2015). RNA viruses replicate in the cytosol, do not integrate into the host genome, and are typically cleared after successful suppression. As a result, they primarily engage cytoplasmic RNA interference pathways and lack a genomic substrate for nuclear silencing mechanisms or chromatin-based repression. In such cases, long-term nuclear maintenance of silencing would provide little selective advantage, as the viral threat does not persist at the level of the genome. Moreover, dsRNA-induced RNAi depends on RNA templates that sustain amplification; as these templates are eliminated, the silencing signal naturally decays.

Together, these observations indicate that dsRNA-triggered RNA interference provides a mode of inheritance that supports short-term regulatory memory across generations without enforcing long-term epigenetic fixation. This balance allows organisms to transmit information about recent RNA-based challenges while preserving the capacity to reset gene expression programs once those challenges have resolved.

3.3 piRNA-Initiated Silencing: Access to Long-Term Epigenetic Maintenance

piRNAs define a distinct mode of small RNA–mediated silencing in *C. elegans* that differs from dsRNA-triggered RNAi in both the stability of the inherited state and the persistence of the initiating signal. Whereas dsRNA-triggered RNAi follows exposure to a transient dsRNA substrate, piRNAs are genomically encoded and expressed in every generation, providing continuous reinitiation of silencing (Batista et al. 2008; Cecere et al. 2012). In specific contexts, this architecture supports long-term or effectively permanent inheritance across generations (Ashe et al. 2012; Luteijn et al. 2012), not because silencing is stronger but because it accesses a regulatory mode compatible with durable epigenetic repression.

A key feature of piRNA-mediated silencing is the restricted nature of its targets. In *C. elegans*, the piRNA pathway operates in the germline as a defense mechanism against invasive DNA sequences (Bagijn et al. 2012). Accordingly, piRNAs preferentially act on sequences excluded from germline expression programs, most prominently transposable elements (Lee et al. 2012; Shirayama et al. 2012).

However, transposons are not the only targets subject to stable piRNA-dependent repression. Pseudogenes and a subset of protein-coding genes that are constitutively silent in the germline are also maintained in a repressed state by piRNA-associated pathways (Gu et al. 2009; Lee et al. 2012; Shirayama et al. 2012; Barucci et al. 2020). In several cases, the expression of such genes is only revealed upon depletion of nuclear Argonaute components, indicating that their silencing reflects active epigenetic maintenance rather than a passive lack of transcription (Rojas et al. 2025).

This behavior contrasts sharply with that of actively expressed germline genes. Although piRNAs can, in principle, target a wide range of germline transcripts, stable silencing of expressed endogenous genes is rarely observed (Gu et al. 2012b; Zhang et al. 2018). In some contexts, factors such as the Argonaute CSR-1 have been shown to counteract piRNA-driven silencing (Seth et al. 2013; Wedeles et al. 2013), most clearly for spermatogenic genes (Cornes et al. 2022). These genes are expressed in the germline only during a brief developmental window, after which piRNA-mediated repression is reinstated in each generation (Cornes et al. 2022). Beyond this specific class of targets, how other actively expressed germline genes avoid piRNA silencing remains poorly understood, suggesting that additional, still unidentified features contribute to resistance (Seth et al. 2018).

Much of the experimental evidence for indefinite piRNA-mediated inheritance derives from reporter systems, most notably single-copy GFP transgenes engineered to contain a piRNA target site. In these contexts, piRNA-initiated silencing can be stably maintained across many generations, even after removal of the initiating piRNA trigger (Ashe et al. 2012; Luteijn et al. 2012). While these experiments provide compelling proof of principle that piRNA pathways can establish self-sustaining epigenetic states, they also highlight the importance of target identity and genomic context. Reporter constructs, by design, lack the regulatory features of endogenous germline genes and may, therefore, access silencing modes that are normally restricted to foreign or excluded sequences.

At the chromatin level, piRNA-mediated silencing is frequently associated with stable repressive domains that persist independently of ongoing transcription (Rojas et al. 2025). Once established, repression can be maintained even when transcription of the target is minimal or absent, suggesting that chromatin-based mechanisms can become self-reinforcing. This long-term maintenance occurs downstream of piRNA-directed initiation and RdRP-dependent amplification, distinguishing piRNA-initiated silencing from dsRNA-triggered RNAi and providing a basis for its extended persistence.

From an evolutionary perspective, the selective access of piRNA-mediated silencing to long-term epigenetic maintenance is consistent with its proposed role in genome defense against DNA-based elements. Transposable elements and related genomic parasites are permanently integrated into the genome and transmitted through the germline. In this context, transient repression would be insufficient, as reactivation at any generation could compromise genome integrity. Stable, self-maintaining silencing, therefore, represents an appropriate regulatory strategy for such targets. This contrasts with dsRNA-triggered RNA interference, which is well-suited to counter RNA-based threats, such as viral infections, that do not integrate into the genome and are typically cleared after suppression.

Together, these observations support a model in which piRNA pathways provide selective access to durable epigenetic maintenance but only for a limited class of targets. Rather than acting as a general silencing system, piRNA-initiated repression enforces germline expression boundaries by stabilizing the silencing of sequences that are permanently excluded from germline transcription. This selectivity explains both the power and the limitations of piRNA-mediated inheritance and underscores why stable, long-term epigenetic repression is the exception rather than the rule.

3.4 pUGylation and Germ Granules: Molecular and Spatial Infrastructure of Inheritance

Small RNA inheritance in *C. elegans* does not rely solely on the transmission of mature small RNAs. Instead, inheritance involves molecular intermediates and cellular structures that together sustain silencing across generational boundaries. Among these, pUG-tailed RNAs (pUG RNAs) and germ granules have emerged as central components of the inheritance process, providing both a molecular substrate and a spatial framework for the maintenance of silencing information (Wan et al. 2018; Shukla et al. 2020).

pUG RNAs are transcripts that acquire characteristic poly(UG) tails through a process of mRNA pUGylation at silenced loci (Shukla et al. 2020). These RNAs serve as preferred templates for RdRPs, with the notable exception of the RdRP EGO-1 associated with the CSR-1 pathway (Roschdi et al. 2022). Rather than representing by-products of silencing, pUG RNAs function as molecular intermediates that enable continued regeneration of secondary small RNAs after the initiating trigger has been removed. Importantly, pUG RNAs can be inherited by embryos, providing a persistent source of templates that facilitates the reestablishment of silencing in the next generation. In this way, inheritance is supported not only by the carryover of mature small RNAs but also by the transmission of an RNA state capable of sustaining amplification.

Equally important is the cellular context in which these RNA species are maintained. In early embryos, inherited small RNAs and pUG RNAs are enriched within germ granules, phase-separated condensates that concentrate RNA and RNA-binding proteins in germline cells. Experimental evidence indicates that intact germ granules are required for the robust maintenance of heritable silencing (Dodson and Kennedy 2019; Lev et al. 2019; Ouyang et al. 2019). Disruption of granule integrity compromises the persistence of small RNA–mediated repression, even when silencing is strong in the parental generation.

A likely consequence of germ granule localization is the protection of inherited RNA species from degradation. RNAs within germ granules appear to be less accessible to nucleases, allowing silencing templates to persist through developmental transitions that would otherwise erase RNA-based regulatory signals (Seydoux and Fire 1994; Quarato et al. 2021). Germ granules may, therefore, function as privileged compartments in which silencing information is preserved, amplified, and selectively transmitted to the germline of the next generation.

The involvement of pUG RNAs and germ granules underscores that small RNA inheritance is an active and organized process rather than a passive molecular carryover. Maintenance requires both RNA substrates capable of sustaining amplification through pUGylation and cellular structures that stabilize and transmit them. Importantly, this infrastructure does not override the constraints described earlier. Even when pUG RNAs and germ granules are present, long-term inheritance remains restricted to specific silencing contexts, reinforcing the idea that persistence reflects regulated access to maintenance mechanisms rather than indiscriminate propagation.

Together, these findings highlight that the heritability of small RNA–mediated silencing depends not only on pathway identity, but also on the physical organization and modification of RNA molecules within the germline. mRNA pUGylation provides a molecular substrate for continued small RNA production, while germ granules provide a cellular environment that promotes stability and transmission. Upstream decisions about target identity and transcriptional context then determine whether this substrate is engaged in a transient or long-term mode of regulation.

An important implication of this framework is that molecular persistence alone should not be conflated with the inheritance of regulatory information. The transmission of pUG RNAs and their protection within germ granules ensure continuity of silencing substrates across generations, but it does not, by itself, determine the regulatory outcome. Indeed, numerous endogenous loci in *C. elegans* produce abundant amplified small RNAs that are heritably maintained across generations without any measurable impact on target gene expression (Gu et al. 2009; Barucci et al. 2020). These cases illustrate that the presence, amplification, and even inheritance of small RNAs can be uncoupled from functional repression. Whether inherited pUG RNAs are engaged in renewed silencing depends on the transcriptional context, pathway accessibility, and downstream regulatory decisions. In this sense, pUGylation and germ granule–mediated protection provide the capacity for inheritance, not its inevitability. This distinction helps explain why robust molecular continuity can coexist with selective, conditional inheritance and why the persistence of silencing intermediates does not guarantee long-term epigenetic fixation.

4 Chromatin Memory, Relic Transposons, and the Persistence of Silencing Over Evolutionary Time

Small RNA–mediated inheritance in *C. elegans* reveals an important distinction between the initiation of silencing and its long-term maintenance. While RNA-based mechanisms such as dsRNA-triggered RNAi and piRNA-initiated repression provide entry points into heritable regulation, they do not, on their own, explain why some silenced states persist across many generations or even over evolutionary timescales. In these cases, chromatin-based mechanisms emerge as central contributors to the durability of repression and can dominate regulatory outcomes long after RNA-mediated initiation has occurred.

Across species, piRNA-mediated silencing is frequently associated with the deposition of repressive chromatin marks, most prominently H3K9me3, at targeted loci (Shirayama et al. 2012; Sienski et al. 2012). Unlike the transient chromatin changes observed following dsRNA-triggered RNAi, these repressive states can be maintained stably even when transcription of the target is minimal or absent. This persistence suggests that, once established, chromatin repression can become self-reinforcing and increasingly decoupled from the RNA triggers that originally initiated silencing (Ahringer and Gasser 2018).

A striking feature of this long-term repression is that it often extends to genomic elements that no longer pose an immediate threat. Over evolutionary time, many transposable elements accumulate mutations that impair their ability to transpose or even be transcribed. Despite this loss of activity, such relic transposons frequently remain embedded within repressive chromatin domains (Ahringer and Gasser 2018). In these cases, small RNA pathways may no longer be actively engaged, as the absence of transcription precludes ongoing RNA-based targeting. Nevertheless, the repressed chromatin state persists.

The maintenance of repression at transcriptionally inert loci raises important questions about the function of chromatin memory. One possibility is that chromatin-based silencing represents a historical record of past genomic conflicts, preserving regulatory decisions made when these elements were active. Alternatively, continued repression of relic elements may serve structural or organizational roles within the nucleus. Repressive chromatin domains contribute to higher-order genome organization, influence the formation of nuclear compartments, and shape the local regulatory environment. From this perspective, the persistence of silencing may reflect constraints imposed by nuclear architecture rather than ongoing selective pressure against element activity.

Importantly, the transition from RNA-based silencing to chromatin-dominated maintenance does not imply a simple handoff or linear pathway. Instead, it suggests a gradual shift in the relative contribution of different regulatory layers. RNA-mediated mechanisms may be essential for recognizing and initiating the repression of new or active targets, whereas chromatin-based mechanisms stabilize and consolidate these decisions over longer timescales. Once transcription ceases, RNA-based pathways naturally disengage, while chromatin states can persist autonomously.

This layered view of silencing provides a useful framework for understanding the limits of small RNA inheritance. RNA-based mechanisms excel at transmitting regulatory information across short evolutionary distances, enabling flexible and reversible responses to environmental and genomic perturbations. Chromatin-based repression, by contrast, supports long-term stability but lacks the responsiveness and specificity of RNA-guided targeting. The persistence of silencing at relic transposons and other germline-excluded loci thus reflects not the indefinite action of small RNA pathways but the enduring imprint left by past regulatory events.

Together, these observations argue against models that treat epigenetic inheritance as a separate evolutionary process independent of genetic context. Instead, small RNA pathways, pUGylation, germ granules, and chromatin modifications form an integrated system in which different layers dominate at different timescales. Heritable variation arises not from a single mechanism, but from the dynamic interplay between transient RNA-based regulation and stable chromatin memory. Understanding this interplay is essential for integrating epigenetic inheritance into evolutionary theory without abandoning the central role of selection and constraint.

5 Tuning the Lifetime of Epigenetic Memory in the Germline

Genetic perturbation studies using dsRNA-triggered RNAi demonstrate that, although small RNA–mediated silencing can be transmitted across generations and persist for extended periods, the duration of heritable RNA-based memory is normally subject to active constraints. These insights derive largely from experimental paradigms in which silencing is forcibly imposed on loci that are already transcriptionally active and licensed in the germline, including both endogenous genes and single-copy transgenes driven by strong germline promoters. In this context, the disruption of the piRNA pathway reveals that RNAi-induced silencing—typically transient and reversible—can become exceptionally long-lived, persisting for years and across hundreds of generations (Shukla et al. 2021). This dramatic stabilization indicates that the piRNA system functions not only to initiate the repression of inappropriate targets but also to limit the persistence of RNAi-driven silencing, preventing amplification and pUGylation from entering a self-sustaining regime. In contrast, perturbation of chromatin-associated silencing pathways results in more modest, quantitative extensions of inheritance, prolonging RNAi memory for additional generations without rendering it irreversible (Houri-Ze'evi et al. 2016). Together, these experiments show that distinct regulatory layers actively shape the lifetime of inherited silencing rather than simply enforcing repression.

A striking implication of these findings is the paradoxical observation that the removal of one repressive pathway can enhance the persistence of another. In the case of piRNA and RNAi pathways, competition for shared molecular components may contribute to this effect. However, such a model is insufficient to explain why weakening chromatin-associated repression—whose effects are primarily transcriptional—can also extend the inheritance of RNAi, even when silencing operates posttranscriptionally. These observations suggest that repressive pathways do not function independently or additively but, instead, form an integrated regulatory network in which distinct modes of silencing actively constrain one another's persistence. How these constraints are implemented at the molecular level—through competition for substrates, modulation of transcriptional flux, spatial organization, or feedback between regulatory layers—remains unresolved.

Beyond experimentally imposed silencing, these considerations raise a more general question: What determines whether an inherited repressive state is transient or becomes stably entrenched? The persistence or decay of small RNA–mediated repression is likely influenced by how effectively a locus is embedded within the regulatory landscape of the germline. While core transcriptional machinery is universally available, regulatory embedding—including cis-regulatory features, chromatin environment, and nuclear context—can vary substantially between loci. When transcriptional engagement is weak, sporadic, or poorly coordinated, inherited silencing may face insufficient and irregular challenge, allowing epigenetic repression to persist and, in some cases, consolidate into long-lasting regulatory states. Conversely, robust and consistent transcriptional activity can actively destabilize inherited repression and promote escape from silencing. Notably, such durable regulatory outcomes do not require fixation through DNA sequence mutation to persist indefinitely; stable epigenetic repression alone can be sufficient to maintain long-term silencing. While genetic mutations may eventually reinforce particular regulatory states, these observations indicate that epigenetic mechanisms themselves can sustain permanent regulatory outcomes and, thereby, contribute to phenotypic persistence without obligatory genetic change.

Finally, the intrinsic stochasticity of RNAi inheritance implies that these processes operate probabilistically rather than deterministically, generating lineage-specific outcomes even in genetically identical populations (Houri-Zeevi et al. 2020). In this view, small RNA inheritance functions as an actively constrained buffering system that modulates the temporal gap between regulatory perturbation and long-term fixation, permitting extended evaluation of regulatory states while preserving mechanisms that normally favor decay and resetting.

Thus, the mechanisms discussed here reveal that epigenetic inheritance is not optimized for persistence but for controllable duration.

6 Open Questions and Evolutionary Implications of Small RNA Inheritance

Despite substantial progress in elucidating the mechanisms and constraints of small RNA inheritance in *C. elegans*, fundamental questions remain regarding its evolutionary significance. Much of the current understanding of RNA-mediated heritable regulation derives from laboratory-based studies conducted in highly controlled environments, often using genetically homogeneous strains. These systems have been essential for revealing the existence of heritable small RNA–mediated silencing and dissecting its molecular logic. At the same time, they necessarily abstract away key features of natural populations, including environmental heterogeneity, genetic diversity, and complex ecological interactions, all of which shape evolutionary dynamics.

A major open question concerns how the modes of inheritance described here—short-term, reversible silencing, and selective long-term epigenetic maintenance—operate outside laboratory conditions. In wild populations, individuals experience fluctuating environments, variable stressors, and diverse genetic backgrounds. How such variability influences the establishment, stability, and erasure of RNA-based regulatory states remains largely unexplored. Interindividual genetic differences may modulate the efficiency of small RNA amplification, pUGylation, germ granule dynamics, or chromatin engagement, thereby shaping strain-specific or population-specific patterns of heritable regulation. Understanding how genetic variation interacts with epigenetic mechanisms will be critical for assessing whether small RNA inheritance contributes to phenotypic diversification, robustness, or canalization in natural populations.

Evidence from studies of natural variation already points to a strong dependence of RNA-based regulatory outcomes on genetic context. Systematic RNA interference targeting essential embryonic genes reveals substantial gene-to-gene variability in phenotypic severity, even under standardized experimental conditions (Paaby et al. 2015). When such perturbations are examined across diverse natural isolates of *C. elegans*, the degree of RNAi-induced embryonic lethality varies widely, reflecting pervasive polygenic modifier effects. These observations indicate that the impact of RNA-based silencing is not an intrinsic property of the RNAi pathway alone but emerges from its interaction with the underlying genetic landscape.

From an evolutionary perspective, these findings challenge strictly mutation-centric models of heredity by revealing that populations harbor preexisting genetic variation that modulates epigenetic regulatory processes. The outcome of environmentally induced regulatory changes—their stability, penetrance, and heritability—is, therefore, in itself, an evolvable trait shaped by genetic architecture. Natural selection acts not only on the direct effects of mutations but also on the regulatory systems that determine how organisms respond to environmental challenges and how those responses are transmitted across generations. Heritable variation, thus, arises not only from genetic mutation and selection acting on coding sequences but also from selection acting on regulatory architectures that shape how environmental information is processed and transmitted across generations.

Several unresolved mechanistic issues further complicate evolutionary extrapolation. One central question is when and how chromatin-based repression becomes independent of RNA-mediated silencing. Although long-term repression is frequently associated with stable chromatin marks, the point at which chromatin maintenance no longer requires ongoing RNA input remains unclear. It is not yet known whether this transition reflects a discrete switch or a gradual decoupling of regulatory layers. Clarifying this relationship will be essential for understanding how transient regulatory signals become consolidated into durable epigenetic states and under what conditions such consolidation remains reversible.

Comparative considerations raise additional questions. The robust amplification-driven inheritance observed in *C. elegans* relies on RdRPs, which are absent in many other animals, including flies and mammals (Castel and Martienssen 2013). In these organisms, RNA-based regulatory signals may be inherently more transient, raising the possibility that chromatin-based mechanisms assume a more immediate or dominant role following RNA-mediated initiation. Whether similar principles of layered regulation apply broadly across species or whether the balance between RNA- and chromatin-based mechanisms differs fundamentally depending on genomic and cellular context remains to be determined. Addressing these issues will require comparative studies that move beyond single-model systems.

The relationship between environmental perturbations and heritable regulatory states also remains an open area of investigation. While experimental paradigms demonstrate that RNA-based mechanisms can transmit regulatory information across generations, it is unclear how frequently such pathways are engaged by environmental inputs in nature and with what consequences. Environmental challenges may not directly induce stable heritable traits, but they could bias regulatory landscapes, modulate developmental trajectories, or alter the probability that particular chromatin states are established or maintained. Over evolutionary timescales, such biases could influence population divergence, strain-specific features, or phenotypic robustness, without implying unrestricted inheritance of acquired characteristics.

Taken together, these considerations indicate that small RNA inheritance represents a regulated system of molecular memory, whose evolutionary relevance depends on the context, genetic background, and timescale. Within defined genetic, cellular, and ecological constraints, small RNA pathways shape how regulatory states are transmitted, stabilized, or erased across generations, thereby influencing the structure of phenotypic variation on which selection acts.

The study of small RNA inheritance in *C. elegans* ultimately reveals that heredity operates through multiple, integrated mechanisms that differ in their temporal dynamics, selectivity, and evolutionary logic. Rather than supporting a simple expansion of heredity to include all environmentally induced changes, these findings demonstrate that heritable regulatory variation is itself subject to sophisticated biological control. The constraints are as important as the capabilities: transient inheritance for temporary challenges, stable maintenance for permanent threats, and regulated transitions between these states. This layered architecture reflects the evolution of regulatory systems designed to process information about environmental and genomic challenges and transmit that information across generations in ways that are adaptive in both duration and specificity. Integrating such mechanisms into evolutionary theory, therefore, requires recognizing that variation arises not only through genetic mutation but also through the operation of evolved regulatory systems that shape the structure and tempo of evolutionary change.

7 Concluding Remarks

This chapter has examined small RNA–mediated inheritance in *C. elegans* as a regulated system of molecular memory, whose defining feature is control over duration rather than unlimited persistence. RNA- and chromatin-based pathways do not simply enable the transmission of regulatory information across generations; they actively constrain when such information persists, when it is erased, and when it is prevented from becoming self-sustaining. The balance between inheritance and erasure is, therefore, not a limitation of epigenetic mechanisms but a central aspect of their biological function.

From this perspective, the evolutionary relevance of small RNA inheritance lies less in its capacity to permanently transmit acquired states than in its ability to buffer the temporal gap between perturbation and fixation. By allowing regulatory states to persist transiently, probabilistically, or conditionally, small RNA pathways enable organisms to evaluate environmental and genomic challenges without prematurely committing them to long-term genetic or chromatin-based outcomes. Selection can, thus, act not only on phenotypes or mutations but on the regulatory architectures that determine how information is processed, stabilized, or discarded across generations.

Viewed in this light, epigenetic inheritance does not weaken mutation-based evolutionary theory, nor does it require its revision. Instead, it refines our understanding of how heritable variation is structured in time by revealing regulatory systems that actively govern when information is retained, when it is erased, and when it is allowed to influence long-term evolutionary outcomes. Integrating small RNA–mediated memory into evolutionary thinking, therefore, requires recognizing that inheritance is shaped both by evolved constraints on persistence and by the capacity to transmit regulatory states. In this sense, the control of duration itself is a fundamental object of selection.

References

Ahringer J, Gasser SM (2018) Repressive chromatin in Caenorhabditis elegans: establishment, composition, and function. Genetics 208(2):491–511. https://doi.org/10.1534/genetics.117.300386

Alcazar RM, Lin R, Fire AZ (2008) Transmission dynamics of heritable silencing induced by double-stranded RNA in Caenorhabditis elegans. Genetics 180(3):1275–1288. https://doi.org/10.1534/genetics.108.089433

Ashe A, Sapetschnig A, Weick E-M, Mitchell J, Bagijn MP, Cording AC, Doebley A-L, Goldstein LD, Lehrbach NJ, Le Pen J, Pintacuda G, Sakaguchi A, Sarkies P, Ahmed S, Miska EA (2012) piRNAs can trigger a multigenerational epigenetic memory in the germline of C. elegans. Cell 150(1):88–99. https://doi.org/10.1016/j.cell.2012.06.018

Ashe A, Sarkies P, Pen JL, Tanguy M, Miska EA (2015) Antiviral RNA interference against Orsay virus is neither systemic nor transgenerational in Caenorhabditis elegans. J Virol 89(23):12035–12046. https://doi.org/10.1128/jvi.03664-14

Bagijn MP, Goldstein LD, Sapetschnig A, Weick E-M, Bouasker S, Lehrbach NJ, Simard MJ, Miska EA (2012) Function, targets, and evolution of Caenorhabditis elegans piRNAs. Science 337(6094):574–578. https://doi.org/10.1126/science.1220952

Barucci G, Cornes E, Singh M, Li B, Ugolini M, Samolygo A, Didier C, Dingli F, Loew D, Quarato P, Cecere G (2020) Small-RNA-mediated transgenerational silencing of histone genes impairs fertility in piRNA mutants. Nat Cell Biol 22(2):235–245. https://doi.org/10.1038/s41556-020-0462-7

Batista PJ, Ruby JG, Claycomb JM, Chiang R, Fahlgren N, Kasschau KD, Chaves DA, Gu W, Vasale JJ, Duan S, Conte D, Luo S, Schroth GP, Carrington JC, Bartel DP, Mello CC (2008) PRG-1 and 21U-RNAs interact to form the piRNA complex required for fertility in C. elegans. Mol Cell 31(1):67–78. https://doi.org/10.1016/j.molcel.2008.06.002

Buckley BA, Burkhart KB, Gu SG, Spracklin G, Kershner A, Fritz H, Kimble J, Fire A, Kennedy S (2012) A nuclear Argonaute promotes multigenerational epigenetic inheritance and germline immortality. Nature 489(7416):447–451. https://doi.org/10.1038/nature11352

Castel SE, Martienssen RA (2013) RNA interference in the nucleus: roles for small RNAs in transcription, epigenetics and beyond. Nat Rev Genet 14(2):100–112. https://doi.org/10.1038/nrg3355

Cecere G (2021) Small RNAs in epigenetic inheritance: from mechanisms to trait transmission. FEBS Lett 595(24):2953–2977. https://doi.org/10.1002/1873-3468.14210

Cecere G, Zheng GXY, Mansisidor AR, Klymko KE, Grishok A (2012) Promoters recognized by Forkhead proteins exist for individual 21U-RNAs. Mol Cell 47(5):734–745. https://doi.org/10.1016/j.molcel.2012.06.021

Cornes E, Bourdon L, Singh M, Mueller F, Quarato P, Wernersson E, Bienko M, Li B, Cecere G (2022) piRNAs initiate transcriptional silencing of spermatogenic genes during C. elegans germline development. Dev Cell 57(2):180–196.e7. https://doi.org/10.1016/j.devcel.2021.11.025

Dodson AE, Kennedy S (2019) Germ granules coordinate RNA-based epigenetic inheritance pathways. Dev Cell 50(6):704–715.e4. https://doi.org/10.1016/j.devcel.2019.07.025

Fire A, Xu S, Montgomery MK, Kostas SA, Driver SE, Mello CC (1998) Potent and specific genetic interference by double-stranded RNA in Caenorhabditis elegans. Nature 391(6669): 806–811. https://doi.org/10.1038/35888

Grishok A, Tabara H, Mello CC (2000) Genetic requirements for inheritance of RNAi in *C. elegans*. Science 287(5462):2494–2497. https://doi.org/10.1126/science.287.5462.2494

Gu W, Shirayama M, Conte D, Vasale J, Batista PJ, Claycomb JM, Moresco JJ, Youngman EM, Keys J, Stoltz MJ, Chen C-CG, Chaves DA, Duan S, Kasschau KD, Fahlgren N, Yates JR, Mitani S, Carrington JC, Mello CC (2009) Distinct Argonaute-mediated 22G-RNA pathways direct genome surveillance in the C. elegans germline. Mol Cell 36(2):231–244. https://doi.org/10.1016/j.molcel.2009.09.020

Gu SG, Pak J, Guang S, Maniar JM, Kennedy S, Fire A (2012a) Amplification of siRNA in Caenorhabditis elegans generates a transgenerational sequence-targeted histone H3 lysine 9 methylation footprint. Nat Genet 44(2):157–164. https://doi.org/10.1038/ng.1039

Gu W, Lee H-C, Chaves D, Youngman EM, Pazour GJ, Conte D, Mello CC (2012b) CapSeq and CIP-TAP identify pol II start sites and reveal capped small RNAs as C. elegans piRNA precursors. Cell 151(7):1488–1500. https://doi.org/10.1016/j.cell.2012.11.023

Houri-Ze'evi L, Korem Y, Sheftel H, Faigenbloom L, Toker IA, Dagan Y, Awad L, Degani L, Alon U, Rechavi O (2016) A tunable mechanism determines the duration of the transgenerational small RNA inheritance in C. elegans. Cell 165(1):88–99. https://doi.org/10.1016/j.cell.2016.02.057

Houri-Zeevi L, Kohanim YK, Antonova O, Rechavi O (2020) Three rules explain transgenerational small RNA inheritance in C. elegans. Cell 182(5):1186–1197.e12. https://doi.org/10.1016/j.cell.2020.07.022

Lee H-C, Gu W, Shirayama M, Youngman E, Conte D, Mello CC (2012) C. elegans piRNAs mediate the genome-wide surveillance of germline transcripts. Cell 150(1):78–87. https://doi.org/10.1016/j.cell.2012.06.016

Lev I, Toker IA, Mor Y, Nitzan A, Weintraub G, Antonova O, Bhonkar O, Shushan IB, Seroussi U, Claycomb JM, Anava S, Gingold H, Zaidel Bar R, Rechavi O (2019) Germ granules govern small RNA inheritance. Curr Biol 29(17):2880–2891.e4. https://doi.org/10.1016/j.cub.2019.07.054

Luteijn MJ, van Bergeijk P, Kaaij LJT, Almeida MV, Roovers EF, Berezikov E, Ketting RF (2012) Extremely stable Piwi-induced gene silencing in Caenorhabditis elegans. EMBO J 31(16): 3422–3430. https://doi.org/10.1038/emboj.2012.213

Ouyang JPT, Folkmann A, Bernard L, Lee C-Y, Seroussi U, Charlesworth AG, Claycomb JM, Seydoux G (2019) P granules protect RNA interference genes from silencing by piRNAs. Dev Cell 50(6):716–728.e6. https://doi.org/10.1016/j.devcel.2019.07.026
Paaby AB, White AG, Riccardi DD, Gunsalus KC, Piano F, Rockman MV (2015) Wild worm embryogenesis harbors ubiquitous polygenic modifier variation. eLife 4:e09178. https://doi.org/10.7554/elife.09178
Pak J, Fire A (2007) Distinct populations of primary and secondary effectors during RNAi in *C. elegans*. Science 315(5809):241–244. https://doi.org/10.1126/science.1132839
Quarato P, Singh M, Cornes E, Li B, Bourdon L, Mueller F, Didier C, Cecere G (2021) Germline inherited small RNAs facilitate the clearance of untranslated maternal mRNAs in C. elegans embryos. Nat Commun 12(1):1441. https://doi.org/10.1038/s41467-021-21691-6
Rojas AM, Chervova A, Bourdon L, Jayaraj J, Morao AK, Feudjio O, Illic S, Capurso S, Quarato P, Ovieve J, Cecere G (2025) Nuclear Argonaute HRDE-1 sustains chromatin-independent transcriptional silencing and epigenetic inheritance. bioRxiv. https://doi.org/10.1101/2025.09.24.678226
Roschdi S, Yan J, Nomura Y, Escobar CA, Petersen RJ, Bingman CA, Tonelli M, Vivek R, Montemayor EJ, Wickens M, Kennedy SG, Butcher SE (2022) An atypical RNA quadruplex marks RNAs as vectors for gene silencing. Nat Struct Mol Biol 29(11):1113–1121. https://doi.org/10.1038/s41594-022-00854-z
Seth M, Shirayama M, Gu W, Ishidate T, Conte D, Mello CC (2013) The C. elegans CSR-1 Argonaute pathway counteracts epigenetic silencing to promote germline gene expression. Dev Cell 27(6):656–663. https://doi.org/10.1016/j.devcel.2013.11.014
Seth M, Shirayama M, Tang W, Shen E-Z, Tu S, Lee H-C, Weng Z, Mello CC (2018) The coding regions of germline mRNAs confer sensitivity to Argonaute regulation in C. elegans. Cell Rep 22(9):2254–2264. https://doi.org/10.1016/j.celrep.2018.02.009
Seydoux G, Fire A (1994) Soma-germline asymmetry in the distributions of embryonic RNAs in Caenorhabditis elegans. Dev Camb Engl 120(10):2823–2834
Shirayama M, Seth M, Lee H-C, Gu W, Ishidate T, Conte D, Mello CC (2012) piRNAs initiate an epigenetic memory of nonself RNA in the C. elegans germline. Cell 150(1):65–77. https://doi.org/10.1016/j.cell.2012.06.015
Shukla A, Yan J, Pagano DJ, Dodson AE, Fei Y, Gorham J, Seidman JG, Wickens M, Kennedy S (2020) poly(UG)-tailed RNAs in genome protection and epigenetic inheritance. Nature 582(7811):283–288. https://doi.org/10.1038/s41586-020-2323-8
Shukla A, Perales R, Kennedy S (2021) piRNAs coordinate poly(UG) tailing to prevent aberrant and perpetual gene silencing. Curr Biol. https://doi.org/10.1016/j.cub.2021.07.076
Sienski G, Dönertas D, Brennecke J (2012) Transcriptional silencing of transposons by Piwi and maelstrom and its impact on chromatin state and gene expression. Cell 151(5):964–980. https://doi.org/10.1016/j.cell.2012.10.040
Sijen T, Fleenor J, Simmer F, Thijssen KL, Parrish S, Timmons L, Plasterk RHA, Fire A (2001) On the role of RNA amplification in dsRNA-triggered gene silencing. Cell 107(4):465–476. https://doi.org/10.1016/s0092-8674(01)00576-1
Vastenhouw NL, Brunschwig K, Okihara KL, Müller F, Tijsterman M, Plasterk RHA (2006) Long-term gene silencing by RNAi. Nature 442(7105):882–882. https://doi.org/10.1038/442882a
Wan G, Fields BD, Spracklin G, Shukla A, Phillips CM, Kennedy S (2018) Spatiotemporal regulation of liquid-like condensates in epigenetic inheritance. Nature 557(7707):679–683. https://doi.org/10.1038/s41586-018-0132-0
Wedeles CJ, Wu MZ, Claycomb JM (2013) Protection of germline gene expression by the C. elegans Argonaute CSR-1. Dev Cell 27(6):664–671. https://doi.org/10.1016/j.devcel.2013.11.016
Yigit E, Batista PJ, Bei Y, Pang KM, Chen C-CG, Tolia NH, Joshua-Tor L, Mitani S, Simard MJ, Mello CC (2006) Analysis of the C. elegans Argonaute family reveals that distinct Argonautes act sequentially during RNAi. Cell 127(4):747–757. https://doi.org/10.1016/j.cell.2006.09.033
Zhang D, Tu S, Stubna M, Wu W-S, Huang W-C, Weng Z, Lee H-C (2018) The piRNA targeting rules and the resistance to piRNA silencing in endogenous genes. Science 359(6375):587–592. https://doi.org/10.1126/science.aao2840

Chapter 9
Transposable Elements in Evolution: Our "Frenemy" Within

Sara Esther Ramos-Lorente, Candela Ruiz-Rodriguez, Pablo Macias-Sanchez, and Francisco Jose Sanchez-Luque

Abstract Transposable elements are a polyphyletic group of repeated sequences that constitute a significant fraction of our genome and are able to replicate and introduce new copies through insertional mutagenesis. This mutagenic potential, as well as other deleterious effects derived from their replication and repetitive nature, has led to considering them as selfish DNA or molecular parasites. However, there is mounting evidence of their positive contribution to the adaptation and evolution of their hosts. This chapter aims to describe paradigmatic examples of these *frenemies* that have contributed to very important features in current living organisms, including the coordination of embryonic gene programs, the defense against retroviruses, the capture of endosymbiotic organelles, the adaptive immune system, and even the lack of a tail in humans.

1 Introduction

Transposable elements (TEs), mobile genetic elements or transposons, are a polyphyletic group of sequences with replication potential present in the genomes of living organisms. They have accumulated copies over evolution and account for a significant part of all genomes (i.e., approximately half of the human genome) (Lander et al. 2001). Generally, TE copies can be grouped in families (Sookdeo et al. 2013; Boissinot and Sookdeo 2016) and typically only a few of the most recent copies remain replication-competent. However, the term "TE" is also inclusive of non-competent copies. Barbara McClintock first described TEs with her work on *Ds/Ac* elements in maize (McClintock 1950), which coordinate reversible inactivation

Authors Sara Esther Ramos-Llorente and Candela Ruiz-Rodriguez have equally contributed to this chapter.

S. E. Ramos Lorente · C. Ruiz-Rodriguez · P. Macias-Sanchez · F. J. Sanchez-Luque (✉)
Institute of Parasitology and Biomedicine Lopez-Neyra, Spanish National Research Council, Armilla, Granada, Spain
e-mail: sanchezluquefj@ipb.csic.es

G. Witzany (ed.), *Integrative Theory of Evolution - Aspects and Insights*,
https://doi.org/10.1007/978-3-032-24295-2_9

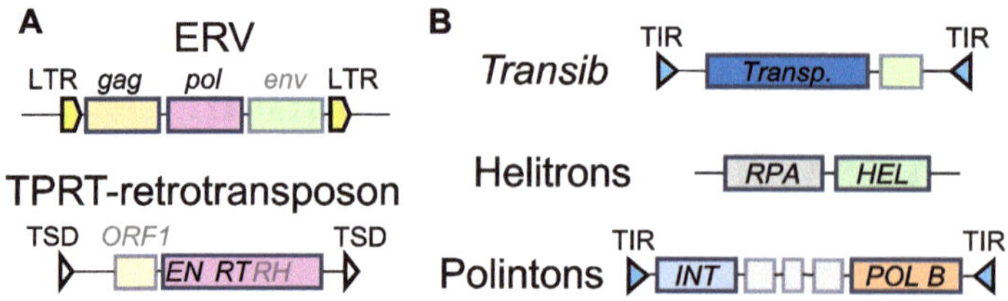

Fig. 9.1 (**a**) Structure of typical retrotransposons. (**b**) Structure of typical DNA transposons. Solid boxes represent consistent features; faded features represent non-conserved features. LTR long terminal repeat, TPRT target-primed reverse transcription, TSD target site duplication, EN endonuclease, RT reverse transcriptase, RH ribonuclease H, TIR terminal inverted repeats, Transp. transposase, RPA replication protein A, HEL helicase, INT integrase, POL B DNA polymerase B

of genes associated with color through TE insertion and excision. The ability to "move" in and out of a *locus* led to naming them as "transposons." Soon, other TEs were described to have a "replicative mobilization," some through an intermediary RNA transcribed from an original copy (donor) and reverse transcribed into a new copy at the target site. These so-called retrotransposons include the long interspersed element-1 (LINE-1 or L1), identified to be active in humans by Dr. Haig H. Kazazian due to *de novo* insertions causing spontaneous hemophilia A cases (Kazazian et al. 1988).

Although their polyphyletic origin and complexity are sources of ambiguity, TEs are typically classified according to the following criteria: replicative mobilization (cut-and-paste vs. copy-and-paste); encoding mobilization machinery (autonomous vs. non-autonomous); and mobilization intermediaries (DNA transposons vs. retrotransposons). Replicative transposons use a copy-and-paste mechanism and include all retrotransposons, as well as some DNA transposons like helitrons and likely polintons. However, other DNA transposons, *a priori* non-replicative, may still achieve limited replication through specific scenarios.

Autonomous TEs encode their replication machinery in one or several open reading frames (ORFs) and require specific features for proper expression (summarized in Fig. 9.1). Conversely, non-autonomous TEs do not encode their replication machinery and are mobilized *in trans* by the machinery of specific autonomous TEs. Their origin may involve processes like the degeneration of autonomous TEs or serendipitous events in which an unrelated sequence gained the ability to hijack the mobilization machinery. Notably, particular cases like HeT-A non-autonomous retrotransposon in *Drosophila* still retain some of the replication machinery, only requiring part of the machinery encoded by TAHRE.

Based on the nature of the molecular intermediate and other specifications of the mobilization mechanism, next, there is a description of groups that include TEs discussed in the examples of positive evolutionary contribution to their hosts in the sections below. Prokaryotic TEs and others with more complex or unclear structure (like penelope-like and tyrosine recombinase TEs) are not included.

1.1 Retrotransposons or Class I TEs

Retrotransposons or Class I TEs replicate through an RNA transcript intermediate that is reverse transcribed into the new copy. They include the following.

Endogenous Retroviruses or Long Terminal Repeats Retrotransposons

They share an evolutionary origin and a similar structure with retroviruses, including *gag* and *pol* open reading frames (ORFs) and long terminal repeats (LTRs) (Mager and Stoye 2015). They frequently have a non-functional or missing *env* ORF, which precludes infectious potential, although some like porcine endogenous retroviruses (PERVs) are positioned in the evolutionary transition between both groups and retain infectious potential. Their intermediary mRNA undergoes the typical reverse transcription cycle of retroviruses.

Target-Primed Reverse Transcription or Non-LTR Retrotransposons

They include autonomous LINEs (encoding one or two ORFs) like human L1 and *Drosophila* TAHRE, and a polyphyletic group of non-autonomous elements only sharing a common replication mechanism by target-primed reverse transcription (TPRT) (Richardson et al. 2015). Typical hallmarks of TPRT include target site duplications (TSD) flanking the insertions and, in many cases, the retrocopied polyA tail in the *de novo* insertion (Luan et al. 1993). Many are transcribed from internal promoters that are included in the mRNA, achieving reconstitution in the new copy. Non-autonomous TEs include small interspersed elements (SINEs) like human *Alu*, as well as larger, approximately 2kb-long, composite TEs like SINE-VNTR-*Alu* (SVA).

1.2 DNA Transposons or Class II TEs

The intermediary for mobilization of Class II TEs is a DNA molecule, although autonomous TEs will still generate mRNAs to synthesize their machinery. They are also a polyphyletic group that includes (Feschotte and Pritham 2007):

Classical "Cut-and-Paste" or Terminal Inverted-Repeat Transposons

They are mobilized by the excision of the donor double-strand copy and its insertion into the target site. This is generally mediated by dimeric transposases that recognize the copies' specific flanking terminal inverted repeats (TIRs). Autonomous elements may contain additional ORFs, while non-autonomous elements only require TIRs. In this group, non-autonomous miniature inverted-repeat TEs (MITES) are remarkably successful in plants (Feschotte and Wessler 2002). The *Ac*/*Ds* pair initially described by McClintock is an autonomous and a non-autonomous TE pair, which achieves limited replication by directional mobilization into unreplicated DNA during the S phase (Ros and Kunze 2001). *Drosophila* P elements also accumulate copies aided by homologous repair of the donor *locus* (Engels et al. 1990).

Helitrons

They achieve a replicative mobilization by transferring one of the DNA strands from the donor copy into the target site, coupled to the synthesis of a new second DNA strand in both *loci*. They include very successful non-autonomous elements, all relying on the typical DNA helicases and replication protein A homologs encoded by the autonomous elements (Kapitonov and Jurka 2001, 2007).

Polintons or Mavericks

They include some of the largest transposons and, like ERVs, are evolutionarily related to viruses, including Sputnik and Mavirus virophages. Their mobilization mechanism remains unclear, although it is speculated to be replicative (Krupovic et al. 2016).

Importantly, TEs have a very disruptive potential either through insertional mutagenesis (Macia et al. 2015; Hancks and Kazazian 2016), the generation of toxic nucleic acids as a by-product of their replication (Thomas et al. 2017; De Cecco et al. 2019; Fukuda et al. 2021), and underlying negative genetic conditions associated with already existing copies (Payer et al. 2017; Horváth et al. 2024). This will have a different impact, particularly concerning insertional mutagenesis, depending on the timeline of the insertion affecting the germline or somatic cells. The negative impact of TEs has been reviewed elsewhere, including their relation to spontaneous genetic disease, cancer, and innate immunity (Hancks and Kazazian 2016; Scott and Devine 2017; Gázquez-Gutiérrez et al. 2021). This has led to them being regarded as molecular parasites or selfish DNA. Nevertheless, this chapter aims to bring up the positive or perhaps necessary contribution of these TEs to their hosts, including fueling adaptive evolution and highlighting them as a *frenemy* within the genome.

TEs are dense in regulatory motifs as a consequence of the presence of functional domains necessary for their replication mechanism, which engage with cellular factors and/or their own machinery. Additionally, some TEs have a complex genomic organization that involves regions prone to forming nucleic acid structures *per se* (i.e., G-quadruplexes) (Sahakyan et al. 2017; Garza et al. 2023; Rosenkrantz et al. 2024) and internal repeated sequences like TSDs, LTRs, and TIRs, and can expand their repertoire of interactions when considering nearby copies of related TEs.

Although inherently disruptive, these sequences represent an important source of genetic diversity and innovation that is hardly achievable by the accumulation of replication-error mutations. Their ultimate success will require specific evolutionary processes, including exaptation, domestication, and pseudogene formation.

In order to illustrate different cases of TE-driven evolution, this chapter is divided into sections based on the types of contributions. A first set of cases will be referred to as insertion-causing mutations when the main selectable effect relies primarily on the disruption of the regulatory landscape and/or the coding sequence due to the location of the copy rather than the inserted sequence itself. Conversely, exaptation will be used for events in which the inserted sequence provides regulatory motifs and/or other features that drive the selectable phenotype, in many cases by repurposing the inserted sequence. Importantly, a specific case is made for insertions introducing TE-unrelated genomic sequences into pseudogenes. Domestication will include examples in which a TE, either in general or as specific copies, has evolved to accomplish a selectable function that confers an evolutionary advantage to the host, mainly through its intrinsically encoded functions. Finally, in a more complex scenario, TEs have promoted biological features that subsequently became the substrate for later evolutionary changes. Thus, although TEs are not directly involved in the evolutionary process itself, such changes would not have been possible without them.

2 Mutations Caused by Transposable Element Insertions

TE insertions can deeply alter genetic functions depending on their target location (Betancourt et al. 2024), causing phenotypes across organisms ranging from prokaryotes to higher eukaryotes (Vale et al. 2022; Benler and Koonin 2022). These effects may occasionally lead to multiple evolution-driving events (Frost et al. 2005; Catlin and Josephs 2022; Galbraith and Hayward 2023), fueling phenotypic diversity and adaptation. A typical mechanism underlaying this involves insertional mutagenesis that disrupts genes and leads to loss of function, such as the paradigmatic appearance of colorless animals, in a manner reminiscent of the original discovery of TEs by Barbara McClintock in maize (McClintock 1950).

The well-studied biological model medaka fish (*Oryzias latipes*) (Koga et al. 1995) displays a wide variety of natural color phenotypes (Moroki et al. 2023), including several independently derived albino mutants, such as albino-i[1] (Kelsh et al. 2004). These mutants present red eyes and amelanotic white skin (Fang et al. 2018) caused by a Tol-1 DNA transposon insertion into exon 1 of the *tyr* (tyrosinase) gene. The insertion introduces a premature stop codon that disrupts the expression of this key gene for melanin synthesis (Koga et al. 1995). Notably, albino-i populations have been observed to spontaneously generate individuals exhibiting mosaic black pigmentation, suggesting restoration of *tyr* gene function following somatic excision of the Tol-1 insertion. This reversion is confirmed to be absent from the germline according to the phenotypes observed in the progeny of melanin-mosaic albino-i individuals. Although the autonomous Tol-1 element responsible for secondary mobilization of the disrupting copy has not been found, this constitutes the first phenotypic evidence of a reversion of a transposon insertion in vertebrates (Tsutsumi et al. 2006). A comparable loss-of-function mechanism underlies the amelanism phenotype in the corn snake (*Pantherophis guttatus*) (Bechtel and Bechtel 1989; Ullate-Agote et al. 2014). This phenotype is characterized by orange dorsal coloration lacking black borders and by pink eyes (Saenko et al. 2015). Saenko and co-workers identified an ERV Copia 6 insertion disrupting the coding sequence of the *oca2* gene, which regulates melanosome pH and is required for proper tyrosinase activity and melanin synthesis (Wang et al. 2025) (Fig. 9.2a).

However, TE-mediated disruption is not limited to loss-of-function mutations and can also act at the regulatory level. The most emblematic example is the adaptive evolution of the British peppered moth (*Biston betularia*) in response to environmental changes driven by industrialization (Wang et al. 2025). During the Industrial Revolution, tree bark exposure favored the melanic (*carbonaria*) form over the ancestral white (*typica*) one through natural selection (Wang et al. 2025). More than a century later, the causal mutation was traced to the insertion of a class II DNA transposon (Hof et al. 2016) within the first intron of the *cortex* gene, which is associated with meiosis regulation and wing color patterning in Lepidoptera (Nadeau et al. 2016). The insertion is not detected in the *cortex* transcripts of carrier individuals, indicating that the phenotypic effect is not related to alteration of the coding sequence (Fig. 9.2b) but likely to the introduction of *cis*-regulatory elements capable of driving rapid phenotypic shifts under selective pressure. Remarkably, the insertion event has been dated to approximately 1819, which is consistent with the first documented appearance of the *carbonaria* phenotype around 1848 and illustrates how the structural phenotype of this mutation conducted a rapid selective replacement. Similar TE-related color disruptions have been described in vertebrates, including irreversible insertions of L1 retrotransposons. In Asian swamp water buffalo (*Bubalus bubalis*), an L1 insertion in sense, upstream of the first coding exon of the agouti signalling protein gene, acts as a strong promoter, causing overexpression of the target gene and subsequent downregulation of melanogenesis (Liang et al. 2021) (Fig. 9.2b).

TE-induced disruption may also occur at the epigenetic level, as illustrated by environmentally sensitive alleles generated by the ONSEN ERV in *Arabidopsis thaliana* (Hayashi and Fernie 2024). Insertion of ONSEN into the first intron of the *FLC* (flowering *locus* C) gene alters its transcriptional response to stress, accelerating flowering and facilitating rapid adaptation to herbicide exposure. Standard DNA methylation of the ONSEN copy promotes intragenic heterochromatin formation and facilitates proper intron splicing coordinated by IRBM2. The adaptive model proposes that ONSEN DNA demethylation leads to chromatin changes that prevent IRBM2 function, leading to splicing alteration and gene disruption. The mechanism behind this may be the differential interference of the DNA methylation profile of the insertion with a nearby H3K27me3 nucleation region. In sum, this creates a system in which the TE-driven gene disruption is conditioned to the stress-regulated TE-epigenetic stage (Raingeval et al. 2024) (Fig. 9.2c).

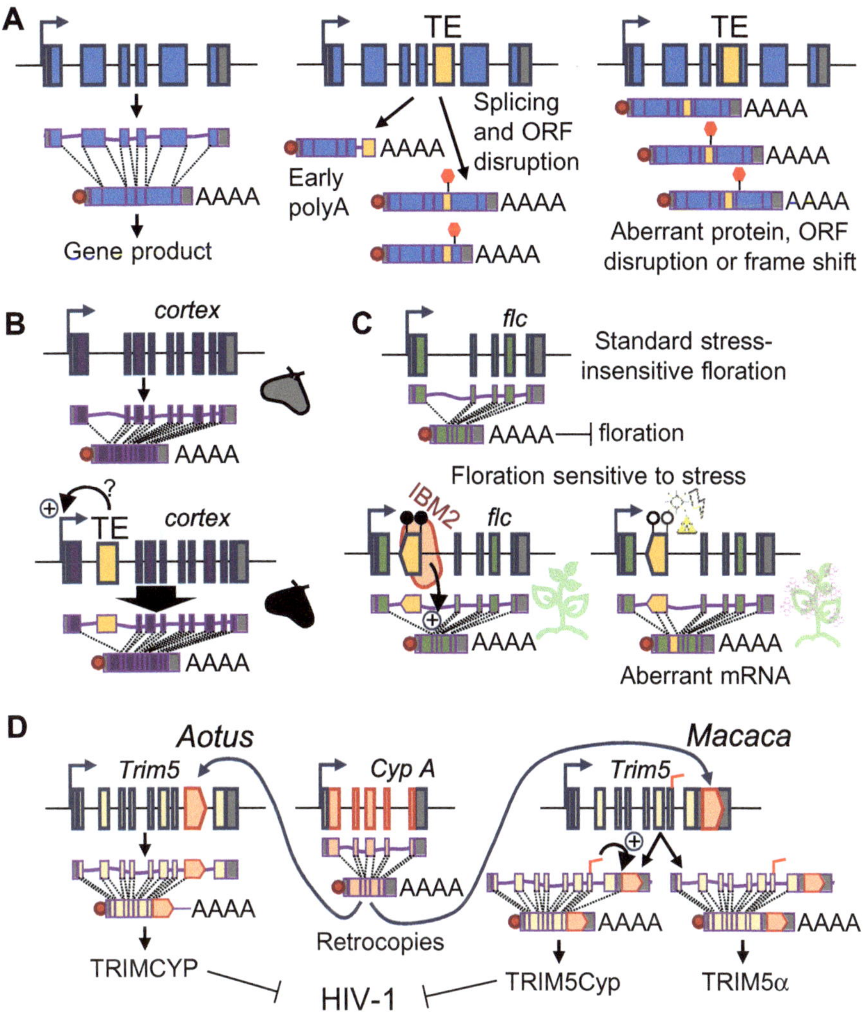

Fig. 9.2 (**a**) Different ways of TE insertions to alter gene expression. Coding exons, blue boxes; non-coding exons, gray boxes; TE insertion, orange boxes; DNA, large boxes; RNA, small boxes; cap, red circle; stop codon, red hexagon. (**b**) Representation of the TE-derived color-associated mutation in *Biston betularia*, purple boxes, coding regions. (**c**) Model for TE-derived stress-modulated floration in *Arabidopsis thaliana*. Green boxes, coding regions; solid circles, DNA methylation marks; empty circles, unmethylated DNA. (**d**) Independent mechanisms for *TRIMCyp* chimera generations in primates originated by Cyclophilin A pseudogenes in the *Trim5* gene. Pale boxes, TRIM5α coding region; red boxes, Cyclophilin A coding regions; point mutation favoring alternative *TRIMCyp* splicing, red flag

Prokaryotic TEs also contribute to the remarkable genetic plasticity in bacteria, which allows them to survive in changing environments. Insertion sequences (ISs) associated with integrons and transposons not only contribute to genetic reorganization but also facilitate the acquisition and dissemination of antibiotic resistance genes (Zhao et al. 2020), particularly in multidrug-resistant strains (Siguier et al. 2015). One illustrative example involves two IS elements flanking the *arn* operon in *Escherichia coli* type B strains, which is responsible for modifying the cellular lipopolysaccharide (LPS). This configuration enables tandem amplification of the operon via IS1 activity, conferring a 10- to 100-fold increase in resistance to polymyxin-B relative to non-amplified strains (Maybin et al. 2024). Notably, this resistance is transient, as the extra copies are lost following removal of antibiotic pressure.

3 Formation of Retrocopied Pseudogenes

Importantly, TEs also have the ability to replicate TE-unrelated genomic sequence through transductions and pseudogene formation. These pseudogenes frequently arise when retrotransposons' machinery binds and reverse-transcribes cellular mRNAs into new genomic locations, producing intron-less gene copies lacking the original promoter and displaying TPRT hallmark (Maestre et al. 1995). Pseudogenes intrinsically lack transcriptional potential and were initially considered "junk DNA" (Pink et al. 2011); however, some have played a role in adaptive evolution. A typical example is the emergence of resistance to retroviral infection in primates through TRIM5α–Cyclophilin A chimeric proteins. In primates, the *TRIM5* gene encodes a restriction factor that binds retroviral capsid during early infection, preventing uncoating and promoting its proteasome-mediated degradation (Stremlau et al. 2004; Burse et al. 2017). In Old World monkeys, this activity is specific to the Human Immunodeficiency Virus type 1 (HIV-1) capsid, and Cyclophilin A appears to have an ambivalent protective and restrictive role depending on the virus, species, and cellular scenario (Towers et al. 2003; Kim et al. 2019). In owl monkeys (*Aotus* spp.), a *CypA* pseudogene retrotransposed into the *TRIM5 locus* and generated a chimeric TRIM5Cyp fusion protein (Sayah et al. 2004) (Fig. 9.2d) that facilitates TRIM5α recognition of assembling HIV-1 particles, rendering infection inefficient (Diaz-Griffero et al. 2006; Zhang et al. 2006). The presence of classical TPRT hallmarks indicates that the *CypA* pseudogene was likely generated by L1, which is the first documented case of a functional chimeric gene by a complete mRNA retrocopy in vertebrates.

Remarkably, an independent event occurred in macaque lineages (*Macaca* spp.), generating a different *CypA* pseudogene insertion at the *TRIM5 locus*. In this case, the retrocopy, together with a single-point mutation, enabled the production of both canonical TRIM5α and TRIM5Cyp isoforms through alternative splicing in lymphocytes (Newman et al. 2008) (Fig. 9.2d). Although the antiviral efficiency and specificity of these isoforms vary among macaque species and against different retroviruses (Liao et al. 2007; Newman et al. 2008), this event, together with the one described in owl monkeys, represents a striking example of convergent evolution. In both cases, TE-mediated retrotransposition independently generated functionally analogous TRIM5Cyp restriction factors in distinct primate lineages lacking a common ancestral fusion.

A similar case of convergent evolution leading to resistance against retroviruses has been described in New World monkeys and mice through independent retrocopies of *CHMP3,* a gene encoding a component of the endosomal sorting complex required for transport (ESCRT) machinery (Scourfield and Martin-Serrano 2017; Rheinemann et al. 2021). The resulting retroCHMP3 ORF is truncated and does not interfere with essential cellular ESCRT functions. Instead, it evolved into a selective inhibitor of the ESCRT-III pathway, thereby impairing the budding of enveloped viruses such as HIV-1 (Rheinemann et al. 2020; Wang et al. 2023).

Finally, pseudogenes can also drive evolutionary innovation on their own when inserted into transcriptionally active genomic contexts. In such cases, retrocopies may acquire expression and subsequently diverge from the ancestral gene, evolving new expression patterns and target preference. A typical example is *GLUD2*, a pseudogene derived from *GLUD1* that is only present in humans and great apes. The ancestral *GLUD1* is a housekeeping gene from the Krebs cycle (glutamate conversion into α-ketoglutarate) expressed in many tissues, including those with high metabolic demand, like liver, kidneys, and muscles (Plaitakis et al. 2017). In contrast, *GLUD2*, which is located on the X chromosome, is specifically expressed in the nervous system, particularly in the brain and retina. The encoded enzyme exhibits biochemical adaptations to neuronal environments, including tolerance to high guanosine triphosphate (GTP) concentrations, rapid activation by adenosine diphosphate (ADP) and L-leucine, and increased activity at lower pH (Andersen et al. 2021, 2022), which confers a high evolutionary advantage. Notably, the adaptive evolution of *GLUD2* coincided temporally with a period of rapid expansion in brain size and complexity in hominins, suggesting a potential contribution to enhanced cognitive capacity and memory in humans and great apes (Burki and Kaessmann 2004).

A similar process has been proposed for YY2 and REX1 pseudogenes, which originated from retrotransposition events involving the gene for the ancestral Yin Yang 1 (YY1) zinc finger transcription factor. These paralogs retained core DNA-binding and regulatory functions of YY1 while diverging in target specificity and tissue expression, with YY2 predominantly active in the germ line and REX1 enriched in embryonic stem cells. This diversification contributed to regulatory innovations associated with pluripotency, early embryonic development, and genomic imprinting (Sui et al. 2004; Luo et al. 2006; Kim et al. 2007; Faulk 2010; Brookes et al. 2022; Ruden et al. 2024; Saha et al. 2024; Li et al. 2025).

4 Exaptation of Transposable Element Insertions

This section addresses how TE insertions have served as a primary resource for evolution through the repurposing of their sequences into new regulatory elements, either directly or together with subsequent fine-tuning and specialization. This has resulted in the selection of specific copies of TEs in discrete locations, which have turned into *cis*-regulatory elements of gene expression, including non-coding regulatory RNAs, three-dimensional genome boundary elements, and protein-coding sequences. This functional diversity is consistent with the intrinsic ability of TEs to act as promoters, enhancers, terminators, silencers, and insulators by providing binding sites for transcription factors. ERVs have been more frequently exapted for gene regulation than other TEs in mammals (Thompson et al. 2016). Their LTRs contain RNA polymerase II promoters, transcription factor binding sites, and, in some cases, splice donors, making them particularly suitable for co-option as regulatory elements. Furthermore, the 5′ and 3′ LTRs are prone to homologous recombination, deleting the internal coding region and leaving behind a functional "solo-LTR" (Belshaw et al. 2005) that can remain as a functional promoter with no ERV coding or replication capacity (Fig. 9.3a). Genome-wide chromatin immunoprecipitation sequencing studies have shown that about 20% of transcription factor binding sites in mammalian genomes are embedded within LTR-derived sequences (Sundaram et al. 2014), many of which have been exapted as tissue- and stage-specific promoters, particularly during early embryogenesis, in germ cells, placenta, and pluripotent stem cells (Thompson et al. 2016).

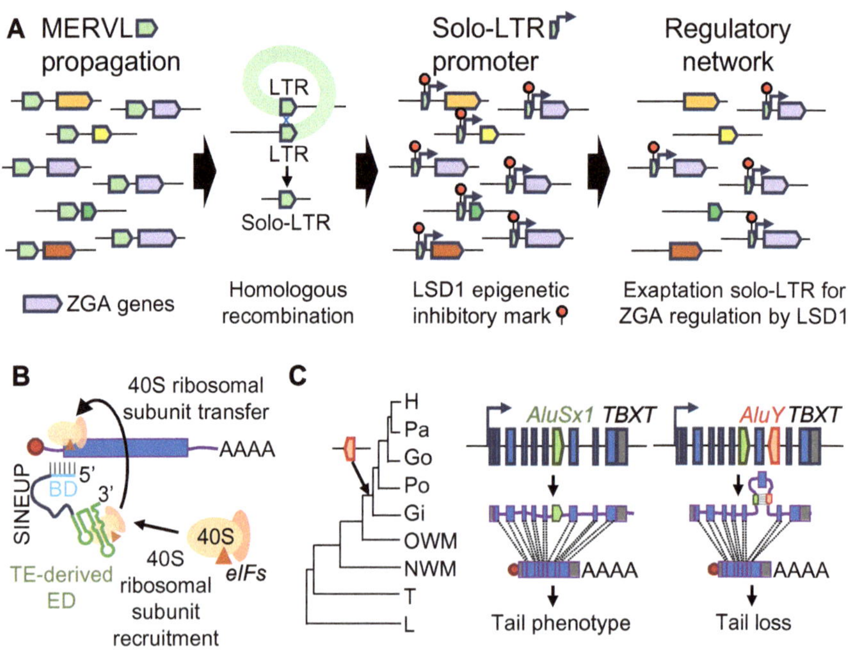

Fig. 9.3 (**a**) Model for a regulatory network based on MERVL solo-LTR exaptation to repress zygotic genome activation (ZGA) genes in early embryonic stages by LSD1. (**b**) SINEUP mechanism of translation enhancement. BD, binding domain; ED, effector domain; eIF, eukaryotic initiation factors. (**c**) Exapted *AluY* insertion mediates exon 6 skipping in the *TBXT* gene in coordination with a preexisting *AluSx1* insertion. The cladogram indicates the tempo of the *AluY* insertion. H human, Pa Pan, Go Gorilla, Gi Gibbon, OWM Old World monkeys, NWM New World monkeys, T Tarsiformes, L Lemurformes

A well-characterized example is provided by LTRs derived from the murine ERV-L (MERVL) (Macfarlan et al. 2011), which function as alternative promoters for a subset of cellular genes whose expression is restricted to zygotic genome activation (ZGA). During early stages of embryonic development, MERVL retrotransposons are silenced by the histone demethylase LSD1/KDM1A, which maintains a repressed chromatin state. KDM1A depletion in embryonic stem cells turns MERVL-derived LTRs' chromatin marks into active, causing their up-regulation and the subsequent activation of ZGA-associated genes in the two-cell stage. This derepression was associated with an altered developmental potential and increased propensity to generate extraembryonic lineages, indicating that MERVL-derived sequences have been exapted for the epigenetic silencing of genes that determine cell fate during early mammalian embryonic development (Macfarlan et al. 2011) (Fig. 9.3a). At a higher regulatory level, mouse L1 elements act in coordination with MERVL solo-LTRs to regulate the entry into the two-cell stage. This process involves L1 RNA acting as a molecular scaffold, binding nucleolin and KAP1, independently repressing MERVLs and promoting ribosome biogenesis (Percharde et al. 2018).

Since TE promoters typically display tropism for the germline and early embryonic stages to ensure their propagation, they appear predisposed for exaptation as *cis*-regulatory elements of genes active during these developmental stages. This model involves waves of TE insertions arising in populations, most of which are subsequently eliminated by negative selection, while productive insertions are retained and ultimately incorporated into host gene regulatory networks. Evidence for this process is found in passeriformes, where large-scale comparative genomics across >300 species revealed recurrent, lineage-specific expansions of ERV-K-derived solo-LTRs coinciding with major episodes of adaptive radiation and speciation (Chen et al. 2024). As an ongoing process, many of these solo-LTRs remain transcriptionally active (particularly in reproductive tissues) and are not yet fixed in populations. Nevertheless, genes harboring Passeriformes-specific ERV-K solo-LTRs exhibit lineage-specific expression patterns compared to their chicken orthologs, consistent with *cis*-regulatory activity (i.e., a solo-LTR influencing expression of *integrin alpha 2* gene, involved in cell adhesion and neural processes) (Chen et al. 2024). Similarly, primate L1s harbor an antisense promoter (ASP) within their 5′ untranslated region (UTR) (Swergold 1990; Speek 2001), which serves as an alternative promoter initiating transcription of approximately 4% of human genes, often in a tissue-dependent manner, including the placenta, brain, and prostate (Criscione et al. 2016). These transcripts incorporate sequences derived from the 5′ UTR of L1 and, occasionally, the coding region, while preserving the native downstream open reading frame. In some cases, ASP becomes the only remaining functional promoter of these genes (Mätlik et al. 2006; Denli et al. 2015). Brain-specific L1 expression has further favored ASP exaptation, contributing to the diversification of the human neuronal transcriptome and potentially increasing brain functional complexity (Jönsson et al. 2019; Garza et al. 2023).

A step further in functional exaptation is exemplified by the murine *Prl3c1 locus*, where TE insertions provide alternative promoters that generate transcripts containing novel coding regions absent from the ancestral gene. *Prl3c1*, a member of the prolactin (PRL) gene family, produces tissue-specific transcripts Prl3c1-v1 and Prl3c1-v2 (Bu et al. 2017). The ancestral Prl3c1-v1 isoform is expressed in the uterine decidua and encodes a secreted cytokine regulating uterine cell growth during pregnancy in rodents (Alam et al. 2008). In *Mus musculus*, Prl3c1-v2 is transcribed from an exapted composite insertion in *Prl3c1* intron 1 that functions as an alternative promoter and consists of two fragments of an ERV (RMER17D_MM) interrupted by an L1 (L1 Mus_3) element. This isoform lacks exon 1, includes an alternative non-coding exon, and is repressed by DNA methylation in non-testicular tissues, while remaining active in Leydig cells. The resulting PLP-J V2 protein has nuclear localization and exerts a cytostatic effect by inhibiting proliferation and promoting cellular maturation, thereby establishing a novel regulatory constraint on testicular growth and androgen production within the otherwise conserved hypothalamic–pituitary–gonadal axis (Bu et al. 2017).

TE exonization has also driven lineage-specific adaptations in the *GPHA locus*, supporting the evolution of hemochorial placentation in anthropoid primates (Roberts et al. 2016). The ancestral *GPHA* gene encodes the common α-subunit of pituitary glycoprotein hormones in vertebrates, is detected in both pituitary and placenta, and heterodimerizes with different subunits to form chorionic gonadotropin (CG), luteinizing hormone, follicle-stimulating hormone, and thyroid-stimulating hormone (Fiddes and Goodman 1981). This multifunctionality strongly constrains *GPHA* diversification, but in the lineage of anthropoid primates, an intron 1 *Alu-J* insertion approximately 35 million years ago introduced cryptic splice sites that exonized part of the *Alu* sequence, generating a 31-amino acid N-terminal extension (Chen et al. 2017). The resulting *Alu*-GPHA isoform is restricted to placental tissues during early pregnancy and produces a chorionic gonadotropin with extended half-life and increased stimulation of trophoblast invasion, which constitutes a selective advantage for successful implantation and vascular remodeling that occurs during hemochorial placentation (Chen et al. 2017).

A hominoid-specific *AluY* insertion into intron 6 of the TBXT gene (Rogers and Gibbs 2014) led to the opposite effect, causing an event of exon skipping responsible for the tail loss in the human lineage. This insertion, together with a preexisting *AluSx1* element in intron 5, forms an inverted-repeat configuration capable of pairing at the RNA level. This arrangement promotes the formation of a stable stem-loop structure in the TBXT pre-mRNA that isolates exon 6, which results in skipping during splicing (Fig. 9.3c). The resulting isoform is associated with tail shortening or complete tail loss in mouse models, supporting *Alu*-mediated exaptation as the mechanism underlying tail loss in hominoids (Xia et al. 2024).

Repurposed TE insertions also frequently function as enhancers (Huda et al. 2011; Wanichnopparat et al. 2013; Xie et al. 2013; Barakat et al. 2018; Roller et al. 2021). One example is a unitary L1M2a insertion within intron 1 of the human *IFNAR1* gene, which acts as an interferon-inducible enhancer in immune cells (Buttler et al. 2023). Indeed, many other enhancers coordinating innate immunity are derived from TEs, including ERV LTRs (Chuong et al. 2016). However, a remarkable example of its complexity is the enhancer controlling *Wnt5a* expression in the developing secondary palate, which arose through the successive integration of AmnSINE1, X6b_DNA, and MER117 at different evolutionary stages, all of them contributing to its function (Nishihara et al. 2016).

One of the most paradigmatic outcomes of TE exaptation is their repurposing into non-coding regulatory RNAs, including small RNAs and long non-coding RNAs (lncRNAs), expanding their impact on the regulation of gene expression at the posttranscriptional level (Chuong et al. 2016). In plants, the ITRs of non-autonomous MITEs favor RNA stem-loop formation that has been repeatedly co-opted as sources of novel microRNAs (miRNAs) (Pegler et al. 2025). A substantial fraction of *de novo* miRNAs originated from MITE insertions belonging to the Mutator, Tc1/Mariner, and PIF/Harbinger superfamilies (Guo et al. 2022). These are typically transcribed as intronic passengers of host genes and subsequently processed into miRNAs targeting genes involved in stress- and environment-responsive pathways, thereby contributing to a wider phenotypic plasticity in the response to changing environmental conditions (Guo et al. 2022).

In mammals, a notable class of TE-derived lncRNAs is SINE-containing translation UP-regulators (SINEUPs), which enhance translation of specific target mRNAs without affecting transcript abundance. Their structure includes a binding domain antisense to the target mRNA (usually overlapping the translation initiation region) and a TE-derived effector domain, most commonly originating from SINEs (Takahashi et al. 2018). The murine AS Uchl1, derived from SINEB2, was the first described SINEUP and enhances Uchl1 translation under cellular stress (Carrieri et al. 2012). Among the human SINEUPs, R12A-AS1 and ITFG2-AS1 enhance translation of PPP1R12A and ITFG2 mRNAs via free right *Alu* monomer (FRAM)–derived and MIRb-derived effector domains, respectively (Schein et al. 2016). The proposed mechanism involves *trans*-internal ribosome entry site (IRES) activity of the effector domain, promoting cap-independent translation by recruiting components of the translation machinery to the paired mRNA (Sharma et al. 2024; D'Agostino et al. 2025).

Finally, human L1 RNA has been shown to function in a non-coding fashion as a structural component of chromatin, potentially through bulk interactions with genomic L1 sequences (Hall et al. 2014). More broadly, TEs are increasingly recognized as integral components of the three-dimensional genome through interactions with transcription factors and chromatin organizers (Lawson et al. 2023). Many TEs (particularly SINEs, but also LINEs and ERVs) harbor binding sites for architectural proteins such as CTCF, YY1, and TFIIIC, enabling them to function as insulators or boundary elements (Schmidt et al. 2012; Wang et al. 2015; Weintraub et al. 2017; Ferrari et al. 2020). Genome-wide chromatin interaction analyses reveal enrichment of TE-derived sequences at topologically associating domain (TAD) and sub-TAD boundaries, where they constrain enhancer-promoter communication and partition the genome into regulatory domains (Dixon et al. 2012; Zhang et al. 2019). Exapted TE-derived architectural elements can preserve conserved chromatin organization while also generating lineage-specific chromatin loops that alter gene regulation (Choudhary et al. 2020, 2023; Diehl et al. 2020). In addition, TEs contribute to higher-order chromatin compartmentalization by organizing heterochromatin macrostructure within B compartments, providing nucleation sites for repressive chromatin marks and promoting large-scale chromatin condensation (Lu et al. 2021).

5 Transposable Elements Domestication

As mentioned in the introduction, domestication represents a specific evolutionary trajectory of TEs in which their intrinsically encoded components are harnessed to perform a positively selectable function for the host while preserving their original biological activity, and either involving multiple copies or a single specialized insertion. One of the earliest identified examples is the co-option of TEs as telomeres in dipteran insects, which resolves the "end replication problem" in the absence of telomerase. In most eukaryotes, telomeric DNA consists of tandem repeats that allow Okazaki fragments generated during lagging-strand synthesis to extend sufficiently into the repeat array, ensuring complete replication of internal chromosomal sequences. These tandem repeats are generated by telomerase, which uses an RNA template to elongate the 3′ end of the lagging strand. Following replication, formation of T-loops and recruitment of telomere-associated proteins stabilize chromosome ends and prevent their recognition as DNA breaks (De Lange 2004). Conversely, *Drosophila melanogaster* has solved this problem through the domestication of three TPRT retrotransposons (HeT-A, TAHRE, and TART) whose new insertions are specifically targeted to chromosome ends and form tandem arrays analogous to canonical telomeres (Zhang and Rong 2012). The autonomous TAHRE element encodes the conventional non-overlapping ORF1 and ORF2, whereas the more abundant HeT-A is non-autonomous but retains ORF1. TART is less frequent and exhibits additional structural features. Notably, rather than using a 5′ UTR internal promoter, these elements harbor an external promoter at their 3′ end. Likely as a consequence of their tandem insertion mode, this arrangement enables the transcription of each copy from the promoter of the preceding one.

The basis of this domestication lies in the highly specific targeting of new insertions to telomeres. Unlike many non-autonomous TPRT retrotransposons, HeT-A conserves ORF1, which encodes a nucleic acid chaperone that binds transposon RNA in *cis*. Large ORF1 protein (ORF1p) assemblies have been detected at telomeres during S phase and require the participation of Verrocchio (a homolog of the conserved telomeric protein Stn1), suggesting a recruitment mechanism for ribonucleoprotein (RNP) complexes that depends on ORF1p (Zhang et al. 2014). Importantly, *Drosophila* must still manage conflicts arising from these potentially deleterious elements, as uncontrolled activity could lead to aberrant telomere elongation. HP1/ORC-associated protein (HOAP) is a telomeric protein that protects chromosome ends from their recognition by the DNA repair machinery. HOAP also participates in silencing telomeric retrotransposons, and comparative studies have shown that it has evolved rapidly in *Drosophila* species, likely adapting to lineage-specific TE repertoires (Zhang et al. 2014). HOAP mediates silencing through interactions with additional proteins, including HP1 and HipHop. Recent evidence further suggests that HipHop plays a dual role, contributing both to RNP recruitment at telomeres and to repression of resident TE copies (Cui et al. 2021). This complex interplay raises the question of whether telomeric elements in Drosophila should be considered fully domesticated or rather tamed ones.

One of the most remarkable examples of unequivocal TE domestication is represented by the *RAG1* and *RAG2* genes, which mediate V(D)J recombination at immune *loci*. This process is essential for generating the diversity of T-cell receptors (TCRs) and B-cell receptors (BCRs) that underlie adaptive immunity in vertebrates. It involves multiple lymphocyte-specific genomic rearrangements catalyzed by a heterodimeric RAG1–RAG2 recombinase, which recognizes highly specific recombination signal sequences (RSSs). RSSs are distributed across V, D, and J gene segments and are organized into two distinct classes that are differentially bound by RAG1 and RAG2. The binding of one RSS from each class brings gene segments into proximity, enabling coordinated DNA cleavage and ligation (Tonegawa 1983). Early on, RSSs were proposed to derive from TIRs of DNA transposons that once mobilized within these *loci*. Structural and evolutionary analyses subsequently revealed that RAG1 and RAG2 evolved from two transposases of the Transib DNA transposon superfamily (Kapitonov and Jurka 2005; Kapitonov and Koonin 2015). Indeed, the discovery of ancestral Transib elements in invertebrates supports a model in which *RAG1* and *RAG2* initially merged within a single transposon that was later domesticated. It is speculated that a former Transib transposon inserted and split an exon of the ancestral TCR gene, originating the proto-V and proto-J segments that would have later expanded together with attached TIRs in tandem and diversified (Fig. 9.4a, b). Next, the coding region of a TransibVDJ copy was domesticated into the modern *RAG1* and *RAG2* genes to recognize the TIRs turned into RSSs (Kapitonov and Koonin 2015; Yakovenko et al. 2021) (Fig. 9.4a, b).

ERVs have also been a source of domesticated elements, profiting from their viral particle formation and fusogenic functions. Arc, a neuronal protein derived from a Gag-type Ty3/gypsy ERV that lost its capacity for autonomous replication, is an interesting example that has been domesticated for intercellular information transfer in the nervous system. Arc proteins preserve hallmark Gag properties, including self-assembly, RNA encapsidation, and release in extracellular vesicles. In mammalian neurons, Arc self-assembles into virus-like capsids, encapsulates RNA, and mediates intercellular transfer of its mRNA and others (Pastuzyn et al. 2018). Thus, in addition to propagating action potentials, neurons became equipped with a system to transmit information in the form of mRNAs. Other Gag domesticated functions have resulted in the emergence of key reproductive innovations associated with mammalian viviparity. Such is the case of *Peg10* and *Peg11*, two paternally expressed imprinted genes that originated from *sushi-ichi*-related retrotransposons and were domesticated in a common ancestor of Theria and Eutheria. Although they share a common origin and retain a functional aspartyl protease domain, Peg10 also retains a functional nucleic acid-binding domain. Peg10 produces two subunits (a short Gag-like protein and a longer Gag-Pol-like protein) that bind RNA and form virus-like particles essential for placental development and protecting the embryo from physical and metabolic stress. Peg11, by contrast, plays a critical role in maintaining placental tissue integrity (Ono et al. 2006; Sekita et al. 2008).

However, perhaps the more paradigmatic examples of ERV domestication involved in placental innovation and defense are related to the fusogenic *env* genes of ERVs like syncytins and supressyns. In humans, Syncytin-1 and Syncytin-2 retain

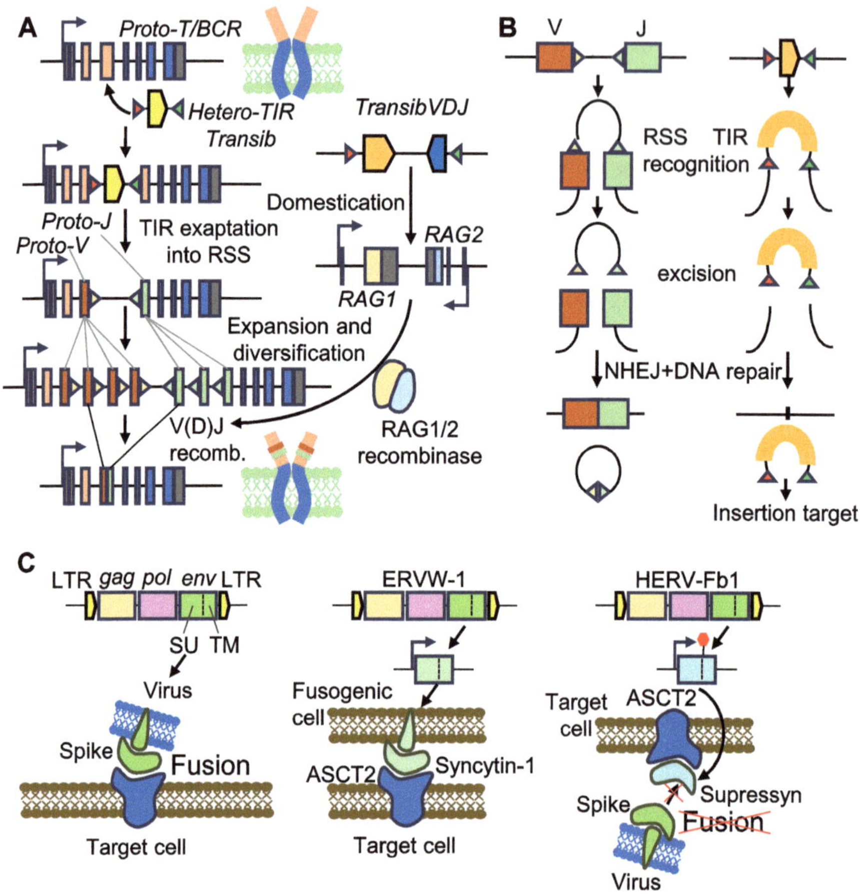

Fig. 9.4 (**a**) Model for V(D)J recombination through exaptation of TIRs and domestication of Transib elements. Gray boxes, non-coding exons; colored boxes, coding exons according to the receptor domains; triangles, TIR and RSS. (**b**) Parallel reaction catalyzed by RAG1/2 recombinase and Transib transposase. Triangles, RSS, and TIR; large boxes, coding regions. (**c**) Domestication of both Syncytin-1 and SUPYN from ERVs and the mechanism of action

the fusogenic properties of retroviral Env proteins and mediate trophoblast cell-cell fusion during placental development, enabling the formation of the syncytiotrophoblast, a multinucleated layer essential for maternal-fetal exchange of nutrients and gases (Chen et al. 2008). Syncytin-1, encoded by the *ERVW-1* gene on chromosome 7q21.2, was acquired by the primate lineage ~25 million years ago (Gimenez et al. 2009). It is a transmembrane protein processed into a surface subunit (SU) and a transmembrane subunit (TM). The SU mediates binding to the cellular receptor ASCT2 (alanine–serine–cysteine transporter 2; SLC1A5), while the TM contains the fusion peptide, immunosuppressive domain, and transmembrane region required for membrane fusion (Tolosa et al. 2012; Toudic et al. 2017) (Fig. 9.4c). The interaction of the SU subunit with ASCT2 induces a conformational rearrangement of the TM subunit that exposes the fusion peptide and triggers membrane fusion (Gong et al. 2005). On the other hand, Syncytin-2 is encoded by an *env* gene of the HERV-FRD (ERVFRD) family and has been conserved in primates for >40 million years (Blaise et al. 2003). While sharing a similar fusogenic mechanism with Syncytin-1, Syncytin-2 has been implicated in establishing maternal immune tolerance, likely reflecting ancestral retroviral strategies to modulate host immune responses (Mangeney et al. 2007). Notably, syncytins have arisen independently in multiple eutherian mammals, marsupials, and even viviparous lizards, indicating repeated and convergent domestication of ERV *env* genes (Lavialle et al. 2013; Cornelis et al. 2015, 2017).

Conversely, suppressyn (SUPYN) is another placental protein that antagonizes the fusogenic activity of Syncytin-1 in trophoblast cells (Priščáková et al. 2023), as a secondary, independently co-opted ERV domesticated to inhibit Syncytin-1 fusogenic activity. The SUPYN coding sequence originated from the *env* gene extracellular subunit of an HERV-Fb1 insertion in chromosome 21q22.3 (Sugimoto et al. 2013), which can be tracked to the common ancestor of catarrhine primates >40 million years ago (Frank et al. 2022). Its high conservation across simian species strongly suggests a functional constraint and adaptive significance following its domestication from a retroviral ancestor (Sugimoto et al. 2019). A premature stop codon truncates SUPYN, eliminating its transmembrane and immunosuppressive domains, resulting in its secretion into vesicles and the extracellular space. Nevertheless, SUPYN retains the ability to bind ASCT2, the receptor for Syncytin-1 and multiple exogenous retroviruses (Sugimoto et al. 2019). SUPYN expression is tightly regulated within the placenta, preventing cell fusion in cytotrophoblasts and extravillous trophoblasts while permitting syncytialization in syncytiotrophoblasts (Sugimoto et al. 2013, 2019; Frank et al. 2022). Notably, SUPYN expression is modulated by oxygen tension (Sugimoto et al. 2019), potentially adjusting trophoblast fusion in response to local oxygen availability (Jaremek et al. 2023). In addition to its developmental role, SUPYN inhibits infection by exogenous retroviruses that use ASCT2 (Fig. 9.4c). It is interesting that the domestication pathway involves exogenous retroviruses first assimilated into ERV, which later lead to domesticated antiviral genes in a model of "fighting fire with fire" (Malfavon-Borja and Feschotte 2015).

6 Contribution of TE to the Evolution of Superior Complex Systems

Up to this section, the contribution of TEs to evolution has been described in terms of direct insertions that disrupt regulatory landscapes and/or through their repurposing or recycling into new functions, with the insertion or derived sequence remaining at the site of action. However, TEs have also contributed to or induced the emergence of complex biological systems that subsequently evolved independently of the original TE sequences, leaving no detectable trace of the copies or TEs that were nonetheless essential for the initial evolutionary transition. One of the most consequential examples of such processes is the acquisition of endosymbiotic mitochondria and chloroplasts by early eukaryotes. Horizontal gene transfer from bacteria to eukaryotes is considered fundamental for the establishment and long-term stabilization of these organelles (Husnik and McCutcheon 2018; Kelly 2021), involving the relocation of thousands of prokaryotic genes to the nucleus, where they became functionally "eukaryotized." Reconstruction of the primary mechanisms underlying these transfers is challenging, given that they occurred approximately 1 billion years ago (Soucy et al. 2015). Strikingly, the work by Calatrava and co-workers (Calatrava et al. 2022) has provided insight into the potential role of TEs in this major evolutionary transition by analyzing the genome of *Paulinella micropora*. This photosynthetic amoeba (Zhang et al. 2017) harbors a chromatophore derived from a recently acquired cyanobacterial endosymbiont, estimated to have been incorporated approximately 1.2 million years ago (Lhee et al. 2019, 2021), enabling direct observation of "early stages" of metabolic integration and gene transfer required to sustain photosynthesis, including the involvement of TEs (Fig. 9.5).

Several prokaryotic genes encoding high-light-inducible (HLI) proteins and photosystem I subunits (PSAE and PSAI) were identified in the nuclear genome, where they function in mitigating light-induced stress. Among these, three PSAE/PSAI nuclear copies and multiple HLI genes exhibit clear hallmarks of retrotransposition, while some of them have degenerated. Notably, although these eukaryotized genes have multiple introns and exons, most of the original prokaryotic ORFs remain within a single exon in the transcript, retaining recognizable TPRT hallmarks. Importantly, unlike non-retrocopies, *bona fide* retrocopies respond to high-light conditions, suggesting that retrotransposition facilitated their integration in host regulatory *loci* responsive to light stress. The authors propose that while the initial migration of prokaryotic DNA into the nucleus remains unresolved and may

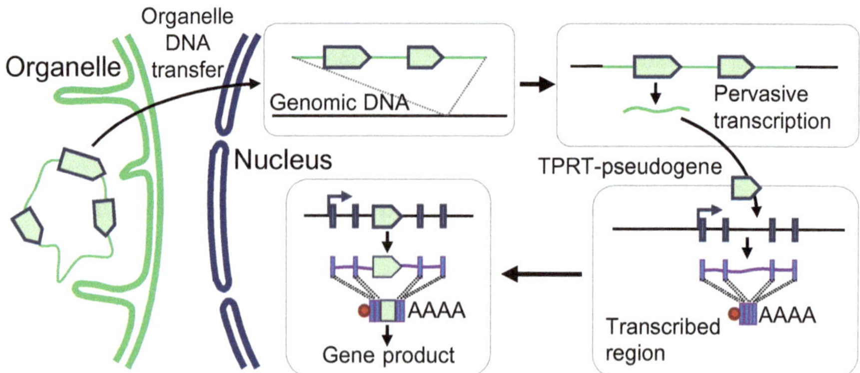

Fig. 9.5 Model for the role of TPRT retrotransposons in the accommodation of organelle genes into suitable gene platforms to be exonized into functional genes. Green boxes, organelle ORFs; blue boxes, eukaryotic coding regions

no longer remain, retrotransposons likely amplified these early fragments across the genome. Over time, one or more retrocopies became exonized within preexisting host transcripts, providing an evolutionary scaffold for the emergence of fully functional nuclear genes and their regulatory adaptations. Indeed, this model is consistent with observations in several plant species, where large organellar DNA fragments recurrently insert into the nuclear genome and segments of them are later redistributed by TE activity (Hazkani-Covo et al. 2010; Michalovova et al. 2013; Husnik and McCutcheon 2018).

Nonetheless, perhaps the most impactful contribution of TEs to complex systems is the creation of regulatory networks controlled by the Krüppel-associated box domain (KRAB) zinc finger proteins (KRAB-ZFP or KZFP). These proteins constitute one of the major host defense systems against TEs. They first appeared approximately 420 million years ago in a common ancestor of coelacanths, birds, and tetrapods. Since then, KZFPs have expanded and diversified via gene duplication to more than 350 and 700 members in the human and murine genomes, respectively, becoming the largest family of transcription factors in vertebrates (Urrutia 2003; Imbeault et al. 2017; Kosuge et al. 2024). KZFPs are composed of two functional domains: an N-terminal KRAB domain and a C-terminal tandem array of C2H2 zinc fingers. The ZFP domain recognizes two to four specific nucleotides with high affinity, allowing the protein to bind to the target DNA sequence (Patel et al. 2018). The KRAB domain mediates recruitment of KAP1 (also known as TRIM28) in most KZFPs (Peng et al. 2000; Bruno et al. 2019), which in turn recruits multiple repressive chromatin modifier factors, including the histone methyltransferase SETDB1, which deposits the repressive H3K9me3 mark (Schultz et al. 2002); the histone deacetylase NuRD complex, which harbors histone deacetylase and nucleosome remodeling activity (Schultz et al. 2001); and heterochromatin protein 1 (HP1) (Ryan et al. 1999).

Approximately two-thirds of human KZFPs recognize and repress the transcriptional activity of TE-derived sequences, including ERVs, L1, SINE, and SVA (Najafabadi et al. 2015; Schmitges et al. 2016; Imbeault et al. 2017). Initially, KZFPs were thought to function exclusively in TE silencing during early embryogenesis, and the wide expansion of KZFPs along the propagation of TE families would be the result of an "arms race." In this model, new TE families emerged and proliferated from individual copies escaping after mutation of the KZFP binding site. Next, KZFP-duplication will allow new genes to be substrates for the evolution of new repressors of these unchecked TE families (Thomas and Schneider 2011). However, accumulating evidence contradicts a strictly antagonistic view. Many KZFPs display highly cell type-specific expression patterns in adult tissues, where they regulate host gene expression by modulating TE-derived *cis*-regulatory elements (Ecco et al. 2016; Imbeault et al. 2017). This supports an alternative (and not mutually exclusive) model in which KZFPs arise after TE dispersion and subsequently exploit TE insertions for gene regulation, enabling positive selection of advantageous copies while facilitating removal of deleterious ones. An example that illustrates this phenomenon is the KZFP-mediated control of ERV-derived enhancers that sculpt germline transcriptomes during spermatogenesis in mice and humans (Otsuka et al. 2025). ERVs function as species-specific enhancers during the transition from mitotic spermatogonia to meiotic spermatocytes (Sakashita et al. 2020). In spermatogonia, high KZFP expression enforces repression of these ERVs through H3K9me3 deposition. Loss of KZFP-mediated repression leads to enhancer activation and induction of adjacent genes in pachytene spermatocytes. This mechanism extends solo-LTR-mediated regulation based on their transcriptional potential by adding an additional layer of coordination imposed by KZFPs originally evolved for TE control. Furthermore, the wide KZFP repertoire also enables the emergence of TE-independent binding sites throughout the genome, adding additional genes into regulatory networks with positive selective value.

The regulatory interplay between KZFPs and TEs is particularly evident in human brain development (Turelli et al. 2020; Playfoot et al. 2021). The primate-specific KZFPs ZNF417 and ZNF587 selectively repress sequences derived from evolutionarily recent retrotransposons, especially human ERV-K and SVA elements. Phylogenetic analyses indicate that ZNF417 and ZNF587, respectively, emerged approximately 43 and 19 million years ago within the primate lineage. Transcriptomic analyses reveal region-specific expression of ZNF417 and ZNF587 in the developing and adult human brain. Their depletion results in derepression of >8 TE copies that act as enhancers regulating neuronal genes involved in GABAergic neurotransmission, ion channel activity, and neuronal differentiation (Turelli et al. 2020). Among the affected genes are AADAT, required for the synthesis of the neuroprotective metabolite kynurenic acid (Campbell et al. 2014), and PRODH, implicated in GABAergic neurotransmission (Kempf et al. 2008), both presenting HERV-K copies recognized by ZNF417/587 upstream of their promoters (Turelli et al. 2020).

Notably, one-third of the human KZFPs do not recognize TEs, but instead target other sites such as promoters, simple repeats, and polyzinc finger protein genes (Imbeault et al. 2017). These KZFPs are older and more evolutionarily conserved than TE-binding KZFPs and, in most cases, do not recruit KAP1 (Schmitges et al. 2016; Imbeault et al. 2017). Nevertheless, these ancient KZFPs play essential roles in mammal-specific processes such as genomic imprinting (Juan and Bartolomei 2019; Takahashi et al. 2019). Whether this is an evolutionary sideline of some of the family members derived from TE-targeting proteins or they are the predecessors of TE-targeting proteins is an open debate.

7 Conclusion

Despite their well-known deleterious potential through aberrant activation and insertional mutagenesis, the sections above illustrate the significant contribution of TEs to evolution, largely dependent on their capacity to generate insertional mutagenesis. Indeed, from the deleterious *vs.* selectable perspective, insertional mutagenesis is not *a priori* different from replication-error mutagenesis, as both of them provide genetic diversity susceptible to natural selection according to their phenotype. However, there is an important aspect to consider: unlike replication error, insertional mutagenesis is catalyzed by a specific machinery encoded by TEs, and, equally important, this machinery is itself subject to multiple layers of host-encoded regulatory control. In fact, according to current models of KZFP evolution, the most recently emerged and transcriptionally active TE families are often not directly regulated by KZFPs. Instead, they are primarily controlled through DNA methylation of their promoters (Muotri et al. 2010; Castro-Diaz et al. 2014; Greenberg and Bourc'his 2019). In specific biological contexts, such as mouse spermatogenesis, this repression has been shown to rely on an adaptive mechanism triggered by TE transcriptional activation and the production of RNA species that give rise to PIWI-interacting RNAs (piRNAs) (Schöpp et al. 2020; Zoch et al. 2024). Some of these piRNAs help guide the silencing machinery to active *loci* through binding the nascent TE transcripts. Notably, additional TE-silencing systems, such as the HUSH complex, also appear to rely on nascent transcription to detect active TE *loci* and tether the silencing machinery to deposit repressive chromatin marks distinct from DNA methylation.

Nevertheless, productive DNA methylation-mediated silencing requires CpG-rich regions in the target (CpG islands). The observation that modern, replication-competent TE families of polyphyletic origin (including LINEs, SVAs, and ERVs) harbor CpG islands (Muotri et al. 2010; Rowe et al. 2013; Horváth et al. 2024) suggests that even TE families unchecked by KZFPs are subject to evolutionary constraints to maintain their activity within tolerable limits. Consequently, ongoing insertional mutagenesis (which perpetuates new TE insertions and contributes to genetic diversity) is restricted to levels compatible with host survival and population viability. This view is supported by the observation that regulatory sequences required for DNA methylation are conserved throughout L1 evolution (Khan et al. 2006; Sanchez-Luque et al. 2019). Indeed, it has been proposed that DNA methylation, now widely employed for epigenetic regulation of host genes, originally evolved as a defense mechanism to control TEs (Greenberg and Bourc'his 2019), positioning it as an additional example of TE-driven innovation discussed in Sect. 5. Altogether, the genomes of living organisms harbor autonomous TEs whose encoded machinery retains the potential to act as powerful evolutionary engines, while the host mechanisms that constrain TE expansion are simultaneously regulating this evolutionary potential. Many of these control systems (including epigenetic mechanisms) are sensitive to factors including stress, aging, and environmental conditions. It results in an interesting conclusion speculating that this could represent a complex evolutionary feedback loop in which the generation of insertional mutagenesis would have somehow become selected to be regulated, to an extent, by the necessity of genetic diversity fueling evolution according to particular environmental pressures.

Acknowledgments The authors thank Dr. Marie-Jeanne H. C. Kempen and Dr. Francisco Macias for reviewing the manuscript. F.J.S-L. acknowledges funding from the Spanish National Research Agency (AEI) grants RYC2021-031920-I, PID2022-143185NA-I00, and CNS2022-136033. P. M-S. acknowledges funding from the AEI PREP2022-000692 PhD fellowship.

References

Alam SMK, Konno T, Sahgal N et al (2008) Decidual cells produce a heparin-binding prolactin family cytokine with putative intrauterine regulatory actions. J Biol Chem 283:18957–18968. https://doi.org/10.1074/jbc.M801826200

Andersen JV, Markussen KH, Jakobsen E et al (2021) Glutamate metabolism and recycling at the excitatory synapse in health and neurodegeneration. Neuropharmacology 196. https://doi.org/10.1016/j.neuropharm.2021.108719

Andersen JV, Schousboe A, Verkhratsky A (2022) Astrocyte energy and neurotransmitter metabolism in Alzheimer's disease: integration of the glutamate/GABA-glutamine cycle. Prog Neurobiol 217. https://doi.org/10.1016/j.pneurobio.2022.102331

Barakat TS, Halbritter F, Zhang M et al (2018) Functional dissection of the enhancer repertoire in human embryonic stem cells. Cell Stem Cell 23:276–288.e8. https://doi.org/10.1016/j.stem.2018.06.014

Bechtel HB, Bechtel E (1989) Color mutations in the corn Snake (Elaphe guttata guttata): review and additional breeding data. J Hered 80:272–276. https://doi.org/10.1093/OXFORDJOURNALS.JHERED.A110853

Belshaw R, Dawson ALA, Woolven-Allen J et al (2005) Genomewide screening reveals high levels of insertional polymorphism in the human endogenous retrovirus family HERV-K(HML2): implications for present-Day activity. J Virol. https://doi.org/10.1128/jvi.79.19.12507-12514.2005

Benler S, Koonin EV (2022) Recruitment of mobile genetic elements for diverse cellular functions in prokaryotes. Front Mol Biosci 9:821197. https://doi.org/10.3389/FMOLB.2022.821197/FULL

Betancourt AJ, Wei KHC, Huang Y, Lee YCG (2024) Causes and consequences of varying transposable element activity: an evolutionary perspective. Annu Rev Genomics Hum Genet 25:1–25. https://doi.org/10.1146/ANNUREV-GENOM-120822-105708

Blaise S, De Parseval N, Bénit L, Heidmann T (2003) Genomewide screening for fusogenic human endogenous retrovirus envelopes identifies syncytin 2, a gene conserved on primate evolution. Proc Natl Acad Sci USA 100:13013–13018. https://doi.org/10.1073/PNAS.2132646100;PAGE:STRING:ARTICLE/CHAPTER

Boissinot S, Sookdeo A (2016) The evolution of LINE-1 in vertebrates. Genome Biol Evol. https://doi.org/10.1093/gbe/evw247

Brookes O, Thorpe SD, Rigby Evans O et al (2022) Covariation of pluripotency markers and biomechanical properties in mouse embryonic stem cells. Front Cell Dev Biol 10:858884. https://doi.org/10.3389/FCELL.2022.858884/BIBTEX

Bruno M, Mahgoub M, Macfarlan TS (2019) The arms race between KRAB-zinc finger proteins and endogenous retroelements and its impact on mammals. Annu Rev Genet 53:393–416. https://doi.org/10.1146/ANNUREV-GENET-112618-043717

Bu P, Yagi S, Shiota K et al (2017) Origin of a rapidly evolving homeostatic control system programming testis function. J Endocrinol 234:217. https://doi.org/10.1530/JOE-17-0250

Burki F, Kaessmann H (2004) Birth and adaptive evolution of a hominoid gene that supports high neurotransmitter flux. Nat Genet 36:1061–1063. https://doi.org/10.1038/NG1431

Burse M, Shi J, Aiken C (2017) Cyclophilin A potentiates TRIM5α inhibition of HIV-1 nuclear import without promoting TRIM5α binding to the viral capsid. PLoS One 12. https://doi.org/10.1371/JOURNAL.PONE.0182298

Buttler CA, Ramirez D, Dowell RD, Chuong EB (2023) An intronic LINE-1 regulates IFNAR1 expression in human immune cells. Mob DNA 14. https://doi.org/10.1186/S13100-023-00308-3

Calatrava V, Stephens TG, Gabr A et al (2022) Retrotransposition facilitated the establishment of a primary plastid in the thecate amoeba Paulinella. Proc Natl Acad Sci USA 119. https://doi.org/10.1073/PNAS.2121241119/-/DCSUPPLEMENTAL

Campbell BM, Charych E, Lee AW, Möller T (2014) Kynurenines in CNS disease: regulation by inflammatory cytokines. Front Neurosci 8. https://doi.org/10.3389/FNINS.2014.00012
Carrieri C, Cimatti L, Biagioli M et al (2012) Long non-coding antisense RNA controls Uchl1 translation through an embedded SINEB2 repeat. Nature 491(7424):454–457. https://doi.org/10.1038/nature11508
Castro-Diaz N, Ecco G, Coluccio A et al (2014) Evolutionally dynamic L1 regulation in embryonic stem cells. Genes Dev. https://doi.org/10.1101/gad.241661.114
Catlin NS, Josephs EB (2022) The important contribution of transposable elements to phenotypic variation and evolution. Curr Opin Plant Biol 65. https://doi.org/10.1016/j.pbi.2021.102140
Chen CP, Chen LF, Yang SR et al (2008) Functional characterization of the human placental Fusogenic membrane protein Syncytin 2. Biol Reprod 79:815–823. https://doi.org/10.1095/BIOLREPROD.108.069765
Chen J, Glémin S, Lascoux M (2017) Genetic diversity and the efficacy of purifying selection across plant and animal species. Mol Biol Evol 34:1417–1428. https://doi.org/10.1093/MOLBEV/MSX088
Chen G, Yu D, Yang Y et al (2024) Adaptive expansion of ERVK solo-LTRs is associated with Passeriformes speciation events. Nat Commun 15(1):3151. https://doi.org/10.1038/s41467-024-47501-3
Choudhary MNK, Friedman RZ, Wang JT et al (2020) Co-opted transposons help perpetuate conserved higher-order chromosomal structures. Genome Biol 21(1):16. https://doi.org/10.1186/S13059-019-1916-8
Choudhary MNK, Quaid K, Xing X et al (2023) Widespread contribution of transposable elements to the rewiring of mammalian 3D genomes. Nat Commun 14:634. https://doi.org/10.1038/S41467-023-36364-9
Chuong EB, Elde NC, Feschotte C (2016) Regulatory activities of transposable elements: from conflicts to benefits. Nat Rev Genet 18(2):71–86. https://doi.org/10.1038/nrg.2016.139
Cornelis G, Vernochet C, Carradec Q et al (2015) Retroviral envelope gene captures and syncytin exaptation for placentation in marsupials. Proc Natl Acad Sci USA 112:E487–E496. https://doi.org/10.1073/PNAS.1417000112
Cornelis G, Funk M, Vernochet C et al (2017) An endogenous retroviral envelope syncytin and its cognate receptor identified in the viviparous placental Mabuya lizard. Proc Natl Acad Sci USA 114:E10991–E11000. https://doi.org/10.1073/PNAS.1714590114
Criscione SW, Theodosakis N, Micevic G et al (2016) Genome-wide characterization of human L1 antisense promoter-driven transcripts. BMC Genomics 17. https://doi.org/10.1186/S12864-016-2800-5
Cui M, Bai Y, Li K, Rong YS (2021) Taming active transposons at Drosophila telomeres: the interconnection between HipHop's roles in capping and transcriptional silencing. PLoS Genet 17:e1009925. https://doi.org/10.1371/JOURNAL.PGEN.1009925
D'Agostino S, Tettey-Matey A, Volpe M et al (2025) Internal ribosome entry sites enhance translation in trans in antisense non-coding SINEUP and circular RNAs. Nucleic Acids Res 53. https://doi.org/10.1093/NAR/GKAF788
De Cecco M, Ito T, Petrashen AP et al (2019) L1 drives IFN in senescent cells and promotes age-associated inflammation. Nature 566:73–78. https://doi.org/10.1038/s41586-018-0784-9
De Lange T (2004) T-loops and the origin of telomeres. Nat Rev Mol Cell Biol 5:323–329. https://doi.org/10.1038/NRM1359
Denli AM, Narvaiza I, Kerman BE et al (2015) Primate-specific ORF0 contributes to retrotransposon-mediated diversity. Cell. https://doi.org/10.1016/j.cell.2015.09.025
Diaz-Griffero F, Vandegraaff N, Li Y et al (2006) Requirements for capsid-binding and an effector function in TRIMCyp-mediated restriction of HIV-1. Virology 351:404–419. https://doi.org/10.1016/j.virol.2006.03.023
Diehl AG, Ouyang N, Boyle AP (2020) Transposable elements contribute to cell and species-specific chromatin looping and gene regulation in mammalian genomes. Nat Commun 11(1):1796. https://doi.org/10.1038/s41467-020-15520-5

Dixon JR, Selvaraj S, Yue F et al (2012) Topological domains in mammalian genomes identified by analysis of chromatin interactions. Nature 485(7398):376–380. https://doi.org/10.1038/nature11082

Ecco G, Cassano M, Kauzlaric A et al (2016) Transposable elements and their KRAB-ZFP controllers regulate gene expression in adult tissues. Dev Cell 36:611–623. https://doi.org/10.1016/j.devcel.2016.02.024

Engels WR, Johnson-Schlitz DM, Eggleston WB, Sved J (1990) High-frequency P element loss in Drosophila is homolog dependent. Cell 62:515–525. https://doi.org/10.1016/0092-8674(90)90016-8

Fang J, Chen T, Pan Q, Wang Q (2018) Generation of albino medaka (Oryzias latipes) by CRISPR/Cas9. J Exp Zool B Mol Dev Evol 330:242–246. https://doi.org/10.1002/JEZ.B.22808

Faulk CD (2010) Evolution of YY1, YY2, REX1 and DNA-binding motifs in vertebrate genomes. LSU Doctoral Dissertations 2042. https://doi.org/10.31390/gradschool_dissertations.2042

Ferrari R, de Llobet Cucalon LI, Di Vona C et al (2020) TFIIIC binding to Alu elements controls gene expression via chromatin looping and histone acetylation. Mol Cell 77:475–487.e11. https://doi.org/10.1016/j.molcel.2019.10.020

Feschotte C, Pritham EJ (2007) DNA transposons and the evolution of eukaryotic genomes. Annu Rev Genet 41:331–368. https://doi.org/10.1146/ANNUREV.GENET.40.110405.090448

Feschotte C, Wessler SR (2002) Mariner-like transposases are widespread and diverse in flowering plants. Proc Natl Acad Sci USA 99:280–285. https://doi.org/10.1073/pnas.022626699

Fiddes JC, Goodman HM (1981) The gene encoding the common alpha subunit of the four human glycoprotein hormones. J Mol Appl Genet 1:3–18

Frank JA, Singh M, Cullen HB et al (2022) Evolution and antiviral activity of a human protein of retroviral origin. Science 378:422–428. https://doi.org/10.1126/SCIENCE.ABQ7871;JOURNAL:JOURNAL:SCIENCE;WGROUP:STRING:PUBLICATION

Frost LS, Leplae R, Summers AO, Toussaint A (2005) Mobile genetic elements: the agents of open source evolution. Nat Rev Microbiol 3:722–732. https://doi.org/10.1038/NRMICRO1235;KWRD

Fukuda S, Varshney A, Fowler BJ et al (2021) Cytoplasmic synthesis of endogenous Alu complementary DNA via reverse transcription and implications in age-related macular degeneration. Proc Natl Acad Sci USA 118. https://doi.org/10.1073/pnas.2022751118

Galbraith JD, Hayward A (2023) The influence of transposable elements on animal colouration. Trends Genet 39:624–638. https://doi.org/10.1016/j.tig.2023.04.005

Garza R, Atacho DAM, Adami A et al (2023) LINE-1 retrotransposons drive human neuronal transcriptome complexity and functional diversification. Sci Adv 9. https://doi.org/10.1126/SCIADV.ADH9543

Gázquez-Gutiérrez A, Witteveldt J, Heras SR, Macias S (2021) Sensing of transposable elements by the antiviral innate immune system. RNA 27:735–752

Gimenez J, Cile Montgiraud CÉ, Oriol G et al (2009) Comparative methylation of ERVWE1/Syncytin-1 and other human endogenous retrovirus LTRs in placenta tissues. DNA Res 16(4):195–211. https://doi.org/10.1093/dnares/dsp011

Gong R, Peng X, Kang S et al (2005) Structural characterization of the fusion core in syncytin, envelope protein of human endogenous retrovirus family W. Biochem Biophys Res Commun 331:1193–1200. https://doi.org/10.1016/j.bbrc.2005.04.032

Greenberg MVC, Bourc'his D (2019) The diverse roles of DNA methylation in mammalian development and disease. Nat Rev Mol Cell Biol 20:590–607

Guo Z, Kuang Z, Tao Y et al (2022) Miniature inverted-repeat transposable elements drive rapid MicroRNA diversification in angiosperms. Mol Biol Evol 39. https://doi.org/10.1093/MOLBEV/MSAC224

Hall LL, Carone DM, Gomez AV et al (2014) Stable C0T-1 repeat RNA is abundant and is associated with euchromatic interphase chromosomes. Cell 156:907–919. https://doi.org/10.1016/j.cell.2014.01.042

Hancks DC, Kazazian HH (2016) Roles for retrotransposon insertions in human disease. Mob DNA 7:9

Hayashi K, Fernie AR (2024) Genomic evidence of a transposon-conferred adaptive response. Cell Rep 43. https://doi.org/10.1016/j.celrep.2024.115009

Hazkani-Covo E, Zeller RM, Martin W (2010) Molecular poltergeists: mitochondrial DNA copies (numts) in sequenced nuclear genomes. PLoS Genet 6:e1000834. https://doi.org/10.1371/JOURNAL.PGEN.1000834

Hof AEV t, Campagne P, Rigden DJ et al (2016) The industrial melanism mutation in British peppered moths is a transposable element. Nature 534:102–105. https://doi.org/10.1038/NATURE17951

Horváth V, Garza R, Jönsson ME et al (2024) Mini-heterochromatin domains constrain the cis-regulatory impact of SVA transposons in human brain development and disease. Nat Struct Mol Biol 31:1543–1556. https://doi.org/10.1038/S41594-024-01320-8

Huda A, Tyagi E, Mariño-Ramírez L et al (2011) Prediction of transposable element derived enhancers using chromatin modification profiles. PLoS One 6:e27513. https://doi.org/10.1371/JOURNAL.PONE.0027513

Husnik F, McCutcheon JP (2018) Functional horizontal gene transfer from bacteria to eukaryotes. Nat Rev Microbiol 16:67–79. https://doi.org/10.1038/NRMICRO.2017.137

Imbeault M, Helleboid PY, Trono D (2017) KRAB zinc-finger proteins contribute to the evolution of gene regulatory networks. Nature 543(7646):550–554. https://doi.org/10.1038/nature21683

Jaremek A, Shaha S, Jeyarajah MJ et al (2023) Genome-wide analysis of hypoxia-inducible factor binding reveals targets implicated in impaired human placental Syncytiotrophoblast formation under low oxygen. Am J Pathol 193:846–865. https://doi.org/10.1016/j.ajpath.2023.03.006

Jönsson ME, Ludvik Brattås P, Gustafsson C et al (2019) Activation of neuronal genes via LINE-1 elements upon global DNA demethylation in human neural progenitors. Nat Commun. https://doi.org/10.1038/s41467-019-11150-8

Juan AM, Bartolomei MS (2019) Evolving imprinting control regions: KRAB zinc fingers hold the key. Genes Dev 33:1–3. https://doi.org/10.1101/GAD.322990.118

Kapitonov VV, Jurka J (2001) Rolling-circle transposons in eukaryotes. Proc Natl Acad Sci USA 98:8714–8719. https://doi.org/10.1073/PNAS.151269298;PAGE:STRING:ARTICLE/CHAPTER

Kapitonov VV, Jurka J (2005) RAG1 core and V(D)J recombination signal sequences were derived from Transib transposons. PLoS Biol 3:0998–1011. https://doi.org/10.1371/JOURNAL.PBIO.0030181

Kapitonov VV, Jurka J (2007) Helitrons on a roll: eukaryotic rolling-circle transposons. Trends Genet 23:521–529. https://doi.org/10.1016/j.tig.2007.08.004

Kapitonov VV, Koonin EV (2015) Evolution of the RAG1-RAG2 locus: both proteins came from the same transposon. Biol Direct 10:20. https://doi.org/10.1186/S13062-015-0055-8

Kazazian HH, Wong C, Youssoufian H et al (1988) Haemophilia A resulting from de novo insertion of L1 sequences represents a novel mechanism for mutation in man. Nature. https://doi.org/10.1038/332164a0

Kelly S (2021) The economics of organellar gene loss and endosymbiotic gene transfer. Genome Biol 22:345. https://doi.org/10.1186/S13059-021-02567-W

Kelsh RN, Inoue C, Momoi A et al (2004) The Tomita collection of medaka pigmentation mutants as a resource for understanding neural crest cell development. Mech Dev 121:841–859. https://doi.org/10.1016/j.mod.2004.01.004

Kempf L, Nicodemus KK, Kolachana B et al (2008) Functional polymorphisms in PRODH are associated with risk and protection for schizophrenia and Fronto-striatal structure and function. PLoS Genet 4:e1000252. https://doi.org/10.1371/JOURNAL.PGEN.1000252

Khan H, Smit A, Boissinot S (2006) Molecular evolution and tempo of amplification of human LINE-1 retrotransposons since the origin of primates. Genome Res. https://doi.org/10.1101/gr.4001406

Kim JD, Faulk C, Kim J (2007) Retroposition and evolution of the DNA-binding motifs of YY1, YY2 and REX1. Nucleic Acids Res 35:3442–3452. https://doi.org/10.1093/NAR/GKM235

Kim K, Dauphin A, Komurlu S et al (2019) Cyclophilin A protects HIV-1 from restriction by human TRIM5α. Nat Microbiol 4:2044–2051. https://doi.org/10.1038/S41564-019-0592-5

Koga A, Inagaki H, Bessho Y, Hori H (1995) Insertion of a novel transposable element in the tyrosinase gene is responsible for an albino mutation in the medaka fish, Oryzias latipes. Mol Gen Genet 249:400–405. https://doi.org/10.1007/BF00287101

Kosuge M, Ito J, Hamada M (2024) Landscape of evolutionary arms races between transposable elements and KRAB-ZFP family. Sci Rep 14(1):23358. https://doi.org/10.1038/s41598-024-73752-7

Krupovic M, Yutin N, Koonin EV (2016) Fusion of a superfamily 1 helicase and an inactivated DNA polymerase is a signature of common evolutionary history of Polintons, polinton-like viruses, Tlr1 transposons and transpovirons. Virus Evol 2. https://doi.org/10.1093/VE/VEW019

Lander ES, Linton LM, Birren B et al (2001) Initial sequencing and analysis of the human genome. Nature. https://doi.org/10.1038/35057062

Lavialle C, Cornelis G, Dupressoir A et al (2013) Paleovirology of "syncytins", retroviral env genes exapted for a role in placentation. Philos Trans R Soc Lond Ser B Biol Sci 368. https://doi.org/10.1098/RSTB.2012.0507

Lawson HA, Liang Y, Wang T (2023) Transposable elements in mammalian chromatin organization. Nat Rev Genet 24:712–723. https://doi.org/10.1038/S41576-023-00609-6

Lhee D, Ha JS, Kim S et al (2019) Evolutionary dynamics of the chromatophore genome in three photosynthetic Paulinella species. Sci Rep 9(1):2560. https://doi.org/10.1038/s41598-019-38621-8

Lhee D, Lee J, Ettahi K et al (2021) Amoeba genome reveals dominant host contribution to plastid endosymbiosis. Mol Biol Evol 38:344–357. https://doi.org/10.1093/MOLBEV/MSAA206

Li Y, Tang L, Jia X, Wu S (2025) Review article: YY2 in tumor regulation: core mechanisms and functional networks. Crit Rev Oncog 30. https://doi.org/10.1615/CRITREVONCOG.2025060905

Liang D, Zhao P, Si J et al (2021) Genomic analysis revealed a convergent evolution of LINE-1 in coat color: a case study in water buffaloes (Bubalus bubalis). Mol Biol Evol 38:1122–1136. https://doi.org/10.1093/MOLBEV/MSAA279

Liao CH, Kuang YQ, Liu HL et al (2007) A novel fusion gene, TRIM5-cyclophilin a in the pig-tailed macaque determines its susceptibility to HIV-1 infection. AIDS 21. https://doi.org/10.1097/01.AIDS.0000304692.09143.1B

Lu JY, Chang L, Li T et al (2021) Homotypic clustering of L1 and B1/Alu repeats compartmentalizes the 3D genome. Cell Res 31(6):613–630. https://doi.org/10.1038/s41422-020-00466-6

Luan DD, Korman MH, Jakubczak JL, Eickbush TH (1993) Reverse transcription of R2Bm RNA is primed by a nick at the chromosomal target site: a mechanism for non-LTR retrotransposition. Cell 72:595–605. https://doi.org/10.1016/0092-8674(93)90078-5

Luo C, Lu X, Stubbs L, Kim J (2006) Rapid evolution of a recently retroposed transcription factor YY2 in mammalian genomes. Genomics 87:348–355. https://doi.org/10.1016/J.YGENO.2005.11.001

Macfarlan TS, Gifford WD, Agarwal S et al (2011) Endogenous retroviruses and neighboring genes are coordinately repressed by LSD1/KDM1A. Genes Dev 25:594–607. https://doi.org/10.1101/GAD.2008511

Macia A, Blanco-Jimenez E, García-Pérez JL (2015) Retrotransposons in pluripotent cells: impact and new roles in cellular plasticity. Biochim Biophys Acta Gene Regul Mech 1849:417

Maestre J, Tchénio T, Dhellin O, Heidmann T (1995) mRNA retroposition in human cells: processed pseudogene formation. EMBO J 14:6333. https://doi.org/10.1002/J.1460-2075.1995.TB00324.X

Mager DL, Stoye JP (2015) Mammalian endogenous retroviruses. Microbiol Spectrum 3. https://doi.org/10.1128/MICROBIOLSPEC.MDNA3-0009-2014

Malfavon-Borja R, Feschotte C (2015) Fighting fire with fire: endogenous retrovirus envelopes as restriction factors. J Virol 89:4047–4050. https://doi.org/10.1128/JVI.03653-14
Mangeney M, Renard M, Schlecht-Louf G et al (2007) Placental syncytins: genetic disjunction between the fusogenic and immunosuppressive activity of retroviral envelope proteins. Proc Natl Acad Sci USA 104:20534–20539. https://doi.org/10.1073/PNAS.0707873105;WEBSITE:WEBSITE:PNAS-SITE;WGROUP:STRING:PUBLICATION
Mätlik K, Redik K, Speek M (2006) L1 antisense promoter drives tissue-specific transcription of human genes. Biomed Res Int 2006:071753. https://doi.org/10.1155/JBB/2006/71753
Maybin M, Ranade AM, Schombel U et al (2024) IS 1-mediated chromosomal amplification of the arn operon leads to polymyxin B resistance in Escherichia coli B strains. MBio 15. https://doi.org/10.1128/MBIO.00634-24
McCLINTOCK B (1950) The origin and behavior of mutable loci in maize. Proc Natl Acad Sci USA 36:344–355. https://doi.org/10.1073/pnas.36.6.344
Michalovova M, Vyskot B, Kejnovsky E (2013) Analysis of plastid and mitochondrial DNA insertions in the nucleus (NUPTs and NUMTs) of six plant species: size, relative age and chromosomal localization. Heredity (Edinb) 111:314–320. https://doi.org/10.1038/HDY.2013.51
Moroki Y, Komori M, Ogawa Y et al (2023) An attempt to identify the medaka receptor for somatolactin alpha using a reverse genetics approach. Genes (Basel) 14. https://doi.org/10.3390/GENES14040796
Muotri AR, Marchetto MCN, Coufal NG et al (2010) L1 retrotransposition in neurons is modulated by MeCP2. Nature. https://doi.org/10.1038/nature09544
Nadeau NJ, Pardo-Diaz C, Whibley A et al (2016) The gene cortex controls mimicry and crypsis in butterflies and moths. Nature 534:106–110. https://doi.org/10.1038/NATURE17961
Najafabadi HS, Mnaimneh S, Schmitges FW et al (2015) C2H2 zinc finger proteins greatly expand the human regulatory lexicon. Nat Biotechnol 33(5):555–562. https://doi.org/10.1038/nbt.3128
Newman RM, Hall L, Kirmaier A et al (2008) Evolution of a TRIM5-CypA Splice Isoform in Old World Monkeys. PLoS Pathog 4:e1000003. https://doi.org/10.1371/JOURNAL.PPAT.1000003
Nishihara H, Kobayashi N, Kimura-Yoshida C et al (2016) Coordinately co-opted multiple transposable elements constitute an enhancer for wnt5a expression in the mammalian secondary palate. PLoS Genet 12:e1006380. https://doi.org/10.1371/JOURNAL.PGEN.1006380
Ono R, Nakamura K, Inoue K et al (2006) Deletion of Peg10, an imprinted gene acquired from a retrotransposon, causes early embryonic lethality. Nat Genet 38. https://doi.org/10.1038/ng1699
Otsuka K, Sakashita A, Maezawa S et al (2025) KRAB zinc-finger proteins regulate endogenous retroviruses to sculpt germline transcriptomes and genome evolution. Genome Res 35:417–431. https://doi.org/10.1101/GR.279924.124
Pastuzyn ED, Day CE, Kearns RB et al (2018) The neuronal gene arc encodes a repurposed retrotransposon gag protein that mediates intercellular RNA transfer. Cell 172:275–288.e18. https://doi.org/10.1016/j.cell.2017.12.024
Patel A, Yang P, Tinkham M et al (2018) DNA conformation induces adaptable binding by tandem zinc finger proteins. Cell 173:221–233.e12. https://doi.org/10.1016/j.cell.2018.02.058
Payer LM, Steranka JP, Yang WR et al (2017) Structural variants caused by Alu insertions are associated with risks for many human diseases. Proc Natl Acad Sci USA. https://doi.org/10.1073/pnas.1704117114
Pegler JL, Oultram JMJ, Grof CPL, Eamens AL (2025) DRB1 and DRB2 are required for an appropriate miRNA-mediated molecular response to salt stress in Arabidopsis thaliana. Plants 14:924. https://doi.org/10.3390/PLANTS14060924/S1
Peng H, Begg GE, Harper SL et al (2000) Biochemical analysis of the Kruppel-associated box (KRAB) transcriptional repression domain. J Biol Chem 275:18000–18010. https://doi.org/10.1074/jbc.m001499200
Percharde M, Lin CJ, Yin Y et al (2018) A LINE1-Nucleolin partnership regulates early development and ESC identity. Cell. https://doi.org/10.1016/j.cell.2018.05.043

Pink RC, Wicks K, Caley DP et al (2011) Pseudogenes: pseudo-functional or key regulators in health and disease? RNA 17:792. https://doi.org/10.1261/RNA.2658311

Plaitakis A, Kalef-Ezra E, Kotzamani D et al (2017) The glutamate dehydrogenase pathway and its roles in cell and tissue biology in health and disease. Biology (Basel) 6. https://doi.org/10.3390/BIOLOGY6010011

Playfoot CJ, Duc J, Sheppard S et al (2021) Transposable elements and their KZFP controllers are drivers of transcriptional innovation in the developing human brain. Genome Res 31:1531–1545. https://doi.org/10.1101/GR.275133.120/-/DC1

Priščáková P, Svoboda M, Feketová Z et al (2023) Syncytin-1, syncytin-2 and suppressyn in human health and disease. J Mol Med 101(12):1527–1542. https://doi.org/10.1007/S00109-023-02385-6

Raingeval M, Leduque B, Baduel P et al (2024) Retrotransposon-driven environmental regulation of FLC leads to adaptive response to herbicide. Nat Plants 10:1672–1681. https://doi.org/10.1038/S41477-024-01807-8;SUBJMETA

Rheinemann L, Downhour DM, Bredbenner K et al (2020) A duplicated ESCRT-III factor blocks enveloped virus budding without inhibiting cellular ESCRT functions. bioRxiv. https://doi.org/10.1101/2020.08.30.273656

Rheinemann L, Downhour DM, Bredbenner K et al (2021) RetroCHMP3 blocks budding of enveloped viruses without blocking cytokinesis. Cell 184:5419. https://doi.org/10.1016/J.CELL.2021.09.008

Richardson SR, Doucet AJ, Kopera HC et al (2015) The influence of LINE-1 and SINE retrotransposons on mammalian genomes. In: Mobile DNA III. ASM Press, Washington, DC

Roberts RM, Green JA, Schulz LC (2016) The evolution of the placenta. Reproduction 152:R179–R189. https://doi.org/10.1530/REP-16-0325

Rogers J, Gibbs RA (2014) Comparative primate genomics: emerging patterns of genome content and dynamics. Nat Rev Genet 15:347–359. https://doi.org/10.1038/NRG3707

Roller M, Stamper E, Villar D et al (2021) LINE retrotransposons characterize mammalian tissue-specific and evolutionarily dynamic regulatory regions. Genome Biol 22. https://doi.org/10.1186/S13059-021-02260-Y

Ros F, Kunze R (2001) Regulation of activator/dissociation transposition by replication and DNA methylation. Genetics 157:1723–1733. https://doi.org/10.1093/GENETICS/157.4.1723

Rosenkrantz JL, Brandorff JE, Raghib S et al (2024) ZNF91 is an endogenous repressor of the molecular phenotype associated with X-linked dystonia–parkinsonism (XDP). Proc Natl Acad Sci USA 121:e2401217121. https://doi.org/10.1073/PNAS.2401217121

Rowe HM, Friedli M, Offner S et al (2013) De novo DNA methylation of endogenous retroviruses is shaped by KRAB-ZFPs/KAP1 and ESET. Development 140:519–529. https://doi.org/10.1242/DEV.087585

Ruden X, Singh A, Marben T et al (2024) A single cell transcriptomic fingerprint of stressed premature, imbalanced differentiation of embryonic stem cells. Birth Defects Res 116:e2409. https://doi.org/10.1002/BDR2.2409;WGROUP:STRING:PUBLICATION

Ryan RF, Schultz DC, Ayyanathan K et al (1999) KAP-1 corepressor protein interacts and colocalizes with heterochromatic and euchromatic HP1 proteins: a potential role for Krüppel-associated box-zinc finger proteins in heterochromatin-mediated gene silencing. Mol Cell Biol 19:4366–4378. https://doi.org/10.1128/MCB.19.6.4366

Saenko SV, Lamichhaney S, Barrio AM et al (2015) Amelanism in the corn snake is associated with the insertion of an LTR-retrotransposon in the OCA2 gene. Sci Rep 5(1):17118. https://doi.org/10.1038/srep17118

Saha K, Nielsen GI, Nandani R et al (2024) YY1 is a transcriptional activator of the mouse LINE-1 Tf subfamily. Nucleic Acids Res 52:12878–12894. https://doi.org/10.1093/NAR/GKAE949

Sahakyan AB, Murat P, Mayer C, Balasubramanian S (2017) G-quadruplex structures within the 3′ UTR of LINE-1 elements stimulate retrotransposition. Nat Struct Mol Biol 24. https://doi.org/10.1038/nsmb.3367

Sakashita A, Maezawa S, Takahashi K et al (2020) Endogenous retroviruses drive species-specific germline transcriptomes in mammals. Nat Struct Mol Biol 27(10):967–977. https://doi.org/10.1038/s41594-020-0487-4

Sanchez-Luque FJ, Kempen M-JHC, Gerdes P et al (2019) LINE-1 evasion of epigenetic repression in humans. Mol Cell. https://doi.org/10.1016/j.molcel.2019.05.024

Sayah DM, Sokolskaja E, Berthoux L, Luban J (2004) Cyclophilin a retrotransposition into TRIM5 explains owl monkey resistance to HIV-1. Nature 430:569–573. https://doi.org/10.1038/NATURE02777

Schein A, Zucchelli S, Kauppinen S et al (2016) Identification of antisense long noncoding RNAs that function as SINEUPs in human cells. Sci Rep 6(1):33605. https://doi.org/10.1038/srep33605

Schmidt D, Schwalie PC, Wilson MD et al (2012) Waves of retrotransposon expansion remodel genome organization and CTCF binding in multiple mammalian lineages. Cell 148:335–348. https://doi.org/10.1016/j.cell.2011.11.058

Schmitges FW, Radovani E, Najafabadi HS et al (2016) Multiparameter functional diversity of human C2H2 zinc finger proteins. Genome Res 26:1742–1752. https://doi.org/10.1101/GR.209643.116/-/DC1

Schöpp T, Zoch A, Berrens RV et al (2020) TEX15 is an essential executor of MIWI2-directed transposon DNA methylation and silencing. Nat Commun 11. https://doi.org/10.1038/s41467-020-17372-5

Schultz DC, Friedman JR, Rauscher FJ (2001) Targeting histone deacetylase complexes via KRAB-zinc finger proteins: the PHD and bromodomains of KAP-1 form a cooperative unit that recruits a novel isoform of the Mi-2alpha subunit of NuRD. Genes Dev 15:428–443. https://doi.org/10.1101/GAD.869501

Schultz DC, Ayyanathan K, Negorev D et al (2002) SETDB1: a novel KAP-1-associated histone H3, lysine 9-specific methyltransferase that contributes to HP1-mediated silencing of euchromatic genes by KRAB zinc-finger proteins. Genes Dev 16:919–932. https://doi.org/10.1101/GAD.973302

Scott EC, Devine SE (2017) The role of somatic L1 retrotransposition in human cancers. Viruses 9:131

Scourfield EJ, Martin-Serrano J (2017) Growing functions of the ESCRT machinery in cell biology and viral replication. Biochem Soc Trans 45:613–634. https://doi.org/10.1042/BST20160479

Sekita Y, Wagatsuma H, Nakamura K et al (2008) Role of retrotransposon-derived imprinted gene, Rtl1, in the feto-maternal interface of mouse placenta. Nat Genet 40:243–248. https://doi.org/10.1038/NG.2007.51

Sharma H, Valentine MNZ, Toki N et al (2024) Decryption of sequence, structure, and functional features of SINE repeat elements in SINEUP non-coding RNA-mediated post-transcriptional gene regulation. Nat Commun 15. https://doi.org/10.1038/S41467-024-45517-3

Siguier P, Gourbeyre E, Varani A et al (2015) Everyman's guide to bacterial insertion sequences. Microbiol Spectrum 3. https://doi.org/10.1128/MICROBIOLSPEC.MDNA3-0030-2014

Sookdeo A, Hepp CM, McClure MA, Boissinot S (2013) Revisiting the evolution of mouse LINE-1 in the genomic era. Mob DNA. https://doi.org/10.1186/1759-8753-4-3

Soucy SM, Huang J, Gogarten JP (2015) Horizontal gene transfer: building the web of life. Nat Rev Genet 16:472–482. https://doi.org/10.1038/NRG3962

Speek M (2001) Antisense promoter of human L1 retrotransposon drives transcription of adjacent cellular genes. Mol Cell Biol. https://doi.org/10.1128/mcb.21.6.1973-1985.2001

Stremlau M, Owens CM, Perron MJ et al (2004) The cytoplasmic body component TRIM5alpha restricts HIV-1 infection in Old World monkeys. Nature 427:848–853. https://doi.org/10.1038/NATURE02343

Sugimoto J, Sugimoto M, Bernstein H et al (2013) A novel human endogenous retroviral protein inhibits cell-cell fusion. Sci Rep 3(1):1462. https://doi.org/10.1038/srep01462

Sugimoto J, Schust DJ, Kinjo T et al (2019) Suppressyn localization and dynamic expression patterns in primary human tissues support a physiologic role in human placentation. Sci Rep 9(1):19502. https://doi.org/10.1038/s41598-019-55933-x

Sui G, Affar EB, Shi Y et al (2004) Yin Yang 1 is a negative regulator of p53. Cell 117:859–872. https://doi.org/10.1016/j.cell.2004.06.004

Sundaram V, Cheng Y, Ma Z et al (2014) Widespread contribution of transposable elements to the innovation of gene regulatory networks. Genome Res 24:1963–1976. https://doi.org/10.1101/GR.168872.113

Swergold GD (1990) Identification, characterization, and cell specificity of a human LINE-1 promoter. Mol Cell Biol. https://doi.org/10.1128/mcb.10.12.6718

Takahashi H, Kozhuharova A, Sharma H et al (2018) Identification of functional features of synthetic SINEUPs, antisense lncRNAs that specifically enhance protein translation. PLoS One 13:e0183229. https://doi.org/10.1371/JOURNAL.PONE.0183229

Takahashi N, Coluccio A, Thorball CW et al (2019) ZNF445 is a primary regulator of genomic imprinting. Genes Dev 33:49–54. https://doi.org/10.1101/GAD.320069.118

Thomas JH, Schneider S (2011) Coevolution of retroelements and tandem zinc finger genes. Genome Res 21:1800–1812. https://doi.org/10.1101/gr.121749.111

Thomas CA, Tejwani L, Trujillo CA et al (2017) Modeling of TREX1-dependent autoimmune disease using human stem cells highlights L1 accumulation as a source of neuroinflammation. Cell Stem Cell. https://doi.org/10.1016/j.stem.2017.07.009

Thompson PJ, Macfarlan TS, Lorincz MC (2016) Long terminal repeats: from parasitic elements to building blocks of the transcriptional regulatory repertoire. Mol Cell 62:766–776. https://doi.org/10.1016/j.molcel.2016.03.029

Tolosa JM, Schjenken JE, Clifton VL et al (2012) The endogenous retroviral envelope protein syncytin-1 inhibits LPS/PHA-stimulated cytokine responses in human blood and is sorted into placental exosomes. Placenta 33:933–941. https://doi.org/10.1016/j.placenta.2012.08.004

Tonegawa S (1983) Somatic generation of antibody diversity. Nature 302:575–581. https://doi.org/10.1038/302575A0

Toudic C, Elisseeff X, Lokossou AG, Barbeau B (2017) Roles of endogenous retrovirus-encoded syncytins in human placentation. In: Human retrotransposons in health and disease, pp 215–238. https://doi.org/10.1007/978-3-319-48344-3_9/FIGURES/3

Towers GJ, Hatziioannou T, Cowan S et al (2003) Cyclophilin A modulates the sensitivity of HIV-1 to host restriction factors. Nat Med 9(9):1138–1143. https://doi.org/10.1038/nm910

Tsutsumi M, Imai S, Kyono-Hamaguchi Y et al (2006) Color reversion of the albino medaka fish associated with spontaneous somatic excision of the Tol-1 transposable element from the tyrosinase gene. Pigment Cell Res 19:243–247. https://doi.org/10.1111/J.1600-0749.2006.00300.X

Turelli P, Playfoot C, Grun D et al (2020) Primate-restricted KRAB zinc finger proteins and target retrotransposons control gene expression in human neurons. Sci Adv 6. https://doi.org/10.1126/SCIADV.ABA3200

Ullate-Agote A, Milinkovitch MC, Tzika AC (2014) The genome sequence of the corn snake (Pantherophis guttatus), a valuable resource for EvoDevo studies in squamates. Int J Dev Biol 58:881–888. https://doi.org/10.1387/IJDB.150060AT

Urrutia R (2003) KRAB-containing zinc-finger repressor proteins. Genome Biol 4:10.1186/GB-2003-4-10-231

Vale FF, Lehours P, Yamaoka Y (2022) Editorial: the role of mobile genetic elements in bacterial evolution and their adaptability. Front Microbiol 13:849667. https://doi.org/10.3389/FMICB.2022.849667/BIBTEX

Wang J, Vicente-García C, Seruggia D et al (2015) MIR retrotransposon sequences provide insulators to the human genome. Proc Natl Acad Sci USA 112:E4428–E4437. https://doi.org/10.1073/PNAS.1507253112

Wang H, Gallet B, Moriscot C et al (2023) An inducible ESCRT-III inhibition tool to control HIV-1 budding. Viruses 15:2289. https://doi.org/10.3390/V15122289/S1

Wang Y, Wang Z, Lv S et al (2025) OCA2 deficiency enhances TPC2 channel activity to reduce melanosomal pH and pigment production. J Invest Dermatol. https://doi.org/10.1016/J.JID.2025.12.011
Wanichnopparat W, Suwanwongse K, Pin-on P et al (2013) Genes associated with the cis-regulatory functions of intragenic LINE-1 elements. BMC Genomics 14. https://doi.org/10.1186/1471-2164-14-205
Weintraub AS, Li CH, Zamudio AV et al (2017) YY1 is a structural regulator of enhancer-promoter loops. Cell. https://doi.org/10.1016/j.cell.2017.11.008
Xia B, Zhang W, Zhao G et al (2024) On the genetic basis of tail-loss evolution in humans and apes. Nature 626(8001):1042–1048. https://doi.org/10.1038/s41586-024-07095-8
Xie M, Hong C, Zhang B et al (2013) DNA hypomethylation within specific transposable element families associates with tissue-specific enhancer landscape. Nat Genet 45(7):836–841. https://doi.org/10.1038/ng.2649
Yakovenko I, Agronin J, Smith LC, Oren M (2021) Guardian of the genome: an alternative RAG/Transib co-evolution hypothesis for the origin of V(D)J recombination. Front Immunol 12:709165. https://doi.org/10.3389/FIMMU.2021.709165
Zhang L, Rong YS (2012) Retrotransposons at Drosophila telomeres: host domestication of a selfish element for the maintenance of genome integrity. Biochim Biophys Acta 1819:771. https://doi.org/10.1016/J.BBAGRM.2012.01.018
Zhang F, Hatziioannou T, Perez-Caballero D et al (2006) Antiretroviral potential of human tripartite motif-5 and related proteins. Virology 353:396–409. https://doi.org/10.1016/j.virol.2006.05.035
Zhang L, Beaucher M, Cheng Y, Rong YS (2014) Coordination of transposon expression with DNA replication in the targeting of telomeric retrotransposons in D rosophila. EMBO J 33:1148–1158. https://doi.org/10.1002/EMBJ.201386940/FIGURES/6
Zhang R, Nowack ECM, Price DC et al (2017) Impact of light intensity and quality on chromatophore and nuclear gene expression in Paulinella chromatophora, an amoeba with nascent photosynthetic organelles. Plant J 90:221–234. https://doi.org/10.1111/TPJ.13488
Zhang Y, Li T, Preissl S et al (2019) Transcriptionally active HERV-H retrotransposons demarcate topologically associating domains in human pluripotent stem cells. Nat Genet 51:1380–1388. https://doi.org/10.1038/S41588-019-0479-7
Zhao QY, Chen PX, Yang L et al (2020) Transmission of plasmid-borne and chromosomal blaCTX-M-64 among Escherichia coli and Salmonella isolates from food-producing animals via ISEcp1-mediated transposition. J Antimicrob Chemother 75:1424–1427. https://doi.org/10.1093/JAC/DKAA044
Zoch A, Konieczny G, Auchynnikava T et al (2024) C19ORF84 connects piRNA and DNA methylation machineries to defend the mammalian germ line. Mol Cell 84:1021–1035.e11. https://doi.org/10.1016/j.molcel.2024.01.014

Chapter 10
Defense and Immune Systems Evolving from Co-opted Selfish Genetic Elements—From the Origin of Life to the Present

Felix Broecker

Abstract Many immune systems—prokaryotic and eukaryotic—appear to have recruited parts from mobile genetic elements (MGEs) they now defend against. In this chapter, I advance an integrative hypothesis that defense pathways across life emerged repeatedly through co-option of MGEs (viruses, transposons, and viroid-like RNAs), with roots extending to protocells in the ancient RNA world. In this view, early ribozyme communities inevitably faced parasitic replicators; *trans*-cleaving ribozymes and superinfection exclusion-like interference provide plausible models for primordial immunity. Random ligation among catalytic RNAs could have produced a multi-ribozyme "protogenome" that domesticated parasites by coupling their transmission to host replication. This coupling could then have enabled genome expansion and, ultimately, a transition to more stable DNA heredity. In prokaryotes, defense islands, restriction–modification systems, and CRISPR-Cas exemplify how mobile modules are domesticated into immunity. In eukaryotes, RNA silencing (siRNA/piRNA), interferon circuitry shaped by transposable-element-derived regulatory elements, and endogenous retroviral co-options (including Env-mediated receptor interference) illustrate the same principle. In vertebrates, adaptive immunity itself derives from a transposon domestication event, in which a *Transib*-related recombinase became the engine of antibody and T-cell receptor diversification. I synthesize these lines into a continuous evolutionary narrative: host-parasite conflict repeatedly turns "weapons into shields," driving innovation from pre-LUCA (Last Universal Common Ancestor) stages to modern immunity. I close with testable experimental proposals to evaluate key steps in this co-option framework. This perspective reframes immunity as a creative legacy of genetic conflict.

F. Broecker (✉)
Idorsia (Berlin) Pharmaceuticals GmbH, Magnusstr. 11, 12489 Berlin, Germany
e-mail: felixbroecker@gmx.net

G. Witzany (ed.), *Integrative Theory of Evolution - Aspects and Insights*,
https://doi.org/10.1007/978-3-032-24295-2_10

1 Introduction

Across all domains of life, genomes carry signatures of ancient conflict between hosts and "selfish" genetic elements. Viruses and transposable elements (TEs) comprise a substantial fraction of cellular DNA, and their long coevolution with hosts has driven diverse defense mechanisms (Broecker and Moelling 2019). A unifying view is that many immune pathways, from prokaryotic CRISPR to vertebrate adaptive immunity, arose by recruiting components from these selfish elements. Prebiotic synthesis of 'biologically relevant' intermediates is not enough: without heredity to stabilize catalysts and connect steps, pathways dissipate faster than they persist, implying that protometabolism had to co-emerge with compartmentalized heredity to maintain catalytic continuity and enable selection (Negrón-Mendoza et al. 2025). This chapter develops an integrative theory of how simple defenses in prebiotic protocells gave rise to complex innate and adaptive immune systems through co-option of viral and transposon machineries. I trace this trajectory from the ancient RNA world to the present, emphasize parasite-host coevolution as a major driver of genetic innovation (Koonin and Dolja 2013), and propose experiments to test central predictions of the model. Key aspects are summarized in the evolutionary timeline in Fig. 10.1.

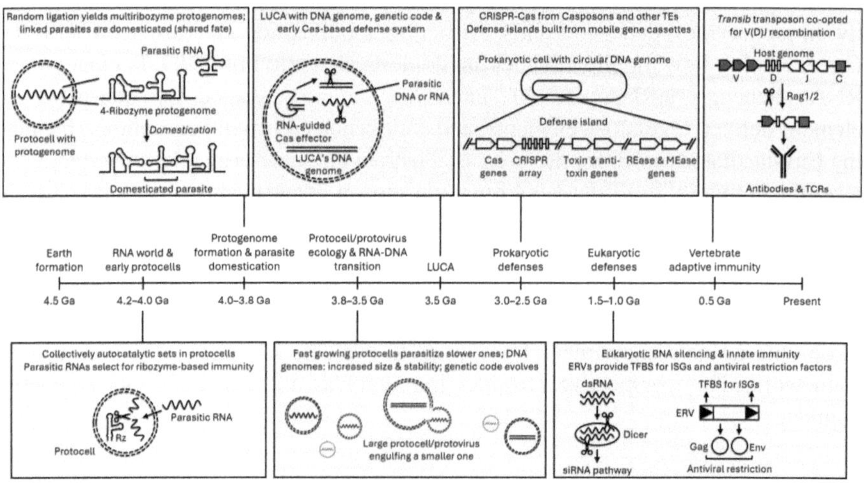

Fig. 10.1 Evolutionary timeline of immune systems. The schematic traces immune evolution from RNA-world protocells containing catalytic and parasitic RNAs, through protogenomes and a primordial virus world, to DNA-based Last Universal Common Ancestor (LUCA) and the subsequent diversification of prokaryotic and eukaryotic immune defenses. At each stage, selfish genetic elements (parasitic RNAs, other protocells, transposons, viruses) are recurrently domesticated into host defense mechanisms, culminating in vertebrate adaptive immunity built on a transposon-derived recombinase. ERV endogenous retrovirus, ISGs interferon-stimulated genes, LUCA last universal common ancestor, MEase methyltransferase, REase restriction endonuclease, Rz ribozyme, TFBS, transcription factor binding site

2 Selfish Genetic Elements as Evolutionary Catalysts

2.1 Viruses, Transposons, and the Host Genome

Selfish genetic elements, including viruses, plasmids, TEs, retroelements, and viroids, are ubiquitous and, in replicator systems, effectively unavoidable (Koonin and Dolja 2013; Koonin et al. 2017a). They enhance their own transmission, in the short term often at the host's expense, yet over evolutionary timescales they have profoundly shaped host genomes and physiology (Chuong et al. 2017). Importantly, it has been shown that host-parasite coevolution in RNA replication systems can be a creative force, sustaining long-term evolutionary dynamics and continually generating new sequence variants, driving the complexity of both hosts and parasites (Furubayashi et al. 2020).

In mammals, viral and transposon-derived sequences comprise roughly half of the genome (Du et al. 2024), or even closer to two-thirds by another estimate (de Koning et al. 2011). Prokaryotic genomes likewise contain substantial mobile DNA; prophages (integrated bacteriophage genomes) and their remnants often account for ~1–5% of bacterial chromosomes (Inglis et al. 2024), in some cases up to 30% (Casjens 2003; Busby et al. 2013). Plant genomes are frequently even more TE-rich (Bennetzen and Wang 2014); for example, ~85% of the maize genome consists of TEs (Schnable et al. 2009).

Hosts, in turn, repeatedly domesticate these sequences, often for immunity (Koonin and Krupovic 2015; tenOever 2016; Broecker and Moelling 2019; Broecker 2021a). Two canonical examples are prokaryotic CRISPR-Cas, whose adaptation machinery derives from MGEs (Krupovic et al. 2014), and vertebrate V(D)J recombination, which repurposes a transposon-derived recombinase to generate antigen-receptor diversity (Kapitonov and Jurka 2005). Recurrent domestications support the broader view that genetic conflict repeatedly supplies molecular parts that can assemble into new host functions (Feschotte 2008; Chuong et al. 2017).

2.2 Parasite–Host Arms Races and Cooperation

Host–parasite coevolution is often framed as a perpetual arms race (Forterre and Prangishvili 2009); hosts evolve defenses, parasites evolve counter-defenses, and this Red Queen dynamic can accelerate evolutionary change and drive complexity (Van Valen 1973; Daugherty and Malik 2012; Furubayashi et al. 2020). Genetic parasites can therefore be major engines of innovation, repeatedly forcing diversification of host-protective strategies (tenOever 2016).

Yet these interactions are not purely antagonistic. They can shift towards cooperation or mutualism when an invading element is tolerated, integrated, or co-opted. A simple example is superinfection exclusion (SIEx), where a preexisting, relatively benign infection reduces susceptibility to a secondary, more virulent strain (discussed below). Such cases blur the boundary between offense and defense. In line with this, Koonin and colleagues argue that both virus–host conflict and alliance have been decisive in evolution, with host genomes repeatedly recruiting viral and other MGE-derived genes for new functions (Koonin and Dolja 2013). Across this chapter, many immune "innovations" can be read as outcomes of this long-running tug-of-war.

3 Origins in the RNA World: Protocells and Primal Immunity

3.1 Prebiotic Protocells and the Rise of Selfish Replicators

I adopt an RNA-world, metabolism-first view in which life began as "protocells": lipid vesicles formed by wet–dry cycling that encapsulated collectively autocatalytic sets of RNA catalysts (ribozymes) (Broecker 2021b). Within such compartments, autocatalytic ribozyme networks could sustain limited heredity-like continuity and evolution before DNA, proteins, or a multi-ribozyme "protogenome" (Damer and Deamer 2020; Adamala and Szostak 2013a; Hordijk et al. 2010; Hordijk and Steel 2017). Short peptides likely entered as early ribozyme cofactors, supported compartmentalization, and may have been precursors of modern protein enzymes (Tagami et al. 2017; Iglesias-Artola et al. 2022; Le Vay et al. 2023). Other origin scenarios remain plausible (e.g., pre-RNA polymers (Yu et al. 2012, Pinheiro et al. 2012) or hydrothermal-vent/iron-sulfur metabolism models (Martin and Russell 2007)), but are not pursued here.

Once RNAs replicate, competition and parasitism emerge. Selfish RNAs that exploit replicases without contributing are theoretically inevitable in replicator ensembles (Koonin et al. 2017a). Classic Qβ replicase experiments showed that short parasitic RNAs can outcompete longer functional templates, but the experiments lacked, for example, compartmentalization and the replicase was externally supplied (Spiegelman et al. 1965; Mills et al. 1967; Kacian et al. 1972). More recent *in vitro* evolution studies, in contrast, demonstrate sustained host-parasite arms-race dynamics that generate novelty in both partners under conditions where compartments are provided (lipid droplets) and the host RNA had to retain its function as it encoded the replicase (Furubayashi et al. 2020). Protocells that restrained parasites, or converted them into useful components, would therefore gain fitness advantages.

This motivates a central premise of this chapter: defense and (proto)genome evolution likely co-emerged, and genetic innovation from life's origin to the present has repeatedly been fueled by host-parasite conflict and parasite domestication. I therefore next consider how an autocatalytic-set protocell might both fend off parasitic RNAs ("ribozyme immunity") and domesticate them.

3.2 *Hypothetical Ribozyme Immunity: "Kill the Parasite"* in Trans

What might a prebiotic immune system look like? One plausible answer is ribozyme-mediated defense (Broecker and Moelling 2019a). Ribozymes can function as sequence-specific endoribonucleases, so an RNA catalyst could function as a molecular "immune" agent by cleaving competing or invading RNAs. Here, modern viroids are an especially informative precedent. Viroids are minimal, circular non-coding RNAs that replicate autonomously in plants, and in the *Avsunviroidae*, replication intermediates are processed by embedded hammerhead ribozymes—a concrete demonstration that small RNAs can couple catalytic activity to successful parasitism (Flores et al. 2004, 2012).

Modern viroids (small circular RNAs infecting plants) illustrate the plausibility of such catalytic RNA motifs: many viroids contain self-cleaving ribozymes, and these *cis*-acting motifs can be engineered to cleave RNA substrates *in trans* (Jimenez et al. 2015; Weinberg et al. 2019). It has therefore been proposed that in an ancient RNA world, a viroid-like replicator could have used a ribozyme to destroy competing RNAs, preventing superinfection by other selfish replicators: a rudimentary immune system (Broecker and Moelling 2019). Consistent with the broader concept, viroid infections can display SIEx in plants, where prior infection reduces establishment of a second strain, often mediated by RNA silencing (Kovalskaya and Hammond 2014; Carbonell 2022; Serra et al. 2023).

3.3 *Random Ligation and the Birth of the Protogenome: How to Domesticate Parasites*

A pivotal step in life's emergence was the formation of the first multigene genome, the protogenome. Building on earlier work (Broecker 2021b), I propose that it arose when an RNA ligase ribozyme randomly ligated members of a collectively autocatalytic RNA set into a single, multi-ribozyme strand. One key advantage of random ligation is domestication: by physically linking parasitic and host functions into one molecule, their transmission becomes coupled, turning genetic conflict into potential utility. In this view, prebiotic chemistry alone cannot sustain stable pathways; catalytic processes must be stabilized and linked by heredity, which ligation-mediated genomic linkage can provide by creating transmissible units (Negrón-Mendoza et al. 2025). Ribozymes capable of ligation have been demonstrated (Lincoln and Joyce 2009), as have polymerase ribozymes that can copy ligated RNAs (Horning and Joyce 2016). Under drying rewetting cycles, ligation could stitch distinct ribozymes into a single RNA that is then replicated, providing an immediate route to hard-link formerly independent—including parasitic—RNAs into a heritable unit (Fig. 10.2). In this framework, coupling parasite and host aligns their inheritance and promotes genomic cooperation, a dynamic expected to be

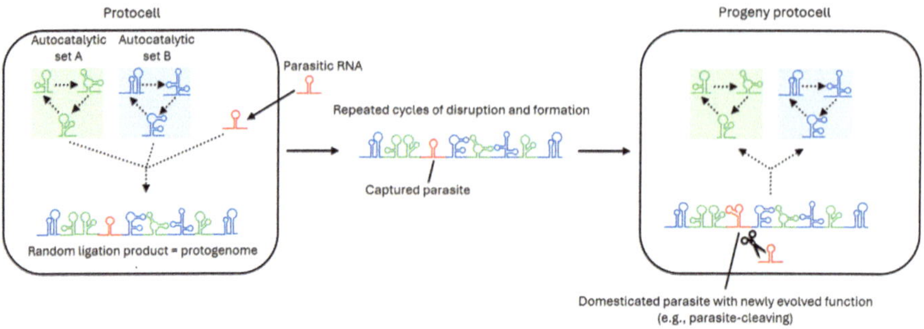

Fig. 10.2 Proposed protogenome in an RNA-world protocell. Left: a fatty-acid protocell (e.g., assembled during wet–dry cycling) encapsulates two collectively autocatalytic ribozyme sets that together constitute a protometabolism (Broecker 2021b; Damer and Deamer 2015). An RNA ligase ribozyme can occasionally generate a longer RNA by template-independent ligation, producing a protogenome that concatenates catalytic domains from both sets; parasitic RNAs (red) may also become incorporated (shown is one protogenome; however, multiple may have coexisted within one protocell, akin to early chromosome-like segments). During repeated wet (vesicle formation/reshuffling) and dry phases (partial dehydration and mixing), the protogenome can be redistributed into newly formed protocells (right), where its retained ribozyme domains can "seed" reconstruction of both autocatalytic sets. By physically linking multiple metabolic functions (and any captured parasite, in this example evolving to an anti-parasite defense) onto a single heritable molecule, this model makes inheritance of a complete catalytic toolkit more likely than stochastic co-encapsulation of many separate short RNAs. The model also mitigates the "first replicator" problem by reducing the requirement that a single long, fully autonomous self-replicating RNA must arise at once from random nucleotide assembly (Robertson and Joyce 2012). (Adapted from Broecker 2021b)

favored by selection as physical linkage aligns the transmission of cooperating replicators (Smith and Szathmáry 1993).

By tethering catalysts and parasites, a protocell forces helpful and harmful RNAs to share the same reproductive fate. However, "free" parasite variants could still arise and siphon resources, so domestication likely required safeguards. One option is to weaponize the incorporated sequence: the protogenome could produce antisense RNAs that sequester or cleave free variants while sparing junction-flanked domesticated copies, turning the co-opted sequence into an anti-parasite module. Sequence-guided antisense silencing is common today, from piRNA-mediated TE control in animals (Weick and Miska 2014) to antiviral siRNA in insects (Marques et al. 2013), supporting the plausibility of early RNA-guided defenses.

A complementary safeguard is self/nonself marking *via* simple chemical tags. A small "marking" ribozyme could install a base or 2′-OH modification on protogenome RNAs (including the domesticated parasite), while an "immune" cleavage ribozyme preferentially targets unmodified RNAs—an RNA-world analog of restriction–modification (Arber and Linn 1969). Modern transcriptomes show that chemical RNA marking is used in biology today: modifications including N6-methyladenosine are widespread and influence RNA fate (Dominissini et al. 2012; Meyer et al. 2012; Saletore et al. 2012).

By seeding multiple functions at once, a protogenome could outcompete lineages that must repeatedly co-encapsulate separate ribozymes (Smith and Szathmáry 1993; Chen et al. 2004). This linkage model promotes genomic cooperation and eases the "first replicator" problem by reducing the need for a single long, autonomous self-replicating RNA to appear in one step (Robertson and Joyce 2012). As a quasispecies with redundancy (e.g., duplicate ribozymes), protogenomes also supply variation for innovation, including early steps towards coding (Eigen 1971; Novozhilov and Koonin 2009). As proteins replaced some ribozymes and addiction modules evolved (Hayes 2003), the protometabolism would become increasingly tied to the protogenome. Rising polymerase fidelity would then permit larger genomes, more complex metabolic pathways, and additional domestication, ultimately enabling expansion and a transition to DNA (Forterre 2005). Overall, genome origins can be viewed partly as an anti-parasite strategy: merging rogue RNAs into a single heritable unit while retaining redundant sequence diversity as raw material for novelty.

What evidence supports multi-ribozyme protogenomes in this framework? Three prerequisites must hold: (i) ribozymes can generate ligation products without strict templating, (ii) ribozyme domains must remain functional when embedded in longer RNAs, and (iii) an RNA polymerase can copy such RNAs with sufficient fidelity to avoid an error catastrophe.

1. *Ligation feasibility*: Template-independent joining of single-stranded nucleic acids is plausible: T4 DNA ligase can ligate ssDNA without a template, and some archaeal RNA ligases show template-independent activity (Kuhn and Frank-Kamenetskii 2005; Zhang and Tripathi 2017). Ribozymes can also be promiscuous in substrate scope (Janzen et al. 2020), and short RNA ligase ribozymes can accept broad substrate combinations (Dhar et al. 2017). Fragmented hairpin ribozymes ligate "a wide variety of fragments with little or no regard for sequence" under freezing/dehydration (Vlassov et al. 2004; Kazakov et al. 2006). Small ligase ribozymes can ligate diverse substrates with minimal sequence constraints (Nomura and Yokobayashi 2019, 2023), and promiscuous ligases tolerating multiple activation chemistries have been evolved (DasGupta et al. 2024). Together, these findings make random side-product formation by ribozymes plausible, and evolutionarily meaningful when linkage couples catalytic functions to heredity, stabilizing multi-step chemical reactions (Negrón-Mendoza et al. 2025).
2. *Activity within longer RNAs*: Many ribozyme domains remain active when embedded in larger RNAs. Hammerhead motifs occur within circular ssRNA genomes (*Avsunviroidae* viroids, satellite circRNAs, hepatitis delta virus) and within retroelements (including SINEs, Penelope-like elements, retrozymes) (Di Serio et al. 2018; de la Peña et al. 2020; Wadkins and Been 2002; Ferbeyre et al. 1998; Cervera and de la Pena 2014; de la Peña and Cervera 2017). Twin-ribozyme introns provide another natural example (Nielsen et al. 2005; Nielsen et al. 2009), and *twister* and related ribozymes retain activity despite extensive flanks (McKinley et al. 2024). Engineered RNAs likewise support multi-domain

activity, including transcripts carrying up to ten active ribozyme domains (Shahi and Banerjea 2002; Kumar and Joyce 2003; Ivanov et al. 2005; Perriman et al. 1993; Cameron and Jennings 1994; Chen et al. 1992; Ohkawa et al. 1993). Collectively, these cases support the feasibility of amalgamated ribozyme domains in a ligated protogenome.

3. *Copy fidelity and the error threshold*: Polymerase fidelity limits protogenome length *via* Eigen's error threshold (Eigen 1971; Tarazona 1992; Domingo et al. 2012), and experimental evolution confirms fidelity is crucial to maintain heritable information in RNA-only systems (Papastavrou et al. 2024). For ribozymes (as opposed to protein enzymes), structural robustness can relax this threshold ~7-8x (Kun et al. 2005). With the best *in vitro* RNA polymerase ribozymes at ~97% per-base fidelity (Horning and Joyce 2016; Tjhung et al. 2020; Attwater et al. 2018; McRae et al. 2024), stable copying is on the order of $\leq$~250 nucleotides, potentially enough for several compact motifs, potentially including ultra-small catalytic cores down to 5 or even 3 nucleotides that have been reported (Turk et al. 2010; Nomura and Yokobayashi 2023). Polymerase activity has also been demonstrated under prebiotically plausible freeze–thaw/pH regimes (Attwater et al. 2025). Larger, more complex ribozymes (e.g., a ~200-nucleotide polymerase) would require higher fidelity, which selection could favor; directed evolution has already raised fidelity from ~91% to >97% (Attwater et al. 2018). At ~99%, the relaxed threshold reaches ~750 nucleotides, sufficient for multiple complex domains, while error-prone copying would also generate diversity, fueling innovation (Domingo et al. 2012).

Having outlined how random ligation could enable stable inheritance of metabolic pathways, domestication, and cooperation, we now follow the consequences: expanding protogenomes, a shift to DNA heredity, concurrent viral emergence, and the road to LUCA.

3.4 Protocells and Protoviruses: Compartment-Level Parasitism in the RNA World

In a pool of fatty-acid protocells, growth is not necessarily symmetric or benign. Model protocell experiments show that vesicles with higher internal solute loads (e.g., encapsulated RNAs) build osmotic pressure, take up membrane amphiphiles from the environment faster, and can even grow at the expense of neighboring vesicles *via* competitive lipid exchange (Chen et al. 2004; Hanczyc et al. 2003). This is reinforced by size-selective retention: small building blocks can exchange readily with the environment, whereas long RNAs are retained, and internally produced protogenomes can accumulate while continuing to draw in resources. Protocells with more active internal chemistry therefore divide more frequently and outcompete less active ones, consistent with theoretical models of lipid competition and growth-division dynamics (Shirt-Ediss et al. 2014; Taneja and Higgs

2025). Encapsulated catalysts amplify this selection: protocells harboring faster ribozymes or metabolic networks grow, divide, and accumulate building blocks more efficiently, enabling compartment-level Darwinian competition (Adamala and Szostak 2013b).

Membranes are also dynamic: fatty-acid vesicles can fuse under prebiotically plausible conditions, allowing contents to mix (Fan et al. 2023; Chen and Walde 2010). For a slow protocell, fusion with a faster one followed by takeover of lipids and contents is functionally similar to infection by a selfish replicator. As shorthand, I refer to especially successful compartments as protoviruses: containing autocatalytic sets that replicate so efficiently that they effectively parasitize other protocells/protoviruses by drawing in lipids, fusing, and overwriting and/or incorporating contents. In this view, protocell ecology already resembles virus-host dynamics: parasitism and cannibalism select for compartments that resist fusion, stabilize membranes, or degrade incoming RNAs, while "invasion-biased" compartments are selected for more efficient takeover, thereby sustaining an arms race. Consistent with this view, Koonin and colleagues proposed a "primordial virus world" in which major viral lineages emerged very early and may predate LUCA (Koonin et al. 2006; Koonin 2009). Building on this perspective, I suggest both LUCA's ancestors and early virus-like lineages arose from mixed protocell populations with frequent fusion, cannibalization, and horizontal exchange. Rather than two discrete categories, protocells and protoviruses form a continuum from self-maintaining chemistries to takeover-prone replicators whose "host" and "parasite" roles are context-dependent. Another distinctive feature might be that protocells evolved the ability to synthesize lipid components themselves (Cho et al. 2025) (as do present-day cells), essentially allowing for autonomous replication outside of wet–dry cycling environments, while protoviruses did not, requiring hijacking the protocells' metabolism for replication (as do modern viruses). (Note usage of *protovirus* differs from Temin's protovirus hypothesis, which concerned provirus/retrovirus origins (Temin 1971).)

4 From Protogenome to LUCA: Transition to DNA and Viral Emergence

As domestication linked parasitic RNAs to host heredity and replication fidelity improved, selection favored larger, more stable genomes, paving the way from RNA protogenomes to DNA genomes and to the lineage termed LUCA. Comparative reconstruction suggests LUCA already had a primordial immune toolkit: it likely encoded multiple Class 1 CRISPR-Cas effector families, consistent with an early RNA-targeting defense that may have lacked the canonical Cas1/Cas2 adaptation module (Moody et al. 2024). The same study infers a LUCA genome of ~2.5–3.0 Mb (encoding ~2600 proteins), far beyond any known RNA genome and consistent with DNA-based heredity.

A long-standing view is that genetic conflict helped drive the transition: DNA is more chemically stable, double-stranded, and compatible with proofreading and repair, making heritable information less vulnerable to parasitic corruption (Forterre 2006a; Koonin and Krupovic 2015). Multiple lines of evidence further implicate viruses as central players, inventing or disseminating DNA replication systems (Forterre 2006b; Koonin 2006; Koonin et al. 2006; Leipe et al. 1999). Notably, core replication machineries differ between bacteria and archaea/eukarya, consistent with multiple invention and/or horizontal transfer events (Leipe et al. 1999; Forterre 2006a; Koonin and Krupovic 2015). More explicit "virus-first" (virocentric) arguments go further than gene shuttling and posit viruses (and their ancestral lineages) as originators of key cellular innovations and identity/immune functions (Villarreal 2009; Koonin and Dolja 2013).

Extant genomes illustrate the size constraints of RNA. Even with proofreading, RNA viruses rarely exceed ~30–32 kb (e.g., coronaviruses; Denison et al. 2011), and the largest reported RNA virus genome (~41 kb nidovirus) relies on proofreading exoribonuclease activity (Saberi et al. 2018). LUCA, by contrast, is reconstructed at the megabase scale, consistent with RNA's higher fragility and limited repair compared with DNA's robust duplex chemistry and layered repair (Moody et al. 2024).

The emergence of the genetic code and translation further separated information storage (nucleic acids) from most catalysis (proteins), enabling a division of labor that expanded catalytic capacity and permitted larger, more regulatable genomes (Koonin and Novozhilov 2009). Mechanistically, the RNA-to-DNA transition need not have required modern protein replisomes; an evolved polymerase ribozyme has been shown to copy RNA into DNA *in vitro*, offering a plausible ribozymatic route to DNA genomes (Samanta and Joyce 2017).

With DNA came new parasite classes (plasmids, DNA viruses) and an entrenched arms race. Plasmids are selfish replicons that reshape host genomes and disseminate traits across lineages (Norman et al. 2009; Smillie et al. 2010), while viruses coevolved with, and helped sculpt, cellular defenses (Koonin and Dolja 2013). Compartmentalization may also contribute to defense: some hypotheses propose the eukaryotic nucleus evolved partly to separate genome replication/transcription from cytoplasmic processes, with some suggesting a virus-related origin of the nucleus (Bell 2001, 2020; Claverie 2006; Forterre 2010; Takemura 2020).

As life diversified, viruses and other selfish elements diversified with it, infiltrating new hosts and exchanging genes horizontally. The next sections examine how each domain of life elaborated distinct immune arsenals by repeatedly co-opting invader-derived nucleic-acid sequences. Before turning to this system, it is worth attempting to define what an immune system is.

5 What Is an Immune System?

A defining step in the emergence of living systems is the ability of an entity to maintain *identity*—to persist as a coherent unit despite continual exposure to foreign molecules and replicators. Classical immunology framed immunity as self/nonself discrimination, most prominently in Burnet's mid-twentieth-century formulation that "non-self" elicits immune responses while "self" is spared (Burnet and Fenner 1949). Modern biology shows this is an oversimplification: immune systems routinely eliminate altered self (e.g., dying or transformed cells) yet tolerate vast amounts of nonself, including dense commensal microbiota. In humans, the intestine alone contains tens of trillions of bacteria, underscoring that effective immunity often involves regulated coexistence, not rejection (Pradeu and Carosella 2006). Accordingly, many contemporary frameworks emphasize harm management and context-dependent regulation over a strict self/nonself binary (Duhamel and Salzet 2025).

Endogenous retroviruses (ERVs) make the same point at the genomic scale: viral sequences can become long-term residents and sometimes contribute to host regulation or defense, even though related exogenous viruses can be pathogenic (Wilkins 2010).

6 Immune Systems in Prokaryotes: Built on Mobile Genetic Elements

Prokaryotes (bacteria and archaea) are under constant attack by viruses (phages and archaeal viruses) and plasmids. They have evolved diverse antiviral and antiplasmid systems, many of which either originate from or are spread by mobile genetic elements (MGEs). Next, I summarize major defense strategies and their selfish-element roots.

6.1 Virus–Virus Warfare: Prophage-Mediated Immunity (Superinfection Exclusion)

One of the most direct antiviral strategies is superinfection exclusion (SIEx), where an existing infection prevents a secondary one (Bucher and Czyż 2024). In bacteria, SIEx often involves prophages encoding host-protective factors (Bondy-Denomy et al. 2016; Bucher and Czyż 2024). In *Pseudomonas aeruginosa*, a filamentous prophage protein (PfsE/Tip) downregulates type IV pili, blocking piliotropic phage attachment or DNA injection (Schmidt et al. 2022). In *Salmonella*, the P22 phage factor SieA provides another model: an inner-membrane exclusion protein prevents DNA translocation across the inner membrane (Leavitt et al. 2024). More generally, prophage carriage can confer broad protection *via* receptor masking and intracellular interference (Bondy-Denomy et al. 2016; Cumby et al. 2012; Samson et al. 2013).

Although prophages can impose fitness costs, surface-modification SIEx can be cost-neutral or context-dependent beneficial (Bondy-Denomy et al. 2016; Bucher et al. 2024). Functionally, SIEx resembles infection-acquired resistance: after "vaccination" by a temperate or mildly virulent phage, cells become resistant to related phages, and vertical transmission (e.g., integrating/temperate phages) makes the resistance heritable and amenable to further refinement (Fischer and Hackl 2016; Bucher et al. 2024).

Evidence for SIEx in archaea is also growing. Virus-borne mini-CRISPR systems can mediate sequence-specific exclusion of competitor viruses (e.g., Portogloboviridae; SNJ1) (Medvedeva et al. 2019; Alarcón-Schumacher et al. 2022; Chen et al. 2020), and chronic/carrier states in *Sulfolobus* rudiviruses (e.g., SIRV3) can bias or block subsequent infections (Papathanasiou et al. 2019). Recent overviews emphasize that archaeal viruses attach to defined surface structures (e.g., S-layer components and filamentous appendages) and enter *via* specific routes, implying SIEx mechanisms analogous to bacteria (Sensevdi et al. 2024).

Cross-domain note: While this section focuses on prokaryotes, SIEx is widespread beyond them, including plants (Zhou et al. 2019), giant amoebal DNA viruses where prior infection downregulates host-surface entry pathways (Aquino et al. 2024), and in animal viruses including SARS-CoV-2, where coinfection with other SARS-CoV-2 particles/lineages can be limited by SIEx-like mechanisms (Sims et al. 2025). A striking eukaryotic example is Mavirus, a virophage that integrates into the genome of protist *Cafeteria roenbergen*sis and can protect it from the lethal giant virus CroV upon reactivation (Fischer and Hackl 2016; Koslová et al. 2024). Together, these cases show that exclusion-like strategies recur across the tree of life.

6.2 *CRISPR-Cas Adaptive Immunity: A Palimpsest of Mobile Genes*

CRISPR-Cas is an adaptive immune system that captures invader fragments (spacers), integrates them into CRISPR arrays, and uses spacer-derived RNAs to guide Cas nucleases to destroy matching elements upon re-encounter (Mosterd et al. 2021; Koonin and Makarova 2024). Comparative genomics suggests that CRISPR-Cas is a mosaic assembled from mobile genetic modules. The adaptation integrase Cas1 is closely linked to casposons, DNA transposons whose casposase is mechanistically and structurally related to Cas1 (Krupovic et al. 2014, 2017; Koonin and Makarova 2017; Hickman et al. 2020). Cas2 is homologous to VapD-like toxins from toxin-antitoxin modules, consistent with recruitment from selfish elements (Kwon et al. 2012; Bertelsen et al. 2021; Anand et al. 2022). Some systems, notably Type III, acquire RNA-derived spacers *via* Cas1-reverse transcriptase fusions likely

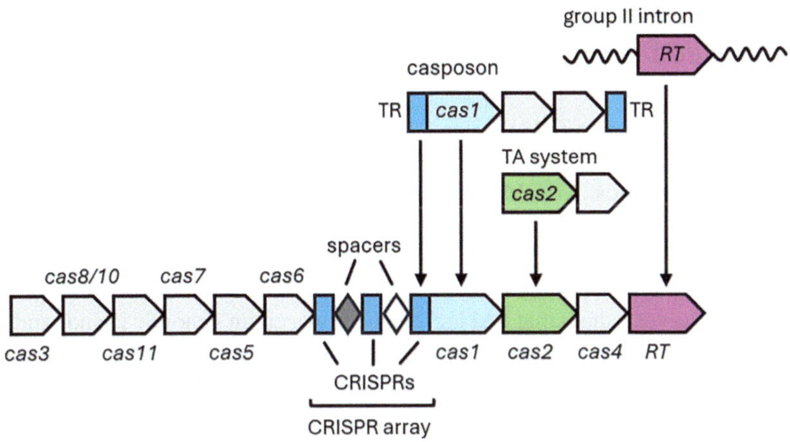

Fig. 10.3 Mobile genetic element origins of CRISPR-Cas systems (illustrated for a Class 1 pathway). Comparative genomics supports a modular origin of CRISPR-Cas *via* domestication of multiple MGEs. Casposons likely contributed the Cas1 integrase ("casposase") and a related DNA-integration mechanism, plausibly first used for inserting casposon terminal repeats and later repurposed for CRISPR spacer acquisition. Cas2 is homologous to VapD-like toxins from toxin–antitoxin (TA) systems, consistent with recruitment from selfish elements. In some Type III systems, a reverse transcriptase (RT) acquired from group II intron retroelements (Novikova and Belfort 2017) enables RT-dependent acquisition of spacers from RNA. Adapted from Koonin and Makarova (2019)

sourced from group II intron retroelements (Silas et al. 2016; Mohr et al. 2018; González-Delgado et al. 2019; Molina-Sánchez et al. 2024).

Effector nucleases also point to mobile ancestry: Cas12 (Type V) appears derived from the TnpB protein of IS200/IS605 transposons, and Cas9 (Type II) is homologous to IscB, both sharing the RuvC-like nuclease fold (Altae-Tran et al. 2023; Nakagawa et al. 2023; Koonin et al. 2023). Mobile elements still shuttle CRISPR in nature: some isolates of the lytic phage ICP1 encode a functional CRISPR-Cas system that disables the phage-inducible chromosomal island-like element (PLE), a subviral element in ICP1's host *Vibrio cholerae* that blocks ICP1 replication, by cleaving PLE DNA as it excises/circularizes and amplifies during infection (Seed et al. 2013; Naser et al. 2017; Barth et al. 2021; Zhang et al. 2024). CRISPR-Cas systems occur in ~40–50% of bacteria and in a majority of archaea (Pinilla-Redondo et al. 2022; Makarova et al. 2025). Overall, CRISPR reads as a palimpsest of domesticated mobile genes, illustrating how transposons and viruses seed new defense innovations (Koonin and Krupovic 2015). Figure 10.3 summarizes this modular origin for a representative Class 1 system: casposons as a likely source of Cas1, toxin–antitoxin modules as a plausible origin for Cas2, and, in some Type III contexts, group II introns contributing reverse transcriptase domains that enable RNA-derived spacer acquisition.

6.3 *Restriction–Modification Systems: Self/Nonself from Mobile DNA*

Restriction–modification (RM) systems are among the most widespread prokaryotic defenses and act as an innate barrier against incoming DNA, including phage genomes and plasmids. An RM system pairs a restriction endonuclease that cleaves specific DNA motifs with a cognate methyltransferase that methylates the same motif on the host genome, thereby marking it as "self." Unmethylated ("nonself") DNA is cut and typically degraded.

Genomic surveys show that RM loci frequently occur in genomic islands and near mobility genes (integrases/transposases), and some are found in transposon-like contexts (e.g., inverted repeats, target-site duplications), consistent with mobility and, in some cases, mobile-element ancestry (Kobayashi 2001; Oliveira et al. 2014; Zheng et al. 2016). For example, plasmid-encoded RM systems with transposon-like organization are found in *Rhizobium* and *Lactococcus* (Rochepeau et al. 1997; Twomey et al. 1998). RM cassettes remain mobile today: they can be carried on conjugative plasmids (Birkholz et al. 2022), prophages (Anton et al. 2025), and defense islands that spread as multigene cassettes between strains (Zhao et al. 2025).

Once established, RM systems can diversify and spread by horizontal gene transfer, providing hosts with a shifting repertoire of defenses while also behaving as selfish modules in their own right (Kusano et al. 1995; Kobayashi 2001; Zheng et al. 2016). RM can even cause "autoimmunity": stochastic cleavage of self-DNA can trigger costly repair mechanisms, imposing a fitness burden (Pleška et al. 2016). A related phenomenon is postsegregational killing: in *Escherichia coli*, plasmid-encoded RM complexes can create a state in which plasmid loss leaves the chromosome insufficiently methylated while the restriction enzyme persists, resulting in lethal self-restriction (Handa and Kobayashi 1999; Handa et al. 2000). Altogether, RM systems exemplify how self/nonself discrimination at the DNA level is shaped by continual exchange and conflict with MGEs.

6.4 *Toxin–Antitoxin Systems: Addiction Modules Turned Abortive Defense*

Beyond RM, many bacteria encode numerous toxin-antitoxin (TA) systems, and a growing subset is now clearly implicated in antiphage defense. TA loci typically pair a stable toxin with a labile antitoxin; when the antitoxin is degraded or its synthesis disrupted, the toxin arrests growth or kills the cell. Several TA systems mediate abortive infection (Abi): during phage attack, infection-driven perturbations of the TA balance activate the toxin, shutting down key host processes needed for phage replication, essentially sacrificing the infected cell for the benefit of surrounding clonemates (Blower et al. 2012; Watson et al. 2019; Lopatina et al. 2020; Arias et al. 2022).

A well-studied example is the Type III ToxIN system of *Pectobacterium atrosepticum*, where the RNase toxin ToxN is normally neutralized by the RNA antitoxin ToxI, but phage infection triggers ToxN activation and rapid cell death, aborting phage production (Fineran et al. 2009; Blower et al. 2012). Other TA systems act through different molecular targets, including toxins that inhibit translation, disrupt membranes, or poison DNA replication (Lopatina et al. 2020). This logic also fits Villarreal's broader "addiction module" framing, in which diverse antiviral defenses, including TA and RM, can be understood as coupled toxin/protection systems that stabilize resident genetic information while excluding competitors (Villarreal 2011, 2012).

TA systems were originally characterized as plasmid-maintenance ("addiction") modules: loss of a plasmid encoding a TA locus kills plasmid-free segregants as the long-lived toxin outlasts the rapidly degraded antitoxin (Hayes 2003; Goeders and Van Melderen 2014). In this view, addiction modules are not merely "selfish" stabilizers: they can be exapted into antiviral Abi because the same kill-switch that enforces element maintenance can also terminate infected cells and block phage spread (Villarreal 2011, 2012). Homologous TA families are now widespread on chromosomes, plasmids, prophages, and other MGEs, and many cluster in defense islands together with other antiviral modules (Goeders and Van Melderen 2014; Lopatina et al. 2020). Functional studies show that some chromosomal TA systems are activated during infection and inhibit diverse phages, consistent with exaptation of addiction modules into dedicated antiphage Abi systems (Song and Wood 2020; Lopatina et al. 2020). Thus, TA systems exemplify a recurring theme: mechanisms that once enforced mobile-element persistence can be repurposed into population-level defenses against viral attack.

6.5 *Other Prokaryotic Systems and Defense Islands*

Beyond CRISPR-Cas, RM and TA systems, prokaryotes encode a rapidly expanding set of antiphage systems, many discovered only in the past decade. Systematic mining of "defense islands," genomic regions enriched for defense genes and MGE signatures, revealed multiple new systems, including bacteriophage exclusion (BREX), defense island system associated with restriction–modification (DISARM), cyclic-oligonucleotide-based antiphage signaling systems (CBASS), among many others (Doron et al. 2018; Koonin et al. 2017b; Millman et al. 2020; Hobbs and Kranzusch 2024). These islands frequently lie within or near prophages, transposons, integrative conjugative elements, and other MGEs, consistent with horizontal dissemination of multigene defense cassettes (Doron et al. 2018; Botelho 2023; Hochhauser et al. 2023; Li et al. 2025; Beamud et al. 2024).

BREX illustrates these recently discovered systems: it is typically a six-gene cassette that uses host methylation patterns as an epigenetic "self" mark and restricts phage development; BREX loci are widespread and often co-occur with other defense systems in composite islands (Goldfarb et al. 2015; Barrangou and van der Oost 2015; Picton et al. 2021; Kelly et al. 2023). DISARM is another multigene system that combines a DNA methylase with helicase- and nuclease-like components to provide broad protection against diverse dsDNA phages, and is frequently found in large defense-associated regions (Ofir et al. 2018). CBASS, by contrast, is signaling-based: a cGAS/DncV-like cyclase senses infection and generates cyclic oligonucleotides, eventually culminating in growth arrest or cell death to halt phage propagation (Millman et al. 2020; Duncan-Lowey and Kranzusch 2022; Wang and Zhang 2023; Hobbs and Kranzusch 2024).

Prokaryotic Argonaute (pAgo) proteins add another layer of nucleic-acid-guided defense. pAgos use short guide DNAs or RNAs to cleave complementary nucleic acids, providing protection against plasmids and phages (Swarts et al. 2014; Kuzmenko et al. 2019; Olovnikov et al. 2013). Their catalytic PIWI domain adopts an RNase H-like nuclease fold, consistent with descent from an ancient nucleic-acid-cleaving enzyme family repeatedly redeployed in mobile elements and host defense (Burroughs et al. 2013; Swarts et al. 2014). Recent work on Asgard archaeal Argonautes with RNA-guided RNA silencing supports an early origin of Argonaute-based defense, potentially predating the last eukaryotic common ancestor (LECA), although vertical inheritance versus extensive horizontal transfer remains debated (Bastiaanssen et al. 2024; Koonin et al. 2017b). In many genomes, pAgo genes occur within defense islands or near other defense loci, again linking them to mobile genetic contexts (Doron et al. 2018; Beamud et al. 2024).

Large-scale genomic surveys further show that transposons can act as vehicles of immune innovation. Tn7-like transposons frequently carry "cargo" defense genes, including diverse antiphage systems and TA modules, mobilizing entire defense cassettes between hosts (Benler et al. 2021).

Together, these examples reinforce a general pattern: defense systems are concentrated in mosaic, mobile regions of prokaryotic genomes rather than being randomly distributed. Comparative genomics indicates that such loci often co-localize with prophages, transposons, and conjugative elements, and are repeatedly found at preferred insertion sites, consistent with assembly and spread through horizontal transfer (Hochhauser et al. 2023; Botelho 2023; Li et al. 2025; Beamud et al. 2024). It is therefore plausible that many defense systems began as mobile elements that conferred a collateral host advantage—phage or plasmid resistance—and were retained and elaborated by selection. As metagenomic surveys and functional screens expand, newly discovered systems frequently bear the imprint of viruses and transposons in their composition or genomic context, underscoring that the prokaryotic "defensome" is largely built from MGEs.

7 Eukaryotic Immune Innovations *via* Element Co-option

The emergence of eukaryotes ~2 billion years ago brought larger, more structured genomes and intensified pressure from genomic parasites, especially due to the expansion and diversification of MGEs. As eukaryotes diversified (and in some lineages evolved multicellularity), defense became layered: intracellular nucleic-acid surveillance was complemented, in animals, by specialized immune cells. Rather than simply inheriting specific prokaryotic defense systems, eukaryotes built on ancient molecular principles and enzyme lineages, and repeatedly domesticated viral and transposon components for new immune functions. This section highlights such co-option events in eukaryotic innate and adaptive immunity.

7.1 RNA Silencing: Transposon-Derived Defense in Eukaryotes

Eukaryotes, including fungi, plants, and animals, share a conserved gene-silencing machinery, RNA interference (RNAi), in which Dicer-like nucleases process double-stranded RNA into small RNAs that load into Argonaute proteins and guide them to complementary targets for cleavage or repression (Fire et al. 1998; Shabalina and Koonin 2008). Although RNAi-derived pathways (siRNA, piRNA, miRNA) now regulate many endogenous genes, comparative genetic analyses strongly support an origin in defense against viruses and MGEs (Obbard et al. 2006, 2009; Shabalina and Koonin 2008).

In animals, the piRNA pathway is specialized for transposon control in the germline. piRNAs are produced from piRNA clusters, heterochromatic “trap” loci enriched in defective TE fragments functioning as genomic archives of past invasions (Brennecke et al. 2007; Girard and Hannon 2008). These loci generate antisense piRNAs that load into PIWI proteins and direct transcriptional and post-transcriptional silencing of active elements, effectively providing an RNA-guided memory of MGEs (Brennecke et al. 2007; Girard and Hannon 2008; Obbard et al. 2009). Conceptually, this parallels CRISPR: invader-derived fragments are stored in the genome and transcribed into guides that later target the same invaders (Koonin and Krupovic 2015).

The RNAi machinery also reflects deep evolutionary—and in part mobile—connections. Phylogenetic reconstruction indicates that core components (Dicer, Argonaute/PIWI, and RNA-dependent RNA polymerases, RdRPs) were already present in LECA (Shabalina and Koonin 2008). Eukaryotic RdRPs belong to the double-ψ β-barrel (“two-barrel”) polymerase superfamily and share a catalytic core with other ancient RNA polymerases, consistent with descent from an early RNA-replicating enzyme lineage later recruited for small-RNA amplification (Iyer et al. 2003; de Farias et al. 2017). Argonaute proteins occur across all three domains: pAgos often function in defense as guide-dependent nucleases, while eukaryotic Argonautes execute RNAi. In both, catalysis relies on a PIWI domain with an RNase H-like fold—part of a broadly distributed nuclease superfamily repeatedly deployed in both MGEs and host defense (Swarts et al. 2014; Majorek et al. 2014; Shabalina and Koonin 2008).

Viruses, in turn, encode diverse suppressors of RNA silencing, and RNAi genes—particularly antiviral arms—often evolve under strong positive selection, consistent with ongoing arms races between host silencing and viral countermeasures (Ding and Voinnet 2017; Burgyán and Havelda 2011; Obbard et al. 2006, 2009). Taken together, eukaryotic RNA silencing is best understood as defense shaped by mobile genetic conflict: hosts use invader-derived fragments (small RNAs) and conserved enzymes (RdRPs, Argonautes) to restrain viruses and transposons. Without this persistent pressure, such an elaborate and conserved silencing immunity is unlikely to have emerged.

7.2 Endogenous Viruses as Antiviral Tools

Eukaryotic genomes, especially in vertebrates, are densely populated by endogenous retroviruses (ERVs), ancient infection fossils that comprise ~5–10% of mammalian DNA. Rather than inert "junk," some ERV-derived sequences have been exapted as antiviral restriction factors or as regulatory modules that tune innate immune-gene expression (Grandi and Tramontano 2018; Srinivasachar Badarinarayan and Sauter 2021).

A classic ERV-based defense is receptor interference: expression of an ERV envelope (Env) glycoprotein can occupy or down-modulate a viral entry receptor, blocking infection by related exogenous viruses—effectively a form of SIEx. In mice, the Fv4 locus encodes an ERV-derived Env that protects against ecotropic murine leukemia virus (MLV) *via* receptor interference (Taylor et al. 2001; Takeda and Matano 2007; Malfavon-Borja and Feschotte 2015). Similar Env-based restriction evolved repeatedly, including Refrex-1 in cats, an ERV-derived Env restriction factor against feline leukemia viruses (Miyake et al. 2022). An endogenous retroviral Env (HERV-R) has also been reported to inhibit SARS-CoV-2 *in vitro*, although the precise mechanism remains unclear (Ansari et al. 2023).

Other ERV-derived restriction factors act post-entry. The mouse factor Fv1, derived from an ERV *gag* gene, binds incoming retroviral capsid proteins and blocks infection post-entry but before integration; *gag*-derived restriction factors show long-term positive selection, consistent with a sustained arms race with retroviruses (Yap et al. 2014; Young et al. 2018).

ERVs also shape immunity through RNA products and regulatory DNA. Bidirectional ERV transcription can generate RNAs (including dsRNA species) that engage pattern-recognition pathways and modulate interferon-stimulated genes (ISGs), linking ERV activity to innate immune tone (Srinivasachar Badarinarayan and Sauter 2021;Chen et al. 2019). At the DNA level, ERV long terminal repeats (LTRs) are potent promoters and enhancers, and comparative epigenomics show ERV/LTR insertions repeatedly rewired interferon-inducible enhancer networks near innate immune genes (Chuong et al. 2016; Ito et al. 2017; Wang et al. 2022; Krchlikova et al. 2025). In primates, immune clusters such as the major histocompatibility (MHC) locus are especially ERV-rich; for example, HERV-K (HML-10) proviruses are biased toward immune-gene neighborhoods and have been implicated in interferon-responsive regulatory effects (Broecker et al. 2016; Grandi and Tramontano 2018).

The youngest human ERV (HERV) family, HERV-K (HML-2), illustrates how ERV activity can intersect with the antiviral state. In preimplantation embryos, HERV-K expression correlates with induction of ISGs such as IFITM1 and an antiviral transcriptional program (Grow et al. 2015). In addition, HERV-K Rec protein can enhance IFITM1 levels and contribute to an antiviral phenotype, and high HERV-K expression contexts are linked to heightened interferon responses (Guo et al. 2022; Bao et al. 2024). Together, these cases underscore a recurring theme: hosts can turn "enemy" modules (Env, Gag, regulatory proteins, LTR-mediated regulation) into restriction factors and immune switches, leaving durable signatures of virus-derived immune innovation (Malfavon-Borja and Feschotte 2015; Wang et al. 2022; Jern and Greenwood 2024). Non-retroviruses have been co-opted as well, albeit less frequently, since genome integration is not part of their lifecycle. For example, an endogenous bornavirus-like nucleoprotein in squirrels can inhibit Borna disease virus replication (Fujino et al. 2014), and in mosquitoes, an endogenous cell-fusing agent virus suppresses replication of cognate RNA viruses *in vivo* (Suzuki et al. 2020).

Finally, koala retrovirus (KoRV) and Jaagsiekte sheep retrovirus (JSRV) provide rare snapshots of recent/ongoing endogenization. KoRV began integrating into the koala germline within the last ~50,000 years and remains geographically polymorphic (Tarlinton et al. 2006; Zheng and Legione 2025). Early anti-KoRV defense dynamics may already be visible: defective recombinant KoRV variants (recKoRVs) formed by recombination between KoRV and older endogenous retroelements, often lack essential coding regions (notably *pol*) and can proliferate, disrupting replication-competent KoRV insertions and contributing to viral decay (Löber et al. 2018; Tarlinton et al. 2022). Endogenous JSRVs (enJSRVs) represent an older, more consolidated system: multiple enJSRV proviruses accumulated over millions of years, and at least two loci (enJS56A1 and enJSRV-20) evolved transdominant restriction of exogenous JSRV *via* receptor interference and/or Gag-mediated restriction (Arnaud et al. 2007; Sistiaga-Poveda and Jugo 2014; Cumer et al. 2019; Moawad et al. 2024). Together, KoRV and enJSRVs illustrate a continuum: endogenization can begin as pathogenic germline invasion, with early restriction emerging as other retroelements disrupt proviruses, and can mature into stable, heritable antiviral functions.

7.3 The Interferon System and "Internal Tripwires"

Multicellular animals, especially vertebrates, evolved a global antiviral alarm system: the interferon (IFN) response. Pattern-recognition receptors (PRRs) such as the RNA sensors RIG-I and MDA5, the cytosolic DNA sensor cGAS, and endosomal Toll-like receptors (TLRs) detect nucleic-acid features that often accompany viral or mobile-element activity (e.g., 5′-triphosphate/uncapped RNA, long dsRNA, cytosolic DNA) (Kell and Gale Jr 2015; Roers et al. 2016; Chow et al. 2015; Wang et al. 2024). PRR signaling induces type I/III interferons, which in turn upregulate hundreds of ISGs that establish an antiviral state.

Many ISGs carry signatures of long-term conflict with viruses and retroelements. APOBEC3 cytidine deaminases restrict retroviruses and endogenous retroelements by mutating their nucleic acids; they expanded in mammals and show recurrent positive selection consistent with prolonged arms races (Duggal and Emerman 2012; Uriu et al. 2021; Ito et al. 2020; Jonathan and Ikeda 2023). TRIM5α, which recognizes retroviral capsids, likewise shows lineage-specific adaptations and repeated positive selection across primates, consistent with historical exposure to diverse retroviruses (Sawyer et al. 2005; Johnson and Sawyer 2009; Newman et al. 2006; Spada et al. 2024). Similar patterns appear in nucleic-acid-sensing pathways, including the PYHIN-family DNA sensors (e.g., AIM2/IFI16), which exhibit adaptive evolution in primates compatible with persistent conflicts involving DNA viruses and intracellular DNA elements (Cridland et al. 2012; Man et al. 2016; Cagliani et al. 2014).

The IFN system can also exploit endogenous mobile elements as "internal tripwires." When epigenetic repression of ERVs or other MGEs is relaxed, by demethylating drugs, genotoxic stress (e.g., during chemotherapy), or developmental reprogramming, these loci can produce RNA species that are sensed as "non-self," triggering type I IFN ("viral mimicry") (Zhao et al. 2018; Clapes et al. 2021; Roers et al. 2016; Cooles et al. 2025). LINE-1 activity can similarly generate endogenous RNA species that engage RNA-sensing pathways; MDA5 and RIG-I signaling can suppress LINE-1 and induce IFN responses (Zhao et al. 2018; Yan et al. 2022). In hematopoietic regeneration, TE-derived RNAs activate MDA5-dependent inflammatory signaling that supports recovery and proliferation (Clapes et al. 2021). Consistently, derepressed endogenous retroelements are now recognized as major interferonogenic stimuli in autoimmune disease and cancer-relevant contexts (Roers et al. 2016; Sauter and Kirchhoff 2021).

Overall, the IFN system reflects layered virus–host coevolution: PRRs are tuned to nucleic-acid signatures typical of mobile elements, ISGs such as APOBEC3s and TRIM5α diversified under viral pressure, and endogenous retroelements can function as built-in alarm wires. Even without direct co-option of a single viral gene, innate immunity is shaped by enduring conflict with selfish genetic elements.

7.4 Adaptive Immunity in Vertebrates: A Transposon's Gift

The most elaborated immune innovation in jawed vertebrates is adaptive immunity, which generates vast repertoires of antibodies and T-cell receptors and supports long-lived immunological memory. Strikingly, its core diversification mechanism, V(D)J recombination, derives from an ancient DNA transposon. The RAG1–RAG2 locus encodes the recombinase that cuts and rejoins V, D, and J segments, and it bears clear hallmarks of *Transib*-like transposons: RAG1 is homologous to *Transib* transposases, and the recombination signal sequences (RSSs) resemble *Transib* terminal inverted repeats (TIRs) (Kapitonov and Jurka 2005; Kapitonov and Koonin 2015). From a virocentric perspective, this system can also be framed as part of a broader history in which persistent viruses and genetic parasites supplied coordinated machinery to vertebrate lineages, rather than single genes in isolation (Villarreal 2009, 2011).

Comparative genomics and structural work support stepwise domestication. Active RAG-like (RAGL) transposons are found in invertebrates (e.g., amphioxus and echinoderms), and across multiple protostome lineages, where they remain mobile or appear partially domesticated (Huang et al. 2016; Martin et al. 2020; Yakovenko et al. 2021). In an early jawed-vertebrate ancestor (~500 Mya), a *ProtoRAG*-like element encoding RAG1L and RAG2L likely invaded the germline and was co-opted: the RAG proteins were repurposed from cut-and-paste transposition into a site-restricted recombination system acting on RSS-flanked antigen-receptor gene segments (Flajnik and Kasahara 2010; Carmona et al. 2016; Carmona and Schatz 2017). In this model, the transposon ends (TIRs) became RSSs, while the DDE/RNase H-like catalytic core was retained in RAG1 (Kapitonov and Jurka 2005; Carmona and Schatz 2017). The result is a tightly regulated program shuffling V, D, and J modules to generate enormous receptor diversity from a finite genome. Domestication required loss of mobility by decoupling RAG genes from TIRs and restricting RAG activity to antigen-receptor loci, consistent with today's architecture in which RSSs flank V/D/J segments while RAG1/2 act *in trans* (with RAG2 as RAG1's cofactor) (Fig. 10.4).

This transposon-derived solution appears specific to jawed vertebrates: jawless vertebrates (lampreys and hagfish) instead use variable lymphocyte receptors assembled by gene conversion of leucine-rich repeat cassettes using lineage-specific cytidine deaminases (Pancer et al. 2004; Boehm et al. 2012; Flajnik and Kasahara 2010). The coexistence of two unrelated systems underscores how transposon capture in the gnathostome lineage redirected a mobile-element machinery into the core of adaptive immunity (Pancer and Cooper 2006; Litman et al. 1999).

Other pillars of adaptive immunity trace to antiviral or retroelement restriction. Activation-induced cytidine deaminase (AID), essential for class switch recombination and somatic hypermutation, belongs to the AID/APOBEC cytidine deaminase family that includes intrinsic restriction factors against retroviruses and retrotransposons (MacDuff and Harris 2006; Koito and Ikeda 2011; Knisbacher et al. 2016). An ancestral anti-retroelement mutator function was redeployed into programmed diversification at immunoglobulin loci (Harris and Dudley 2015; Meshcheryakova et al. 2021). Likewise, the MHC region is densely populated by endogenous retroelements whose LTRs and related sequences contributed regulatory modules shaping antigen-presentation gene expression (Chuong et al. 2016; Grandi et al. 2017; Flajnik and Kasahara 2010). Together, V(D)J recombination, AID-driven diversification, and immune-gene regulatory rewiring exemplify how MGEs repeatedly supplied the raw machinery for vertebrate immune innovation.

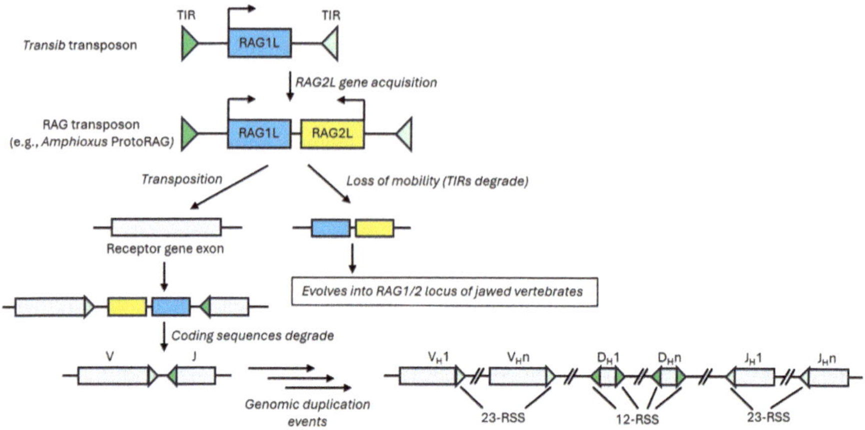

Fig. 10.4 Proposed mobile genetic element origin of jawed-vertebrate adaptive immunity *via* V(D)J recombination. A leading model traces the RAG1 catalytic core of V(D)J recombination to *Transib* transposons, whose transposase is homologous to RAG1/RAG1-like (RAG1L) and whose terminal inverted repeats (TIRs) show sequence/architectural features consistent with an evolutionary relationship to recombination signal sequences (RSS) (Kapitonov and Jurka 2005; Huang et al. 2016). An extant "living fossil" consistent with this scenario is *ProtoRAG*, a RAG1L-RAG2L-encoding transposon discovered in amphioxus, in which RAG1L and RAG2L are linked and flanked by TIRs, supporting the idea that an ancestral mobile-element encoded both components (Huang et al. 2016). During domestication in an early jawed-vertebrate ancestor, the transposase activity was repurposed from transposition to site-restricted recombination, with the transposon ends (TIRs) giving rise to RSSs that flank V, D, and J gene segments (D segments are flanked on both sides; V and J segments on one side), while the RAG genes became fixed at a dedicated RAG1/2 locus and lost autonomous mobility. The characteristic 12/23 rule—joining of a 12-bp spacer RSS with a 23-bp spacer RSS—has been proposed to reflect ancestral asymmetry of transposon ends and ensure productive V(D)J assembly (Kapitonov and Jurka 2005). Expansion of the V, D, and J segment repertoires likely occurred subsequently *via* duplication and diversification of receptor gene segments (Fugmann et al. 2000). Bottom right: schematic organization of a representative immunoglobulin heavy-chain V(D)J locus

8 Experimental Outlook: Testing the Model

The model proposed in this chapter spans events from the origin of life to modern cells. Many steps are not directly observable, but several experimental strategies can test whether protocell-level "immunity," protogenome formation, and mobile-element co-option are mechanistically plausible.

8.1 *Simulating Protogenome Formation*

The random-ligation protogenome hypothesis can be reenacted with a minimal *in vitro* system that combines a collection of ribozymes (e.g., an autocatalytic set) and appropriate RNA ligase and RNA polymerase ribozymes inside simple compartments (fatty-acid vesicles or droplets) while applying wet–dry or freeze–thaw cycling to promote ligation and copying. Cross-replicating ligase ribozymes (Lincoln and Joyce 2009) and relatively processive RNA polymerase ribozymes (Horning and Joyce 2016; Attwater et al. 2013, 2018, 2025) already exist, and ribozymes can operate in protocell-like vesicles (Mansy et al. 2008; Adamala and Szostak 2013a, b). Promiscuous, near sequence-independent ligation can occur under dehydration/freezing conditions (Vlassov et al. 2004; Nomura and Yokobayashi 2019; Nomura and Yokobayashi 2023). The key readouts would be whether multi-ribozyme chimeras arise, become enriched under replication, and retain multiple catalytic activities.

8.2 Recreating Ribozyme "Immunity"

RNA-only defense can be probed by evolving ribozymes that cleave a defined parasitic RNA in a replicating environment (e.g., using Qβ replicase, or a polymerase/ligase ribozyme system). Starting from random RNA pools or known self-cleaving motifs, one can impose selection such that replication is impaired unless the parasite is inactivated. *In vitro* evolution can yield *trans*-cleaving ribozymes with strong sequence specificity (Breaker and Joyce 1994; Jimenez et al. 2015); the critical test here is whether "guard" ribozymes emerge under parasite pressure and whether they can discriminate between closely related RNAs (e.g., free versus junction-flanked variants), providing a tractable proxy for early self/nonself recognition.

8.3 Mining Genomes for New Co-options

Comparative genomics coupled with functional assays can systematically expand the catalog of domesticated MGEs in defense. Large surveys show that diverse defense systems often cluster in "defense islands" enriched for transposases, integrases, and prophage remnants (Makarova et al. 2011, 2020; Doron et al. 2018; Gao et al. 2020). Scaling these analyses across expanded genome and metagenome datasets can identify candidate mobile-derived defense genes based on domain composition and synteny. Machine-learning and context-based prediction pipelines have already accelerated discovery of new antiphage modules (Doron et al. 2018; Millman et al. 2020; Tesson et al. 2022); focusing these approaches on transposase-like domains, RNase H-like folds, and viral structural signatures should reveal additional domestications, which can then be cloned and tested for antiphage/antiplasmid activity in challenge assays (Doron et al. 2018; Gao et al. 2020). A parallel opportunity exists in plants, whose genomes are often extremely TE-rich (e.g., maize and wheat with >80% repeats/TEs) (Schnable et al. 2009; International Wheat Genome Sequencing Consortium 2018). In plants, TEs can supply stress-responsive regulatory elements and shape nucleotide-binding leucine-rich repeat (NLR) resistance-gene clusters (Slotkin and Martienssen 2007; Lisch 2013; Quadrana et al. 2016), while endogenous pararetroviral sequences can generate small RNAs that silence related exogenous viruses, functioning as heritable antiviral memory (Richert-Pöggeler et al. 2022; Mette et al. 2002). TE-derived small RNAs also contribute broadly to RNA-directed DNA methylation and transcriptional silencing programs with direct relevance to defense (Matzke and Mosher 2014; Slotkin and Martienssen 2007). Together, these approaches can uncover transitional forms and test the broad prediction that many immune modules originate as mobile-element "spare parts," later refined by selection (Lisch 2013; Quadrana et al. 2016).

9 Conclusion: An Integrative Evolutionary Scenario

Stepping back, a consistent pattern emerges: genetic conflict can trigger evolutionary creativity. Escalation between host defenses and parasite counter-defenses has driven genomic and molecular novelty in ways that are difficult to explain from a purely host-centric perspective. Selfish genetic elements such as TEs and viruses are widely recognized as major drivers of evolutionary innovation (Werren 2011), and their interaction with hosts is often framed as a Red Queen dynamic in which both sides must keep evolving to maintain relative fitness (McLaughlin Jr and Malik 2017).

This chapter traced that logic from the origin of life to modern immunity. In primordial RNA protocells, the emergence of parasitic replicators would have immediately selected for rudimentary defenses, including catalytic RNAs that cleave other RNAs. Formation of a multigene protogenome by random ligation can be viewed as an early co-option of "rogue" molecules into a cooperative unit, an immune strategy that turned competitors into potential partners (Broecker and Moelling 2019; Broecker 2021b). In this sense, early evolution already exemplifies what Witzany calls *natural genome editing*: genomes as dynamic, negotiated "texts" continuously rewritten through interactions among genetic agents such as viruses, plasmids, and transposons (Witzany 2006, 2011).

As DNA-based cells emerged, they carried this principle forward. LUCA was likely embedded in a virus-rich, horizontally connected community and had to contend with plasmid-like and virus-like elements from the outset. Across prokaryotes, immune systems repeatedly bear the imprint of MGEs: RM modules with mobile dynamics, CRISPR-Cas assembled from transposases, casposons, and plasmid toxins, TA and Abi systems rooted in phage and plasmid genes, and other systems clustered in defense islands. Eukaryotes layered these conflicts into genome management and multicellular physiology: small-RNA pathways to police transposons, interferon systems and sensors tuned to viral signatures, and tissue-level containment strategies such as programmed cell death. They also domesticated parts of viruses, for example, as antiviral restriction factors and as regulatory modules rewiring immune-gene networks. Jawed vertebrates then capped this history by evolving adaptive immunity, whose core diversification mechanism, V(D)J recombination, derives from a *Transib* transposon domesticated into the RAG recombinase machinery (Kapitonov and Jurka 2005; Carmona and Schatz 2017).

These processes likely shaped broader evolutionary transitions. The origin of eukaryotes (and, in some hypotheses, the nucleus) has been linked to the need to segregate host DNA from cytosolic viral replication and mobile elements (Bell 2001; Forterre 2006a; Koonin 2016). Multicellularity may have been favored in part because differentiation, restricted germlines, and altruistic cell death can contain infections at the level of cells or tissues (Iranzo et al. 2016). Viruses and transposons also facilitate horizontal gene transfer and evolutionary "leaps," moving defense systems between species and relocating genes and regulatory elements across genomes (Frost et al. 2005; Koonin and Dolja 2013). Notably, what gets mobilized is itself shaped by conflict: transposon cargo is enriched for anti-MGE defenses, promoting dissemination of immunity as transferable modules (Benler et al. 2021).

Coevolution can also culminate in mutualism. Syncytins (ERV-derived Env proteins essential for placental development in multiple mammalian lineages) have also been found as analogous ERV-Env-based syncytins in viviparous lizards, a striking case of virus-mediated convergent evolution (Mi et al. 2000; Dupressoir et al. 2005; Cornelis et al. 2017). Such examples underscore how "selfish elements" can become "self."

Seen through this lens, parasite–host coevolution is not just a filter that removes the unfit; it is a creative engine that builds new functionalities. Levin and colleagues frame evolution as an expansion of multi-scale problem-solving and information processing across cells, tissues, and organisms, mediated by biochemical and bioelectric signaling networks (Levin 2019; Fields and Levin 2023). Immune systems fit naturally into this broader context: distributed decision-making layers that detect perturbations, store experience, and reshape physiology—often using repurposed viral or transposon "hardware."

In summary, the tug-of-war between genomes and their internal invaders is a central narrative of an integrative theory of evolution. Immune systems repeatedly arose through molecular reuse—building new defenses from parts repurposed from genetic enemies—explaining why immunity is studded with seemingly "odd" components. This unifying perspective links protocells to humans (Koonin and Dolja 2013; Werren 2011; Koonin 2009) and may also guide bioengineering, as we increasingly repurpose viral capsids, transposases, and ERV-derived regulatory modules for therapies and synthetic circuits. Darwin's famous "warm little pond" was likely less a tranquil cradle than an arena of competing replicators, where early protocells first learned to survive by converting genetic threats into heritable safeguards—the same conflict-driven logic that underlies immune innovation until today.

References

Adamala K, Szostak JW (2013a) Nonenzymatic template-directed RNA synthesis inside model protocells. Science 342:1098–1100. https://doi.org/10.1126/science.1241888

Adamala K, Szostak JW (2013b) Competition between model protocells driven by an encapsulated catalyst. Nat Chem 5:495–501. https://doi.org/10.1038/nchem.1650

Alarcón-Schumacher T, Naor A, Gophna U, Erdmann S (2022) Isolation of a virus causing a chronic infection in the archaeal model organism *Haloferax volcanii* reveals antiviral activities of a provirus. Proc Natl Acad Sci USA 119:e2205037119. https://doi.org/10.1073/pnas.2205037119

Altae-Tran H, Shmakov SA, Makarova KS et al (2023) Diversity, evolution, and classification of the RNA-guided nucleases TnpB and Cas12. Proc Natl Acad Sci USA 120:e2308224120. https://doi.org/10.1073/pnas.2308224120

Anand V, Prabhakaran HS, Gogoi P, Kanaujia SP, Kumar M (2022) Structural and functional characterization of Cas2 of CRISPR-Cas subtype I-C lacking the CRISPR component. Front Mol Biosci 9:988569. https://doi.org/10.3389/fmolb.2022.988569

Ansari S, Gupta N, Verma R et al (2023) Antiviral activity of the human endogenous retrovirus-R envelope protein against SARS-CoV-2. EMBO Rep 24:e55900. https://doi.org/10.15252/embr.202255900

Anton BP, Blumenthal R, Eaglesham JB et al (2025) Biology of host-dependent restriction-modification in prokaryotes. EcoSal Plus 13:eesp00142022. https://doi.org/10.1128/ecosalplus.esp-0014-2022

Aquino ILM, Reis ES, Moreira ROAM et al (2024) Giant viruses inhibit superinfection by downregulating phagocytosis in *Acanthamoeba*. J Virol 98:e0104524. https://doi.org/10.1128/jvi.01045-24

Arber W, Linn S (1969) DNA modification and restriction. Annu Rev Biochem 38:467–500. https://doi.org/10.1146/annurev.bi.38.070169.002343

Arias CF, Acosta FJ, Bertocchini F et al (2022) The coordination of anti-phage immunity mechanisms in bacterial cells. Nat Commun 13:7412. https://doi.org/10.1038/s41467-022-35203-7

Arnaud F, Caporale M, Varela M et al (2007) A paradigm for virus-host coevolution: sequential counter-adaptations between endogenous and exogenous retroviruses. PLoS Pathog 3:e170. https://doi.org/10.1371/journal.ppat.0030170

Attwater J, Wochner A, Holliger P (2013) In-ice evolution of RNA polymerase ribozyme activity. Nat Chem 5:1011–1018. https://doi.org/10.1038/nchem.1781

Attwater J, Raguram A, Morgunov AS, Gianni E, Holliger P (2018) Ribozyme-catalysed RNA synthesis using triplet building blocks. Elife 7:e35255. https://doi.org/10.7554/eLife.35255

Attwater J, Augustin TL, Curran JF et al (2025) Trinucleotide substrates under pH-freeze-thaw cycles enable open-ended exponential RNA replication by a polymerase ribozyme. Nat Chem 17:1129–1137. https://doi.org/10.1038/s41557-025-01830-y

Bao C, Gao Q, Xiang H et al (2024) Human endogenous retroviruses and exogenous viral infections. Front Cell Infect Microbiol 14:1439292. https://doi.org/10.3389/fcimb.2024.1439292

Barrangou R, van der Oost J (2015) Bacteriophage exclusion, a new defense system. EMBO J 34:134–135. https://doi.org/10.15252/embj.201490620

Barth ZK, Nguyen MH, Seed KD (2021) A chimeric nuclease substitutes a phage CRISPR-Cas system to provide sequence-specific immunity against subviral parasites. Elife 10:e68339. https://doi.org/10.7554/eLife.68339

Bastiaanssen C, Bobadilla Ugarte P, Kim K et al (2024) RNA-guided RNA silencing by an Asgard archaeal Argonaute. Nat Commun 15:5499. https://doi.org/10.1038/s41467-024-49452-1

Beamud B, Benz F, Bikard D (2024) Going viral: the role of mobile genetic elements in bacterial immunity. Cell Host Microbe 32:804–819. https://doi.org/10.1016/j.chom.2024.05.017

Bell PJ (2001) Viral eukaryogenesis: was the ancestor of the nucleus a complex DNA virus? J Mol Evol 53:251–256. https://doi.org/10.1007/s002390010215

Bell PJL (2020) Evidence supporting a viral origin of the eukaryotic nucleus. Virus Res 289: 198168. https://doi.org/10.1016/j.virusres.2020.198168
Benler S, Faure G, Altae-Tran H, Shmakov S, Zheng F, Koonin E (2021) Cargo genes of Tn7-like transposons comprise an enormous diversity of defense systems, mobile genetic elements, and antibiotic resistance genes. mBio 12:e0293821. https://doi.org/10.1128/mBio.02938-21
Bennetzen JL, Wang H (2014) The contributions of transposable elements to the structure, function, and evolution of plant genomes. Annu Rev Plant Biol 65:505–530. https://doi.org/10.1146/annurev-arplant-050213-035811
Bertelsen MB, Senissar M, Nielsen MH et al (2021) Structural basis for toxin inhibition in the VapXD toxin-antitoxin system. Structure 29:139–150.e3. https://doi.org/10.1016/j.str.2020.10.002
Birkholz N, Jackson S, Fagerlund RD, Fineran PC (2022) A mobile restriction-modification system provides phage defence and resolves an epigenetic conflict with an antagonistic endonuclease. Nucleic Acids Res 50:3348–3361. https://doi.org/10.1093/nar/gkac147
Blower TR, Evans TJ, Przybilski R et al (2012) Viral evasion of a bacterial suicide system by RNA-based molecular mimicry enables infectious altruism. PLoS Genet 8:e1003023. https://doi.org/10.1371/journal.pgen.1003023
Boehm T, McCurley N, Sutoh Y et al (2012) VLR-based adaptive immunity. Annu Rev Immunol 30:203–220. https://doi.org/10.1146/annurev-immunol-020711-075038
Bondy-Denomy J, Qian J, Westra ER et al (2016) Prophages mediate defense against phage infection through diverse mechanisms. ISME J 10:2854–2866. https://doi.org/10.1038/ismej.2016.79
Botelho J (2023) Defense systems are pervasive across chromosomally integrated mobile genetic elements and are inversely correlated to virulence and antimicrobial resistance. Nucleic Acids Res 51:4385–4397. https://doi.org/10.1093/nar/gkad282
Breaker RR, Joyce GF (1994) A DNA enzyme that cleaves RNA. Chem Biol 1:223–229. https://doi.org/10.1016/1074-5521(94)90014-0
Brennecke J, Aravin AA, Stark A et al (2007) Discrete small RNA-generating loci as master regulators of transposon activity in *Drosophila*. Cell 128:1089–1103. https://doi.org/10.1016/j.cell.2007.01.043
Broecker F (2021a) The contribution of viruses to immune systems. In: Hurst CJ (ed) The biological role of a virus, Advances in environmental microbiology, vol 9. Springer, Cham, pp 253–283. https://doi.org/10.1007/978-3-030-85395-2_8
Broecker F (2021b) Genome evolution from random ligation of RNAs of autocatalytic sets. Int J Mol Sci 22:13526. https://doi.org/10.3390/ijms222413526
Broecker F, Moelling K (2019) Evolution of immune systems from viruses and transposable elements. Front Microbiol 10:51. https://doi.org/10.3389/fmicb.2019.00051
Broecker F, Horton R, Heinrich J et al (2016) The intron-enriched HERV-K(HML-10) family suppresses apoptosis, an indicator of malignant transformation. Mob DNA 7:25. https://doi.org/10.1186/s13100-016-0081-9
Bucher MJ, Czyż DM (2024) Phage against the machine: the SIE-ence of superinfection exclusion. Viruses 16:1348. https://doi.org/10.3390/v16091348
Burgyán J, Havelda Z (2011) Viral suppressors of RNA silencing. Trends Plant Sci 16:265–272. https://doi.org/10.1016/j.tplants.2011.02.010
Burnet FM, Fenner F (1949) The production of antibodies, 2nd edn. Macmillan and Co., Ltd, Melbourne
Burroughs AM, Iyer LM, Aravind L (2013) Two novel PIWI families: roles in inter-genomic conflicts in bacteria and Mediator-dependent modulation of transcription in eukaryotes. Biol Direct 8:13. https://doi.org/10.1186/1745-6150-8-13
Busby B, Kristensen DM, Koonin EV (2013) Contribution of phage-derived genomic islands to the virulence of facultative bacterial pathogens. Environ Microbiol 15:307–312. https://doi.org/10.1111/j.1462-2920.2012.02886.x

Cagliani R, Forni D, Biasin M et al (2014) Ancient and recent selective pressures shaped genetic diversity at AIM2-like nucleic acid sensors. Genome Biol Evol 6:830–845. https://doi.org/10.1093/gbe/evu066

Cameron FH, Jennings PA (1994) Multiple domains in a ribozyme construct confer increased suppressive activity in monkey cells. Antisense Res Dev 4:87–94. https://doi.org/10.1089/ard.1994.4.87

Carbonell A (2022) RNAi tools for controlling viroid diseases. Virus Res 313:198729. https://doi.org/10.1016/j.virusres.2022.198729

Carmona LM, Schatz DG (2017) New insights into the evolutionary origins of the recombination-activating gene proteins and V(D)J recombination. FEBS J 284:1590–1605. https://doi.org/10.1111/febs.13990

Carmona LM, Fugmann SD, Schatz DG (2016) Collaboration of RAG2 with RAG1-like proteins during the evolution of V(D)J recombination. Genes Dev 30:909–917. https://doi.org/10.1101/gad.278432.116

Casjens S (2003) Prophages and bacterial genomics: what have we learned so far? Mol Microbiol 49:277–300. https://doi.org/10.1046/j.1365-2958.2003.03580.x

Cervera A, De la Peña M (2014) Eukaryotic penelope-like retroelements encode hammerhead ribozyme motifs. Mol Biol Evol 31:2941–2947. https://doi.org/10.1093/molbev/msu232

Chen IA, Walde P (2010) From self-assembled vesicles to protocells. Cold Spring Harb Perspect Biol 2:a002170. https://doi.org/10.1101/cshperspect.a002170

Chen CJ, Banerjea AC, Harmison GG, Haglund K, Schubert M (1992) Multitarget-ribozyme directed to cleave at up to nine highly conserved HIV-1 env RNA regions inhibits HIV-1 replication--potential effectiveness against most presently sequenced HIV-1 isolates. Nucleic Acids Res 20:4581–4589. https://doi.org/10.1093/nar/20.17.4581

Chen IA, Roberts RW, Szostak JW (2004) The emergence of competition between model protocells. Science 305:1474–1476. https://doi.org/10.1126/science.1100757

Chen S, Hu X, Cui IH et al (2019) An endogenous retroviral element exerts an antiviral innate immune function via the derived lncRNA lnc-ALVE1-AS1. Antivir Res 170:104571. https://doi.org/10.1016/j.antiviral.2019.104571

Chen B, Chen Z, Wang Y et al (2020) ORF4 of the temperate archaeal virus SNJ1 governs the lysis-lysogeny switch and superinfection immunity. J Virol 94:e00841-20. https://doi.org/10.1128/JVI.00841-20

Cho CJ, An T, Lai YC et al (2025) Protocells by spontaneous reaction of cysteine with short-chain thioesters. Nat Chem 17:148–155. https://doi.org/10.1038/s41557-024-01666-y

Chow J, Franz KM, Kagan JC (2015) PRRs are watching you: localization of innate sensing and signaling regulators. Virology 479-480:104–109. https://doi.org/10.1016/j.virol.2015.02.051

Chuong EB, Elde NC, Feschotte C (2016) Regulatory evolution of innate immunity through co-option of endogenous retroviruses. Science 351:1083–1087. https://doi.org/10.1126/science.aad5497

Chuong EB, Elde NC, Feschotte C (2017) Regulatory activities of transposable elements: from conflicts to benefits. Nat Rev Genet 18:71–86. https://doi.org/10.1038/nrg.2016.139

Clapes T, Polyzou A, Prater P et al (2021) Chemotherapy-induced transposable elements activate MDA5 to enhance haematopoietic regeneration. Nat Cell Biol 23:704–717. https://doi.org/10.1038/s41556-021-00707-9

Claverie JM (2006) Viruses take center stage in cellular evolution. Genome Biol 7:110. https://doi.org/10.1186/gb-2006-7-6-110

Cooles FAH, Vidal Pedrola G, Naamane N et al (2025) Endogenous Retroelement activation is implicated in interferon-α production and anti-cyclic citrullinated peptide autoantibody generation in early rheumatoid arthritis. Arthritis Rheumatol 77:646–657. https://doi.org/10.1002/art.43083

Cornelis G, Funk M, Vernochet C et al (2017) An endogenous retroviral envelope syncytin and its cognate receptor identified in the viviparous placental *Mabuya* lizard. Proc Natl Acad Sci USA 114:E10991–E11000. https://doi.org/10.1073/pnas.1714590114

Cridland JA, Curley EZ, Wykes MN et al (2012) The mammalian PYHIN gene family: phylogeny, evolution and expression. BMC Evol Biol 12:140. https://doi.org/10.1186/1471-2148-12-140
Cumby N, Edwards AM, Davidson AR, Maxwell KL (2012) The bacteriophage HK97 gp15 moron element encodes a novel superinfection exclusion protein. J Bacteriol 194:5012–5019. https://doi.org/10.1128/JB.00843-12
Cumer T, Pompanon F, Boyer F (2019) Old origin of a protective endogenous retrovirus (enJSRV) in the Ovis genus. Heredity 122:187–194. https://doi.org/10.1038/s41437-018-0112-z
Damer B, Deamer D (2015) Coupled phases and combinatorial selection in fluctuating hydrothermal pools: a scenario to guide experimental approaches to the origin of cellular life. Life 5:872–887. https://doi.org/10.3390/life5010872
Damer B, Deamer D (2020) The hot spring hypothesis for an origin of life. Astrobiology 20:429–452. https://doi.org/10.1089/ast.2019.2045
DasGupta S, Weiss Z, Nisler C, Szostak JW (2024) Evolution of the substrate specificity of an RNA ligase ribozyme from phosphorimidazole to triphosphate activation. Proc Natl Acad Sci USA 121:e2407325121. https://doi.org/10.1073/pnas.2407325121
Daugherty MD, Malik HS (2012) Rules of engagement: molecular insights from host-virus arms races. Annu Rev Genet 46:677–700. https://doi.org/10.1146/annurev-genet-110711-155522
de Farias ST, Dos Santos Junior AP, Rêgo TG, José MV (2017) Origin and evolution of RNA-dependent RNA polymerase. Front Genet 8:125. https://doi.org/10.3389/fgene.2017.00125
de Koning AP, Gu W, Castoe TA, Batzer MA, Pollock DD (2011) Repetitive elements may comprise over two-thirds of the human genome. PLoS Genet 7:e1002384. https://doi.org/10.1371/journal.pgen.1002384
de la Peña M, Cervera A (2017) Circular RNAs with hammerhead ribozymes encoded in eukaryotic genomes: the enemy at home. RNA Biol 14:985–991. https://doi.org/10.1080/15476286.2017.1321730
de la Peña M, Ceprián R, Cervera A (2020) A singular and widespread group of mobile genetic elements: RNA circles with autocatalytic ribozymes. Cells 9:2555. https://doi.org/10.3390/cells9122555
Denison MR, Graham RL, Donaldson EF, Eckerle LD, Baric RS (2011) Coronaviruses: an RNA proofreading machine regulates replication fidelity and diversity. RNA Biol 8:270–279. https://doi.org/10.4161/rna.8.2.15013
Dhar N, Weinberg MS, Michod RE, Durand PM (2017) Molecular trade-offs in RNA ligases affected the modular emergence of complex ribozymes at the origin of life. R Soc Open Sci 4:170376. https://doi.org/10.1098/rsos.170376
Di Serio F, Li SF, Matoušek J et al (2018) ICTV virus taxonomy profile: Avsunviroidae. J Gen Virol 99:611–612. https://doi.org/10.1099/jgv.0.001045
Ding SW, Voinnet O (2017) Antiviral immunity directed by small RNAs. Cell 130:413–426. https://doi.org/10.1016/j.cell.2007.07.039
Domingo E, Sheldon J, Perales C (2012) Viral quasispecies evolution. Microbiol Mol Biol Rev 76:159–216. https://doi.org/10.1128/MMBR.05023-11
Dominissini D, Moshitch-Moshkovitz S, Schwartz S et al (2012) Topology of the human and mouse m6A RNA methylomes revealed by m6A-seq. Nature 485:201–206. https://doi.org/10.1038/nature11112
Doron S, Melamed S, Ofir G et al (2018) Systematic discovery of antiphage defense systems in the microbial pangenome. Science 359:eaar4120. https://doi.org/10.1126/science.aar4120
Du AY, Chobirko JD, Zhuo X, Feschotte C, Wang T (2024) Regulatory transposable elements in the encyclopedia of DNA elements. Nat Commun 15:7594. https://doi.org/10.1038/s41467-024-51921-6
Duggal NK, Emerman M (2012) Evolutionary conflicts between viruses and restriction factors shape immunity. Nat Rev Immunol 12:687–695. https://doi.org/10.1038/nri3295
Duhamel M, Salzet M (2025) Self or nonself: end of a dogma? Front Immunol 16:1595764. https://doi.org/10.3389/fimmu.2025.1595764

Duncan-Lowey B, Kranzusch PJ (2022) CBASS phage defense and evolution of antiviral nucleotide signaling. Curr Opin Immunol 74:156–163. https://doi.org/10.1016/j.coi.2022.01.002
Dupressoir A, Marceau G, Vernochet C et al (2005) Syncytin-A and syncytin-B, two fusogenic placenta-specific murine envelope genes of retroviral origin conserved in Muridae. Proc Natl Acad Sci USA 102:725–730. https://doi.org/10.1073/pnas.0406509102
Eigen M (1971) Selforganization of matter and the evolution of biological macromolecules. Naturwissenschaften 58:465–523. https://doi.org/10.1007/BF00623322
Fan Z, Deckel Y, Lowe LA et al (2023) Lipid exchange promotes fusion of model protocells. Small Methods 7:e2300126. https://doi.org/10.1002/smtd.202300126
Ferbeyre G, Smith JM, Cedergren R (1998) Schistosome satellite DNA encodes active hammerhead ribozymes. Mol Cell Biol 18:3880–3888. https://doi.org/10.1128/MCB.18.7.3880
Feschotte C (2008) Transposable elements and the evolution of regulatory networks. Nat Rev Genet 9:397–405. https://doi.org/10.1038/nrg2337
Fields C, Levin M (2023) Regulative development as a model for origin of life and artificial life studies. Biosystems 229:104927. https://doi.org/10.1016/j.biosystems.2023.104927
Fineran PC, Blower TR, Foulds IJ et al (2009) The phage abortive infection system, ToxIN, functions as a protein-RNA toxin-antitoxin pair. Proc Natl Acad Sci USA 106:894–899. https://doi.org/10.1073/pnas.0808832106
Fire A, Xu S, Montgomery MK, Kostas SA, Driver SE, Mello CC (1998) Potent and specific genetic interference by double-stranded RNA in *Caenorhabditis elegans*. Nature 391:806–811. https://doi.org/10.1038/35888
Fischer MG, Hackl T (2016) Host genome integration and giant virus-induced reactivation of the virophage mavirus. Nature 540:288–291. https://doi.org/10.1038/nature20593
Flajnik MF, Kasahara M (2010) Origin and evolution of the adaptive immune system: genetic events and selective pressures. Nat Rev Genet 11:47–59. https://doi.org/10.1038/nrg2703
Flores R, Delgado S, Gas ME et al (2004) Viroids: the minimal non-coding RNAs with autonomous replication. FEBS Lett 567:42–48. https://doi.org/10.1016/j.febslet.2004.03.118
Flores R, Serra P, Minoia S, Di Serio F, Navarro B (2012) Viroids: from genotype to phenotype just relying on RNA sequence and structural motifs. Front Microbiol 3:217. https://doi.org/10.3389/fmicb.2012.00217
Forterre P (2005) The two ages of the RNA world, and the transition to the DNA world: a story of viruses and cells. Biochimie 87:793–803. https://doi.org/10.1016/j.biochi.2005.03.015
Forterre P (2006a) The origin of viruses and their possible roles in major evolutionary transitions. Virus Res 117:5–16. https://doi.org/10.1016/j.virusres.2006.01.010
Forterre P (2006b) Three RNA cells for ribosomal lineages and three DNA viruses to replicate their genomes: a hypothesis for the origin of cellular domain. Proc Natl Acad Sci USA 103:3669–3674. https://doi.org/10.1073/pnas.0510333103
Forterre P (2010) Defining life: the virus viewpoint. Orig Life Evol Biosph 40:151–160. https://doi.org/10.1007/s11084-010-9194-1
Forterre P, Prangishvili D (2009) The great billion-year war between ribosome- and capsid-encoding organisms (cells and viruses) as the major source of evolutionary novelties. Ann N Y Acad Sci 1178:65–77. https://doi.org/10.1111/j.1749-6632.2009.04993.x
Frost LS, Leplae R, Summers AO, Toussaint A (2005) Mobile genetic elements: the agents of open source evolution. Nat Rev Microbiol 3:722–732. https://doi.org/10.1038/nrmicro1235
Fugmann SD, Lee AI, Shockett PE, Villey IJ, Schatz DG (2000) The RAG proteins and V(D)J recombination: complexes, ends, and transposition. Annu Rev Immunol 18:495–527. https://doi.org/10.1146/annurev.immunol.18.1.495
Fujino K, Horie M, Honda T, Merriman DK, Tomonaga K (2014) Inhibition of Borna disease virus replication by an endogenous bornavirus-like element in the ground squirrel genome. Proc Natl Acad Sci USA 111:13175–13180. https://doi.org/10.1073/pnas.1407046111
Furubayashi T, Ueda K, Bansho Y et al (2020) Emergence and diversification of a host-parasite RNA ecosystem through Darwinian evolution. Elife 9:e56038. https://doi.org/10.7554/eLife.56038

Gao L, Altae-Tran H, Böhning F et al (2020) Diverse enzymatic activities mediate antiviral immunity in prokaryotes. Science 369:1077–1084. https://doi.org/10.1126/science.aba0372
Girard A, Hannon GJ (2008) Conserved themes in small-RNA-mediated transposon control. Trends Cell Biol 18:136–148. https://doi.org/10.1016/j.tcb.2008.01.004
Goeders N, Van Melderen L (2014) Toxin-antitoxin systems as multilevel interaction systems. Toxins 6:304–324. https://doi.org/10.3390/toxins6010304
Goldfarb T, Sberro H, Weinstock E et al (2015) BREX is a novel phage resistance system widespread in microbial genomes. EMBO J 34:169–183. https://doi.org/10.15252/embj.201489455
González-Delgado A, Mestre MR, Martínez-Abarca F, Toro N (2019) Spacer acquisition from RNA mediated by a natural reverse transcriptase-Cas1 fusion protein associated with a type III-D CRISPR-Cas system in Vibrio vulnificus. Nucleic Acids Res 47:10202–10211. https://doi.org/10.1093/nar/gkz746
Grandi N, Tramontano E (2018) Human endogenous retroviruses are ancient acquired elements still shaping innate immune responses. Front Immunol 9:2039. https://doi.org/10.3389/fimmu.2018.02039
Grandi N, Cadeddu M, Pisano MP et al (2017) Identification of a novel HERV-K(HML10): comprehensive characterization and comparative analysis in non-human primates provide insights about HML10 proviruses structure and diffusion. Mob DNA 8:15. https://doi.org/10.1186/s13100-017-0099-7
Grow EJ, Flynn RA, Chavez SL et al (2015) Intrinsic retroviral reactivation in human preimplantation embryos and pluripotent cells. Nature 522:221–225. https://doi.org/10.1038/nature14308
Guo Y, Yang C, Liu Y et al (2022) High expression of HERV-K (HML-2) might stimulate interferon in COVID-19 patients. Viruses 14:996. https://doi.org/10.3390/v14050996
Hanczyc MM, Fujikawa SM, Szostak JW (2003) Experimental models of primitive cellular compartments: encapsulation, growth, and division. Science 302:618–622. https://doi.org/10.1126/science.1089904
Handa N, Kobayashi I (1999) Post-segregational killing by restriction modification gene complexes: observations of individual cell deaths. Biochimie 81:931–938. https://doi.org/10.1016/s0300-9084(99)00201-1
Handa N, Ichige A, Kusano K, Kobayashi I (2000) Cellular responses to postsegregational killing by restriction-modification genes. J Bacteriol 182:2218–2229. https://doi.org/10.1128/JB.182.8.2218-2229.2000
Harris RS, Dudley JP (2015) APOBECs and virus restriction. Virology 479-480:131–145. https://doi.org/10.1016/j.virol.2015.03.012
Hayes F (2003) Toxins-antitoxins: plasmid maintenance, programmed cell death, and cell cycle arrest. Science 301:1496–1499. https://doi.org/10.1126/science.1088157
Hickman AB, Kailasan S, Genzor P, Haase AD, Dyda F (2020) Casposase structure and the mechanistic link between DNA transposition and spacer acquisition by CRISPR-Cas. Elife 9:e50004. https://doi.org/10.7554/eLife.50004
Hobbs SJ, Kranzusch PJ (2024) Nucleotide immune signaling in CBASS, Pycsar, Thoeris, and CRISPR antiphage defense. Ann Rev Microbiol 78:255–276. https://doi.org/10.1146/annurev-micro-041222-024843
Hochhauser D, Millman A, Sorek R (2023) The defense Island repertoire of the *Escherichia coli* pan-genome. PLoS Genet 19:e1010694. https://doi.org/10.1371/journal.pgen.1010694
Hordijk W, Steel M (2017) Chasing the tail: the emergence of autocatalytic networks. Biosystems 152:1–10. https://doi.org/10.1016/j.biosystems.2016.12.002
Hordijk W, Hein J, Steel M (2010) Autocatalytic sets and the origin of life. Entropy 12:1733–1742. https://doi.org/10.3390/e12071733
Horning DP, Joyce GF (2016) Amplification of RNA by an RNA polymerase ribozyme. Proc Natl Acad Sci USA 113:9786–9791. https://doi.org/10.1073/pnas.1610103113
Huang S, Tao X, Yuan S et al (2016) Discovery of an active RAG transposon illuminates the origins of V(D)J recombination. Cell 166:102–114. https://doi.org/10.1016/j.cell.2016.05.032

Iglesias-Artola JM, Drobot B, Kar M et al (2022) Charge-density reduction promotes ribozyme activity in RNA-peptide coacervates via RNA fluidization and magnesium partitioning. Nat Chem 14:407–416. https://doi.org/10.1038/s41557-022-00890-8

Inglis LK, Roach MJ, Edwards RA (2024) Prophages: an integral but understudied component of the human microbiome. Microb Genom 10:001166. https://doi.org/10.1099/mgen.0.001166

International Wheat Genome Sequencing Consortium (IWGSC) (2018) Shifting the limits in wheat research and breeding using a fully annotated reference genome. Science 361:eaar7191. https://doi.org/10.1126/science.aar7191

Iranzo J, Puigbò P, Lobkovsky AE, Wolf YI, Koonin EV (2016) Inevitability of Genetic Parasites. Genome Biol Evol 8:2856–2869. https://doi.org/10.1093/gbe/evw193

Ito J, Sugimoto R, Nakaoka H et al (2017) Systematic identification and characterization of regulatory elements derived from human endogenous retroviruses. PLoS Genet 13:e1006883. https://doi.org/10.1371/journal.pgen.1006883

Ito J, Gifford RJ, Sato K (2020) Retroviruses drive the rapid evolution of mammalian APOBEC3 genes. Proc Natl Acad Sci USA 117:610–618. https://doi.org/10.1073/pnas.1914183116

Ivanov SA, Vauléon S, Müller S (2005) Efficient RNA ligation by reverse-joined hairpin ribozymes and engineering of twin ribozymes consisting of conventional and reverse-joined hairpin ribozyme units. FEBS J 272:4464–4474. https://doi.org/10.1111/j.1742-4658.2005.04865.x

Iyer LM, Koonin EV, Aravind L (2003) Evolutionary connection between the catalytic subunits of DNA-dependent RNA polymerases and eukaryotic RNA-dependent RNA polymerases and the origin of RNA polymerases. BMC Struct Biol 3:1. https://doi.org/10.1186/1472-6807-3-1

Janzen E, Blanco C, Peng H, Kenchel J, Chen IA (2020) Promiscuous ribozymes and their proposed role in prebiotic evolution. Chem Rev 120:4879–4897. https://doi.org/10.1021/acs.chemrev.9b00620

Jern P, Greenwood AD (2024) Wildlife endogenous retroviruses: colonization, consequences, and cooption. Trends Genet 40:149–159. https://doi.org/10.1016/j.tig.2023.10.014

Jimenez RM, Polanco JA, Lupták A (2015) Chemistry and biology of self-cleaving ribozymes. Trends Biochem Sci 40:648–661. https://doi.org/10.1016/j.tibs.2015.09.001

Johnson WE, Sawyer SL (2009) Molecular evolution of the antiretroviral TRIM5 gene. Immunogenetics 61:163–176. https://doi.org/10.1007/s00251-009-0358-y

Jonathan M, Ikeda T (2023) APOBEC3 family proteins as drivers of virus evolution. Front Virol 3:1332010. https://doi.org/10.3389/fviro.2023.1332010

Kacian DL, Mills DR, Kramer FR, Spiegelman S (1972) A replicating RNA molecule suitable for a detailed analysis of extracellular evolution and replication. Proc Natl Acad Sci USA 69:3038–3042. https://doi.org/10.1073/pnas.69.10.3038

Kapitonov VV, Jurka J (2005) RAG1 core and V(D)J recombination signal sequences were derived from *Transib* transposons. PLoS Biol 3:e181. https://doi.org/10.1371/journal.pbio.0030181

Kapitonov VV, Koonin EV (2015) Evolution of the RAG1-RAG2 locus: both proteins came from the same transposon. Biol Direct 10:20. https://doi.org/10.1186/s13062-015-0055-8

Kazakov SA, Balatskaya SV, Johnston BH (2006) Ligation of the hairpin ribozyme in cis induced by freezing and dehydration. RNA 12:446–456. https://doi.org/10.1261/rna.2123506

Kell AM, Gale M Jr (2015) RIG-I in RNA virus recognition. Virology 479-480:110–121. https://doi.org/10.1016/j.virol.2015.02.017

Kelly A, Went SC, Mariano G et al (2023) Diverse Durham collection phages demonstrate complex BREX defense responses. Appl Environ Microbiol 89:e0062323. https://doi.org/10.1128/aem.00623-23

Knisbacher BA, Gerber D, Levanon EY (2016) DNA editing by APOBECs: a genomic preserver and transformer. Trends Genet 32:16–28. https://doi.org/10.1016/j.tig.2015.10.005

Kobayashi I (2001) Behavior of restriction-modification systems as selfish mobile elements and their impact on genome evolution. Nucleic Acids Res 29:3742–3756. https://doi.org/10.1093/nar/29.18.3742

Koito A, Ikeda T (2011) Intrinsic restriction activity by AID/APOBEC family of enzymes against the mobility of retroelements. Mob Genet Elements 1:197–202. https://doi.org/10.4161/mge.1.3.17430
Koonin EV (2006) Temporal order of evolution of DNA replication systems inferred by comparison of cellular and viral DNA polymerases. Biol Direct 1:39. https://doi.org/10.1186/1745-6150-1-39
Koonin EV (2009) On the origin of cells and viruses: primordial virus world scenario. Ann N Y Acad Sci 1178:47–64. https://doi.org/10.1111/j.1749-6632.2009.04992.x
Koonin EV (2016) Viruses and mobile elements as drivers of evolutionary transitions. Philos Trans R Soc Lond Ser B Biol Sci 371:20150442. https://doi.org/10.1098/rstb.2015.0442
Koonin EV, Dolja VV (2013) A virocentric perspective on the evolution of life. Curr Opin Virol 3: 546–557. https://doi.org/10.1016/j.coviro.2013.06.008
Koonin EV, Krupovic M (2015) Evolution of adaptive immunity from transposable elements combined with innate immune systems. Nat Rev Genet 16:184–192. https://doi.org/10.1038/nrg3859
Koonin EV, Makarova KS (2017) Mobile genetic elements and evolution of CRISPR-Cas systems: all the way there and Back. Genome Biol Evol 9:2812–2825. https://doi.org/10.1093/gbe/evx192
Koonin EV, Makarova KS (2019) Origins and evolution of CRISPR-Cas systems. Philos Trans R Soc Lond Ser B Biol Sci 374:20180087. https://doi.org/10.1098/rstb.2018.0087
Koonin EV, Makarova KS (2024) CRISPR in mobile genetic elements: counter-defense, inter-element competition and RNA-guided transposition. BMC Biol 22:295. https://doi.org/10.1186/s12915-024-02090-x
Koonin EV, Novozhilov AS (2009) Origin and evolution of the genetic code: the universal enigma. IUBMB Life 61:99–111. https://doi.org/10.1002/iub.146
Koonin EV, Senkevich TG, Dolja VV (2006) The ancient virus world and evolution of cells. Biol Direct 1:29. https://doi.org/10.1186/1745-6150-1-29
Koonin EV, Wolf YI, Katsnelson MI (2017a) Inevitability of the emergence and persistence of genetic parasites caused by evolutionary instability of parasite-free states. Biol Direct 12:31. https://doi.org/10.1186/s13062-017-0202-5
Koonin EV, Makarova KS, Wolf YI (2017b) Evolutionary genomics of defense systems in archaea and bacteria. Ann Rev Microbiol 71:233–261. https://doi.org/10.1146/annurev-micro-090816-093830
Koonin EV, Gootenberg JS, Abudayyeh OO (2023) Discovery of diverse CRISPR-Cas systems and expansion of the genome engineering toolbox. Biochemistry 62:3465–3487. https://doi.org/10.1021/acs.biochem.3c00159
Koslová A, Hackl T, Bade F et al (2024) Endogenous virophages are active and mitigate giant virus infection in the marine protist *Cafeteria burkhardae*. Proc Natl Acad Sci USA 121: e2314606121. https://doi.org/10.1073/pnas.2314606121
Kovalskaya N, Hammond RW (2014) Molecular biology of viroid-host interactions and disease control strategies. Plant Sci 228:48–60. https://doi.org/10.1016/j.plantsci.2014.05.006
Krchlikova V, Lu Y, Sauter D (2025) Viral influencers: deciphering the role of endogenous retroviral LTR12 repeats in cellular gene expression. J Virol 99:e0135124. https://doi.org/10.1128/jvi.01351-24
Krupovic M, Makarova KS, Forterre P, Prangishvili D, Koonin EV (2014) Casposons: a new superfamily of self-synthesizing DNA transposons at the origin of prokaryotic CRISPR-Cas immunity. BMC Biol 12:36. https://doi.org/10.1186/1741-7007-12-36
Krupovic M, Béguin P, Koonin EV (2017) Casposons: mobile genetic elements that gave rise to the CRISPR-Cas adaptation machinery. Curr Opin Microbiol 38:36–43. https://doi.org/10.1016/j.mib.2017.04.004
Kuhn H, Frank-Kamenetskii MD (2005) Template-independent ligation of single-stranded DNA by T4 DNA ligase. FEBS J 272:5991–6000. https://doi.org/10.1111/j.1742-4658.2005.04954.x

Kumar RM, Joyce GF (2003) A modular, bifunctional RNA that integrates itself into a target RNA. Proc Natl Acad Sci USA 100:9738–9743. https://doi.org/10.1073/pnas.1334190100
Kun A, Santos M, Szathmáry E (2005) Real ribozymes suggest a relaxed error threshold. Nat Genet 37:1008–1011. https://doi.org/10.1038/ng1621
Kusano K, Naito T, Handa N, Kobayashi I (1995) Restriction-modification systems as genomic parasites in competition for specific sequences. Proc Natl Acad Sci USA 92:11095–11099. https://doi.org/10.1073/pnas.92.24.11095
Kuzmenko A, Yudin D, Ryazansky S, Kulbachinskiy A, Aravin AA (2019) Programmable DNA cleavage by Ago nucleases from mesophilic bacteria *clostridium butyricum* and *Limnothrix rosea*. Nucleic Acids Res 47:5822–5836. https://doi.org/10.1093/nar/gkz379
Kwon AR, Kim JH, Park SJ et al (2012) Structural and biochemical characterization of HP0315 from *Helicobacter pylori* as a VapD protein with an endoribonuclease activity. Nucleic Acids Res 40:4216–4228. https://doi.org/10.1093/nar/gkr1305
Le Vay KK, Salibi E, Ghosh B, Tang TYD, Mutschler H (2023) Ribozyme activity modulates the physical properties of RNA-peptide coacervates. elife 12:e83543. https://doi.org/10.7554/eLife.83543
Leavitt JC, Woodbury BM, Gilcrease EB et al (2024) Bacteriophage P22 SieA-mediated superinfection exclusion. MBio 15:e0216923. https://doi.org/10.1128/mbio.02169-23
Leipe DD, Aravind L, Koonin EV (1999) Did DNA replication evolve twice independently? Nucleic Acids Res 27:3389–3401. https://doi.org/10.1093/nar/27.17.3389
Levin M (2019) The computational boundary of a "Self": developmental bioelectricity drives multicellularity and scale-free cognition. Front Psychol 10:2688. https://doi.org/10.3389/fpsyg.2019.02688
Li P, Wu Z, Liu T et al (2025) The defensome of prokaryotes in aquifers. Nat Commun 16:6482. https://doi.org/10.1038/s41467-025-61467-w
Lincoln TA, Joyce GF (2009) Self-sustained replication of an RNA enzyme. Science 323:1229–1232. https://doi.org/10.1126/science.1167856
Lisch D (2013) How important are transposons for plant evolution? Nat Rev Genet 14:49–61. https://doi.org/10.1038/nrg3374
Litman GW, Anderson MK, Rast JP (1999) Evolution of antigen binding receptors. Annu Rev Immunol 17:109–147. https://doi.org/10.1146/annurev.immunol.17.1.109
Löber U, Hobbs M, Dayaram A et al (2018) Degradation and remobilization of endogenous retroviruses by recombination during the earliest stages of a germ-line invasion. Proc Natl Acad Sci USA 115:8609–8614. https://doi.org/10.1073/pnas.1807598115
Lopatina A, Tal N, Sorek R (2020) Abortive infection: bacterial suicide as an antiviral immune strategy. Annu Rev Virol 7:371–384. https://doi.org/10.1146/annurev-virology-011620-040628
Macduff DA, Harris RS (2006) Directed DNA deamination by AID/APOBEC3 in immunity. Curr Biol 16:R186–R189. https://doi.org/10.1016/j.cub.2006.02.035
Majorek KA, Dunin-Horkawicz S, Steczkiewicz K et al (2014) The RNase H-like superfamily: new members, comparative structural analysis and evolutionary classification. Nucleic Acids Res 42: 4160–4179. https://doi.org/10.1093/nar/gkt1414
Makarova KS, Wolf YI, Snir S, Koonin EV (2011) Defense islands in bacterial and archaeal genomes and prediction of novel defense systems. J Bacteriol 193:6039–6056. https://doi.org/10.1128/JB.05535-11
Makarova KS, Wolf YI, Iranzo J et al (2020) Evolutionary classification of CRISPR-Cas systems: a burst of class 2 and derived variants. Nat Rev Microbiol 18:67–83. https://doi.org/10.1038/s41579-019-0299-x
Makarova KS, Shmakov SA, Wolf YI et al (2025) An updated evolutionary classification of CRISPR-Cas systems including rare variants. Nat Microbiol. https://doi.org/10.1038/s41564-025-02180-8
Malfavon-Borja R, Feschotte C (2015) Fighting fire with fire: endogenous retrovirus envelopes as restriction factors. J Virol 89:4047–4050. https://doi.org/10.1128/JVI.03653-14

Man SM, Karki R, Kanneganti TD (2016) AIM2 inflammasome in infection, cancer, and autoimmunity: role in DNA sensing, inflammation, and innate immunity. Eur J Immunol 46:269–280. https://doi.org/10.1002/eji.201545839

Mansy SS, Schrum JP, Krishnamurthy M et al (2008) Template-directed synthesis of a genetic polymer in a model protocell. Nature 454:122–125. https://doi.org/10.1038/nature07018

Marques JT, Wang JP, Wang X et al (2013) Functional specialization of the small interfering RNA pathway in response to virus infection. PLoS Pathog 9:e1003579. https://doi.org/10.1371/journal.ppat.1003579

Martin W, Russell MJ (2007) On the origin of biochemistry at an alkaline hydrothermal vent. Philos Trans R Soc Lond Ser B Biol Sci 362:1887–1925. https://doi.org/10.1098/rstb.2006.1881

Martin EC, Vicari C, Tsakou-Ngouafo L et al (2020) Identification of RAG-like transposons in protostomes suggests their ancient bilaterian origin. Mob DNA 11:17. https://doi.org/10.1186/s13100-020-00214-y

Matzke MA, Mosher RA (2014) RNA-directed DNA methylation: an epigenetic pathway of increasing complexity. Nat Rev Genet 15:394–408. https://doi.org/10.1038/nrg3683

McKinley LN, Meyer MO, Sebastian A et al (2024) Direct testing of natural twister ribozymes from over a thousand organisms reveals a broad tolerance for structural imperfections. Nucleic Acids Res 52:14133–14153. https://doi.org/10.1093/nar/gkae908

McLaughlin RN Jr, Malik HS (2017) Genetic conflicts: the usual suspects and beyond. J Exp Biol 220(Pt 1):6–17. https://doi.org/10.1242/jeb.148148

McRae EKS, Wan CJK, Kristoffersen EL et al (2024) Cryo-EM structure and functional landscape of an RNA polymerase ribozyme. Proc Natl Acad Sci USA 121:e2313332121. https://doi.org/10.1073/pnas.2313332121

Medvedeva S, Liu Y, Koonin EV et al (2019) Virus-borne mini-CRISPR arrays are involved in interviral conflicts. Nat Commun 10:5204. https://doi.org/10.1038/s41467-019-13205-2

Meshcheryakova A, Pietschmann P, Zimmermann P et al (2021) AID and APOBECs as multifaceted intrinsic virus-restricting factors: emerging concepts in the light of COVID-19. Front Immunol 12:690416. https://doi.org/10.3389/fimmu.2021.690416

Mette MF, Kanno T, Aufsatz W et al (2002) Endogenous viral sequences and their potential contribution to heritable virus resistance in plants. EMBO J 21:461–469. https://doi.org/10.1093/emboj/21.3.461

Meyer KD, Saletore Y, Zumbo P et al (2012) Comprehensive analysis of mRNA methylation reveals enrichment in 3′ UTRs and near stop codons. Cell 149:1635–1646. https://doi.org/10.1016/j.cell.2012.05.003

Mi S, Lee X, Li X et al (2000) Syncytin is a captive retroviral envelope protein involved in human placental morphogenesis. Nature 403:785–789. https://doi.org/10.1038/35001608

Millman A, Melamed S, Amitai G, Sorek R (2020) Diversity and classification of cyclic-oligonucleotide-based anti-phage signalling systems. Nat Microbiol 5:1608–1615. https://doi.org/10.1038/s41564-020-0777-y

Mills DR, Peterson RL, Spiegelman S (1967) An extracellular Darwinian experiment with a self-duplicating nucleic acid molecule. Proc Natl Acad Sci USA 58:217–224. https://doi.org/10.1073/pnas.58.1.217

Miyake A, Ngo MH, Wulandari S et al (2022) Convergent evolution of antiviral machinery derived from endogenous retrovirus truncated envelope genes in multiple species. Proc Natl Acad Sci USA 119:e2114441119. https://doi.org/10.1073/pnas.2114441119

Moawad AS, Wang F, Zheng Y et al (2024) Evolution of endogenous retroviruses in the subfamily of *Caprinae*. Viruses 16:398. https://doi.org/10.3390/v16030398

Mohr G, Silas S, Stamos JL et al (2018) A reverse transcriptase-Cas1 fusion protein contains a Cas6 domain required for both CRISPR RNA biogenesis and RNA spacer acquisition. Mol Cell 72:700–714.c8. https://doi.org/10.1016/j.molcel.2018.09.013

Molina-Sánchez MD, Martínez-Abarca F, Millán V et al (2024) Adaptive immunity of type VI CRISPR-Cas systems associated with reverse transcriptase-Cas1 fusion proteins. Nucleic Acids Res 52:14229–14243. https://doi.org/10.1093/nar/gkae1154

Moody ERR, Álvarez-Carretero S, Mahendrarajah TA et al (2024) The nature of the last universal common ancestor and its impact on the early Earth system. Nat Ecol Evol 8:1654–1666. https://doi.org/10.1038/s41559-024-02461-1

Mosterd C, Rousseau GM, Moineau S (2021) A short overview of the CRISPR-Cas adaptation stage. Can J Microbiol 67:1–12. https://doi.org/10.1139/cjm-2020-0212

Nakagawa R, Hirano H, Omura SN et al (2023) Cryo-EM structure of the transposon-associated TnpB enzyme. Nature 616:390–397. https://doi.org/10.1038/s41586-023-05933-9

Naser IB, Hoque MM, Nahid MA et al (2017) Analysis of the CRISPR-Cas system in bacteriophages active on epidemic strains of Vibrio cholerae in Bangladesh. Sci Rep 7:14880. https://doi.org/10.1038/s41598-017-14839-2

Negrón-Mendoza A, Hernández-Morales R, Lazcano A (2025) Can the origin of biosynthetic routes be explained by a Frankenstein's monster-like spontaneous assembly of prebiotic reactants? Philos Trans R Soc Lond Ser B Biol Sci 380:20240289. https://doi.org/10.1098/rstb.2024.0289

Newman RM, Hall L, Connole M et al (2006) Balancing selection and the evolution of functional polymorphism in old world monkey TRIM5alpha. Proc Natl Acad Sci USA 103:19134–19139. https://doi.org/10.1073/pnas.0605838103

Nielsen H, Westhof E, Johansen S (2005) An mRNA is capped by a 2′, 5′ lariat catalyzed by a group I-like ribozyme. Science 309:1584–1587. https://doi.org/10.1126/science.1113645

Nielsen H, Einvik C, Lentz TE, Hedegaard MM, Johansen SD (2009) A conformational switch in the DiGIR1 ribozyme involved in release and folding of the downstream I-DirI mRNA. RNA 15:958–967. https://doi.org/10.1261/rna.669209

Nomura Y, Yokobayashi Y (2019) Systematic minimization of RNA ligase ribozyme through large-scale design-synthesis-sequence cycles. Nucleic Acids Res 47:8950–8960. https://doi.org/10.1093/nar/gkz729

Nomura Y, Yokobayashi Y (2023) RNA ligase ribozymes with a small catalytic core. Sci Rep 13:8584. https://doi.org/10.1038/s41598-023-35584-9

Norman A, Hansen LH, Sørensen SJ (2009) Conjugative plasmids: vessels of the communal gene pool. Philos Trans R Soc Lond Ser B Biol Sci 364:2275–2289. https://doi.org/10.1098/rstb.2009.0037

Novikova O, Belfort M (2017) Mobile group II introns as ancestral eukaryotic elements. Trends Genet 33:773–783. https://doi.org/10.1016/j.tig.2017.07.009

Novozhilov AS, Koonin EV (2009) Exceptional error minimization in putative primordial genetic codes. Biol Direct 4:44. https://doi.org/10.1186/1745-6150-4-44

Obbard DJ, Jiggins FM, Halligan DL, Little TJ (2006) Natural selection drives extremely rapid evolution in antiviral RNAi genes. Curr Biol 16:580–585. https://doi.org/10.1016/j.cub.2006.01.065

Obbard DJ, Gordon KH, Buck AH, Jiggins FM (2009) The evolution of RNAi as a defence against viruses and transposable elements. Philos Trans R Soc Lond Ser B Biol Sci 364:99–115. https://doi.org/10.1098/rstb.2008.0168

Ofir G, Melamed S, Sberro H et al (2018) DISARM is a widespread bacterial defence system with broad anti-phage activities. Nat Microbiol 3:90–98. https://doi.org/10.1038/s41564-017-0051-0

Ohkawa J, Yuyama N, Takebe Y, Nishikawa S, Taira K (1993) Importance of independence in ribozyme reactions: kinetic behavior of trimmed and of simply connected multiple ribozymes with potential activity against human immunodeficiency virus. Proc Natl Acad Sci USA 90:11302–11306. https://doi.org/10.1073/pnas.90.23.11302

Oliveira PH, Touchon M, Rocha EP (2014) The interplay of restriction-modification systems with mobile genetic elements and their prokaryotic hosts. Nucleic Acids Res 42:10618–10631. https://doi.org/10.1093/nar/gku734

Olovnikov I, Chan K, Sachidanandam R, Newman DK, Aravin AA (2013) Bacterial argonaute samples the transcriptome to identify foreign DNA. Mol Cell 51:594–605. https://doi.org/10.1016/j.molcel.2013.08.014

Pancer Z, Cooper MD (2006) The evolution of adaptive immunity. Annu Rev Immunol 24:497–518. https://doi.org/10.1146/annurev.immunol.24.021605.090542

Pancer Z, Amemiya CT, Ehrhardt GR et al (2004) Somatic diversification of variable lymphocyte receptors in the agnathan sea lamprey. Nature 430:174–180. https://doi.org/10.1038/nature02740

Papastavrou N, Horning DP, Joyce GF (2024) RNA-catalyzed evolution of catalytic RNA. Proc Natl Acad Sci USA 121:e2321592121. https://doi.org/10.1073/pnas.2321592121

Papathanasiou P, Erdmann S, Leon-Sobrino C et al (2019) Stable maintenance of the rudivirus SIRV3 in a carrier state in *Sulfolobus islandicus* despite activation of the CRISPR-Cas immune response by a second virus SMV1. RNA Biol 16:557–565. https://doi.org/10.1080/15476286.2018.1511674

Perriman R, Graf L, Gerlach WL (1993) A ribozyme that enhances gene suppression in tobacco protoplasts. Antisense Res Dev 3:253–263. https://doi.org/10.1089/ard.1993.3.253

Picton DM, Luyten YA, Morgan RD et al (2021) The phage defence Island of a multidrug resistant plasmid uses both BREX and type IV restriction for complementary protection from viruses. Nucleic Acids Res 49:11257–11273. https://doi.org/10.1093/nar/gkab906

Pinheiro VB, Taylor AI, Cozens C et al (2012) Synthetic genetic polymers capable of heredity and evolution. Science 336:341–344. https://doi.org/10.1126/science.1217622

Pinilla-Redondo R, Russel J, Mayo-Muñoz D et al (2022) CRISPR-Cas systems are widespread accessory elements across bacterial and archaeal plasmids. Nucleic Acids Res 50:4315–4328. https://doi.org/10.1093/nar/gkab859

Pleška M, Qian L, Okura R et al (2016) Bacterial autoimmunity due to a restriction-modification system. Curr Biol 26:404–409. https://doi.org/10.1016/j.cub.2015.12.041

Pradeu T, Carosella ED (2006) The self model and the conception of biological identity in immunology. Biol Philos 21:235–252. https://doi.org/10.1007/s10539-005-8621-6

Quadrana L, Bortolini Silveira A, Mayhew GF et al (2016) The *Arabidopsis thaliana* mobilome and its impact at the species level. Elife 5:e15716. https://doi.org/10.7554/eLife.15716

Richert-Pöggeler KR, Iskra-Caruana ML, Kishima Y (2022) Editorial: DNA virus and host plant interactions from antagonism to endogenization. Front Plant Sci 13:1014516. https://doi.org/10.3389/fpls.2022.1014516

Robertson MP, Joyce GF (2012) The origins of the RNA world. Cold Spring Harb Perspect Biol 4:a003608. https://doi.org/10.1101/cshperspect.a003608

Rochepeau P, Selinger LB, Hynes MF (1997) Transposon-like structure of a new plasmid-encoded restriction-modification system in *Rhizobium leguminosarum* VF39SM. Mol Gen Genet 256:387–396. https://doi.org/10.1007/s004380050582

Roers A, Hiller B, Hornung V (2016) Recognition of endogenous nucleic acids by the innate immune system. Immunity 44:739–754. https://doi.org/10.1016/j.immuni.2016.04.002

Saberi A, Gulyaeva AA, Brubacher JL, Newmark PA, Gorbalenya AE (2018) A planarian nidovirus expands the limits of RNA genome size. PLoS Pathog 14:e1007314. https://doi.org/10.1371/journal.ppat.1007314

Saletore Y, Meyer K, Korlach J et al (2012) The birth of the Epitranscriptome: deciphering the function of RNA modifications. Genome Biol 13:175. https://doi.org/10.1186/gb-2012-13-10-175

Samanta B, Joyce GF (2017) A reverse transcriptase ribozyme. Elife 6:e31153. https://doi.org/10.7554/eLife.31153

Samson JE, Magadán AH, Sabri M, Moineau S (2013) Revenge of the phages: defeating bacterial defences. Nat Rev Microbiol 11:675–687. https://doi.org/10.1038/nrmicro3096

Sauter D, Kirchhoff F (2021) Evolutionary conflicts and adverse effects of antiviral factors. Elife 10:e65243. https://doi.org/10.7554/eLife.65243

Sawyer SL, Wu LI, Emerman M, Malik HS (2005) Positive selection of primate TRIM5alpha identifies a critical species-specific retroviral restriction domain. Proc Natl Acad Sci USA 102:2832–2837. https://doi.org/10.1073/pnas.0409853102

Schmidt AK, Fitzpatrick AD, Schwartzkopf CM et al (2022) A filamentous bacteriophage protein inhibits type IV pili to prevent superinfection of *Pseudomonas aeruginosa*. MBio 13:e0244121. https://doi.org/10.1128/mbio.02441-21

Schnable PS, Ware D, Fulton RS et al (2009) The B73 maize genome: complexity, diversity, and dynamics. Science 326:1112–1115. https://doi.org/10.1126/science.1178534
Seed KD, Lazinski DW, Calderwood SB, Camilli A (2013) A bacteriophage encodes its own CRISPR/Cas adaptive response to evade host innate immunity. Nature 494:489–491. https://doi.org/10.1038/nature11927
Sensevdi ER, Sourrouille ZA, Quax TE (2024) Host range and cell recognition of archaeal viruses. Curr Opin Microbiol 77:102423. https://doi.org/10.1016/j.mib.2023.102423
Serra P, Navarro B, Forment J et al (2023) Expression of symptoms elicited by a hammerhead viroid through RNA silencing is related to population bottlenecks in the infected host. New Phytol 239:240–254. https://doi.org/10.1111/nph.18934
Shabalina SA, Koonin EV (2008) Origins and evolution of eukaryotic RNA interference. Trends Ecol Evol 23:578–587. https://doi.org/10.1016/j.tree.2008.06.005
Shahi S, Banerjea AC (2002) Multitarget ribozyme against the S1 genome segment of reovirus possesses novel cleavage activities and is more efficacious than its constituent mono-ribozymes. Antivir Res 55:129–140. https://doi.org/10.1016/s0166-3542(02)00012-8
Shirt-Ediss B, Ruiz-Mirazo K, Mavelli F, Solé RV (2014) Modelling lipid competition dynamics in heterogeneous protocell populations. Sci Rep 4:5675. https://doi.org/10.1038/srep05675
Silas S, Mohr G, Sidote DJ et al (2016) Direct CRISPR spacer acquisition from RNA by a natural reverse transcriptase-Cas1 fusion protein. Science 351:aad4234. https://doi.org/10.1126/science.aad4234
Sims A, Weir DJ, Cole SJ, Hutchinson E (2025) SARS-CoV-2 cellular coinfection is limited by superinfection exclusion. J Virol 99:e0207724. https://doi.org/10.1128/jvi.02077-24
Sistiaga-Poveda M, Jugo BM (2014) Evolutionary dynamics of endogenous Jaagsiekte sheep retroviruses proliferation in the domestic sheep, mouflon and Pyrenean chamois. Heredity 112:571–578. https://doi.org/10.1038/hdy.2013.136
Slotkin RK, Martienssen R (2007) Transposable elements and the epigenetic regulation of the genome. Nat Rev Genet 8:272–285. https://doi.org/10.1038/nrg2072
Smillie C, Garcillán-Barcia MP, Francia MV, Rocha EP, de la Cruz F (2010) Mobility of plasmids. Microbiol Mol Biol Rev 74:434–452. https://doi.org/10.1128/MMBR.00020-10
Smith JM, Szathmáry E (1993) The origin of chromosomes. I. Selection for linkage. J Theor Biol 164:437–446. https://doi.org/10.1006/jtbi.1993.1165
Song S, Wood TK (2020) A primary physiological role of toxin/antitoxin systems is phage inhibition. Front Microbiol 11:1895. https://doi.org/10.3389/fmicb.2020.01895
Spada SJ, Grigg ME, Bouamr F, Best SM, Zhang P (2024) TRIM5α: a protean architect of viral recognition and innate immunity. Viruses 16:997. https://doi.org/10.3390/v16070997
Spiegelman S, Haruna I, Holland IB, Beaudreau G, Mills D (1965) The synthesis of a self-propagating and infectious nucleic acid with a purified enzyme. Proc Natl Acad Sci USA 54:919–927. https://doi.org/10.1073/pnas.54.3.919
Srinivasachar Badarinarayan S, Sauter D (2021) Switching sides: how endogenous retroviruses protect us from viral infections. J Virol 95:e02299–e02220. https://doi.org/10.1128/JVI.02299-20
Suzuki Y, Baidaliuk A, Miesen P et al (2020) Non-retroviral endogenous viral element limits cognate virus replication in *Aedes aegypti* Ovaries. Curr Biol 30:3495–3506.e6. https://doi.org/10.1016/j.cub.2020.06.057
Swarts DC, Makarova K, Wang Y et al (2014) The evolutionary journey of Argonaute proteins. Nat Struct Mol Biol 21:743–753. https://doi.org/10.1038/nsmb.2879
Tagami S, Attwater J, Holliger P (2017) Simple peptides derived from the ribosomal core potentiate RNA polymerase ribozyme function. Nat Chem 9:325–332. https://doi.org/10.1038/nchem.2739
Takeda A, Matano T (2007) Inhibition of infectious murine leukemia virus production by Fv-4 env gene products exerting dominant negative effect on viral envelope glycoprotein. Microbes Infect 9:1590–1596. https://doi.org/10.1016/j.micinf.2007.09.012

Takemura M (2020) Medusavirus ancestor in a proto-eukaryotic cell: updating the hypothesis for the viral origin of the nucleus. Front Microbiol 11:571831. https://doi.org/10.3389/fmicb.2020.571831
Taneja J, Higgs PG (2025) Protocell dynamics: modelling growth and division of lipid vesicles driven by an autocatalytic reaction. Life 15:724. https://doi.org/10.3390/life15050724
Tarazona P (1992) Error thresholds for molecular quasispecies as phase transitions: from simple landscapes to spin-glass models. Phys Rev A 45:6038–6050. https://doi.org/10.1103/physreva.45.6038
Tarlinton RE, Meers J, Young PR (2006) Retroviral invasion of the koala genome. Nature 442:79–81. https://doi.org/10.1038/nature04841
Tarlinton RE, Legione AR, Sarker N et al (2022) Differential and defective transcription of koala retrovirus indicates the complexity of host and virus evolution. J Gen Virol 103. https://doi.org/10.1099/jgv.0.001749
Taylor GM, Gao Y, Sanders DA (2001) Fv-4: identification of the defect in Env and the mechanism of resistance to ecotropic murine leukemia virus. J Virol 75:11244–11248. https://doi.org/10.1128/JVI.75.22.11244-11248.2001
Temin HM (1971) The protovirus hypothesis: speculations on the significance of RNA-directed DNA synthesis for normal development and for carcinogenesis. J Natl Cancer Inst 46:3–7
tenOever BR (2016) The evolution of antiviral defense systems. Cell Host Microbe 19:142–149. https://doi.org/10.1016/j.chom.2016.01.006
Tesson F, Hervé A, Mordret E et al (2022) Systematic and quantitative view of the antiviral arsenal of prokaryotes. Nat Commun 13:2561. https://doi.org/10.1038/s41467-022-30269-9
Tjhung KF, Shokhirev MN, Horning DP, Joyce GF (2020) An RNA polymerase ribozyme that synthesizes its own ancestor. Proc Natl Acad Sci USA 117:2906–2913. https://doi.org/10.1073/pnas.1914282117
Turk RM, Chumachenko NV, Yarus M (2010) Multiple translational products from a five-nucleotide ribozyme. Proc Natl Acad Sci USA 107:4585–4589. https://doi.org/10.1073/pnas.0912895107
Twomey DP, McKay LL, O'Sullivan DJ (1998) Molecular characterization of the Lactococcus lactis LlaKR2I restriction-modification system and effect of an IS982 element positioned between the restriction and modification genes. J Bacteriol 180:5844–5854. https://doi.org/10.1128/JB.180.22.5844-5854.1998
Uriu K, Kosugi Y, Ito J, Sato K (2021) The battle between retroviruses and APOBEC3 genes: its past and present. Viruses 13:124. https://doi.org/10.3390/v13010124
Van Valen L (1973) A new evolutionary law. Evol Theory 1:1–30
Villarreal LP (2009) The source of self: genetic parasites and the origin of adaptive immunity. Ann N Y Acad Sci 1178:194–232. https://doi.org/10.1111/j.1749-6632.2009.05020.x
Villarreal LP (2011) Viral ancestors of antiviral systems. Viruses 3:1933–1958. https://doi.org/10.3390/v3101933
Villarreal LP (2012) The addiction module as a social force. In: Witrany G (ed) Viruses: essential agents of life. Springer, Dordrecht. https://doi.org/10.1007/978-94-007-4899-6_6
Vlassov AV, Johnston BH, Landweber LF, Kazakov SA (2004) Ligation activity of fragmented ribozymes in frozen solution: implications for the RNA world. Nucleic Acids Res 32:2966–2974. https://doi.org/10.1093/nar/gkh601
Wadkins TS, Been MD (2002) Ribozyme activity in the genomic and antigenomic RNA strands of hepatitis delta virus. Cell Mol Life Sci 59:112–125. https://doi.org/10.1007/s00018-002-8409-7
Wang L, Zhang L (2023) The arms race between bacteria CBASS and bacteriophages. Front Immunol 14:1224341. https://doi.org/10.3389/fimmu.2023.1224341
Wang Y, Liu M, Guo X et al (2022) Endogenous retrovirus elements are co-expressed with IFN stimulation genes in the JAK-STAT pathway. Viruses 15:60. https://doi.org/10.3390/v15010060

Wang J, Dong Y, Zheng X et al (2024) Host factors modulate virus-induced IFN production via pattern recognition receptors. J Inflamm Res 17:3737–3752. https://doi.org/10.2147/JIR.S455035
Watson BNJ, Vercoe RB, Salmond GPC et al (2019) Type I-F CRISPR-Cas resistance against virulent phages results in abortive infection and provides population-level immunity. Nat Commun 10:5526. https://doi.org/10.1038/s41467-019-13445-2
Weick EM, Miska EA (2014) piRNAs: from biogenesis to function. Development 141:3458–3471. https://doi.org/10.1242/dev.094037
Weinberg CE, Weinberg Z, Hammann C (2019) Novel ribozymes: discovery, catalytic mechanisms, and the quest to understand biological function. Nucleic Acids Res 47:9480–9494. https://doi.org/10.1093/nar/gkz737
Werren JH (2011) Selfish genetic elements, genetic conflict, and evolutionary innovation. Proc Natl Acad Sci USA 108(Suppl 2):10863–10870. https://doi.org/10.1073/pnas.1102343108
Wilkins AS (2010) The enemy within: an epigenetic role of retrotransposons in cancer initiation. BioEssays 32:856–865. https://doi.org/10.1002/bies.201000008
Witzany G (2006) Natural genome-editing competences of viruses. Acta Biotheor 54:235–253. https://doi.org/10.1007/s10441-006-9000-7
Witzany G (2011) The agents of natural genome editing. J Mol Cell Biol 3:181–189. https://doi.org/10.1093/jmcb/mjr005
Yakovenko I, Agronin J, Smith LC, Oren M (2021) Guardian of the genome: an alternative RAG/*Transib* co-evolution hypothesis for the origin of V(D)J recombination. Front Immunol 12:709165. https://doi.org/10.3389/fimmu.2021.709165
Yan J, Zhao Y, Du J et al (2022) RNA sensor MDA5 suppresses LINE-1 retrotransposition by regulating the promoter activity of LINE-1 5′-UTR. Mob DNA 13:10. https://doi.org/10.1186/s13100-022-00268-0
Yap MW, Colbeck E, Ellis SA, Stoye JP (2014) Evolution of the retroviral restriction gene Fv1: inhibition of non-MLV retroviruses. PLoS Pathog 10:e1003968. https://doi.org/10.1371/journal.ppat.1003968
Young GR, Yap MW, Michaux JR, Steppan SJ, Stoye JP (2018) Evolutionary journey of the retroviral restriction gene Fv1. Proc Natl Acad Sci USA 115:10130–10135. https://doi.org/10.1073/pnas.1808516115
Yu H, Zhang S, Chaput JC (2012) Darwinian evolution of an alternative genetic system provides support for TNA as an RNA progenitor. Nat Chem 4:183–187. https://doi.org/10.1038/nchem.1241
Zhang L, Tripathi A (2017) Archaeal RNA ligase from *Thermoccocus kodakarensis* for template dependent ligation. RNA Biol 14:36–44. https://doi.org/10.1080/15476286.2016.1239688
Zhang L, Wang H, Zeng J et al (2024) Cas1 mediates the interference stage in a phage-encoded CRISPR-Cas system. Nat Chem Biol 20:1471–1481. https://doi.org/10.1038/s41589-024-01659-5
Zhao K, Du J, Peng Y et al (2018) LINE1 contributes to autoimmunity through both RIG-I- and MDA5-mediated RNA sensing pathways. J Autoimmun 90:105–115. https://doi.org/10.1016/j.jaut.2018.02.007
Zhao Y, Ma Y, Vasileiou C et al (2025) Jumbo phage-mediated transduction of genomic islands. Proc Natl Acad Sci USA 122:e2512465122. https://doi.org/10.1073/pnas.2512465122
Zheng L, Legione AR (2025) A high frequency of detection of koala retrovirus fragments in Victorian koalas suggests historic integration of KoRV. J Gen Virol 106:002097. https://doi.org/10.1099/jgv.0.002097
Zheng H, Dietrich C, Hongoh Y, Brune A (2016) Restriction-modification systems as mobile genetic elements in the evolution of an intracellular symbiont. Mol Biol Evol 33:721–725. https://doi.org/10.1093/molbev/msv264
Zhou X, Sun K, Zhou X, Jackson AO, Li Z (2019) The matrix protein of a plant Rhabdovirus mediates superinfection exclusion by inhibiting viral transcription. J Virol 93:e00680–e00619. https://doi.org/10.1128/JVI.00680-19

Chapter 11
Non-mendelian Inheritance and Evolution in Ciliates with a Focus on Transposons

Sarah E. Allen and Mariusz Nowacki

Abstract Transposable elements (TEs) have driven the evolution of complex traits in eukaryotes via a range of different mechanisms. Ciliated protozoa (ciliates) are a valuable model eukaryote in which several groundbreaking discoveries have been made over the years. Their unusual life cycle and genome organisation have led to a unique co-evolution with transposable elements (TEs), making them a fascinating system in which to study TE insertion, silencing and domestication. In this chapter, we discuss non-Mendelian inheritance and evolution in ciliates, with a focus on TEs.

1 Introduction

In the traditional Darwinian model of inheritance, evolutionary pressure is exerted at the level of the germline. Evolutionary fitness is determined by the genes inherited from the parental germlines and manifests as the ability to pass on those same genes to the next generation. In this model, the 'soma', or non-germline cells (or genomes), dies off at every generation and is thus disposable—merely a vehicle for the transmission of genes to the generations to come. In general, the information flow between germline and soma is considered to be strictly unidirectional, passing from germline to soma but never the other way, in accordance with August Weismann's theory of germ plasm. This maxim is termed the 'Weismann barrier'. Thus, an organism's evolutionary fitness in a given environment is determined at conception and cannot be altered during its lifetime.

As time and science have moved on, exceptions to this model have been studied, and of course, the truth is more nuanced than previously thought. While the genome is determined at conception, organisms show a remarkable ability to adapt to their environments, thus increasing the number and survivability of offspring that they produce. Adaptability itself is emerging as a heritable genetic trait (Jerison et al. 2017), indicating that the ability to alter one's phenotype is an important and

S. E. Allen · M. Nowacki (✉)
Institute of Cell Biology, University of Bern, Bern, Switzerland
e-mail: mariusz.nowacki@unibe.ch

G. Witzany (ed.), *Integrative Theory of Evolution - Aspects and Insights*,
https://doi.org/10.1007/978-3-032-24295-2_11

selectable attribute. This adaptability may be on the phenotypic level, over the lifespan of an individual organism, but may also be at the genetic level, over generations of natural selection. In the traditional model of inheritance, genetic diversity arises through mutations in individuals, and most mutations are deleterious. There must therefore be a balance between protecting the organism from deleterious mutations and allowing evolution to progress.

A defence against deleterious mutations and a source of adaptability for an organism is increased complexity. In a pathway with a few genes, each one will have a crucial role, and any change is likely to destroy the function of the pathway and be lethal for the cell. But in a pathway with dozens of genes and functional redundancy at one or more points, the scope for evolution of new functions for genes is much greater. Therefore, arguably, complexity itself is a favourable trait in evolution as it leads to greater adaptability and therefore the possibility of occupying more evolutionary niches.

2 Non-mendelian Inheritance and Evolution in Ciliates

2.1 TEs in Evolution: A Background

Transposable elements (TEs) are mobile genetic elements that reproduce and insert themselves into host genomes. There are many different types of TE, but they are broadly divided into two categories, DNA transposons and retrotransposons (RTs). The main differentiator between the two classes is their mechanism of transposition—how they relocate themselves within a genome. DNA TEs have the capacity to cut themselves out and then re-ligate themselves into a new genomic locus. RTs, on the other hand, make a copy of themselves using RNA, then reverse transcribe the RNA into cDNA. The cDNA is then inserted into a new genomic locus (Hayward and Gilbert 2022). RTs therefore increase in number with each transposition event, whereas DNA transposons only increase in number if they transpose during DNA replication (a transposon that has already been replicated may jump to an unreplicated region). TEs are typically in the order of a few hundred base pairs (bp) up to several kilobase pairs (kbp) long, but can be as long as 40 kbp (Arkhipova and Yushenova 2019; Wells and Feschotte 2020). A TE insertion is therefore a major event in any given genomic locus, and can have hugely disruptive consequences. TEs have often been called 'selfish' or 'parasitic' DNA, as they rely on the transcription and translation machinery of their hosts, which they co-opt solely in order to replicate themselves. Their success in this endeavour over evolutionary time is attested to by the fact that TEs are present in almost every eukaryotic genome, where they make up a greater or smaller proportion of the genome depending on the species. In yeast, the proportion of TEs and TE-derived sequences is 3%, while in humans it is two thirds and in maize it is 84% (Hayward and Gilbert 2022). The reason why some organisms tolerate TEs at higher levels than others is not known with certainty.

Historically, most discussions of TE–host relationships have assumed that TEs are purely parasitic, at best being a small net energy drain on their host and at worst inserting themselves into important protein-coding or regulatory regions, disrupting endogenous sequences and killing their host. However, recently it has become clear that the relationship between TEs and host genome evolution is tightly linked, with TEs driving the evolution of a myriad of functions that are beneficial to their hosts, in what could be described as more of a symbiosis than a host-parasite relationship. The pressure to prevent unchecked TE invasion has driven hosts to evolve strategies to repress TEs. These strategies may then be co-opted by the host for its own purposes. An example of this is heterochromatinisation, which is thought to have evolved as a means of silencing TEs and thereby preventing their spread (Déléris et al. 2021; Almeida et al. 2022). Heterochromatin is an important feature of eukaryotic genomes, where it is both regulatory and structural, and many pathways and mechanisms have evolved to aid in its establishment, spreading and restriction (Allshire and Madhani 2018; McKinley and Cheeseman 2016; Scott et al. 2006; Sullivan et al. 2016; Uhlmann 2016). TEs have also contributed to eukaryote evolution directly, through incorporation into the host genome, where they may become domesticated and gain new functions (Almeida et al. 2022). There are many such examples, including the RAG1 and RAG2 genes necessary for V(D)J recombination in lymphocyte development in jawed vertebrates (Kapitonov and Jurka 2005; Zhou et al. 2004). RAG1 and RAG2 are ancient TEs of the *Transib* superfamily, and their ability to cleave and rejoin DNA was co-opted by their host to provide diversity in its immune system. Domesticated transposons have also been implicated in the evolution of sex determination in mice (Miyawaki et al. 2020) and placental development in mammals (Shiura et al. 2021; Mangeney et al. 2007; Kitazawa et al. 2017), among other features (Modzelewski et al. 2022). In addition to the domestication of TEs into protein-coding genes, many examples exist in nature of promoters and other regulatory regions originating from TE insertions. For example, the transition from embryonic stem cells to the 2-cell embryo in mice appears to be regulated by a domesticated retrotransposon, which functions as an alternative promoter to a swathe of host genes involved in cell fate determination (Torres-Padilla 2020; Evsikov et al. 2004; Peaston et al. 2004). Tail loss in the primate lineage, including humans, is also the result of a transposition event leading to alternative splicing of the transcription factor *TBXT* in early embryonic development (Xia et al. 2024). Transposition events may also multiply and move endogenous host sequences into new genomic loci if the TE excision is imprecise. Importantly, transposition can occur not only within an organism but also through horizontal gene transfer across organisms or even species. This may take place via a viral or other pathogenic intermediate, which can pick up TEs from a host and pass them on, or directly via sexual reproduction between two members of the same or closely related species. This horizontal TE transfer may also include host genomic sequences and therefore provides a mechanism for sequences to jump between two unrelated host organisms (Gilbert and Feschotte 2018; Cosby et al. 2019). These examples demonstrate the remarkable power that TEs have to rapidly drive the evolution of host organisms, moving entire genes and regulatory regions both within and between genomes.

2.2 Ciliates Have an Unusual Life Cycle in which TEs Play a Vital Role

Ciliates provide a fascinating model for the study of non-Mendelian inheritance for a number of reasons. Ciliates are protozoa that separate their germline from their soma by containing them in two distinct nuclei within a single cell. The two nuclei are not only functionally distinct but are also genetically distinct, with the germline nucleus containing many sequences that are not present in the somatic nucleus (Duret et al. 2008; Arnaiz et al. 2012; Chen and Landweber 2016; Eisen et al. 2006). These germline-limited sequences consist of transposable elements (TEs) and transposon-derived sequences called internal eliminated sequences (IESs). Like in multicellular organisms, the soma is disposable and is renewed at each sexual reproduction, while the germline is immortal and provides the genetic information for building the soma. The germline nucleus is usually diploid, and a ciliate will have at least one but may have several, depending on its species. The somatic nucleus is highly polyploid, 45n in *Tetrahymena* or 800n in *Paramecium*, and a cell will usually have only one (Allen and Nowacki 2020). Due to their relative difference in ploidy, the two nuclei are of vastly different sizes, which gives them their names, ‘micronucleus’ (germline) and ‘macronucleus’ (somatic nucleus). Ciliates have two stages in their life cycle: vegetative growth and sexual reproduction. During vegetative growth, the diploid micronucleus divides by mitosis, while the polyploid macronucleus divides by a poorly understood process called amitosis, in which there is no mitotic spindle and no apparent chromosome segregation (Fu et al. 2022; Tucker et al. 1980) (see Fig. 11.1). Due to the high copy number of the genome, each daughter macronucleus receives at least one copy of each chromosome, which is then amplified up to the correct amount following cell division. A certain number of vegetative divisions must take place before a ciliate can undergo sexual reproduction. During sexual reproduction, a new somatic nucleus is formed from the germline. Briefly, the micronuclei undergo meiosis to form eight haploid pronuclei, seven of which degrade and one of which divides by mitosis, to form two haploid pronuclei. One pronucleus is exchanged with another cell of the opposite mating type in a process termed ‘conjugation’, in which two cells fuse and a cytoplasmic bridge is formed between them, allowing transfer of a nucleus. The two pronuclei in each mating pair then fuse and undergo two more mitotic divisions, resulting in four diploid nuclei (Phadke and Zufall 2009). Two of these will continue the germline as the two new micronuclei, and the other two will undergo programmed genome rearrangement to form two new macronuclei, in a process termed ciliate ‘development’. Meanwhile, the old somatic nucleus—the maternal macronucleus—fragments and is degraded. Programmed genome rearrangement comprises three main events: chromosome fragmentation, genome amplification and the removal of all TEs, IESs and other repetitive or unwanted DNA. Chromosome fragmentation occurs via the formation of double-stranded breaks at certain loci in the micronuclear chromosomes, followed by de novo telomere addition. The result is a larger number of shorter chromosomes: in *Tetrahymena*, around 200 macronuclear chromosomes are formed from five

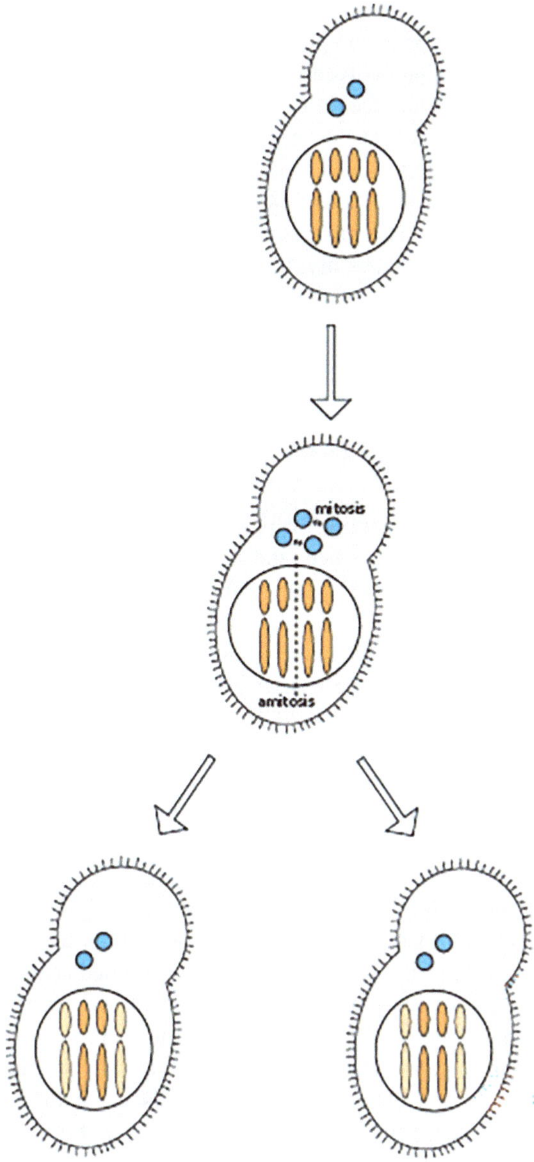

Fig. 11.1 Amitosis. The two different types of nuclei in *Paramecium* are shown in blue (micronucleus, germline) and cream (macronucleus, soma). Chromosomes are shown in orange and in 4n as opposed to 800n for simplicity. During mitosis, chromosome segregation in the micronucleus proceeds as in other organisms, with replication occurring prior to chromosome condensation, spindle formation and equal division of chromatids into daughter nuclei. In amitosis, the macronucleus is roughly divided in half, and the chromosome copy number is increased up to 800n following nuclear division. Newly synthesised chromosomes are represented by a lighter shade of orange in the new macronucleus

micronuclear chromosomes (Hamilton et al. 2005, 2016; Fass et al. 2011). The removal of TEs and IESs is carried out by one or more domesticated transposons, which make double-stranded breaks in the DNA flanking the sequences that are to be removed (Baudry et al. 2009; Cheng et al. 2010; Vogt and Mochizuki 2013; Doak et al. 1994; Nowacki et al. 2009). This process is hugely complex and involves distinct mechanisms to identify IESs and TEs, find their ends, mark them for excision, find and reanneal the correct ends following excision and then process the excised DNA. This is a fascinating example of how TEs have driven the evolution of a complex major event in cellular development, with a myriad of genes having evolved to ensure the correct and timely removal of unwanted DNA. The fact that the key players in this pathway—the excisases themselves—are domesticated transposons, illustrates the dual-faceted nature of transposon invasion into host genomes. The cell, needing to remove TEs and their derivatives from its somatic nucleus, co-opts another TE to perform the removal. The result is not only a TE-free somatic nucleus but also an additional large and elaborate pathway with hundreds of genes involved, all of which have the scope to further evolve and add more layers of complexity to the cell's functions.

2.3 *Small RNAs Help Ciliates to Distinguish Self from Non-self DNA*

How the cell identifies which sequences to keep and which to excise during genome rearrangement has been the subject of much research over the last decades. In three well-studied species, *Oxytricha*, *Paramecium* and *Tetrahymena*, it is achieved with the help of piwi-binding small RNAs (piRNAs). piRNAs were first discovered as a host defence mechanism against transposable elements in *Drosophila*, where they bind to and silence transcripts from TEs in the germline, preventing their spread (Malone and Hannon 2009). They have subsequently been discovered across the entire animal kingdom, where they play a similar role in TE silencing. In the ciliates *Tetrahymena* and *Paramecium*, 25 nt piRNAs are formed genome-wide in the micronucleus at the start of sexual reproduction, and are termed scan RNAs (scnRNAs) (Mochizuki and Gorovsky 2004; Lepère et al. 2008; Sandoval et al. 2014). They are bound by a piwi protein and transported to the maternal macronucleus, where they are compared ('scanned') against the entire maternal genome to find their matching sequence. Those that fail to find a match in the maternal genome must logically derive from germline-limited sequences in the micronucleus, which are to be removed during programmed genome rearrangement. They are therefore sent to the developing macronucleus, which at this stage is newly formed from meiotic products of the germline and thus still contains all TEs, IESs and other unwanted sequences. Here, the piRNAs identify the germline-limited sequences, which can then be excised by a domesticated transposase, and the chromosome ends are re-annealed. In *Oxytricha*, a related but opposite pathway exists, where piRNAs are produced in the maternal macronucleus and therefore by definition do not match any germline-limited sequences (Fang et al. 2012). Their role in developing a new macronucleus is to protect macronuclear sequences against excision and degradation.

This genome comparison event, in which small RNAs help the daughter macronucleus develop as a copy of its parental macronucleus, has benefits for both the host and the invading TEs. It protects the host against TEs and their remnants, preventing them from disrupting genes and other important sequences. It also mitigates much of the possible damage from TE proliferation in the germline, which is advantageous to TEs themselves as they can spread further without destroying their host.

2.4 Epigenetic Trans-generational Inheritance in Ciliates

The small RNA pathways described above use the maternal soma as a template for making a new daughter soma, describing what should be contained therein and how it should be organised and using RNA as its intermediary. Importantly, it does not affect the germline, which retains all its sequences and transmits them faithfully through the generations, but it does affect which germline sequences get expressed in the newly developed organism. It thus provides an intriguing possibility for the trans-generational inheritance of acquired characteristics, which is antithetical to the traditional view of Darwinian evolution. Theoretically, deletions or rearrangements that occur in the maternal macronucleus during vegetative growth and that confer an advantage in a given environment could be transmitted to the developing daughter macronucleus via the small RNA pathways activated during development. This type of inheritance would not break the Weismann barrier or contradict the laws of Darwinian evolution, but would rather constitute another level of selection.

2.5 IESs Evolve Towards Being Readily Excised from Ciliate Genomes

It is clear that the evolution of ciliate genomes and TEs is tightly linked. In the *Paramecium* germline, 45,000 unique IESs are present, all of which derive from transposon insertions. The majority of them are located within protein-coding genes, meaning that their precise (to the single nucleotide) excision is crucial for gene function and hence cell survival (Allen and Nowacki 2020). It might seem counter-intuitive that the ciliate genome should have retained all these IESs over evolutionary time. In other organisms, a transposon insertion within a crucial gene would be an evolutionary dead end as it would be immediately lethal—but in ciliates, it is tolerated at extremely high levels. Interestingly, once inserted, it is clear that there is a selective pressure on IESs to be 'exciseable', that is, to be readily excised by the existing cellular machinery that performs the genome rearrangements. 'Exciseability' as a feature of IESs has several aspects; in *Paramecium*, these include obligate TA dinucleotides flanking the IES, a length preference and end recognition sequence preferences. The length preference for *Paramecium* IESs is particularly interesting, as it follows a sinusoidal distribution that peaks at 26–28 bp, then has peaks every ~10 bp, but not including 36–38 bp, which is strongly unfavourable. Interestingly, newer IESs (more recently inserted) tend to be longer and older IESs

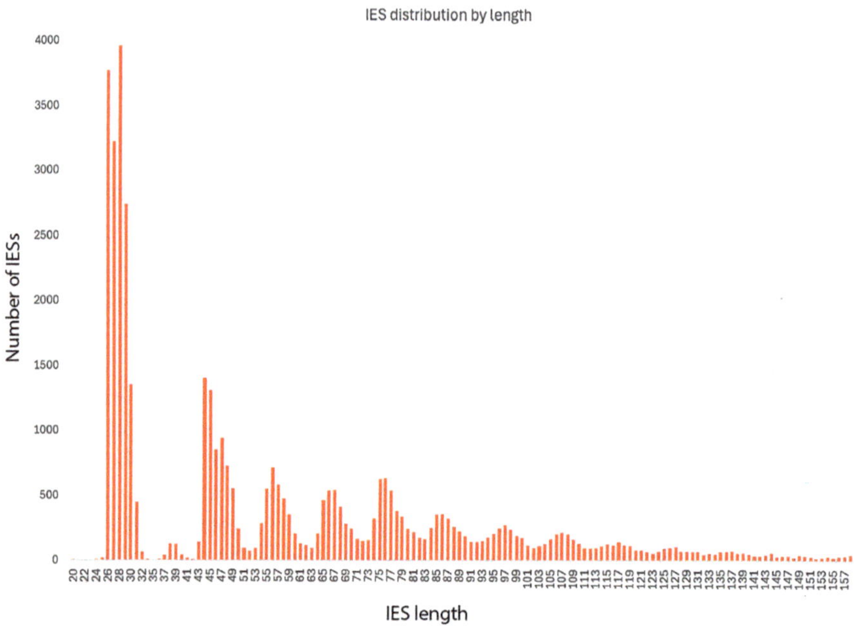

Fig. 11.2 IES distribution according to length. The x-axis shows IES length, cut off at 158 bp to better show the majority of IESs. Histogram bar lengths (y-axis) correspond to the number of IESs for a given length. A striking periodicity is seen, peaking at 26–28 bp and with peaks roughly every 10 bp

shorter, implying that they evolve to become shorter over time (Arnaiz et al. 2012; Guérin et al. 2017). However, 26 bp is the shortest length possible for *Paramecium* IESs, and below 26 bp, there are no IESs at all. This distribution (see Fig. 11.2) provides clues as to how the excision machinery works, although as yet the mechanism has not been clearly elucidated. It also provides insight into the evolution of IESs from TEs, and raises several interesting questions. Clearly, there is pressure on IESs to grow smaller over evolutionary time, but this shrinking is non-uniform: smaller is only preferred within the confines of the excision machinery. In all, 46 bp is preferred over 56 bp, but 36 bp is not preferred over 46 bp, nor is 42 bp. What is being selected for is exciseability rather than length per se.

Another measure of 'exciseability' is the dependence of an IES on the small RNA pathways that identify non-host DNA and target it for excision (see Sect. 2.3). In *Paramecium*, around 85% of IESs do not require the scnRNA pathway in order to be excised (Swart et al. 2017), and are recognised and removed independently of their presence or absence in the maternal macronucleus. They are thus termed 'non-maternally controlled' IESs, and their recognition is as yet not entirely understood. What is clear, however, is that newer IESs are more heavily dependent on the scnRNA pathway, and older IESs tend towards being non-maternally controlled. It therefore seems that over time, IESs evolve towards being excised automatically, independently of the small RNA pathways put there to help identify them.

2.6 Host-IES Symbiosis in Ciliates

Why do IESs evolve to be more and more easily excised, but still present in the genome? What, if any, advantage do they provide the cell? One possible answer is that there is no advantage at all, and that IESs are simply under pressure to shrink, become more and more unobtrusive, and eventually disappear from the genome. In this 'evolutionary arms race' model, the IESs we observe today are in the process of being excluded from the genome, but are too recently inserted to have been eliminated yet. The evolution towards easy excision is explained as simply a temporary measure to mitigate the harm caused by IESs, before they are eventually selected against and disappear. A second possibility is that the reason that IESs persist in all ciliate genomes is that they confer some advantage to the cell. We may call this the 'host-IES symbiosis model'. If true, the most plausible potential advantage is increased adaptability of the genome, as seen in other organisms (see Sect. 2.1). The programmed genome rearrangement process is by nature error-prone, with failure to excise IESs, IESs being excised but with shifted boundaries, and erroneous excision of macronuclear sequences flanked by TAs ('cryptic' IES excision), all occurring at a low but steady rate (Duret et al. 2008; Arnaiz et al. 2012). This type of error would be strongly avoided in any species with only two copies of each chromosome, such as animals or most plants, but in the highly polyploid ciliate, it can be tolerated, as for any chromosome carrying a fatal error, there are hundreds of correct versions. It may only become a serious problem for a cell if random chromosome assortment during amitosis leads to higher proportions of errors in a proportion of vegetative offspring, but even then, the affected cell will be genetically identical to its unaffected sister cells on the germline level, and so evolutionarily speaking, the fitness cost is small (see Sect. 2.8). In the host-IES symbiosis model, this low but constant error rate is not a bug, but a feature.

2.7 Co-evolution of Ciliate Genomes and TEs

Whether IESs and their excision machinery confer an advantage to ciliate genomes or are simply the fallout of an evolutionary arms race between ciliates and TEs, it is clear that their presence has led to the evolution of a number of other genes and processes. As described in the introductory section on 'TEs in evolution', many known examples exist in biology of traits whose evolution was driven by TE insertion or the need to silence inserted TEs. Ciliates are no exception to this. An elegant example is the mechanism by which the *Paramecium* mating type is determined. Two mating types exist in *Paramecium*, termed E and O. Mating type E expresses a transmembrane surface protein called MtA, while O does not, and cells of opposite mating types recognise each other with the help of the MtA protein. MtA is therefore necessary for conjugation in *Paramecium*. Mating type is inherited maternally via the scnRNA pathway, meaning that a daughter cell will have the same mating type as its cytoplasmic parent (see Sect. 2.3). In the species *Paramecium tetraurelia*, the promoter for the *MtA* gene is excised by the IES excision machinery in mating type O, but retained in mating type E. Strikingly, the *MtA* gene does not appear to be TE-derived, meaning that in mating type determination, the IES excision pathway has been co-opted to excise a bona fide piece of coding genome. The 195 bp section of promoter that is removed is flanked by TAs and has loose IES consensus inverted repeats at its ends. This demonstrates that the genome rearrangement machinery can be used to excise any piece of DNA from the genome that is of a suitable length and is flanked by TA dinucleotides—and that this is not just a theoretical possibility but has genuinely occurred to control an important cellular function. Even more strikingly, in the sister species *Paramecium septaurelia*, the same *MtA* gene controls mating type determination, which is also maternally inherited via the scnRNA pathway. However, in this species, it is not the promoter of *MtA* that is excised in mating type O, but sequences within the transcription factor *MtB*, which controls the expression of *MtA*. The *MtA* gene and its promoter are intact, but it is simply not expressed in the O mating type, as its transcription factor *MtB* is non-functional due to deletions. These deletions also occur between flanking TA dinucleotides in non-TE, non-IES sequences. The two mechanisms of mating type determination, therefore, both involve co-option of the IES excision machinery but appear to have evolved separately, demonstrating the practicality of the genome rearrangement processes for the evolution of new traits and mechanisms (Singh et al. 2014).

Domesticated transposons have, as described, important functions in nearly every eukaryote studied. As mentioned previously, ciliates use domesticated transposons for a number of different functions, not least the genome rearrangement pathway itself. This domestication has also occurred independently in different ciliate species: in *Paramecium*, the IES and TE excision is carried out by a PiggyBac-like domesticated transposase, PiggyMac (Baudry et al. 2009), while in *Tetrahymena*, it is done by three PiggyBac-like transposases, TPB2, TPB1 and TPB6 (Vogt and Mochizuki 2013; Cheng et al. 2016). In *Oxytricha*, rather than a single or small number of transposon domestication events having occurred, the genome rearrangement is carried out by four families of Tc1/Mariner-type transposases, comprising thousands of active transposases (Nowacki et al. 2009). These are expressed from germline-limited sequences during early development, and they then excise their own genomic sequences along with all IESs and TE remnants in the genome. This convergent evolution between three distantly related ciliate species is evidence of the potential of TEs in the germline for adaptation into genes useful to their hosts, and is also an interesting illustration of how TEs can be co-opted in their own defence. This elaborate system, from the outside appearing to be a zero-sum game, in fact adds a vast new level of complexity and therefore adaptability to the species.

Apart from the IES excision machinery being co-opted for the ciliate hosts' own use, IESs themselves are frequently assimilated into host genomes. Comparison of the macronuclear genomes in different strains of *Paramecium tetraurelia* reveals many instances of IESs being excised in one strain, but retained in another (Arnaiz et al. 2020). This differential excision has evolved subsequent to the IES insertion into the genome, as evidenced by the fact that the same IES sequence is present in both strains but is only excised in one of them. Therefore, it is clear that there is ongoing integration of IESs into host genomes. IES integration is another means by which rapid evolution may occur, if the IES in question confers (or evolves) a beneficial function to the cell.

2.8 'Somatic Evolution' in Ciliates

Unlike many organisms that separate their germline from their soma, ciliates do not differentiate specific cell types to contain their germline, as distinct from their somatic cell types. In a mammal, for example, only the gonads produce germ cells, and all other cell types of the body are genetic dead ends. In ciliates, any cell can undergo sexual development and contribute genetically to the next generation. In most cases, this requires a partner with whom to exchange a meiotic pronucleus, but some species are also capable of reproducing without a sexual partner. An example is *Paramecium*, one of the more well-studied species of ciliate, which can self-fertilise in a process termed 'autogamy'. Here, the meiotic pronucleus is not exchanged with another cell, but simply fuses with the second pronucleus within the same cell. After this, macronuclear development continues as after conjugation. Autogamy can only occur after a *Paramecium* has undergone roughly 20 vegetative divisions, as compared to conjugation, which can take place after a minimum of three to four divisions (Berger 1986). In other words, a new zygotic Paramecium cell will theoretically produce at a minimum of 2 (Evsikov et al. 2004) vegetative daughter cells before any of them undergo meiosis and self-fertilise, or 16 vegetative daughter cells before conjugation can take place. While each of these vegetative offspring is genetically identical in terms of their germline, their somatic nuclei may start to vary significantly during the vegetative life cycle. Amitosis does not segregate chromosomes equally or consistently, as described previously; rather, the number of chromosomes inherited into each daughter cell is random, and the ploidy is made up in the daughter cell through DNA replication. Therefore, with each division, biases in DNA content may arise quickly and may have selective advantages or disadvantages in a given environment. As described in the 'Host-IES symbiosis in ciliates' section, the genome rearrangement pathway consistently makes excision errors at a low level following each sexual reproduction event, introducing variability in the form of indels in around 0.1% of chromosomes at a given IES locus. Additionally, the chromosome fragmentation that occurs during genome rearrangement is, by nature, variable, with certain regions being broken and telomerised in some instances but remaining joined in others, leading to variability in chromosome length and number. Including this phenomenon, and other random genetic recombination events that may occur in an actively transcribed genome with high ploidy, the scope for alternative versions of the Paramecium somatic genomes to arise in the 20 divisions between two autogamies is huge. As these genomes are also all under selective pressure, there could arise a sort of rapid 'somatic evolution', in which very fast adaptation of the somatic nucleus occurs in response to environmental pressures (see Fig. 11.3). This adaptation would have no effect on the germline and therefore be low risk for the species, and indeed, low risk for the rapidly somatically adapting cells themselves, genetically identical as they are (on the germline level) to every other cell that has emerged from a single zygote following the last sexual reproduction event. As described above, the majority of IESs are excised independently of the maternal macronucleus or scnRNA pathway, and it is therefore reasonable to assume that in most cases, a change in the somatic genome will not be inherited into the daughter cell. However, as evidenced by the mating type IES, there is nonetheless a little possibility of inheriting somatic changes via the scnRNA pathway into the daughter macronucleus. Combined with rapid cell division rates and large populations of ciliates, the scope for rapid and heritable adaptation to changing environments is large.

2.9 Whole-Genome Duplications Provide Complexity and Adaptability to Ciliate Genomes

A special feature of ciliates is their propensity towards whole-genome duplications. The *Paramecium aurelia* species has undergone at least three whole-genome duplications in its history, with the most recent having led to the speciation of the *P. aurelia* complex, which consists of 15 sibling species. A whole-genome duplication can lead to the rapid evolution of novel functions of genes, which will suddenly be present in duplicate and hence have the flexibility to mutate without compromising their original function. In ciliates, whole-genome duplications are tolerated more readily than in many species, due to the nuclear dimorphism and constraints on the ratio of macronuclear DNA to cytoplasm (Berger and Schmidt 1978). These constraints mean that the germline may duplicate without altering the ploidy of the macronucleus, and hence gene dosage will not be affected.

2.10 Other Non-mendelian Inheritance in Ciliates: Cortical Inheritance

A feature of ciliates is that following sexual reproduction, the new nucleus develops within the maternal cell. The new zygotic ciliate after genome rearrangement is, therefore, cytoplasmically and cytoskeletally speaking the same cell as its parent, only the nuclei have been replaced. However, as discussed, a newly developed ciliate behaves differently from a ciliate that has undergone several vegetative divisions, not least in its ability to undergo another round of sexual reproduction. This indicates that maturation and ageing are inherent parts of the ciliate life cycle,

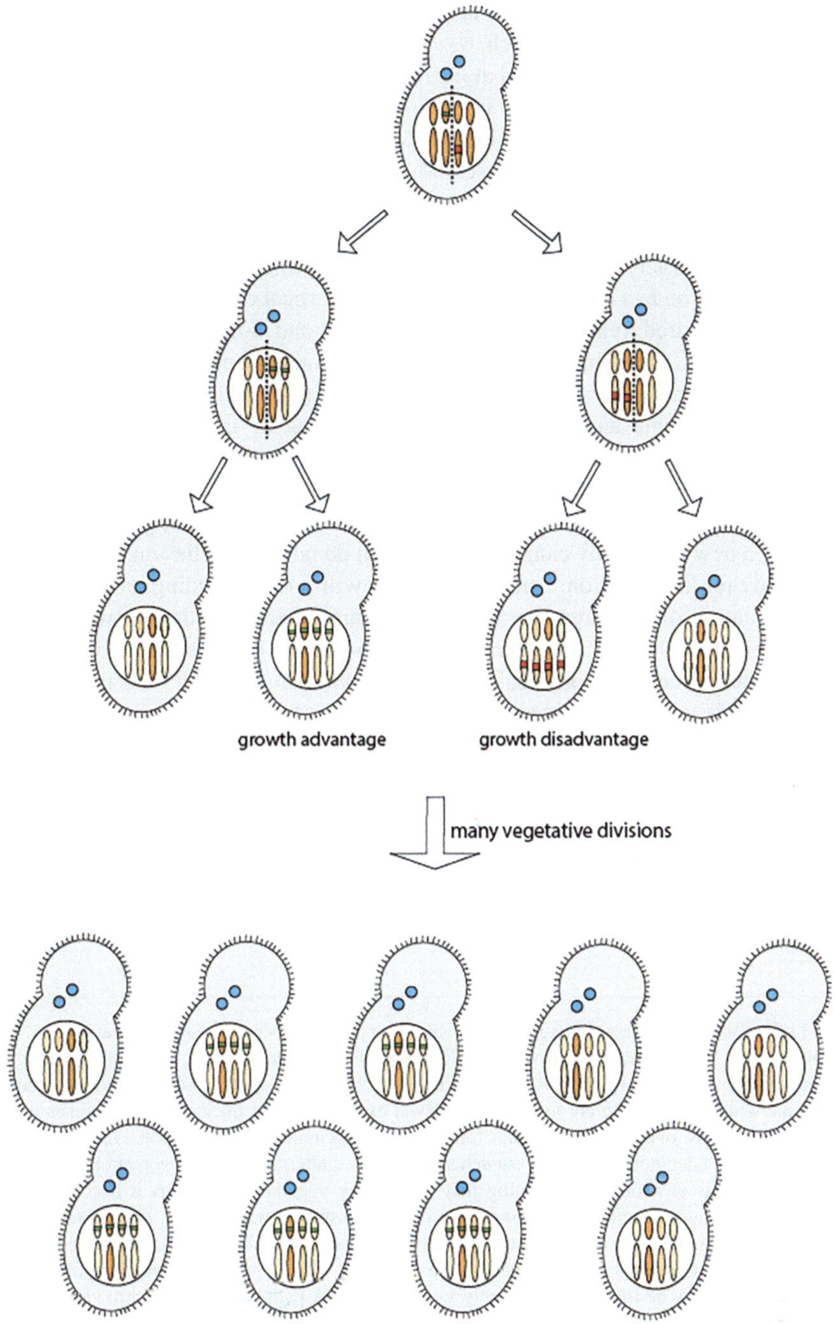

Fig. 11.3 Somatic assortment in *Paramecium* vegetative growth, a simplified scheme. The top cell represents a newly developed cell following sexual reproduction. The 800n copy number in the macronucleus is represented by four chromosomes of each type. The inherent error-proneness of the genome rearrangement process is represented by two alternative excision events, one red and one

just as in multicellular organisms, and the activation of zygotic genes must have profound consequences for the cell.

A fascinating phenomenon that is readily observable in ciliates is 'cortical inheritance'. It was discovered in the 1930s by the ciliate researcher Tracy Sonneborn, who observed that changes to the cortical structure of Paramecium were inherited in vegetative and sexual offspring, independently of the genetic makeup of the cell. These changes include grafts of cortex from one ciliate to another, residual pieces of the cytoplasmic bridge that broke off during conjugation, or even microsurgery performed on the cortex, in which a piece was cut out and re-grafted back on in a different orientation. Such cortical changes could be inherited for several hundred vegetative divisions (Beisson and Sonneborn 1965; Soh and Pearson 2022). Cortical inheritance falls under the category of structural inheritance, in which information is transmitted via spatial structures that reproduce themselves in three dimensions and therefore produce replicas of themselves, even when manipulated or altered in other ways. In *Paramecium*, cortical structural inheritance originates from the fact that basal bodies on the cell surface are produced in a templated manner from 'parent' basal bodies. Therefore, like a living cell or a strand of DNA, a new basal body cannot be assembled de novo but relies on a pre-existing basal body for its formation. A new basal body will form depending on the structure and orientation of its parent basal body, and any alterations thereto will thus be inherited with each replication event. This phenomenon emphasises the importance of the maternal cell as a physical template for the daughter cell, and the fact that information is transmitted physically and spatially, as well as digitally, through the DNA code. Prions, such as those that transmit the disease BSE ('mad cow disease'), are another example of structural inheritance. *Tetrahymena* also transmits variations in its surface structures, termed 'handedness', in a non-genetic, structural manner (Nelsen et al. 1989).

Fig. 11.3 (continued) green. In reality, these would be present in very low numbers and not 1/4 of chromosomes. The first two divisions illustrate how amitosis can segregate alternative rearrangements into daughter cells, where they will become over-represented over vegetative divisions, which amplify the chromosomes received by the daughter nuclei. In this representation, after two divisions, one daughter cell has four copies of each alternative excision event, though in reality, it would take much longer. In our scheme, the green alternative rearrangement has a growth advantage in this environment, meaning that over many vegetative divisions it becomes over-represented in the population, as it competes more successfully for resources. The red alternative rearrangement, on the other hand, has a growth disadvantage and is eliminated from the population. Importantly, these cells are all clones on the germline level. At low frequency, alternative rearrangements may be inherited maternally via the scnRNA pathway and thus provide a growth advantage for future sexual offspring

3 Conclusions

3.1 *Where Does 'Parasite' End and 'Host' Begin?*

As our understanding of biology increases, our assumptions and hypotheses must be updated in keeping with new information. Since their discovery, TEs have been seen as parasitic DNA, but this view is having to adapt in light of how intricately connected their evolution is to their hosts'. It is no exaggeration to say that eukaryotic hosts would look nothing like their modern selves were it not for millions of years of co-evolution with TEs. Nowhere is this more clear than in ciliates, whose genomes are literally peppered with TEs that flit between 'host' and 'invader' status on a strain-by-strain basis, and who orchestrate their own removal from the genome. One can argue that ciliates have evolved complex systems to defend themselves against TEs, but one could equally well argue that TEs have evolved an ideal niche in which to reside, and we have named it 'ciliates' (or any other eukaryote, for that matter). The origins of life on Earth are believed to be self-replicating nucleic acids, the very definition of what we now term 'selfish'. Therefore, the idea that life evolved into cells with genomes, which were only later populated by 'selfish' DNA elements and have been waging war against these ever since, is nonsensical. The theory of evolution is based on the premise that every gene, sequence, or combination thereof is selfish and will evolve in the direction of perpetuating itself the most efficiently. Describing the relationship between TEs and their hosts as an 'arms race' may be true from a certain perspective, but it is misleading in the sense that it implies agency and anthropomorphises the relationship between different DNA sequences that all originated from the same place—the first universal common ancestor (FUCA)—and continue to co-evolve to this day. A TE that has lost its mobility and been co-opted into an important gene in a highly successful eukaryotic species cannot be said to have 'lost' an 'arms race' simply because it is no longer able to actively transpose itself—it is still present in countless copies in genomes worldwide and is therefore arguably a huge success story. Likewise, ciliates, whose germline genomes are anything up to 96% TE-derived (Yerlici and Landweber 2014), comprise a biomass several times higher than any terrestrial animal (Bar-On et al. 2018), and cannot be said to be 'losing' a war on TEs.

3.2 *Conclusions: Ciliates and Non-mendelian Inheritance*

As discussed in this chapter, ciliates provide a fascinating model for studying many aspects of evolution that do not fit into the traditional error replication model of inheritance. The co-evolution of TEs and ciliate genomes demonstrates the remarkable power of mobile genetic elements to drive the evolution of complex traits, not only by inserting themselves into genomes, through horizontal gene transfer within and between species, and by necessitating genome defence pathways that mitigate TE damage, but also provides scope for evolving new features in the host.

Ciliates also display a remarkable two-tier environmental adaptation process, thanks to their nuclear dimorphism. Apart from the possibility of germline mutations leading to Darwinian evolution over sexual generations, the somatic macronucleus with its amitotic divisions and phenotypic assortment can also adapt rapidly between sexual cycles. The epigenetic inheritance from mother to daughter cells makes it possible to inherit such adaptations in rare cases.

Furthermore, ciliates demonstrate the phenomenon of structural inheritance in their cortex, demonstrating the inheritance of three-dimensional structures and orientations independently of their digital DNA code.

Many of the principles that become clear when we study diverse organisms, such as ciliates, give us insight into our own biology as well. Therefore, ciliates exemplify the need to look beyond ourselves and our closest relatives if we wish to gain a complete understanding of biology.

References

Allen SE, Nowacki M (2020) Roles of noncoding RNAs in ciliate genome architecture. J Mol Biol 432:4186–4198. https://doi.org/10.1016/j.jmb.2019.12.042

Allshire RC, Madhani HD (2018) Ten principles of heterochromatin formation and function. Nat Rev Mol Cell Biol 19:229–244. https://doi.org/10.1038/nrm.2017.119

Almeida MV, Vernaz G, Putman ALK, Miska EA (2022) Taming transposable elements in vertebrates: from epigenetic silencing to domestication. Trends Genet 38:529–553. https://doi.org/10.1016/j.tig.2022.02.009

Arkhipova IR, Yushenova IA (2019) Giant transposons in eukaryotes: is bigger better? Genome Biol Evol 11:906–918. https://doi.org/10.1093/gbe/evz041

Arnaiz O, Mathy N, Baudry C, Malinsky S, Aury J-M, Denby Wilkes C, Garnier O, Labadie K, Lauderdale BE, Le Mouël A et al (2012) The Paramecium germline genome provides a niche for intragenic parasitic DNA: evolutionary dynamics of internal eliminated sequences. PLoS Genet 8:e1002984. https://doi.org/10.1371/journal.pgen.1002984

Arnaiz O, Meyer E, Sperling L (2020) ParameciumDB 2019: integrating genomic data across the genus for functional and evolutionary biology. Nucleic Acids Res 48:D599–D605. https://doi.org/10.1093/nar/gkz948

Bar-On YM, Phillips R, Milo R (2018) The biomass distribution on earth. Proc Natl Acad Sci 115: 6506–6511. https://doi.org/10.1073/pnas.1711842115

Baudry C, Malinsky S, Restituito M, Kapusta A, Rosa S, Meyer E, Bétermier M (2009) PiggyMac, a domesticated piggyBac transposase involved in programmed genome rearrangements in the ciliate Paramecium tetraurelia. Genes Dev 23:2478–2483. https://doi.org/10.1101/gad.547309

Beisson J, Sonneborn TM (1965) Cytoplasmic inheritance of the organization of the cell cortex in Paramecium Aurelia*. Proc Natl Acad Sci 53:275–282. https://doi.org/10.1073/pnas.53.2.275

Berger JD (1986) Autogamy in Paramecium. Cell cycle stage-specific commitment to meiosis. Exp Cell Res 166:475–485. https://doi.org/10.1016/0014-4827(86)90492-1

Berger JD, Schmidt HJ (1978) Regulation of macronuclear DNA content in Paramecium tetraurelia. J Cell Biol 76:116–126. https://doi.org/10.1083/jcb.76.1.116

Chen X, Landweber LF (2016) Phylogenomic analysis reveals genome-wide purifying selection on TBE transposons in the ciliate Oxytricha. Mob DNA 7:2–2. https://doi.org/10.1186/s13100-016-0057-9

Cheng C-Y, Vogt A, Mochizuki K, Yao M-C (2010) A domesticated piggyBac transposase plays key roles in heterochromatin dynamics and DNA cleavage during programmed DNA deletion in

Tetrahymena thermophila. Mol Biol Cell 21:1753–1762. https://doi.org/10.1091/mbc.E09-12-1079
Cheng C-Y, Young JM, Lin C-YG, Chao J-L, Malik HS, Yao M-C (2016) The piggyBac transposon-derived genes TPB1 and TPB6 mediate essential transposon-like excision during the developmental rearrangement of key genes in Tetrahymena thermophila. Genes Dev 30: 2724–2736. https://doi.org/10.1101/gad.290460.116
Cosby RL, Chang N-C, Feschotte C (2019) Host–transposon interactions: conflict, cooperation, and cooption. Genes Dev 33:1098–1116. https://doi.org/10.1101/gad.327312.119
Déléris A, Berger F, Duharcourt S (2021) Role of Polycomb in the control of transposable elements. Trends Genet 37:882–889. https://doi.org/10.1016/j.tig.2021.06.003
Doak TG, Doerder FP, Jahn CL, Herrick G (1994) A proposed superfamily of transposase genes: transposon-like elements in ciliated protozoa and a common "D35E" motif. Proc Natl Acad Sci USA 91:942–946. https://doi.org/10.1073/pnas.91.3.942
Duret L, Cohen J, Jubin C, Dessen P, Goût J-F, Mousset S, Aury J-M, Jaillon O, Noël B, Arnaiz O et al (2008) Analysis of sequence variability in the macronuclear DNA of Paramecium tetraurelia: a somatic view of the germline. Genome Res 18:585–596. https://doi.org/10.1101/gr.074534.107
Eisen JA, Coyne RS, Wu M, Wu D, Thiagarajan M, Wortman JR, Badger JH, Ren Q, Amedeo P, Jones KM et al (2006) Macronuclear genome sequence of the ciliate Tetrahymena thermophila, a model eukaryote. PLoS Biol 4:e286. https://doi.org/10.1371/journal.pbio.0040286
Evsikov AV, de Vries WN, Peaston AE, Radford EE, Fancher KS, Chen FH, Blake JA, Bult CJ, Latham KE, Solter D et al (2004) Systems biology of the 2-cell mouse embryo. Cytogenet Genome Res 105:240–250. https://doi.org/10.1159/000078195
Fang W, Wang X, Bracht JR, Nowacki M, Landweber LF (2012) Piwi-interacting RNAs protect DNA against loss during Oxytricha genome rearrangement. Cell 151:1243–1255. https://doi.org/10.1016/j.cell.2012.10.045
Fass JN, Joshi NA, Couvillion MT, Bowen J, Gorovsky MA, Hamilton EP, Orias E, Hong K, Coyne RS, Eisen JA et al (2011) Genome-scale analysis of programmed DNA elimination sites in Tetrahymena thermophila. G3: Genes|Genomes|Genetics 1:515–522. https://doi.org/10.1534/g3.111.000927
Fu Y-X, Wang G, Chen K, Ma X, Liu S-Q, Miao W (2022) Amitosis as a strategy of cell division—insight from the proliferation of Tetrahymena thermophila macronuclei. Theor Popul Biol 145: 52–62. https://doi.org/10.1016/j.tpb.2022.03.004
Gilbert C, Feschotte C (2018) Horizontal acquisition of transposable elements and viral sequences: patterns and consequences. Curr Opin Genet Dev 49:15–24. https://doi.org/10.1016/j.gde.2018.02.007
Guérin F, Arnaiz O, Boggetto N, Denby Wilkes C, Meyer E, Sperling L, Duharcourt S (2017) Flow cytometry sorting of nuclei enables the first global characterization of Paramecium germline DNA and transposable elements. BMC Genomics 18:327. https://doi.org/10.1186/s12864-017-3713-7
Hamilton E, Bruns P, Lin C, Merriam V, Orias E, Vong L, Cassidy-Hanley D (2005) Genome-wide characterization of Tetrahymena thermophila chromosome breakage sites. I. Cloning and identification of functional sites. Genetics 170:1611–1621. https://doi.org/10.1534/genetics.104.031401
Hamilton EP, Kapusta A, Huvos PE, Bidwell SL, Zafar N, Tang H, Hadjithomas M, Krishnakumar V, Badger JH, Caler EV et al (2016) Structure of the germline genome of Tetrahymena thermophila and relationship to the massively rearranged somatic genome. eLife 5:e19090. https://doi.org/10.7554/eLife.19090
Hayward A, Gilbert C (2022) Transposable elements. Curr Biol 32:R904–R909. https://doi.org/10.1016/j.cub.2022.07.044
Jerison ER, Kryazhimskiy S, Mitchell JK, Bloom JS, Kruglyak L, Desai MM (2017) Genetic variation in adaptability and pleiotropy in budding yeast. eLife 6:e27167. https://doi.org/10.7554/eLife.27167

Kapitonov VV, Jurka J (2005) RAG1 core and V (D) J recombination signal sequences were derived from Transib transposons. PLoS Biol 3:e181

Kitazawa M, Tamura M, Kaneko-Ishino T, Ishino F (2017) Severe damage to the placental fetal capillary network causes mid-to late fetal lethality and reduction in placental size in Peg11/Rtl1 KO mice. Genes Cells 22:174–188

Lepère G, Bétermier M, Meyer E, Duharcourt S (2008) Maternal noncoding transcripts antagonize the targeting of DNA elimination by scanRNAs in Paramecium tetraurelia. Genes Dev 22:1501–1512. https://doi.org/10.1101/gad.473008

Malone CD, Hannon GJ (2009) Small RNAs as guardians of the genome. Cell 136:656–668. https://doi.org/10.1016/j.cell.2009.01.045

Mangeney M, Renard M, Schlecht-Louf G, Bouallaga I, Heidmann O, Letzelter C, Richaud A, Ducos B, Heidmann T (2007) Placental syncytins: genetic disjunction between the fusogenic and immunosuppressive activity of retroviral envelope proteins. Proc Natl Acad Sci 104:20534–20539

McKinley KL, Cheeseman IM (2016) The molecular basis for centromere identity and function. Nat Rev Mol Cell Biol 17:16–29

Miyawaki S, Kuroki S, Maeda R, Okashita N, Koopman P, Tachibana M (2020) The mouse Sry locus harbors a cryptic exon that is essential for male sex determination. Science 370:121–124

Mochizuki K, Gorovsky MA (2004) Conjugation-specific small RNAs in Tetrahymena have predicted properties of scan (scn) RNAs involved in genome rearrangement. Genes Dev 18: 2068–2073. https://doi.org/10.1101/gad.1219904

Modzelewski AJ, Gan Chong J, Wang T, He L (2022) Mammalian genome innovation through transposon domestication. Nat Cell Biol 24:1332–1340. https://doi.org/10.1038/s41556-022-00970-4

Nelsen EM, Frankel J, Jenkins LM (1989) Non-genic inheritance of cellular handedness. Development 105:447–456. https://doi.org/10.1242/dev.105.3.447

Nowacki M, Higgins BP, Maquilan GM, Swart EC, Doak TG, Landweber LF (2009) A functional role for transposases in a large eukaryotic genome. Science (New York, NY) 324:935–938. https://doi.org/10.1126/science.1170023

Peaston AE, Evsikov AV, Graber JH, de Vries WN, Holbrook AE, Solter D, Knowles BB (2004) Retrotransposons regulate host genes in mouse oocytes and preimplantation embryos. Dev Cell 7:597–606. https://doi.org/10.1016/j.devcel.2004.09.004

Phadke SS, Zufall RA (2009) Rapid diversification of mating systems in ciliates. Biol J Linn Soc 98:187–197. https://doi.org/10.1111/j.1095-8312.2009.01250.x

Sandoval PY, Swart EC, Arambasic M, Nowacki M (2014) Functional diversification of dicer-like proteins and small RNAs required for genome sculpting. Dev Cell 28:174–188. https://doi.org/10.1016/j.devcel.2013.12.010

Scott KC, Merrett SL, Willard HF (2006) A heterochromatin barrier partitions the fission yeast centromere into discrete chromatin domains. Curr Biol 16:119–129

Shiura H, Ono R, Tachibana S, Kohda T, Kaneko-Ishino T, Ishino F (2021) PEG10 viral aspartic protease domain is essential for the maintenance of fetal capillary structure in the mouse placenta. Development 148:dev199564. https://doi.org/10.1242/dev.199564

Singh DP, Saudemont B, Guglielmi G, Arnaiz O, Gout J-F, Prajer M, Potekhin A, Przybos E, Aubusson-Fleury A, Bhullar S et al (2014) Genome-defence small RNAs exapted for epigenetic mating-type inheritance. Nature 509:447–452

Soh AWJ, Pearson CG (2022) Ciliate cortical organization and dynamics for cell motility: comparing ciliates and vertebrates. J Eukaryot Microbiol 69:e12880. https://doi.org/10.1111/jeu.12880

Sullivan LL, Maloney KA, Towers AJ, Gregory SG, Sullivan BA (2016) Human centromere repositioning within euchromatin after partial chromosome deletion. Chromosom Res 24: 451–466

Swart E, Denby Wilkes C, Sandoval P, Hoehener C, Singh A, Furrer D, Arambasic M, Ignarski M, Nowacki M (2017) Identification and analysis of functional associations among natural

eukaryotic genome editing components [version 1; peer review: 1 approved, 1 approved with reservations]. F1000Research 6. https://doi.org/10.12688/f1000research.12121.1
Torres-Padilla M-E (2020) On transposons and totipotency. Philos Trans R Soc B Biol Sci 375: 20190339. https://doi.org/10.1098/rstb.2019.0339
Tucker JB, Beisson J, Roche DL, Cohen J (1980) Microtubules and control of macronuclear 'amitosis' in Paramecium. J Cell Sci 44:135
Uhlmann F (2016) SMC complexes: from DNA to chromosomes. Nat Rev Mol Cell Biol 17:399–412
Vogt A, Mochizuki K (2013) A domesticated PiggyBac transposase interacts with heterochromatin and catalyzes reproducible DNA elimination in Tetrahymena. PLoS Genet 9:e1004032. https://doi.org/10.1371/journal.pgen.1004032
Wells JN, Feschotte C (2020) A field guide to eukaryotic transposable elements. Annu Rev Genet 54:539–561. https://doi.org/10.1146/annurev-genet-040620-022145
Xia B, Zhang W, Zhao G, Zhang X, Bai J, Brosh R, Wudzinska A, Huang E, Ashe H, Ellis G et al (2024) On the genetic basis of tail-loss evolution in humans and apes. Nature 626:1042–1048. https://doi.org/10.1038/s41586-024-07095-8
Yerlici, V.T., and Landweber, L.F. (2014). Programmed genome rearrangements in the ciliate Oxytricha. Microbiol Spectrum 2, 10.1128/microbiolspec.MDNA3-0025–2014
Zhou L, Mitra R, Atkinson PW, Burgess Hickman A, Dyda F, Craig NL (2004) Transposition of hAT elements links transposable elements and V (D) J recombination. Nature 432:995–1001

Chapter 12
Structural Neutrality and Accommodation Shape RNA Evolution

Eric Westhof

Abstract RNA molecules maintain conserved three-dimensional (3D) structures despite extensive sequence variability, underpinning diverse biological functions. The mechanisms enabling such structural conservation amid sequence diversity remain incompletely understood. In this chapter, we analyze RNA tertiary interactions across homologous sequences using structural alignments and reveal pervasive 3D molecular neutrality or a tolerance of local sequence variations through isosteric and conformational accommodations. These RNA molecular properties allow many sequences to fold into similar architectures with minimal conserved nucleotides. Further, molecular neutrality facilitates evolutionary innovations via alternative conformations, epistasis, and allosteric effects. Our findings highlight how structure-based molecular neutrality shapes RNA evolvability, providing a framework to interpret RNA structural diversity and the emergence of novel functions. It is hoped that the deep links between structural sequence alignments and RNA folding principles will help the design of RNA prediction tools.

Keywords RNA molecular neutrality · RNA structural homology · RNA evolvability · Isosteric base pairs · RNA tertiary structures · RNA 3D predictions

1 Structural Homology: Many Sequences, One Architecture

The origins of molecular evolution of life are rooted in the evolution of RNA molecules, which are now recognized to participate in a wide variety of biological functions (Cech 2024; Mattick and Amaral 2022). Extant RNA molecules form remarkable molecular architectures, sometimes referred to as nanoobjects, either on

E. Westhof (✉)
Université de Strasbourg, Institut de biologie moléculaire et cellulaire du CNRS, Strasbourg, France

Wenzhou Institute, Chinese Academy of Sciences, Wenzhou, China
e-mail: eric.westhof@ibmc-cnrs.unistra.fr

G. Witzany (ed.), *Integrative Theory of Evolution - Aspects and Insights*,
https://doi.org/10.1007/978-3-032-24295-2_12

their own or in association with protein or metabolite cofactors (Berman et al. 2000; Burley et al. 2019). These RNA nanoobjects rarely derive from a single sequence; rather, many similar RNA sequences are commonly retrieved from sequencing projects within or across species from all three kingdoms of life (Consortium et al. 2017; Ontiveros et al. 2024). Functionally, these RNA sequences are considered homologous, and it is understood that the entire set of homologous sequences folds into a consensus, functional three-dimensional architecture (Levitt 1969; Noller 2005; Wuyts et al. 2001). This shared three-dimensional (3D) architecture can accommodate sequence variations and the resulting local changes. It is suggested here that the collection of structures associated with a functional RNA architecture be considered as forming a *structural quasi-species*,[1] a population of related molecules sharing a common structure despite sequence diversity. In large molecules, the homologous architecture is often restricted to a conserved core, onto which various appendices can be grafted depending on the subset of sequences (see, for example, the two main families of RNase P (Massire et al. 1998; Westhof and Massire 2004). Strikingly, the number of conserved residues in homologous sequences is only a fraction of the total number of nucleotides in the aligned RNA sequences (Singhal and Fallis 1979; Schultes and Bartel 2000). For example, all transfer RNA sequences fold into a common secondary structure with four helical stems (the cloverleaf) and a common three-dimensional L-shaped architecture. However, in a small subset of transfer RNA (tRNA) sequences, a variable loop linking two helical stems expands, resulting in a fifth helical segment. Searches across all three kingdoms of life generally reveal functional tRNAs with mutations at conserved positions. More restricted searches and alignments (e.g., within mammals) show that the number of invariant residues is approximately 7 (among around 76 nucleotides) (Westhof et al. 2022). The self-splicing group I introns have a complex common core onto which various local structures are grafted, leading to several sub-groups (Michel and Dujon 1983; Lehnert et al. 1996; Michel and Westhof 1990). However, the key intramolecular interactions maintaining the core are either preserved or accommodated despite the presence of additional external elements. The folding processes and pathways are not discussed here, but it has been shown that some of the conserved interactions occur early in the folding process, leading to the emergence of cooperative interaction networks that guide the RNA to its native fold (Behrouzi et al. 2012). The number of invariant residues is around 7 (over more than 400 nucleotides) (Lisacek et al. 1994), but several of them can be mutated if other compensatory mutations are present (Been and Perrotta 1991).

[1]This is an extrapolation of the concept introduced by Eigen and Schuster for viral species, especially RNA viruses (Eigen 1971; Eigen and Schuster 1977; Biebricher and Eigen 2005; Sardanyes et al. 2024). In viruses, there is no static single genome, but rather dynamic distributions of closely related genomic variants that form a quasispecies. The sets of sequences in sequence space build a landscape of fitness that depends on the viral type and the environmental conditions. Catalytic RNA sequences, such as group I/II introns, can be grouped into groups and subgroups that form the functional landscape.

Many of the concepts discussed here, such as neutrality and evolvability, have been extensively and insightfully analyzed using secondary structure prediction algorithms (Fontana and Schuster 1998; Ancel and Fontana 2000; Huynen et al. 1996). In this work, I aim to illustrate the concept of molecular neutrality as it applies to the interactions between components of RNA architectures. Here, 3D molecular neutrality refers to the ability of local interactions within and between RNA elements to tolerate variations in sequence and length. While 3D molecular neutrality does not directly determine specific folds, it enables their maintenance throughout evolution. As a result, some folds may act as passive bystanders within the overall functional architecture. However, after one or more mutations, these folds can opportunistically promote changes in the architecture, potentially leading to alternative conformations that may result in more efficient function, through mechanisms such as epistasis (interactions between mutations), allostery (regulation at a distance), or even new functions via exaptation (the co-option of existing structures for new purposes). Not all intra- and inter-molecular interactions are necessarily always optimized and fully satisfied, giving rise to potential 'frustrated states', local alternative micro-states in dynamic equilibrium. Importantly, because the structure of a molecule precedes its evolution (Geoffroy St Hilaire), such structural quasi-species may facilitate the emergence of evolutionary innovations in both structure and function.

In summary, molecular neutrality, both at the 2D and at the 3D levels, implies that, during molecular evolution and in conjunction with mutations, new states can be reached that may increase stability or enable alternative foldings conducive to novel architectures or functions. The physico-chemical basis for RNA molecular neutrality includes, among other factors, the ability of base and ribose atoms to form hydrogen bonds, the various binding modes of water molecules and ions, and the diversity of interactions between base pairs. In this work, I will emphasize the power of structural alignments for organizing structural quasi-species. Structural alignments link experimentally determined structures to the set of homologous sequences that can fold into a conserved RNA architecture. Selected molecular examples based on tRNA structures will demonstrate how RNA molecules are particularly prone to interactions that accommodate molecular neutrality. Although not discussed in detail here, RNA molecular neutrality plays a significant role in the efficacy of in vitro Darwinian evolution, such as in SELEX experiments (Tuerk and Gold 1990; Ellington and Szostak 1990; Robertson and Joyce 1990). Because of molecular neutrality, RNA elements can arise that do not disrupt an existing structural state but may allow for sudden changes or new conformations to emerge during subsequent cycles of evolution.

2 Methodology

We now have access to a vast number of RNA sequences across the phylogeny, which are organized in databases for easy retrieval, such as RFAM (Ontiveros et al. 2024) and RNACentral (Consortium et al. 2017). In addition, the number of relevant

RNA structures is steadily increasing, thanks to advances in cryogenic techniques in X-ray crystallography and the development of highly efficient detectors in electron microscopy (Chua et al. 2022). These structures are collected in the Protein Data Bank (Berman et al. 2000; Burley et al. 2019). This is particularly true for ribosomes, for which hundreds of structures from *Eubacteria*, *Archaea*, and *Eukaryota* are now available. As a result, precise structural alignments can be derived, compared, and analyzed across the whole set of sequences and throughout whole molecules, or by considering only a subset of sequences or a partial architectural fold, to follow the molecular contacts and interactions underlying molecular adaptations and accommodations within the common RNA architecture.

Here, I will illustrate the approach and concepts of structural sequence alignments using transfer RNA (tRNA) as an example that is easily transferable and applicable to other structured RNA molecules. Transfer RNAs are ubiquitous RNA molecules in biological cells. They carry a specific amino acid to the ribosome for protein translation. Each tRNA has a defined anticodon, complementary to the codon corresponding to the amino acid according to the genetic code, and each tRNA is charged, at its 3′-end, with the proper amino acid by one of the 20 aminoacyl tRNA synthetases of each cell (Quigley and Rich 1976; Biela et al. 2023) (see Figs. 12.1 and 12.2).

For annotating base pairs, the following conventions are used. Regular Watson–Crick pairs will be denoted as A–U, G = C, or by their positions, such as 13–22. The usual wobble pair will be written as G•U. In the text, non-Watson–Crick pairs will be denoted by a "/" (e.g., G15/C48 or 54/58) to emphasize that they are not regular Watson–Crick pairs. In the figures and tables, I will use the L–W notation[2] (Leontis and Westhof 2001).

[2]The L–W notation annotates interactions between two nucleotides that are mediated by H-bonds between atoms belonging to the two interacting bases of nucleic acid structures. The emphasis in the L–W notation is on the bases because, in nucleic acids, the bases characterize the sequence, and because the aim of the notation is to follow how the interacting residues behave in homologous sequences. The L-W notation can also be used to annotate schematic drawings representing experimental structures. The names of the edges come naturally to the mind (Watson–Crick edge, Hoogsteen edge, and the third one, with O2 or N3 of the base involved, the sugar edge, since it is on that side of a nucleotide). In RNA, the O2′ of the sugar is included in the description of the H-bonds (but not in the classification per se) because, in some interactions, it does clearly contribute to the orientation and strength of the pair and, thus, does help in distinguishing base pairs. The names given to a specific interaction between two residues are thus not arbitrary; they describe directly the edges involved in the hydrogen bonding contacts between the base atoms (in analogy to the Latin phrase *nomen omen*, suggesting that a name is not just an arbitrary label, but carries a deeper meaning). The *cis* and *trans* distinction is necessary because the sugar ring is asymmetrically linked with respect to the bases, and it is directly related to the $5' > 3'$ orientation of the strands.

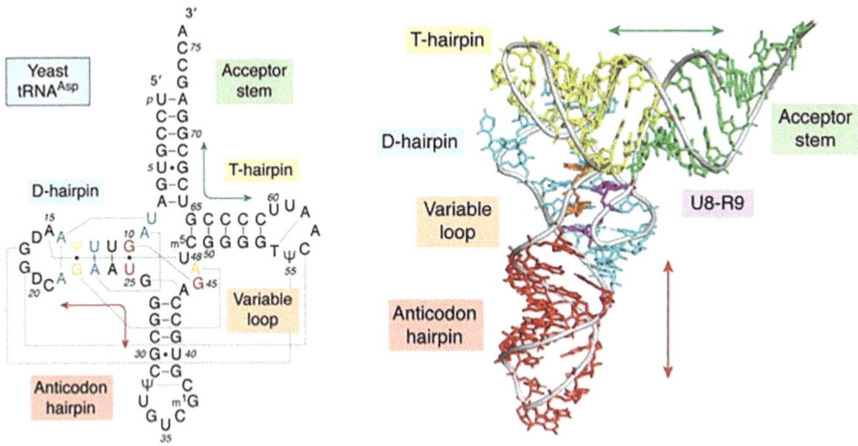

Fig. 12.1 The secondary structure or 2D (left drawing) and the three-dimensional structure or 3D (right drawing) representations of a transfer RNA (tRNA). The sequence (Gangloff et al. 1971) and the refined three-dimensional structure (Westhof et al. 1985) (PDB number 2TRA) correspond to yeast tRNA-Asp. Here, we focus on cytoplasmic tRNAs in eukaryotes, but their characteristics extend into the other phylogenetic phyla. The secondary structure or 2D can be drawn on a planar drawing and represents the Watson–Crick (G = C and A-U) with the G•U pairs that build the helices. The 2D structure is made of four helices: the acceptor stem (A-stem), the D-hairpin, the Anticodon hairpin, and the T-hairpin. A hairpin is made of a helix capped by an apical loop. The A-stem has always seven base pairs, and both the AC- and T-stems have five base pairs. The D-stem has generally four base pairs, and it is framed by the quasi-invariant U8/A14/A21 triple; A14 is followed by the quasi-invariant R15/Y48 trans Watson–Crick/Watson–Crick pair. The D-loop has two variable regions: after R15 and before the invariant GG, and after the invariant GG and before the invariant A21. The variable loop has three residues involved in conserved contacts: one forming a capping non-WC pair on top of the AC-stem, R15/Y48 (see Fig. 12.3b), and two others interacting with the D-stem. Additional residues occur before Y48. Some tRNAs, called long-arm tRNAs, have an additional helix, but 44 and 48 are maintained. Both the T- and AC-loops have seven nucleotides but display very different conservation and variations. Because of its characteristic 2D structure, it was named the cloverleaf structure of tRNA. All tRNAs, except some mitochondrial ones, display such a structure (some tRNAs have a long variable loop that can form an additional helix). Names are given by the modified bases present; the D-hairpin often contains a dihydrouridine, and the T-hairpin a thymine instead of a uracil. tRNA molecules are heavily modified. In the present example, there are three additional modifications: a pseudouridine at positions 32 and 55, a G modified by a methyl group on N1 (m1G), and a C modified by a methyl group on C5 (m5C). These modifications have many roles; those in the core contribute mainly to stability and avoidance of degradation, and those in the anticodon loop are key for the fidelity of the decoding process. The dotted lines between nucleotides indicate the tertiary contacts that are observed in the 3D structure. The red and green double arrows indicate that the helices stack on top of each other to build up the two main arms of the L structure of the 3D structure (as shown at the right). The coaxial stacking of the helices allows for the formation of loop-loop tertiary pairs between the T- and D-loops typical of tRNA architecture. The linking single-stranded segments (U8R9, with R being a purine, either A or G, and the variable loop bases 45–48) interact with the bases of the D-hairpin. The amino acid is attached at the 3′-end, and the anticodon triplet is at the other end (about 90 Å away). (The drawings are reproduced from the reference (Westhof and Leontis 2021) with permission (CC-BY License))

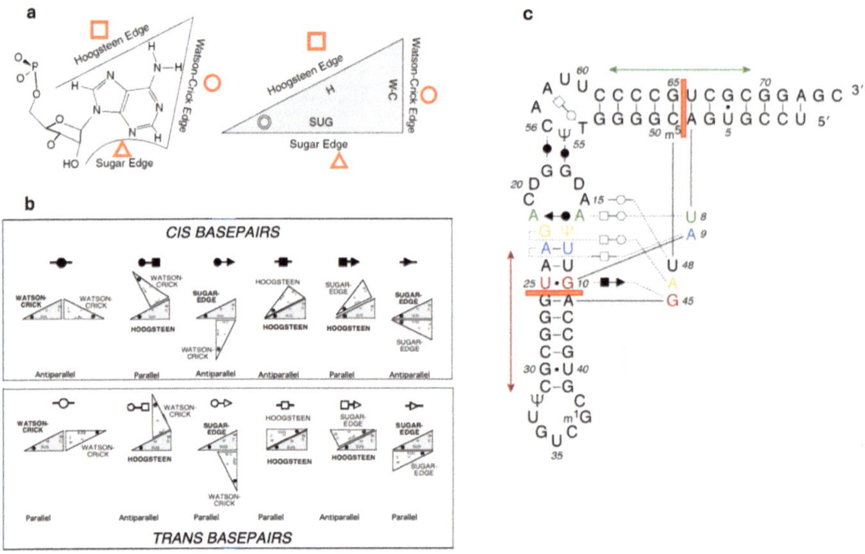

Fig. 12.2 (**a**) The Leontis–Westhof nomenclature annotates all pairs formed between any of the four nucleic acid bases (see below) (Leontis and Westhof 2001). Any nucleotide can be represented by a triangle where each edge is given the name of the type of interactions made by the planar bases. The Watson–Crick edge forms the Watson–Crick interactions, the Hoogsteen edge the Hoogsteen interactions, and the Sugar edge, characteristic of RNA, the interactions involving the hydroxyl group O2′. These interactions are mediated by H-bonds between the nucleosides (base + ribose sugar). The edges are annotated, respectively, by a circle, a square, and a triangle. (**b**) Simple combinatorics leads to six types of pairs with the ribose rings on the same side of the interacting H-bonds (in *cis*) and six additional types with the sugar rings on opposite sides of the interacting H-bonds (in *trans*). The name of each type of pair is given by the name of the interacting edges. Thus, in an A/G pair in the *trans* Hoogsteen/Sugar edge family, the A gives its Hoogsteen edge to the Sugar edge of the G. Assuming standard conformations of the nucleotides (base + ribose + phosphate group), the choice of edge with cis or trans orientations leads to a parallel or an antiparallel orientation of the polynucleotide strands (Westhof 1992). (**c**) A drawing of the tertiary structure of tRNA following the representation suggested by Kim (1978). The L-shaped representation highlights the coaxial stacks between the acceptor and the thymine stems as well as between the D and the anticodon stems (*indicated by double arrows*). The red rectangles show the stacking pairs between the coaxial helices (G10•U25/G26•A44 and A7-U66/G65 = m^{5}C49, where m^5 indicates that residue C49 carries a methyl group on position C5), illustrating the absence of gaps in the D-AC strand and in the T-AA strand (see Fig. 12.4). The pairs formed in the three-dimensional structure are shown with the L-W nomenclature

2.1 Geometric Isostericity

The geometric concept of isostericity is commonly used for comparing nucleic acid base pairs (Topal and Fresco 1976; Westhof 2014; Stombaugh et al. 2009). Base pairs are considered isosteric when the glycosyl bonds of the two sets of paired nucleotides can be superimposed, meaning that structurally equivalent atoms form hydrogen-bonding contacts in both pairs. In standard Watson–Crick pairs, this

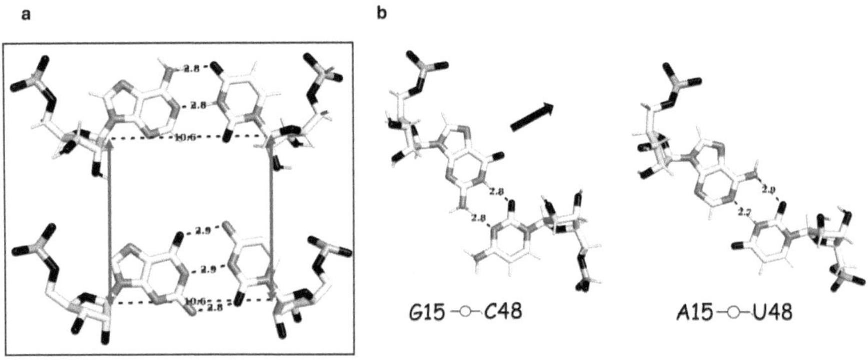

Fig. 12.3 (**a**) All combinations of the classical *cis* Watson–Crick/Watson–Crick base pairs are isosteric between them (or can be superimposed upon each other). The corresponding distances between the ribose C1′ atoms are around 10.6 Å. (**b**) The *trans* Watson–Crick/Watson–Crick G/C and A/U base pairs are not isosteric with each other. Clearly, the atoms forming the H-bonds do not correspond in the two pairs; the G moves upwards (*indicated by the thick black arrow*) so that N1 (G) interacts with O2(C), while N6 interacts with O2(U) in the AU pair. To be isosteric, the corresponding atom in G would have to be O6, but it does not carry a H atom necessary to form an H-bond. The examples shown come from tRNA structures: 1EHZ (Shi and Moore 2000) for G15/C48 and 2TRA (Westhof et al. 1985) for A15/U48. The distances between the C1′ atoms are around 10.9 Å. To help with visualization, hydrogen atoms are shown in all molecular drawings. They were added automatically and were not experimentally detected. For the hydroxyl groups, their positions are arbitrary. (All molecular drawings were generated using *The PyMOL Molecular Graphics System, Version 3.0 Schrödinger, LLC*)

results in similar shapes of the base pairs with a C1′–C1′ distance of approximately 10.6 Å (Fig. 12.3a). Complementary Watson–Crick pairs are isosteric with each other, as well as upon their reversal (G = C vs C = G) (Fig. 12.3a). However, the wobble G•U pairs are not isosteric upon reversal; a G•U pair is not superimposable with a U•G pair (Masquida and Westhof 2000). Isosteric base pairs also occur among the non-Watson–Crick pairs (Stombaugh et al. 2009), but these do not respect complementarity in the Watson–Crick sense (Chargaff 1950; Elson and Chargaff 1954). For example, the *trans* Watson–Crick/Watson–Crick pairs G/U and A/C are isosteric between them (see Fig. 12.2 of (Westhof 2014)). However, the *trans* Watson–Crick/Watson–Crick pairs G/C and A/U are not (Fig. 12.3b). Geometric isostericity forms the basis of structural sequence alignments.

3 From Multiple Sequence Alignments (MSA) to Structural Sequence Alignments

Sequence alignments demonstrate that biological macromolecular structures evolve at a slower pace than sequences do (Fox and Woese 1975). Therefore, the structural elements and their intramolecular contacts must possess invariant relationships that

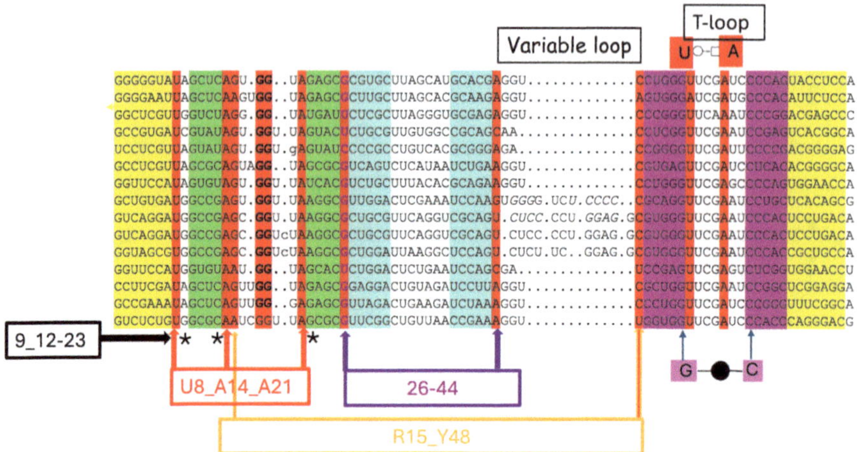

Fig. 12.4 An illustrative example of a structural alignment based on an arbitrary set of human tRNA sequences (Chan and Lowe 2016). The covarying helical strands forming the helices of the 2D structure are color-coded for ease of recognition (yellow: A-stem, green: D-stem, cyan: AC-stem, and magenta: T-stem). The tertiary pairs of the 3D structure are highlighted in red. The single-stranded and loop regions are in white. The T-loop is marked, and two conserved pairs are shown: the Watson–Crick pair G53 = C61 closing the T-stem (below) and the *trans* Watson–Crick/Hoogsteen U54/A58 (top)

are maintained throughout the manifold of sequences forming the structural quasi-species. The sequences accommodate and transform from one to another through geometrical and energetic equivalences. These sequence-to-sequence transformations result from a selected mixture of isosteric and isoenergetic geometric interactions governed by equilibria, compatible with the highly charged environment of RNA molecules, between base stacking, hydrogen bonds, and solvent contacts.

Unlike standard sequence alignments, which focus on the vertical alignment of letters, structural sequence alignments emphasize the vertical alignment of sequence elements that are structurally equivalent. Structural equivalence is determined by observing that sequences perform the same function, even in different cells or organisms under varying physiological conditions, and by analyzing similarities between sequences, particularly the distribution of critical sequence elements. The underlying hypothesis is that aligning multiple homologous sequences with reference to a high-quality experimental three-dimensional structure provides a more integrated understanding of RNA architecture.

4 What Can We Capture from Structural Sequence Alignments?

The interactions between the nucleotide bases (the side chains) dominate RNA architectures and, consequently, they are usually characterized and described in terms of the base pairs they contain. The base pairs are commonly partitioned into pairs of the secondary structure and those of the tertiary structure. The Watson–Crick pairs maintaining the helices of the secondary structure contain a large chunk of the folding energy of an RNA sequence, and their determinations constitute the first step toward the construction of a folded three-dimensional architecture (Westhof et al. 2011). The non-Watson–Crick pairs are characteristic of the three-dimensional structure. They organize hairpin loops and single-stranded regions that interact with double-stranded helices or other single-stranded segments (Westhof and Fritsch 2000; Hermann and Westhof 1999).

4.1 The Secondary Structure Helices Form the Architectural Skeleton

Standard multiple sequence alignments (MSAs) primarily highlight regions that covary, reflecting variations compatible with the underlying set of Watson–Crick base-paired helices that form the secondary structure (Michel et al. 2000). Covariations between complementary nucleotides in Watson–Crick pairs maintain the secondary structure by replacing one Watson–Crick pair with another isosterically (e.g., A–U, U–A, G = C, C = G). These complementary regions can form helical stems. In the absence of an experimentally determined structure, the presence of a helix is inferred from the number of observed base covariations in the helical strands (Noller and Woese 1981; Michel and Dujon 1983). However, the helical strands may be fully complementary without any covariation. The absence of covariation does not necessarily mean the helix does not form. It simply indicates that sequence analysis does not provide evidence for such a helix (Rivas 2023; Rivas et al. 2017). Pseudoknots are three-dimensional elements in which a stretch of the loop of a hairpin forms a Watson–Crick base-paired helix with a single-stranded region beyond the hairpin stem. The observation of complementarity between the hairpin loop and the single-stranded region suggests the potential presence of a pseudoknot. Pseudoknots tend to deviate from normal two-dimensional helices and can occur without covariation, often involving conserved nucleotides and non-Watson–Crick pairs.

4.2 The Helical Junctions and the Relative Positions of Helices

Structural alignments display, among the sequences forming a structural quasi-species, the variations that can fit within or conform to an overall three-dimensional architecture (Fig. 12.4). I will now describe how careful structural alignments yield valuable information on the 3D maintenance and its evolution. Between structurally aligned elements, the number of empty spaces or gaps that have no structural basis or are only present for the convenience of representation should be minimized systematically. The enforcement of a stringent control of the number of gaps and the avoidance of misleading or unnecessary insertions highlight precisely the start and end of the helical base pairing strands. Such positions contain very basic critical three-dimensional information. Indeed, when the strands form helices, the precise position of the junction is often dictated by covariations between the residues flanking the junctions and their helical partners. The absence of a gap between two segments of sequences, especially when involved in the formation of helices, indicates that the two helical regions in which they participate have their helical axes not far from being parallel, leading to a coaxial stacked arrangement of the helices (Kim and Cech 1987). In ligand-binding RNAs, like aptamers, the interacting ligand can disrupt this folding rule (as in the SMK box complexed with S-adenosylmethionine (Lu et al. 2008)). On the other hand, the presence of a variable number of residues between helical segments points to diverging strands and the potential involvement of non-Watson–Crick pairs.

4.3 The Absence of Gaps Helps to Identify Key Conserved Residues

In human tRNAs, the two variable regions flanking the -GG- sequence in the D loop have, respectively, at most two and three residues. At the same time, the structural alignment indicates that the -GG- sequence is invariant and occupies similar positions in space. However, the variable loop preceding the T-hairpin contains, in class I tRNAs, two to four residues and, in class II tRNAs, a larger number of nucleotides that form an additional helix. It thus reflects a variable and open space in the overall tRNA architecture. Secondly, an absence of a gap highlights the conservation of nucleotides at the ends of helices or at the interfaces, such as residues 8, 14, and 21, which all border a helical strand and are involved in a conserved triple interaction. Similarly, residues 15 and 54 have specific conservation patterns. In contrast, nucleotides 26 and 44 are variable, indicating an interfacing role between the AC-hairpin and the rest of the tRNA body.

In the alignments of tRNAs, there is never a residue between the last covarying nucleotide of the second strand of the T-stem and the first covarying nucleotide of the second strand of the AA stem. Structures show that these two stems are coaxially stacked. Another related observation is the following. The base pair of the T-hairpin before the loop is always a G (G53, which is the last nucleotide on the 5′-strand, the first strand encountered going from the 5′ to the 3′-end of the RNA sequence) paired with C (C61, the first nucleotide on the 3′-strand, the second strand encountered going from the 5′ to the 3′-end of the RNA sequence). Such a conservation indicates a stabilizing constraint on the tertiary fold generally mediated by a non-Watson–Crick base pair. In the case of G53 = C61, a stabilizing H-bond between the amino group of C61 and an anionic phosphate oxygen atom of the preceding residue. The amino groups of the nucleic acid bases are regularly involved in H-bonding contacts with negative phosphate oxygen atoms. Structurally, the most important ones are first the amino group of G (the G-phosphate interactions (Mokdad et al. 2006)) followed by C (C-phosphate interactions occur in 4-way junctions (Timsit et al. 1989; Mayer-Jung et al. 1997)) and then A.

4.4 Covariations Versus Neutral Networks

The fundamental difference between Watson–Crick and non-Watson–Crick pairs lies in their structurally-based evolutionary behaviors. As discussed in Sect. 2.1, the bases in Watson–Crick pairs covary following a precise pattern that maintains isostericity and similarity in shape. The non-Watson–Crick base pairs do not, *stricto* sensu, covary. Structural comparisons show instead that they exchange within neutral networks of interactions where either both residues change during evolution or only one of the two (Dutheil et al. 2010). In other words, for a given base type, many other bases can lead to structurally isosteric base pairs. Thus, the underlying signals in the sequence alignments are blurred and require additional considerations to be revealed. This phenomenon is best illustrated in the isostericity matrices of the twelve families of base pairs (Leontis and Westhof 2001; Leontis et al. 2002). An illustration of how non-Watson–Crick pairs evolve on neutral networks is shown in Fig. 12.5. In the *trans* Hoogsteen/Sugar Edge family pair, two networks can be observed; in one, the base giving its Hoogsteen belongs to the amino bases (C and A), and in the second one, it belongs to the keto family (U and G). The first network is more populated than the second one, because in such arrangements, two H-bonds can form. Furthermore, often the nucleotides involved in non-Watson–Crick pairs are invariant in the structural species ensemble. For example, in pseudoknots that are formed by a single Watson–Crick base pair creating a three-dimensional interaction, the nucleotides involved are very often invariant and do not covary. In tRNAs, an example is the 3D Watson–Crick pair between the invariant G19 and C56 of loop D and loop T, respectively.

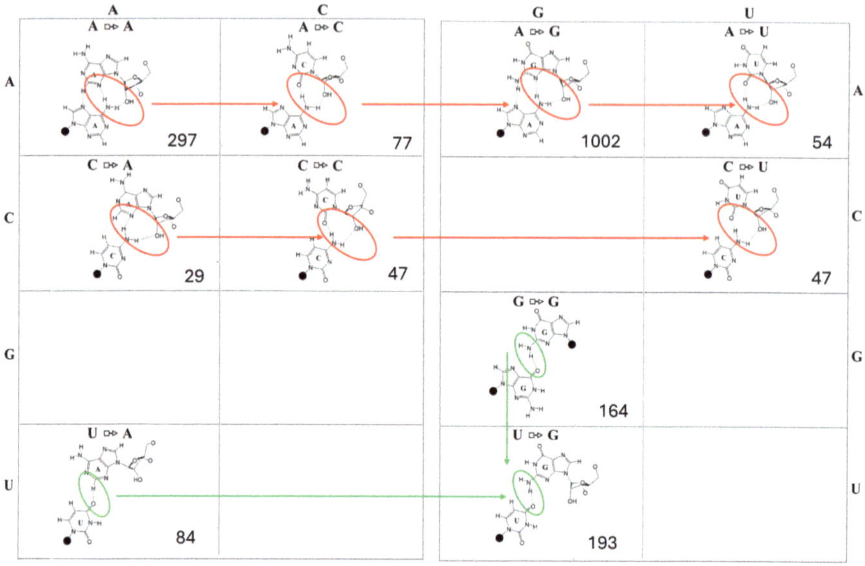

Fig. 12.5 Representations of the family of the *trans* Hoogsteen/SugarEdge base pairs. The H-bonds involved are circled and linked (in red for the largest neutral network and in green for the small neutral network). The number of occurrences in 1297 PDB structures below 2.5 Å are indicated. (These were obtained using WebFR3D (Petrov et al. 2011))

The Non-Watson–Crick Exchange Within Neutral Networks

Another example can be found in the R15/Y48 pairs of tRNAs (with a purine R always at position 15 and a pyrimidine Y at position 48) (Fig. 12.3b). Most commonly, R15 and Y48 form *trans* Watson–Crick/Watson–Crick A/U and G/C pairs. There are exceptions in some tRNAs where R15/R48 pairs occur, as in some bacterial Cys-tRNAs (Hou et al. 1993) and some eukaryotic Ala-tRNAs (Westhof et al. 2020). However, in general, the presence of a pyrimidine at position 15 or of a G15/U48 pair is unusual and may indicate an error in sequencing, a pseudogene of tRNA, or a mutant tRNA. An example where only one of the two bases of a tertiary pair changes is the *trans* Watson–Crick/Hoogsteen 54/58; it is overwhelmingly U/A, but some specific tRNAs present an A/A pair, as observed in eukaryotic initiator Met-tRNAs and some isodecoders of eukaryotic ACG-Ala-tRNAs (Westhof et al. 2020). Exceptions in a conserved tertiary pair can occur together with other changes; for example, the presence of R15/R48 pairs in Ala-tRNAs occurs together with A54/A58 (Westhof et al. 2020). Similarly, the presence of G15/G48 in bacterial Cys-tRNA occurs with a change in a triple pair where residue 9 contacts 13–22 in the D-stem (like in long-arm tRNAs) and not the usual 12–23 (see below) (Hauenstein et al. 2004).

4.5 The Interacting Edges of Non-Watson–Crick Pairs Are Conserved

Analyses of RNA three-dimensional structures show that the concept of isostericity should not be applied too strictly to tertiary interactions or non-Watson–Crick pairs. The main advantage of the Leontis and Westhof (LW) nomenclature (Leontis and Westhof 2001) is that the base pair name contains the base edges that interact with each other. A realistic application of isostericity is that, throughout the aligned sequences, the same base edges can approach each other without imposing that equivalent H-bonds be formed. This is particularly clear when the nucleotides form more than one base pairing contact, as in triple and quadruple assemblies of nucleotides (Gautheret et al. 1995). Indeed, nucleic acid bases can form base interactions between three or four nucleotides, not just two. These multiple base-base interactions can be decomposed into one or more base pair types. For example, in G-quadruplexes, each G nucleotide forms *cis* Watson–Crick/Hoogsteen pairs with two adjacent Gs.

In tRNAs, there is a recurrent triple interaction involving the conserved residues U8, A14, and A21 (Fig. 12.6). The nucleotides U8 and A14 form a *trans* Watson–Crick/Hoogsteen pair. However, in yeast tRNA-Phe, A21 presents its Watson–Crick edge to the sugar edge of U8 and, in yeast tRNA-Asp, A21 presents its sugar edge to the Watson–Crick edge of A14. In both cases, A21 occupies the same space around the U8/A14 pair. Besides, slightly different positions of A21 can be observed.

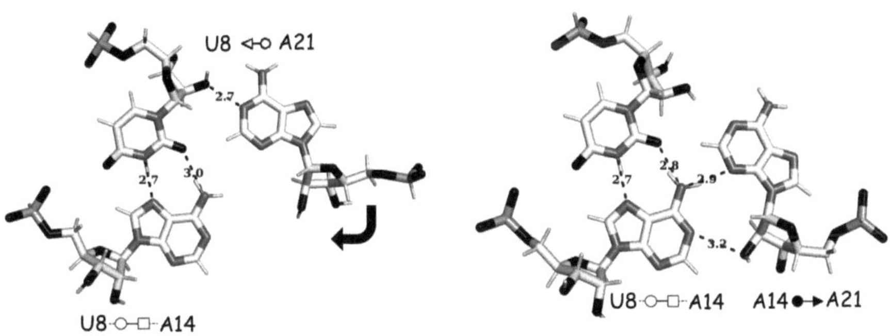

Fig. 12.6 Two different molecular arrangements of a conserved triple in tRNAs. (*Left*) In yeast Phe-tRNA (1EHZ) (Shi and Moore 2000) A21 interacts via its Watson–Crick edge with the Sugar edge of U8. (*Right*) In yeast Asp-tRNA (2TRA) (Westhof et al. 1985) A21 interacts via its Sugar edge with the Watson–Crick of A14. The *black arrow* indicates the slight rotation undergone by A21 between the two structures

4.6 *Tertiary Interactions Vary with the Set of Secondary Helices*

In addition, variations in secondary structure elements can induce changes in interacting residues. There are two classes of common elongator tRNAs; the most frequent ones (class I) have four helical stems, and the second class (class II or long-arm tRNAs) has an additional helix that results from an expansion of the short variable segment of type I tRNAs. In class I tRNAs, there is a triple between residue 9 and base pair 12–23, while in class II tRNAs, the triple occurs between residue 9 and base pair 13–22 (Westhof and Auffinger 2012) (see Table 12.1 for the number of instances) (Figs. 12.7 and 12.8). The examples shown in Figs. 12.7 and 12.8 illustrate that when a third base interacts with a helical base pair, deviations from strict geometric isostericity are observed in the sequences. However, the key point is that the types of base edges involved are generally maintained. Throughout the three kingdoms of life, one can observe specific variations in tRNA secondary structures, like in the selenocysteine-tRNAs (Sturchler et al. 1993) or the mitochondrial tRNAs (Suzuki et al. 2011). Structural sequence alignments should, therefore, best be made separately for each given set of 2D helical stems.

Table 12.1 Nucleotide occurrences in the triples formed between the D-stem 12–23 base pair and nucleotide 9 in 745 sequences of class I tRNAs (top) and between the D-stem 13–22 base pair and nucleotide 9 in 187 sequences of class II tRNAs (below)

	Residue 9			
Class I Base pair 12/23	A	U	G	C
A–U	21	–	–	4
U–A	**412**	–	34	–
G = C	39	4	**104**	–
C = G	23	2	64	13
Others	9	1	10	1
Class II Base pair 13/22				
A ◁-□ A	**32**	–	8	3
G ◁-□ A	–	4	**115**	–
G ◁-□ U	–	–	11	–
Others	7	1	4	2

Adapted from (Westhof and Auffinger 2012). The most frequent occurrences (in bold) are illustrated in Figs. 12.7 and 12.8

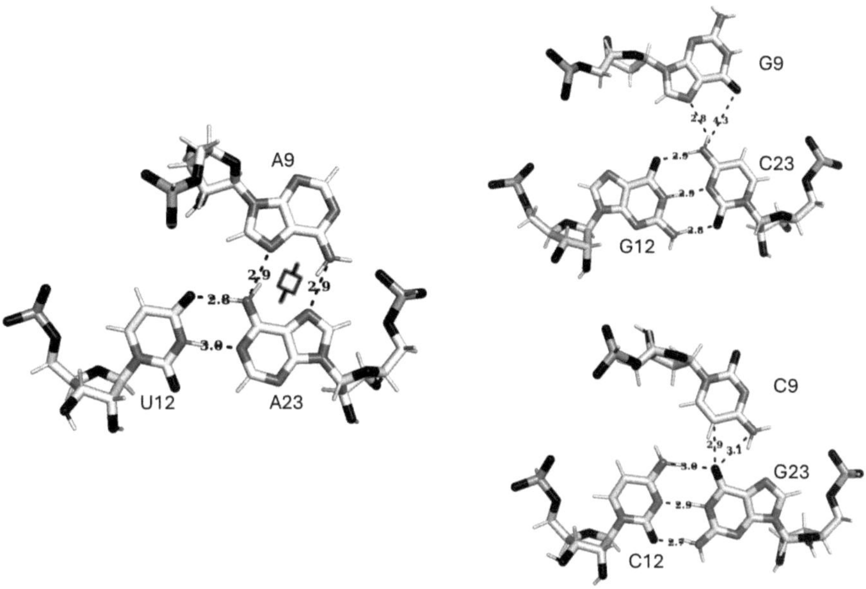

Fig. 12.7 Examples of contacts made by the variable residue at position 9 with the variable Watson–Crick pair 12–23 in the D-stem of some class I tRNAs. At the *left* with A9 (1EHZ) (Shi and Moore 2000), at the *top right* G9 (8RQK) (Bartuli et al. 2025) and at the *bottom right* C9 (1G59) (Sekine et al. 2001). However, in each case, the Hoogsteen edge of residue 9 interacts with the Hoogsteen edge of residue 23 with a *trans* orientation of the glycosyl bonds with respect to the line of contact between the two bases. The LW symbols are indicated between the H-bonds linking the pairs

5 The Intrinsic Processes of Molecular Accommodations in RNA Architectures

The examples described previously illustrate the diverse potential for the local molecular accommodations accessible to RNA molecules. These molecular accommodations arise from their atomic arrangements, stereochemistry, and charges. The main factors contributing to the molecular accommodations are the following.

5.1 The Sugar–Phosphate Backbone

It spans seven torsion angles, one of which is constrained by the sugar pucker type. Although they possess highly preferred ranges, the torsion angles can vary, in concert or not with neighboring torsion angles (Sundaralingam 1975; Saenger 1984). Minor variations in the C3′-*endo* region lead to significant displacements of the linking oxygen atoms O5′ and O3′ (Sundaralingam and Westhof 1980). Similarly, minor variations in the C2′-*endo* region lead to reorientations of the

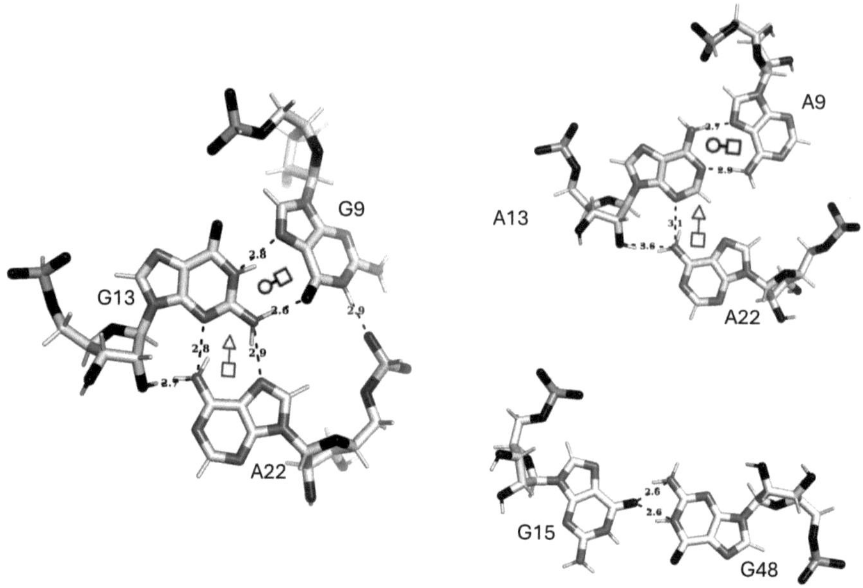

Fig. 12.8 Examples of contacts made by the variable residue at position 9 with the last base pair of the D-stem 13/22. In human tRNAs, that base pair is Watson–Crick, except in class II tRNAs, where it is a *trans* Sugar edge/Hoogsteen G/A pair. At the *left*, the triple interaction in the class II yeast Ser-tRNA (1SER) (Biou et al. 1994). At the *top right*, a triple isosteric to the one shown at the *left*, with only A residues from a class I bacterial Cys-tRNA (1U0B) (Hauenstein et al. 2004). At the *bottom right*, the simultaneous presence of the rare G15/G48 bifurcated pair (Hou et al. 1993). The LW symbols are indicated between the H-bonds linking the pairs

nucleotide base (Amerein et al. 1987). While variations in torsion angles do occur in helical nucleotides, they are key in single-stranded segments for leaving helices, forming loops, and structured or unstructured linking regions. For example, instead of the common C3′-*endo* sugar pucker in helical stems, when the last 3′ nucleotide of a helical strand branches off, it often adopts the C2′-*endo* conformation (as in nucleotide 7 in the acceptor stem of tRNAs).

5.2 *Stacking Interactions*

Although of variable strength (Sponer et al. 1996), the stacking interactions are not highly constrained in geometry (Bugg et al. 1971; Sponer et al. 2008). However, stacking interactions contribute significantly to the stabilization of nucleic acids, and a loss of some inter-base H-bonding can be easily compensated by a continuous stack between bases.

5.3 *The Hydroxyl Group*

The 2′-hydroxyl group of the ribose can give or receive H-bonds in three directions, with preferred ranges (Auffinger and Westhof 1997).

5.4 *The Solvent*

Water molecules can easily complement H-bonding losses, interact with the exocyclic groups of the bases, or fill molecular voids (Westhof 1988; Auffinger and Westhof 2001).

5.5 *The Ions*

Ions from the solvent shell can compensate for electrostatic repulsion either in a delocalized fashion or by localized binding to RNA nucleotides or folds with a variable number of liganded water molecules (Leonarski et al. 2025).

These processes are based on weak forces, but together they enable the molecular accommodations necessary to understand the molecular neutrality underlying the structural sequence alignments linking homologous sequences and structures. And, beyond, they may help explain the efficiency of in vitro molecular Darwinian evolution, where it is regularly observed that only 5–8 cycles of amplification and selection are necessary to observe targeted functional molecules.

6 Molecular Neutrality or How to Draw on Errors in RNA Evolution

All processes in living cells, within organisms and between organisms, rely on molecules and their molecular interactions and exchanges. Molecules are subject to errors, and all macromolecular processes have evolved sophisticated mechanisms to keep error rates within ranges adapted to the environmental conditions of the living cells in which they occur. In RNA, as in all other biological macromolecules, the molecular errors or mutations do not have all the same biological impact. RNA polymerases are notoriously error-prone (Lynch 2010). The errors are not only point mutations but also sequence insertions or deletions (Gout et al. 2017). Here, examples will be given illustrating how such sequence variations due to error propagation can be positively incorporated, leading to novel structural elements.

In RNA, the best-known sequence variations are those that maintain secondary structure, and in some cases, tertiary elements such as pseudoknots. These are achieved through covariations between complementary nucleotides in Watson–Crick pairs, where one Watson–Crick pair (A–U, U–A, G = C, C = G) is replaced by another. However, in three-dimensional folds where helices stack and pack together to form the architecture, some helical base pairs form specific stabilizing contacts with neighboring nucleotides, which restrict nucleotide variation. This occurs frequently at the ends of helices, as described above. There are also several other tertiary contacts that restrict variations in helical pairs.

In helices, other types of non-Watson–Crick occur; the most frequent one is the G•U pair within or at the ends of helices, followed by the G/A pair, especially at the ends of helices. Both the G•U and G•A pairs are easily accommodated within helical regions and can be considered as intermediate states between various standard Watson–Crick pairs during molecular evolution:

G=C > < G•U > < A-U
C=G > < U•G > < U-A
A-U > < A•G > < C=G
G=C > < G•A > < U-A

However, both G/U and G/A pairs can exist in different geometrical arrangements. While within helices, the G•U pairs adopt in most cases the standard wobble geometry, at the ends of helices, many different types of relative positions of the G and U nucleotides are observed. Similarly, within helices, the G/A pairs form extended Watson–Crick G•A pairs, but together with other non-Watson–Crick pairs or at the ends of helices, they can also adopt different pairing geometries such as the *trans* SugarEdge/Hoogsteen G/A pair. Thus, those types of pairs can accommodate their geometries to the local environment, remaining passive bystanders during molecular evolution. Or, following the emergence of new elements, they change more severely their conformations to allow the formation of a new fold. During their interactions, G and U have the propensity to change their electronic structure and form tautomeric or anionic pairs (Westhof et al. 2019, 2023; Demeshkina et al. 2012; Rozov et al. 2018), especially at the ends of helices where the diversity of interactions between G and U increases sharply. Some examples are presented in Fig. 12.9. Recently, it was shown that the replacement of a Watson–Crick pair by a G•U pair around the middle of the first stem of an RNA structure led to crystals of better resolution (Ren et al. 2025). In RNA structures, helices tend to coaxially stack two-by-two, and, clearly, in multiple junctions where more than two helices merge, the ends of one or more helices will be exposed to connect with single-stranded segments. In such regions, a larger conformational space of the terminal nucleotides is an advantage for folding into a compact structure.

The construction of complex RNA architectures is characterized by the recurrent use of structurally similar units or modules that cap helical fragments or link them with variable torsions or kinking. These structural RNA modules are assembled from various sets of non-Watson–Crick pairs in a precise order, are limited in the number of families, and occur in all kinds of functional RNAs. Figure 12.10 shows one example of an RNA module together with one of its variants: an internal loop fully base-paired in a non-Watson–Crick fashion is converted into a Kink-turn by a trinucleotide sequence repetition. In three-dimensional modules, the set of

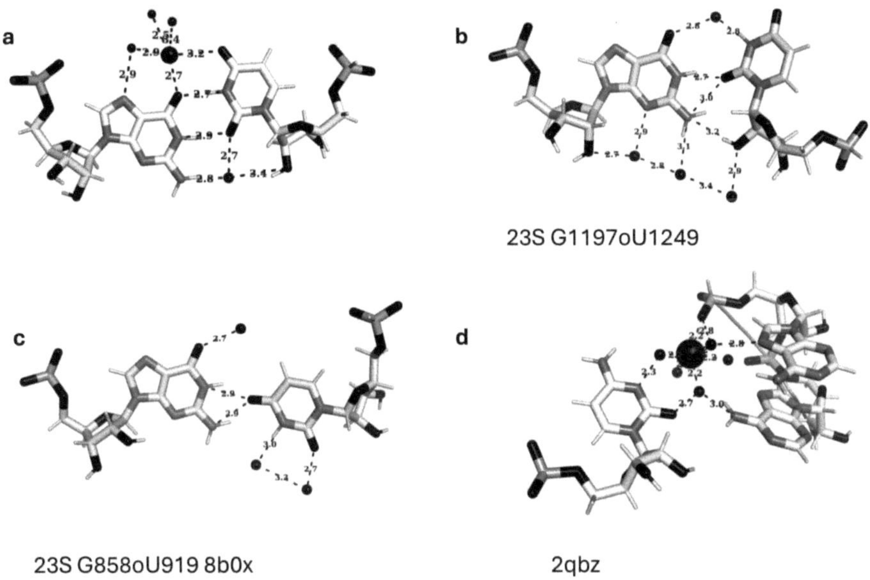

Fig. 12.9 Illustrations of some of some G•U base pairs in RNA structures. (**a**) A standard G•U pair. The positions of the H atoms are only representative (from PDB 1HQ1). A hydrated potassium ion binds in the major groove and contacts directly O6(G) and O4(U). A water molecule binds in the minor groove and links N2(G) to O2(U). Additional water molecules, not shown, interact with the pair. (**b**) A partially opened G•U pair with the O2(U) pointing to both N1(G) and N2(G) with an insertion of a water molecule between O6(G) and O4(U) (from the 23S rRNA G1197•U1249 of 8B0X). (**c**) A fully bifurcated G•U pair with O4(U) pointing to both N1(G) and N2(G) and water molecules binding to O6(G) as well as to N3(U) and O2(U) (from the 23S rRNA G858•U919 of 8B0X). (**d**) A fully opened G•U pair with a strong interaction with a hydrated magnesium ion (from 2QBZ)

non-Watson–Crick has a defined order, but the order can be reversed. In other words, a loop-E module is characterized by two sets of sequences 5′GAA3′ and 5′AGUA3′, but the order in which those two sequences appear along the one-dimensional sequence is irrelevant, and such strand exchanges can occur in homologous sequences (Westhof 2010; Weinberg et al. 2010).

An analysis of three-way junctions illustrates the diversity of contacts and non-Watson–Crick pairs at the junction core that can accommodate the different choices of coaxiality of helices in three-way junctions (Lescoute and Westhof 2006). Although there is a range of angles between the helical axes, in a three-way junction, generally two helical stems tend to be coaxial with the third one at an angle (if helices are numbered I, II, and III, three broad families can be observed: either with I & II stacking, or I & III, or II & III). The lengths of the single-stranded segments linking covalently to the helical strands participate structurally in the final choice. Some internal arrangements are preferred, but there is a clear diversity of contacts in each of the three families (Lescoute and Westhof 2006). Figure 12.11 illustrates the case for one of the three main families, where four variations in the central non-Watson–Crick pairing between helices P1 and P2 are shown.

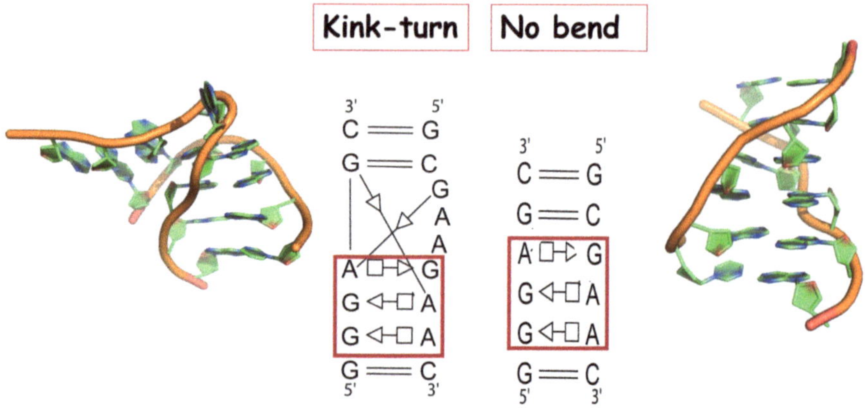

Fig. 12.10 Two examples of recurrent RNA modules. At the left, a Kink-turn that links at 90° to helical stems. At the right, an internal loop between helical pairs, with the same non-Watson–Crick pairs as in the Kink-turn (red boxes). The Kink-turn presents a duplication of the three nucleotides GAA, with additional non-Watson–Crick pairs between the sugar edges of the helical strands

7 Concluding Comments on RNA Modifications, Weak Forces, and RNA 3D Predictions

Here, we have considered mainly the four natural bases A, G, C, and U to highlight the richness in the variety of contacts and recognition modes. However, it is now well appreciated that RNA is often chemically modified by dedicated enzymes that modify specific nucleotides, often in a specific context (see Fig. 12.1). These modifications generally amplify the inherent interaction modalities of RNA molecules, rather than introducing additional ones. Indeed, the most frequent modifications stabilize or restrict nucleotides to their most preferred conformations of nucleotides (C3′-*endo* and the *anti* orientation of the base), like pseudouridylation or 2′-O-methylation and methylation or acetylation on the C5 atom of pyrimidines. The modifications influence conformational preferences by slight variations in partial atomic charges within the base and by reorganizing the water network between the base and the phosphate groups (Auffinger and Westhof 1998, 2001; Garcia-Campos et al. 2025; Rozman et al. 2026). As described in Sects. 5.4 and 5.5, the molecular interactions at play rely on weak forces acting at short distances within the RNA solute and with the solvent molecules. The modifications can occur at many places throughout the RNA, as in ribosomal RNAs of hyperthermophilic archaeal species, where the amount of methylation and acetylation increases very significantly in the RNA helices to cope with boiling water temperatures (Garcia-Campos et al. 2025). The modifications can also occur only at some specific positions that are key for the folding and maintenance of the three-dimensional architecture. For example (see Fig. 12.2c), the first residue of the T-stem, C49, that stacks with the last residue of the AA stem, A7, is very frequently methylated on C5,

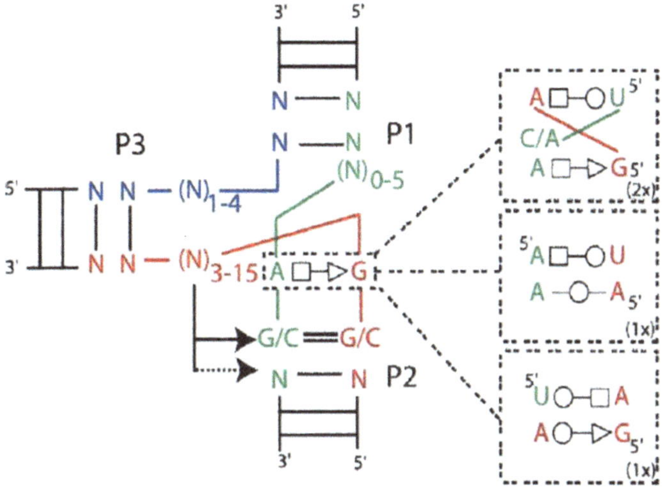

Fig.12.11 A consensus for family A of the three-way junctions. In family A, helix P3 is roughly perpendicular to the coaxial stack between helices P1 and P2. The junction J12 between stems 1 and 2 comprises 0–5 nts; the junction J23 between helices 2 and 3, 3–15 nts; and the junction J31 between stems 3 and 1, 1–4 nts. Nucleotides at position 3, 4, or 5 of J23 make tertiary interactions with the shallow groove of the first one or the first two Watson–Crick base pairs of P2 (indicated by black arrows). The closing base pair of P2 is always in the *trans* orientation. To family A belong the three-way junctions between h20–h21–h22 in 16S rRNA or H49–H59–1–Hx in 23S rRNA (Lescoute and Westhof 2006). These correspond to the three-way junctions depicted, respectively, in Fig. 12.1 (b) and (e) of (Leontis et al. 2006). These two three-way junctions incorporate a Ω-turn sequence motif (Wadley and Pyle 2004) in which a *syn* base is extruded from the helix to form a *cis* Watson–Crick base pair with a base often engaged in a *trans* Hoogsteen pair. Interestingly, the Ω-turn sequence motif also occurs in internal loops linking only two helical segments (Wadley and Pyle 2004). (Figure adapted from Lescoute and Westhof 2006)

and the methyl group contacts phosphate groups key for the tRNA fold. Analyses of specific RNA modulations show that a stabilization of selected tertiary positions, based on weak interactions, can propagate folding and the maintenance of the architecture. In addition, 2′-O-methylation of ribose specifically protects nucleotides from hydrolytic degradation.

Finally, RNA, like all other macromolecules, undergoes constant molecular dynamics depending on external factors like temperature, even in structured RNAs. Many RNA sequences have a great potential for global alternative pairing and conformational folds that do not result solely from the molecular dynamics intrinsic to any molecule (Bussi et al. 2024). In many instances, these alternative states participate in biological regulation or underlying functional mechanisms. The above descriptions, based on structured RNAs, are not intended to mean that RNA is generally a more fluid molecule, but, instead, that even in structured RNAs, regular molecular dynamics together with structural accommodations that are mediated mainly by weak forces can lead to stable folds despite sequence variations.

The structural phenomena described here contribute to the difficulty in predicting the three-dimensional structures of RNA (Bu et al. 2025). Generally, although global arrangements of helices can be reached, the correct predictions of non-Watson–Crick and the assignments of conservation in sequence alignments are very poorly predicted (Westhof et al. 2026; Kretsch et al. 2026). As shown by many assessment scores (Bu et al. 2025; Westhof et al. 2026; Kretsch et al. 2026), it is apparent that global fold is not sufficient to capture the key 3D features, maintaining the architecture and explaining the sequence alignments and the biological function. In other words, prediction tools somehow manage to accommodate single-stranded segments linking helices in various ways. However, these conformational pathways, even in the absence of conflict with stereochemistry or geometry, do not reflect or rationalize sequence alignments. In previous modeling efforts (Michel and Westhof 1990; Lehnert et al. 1996; Massire et al. 1998), tertiary anchor contacts between secondary elements derived from sequence alignments constrained the progressive assembly of the final RNA architecture successfully. Hopefully, more evolution-based modeling with a systematic use of the deep links between multiple sequence alignments and RNA architectural principles will form the core of future developments in RNA 3D structure prediction.

Acknowledgments I wish to thank Craig Zirbel for help with FR3D and numerous discussions on RNA base pairs and Sarah Woodson for reading the manuscript and suggesting several clarifying points. I acknowledge funding by the Wenzhou Institute, University of Chinese Academy of Sciences (WIUCASQD2024002).

References

Amerein B, Ripp R, Dumas P (1987) PUCK: a real-time modification of sugar pucker on a PS300. J Mol Graph 5:5

Ancel LW, Fontana W (2000) Plasticity, evolvability, and modularity in RNA. J Exp Zool 288: 242–283

Auffinger P, Westhof E (1997) Rules governing the orientation of the 2′-hydroxyl group in RNA. J Mol Biol 274:54–63

Auffinger P, Westhof E (1998) Hydration of RNA base pairs. J Biomol Struct Dyn 16:693–707

Auffinger P, Westhof E (2001) Hydrophobic groups stabilize the hydration Shell of 2′-O-methylated RNA duplexes. Angew Chem Int Ed Engl 40:4648–4650

Bartuli J, Jungwirth S, Dixit M, Okuda T, Zimmermann JP, Erlacher M, Pan T, Volz A, Huttenhofer A, Warscheid B, Hobartner C, Grimm C, Fischer U (2025) tRNA as an assembly chaperone for a macromolecular transcription-processing complex. Nat Struct Mol Biol 32:2349

Been MD, Perrotta AT (1991) Group I intron self-splicing with adenosine: evidence for a single nucleoside-binding site. Science 252:434–437

Behrouzi R, Roh JH, Kilburn D, Briber RM, Woodson SA (2012) Cooperative tertiary interaction network guides RNA folding. Cell 149:348–357

Berman HM, Westbrook J, Feng Z, Gilliland G, Bhat TN, Weissig H, Shindyalov IN, Bourne PE (2000) The protein data bank. Nucleic Acids Res 28:235–242

Biebricher CK, Eigen M (2005) The error threshold. Virus Res 107:117–127

Biela A, Hammermeister A, Kaczmarczyk I, Walczak M, Koziej L, Lin TY, Glatt S (2023) The diverse structural modes of tRNA binding and recognition. J Biol Chem 299:104966

Biou V, Yaremchuk A, Tukalo M, Cusack S (1994) The 2.9 A crystal structure of T. thermophilus seryl-tRNA synthetase complexed with tRNA(Ser). Science 263:1404–1410

Bu F, Adam Y, Adamiak RW, Antczak M, de Aquino BRH, Badepally NG, Batey RT, Baulin EF, Boinski P, Boniecki MJ, Bujnicki JM, Carpenter KA, Chacon J, Chen SJ, Chiu W, Cordero P, Das NK, Das R, Dawson WK, DiMaio F, Ding F, Dock-Bregeon AC, Dokholyan NV, Dror RO, Dunin-Horkawicz S, Eismann S, Ennifar E, Esmaeeli R, Farsani MA, Ferre-D'Amare AR, Geniesse C, Ghanim GE, Guzman HV, Hood IV, Huang L, Jain DS, Jaryani F, Jin L, Joshi A, Karelina M, Kieft JS, Kladwang W, Kmiecik S, Koirala D, Kollmann M, Kretsch RC, Kurcinski M, Li J, Li S, Magnus M, Masquida B, Moafinejad SN, Mondal A, Mukherjee S, Nguyen THD, Nikolaev G, Nithin C, Nye G, Pandaranadar Jeyeram IPN, Perez A, Pham P, Piccirilli JA, Pilla SP, Pluta R, Poblete S, Ponce-Salvatierra A, Popenda M, Popenda L, Pucci F, Rangan R, Ray A, Ren A, Sarzynska J, Sha CM, Stefaniak F, Su Z, Suddala KC, Szachniuk M, Townshend R, Trachman RJ 3rd, Wang J, Wang W, Watkins A, Wirecki TK, Xiao Y, Xiong P, Xiong Y, Yang J, Yesselman JD, Zhang J, Zhang Y, Zhang Z, Zhou Y, Zok T, Zhang D, Zhang S, Zyla A, Westhof E, Miao Z (2025) RNA-puzzles round V: blind predictions of 23 RNA structures. Nat Methods 22:399–411

Bugg CE, Thomas JM, Sundaralingam M, Rao ST (1971) Stereochemistry of nucleic acids and their constituents. X. Solid-state base-stacking patterns in nucleic acid constituents and polynucleotides. Biopolymers 10:175–219

Burley SK, Berman HM, Bhikadiya C, Bi C, Chen L, Di Costanzo L, Christie C, Dalenberg K, Duarte JM, Dutta S, Feng Z, Ghosh S, Goodsell DS, Green RK, Guranovic V, Guzenko D, Hudson BP, Kalro T, Liang Y, Lowe R, Namkoong H, Peisach E, Periskova I, Prlic A, Randle C, Rose A, Rose P, Sala R, Sekharan M, Shao C, Tan L, Tao YP, Valasatava Y, Voigt M, Westbrook J, Woo J, Yang H, Young J, Zhuravleva M, Zardecki C (2019) RCSB protein data bank: biological macromolecular structures enabling research and education in fundamental biology, biomedicine, biotechnology and energy. Nucleic Acids Res 47:D464–DD74

Bussi G, Bonomi M, Gkeka P, Sattler M, Al-Hashimi HM, Auffinger P, Duca M, Foricher Y, Incarnato D, Jones AN, Kirmizialtin S, Krepl M, Orozco M, Palermo G, Pasquali S, Salmon L, Schwalbe H, Westhof E, Zacharias M (2024) RNA dynamics from experimental and computational approaches. Structure 32:1281–1287

Cech TR (2024) The catalyst, RNA and the quest to unlock life's deepest secrets. W.W. Norton

Chan PP, Lowe TM (2016) GtRNAdb 2.0: an expanded database of transfer RNA genes identified in complete and draft genomes. Nucleic Acids Res 44:D184–D189

Chargaff E (1950) Chemical specificity of nucleic acids and mechanism of their enzymatic degradation. Experientia 6:201–209

Chua EYD, Mendez JH, Rapp M, Ilca SL, Tan YZ, Maruthi K, Kuang H, Zimanyi CM, Cheng A, Eng ET, Noble AJ, Potter CS, Carragher B (2022) Better, faster, cheaper: recent advances in cryo-electron microscopy. Annu Rev Biochem 91:1–32

Consortium, The RNAcentral, Petrov AI, Kay SJE, Kalvari I, Howe KL, Gray KA, Bruford EA, Kersey PJ, Cochrane G, Finn RD, Bateman A, Kozomara A, Griffiths-Jones S, Frankish A, Zwieb CW, Lau BY, Williams KP, Chan PP, Lowe TM, Cannone JJ, Gutell R, Machnicka MA, Bujnicki JM, Yoshihama M, Kenmochi N, Chai B, Cole JR, Szymanski M, Karlowski WM, Wood V, Huala E, Berardini TZ, Zhao Y, Chen R, Zhu W, Paraskevopoulou MD, Vlachos IS, Hatzigeorgiou AG, Ma L, Zhang Z, Puetz J, Stadler PF, McDonald D, Basu S, Fey P, Engel SR, Cherry JM, Volders PJ, Mestdagh P, Wower J, Clark MB, Quek XC, Dinger ME (2017) RNAcentral: a comprehensive database of non-coding RNA sequences. Nucleic Acids Res 45:D128–DD34

Demeshkina N, Jenner L, Westhof E, Yusupov M, Yusupova G (2012) A new understanding of the decoding principle on the ribosome. Nature 484:256–259

Dutheil JY, Jossinet F, Westhof E (2010) Base pairing constraints drive structural epistasis in ribosomal RNA sequences. Mol Biol Evol 27:1868–1876

Eigen M (1971) Selforganization of matter and the evolution of biological macromolecules. Naturwissenschaften 58:465–523

Eigen M, Schuster P (1977) The hypercycle. A principle of natural self-organization. Part a: emergence of the hypercycle. Naturwissenschaften 64:541–565
Ellington AD, Szostak JW (1990) In vitro selection of RNA molecules that bind specific ligands. Nature 346:818–822
Elson D, Chargaff E (1954) Regularities in the composition of pentose nucleic acids. Nature 173:1037–1038
Fontana W, Schuster P (1998) Continuity in evolution: on the nature of transitions. Science 280: 1451–1455
Fox GE, Woese CR (1975) 5S RNA secondary structure. Nature 256:505–507
Gangloff J, Keith G, Ebel JP, Dirheimer G (1971) Structure of aspartate-tRNA from brewer's yeast. Nat New Biol 230:125–126
Garcia-Campos MA, Georgeson J, Nir R, Reichelt R, Fluke KA, Matzov D, Iyer V, Burkhart BW, Lui L, Kustanovich A, Grunberger F, Thalalla-Gamage S, Howpay-Manage SA, Gerovac M, Alexandre N, Nobe Y, Nowak JS, Perera M, Apostle A, Fang S, Glatt S, Jona G, Ferreira-Cerca S, Vogel J, Taoka M, Meier JL, Westhof E, Santangelo TJ, Grohmann D, Shalev-Benami M, Schwartz S (2025) Pan-modification profiling facilitates a cross-evolutionary dissection of the thermoregulated ribosomal epitranscriptome. Cell 188:6825–6844. e28
Gautheret D, Damberger SH, Gutell RR (1995) Identification of base-triples in RNA using comparative sequence analysis. J Mol Biol 248:27–43
Gout JF, Li W, Fritsch C, Li A, Haroon S, Singh L, Hua D, Fazelinia H, Smith Z, Seeholzer S, Thomas K, Lynch M, Vermulst M (2017) The landscape of transcription errors in eukaryotic cells. Sci Adv 3:e1701484
Hauenstein S, Zhang CM, Hou YM, Perona JJ (2004) Shape-selective RNA recognition by cysteinyl-tRNA synthetase. Nat Struct Mol Biol 11:1134–1141
Hermann T, Westhof E (1999) Non-Watson-Crick base pairs in RNA-protein recognition. Chem Biol 6:R335–R343
Hou YM, Westhof E, Giege R (1993) An unusual RNA tertiary interaction has a role for the specific aminoacylation of a transfer RNA. Proc Natl Acad Sci USA 90:6776–6780
Huynen MA, Stadler PF, Fontana W (1996) Smoothness within ruggedness: the role of neutrality in adaptation. Proc Natl Acad Sci USA 93:397–401
Kim SH (1978) Three-dimensional structure of transfer RNA and its functional implications. Adv Enzymol Relat Areas Mol Biol 46:279–315
Kim SH, Cech TR (1987) Three-dimensional model of the active site of the self-splicing rRNA precursor of Tetrahymena. Proc Natl Acad Sci USA 84:8788–8792
Kretsch RC, Hummer AM, He S, Yuan R, Zhang J, Karagianes T, Cong Q, Kryshtafovych A, Das R (2026) Assessment of nucleic acid structure prediction in CASP16. Proteins 94:192–217
Lehnert V, Jaeger L, Michel F, Westhof E (1996) New loop-loop tertiary interactions in self-splicing introns of subgroup IC and ID: a complete 3D model of the Tetrahymena thermophila ribozyme. Chem Biol 3:993–1009
Leonarski F, Henning-Knechtel A, Kirmizialtin S, Ennifar E, Auffinger P (2025) Principles of ion binding to RNA inferred from the analysis of a 1.55 A resolution bacterial ribosome structure – Part I: Mg2. Nucleic Acids Res 53
Leontis NB, Westhof E (2001) Geometric nomenclature and classification of RNA base pairs. RNA 7:499–512
Leontis NB, Stombaugh J, Westhof E (2002) The non-Watson-Crick base pairs and their associated isostericity matrices. Nucleic Acids Res 30:3497–3531
Leontis NB, Lescoute A, Westhof E (2006) The building blocks and motifs of RNA architecture. Curr Opin Struct Biol 16:279–287
Lescoute A, Westhof E (2006) Topology of three-way junctions in folded RNAs. RNA 12:83–93
Levitt M (1969) Detailed molecular model for transfer ribonucleic acid. Nature 224:759–763
Lisacek F, Diaz Y, Michel F (1994) Automatic identification of group I intron cores in genomic DNA sequences. J Mol Biol 235:1206–1217

Lu C, Smith AM, Fuchs RT, Ding F, Rajashankar K, Henkin TM, Ke A (2008) Crystal structures of the SAM-III/S(MK) riboswitch reveal the SAM-dependent translation inhibition mechanism. Nat Struct Mol Biol 15:1076–1083

Lynch M (2010) Evolution of the mutation rate. Trends Genet 26:345–352

Masquida B, Westhof E (2000) On the wobble GoU and related pairs. RNA 6:9–15

Massire C, Jaeger L, Westhof E (1998) Derivation of the three-dimensional architecture of bacterial ribonuclease P RNAs from comparative sequence analysis. J Mol Biol 279:773–793

Mattick J, Amaral P (2022) RNA, the epicenter of genetic information. CRC Press, Boca Raton

Mayer-Jung C, Moras D, Timsit Y (1997) Effect of cytosine methylation on DNA-DNA recognition at CpG steps. J Mol Biol 270:328–335

Michel F, Dujon B (1983) Conservation of RNA secondary structures in two intron families including mitochondrial-, chloroplast- and nuclear-encoded members. EMBO J 2:33–38

Michel F, Westhof E (1990) Modelling of the three-dimensional architecture of group I catalytic introns based on comparative sequence analysis. J Mol Biol 216:585–610

Michel F, Costa M, Massire C, Westhof E (2000) Modeling RNA tertiary structure from patterns of sequence variation. Methods Enzymol 317:491–510

Mokdad A, Krasovska MV, Sponer J, Leontis NB (2006) Structural and evolutionary classification of G/U wobble basepairs in the ribosome. Nucleic Acids Res 34:1326–1341

Noller HF (2005) RNA structure: reading the ribosome. Science 309:1508–1514

Noller HF, Woese CR (1981) Secondary structure of 16S ribosomal RNA. Science 212:403–411

Ontiveros N, Cooke E, Nawrocki EP, Triebel S, Marz M, Rivas E, Griffiths-Jones S, Petrov AI, Bateman A, Sweeney B (2024) Rfam 15: RNA families database in 2025. bioRxiv

Petrov AI, Zirbel CL, Leontis NB (2011) WebFR3D–a server for finding, aligning and analyzing recurrent RNA 3D motifs. Nucleic Acids Res 39:W50–W55

Quigley GJ, Rich A (1976) Structural domains of transfer RNA molecules. Science 194:796–806

Ren Y, Lin X, Liao W, Peng X, Deng J, Zhang Z, Zhan J, Zhou Y, Westhof E, Lilley DMJ, Wang J, Huang L (2025) A general strategy for engineering GU base pairs to facilitate RNA crystallization. Nucleic Acids Res 53

Rivas E (2023) RNA covariation at helix-level resolution for the identification of evolutionarily conserved RNA structure. PLoS Comput Biol 19:e1011262

Rivas E, Clements J, Eddy SR (2017) A statistical test for conserved RNA structure shows lack of evidence for structure in lncRNAs. Nat Methods 14:45–48

Robertson DL, Joyce GF (1990) Selection in vitro of an RNA enzyme that specifically cleaves single-stranded DNA. Nature 344:467–468

Rozman B, Broennimann K, Rajan KS, Nachshon A, Saha C, Arazi T, Mohan V, Geiger T, Wollner CJ, Richner JM, Westhof E, Yonath A, Bashan A, Stern-Ginossar N (2026) N(1)-Methylpseudouridine directly modulates translation dynamics. Nature 651:533

Rozov A, Wolff P, Grosjean H, Yusupov M, Yusupova G, Westhof E (2018) Tautomeric G*U pairs within the molecular ribosomal grip and fidelity of decoding in bacteria. Nucleic Acids Res 46: 7425–7435

Saenger W (1984) Principles of nucleic acid structure. Springer Verlag

Sardanyes J, Perales C, Domingo E, Elena SF (2024) Quasispecies theory and emerging viruses: challenges and applications. NPJ Viruses 2:54

Schultes EA, Bartel DP (2000) One sequence, two ribozymes: implications for the emergence of new ribozyme folds. Science 289:448–452

Sekine S, Nureki O, Shimada A, Vassylyev DG, Yokoyama S (2001) Structural basis for anticodon recognition by discriminating glutamyl-tRNA synthetase. Nat Struct Biol 8:203–206

Shi H, Moore PB (2000) The crystal structure of yeast phenylalanine tRNA at 1.93 Å resolution: a classic structure revisited. RNA 6:1091–1105

Singhal RP, Fallis PA (1979) Structure, function, and evolution of transfer RNAs (with appendix giving complete sequences of 178 tRNAs). Prog Nucleic Acid Res Mol Biol 23:227–290

Sponer J, Leszczynski J, Hobza P (1996) Nature of nucleic Acid-Base stacking: nonempirical ab initio and empirical potential characterization of 10 stacked base dimers. Comparison of stacked and H-bonded base pairs. J Phys Chem 100:5590–5596

Sponer J, Riley KE, Hobza P (2008) Nature and magnitude of aromatic stacking of nucleic acid bases. Phys Chem Chem Phys 10:2595–2610
Stombaugh J, Zirbel CL, Westhof E, Leontis NB (2009) Frequency and isostericity of RNA base pairs. Nucleic Acids Res 37:2294–2312
Sturchler C, Westhof E, Carbon P, Krol A (1993) Unique secondary and tertiary structural features of the eucaryotic selenocysteine tRNA(Sec). Nucleic Acids Res 21:1073–1079
Sundaralingam M (1975) Structure and conformation of nucleosides and nucleotides and their analogs as determined by x-ray diffraction. Ann NY Acad Sci 255:3–42
Sundaralingam M, Westhof E (1980) Correlated motions in DNA. Biophys J 32:250–252
Suzuki T, Nagao A, Suzuki T (2011) Human mitochondrial tRNAs: biogenesis, function, structural aspects, and diseases. Annu Rev Genet 45:299–329
Timsit Y, Westhof E, Fuchs RP, Moras D (1989) Unusual helical packing in crystals of DNA bearing a mutation hot spot. Nature 341:459–462
Topal MD, Fresco JR (1976) Base pairing and fidelity in codon-anticodon interaction. Nature 263: 289–293
Tuerk C, Gold L (1990) Systematic evolution of ligands by exponential enrichment: RNA ligands to bacteriophage T4 DNA polymerase. Science 249:505–510
Wadley LM, Pyle AM (2004) The identification of novel RNA structural motifs using COMPADRES: an automated approach to structural discovery. Nucleic Acids Res 32:6650–6659
Weinberg Z, Wang JX, Bogue J, Yang J, Corbino K, Moy RH, Breaker RR (2010) Comparative genomics reveals 104 candidate structured RNAs from bacteria, archaea, and their metagenomes. Genome Biol 11:R31
Westhof E (1988) Water: an integral part of nucleic acid structure. Annu Rev Biophys Biophys Chem 17:125–144
Westhof E (1992) Westhof's rule. Nature 358:459–460
Westhof E (2010) The amazing world of bacterial structured RNAs. Genome Biol 11:108
Westhof E (2014) Isostericity and tautomerism of base pairs in nucleic acids. FEBS Lett 588:2464–2469
Westhof E, Auffinger P (2012) Transfer RNA structure. In: Encyclopedia of life sciences. Wiley, Chichester
Westhof E, Fritsch V (2000) RNA folding: beyond Watson-Crick pairs. Structure 8:R55–R65
Westhof E, Leontis NB (2021) An RNA-centric historical narrative around the protein data bank. J Biol Chem 296:100555
Westhof E, Massire C (2004) Structural biology. Evolution of RNA architecture. *Science* 306:62–63
Westhof E, Dumas P, Moras D (1985) Crystallographic refinement of yeast aspartic acid transfer RNA. J Mol Biol 184:119–145
Westhof E, Masquida B, Jossinet F (2011) Predicting and modeling RNA architecture. Cold Spring Harb Perspect Biol 3
Westhof E, Yusupov M, Yusupova G (2019) The multiple flavors of GoU pairs in RNA. J Mol Recognit 32:e2782
Westhof E, Liang S, Tong X, Ding X, Zheng L, Dai F (2020) Unusual tertiary pairs in eukaryotic tRNA(Ala). RNA 26:1519–1529
Westhof E, Thornlow B, Chan PP, Lowe TM (2022) Eukaryotic tRNA sequences present conserved and amino acid-specific structural signatures. Nucleic Acids Res 50:4100–4112
Westhof E, Watson ZL, Zirbel CL, Cate JHD (2023) Anionic G*U pairs in bacterial ribosomal rRNAs. RNA 29:1069–1076
Westhof E, Sun H, Bu F, Miao Z (2026) The RNA-puzzles assessments of RNA-only targets in CASP16. Proteins 94:218–229
Wuyts J, Van de Peer Y, De Wachter R (2001) Distribution of substitution rates and location of insertion sites in the tertiary structure of ribosomal RNA. Nucleic Acids Res 29:5017–5028

Chapter 13
RNA Dynamics in the Evolution of RNA Function

Constanze Ebermann and Sabine Müller

Abstract RNA shows significantly higher folding dynamics than DNA because it predominantly occurs in its single-stranded form. Depending on sequence, length, and possible interaction partners, a single RNA strand can adopt a variety of folds, leading to a broad spectrum of distinct conformations with characteristic functions. Due to its folding versatility, RNA has the potential to perform a wide variety of functions, including regulation of gene expression and catalysis. Throughout evolution, numerous functional RNAs have emerged, and this process continues today. It is accompanied by significant efforts in in vitro RNA engineering, adding an "artificial evolutionary" aspect. In this chapter, we describe how different functionalities have been engineered into a small self-cleaving RNA. We demonstrate the power of this RNA in supporting highly dynamic processes, such as RNA recombination and splicing. These processes require cascades of conformational changes, and RNA folds to perform individual processing steps and to generate functional novelty without sequence change.

Keywords Allosteric regulation · Ribozyme · Recombination · RNA folding · Splicing

Abbreviations

CHR	chameleon ribozyme
DNA	deoxyribonucleic acid
FMN	flavine mononucleotide
FWJ	four-way junction
GOI	gene of interest
HDV	Hepatitis Delta Virus
HHR	hammerhead ribozyme

C. Ebermann · S. Müller (✉)
University of Greifswald, Institute of Biochemistry, Greifswald, Germany
e-mail: sabine.mueller@uni-greifswald.de

G. Witzany (ed.), *Integrative Theory of Evolution - Aspects and Insights*,
https://doi.org/10.1007/978-3-032-24295-2_13

HPR	hairpin ribozyme
IRES	internal ribosome entry site
mRNA	messenger RNA
ORF	open reading frame
RBS	ribosome binding site
RNA	ribonucleic acid
RNAT	RNA thermometer
SF	supporting factor
siRNA	small interfering RNA
snRNPs	small nuclear ribonucleoproteins
TRAP	trp-RNA-binding attenuation protein
TWJ	three-way junction
UTR	untranslated region
VS	Varkud Satellite

1 Small Self-Cleaving Ribozymes

As the name suggests, this class of ribozymes comprises RNAs with a relatively small number of nucleotides that nonetheless demonstrate high functional diversity and agility. This combination makes these ribozymes ideal subjects for studying RNA function and folding characteristics. Due to their easier handling and higher stability compared to longer RNA strands, small ribozymes can be easily manipulated and modified in vitro. This allows for the engineering of new ribozyme variants through evolutionary and/or rational design in vitro (Ebermann and Müller 2025).

The hairpin ribozyme (HPR) and the hammerhead ribozyme (HHR) are prime examples of small ribozymes. They are the most well-studied and well-understood catalysts in the class of small, self-cleaving ribozymes, and will therefore be discussed in more detail in Sect. 2. The structures and mechanisms of other small self-cleaving ribozymes, such as the twister, twister sister, pistol, hatchet, hepatitis delta virus (HDV), and Varkud satellite (VS) ribozymes, have been extensively reviewed (for a more recent addition, see Peng et al. 2021). In addition, further information on ribozymes can be found on the online platform Ribocentre (https://www.ribocentre.org/), where comprehensive information on all natural ribozymes is compiled (Deng et al. 2023).

Despite variations in structure and catalytic strategy, the overall mechanism of small self-cleaving ribozymes remains the same. Notably, such ribozymes often occur in or are associated with viral RNA, as well as highly conserved, ancient sections of bacterial and eukaryotic RNA (Peng et al. 2021). The high degree of conservation of the ribozyme sequences and their independence from proteins suggest that these functional RNAs may have played a significant role in the evolution of life and the emergence of the RNA world. In their natural environment, small self-cleaving ribozymes function as catalysts for RNA self-processing by

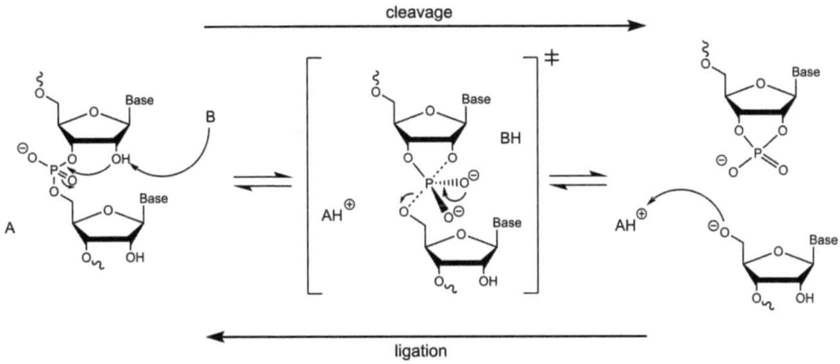

Fig. 13.1 General mechanism of small self-cleaving ribozymes by general acid-base catalysis

means of reversible cleavage of an RNA strand in *cis* (i.e., the same RNA strand). However, ribozymes can also be engineered to function as true catalysts, cleaving or ligating a substrate strand in a *trans* fashion with multiple turnovers. In both cases, the ribozymes execute a transesterification reaction through general acid-base catalysis. In this reaction, a nucleobase within the ribozyme structure, hydrated magnesium ions, or external cofactors function as the acid or base (Peng et al. 2021; Deng et al. 2023). The general base abstracts a proton from a specific 2′-OH group, thereby activating the oxygen for nucleophilic attack on the adjacent phosphorus. This results in the formation of a trigonal bipyramidal intermediate. Subsequently, the 5′-OH group is protonated by a general acid, resulting in the formation of a suitable leaving group. Subsequent to cleavage, the typical products are a fragment with a free 5′-OH group and the other with a 2′,3′-cyclic phosphate (Fig. 13.1). Ligation progresses along the same pathway in the opposite direction, thereby reconnecting two fragments with the aforementioned end functionalities (5′-OH and 2′, 3′-cyclic phosphate; Lilley 2011).

The majority of these ribozymes depend on magnesium or other divalent cations for optimal functionality (Schnabl and Sigel 2010). The presence of Mg^{2+} is crucial for facilitating the appropriate folding of the ribozyme into its active conformation by counteracting the negative charge of the phosphate backbone (Misra and Draper 1998; Hanna and Doudna 2000; Yamagami et al. 2021). However, as previously mentioned, hydrated magnesium ions can also directly participate in the catalytic mechanism itself, acting as a general acid or base or as a Lewis acid. In this case, they interact with specific residues in the catalytic core and thus stabilize products or intermediates of the reaction (Hanna and Doudna 2000).

2 Two Examples: The Hairpin and the Hammerhead Ribozyme

2.1 Hairpin Ribozyme

The hairpin ribozyme (HPR) was discovered in the negative strand of tobacco ringspot virus satellite RNA, where it plays a critical role in processing intermediates of rolling circle replication (Buzayan et al. 1986). Over the years, extensive research on the HPR has yielded solved crystal structures of several variants (Rupert and Ferré-D'Amaré 2001; Alam et al. 2005), along with characterization of its catalytic properties and sequence requirements (Fedor 1999, 2000; Müller et al. 2012). In addition, a variety of modified versions of the natural HPR were optimized for applications that extend beyond its natural function.

The wild-type HPR forms a four-way junction (FWJ) with four partially double-stranded arms (Butcher and Burke 1994; Fig. 13.2). However, the minimally functional HPR consists of only two of these arms connected by a hinge region, thereby forming a two-way junction structure. The two arms each consist of two helical regions interrupted by a loop. Loop A contains the cleavage site between the guanosine at position +1 (G^{+1}) and the upstream base (B^{-1}) within the consensus sequence N*GUC (Joseph et al. 1993; the star marks the cleavage site). The sequence of the helical regions exhibits significant variability, provided that base

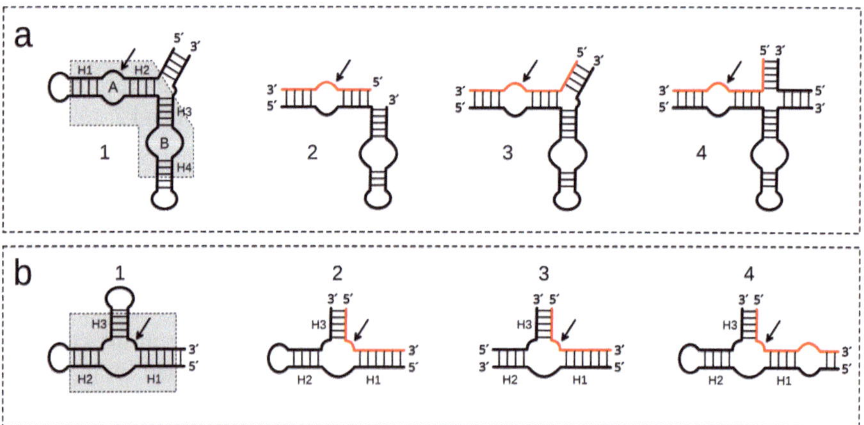

Fig. 13.2 Schematics of HPR and HHR structures. (**a**) Structural variants of the hairpin ribozyme: 1: General structure of a TWJ hairpin ribozyme with loops A and B and helices (H) 1–4; the minimal motif highlighted in grey. The cleavage site is indicated by an arrow. The HPR as a minimal motif (two-way junction), TWJ, and FWJ (2–4). The substrate strand is indicated in red and the HPR strand in black. (**b**) Structural variants of the hammerhead ribozyme: 1: General structure of HHR with stems/helices (H) 1–3; the minimal motif highlighted in grey. The cleavage site is indicated by an arrow. HHR variants with open or closed stems (2–4). The substrate strand is indicated in red, the HHR strand(s) in black

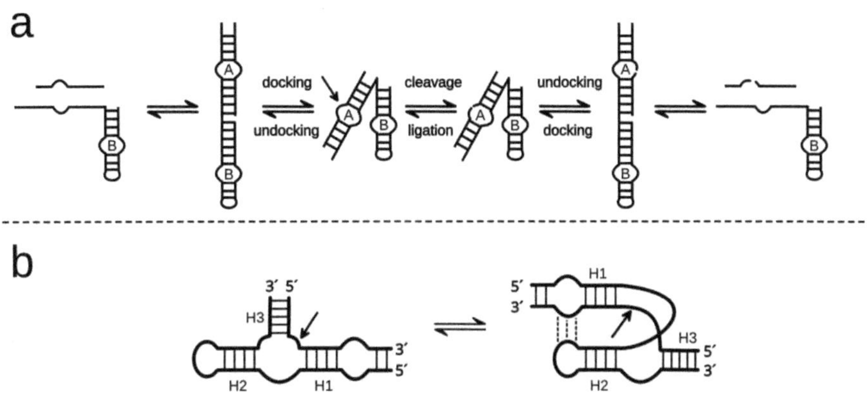

Fig. 13.3 (**a**) Conformational dynamics of the HPR. Note, substrate/product binding and release can occur only from the extended state, while reaction proceeds only from the docked state. (**b**) Folding of the wild-type HHR with an additional loop in H1 supported by loop–loop interaction. The cleavage site is indicated by an arrow

pairing is maintained. In contrast, nucleobases in loops A and B are strongly conserved (Berzal-Herranz et al. 1993; Pinard et al. 1999; Fedor 2000).

For proper folding and activity, helix 1 is required to span a minimum of four base pairs, while helix 2 is required to have an exact length of four base pairs. The high conservation of the two loops can be attributed to their critical role in the folding of the HPR into its catalytically active structure and formation of the catalytic core. The interaction of bases in the loops, particularly through a ribose zipper and an interdomain Watson–Crick base pair, leads to the close proximity of the loops, a process referred to as docking (Fig. 13.3; Earnshaw et al. 2000). In the docked state, the ribozyme exhibits a bent conformation (Fedor 1999). Cleavage and ligation of the HPR can only occur from the docked state, while substrate/product binding and dissociation are only possible from an extended conformation (Zhuang et al. 2002). It is important to note that, in order to function properly, the ribozyme must undergo a dynamic transition between extended and docked states. Ligation requires a stable docked state, whereas cleavage, in contrast, can also occur from a transiently docked conformation. If the lifetime of the docked state is too short, dissociation of the substrates is likely to happen too rapidly for productive ligation. Consequently, ligation of two strands is augmented by structural elements or environmental conditions that stabilize the docked state (Paudel and Rueda 2014). A series of experiments demonstrated that the rates of hairpin ribozyme cleavage and ligation are strongly dependent on pH, which is consistent with the hypothesis that proton transfer occurs within the transition state during the catalytic process (Pinard 2001; Nahas et al. 2004). It is noteworthy that an internal equilibrium constant between cleavage and ligation $K_{\mathrm{int}} = K_{\mathrm{lig}}/K_{\mathrm{cleav}}$, of 34 was determined, indicating that the reaction is substantially biased toward ligation (Nahas et al. 2004).

As aforementioned, the docked state is stabilized by tertiary interactions within the folded structure. Stabilization can additionally arise from the specific junction structure, which is comprised of two arms (minimal motif), a three-way junction (TWJ), or a four-way junction (FWJ; Fig. 13.2). The wild-type hairpin ribozyme has been shown to form a FWJ, which facilitates folding in the docked state and close contact between loops A and B.

In contrast, in the minimal two-way junction HPR, it functions more akin to a hinge, thereby rendering its stabilizing contribution virtually negligible. A three-way junction (TWJ) structure is situated between these two extremes. The presence of an additional arm and a short bulge at the appropriate position assists folding and stabilizes the docked state (Walter et al. 1999; Welz et al. 2001). The sequence around the junction and of the bulge is of particular importance to achieve these structural effects. The HPR, like all small self-cleaving ribozymes, is dependent on magnesium ions for activity. However, magnesium ions do not constitute an integral component of the catalytic mechanism itself. The metal ions function as a charge buffer, thereby facilitating proper folding through the neutralization of the ribozymes' negatively charged phosphate backbone. Consequently, the retention of functional HPR catalysis can be achieved in the absence of magnesium ions under various conditions (Murray et al. 1998; Welz et al. 2001; Vlassov et al. 2005; Mutschler et al. 2015). Moreover, the facilitation of this process can be accomplished by other divalent cations (Schnabl and Sigel 2010). As the stability of the docked state increases with the number of arms involved in the junction, FWJ HPRs exhibit higher activity and are active at lower magnesium ion concentrations than TWJ HPRs or the minimal two-way junction motif (Walter et al. 1999; Wilson et al. 2007). Consequently, FWJ HPRs have been observed to promote ligation, while the equilibrium shifts toward cleavage as the number of junction arms decreases.

2.2 Hammerhead Ribozyme

The hammerhead ribozyme (HHR) was the first small self-cleaving ribozyme to be discovered, and it has been the subject of extensive research since then (Usman et al. 1996). Similar to the HPR, it has been found in satellite RNAs of plant viruses and plant viroids, playing a functional role in rolling circle replication (Prody et al. 1986). Crystallography studies have revealed that the HHR structure features a junction region with three helices/stems, with helix 2 being an open or hairpin stem structure (Pley et al. 1994; Fig. 13.2). G8 and G12 in the catalytic strand are indispensable for catalytic activity, and are therefore highly conserved. However, the sequence of the helical regions is completely variable (Birikh et al. 1997). The cleavage site is located in the junction region between stems I and III within the sequence UC*N. In the full-length HHR, a loop–loop interaction between stems I and II assists in the folding of the ribozyme (Fig. 13.3; La Peña et al. 2003; Martick and Scott 2006). These tertiary interactions are critical for optimal activity, but not essential for catalysis. The minimal motif lacks the loops in stems I and II, resulting in the absence of loop–loop interactions in the minimal structure. Nevertheless, this minimal HHR exhibits cleavage activity, but has a significantly lower turnover rate and requires a higher concentration of magnesium ions than the wild type (Hammann et al. 2012; Scott et al. 2013).

Unlike the HPR, which favors ligation, the HHR exhibits poor ligation activity compared to cleavage (Hertel et al. 1994; Hertel and Uhlenbeck 1995; Stage-Zimmermann and Uhlenbeck 1998). This strong preference for cleavage may be due to the flexibility of the hammerhead ribozyme complex with two ligation substrates, resulting in an entropic advantage for cleavage (Hertel and Uhlenbeck 1995). Because of this, hammerhead ribozymes have primarily been used for RNA cleavage. The example of the HHR demonstrates that, although folding flexibility can be advantageous for the evolution of functionality, the development of other RNA activities may benefit from a more rigid tertiary structure. Additionally, the evolutionary trajectory of each nucleotide in the HHR's catalytic core can be directly correlated with its functional importance, a feature that probably applies also to other functional nucleic acids (Brill et al. 2024).

3 Functional Dynamics of Small Ribozymes Beyond Cleavage and Ligation

3.1 Twin Ribozymes

The concept of twin ribozymes, which mediate an RNA fragment exchange reaction (Welz et al. 2003), is one approach that clearly demonstrates the potential of the characteristic cleavage/ligation behavior of the hairpin ribozyme (HPR) caused by its folding dynamics. In this system, two HPRs are fused together by tandem duplication, creating two cleavage/ligation sites upon binding a suitable substrate (Fig. 13.4). A predefined section of the RNA substrate is excised and replaced with

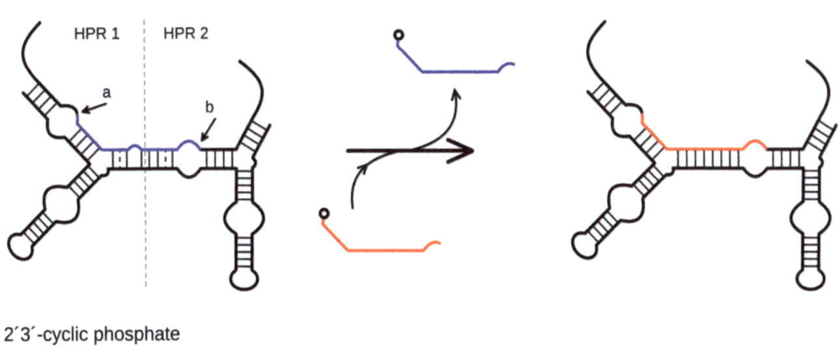

Fig. 13.4 Principle of twin ribozyme-mediated RNA recombination/strand exchange. Blue: substrate patch (with mismatches and/or bulge[s]) to be exchanged; red: substrate patch to be inserted, unfilled circle: 2′,3′-cyclic phosphate. The twin ribozyme consists of two HPR units, 1 and 2, each providing one cleavage site (**a**, **b**). Cleavage sites are indicated by arrows. Mismatches are displayed as dashed lines

an externally added oligonucleotide. This process requires two cleavages and two ligation events that have to occur in a strictly controlled manner (Fig. 13.4).

Thus, in a twin ribozyme reaction, the substrate is first cleaved at two predetermined sites. This results in a central fragment that contains destabilizing features, such as mismatches or loops, and can therefore dissociate relatively easily. In contrast, the two RNA fragments that flank the cleavage sites form stable duplexes with the ribozyme, causing them to preferentially remain bound to the ribozyme. An oligonucleotide is added externally after cleavage and is designed to bind strongly to the gap left by dissociation of the excised fragment. Together with a slight molar excess of this oligonucleotide, the reaction shifts towards ligation and the formation of the recombined end product. Depending on the design, fragments of similar lengths can be exchanged (Vauléon et al. 2005) or a shorter strand replaced by a longer one (Welz et al. 2003; Ivanov et al. 2005; Vauléon et al. 2005; Balke et al. 2014, 2016) and vice versa (Drude et al. 2007). Ultimately, this system enables insertions, base substitutions, and the repair of short deletions within a target RNA strand (Balke et al. 2016). This approach can also be useful for the site-specific modification of longer RNAs. We have demonstrated that twin ribozyme-mediated recombination is compatible with site-specifically functionalized or modified oligonucleotides. This allows for the internal modification of RNA strands whose length exceeds the practical range of solid-phase synthesis (Vauléon et al. 2005).

3.2 Recombination

Based on the two fundamental reactions of cleavage and ligation, which are supported by the HPR, in particular, it is evident that the functional scope can be expanded by carefully adjusting their properties. The most obvious example is the process of recombination, which was already described in the previous section with the twin ribozymes. Likewise, a single HPR can be used for supporting recombination (Hieronymus et al. 2016, 2022; Hieronymus and Müller 2019, 2021; Zhu et al. 2023). Recombination is a crucial evolutionary principle because it allows two previously unassociated sequences to combine into a new molecule. Thus, the potential alteration of properties and functionalities by a change in sequence expands beyond random mutations of single nucleotides. Regarding early RNA evolution, possibly in the absence of proteins, it is of special interest to investigate whether the relatively small and simple HPR can perform such reactions. Due to its ability to catalyze cleavage and ligation to similar extents, the HPR is a promising candidate for such applications. Careful manipulation of the HPR structure can shift the internal equilibrium between the two reactions towards either one. The stability of the ribozyme-substrate complex is an important factor.

In general, successful recombination by the HPR requires cleavage of an RNA substrate, yielding two products: one with a 5′-OH group and the other with a 2′,3′-cyclic phosphate. The product bearing the 5′-OH group is readily released due to its relatively low binding affinity to the ribozyme. The second fragment, which carries the 2′,3′-cyclic phosphate, remains bound to the ribozyme and

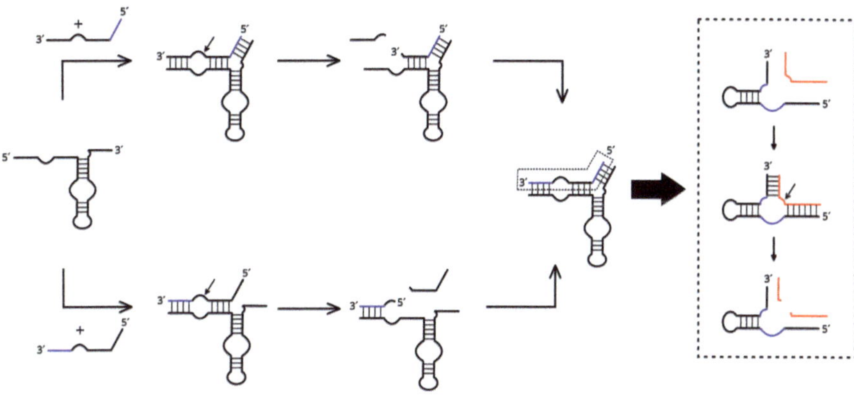

Fig. 13.5 Reaction scheme of HPR-supported recombination to yield a functional hammerhead ribozyme. Conserved nucleotides of the HHR are marked in blue. The HHR substrate is marked in red

becomes the substrate for the subsequent ligation with an added ligation substrate. This ligation substrate binds more tightly to the ribozyme and can easily replace the shorter fragment to become ligated. This results in the recombination product. This type of recombination reaction can be expanded beyond simple recombination to generate RNAs with new functions (Hieronymus et al. 2016). We previously designed a variant of the HPR that could recombine two RNA molecules, yielding a functional hammerhead ribozyme as the recombination product (Fig. 13.5). The two educt RNA strands themselves exhibited no catalytic activity. However, each consists of a nonfunctional and a profunctional sequence patch. After cleavage, the nonfunctional parts are released, leaving behind the profunctional sequences. Due to their preferred binding to the HPR, these sequences can be efficiently ligated to yield the recombined functional RNA strand. Since the recombination product is an active hammerhead ribozyme, the success of the recombination reaction can be verified using an HHR functionality assay with a compatible substrate. After optimizing the experimental conditions, yields of up to 76% were achieved (Hieronymus et al. 2016).

Similar to the aforementioned scheme, HPR-mediated recombination was utilized to assemble an RNA strand with two functionalities: an aptazyme consisting of an FMN or theophylline aptamer unit, as well as a hammerhead ribozyme unit (Hieronymus and Müller 2021; Fig. 13.6). The HPR used could cleave two RNA strands carrying either the aptamer or the hammerhead ribozyme sequence, respectively.

After cleavage, the resulting fragments were ligated to form an allosterically controlled hammerhead aptazyme. The activity of this aptazyme can then be regulated by externally added FMN or theophylline ligands. Yields of up to 34% of these functional aptazymes were obtained. Also, other long RNA molecules, such as an RNA polymerase ribozyme, can be produced by HPR recombination in similar assays (Mutschler et al. 2015; Le Vay et al. 2021).

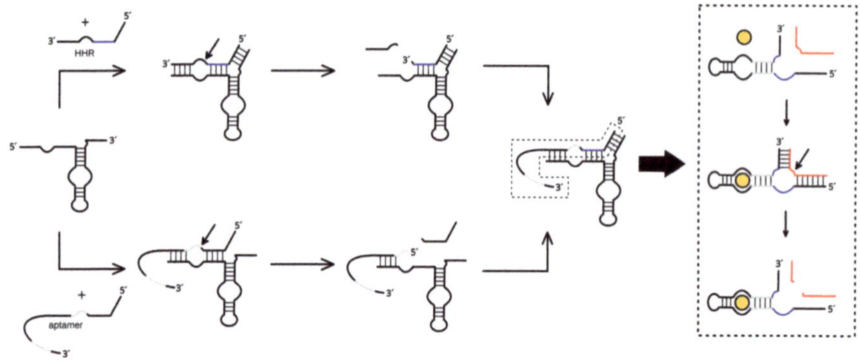

Fig. 13.6 Reaction scheme of HPR-supported recombination to yield a functional aptazyme consisting of an HHR unit and a theophylline or FMN aptamer unit. Conserved sequence parts of the HHR are marked in blue. Essential sequence parts for the formation of the aptamer are marked in grey. The HHR substrate is marked in red. The cognate ligand, theophylline or FMN, is represented by a yellow circle

3.3 *Splicing and Circularization*

Splicing is essentially a type of recombination. In modern biological systems, this event is executed by either group I or II self-splicing introns, which are self-cleaving/self-ligating ribozymes, or by the spliceosomal complex. These RNA structures support transesterifications in the sugar-phosphate backbone using a common reaction mechanism. However, these ribozymes are rather large and complex compared to the small HPR (Cech 1990; Lehmann and Schmidt 2003; Will and Luhrmann 2011). The spliceosome depends on proteins for structural stabilization and regulation. Regarding an evolving RNA world, in which neither proteins nor very long RNAs existed in high abundance, it is interesting to consider whether smaller ribozymes could perform splicing reactions. Using the hairpin ribozyme, we demonstrated its ability to mimic the splicing process, splice out an intron from between two exons, resulting in the ligation of the exons and release of the intron sequence (Hieronymus et al. 2022; Zhu et al. 2023; Figs. 13.7 and 13.8). In analogy to the exon-intron-exon composition of an mRNA before splicing, a substrate to be spliced by an HPR variant must carry two HPR cleavage sites. The two sites are cleaved consecutively. The sequence in between, mimicking the intron, is spliced out, and the two flanking patches, mimicking the exons, are ligated. To successfully catalyze this reaction sequence, the HPR must fold precisely in response to the bound substrate. It has to adopt two alternative folds that position one of the substrate's two cleavage sites in the correct location. After the first cleavage event, the ribozyme refolds to enable the second cleavage. After dissociation of the intron, the remaining two exons comprise the terminal functionalities required for subsequent ligation.

For the described splicing reaction to be successful, it is essential that the intron sequence can easily dissociate from the HPR to prevent religation to the exons.

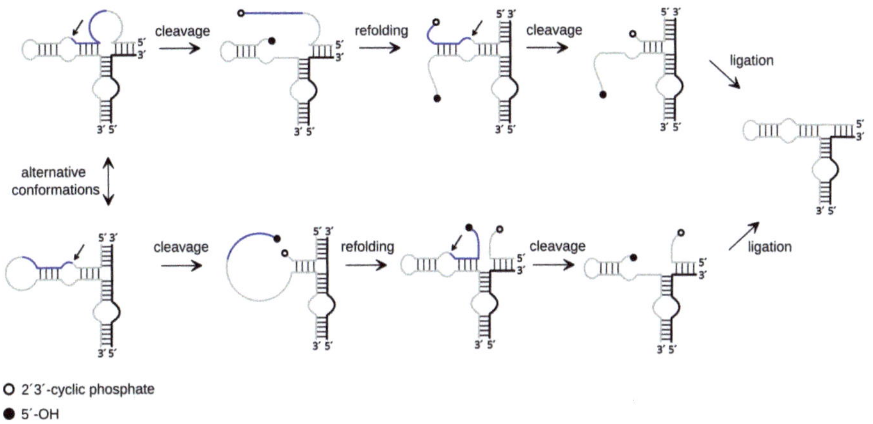

Fig. 13.7 Reaction scheme of HPR-based RNA splicing. Blue: intron to be spliced out, grey: exon; black: supporting factor; unfilled circle: 2′,3′-cyclic phosphate, filled circle: 5′-OH. The cleavage sites are indicated by arrows

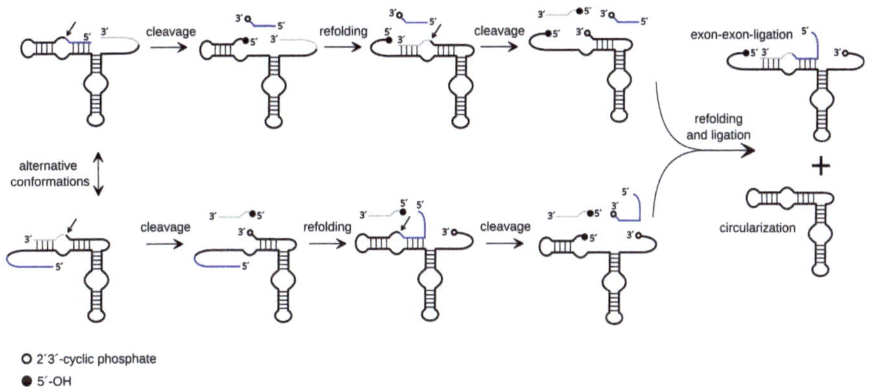

Fig. 13.8 Reaction scheme of HPR-based RNA self-splicing. Blue: 5′-exon, grey: 3′-exon; unfilled circle: 2′,3′-cyclic phosphate, filled circle: 5′-OH. The cleavage sites are indicated by arrows

Second, after intron release, the remaining HPR-substrate complex must fold into a stable docked conformation that enables ligation. To develop an HPR-based splicing system, we carefully investigated the possible folding conformations of all reaction intermediates and implemented two different approaches.

In a first design, the ribozyme variant, which we designated the hairpin ribozyme-derived spliceozyme, the HPR sequence was partially included in the exon sequence (Fig. 13.7). To enable better control of the reaction, the active ribozyme was divided into two parts: one containing only the exon-intron-exon sequence (shown in grey and blue in Fig. 13.7), which requires reconstitution into an active HPR by the addition of another RNA strand, called the supporting factor (SF). Thus, the presence or absence of the SF determines whether splicing occurs. Successful splicing was demonstrated by fragment length analysis (Zhu et al. 2023). Furthermore, the design included that the excised intron has a downstream function to act as a positive regulator of the activity of a bipartite DNAzyme. Thus, the success of the splicing reaction could be validated not only by fragment length analysis of the resulting products but also by a downstream functional assay based on DNAzyme activity (Zhu et al. 2023).

Another approach to a self-splicing HPR is presented in Fig. 13.8 (Hieronymus et al. 2022). Similar to the aforementioned system, the ribozyme adopts two alternative conformations, in each of which one of the two possible cleavage sites is cleaved. However, unlike the scheme described above, the catalytic activity is executed by the intron itself, as in naturally occurring self-splicing introns, rather than being located in the exonic regions. After the first cleavage event in one of the two conformations, one exon is released. Then the HPR refolds into the alternative conformation, which permits the second cleavage to occur. After both cleavage reactions are completed and the products are released, there are two possible outcomes. The exons can rebind to the HPR and be ligated, resulting in the desired splicing product. Or, in an alternative reaction, the intron/HPR refolds to bring its termini with the functionalities required for ligation (5′-OH and 2′,3′-cyclic phosphate) to the catalytic center, leading to intramolecular ligation and thereby circularization of the intron (Fig. 13.8).

The described examples of small self-splicing ribozymes emphasize the importance of precise folding and exact recognition of splice sites through the correct hybridization of the ribozyme and substrate strands. The dynamics in the folding of RNA strands are essential for the evolution of a complex process like splicing in higher organisms. The rather sophisticated spliceosome complex, which cannot act without the small nuclear ribonucleoproteins (snRNPs), probably emerged from group II self-splicing introns that do not depend on proteins for activity (Harris and Rogers 2008; Rogozin et al. 2012; Müller 2017; Vosseberg and Snel 2017; Galej et al. 2018; Borao et al. 2021). Thus, it is also likely that these self-splicing introns originated from smaller ribozyme structures capable of performing a primordial splicing reaction (Harris and Rogers 2008), a hypothesis that has been experimentally investigated also by other research groups (Müller 2017).

In addition to the regular splicing process discussed above, back-splicing can occur in biological systems. In this process, rather than joining an upstream 5′-splice site to a downstream 3′-splice site, a downstream 5′-splice site ligates to an upstream 3′-splice site. This results in the formation of a circular RNA (Salzman et al. 2012). As shown by the example of our self-splicing HPR, circularization can occur as a side reaction to the intended splicing process (Hieronymus et al. 2022). Based on this finding, the design principle for a circularization system is clear, and intended circularization by a self-splicing HPR variant has been demonstrated experimentally (Pieper et al. 2007; Petkovic and Müller 2013; Petkovic et al. 2015). As in regular splicing, two cleavage sites must be present, and the HPR must fold into two alternative conformations, each allowing cleavage at one of the sites. After the first cleavage reaction, the ribozyme refolds to enable the second cleavage reaction. The resulting free intron ends are then ligated to yield a circular RNA (Petkovic et al.

2015). However, this circular product still contains the active HPR sequence. Therefore, a notable disadvantage of this approach is the relative instability of the circular products due to their intrinsic self-cleavage activity.

3.4 Chameleon Ribozyme

As shown in previous sections, a single ribozyme can exhibit significant functional diversity based solely on its dynamic folding properties. However, a ribozyme can be prompted to prefer one cleavage site over another, while retaining its characteristic identity (i.e., the HPR). It is also possible to transition a ribozyme between two distinct folds, each of which represents a different activity mode (Schultes and Bartel 2000; DasGupta et al. 2021; Zhu et al. 2024). We recently synthesized a ribozyme capable of folding into either an active HPR or an active HHR, depending on the substrate added to the reaction. Because this ribozyme changes its appearance in response to its environment, we designated it the chameleon ribozyme (CHR) (Zhu et al. 2024; Fig. 13.9a). To enable this functional dynamic, the sequence was very carefully designed. Unlike the aforementioned approaches, the structural and sequence requirements of two ribozymes had to be considered simultaneously. Sequence changes that stabilize one structure must not interfere with the structure of the other. Investigating the minimal conserved sequences of the HPR and HHR revealed that the two ribozymes' conserved regions were compatible (DasGupta et al. 2021).

Positions conserved in the HPR are not conserved in the HHR, and vice versa. This observation enabled us to combine the minimal sequences of both ribozymes into a single RNA strand. After several steps of sequence optimization involving mutations and extensions of the ribozymes' minimal motifs to facilitate more stable folding, the final chameleon ribozyme exhibited cleavage and ligation activities with the respective substrates in separate reactions and with different substrates present in a one-pot reaction (Zhu et al. 2024). The first design, CHR1 (Fig. 13.9b), had a long binding arm for the substrate in the HPR fold. This led to a high probability of back-ligation and relatively low cleavage yields of the hairpin ribozyme fold. To adjust the two activities of the CHR- cleavage by the HHR and ligation by the HPR-, the ribozyme sequence was extended in the 3′-direction to provide a longer binding arm for the 5′-ligation substrate of the HPR (CHR2). This stabilizes the docked state of the HPR fold, thereby promoting ligation activity. To further stabilize the docked state, we elongated helix 4 to stabilize loop B and introduced additional mutations to the sequence, ultimately yielding CHR3 (Fig. 13.9b). All three variants exhibited high cleavage activity with the hammerhead substrate, suggesting that the hammerhead fold remained unaffected by the modifications made to promote HPR docking. Variants CHR2 and CHR3 exhibited satisfying ligation activity with substrates for the hairpin fold. As expected, the CHR3 variant had higher ligation yields due to stronger stabilization of the docked state (Zhu et al. 2024).

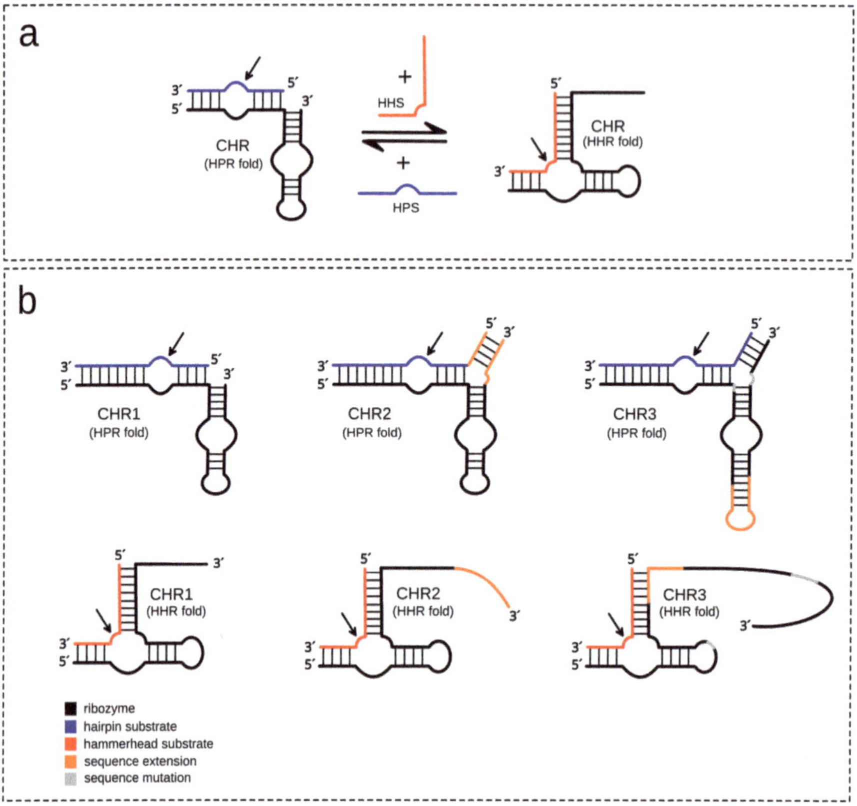

Fig. 13.9 (**a**) General principle of the chameleon ribozyme (CHR), which adopts either the HPR (left) or the HHR (right) fold, depending on the added substrate. Blue: HPR substrate, red: HHR substrate. The cleavage sites are indicated by arrows. (**b**) Alterations of CHR sequence for activity optimization

4 Allosteric Control of Small Self-Cleaving Ribozymes

In nature, enzymes are often regulated by external cofactors, molecules that bind to the enzyme, causing it to alter its structure. Similar to the principle of allosteric control of enzymatic activity in biological systems, ribozymes can also be regulated by external cofactors. This means that when an effector molecule binds, the ribozyme's structure and, consequently, its activity change. Such an effector can be a small molecule (true allosteric control) or an accessory RNA strand called an effector oligonucleotide (allosteric-like control).

4.1 Allosteric Control by Effector Oligonucleotides

RNA conformation can change with the addition of external RNA or DNA strands (effector oligonucleotides) that bind to a specific ribozyme sequence segment via complementary base pairing. This binding either facilitates (positive control/induction) or prevents (negative control/repression) the ribozymes' ability to fold correctly and catalyze their respective reaction.

Several approaches have been developed for inducing or repressing HPR activity (Komatsu et al. 2002; Vauléon and Müller 2003; Najafi-Shoushtari 2004). In all cases, the applied effector oligonucleotide reconstituted the active tertiary structure of a misfolded hairpin ribozyme. The system previously presented by us utilized a mutated HPR that showed no activity due to a mutated essential nucleotide necessary for docking, as well as further mismatches that destabilized the catalytic domain (Vauléon and Müller 2003). The effector oligonucleotide was designed to contain the correct nucleotide in the right position and hybridize with the HPR in helical regions 3 and 4. Due to the effector's preferred complementary binding, the mutated part of the HPR sequence was replaced by the external RNA strand. This complementation yielded the correct HPR fold, restoring activity (Vauléon and Müller 2003; Fig. 13.10a). A similar approach was used by Komatsu et al. (2002). Nonfunctional HPRs due to a loss of base pairing in helix 4 were activated again by the addition of a short oligonucleotide complementary to the nucleotides in the proposed helix. Upon binding of the oligonucleotide effector, helix 4 formed, allowing ribozyme catalysis to proceed (Komatsu et al. 2002; Fig. 13.10b).

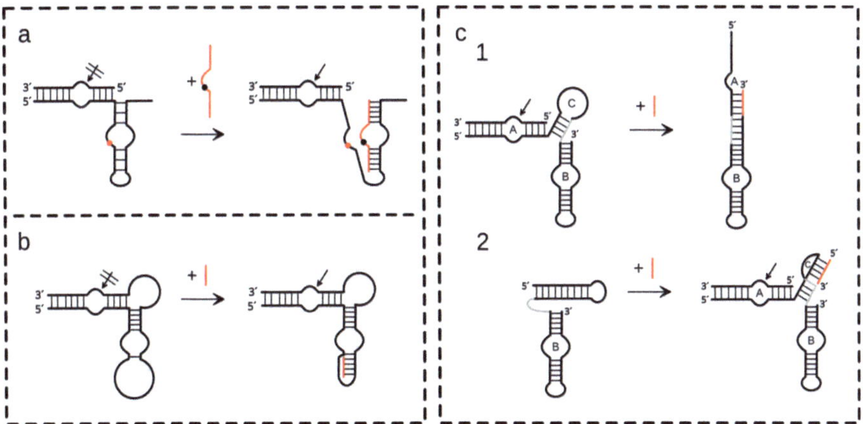

Fig. 13.10 Positive allosteric control of HPR cleavage activity by an effector oligonucleotide. (**a**) Restored activity of HPR with a mismatch (red dot) after addition of an effector oligonucleotide (red) with a correct base (black dot). (**b**) Addition of a small effector oligonucleotide (red) restores the structure of helix 4, and activity is upregulated. (**c**) Design of HPRs that are either repressible (1) or inducible (2) by the addition of an oligonucleotide effector binding to the introduced regulatory domain C. Red: effector oligonucleotide; Grey: base-pairing sequence for domain C

Even more sophisticated is the work by Famulok and coworkers (Najafi-Shoushtari 2004). Their approach includes negative control by stabilizing the ribozyme in an inactive conformation with an oligonucleotide effector. The introduction of a regulatory domain C via a pseudo-three-way junction between helices 2 and 3 resulted in HPRs that, dependent on the specific design, could be either induced or repressed by an external oligonucleotide. The sequence of domain C can form stable alternative structures within the ribozyme, such as an internal stem-loop structure (Fig. 13.10c1) or a pseudo-half-knot (Fig. 13.10c2). The addition of an effector oligonucleotide complementary to domain C of a repressible HPR results in a stable stretched conformation, thereby causing a loss of activity (Fig. 13.10c1). The same effector added to an inducible HPR variant restored activity through the formation of a pseudo-half-knot (Fig. 13.10c2). The design of domain C included complementarity to part of the Trp leader mRNA. Thus, induction or repression of ribozyme activity could sense and monitor the presence of that mRNA and, consequently, its interaction with the protein that binds to and sequesters the ribozyme binding site (the L-tryptophan-activated Trp-RNA-binding attenuation protein, short TRAP).

The effector oligonucleotide can act not only as a simple regulator, but also as a substrate. An example is shown in Sect. 3.4 with the chameleon ribozyme (Zhu et al. 2024). The CHR can be viewed as a ribozyme regulated by an effector oligonucleotide. In this special case, however, rather than being switched on or off, the conformation shifts between two active states, which are triggered by the respective oligonucleotide. First, the oligonucleotide induces the conformational change of the ribozyme, and then it is cleaved as the substrate of the ribozyme reaction.

4.2 Allosteric Control by Small Molecules

In addition to controlling ribozyme reactions through complementary binding of oligonucleotides, the addition of smaller molecules can influence catalytic activity, providing true allosteric control. This requires two different types of functional RNAs—a ribozyme and an aptamer—to be combined into a single molecule. In the resulting RNA, the aptamer unit binds a specific small molecule, and the ribozyme unit acts as the catalytic entity. When the small molecule binds to the aptamer unit, this triggers a conformational change in the ribozyme unit, resulting in the up- or downregulation of its activity. Depending on the specific design, the ribozyme can be allosterically controlled to only be active when a certain molecule is present (positive control) or absent (negative control; Tang and Breaker 1997).

Such a system can be established by fusing the two functional domains either without intersection or with overlap, resulting in an aptazyme or riboswitch. Early work in this field was done by Breaker and coworkers (Tang and Breaker 1997; Soukup and Breaker 1999). Since then, several aptazymes have been developed, most of which are based on the hammerhead ribozyme fused to various aptamers (Araki et al. 1998; Frommer et al. 2015; Felletti and Hartig 2017; Kaloudas et al. 2023). However, other ribozymes have also been used, such as a Diels–Alder

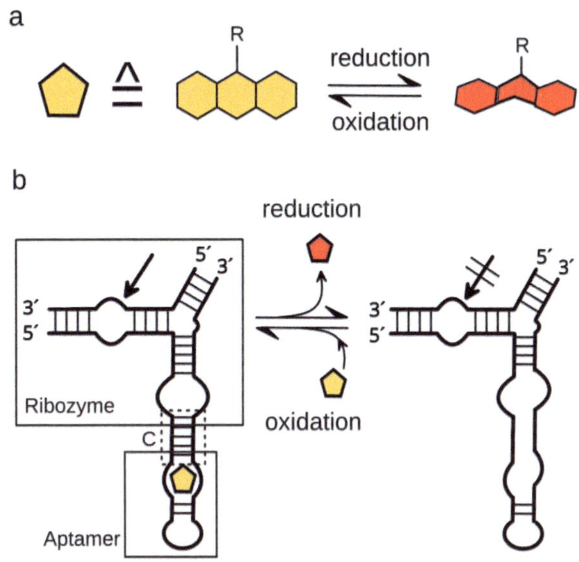

Fig. 13.11 FMN-responsive hairpin ribozyme. (**a**) Change of the FMN shape upon reduction/oxidation. (**b**) Schematic representation of the secondary structure of an FMN-responsive hairpin aptazyme. The binding of oxidized FMN (left) enables cleavage. Upon reduction, FMN cannot bind to the aptamer, and cleavage is repressed. (**c**) communication module connecting the two units. The cleavage site is indicated by an arrow; yellow pentagon: FMN

ribozyme that is allosterically controlled by theophylline (Helm et al. 2005). Most of these ribozymes were discovered through in vitro selection. While these examples are artificial constructs that did not evolve naturally, they demonstrate the potential of RNA diversification in biological systems.

In our laboratory, we used a rational approach to develop a hairpin ribozyme sensitive to flavin mononucleotide (FMN) by integrating the FMN aptamer into the hairpin ribozyme structure (Fig. 13.11). The HPR unit and the aptamer unit are connected by a communication module that transfers the aptamer's structural alteration upon FMN binding into a conformational change in the ribozyme unit. This results in an increase in activity. Since only the oxidized form of FMN can bind to the aptamer domain, the aptazyme can sense the oxidation state of its environment (Strohbach et al. 2006). The aptamer's much lower binding affinity to the reduced form of FMN than the oxidized form derives from the significantly divergent shapes of the two states. The oxidized form of FMN has planar geometry due to its aromatic characteristics. Reduction, however, leads to loss of aromaticity, resulting in a bent shape of the isoalloxazine ring. Since FMN binding by the aptamer is primarily mediated by hydrophobic stacking interactions between FMN and the aptamer's nucleobases, the loss of planarity due to reduction impedes these interactions, preventing effective binding of FMN. We have demonstrated that FMN reduction can be achieved through chemical and electrochemical means. Additionally, we demonstrated electron transport through the aptamer region, marking the inception point of a novel approach that couples FMN reduction with RNA charge transfer (Frommer and Müller 2020).

Alternatively, the ribozyme and aptamer sequences can be fused in tandem as a means of gene regulation. In this scenario, the two sequences must be incorporated into the untranslated region (UTR) of the respective mRNA. In the case of an OFF riboswitch, the aptamer without the ligand exhibits a conformation that impedes the correct folding of the ribozyme unit. This results in inhibited cleavage and, consequently, successful mRNA translation. However, in the presence of the ligand, the ribozyme folds into its correct conformation, cleaving the mRNA in *cis* and disabling translation. For an ON riboswitch, the aforementioned mechanism must function in the opposite direction. The use of different ribozymes for these purposes has been demonstrated in many scenarios (Ogawa and Maeda 2008; Nomura et al. 2013; Strobel et al. 2015; Zhong et al. 2016; Takahashi and Yokobayashi 2019; Yokobayashi 2019; Mustafina et al. 2020, 2021; Shanidze et al. 2020; Spöring et al. 2020a, b; Fang et al. 2023; Kläge et al. 2024).

5 RNA Dynamics in Modern Biochemistry: Riboswitches and RNA Thermometer

The ribozymes mentioned in this chapter represent a fundamental, recurrent principle found in artificial and natural RNAs of different functionalities, sequences, and lengths. The versatility of distinct folding behaviors has great potential to act as a critical regulator in different biological processes. Due to the sensitivity of RNA conformation to its environment, including concentration and species of metal ions, osmotic pressure of the universal solvent water, presence of small metabolites, complementary RNA strands, and temperature, RNA unifies sensing and reacting to a specific stimulus in a single molecule that, in some cases, is only a few nucleotides long. Clearly, in early evolution, this minimalistic system would have been advantageous. Dynamic RNA folding offers rapid, direct reactions in different regulatory dimensions because it unifies sensory and coding components in one molecule. Since RNA is part of both transcription and translation, the two processes of gene expression, it indirectly regulates metabolic processes in living systems without taking part in them. Genes encoding metabolic proteins can only be efficiently expressed when the respective mRNA is correctly transcribed. This is only possible if the mRNA does not fold into a tertiary structure that would impede the RNA polymerase. After transcription, the mRNA must be translated at the ribosome, which is a ribozyme that performs the peptidyl transferase reaction (Lilley 2001; Steitz and Moore 2003). For the effective translation of desired genes, the mRNA must be spliced beforehand, a process catalyzed by RNAs with distinct folding behaviors. After processing, the mRNA must not exhibit repressive tertiary structures that hinder ribosome binding or processivity. Conversely, some distinctively folded conformations can enhance translation of mRNAs that would not otherwise be translated, as seen with viral RNA containing IRESs (Pelletier and Sonenberg 1988; Martinez-Salas et al. 2018). Of course, the ribosomal RNA itself must also fold in a specific way to enable translation. During evolution, ribosomal RNA was extended by expansion segments, which added new layers of regulatory potential (Knorr et al. 2019; Shankar et al. 2020; Krauer et al. 2021; Hariharan et al. 2023;

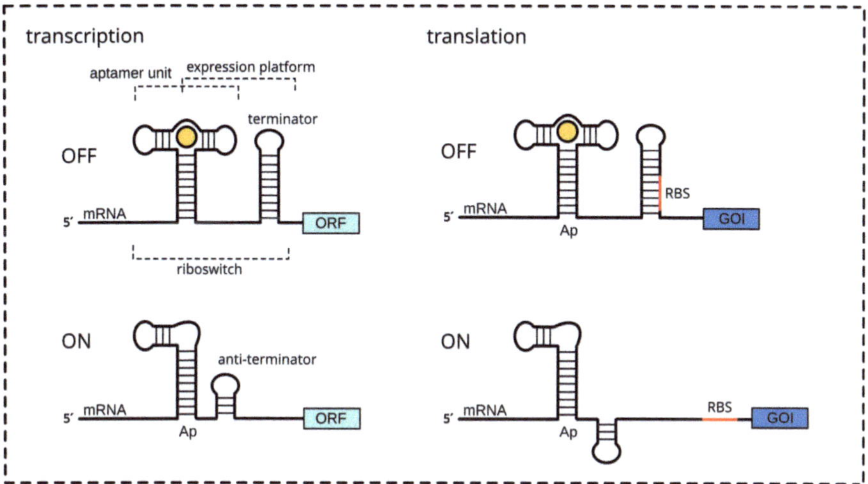

Fig. 13.12 Functional principle of an OFF riboswitch that inhibits transcription (left) or translation (right) in response to binding of its respective ligand. Ap aptamer unit, ORF open reading frame, RBS ribosome binding site, GOI gene of interest

Rauscher et al. 2024; Rauscher and Polacek 2024; Nagai et al. 2025). This process left the core region of the ribosome, which contains the peptidyl transferase center, unperturbed (Petrov et al. 2014, 2015).

Regulatory RNAs can act in *cis,* meaning they act nearby on the same RNA strand, or in *trans,* meaning they act at a greater distance or on another RNA strand. Here, we will not discuss *trans*-acting RNAs, such as antisense RNAs, microRNAs, and siRNAs. For further information, readers can refer to Andersen et al. (2019), Gil and Ulitsky (2020), Shang et al. (2023), Zhang et al. (2023), and Rehman et al. (2024). More interesting with regard to RNA folding dynamics are regulatory RNAs that act in *cis*. These elements are located in the 5′-UTR of the respective mRNA, and each consists of two functional domains: the sensory domain and the expression platform (Fig. 13.12). The sequences of these two domains overlap, resulting in the translation of a conformational change in the sensory domain into a shift in the folding of the expression platform. Depending on the stimulus, the regulatory domain can be an RNA thermometer (RNAT), for example, or an aptamer that binds a specific ligand, such as a metabolite. Binding of the ligand or an alteration in temperature causes the sensory domain to refold into a different tertiary structure, which also affects the expression platform. This results in altered gene expression through transcription termination or progression, or translation initiation or inhibition. Control of splicing has also been observed (Kubodera et al. 2003). The exact process of regulation depends on the type of riboswitch. An OFF riboswitch turns off gene expression in the presence of a stimulus, while an ON riboswitch turns it on (Fig. 13.12).

Riboswitches were briefly introduced in Sect. 4.2, where the general principle of their function was explained in terms of translation regulation by allosteric control.

A riboswitch is an RNA that can specifically bind a small molecule in its aptamer domain. This induces a conformational shift and, consequently, alters transcription or translation activity of the respective mRNA (Serganov and Patel 2007; Breaker 2012, 2018). Many such riboswitches have been discovered and described (Mellin and Cossart 2015; Mehdizadeh Aghdam et al. 2016; Sinumvayo et al. 2018; Kavita and Breaker 2023). For transcription regulation, the mRNA containing the riboswitch must fold into a stem-like terminator structure, which inhibits transcription progression. To enable productive transcription, the RNA must adopt an alternative fold: the anti-terminator structure. Switching between these two states is facilitated by the two riboswitch structures, either with or without the respective ligand (Fig. 13.12, left). The principle of translational control is quite similar. In the repressive structure, the ribosome binding site (RBS) is included in a hybridized stem structure, which renders it inaccessible to the ribosome. Therefore, translation is only possible if the RBS is released from the stem structure (Fig. 13.12, right). The elaborate folding behaviors and dynamics of riboswitches in different states are the subject of thorough investigation (Steinert et al. 2017; Gong et al. 2017; Liao et al. 2023; Bushhouse et al. 2025).

Allosterically controlled ribozymes that act similarly to the described riboswitches are rarely found in nature (Lee et al. 2010; Panchapakesan and Breaker 2021; Kavita and Breaker 2023). One thoroughly studied example is the *glmS* ribozyme (Winkler et al. 2004; Collins et al. 2007).

RNA thermometers (RNATs) demonstrate how RNA folding can influence the evolutionary process (Narberhaus et al. 2006; Shah and Gilchrist 2010; Sharma et al. 2022). At low temperatures, mRNA containing an RNAT folds into a stem-loop structure that occludes the ribosome binding site (RBS). This repressive structure inhibits the expression of some virulence factors, which inhibits the pathogenicity of bacteria at low temperatures (e.g., outside a host organism; Fig. 13.13). However, when the bacteria reach their host and are exposed to its elevated body temperature, the stem-loop structure melts, rendering the RBS accessible. Gene expression of virulence factors then proceeds, enabling infection of the host by the bacteria. The infection, in turn, triggers a host response, thereby leading to evolutionary competition between the host and the bacteria.

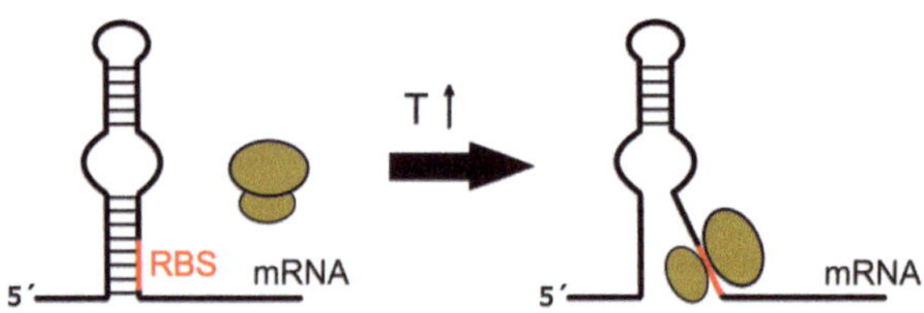

Fig. 13.13 Principle of RNA thermometer function in translation regulation. At low temperatures (left), the ribosome binding site (RBS) is occluded, inhibiting binding of the ribosome. At higher temperatures (~37 °C, right), the stem structure melts and the RBS is accessible

6 Conclusion

The functional and regulatory RNA elements described in this chapter demonstrate how precisely these RNAs fold to efficiently fulfill their functions and control critical mechanisms in living cells. RNA folding has certainly played a crucial role in evolution. Since many RNAs do not depend on proteins, their functional performance probably represents an ancient principle that has persisted throughout evolution. Understanding RNA dynamics is crucial because it demonstrates that evolutionary innovation does not primarily result from error; it can also arise from RNA molecules that fold into multiple functional conformations, change behavior in response to their environment, and interact in networks. Therefore, RNA function can change without a change in sequence. Mutation and selection remain essential, but evolutionary novelty can also emerge from the context-dependent, dynamic behavior that is intrinsic to RNA molecules.

References

Alam S, Grum-Tokars V, Krucinska J, Kundracik ML, Wedekind JE (2005) Conformational heterogeneity at position U37 of an All-RNA hairpin ribozyme with implications for metal binding and the catalytic structure of the S-Turn. Biochemistry 44(44):14396–14408

Andersen RE, Hong SJ, Lim JJ, Cui M, Harpur BA, Hwang E, Delgado RN, Ramos AD, Liu SJ, Blencowe BJ, Lim DA (2019) The long noncoding RNA Pnky is a trans-acting regulator of cortical development in vivo. Dev Cell 49(4):632–642.e7

Araki M, Okuno Y, Hara Y, Sugiura Y (1998) Allosteric regulation of a ribozyme activity through ligand-induced conformational change. Nucleic Acids Res 26(14):3379–3384

Balke D, Zieten I, Strahl A, Müller O, Müller S (2014) Design and characterization of a twin ribozyme for potential repair of a deletion mutation within the oncogenic CTNNB1-ΔS45 mRNA. ChemMedChem 9(9):2128–2137

Balke D, Becker A, Müller S (2016) In vitro repair of a defective EGFP transcript and translation into a functional protein. Org Biomol Chem 14(28):6729–6737

Berzal-Herranz A, Joseph S, Chowrira BM, Butcher SE, Burke JM (1993) Essential nucleotide sequences and secondary structure elements of the hairpin ribozyme. EMBO J 12(6):2567–2573

Birikh KR, Heaton PA, Eckstein F (1997) The structure, function and application of the hammerhead ribozyme. Eur J Biochem 245(1):1–16

Borao S, Ayté J, Hümmer S (2021) Evolution of the early spliceosomal complex—from constitutive to regulated splicing. IJMS 22(22):12444

Breaker RR (2012) Riboswitches and the RNA world. Cold Spring Harb Perspect Biol 4(2): a003566

Breaker RR (2018) Riboswitches and translation control. Cold Spring Harb Perspect Biol 10(11): a032797

Brill J, Nurmi C, Li Y (2024) Elucidating evolutionary mechanisms and variants of the hammerhead ribozyme using *in vitro* selection. Chembiochem 25(21):e202400432

Bushhouse DZ, Fu J, Lucks JB (2025) RNA folding kinetics control riboswitch sensitivity in vivo. Nat Commun 16(1):953

Butcher SE, Burke JM (1994) Structure-mapping of the hairpin ribozyme. J Mol Biol 244(1):52–63

Buzayan JM, Gerlach WL, Bruening G (1986) Non-enzymatic cleavage and ligation of RNAs complementary to a plant virus satellite RNA. Nature 323(6086):349–353

Cech TR (1990) Self-splicing of group I introns. Annu Rev Biochem 59:543–568
Collins JA, Irnov I, Baker S, Winkler WC (2007) Mechanism of mRNA destabilization by the *glmS* ribozyme. Genes Dev 21(24):3356–3368
DasGupta S, Nykiel K, Piccirilli JA (2021) The hammerhead self-cleaving motif as a precursor to complex endonucleolytic ribozymes. RNA 27(9):1017–1024
Deng J, Shi Y, Peng X, He Y, Chen X, Li M, Lin X, Liao W, Huang Y, Jiang T, Lilley DMJ, Miao Z, Huang L (2023) Ribocentre: a database of ribozymes. Nucleic Acids Res 51(D1): D262–D268
Drude I, Vauléon S, Müller S (2007) Twin ribozyme mediated removal of nucleotides from an internal RNA site. Biochem Biophys Res Commun 363(1):24–29
Earnshaw DJ, Hamm ML, Piccirilli JA, Karpeisky A, Beigelman L, Ross BS, Manoharan M, Gait MJ (2000) Investigation of the proposed interdomain ribose zipper in hairpin ribozyme cleavage using 2′-modified nucleosides. Biochemistry 39(21):6410–6421
Ebermann C, Müller S (2025) Engineering of small ribozymes acting on RNA: what is needed to make a new function work with an existing catalyst? Chembiochem 26(10):e202500213
Fang J, Wang J, Wang Y, Liu X, Chen B, Zou W (2023) Ribo-On and Ribo-Off tools using a self-cleaving ribozyme allow manipulation of endogenous gene expression in C. elegans. Commun Biol 6(1):e00471
Fedor MJ (1999) Tertiary structure stabilization promotes hairpin ribozyme ligation. Biochemistry 38(34):11040–11050
Fedor MJ (2000) Structure and function of the hairpin ribozyme. J Mol Biol 297(2):269–291
Felletti M, Hartig JS (2017) Ligand-dependent ribozymes. WIREs RNA 8(2):e1395
Frommer J, Müller S (2020) Reductive charge transfer through an RNA Aptamer. Angew Chem Int Ed 59(51):22999–23004
Frommer J, Appel B, Müller S (2015) Ribozymes that can be regulated by external stimuli. Curr Opin Biotechnol 31:35–41
Galej WP, Toor N, Newman AJ, Nagai K (2018) Molecular mechanism and evolution of nuclear pre-mRNA and group II intron splicing: insights from cryo-electron microscopy structures. Chem Rev 118(8):4156–4176
Gil N, Ulitsky I (2020) Regulation of gene expression by cis-acting long non-coding RNAs. Nat Rev Genet 21(2):102–117
Gong S, Wang Y, Wang Z, Zhang W (2017) Co-transcriptional folding and regulation mechanisms of riboswitches. Molecules 22(7):1169
Hammann C, Luptak A, Perreault J, La Peña M (2012) The ubiquitous hammerhead ribozyme. RNA 18(5):871–885
Hanna R, Doudna JA (2000) Metal ions in ribozyme folding and catalysis. Curr Opin Chem Biol 4(2):166–170
Hariharan N, Ghosh S, Palakodeti D (2023) The story of rRNA expansion segments: finding functionality amidst diversity. WIREs RNA 14(1):e1732
Harris L, Rogers SO (2008) Splicing and evolution of an unusually small group I intron. Curr Genet 54(4):213–222
Helm M, Petermeier M, Ge B, Fiammengo R, Jäschke A (2005) Allosterically activated Diels−Alder catalysis by a ribozyme. J Am Chem Soc 127(30):10492–10493
Hertel KJ, Uhlenbeck OC (1995) The internal equilibrium of the hammerhead ribozyme reaction. Biochemistry 34(5):1744–1749
Hertel KJ, Herschlag D, Uhlenbeck OC (1994) A kinetic and thermodynamic framework for the hammerhead ribozyme reaction. Biochemistry 33(11):3374–3385
Hieronymus R, Müller S (2019) Engineering of hairpin ribozyme variants for RNA recombination and splicing. Ann N Y Acad Sci 1447(1):135–143
Hieronymus R, Müller S (2021) Towards higher complexity in the RNA world: hairpin ribozyme supported RNA recombination. ChemSystemsChem 3(4):e2100003
Hieronymus R, Godehard SP, Balke D, Müller S (2016) Hairpin ribozyme mediated RNA recombination. Chem Commun (Camb) 52(23):4365–4368

Hieronymus R, Zhu J, Müller S (2022) RNA self-splicing by engineered hairpin ribozyme variants. Nucleic Acids Res 50(1):368–377

Ivanov SA, Vauléon S, Müller S (2005) Efficient RNA ligation by reverse-joined hairpin ribozymes and engineering of twin ribozymes consisting of conventional and reverse-joined hairpin ribozyme units. FEBS J 272(17):4464–4474

Joseph S, Berzal-Herranz A, Chowrira BM, Butcher SE, Burke JM (1993) Substrate selection rules for the hairpin ribozyme determined by in vitro selection, mutation, and analysis of mismatched substrates. Genes Dev 7(1):130–138

Kaloudas D, Pavlova N, Penchovsky R (2023) GHOST-NOT and GHOST-YES: two programs for generating high-speed biosensors with randomized oligonucleotide binding sites with NOT or YES Boolean logic functions based on experimentally validated algorithms. J Biotechnol 373: 82–89

Kavita K, Breaker RR (2023) Discovering riboswitches: the past and the future. Trends Biochem Sci 48(2):119–141

Kläge D, Müller E, Hartig JS (2024) A comparative survey of the influence of small self-cleaving ribozymes on gene expression in human cell culture. RNA Biol 21(1):1–11

Knorr AG, Schmidt C, Tesina P, Berninghausen O, Becker T, Beatrix B, Beckmann R (2019) Ribosome–NatA architecture reveals that rRNA expansion segments coordinate N-terminal acetylation. Nat Struct Mol Biol 26(1):35–39

Komatsu Y, Nobuoka K, Karino-Abe N, Matsuda A, Ohtsuka E (2002) In vitro selection of hairpin ribozymes activated with short oligonucleotides. Biochemistry 41(29):9090–9098

Krauer N, Rauscher R, Polacek N (2021) tRNA synthetases are recruited to yeast ribosomes by rRNA expansion segment 7L but do not require association for functionality. ncRNA 7(4):73

Kubodera T, Watanabe M, Yoshiuchi K, Yamashita N, Nishimura A, Nakai S, Gomi K, Hanamoto H (2003) Thiamine-regulated gene expression of *Aspergillus oryzae thiA* requires splicing of the intron containing a riboswitch-like domain in the 5′-UTR. FEBS Lett 555(3):516–520

La Peña M, Gago S, Flores R (2003) Peripheral regions of natural hammerhead ribozymes greatly increase their self-cleavage activity. EMBO J 22(20):5561–5570

Le Vay K, Song EY, Ghosh B, Tang T-YD, Mutschler H (2021) Enhanced ribozyme-catalyzed recombination and oligonucleotide assembly in peptide-RNA condensates. Angew Chem Int Ed 60(50):26096–26104

Lee ER, Baker JL, Weinberg Z, Sudarsan N, Breaker RR (2010) An allosteric self-splicing ribozyme triggered by a bacterial second messenger. Science 329(5993):845–848

Lehmann K, Schmidt U (2003) Group II introns: structure and catalytic versatility of large natural ribozymes. Crit Rev Biochem Mol Biol 38(3):249–303

Liao T-W, Huang L, Wilson TJ, Ganser LR, Lilley DMJ, Ha T (2023) Linking folding dynamics and function of SAM/SAH riboswitches at the single molecule level. Nucleic Acids Res 51(17): 8957–8969

Lilley DMJ (2001) The ribosome functions as a ribozyme. Chembiochem 2(1):31–35

Lilley DMJ (2011) Mechanisms of RNA catalysis. Philos Trans R Soc B 366(1580):2910–2917

Martick M, Scott WG (2006) Tertiary contacts distant from the active site prime a ribozyme for catalysis. Cell 126(2):309–320

Martinez-Salas E, Francisco-Velilla R, Fernandez-Chamorro J, Embarek AM (2018) Insights into structural and mechanistic features of viral IRES elements. Front Microbiol 8:2629

Mehdizadeh Aghdam E, Hejazi MS, Barzegar A (2016) Riboswitches: from living biosensors to novel targets of antibiotics. Gene 592(2):244–259

Mellin JR, Cossart P (2015) Unexpected versatility in bacterial riboswitches. Trends Genet 31(3): 150–156

Misra VK, Draper DE (1998) On the role of magnesium ions in RNA stability. Biopolymers 48(2–3):113–135

Müller U (2017) Design and experimental evolution of trans-splicing group I intron ribozymes. Molecules 22(1):75

Müller S, Appel B, Krellenberg T, Petkovic S (2012) The many faces of the hairpin ribozyme. IUBMB Life 64(1):36–47
Murray JB, Seyhan AA, Walter NG, Burke JM, Scott WG (1998) The hammerhead, hairpin and VS ribozymes are catalytically proficient in monovalent cations alone. Chem Biol 5(10):587–595
Mustafina K, Fukunaga K, Yokobayashi Y (2020) Design of Mammalian ON-riboswitches based on tandemly fused Aptamer and ribozyme. ACS Synth Biol 9(1):19–25
Mustafina K, Nomura Y, Rotrattanadumrong R, Yokobayashi Y (2021) Circularly-permuted pistol ribozyme. ACS Synth Biol 10(8):2040–2048
Mutschler H, Wochner A, Holliger P (2015) Freeze–thaw cycles as drivers of complex ribozyme assembly. Nature Chem 7(6):502–508
Nagai R, Milam OL, Niwa T, Howell WJ, Best JA, Yoshida H, Freeburg CD, Koomen JM, Fujii K (2025) Ribosomal expansion segment contributes to translation fidelity via N-terminal processing of ribosomal proteins. Nucleic Acids Res 53(10):gkaf448
Nahas MK, Wilson TJ, Hohng S, Jarvie K, Lilley DMJ, Ha T (2004) Observation of internal cleavage and ligation reactions of a ribozyme. Nat Struct Mol Biol 11(11):1107–1113
Najafi-Shoushtari SH (2004) Sensing complex regulatory networks by conformationally controlled hairpin ribozymes. Nucleic Acids Res 32(10):3212–3219
Narberhaus F, Waldminghaus T, Chowdhury S (2006) RNA thermometers. FEMS Microbiol Rev 30(1):3–16
Nomura Y, Zhou L, Miu A, Yokobayashi Y (2013) Controlling mammalian gene expression by allosteric hepatitis delta virus ribozymes. ACS Synth Biol 2(12):684–689
Ogawa A, Maeda M (2008) An artificial aptazyme-based riboswitch and its cascading system in E. Coli. Chembiochem 9(2):206–209
Panchapakesan SSS, Breaker RR (2021) The case of the missing allosteric ribozymes. Nat Chem Biol 17(4):375–382
Paudel BP, Rueda D (2014) Molecular crowding accelerates ribozyme docking and catalysis. J Am Chem Soc 136(48):16700–16703
Pelletier J, Sonenberg N (1988) Internal initiation of translation of eukaryotic mRNA directed by a sequence derived from poliovirus RNA. Nature 334(6180):320–325
Peng H, Latifi B, Müller S, Lupták A, Chen IA (2021) Self-cleaving ribozymes: substrate specificity and synthetic biology applications. RSC Chem Biol 2(5):1370–1383
Petkovic S, Müller S (2013) RNA self-processing. FEBS Lett 587(15):2435–2440
Petkovic S, Badelt S, Block S, Flamm C, Delcea M, Hofacker I, Müller S (2015) Sequence-controlled RNA self-processing. RNA 21(7):1249–1260
Petrov AS, Bernier CR, Hsiao C, Norris AM, Kovacs NA, Waterbury CC, Stepanov VG, Harvey SC, Fox GE, Wartell RM, Hud NV, Williams LD (2014) Evolution of the ribosome at atomic resolution. Proc Natl Acad Sci USA 111(28):10251–10256
Petrov AS, Gulen B, Norris AM, Kovacs NA, Bernier CR, Lanier KA, Fox GE, Harvey SC, Wartell RM, Hud NV, Williams LD (2015) History of the ribosome and the origin of translation. Proc Natl Acad Sci USA 112(50):15396–15401
Pieper S, Vauléon S, Müller S (2007) RNA self-processing towards changed topology and sequence oligomerization. bchm 388(7):743–746
Pinard R (2001) Functional involvement of G8 in the hairpin ribozyme cleavage mechanism. EMBO J 20(22):6434–6442
Pinard R, Lambert D, Walter NG, Heckman JE, Major F, Burke JM (1999) Structural basis for the guanosine requirement of the hairpin ribozyme. Biochemistry 38(49):16035–16039
Pley HW, Flaherty KM, McKay DB (1994) Three-dimensional structure of a hammerhead ribozyme. Nature 372(6501):68–74
Prody GA, Bakos JT, Buzayan JM, Schneider IR, Bruening G (1986) Autolytic processing of dimeric plant virus satellite RNA. Science 231(4745):1577–1580
Rauscher R, Polacek N (2024) Ribosomal RNA expansion segments and their role in ribosome biology. Biochem Soc Trans 52(3):1317–1325

Rauscher R, Eggers C, Dimitrova-Paternoga L, Shankar V, Rosina A, Cristodero M, Paternoga H, Wilson DN, Leidel SA, Polacek N (2024) Evolving precision: rRNA expansion segment 7S modulates translation velocity and accuracy in eukaryal ribosomes. Nucleic Acids Res 52(7): 4021–4036

Rehman SU, Ullah N, Zhang Z, Zhen Y, Din A-U, Cui H, Wang M (2024) Recent insights into the functions and mechanisms of antisense RNA: emerging applications in cancer therapy and precision medicine. Front Chem 11:1335330

Rogozin IB, Carmel L, Csuros M, Koonin EV (2012) Origin and evolution of spliceosomal introns. Biol Direct 7(1):11

Rupert PB, Ferré-D'Amaré AR (2001) Crystal structure of a hairpin ribozyme–inhibitor complex with implications for catalysis. Nature 410(6830):780–786

Salzman J, Gawad C, Wang PL, Lacayo N, Brown PO (2012) Circular RNAs are the predominant transcript isoform from hundreds of human genes in diverse cell types. PLoS One 7(2):e30733

Schnabl J, Sigel RKO (2010) Controlling ribozyme activity by metal ions. Curr Opin Chem Biol 14(2):269–275

Schultes EA, Bartel DP (2000) One sequence, two ribozymes: implications for the emergence of new ribozyme folds. Science 289(5478):448–452

Scott WG, Horan LH, Martick M (2013) The hammerhead ribozyme. In: Progress in molecular biology and translational science. https://doi.org/10.1016/b978-0-12-381286-5.00001-9

Serganov A, Patel DJ (2007) Ribozymes, riboswitches and beyond: regulation of gene expression without proteins. Nat Rev Genet 8(10):776–790

Shah P, Gilchrist MA (2010) Is thermosensing property of RNA thermometers unique? PLoS One 5(7):e11308

Shang R, Lee S, Senavirathne G, Lai EC (2023) microRNAs in action: biogenesis, function and regulation. Nat Rev Genet 24(12):816–833

Shanidze N, Lenkeit F, Hartig JS, Funck D (2020) A theophylline-responsive riboswitch regulates expression of nuclear-encoded genes. Plant Physiol 182(1):123–135

Shankar V, Rauscher R, Reuther J, Gharib WH, Koch M, Polacek N (2020) rRNA expansion segment 27Lb modulates the factor recruitment capacity of the yeast ribosome and shapes the proteome. Nucleic Acids Res 48(6):3244–3256

Sharma A, Alajangi HK, Pisignano G, Sood V, Singh G, Barnwal RP (2022) RNA thermometers and other regulatory elements: diversity and importance in bacterial pathogenesis. WIREs RNA 13(5):e1711

Sinumvayo JP, Zhao C, Tuyishime P (2018) Recent advances and future trends of riboswitches: attractive regulatory tools. World J Microbiol Biotechnol 34(11):171

Soukup GA, Breaker RR (1999) Engineering precision RNA molecular switches. Proc Natl Acad Sci USA 96(7):3584–3589

Spöring M, Boneberg R, Hartig JS (2020a) Aptamer-mediated control of polyadenylation for gene expression regulation in mammalian cells. ACS Synth Biol 9(11):3008–3018

Spöring M, Finke M, Hartig JS (2020b) Aptamers in RNA-based switches of gene expression. Curr Opin Biotechnol 63:34–40

Stage-Zimmermann TK, Uhlenbeck OC (1998) Hammerhead ribozyme kinetics. RNA 4(8): 875–889

Steinert H, Sochor F, Wacker A, Buck J, Helmling C, Hiller F, Keyhani S, Noeske J, Grimm S, Rudolph MM, Keller H, Mooney RA, Landick R, Suess B, Fürtig B, Wöhnert J, Schwalbe H (2017) Pausing guides RNA folding to populate transiently stable RNA structures for riboswitch-based transcription regulation. elife 6:e21297

Steitz TA, Moore PB (2003) RNA, the first macromolecular catalyst: the ribosome is a ribozyme. Trends Biochem Sci 28(8):411–418

Strobel B, Klauser B, Hartig JS, Lamla T, Gantner F, Kreuz S (2015) Riboswitch-mediated attenuation of transgene cytotoxicity increases adeno-associated virus vector yields in HEK-293 cells. Mol Ther 23(10):1582–1591

Strohbach D, Novak N, Müller S (2006) Redox-Active Riboswitching. Angew Chem Int Ed 45(13): 2127–2129

Takahashi K, Yokobayashi Y (2019) Reversible gene regulation in mammalian cells using riboswitch-engineered vesicular stomatitis virus vector. ACS Synth Biol 8(9):1976–1982

Tang J, Breaker RR (1997) Rational design of allosteric ribozymes. Chem Biol 4(6):453–459

Usman N, Beigelman L, McSwiggen JA (1996) Hammerhead ribozyme engineering. Curr Opin Struct Biol 6(4):527–533

Vauléon S, Müller S (2003) External regulation of hairpin ribozyme activity by an oligonucleotide effector. Chembiochem 4(2–3):220–224

Vauléon S, Ivanov SA, Gwiazda S, Müller S (2005) Site-specific fluorescent and affinity labelling of RNA by using a small engineered twin ribozyme. Chembiochem 6(12):2158–2162

Vlassov AV, Johnston BH, Kazakov SA (2005) Hairpin ribozyme-catalyzed ligation in water-alcohol solutions. Oligonucleotides 15(4):303–309

Vosseberg J, Snel B (2017) Domestication of self-splicing introns during eukaryogenesis: the rise of the complex spliceosomal machinery. Biol Direct 12(1):30

Walter NG, Burke JM, Millar DP (1999) Stability of hairpin ribozyme tertiary structure is governed by the interdomain junction. Nat Struct Biol 6(6):544–549

Welz R, Schmidt C, Müller S (2001) Spermine supports catalysis of hairpin ribozyme variants to differing extents. Biochem Biophys Res Commun 283(3):648–654

Welz R, Bossmann K, Klug C, Schmidt C, Fritz H, Müller S (2003) Site-directed alteration of RNA sequence mediated by an engineered twin ribozyme. Angew Chem Int Ed 42(21):2424–2427

Will CL, Luhrmann R (2011) Spliceosome structure and function. Cold Spring Harb Perspect Biol 3(7):a003707

Wilson TJ, Nahas M, Araki L, Harusawa S, Ha T, Lilley DMJ (2007) RNA folding and the origins of catalytic activity in the hairpin ribozyme. Blood Cells Mol Dis 38(1):8–14

Winkler WC, Nahvi A, Roth A, Collins JA, Breaker RR (2004) Control of gene expression by a natural metabolite-responsive ribozyme. Nature 428(6980):281–286

Yamagami R, Sieg JP, Bevilacqua PC (2021) Functional roles of chelated magnesium ions in RNA folding and function. Biochemistry 60(31):2374–2386

Yokobayashi Y (2019) Aptamer-based and aptazyme-based riboswitches in mammalian cells. Curr Opin Chem Biol 52:72–78

Zhang J, Chen B, Gan C, Sun H, Zhang J, Feng L (2023) A comprehensive review of small interfering RNAs (siRNAs): mechanism, therapeutic targets, and delivery strategies for cancer therapy. IJN 18:7605–7635

Zhong G, Wang H, Bailey CC, Gao G, Farzan M (2016) Rational design of aptazyme riboswitches for efficient control of gene expression in mammalian cells. elife 5

Zhu J, Hieronymus R, Müller S (2023) A hairpin ribozyme derived spliceozyme. ChemBioChem 24(13):475

Zhu J, Dierks D, Möller C, Balke D, Müller S (2024) Two RNA folds from one sequence. Angew Chem Int Ed Engl 63(37):e202409047

Zhuang X, Kim H, Pereira MJB, Babcock HP, Walter NG, Chu S (2002) Correlating structural dynamics and function in single ribozyme molecules. Science 296(5572):1473–1476

Chapter 14
'Location, Location, Location': Cis-Acting Antisense Transcripts in Evolution and Disease

Surar Al-Hashimi, Suwalak Chitchorean, Mouna Alhouli, Shaymaa Sadeq, and Andreas Werner

Abstract Natural antisense transcripts (NATs) are noncoding, processed RNAs that originate from the opposite strand of protein-coding genes. They can be considered as long noncoding RNAs (lncRNAs) that originate from the same genomic locus as their potential regulatory target and are therefore particularly versatile regulators of gene expression. Their potential to form structures such as hairpins, R-loops, triple helices, and RNA-RNA hybrids with the related sense transcript provides a myriad of opportunities to modulate the processing of their cognate protein-coding transcript and to bind to chromatin-modifying enzymes and transcription factors that eventually shape the activity of affected genes. Moreover, the close proximity of the effector (NAT) and its target (the related protein-coding sense gene) minimizes off-target effects and enables condensate formation to regulate sense transcription. The fact that NATs are generally processed and modified with mRNA-like features expands their regulatory reach to the cytoplasm, where examples of NATs regulating mRNA stability and translation efficiency have been reported. Considering their versatility, it comes as no surprise that an ever-increasing number of NATs are associated with development and disease. Conversely, bidirectional transcription can have detrimental effects on adjacent genes. For example, double-stranded RNA (dsRNA) is highly immunogenic or, as in *Caenorhabditis elegans*, may lead to transgenerational gene silencing. These challenges may constitute evolutionary pressure to develop mitigating strategies.

S. Al-Hashimi
Department of Internal Medicine, College of Medicine, University of Misan, Misan, Iraq

S. Chitchorean
Department of Microbiology, Faculty of Medicine, Khon Kaen University, Khon Kaen, Thailand

M. Alhouli
Department of Biological Sciences, College of Science, Kuwait University, Safat, Kuwait

S. Sadeq · A. Werner (✉)
Biosciences Institute, Newcastle University, Newcastle upon Tyne, UK
e-mail: andreas.werner@newcastle.ac.uk

G. Witzany (ed.), *Integrative Theory of Evolution - Aspects and Insights*,
https://doi.org/10.1007/978-3-032-24295-2_14

1 Introduction

The occurrence of antisense transcripts was established more than 40 years ago in prokaryotes and eukaryotes (Inouye 1988). The early reports were met with skepticism, especially when potential biological functions were questioned, and NATs were often perceived as transcriptional noise. Even after transcriptomics approaches revealed the genome-wide scale of bidirectional transcription, reservations persisted concerning the functional relevance of NATs (Timmons and Good 2006). Only recently has the perception of long noncoding RNAs (lncRNAs) from "transcriptional noise" changed to "versatile and essential regulators of gene expression." In fact, NATs are nowadays often characterized and reported as a subclass of lncRNAs without emphasizing the genomic configuration antisense to a protein-coding sense transcript (Mattick et al. 2023).

Bidirectional transcription that gives rise to NATs is prevalent in all cells and occurs at a significant scale. NATs are usually expressed at lower levels than the corresponding sense transcripts; generally, the difference is about ten-fold. There are species-specific variations in how the resulting NATs affect the expression of the related protein-coding sense genes, as well as how specific cells or entire organisms mitigate potentially harmful consequences of NATs (Werner et al. 2024).

Significant differences in the biological roles of NATs and the related regulatory mechanisms are observed between bacterial and eukaryotic cells. For example, antisense transcripts in bacteria regulate plasmid numbers and transposon mobility and improve fitness through resistance against bacteriophages and nutrient acquisition, among other functions (Millar and Raghavan 2020). Mechanistically, this is mostly achieved through kissing oligonucleotide complexes or double-stranded RNA (dsRNA) formation that affects transcription or translation of the target gene (Georg and Hess 2011). The biology of regulatory NATs in eukaryotes, the main focus of this article, is more complex and generally better understood than in prokaryotes.

2 Definition and Nomenclature of NATs

Natural antisense transcripts (NATs) are non-protein-coding RNAs transcribed from the strand opposite protein-coding genes. Descriptors to specify the architecture of such genes are often added, including convergent/divergent, head-to-head/tail-to-tail/full overlap, or with/without exonic overlap (Fig. 14.1). Importantly, these various subgroups potentially share common regulatory mechanisms and/or biological roles (Werner et al. 2024). For example, head-to-head and divergent sense/antisense transcripts often derive from a shared promoter, and transcription of sense and antisense RNAs is co-regulated (Statello et al. 2021). Conversely, tail-to-tail arrangements and convergent transcription are likely associated with posttranscriptional regulation and may predominantly affect splicing, localization, or stability of

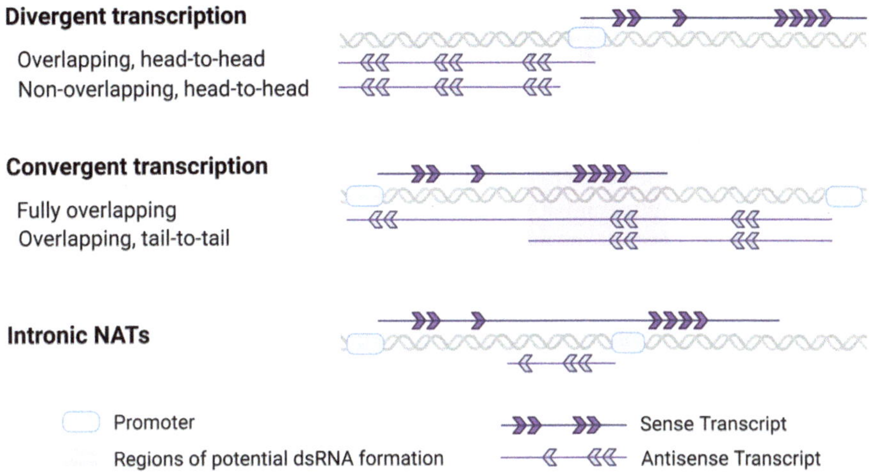

Fig. 14.1 Different types of natural antisense transcripts. Sense and antisense transcripts can be driven from a joint bidirectional promoter (blue shape, top panel). The transcripts either share complementary sequences (overlapping, pink rectangles) or do not (non-overlapping). Divergent transcription results in a so-called head-to-head conformation of sense and antisense RNA. Antisense expression causes transcriptional interference at the common promoter that results in up- or downregulated expression of the sense RNA. The middle panel represents convergent transcription, where sense- and antisense transcripts are driven by separate promoters localized downstream of each other (blue shapes). Depending on the length of the NAT, it can be fully overlapping or complementary is limited to the 3′ end (tail-to-tail). Frequently, NATs can be embedded in a single intron of the sense gene (intronic NAT, lower panel)

the related sense transcript. Genome-wide studies assessing transcription factor binding sites in human cells (Cawley et al. 2004) and antisense transcripts in mice (Katayama et al. 2005) both indicate that antisense transcription is about equally likely to originate at the 5′ or the 3′ end of protein-coding genes. NATs that share an exonic overlap with the related sense transcript may involve an essential dsRNA intermediate in their regulatory mechanism. However, dsRNA is highly immunogenic, and sense-antisense hybridization at a significant scale may only occur in immune-privileged cells, for example, in germ cells (see below) (Sadeq et al. 2021).

3 The Scale of Antisense Transcription in Eukaryotes

First efforts to characterize the antisense transcriptome of eukaryotic cells in the early phase of the genomic era involved data mining for complementary RNAs in the rapidly growing sequence repositories (Shendure and Church 2002). Soon afterwards, the assembly of various reference genomes allowed a more comprehensive and accurate description of antisense transcription in various model organisms (Katayama et al. 2005). A consensus emerged that anywhere between 15% and

70% of genes express transcripts that map to the opposite strand of the protein-coding strand, with only few exceptions (see below). The specific scale essentially depends on the interrogated tissue, experimental conditions, and the bioinformatic pipeline used to analyze the sequencing data. For example, in yeast, 15% (*Saccharomyces cerevisiae*) and 32% (*Schizosaccharomyces pombe*) (Yassour et al. 2010; Ni et al. 2010) of genes were found to be bi-directionally transcribed, whereas up to 72% were found in mice and 40–50% in humans (Katayama et al. 2005; Kiyosawa et al. 2003). Importantly, in more complex organisms, the most comprehensive expression of NATs is found in the testis, specifically in developing germ cells (Werner et al. 2007).

Whether NATs are phylogenetically conserved has long been debated, and if only nucleotide sequence similarities are concerned, conservation is indeed minimal (Veeramachaneni et al. 2004; Wood et al. 2013). More refined strategies that include genomic position, synteny, and expression patterns (using pipelines established to assess the phylogenicity of lncRNAs) revealed lineage specificity and expression patterns that matched the conservation of protein-coding genes (Hezroni et al. 2015). Remarkably, short stretches of highly conserved nucleotides are often found in lncRNAs (and NATs), pointing to the functional or structural relevance of sequence motifs within a particular transcript (Ruiz-Orera and Albà 2019).

4 Transcriptional Regulation

Since the discovery of NATs, a myriad of studies have investigated their potential biological roles and established mechanisms by which NATs regulate the expression of the related sense transcripts (Faghihi and Wahlestedt 2009; Werner et al. 2024). The focus of this paragraph is on antisense transcription initiating in the broader promoter region of the protein-coding sense transcript, resulting in divergent, head-to-head sense-antisense RNAs. The proximity of the transcription start sites in the sense and antisense directions leads to transcriptional interference, where transcriptional activity in one direction affects the output from the opposite strand. The regulatory impact of NATs can manifest as altered chromatin marks, structural features such as R-loops or dsRNA, or condensate formation as detailed below (Fig. 14.2). The result of antisense transcription consists of reduced or increased expression of the sense gene, depending on the nature of the altered chromatin marks and the recruitment of repressing or enhancing transcription factors (He et al. 2008).

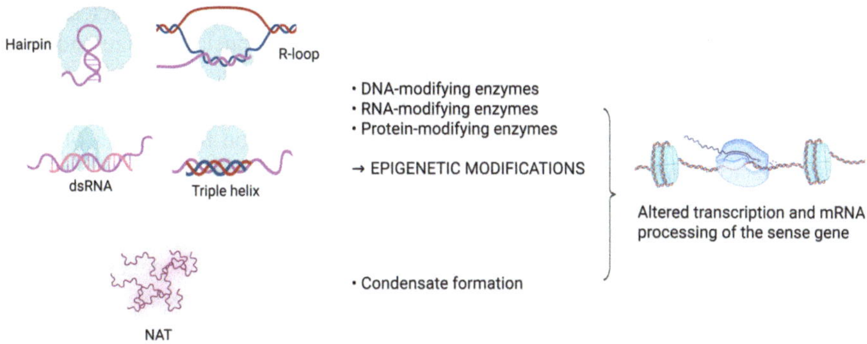

Fig. 14.2 Transcriptional interference. NATs can adopt distinct structural features, hairpins, R-loops, dsRNA, or triple helices that recruit protein and chromatin-modifying enzymes. Altered DNA methylation and histone modifications affect the transcription of the cognate sense gene; both up- and downregulation can be observed. In addition, NAT expression can directly modulate Pol II processivity. This can either be achieved by modifying Pol II or by inducing condensate formation to sequester transcription factors (lower panel)

5 Transcriptional Interference in Yeast

The mechanisms of antisense transcription and its interference with the expression of the sense gene have been extensively studied in the two model yeast species *S. cerevisiae* and *S. pombe*. The general principles can potentially be extrapolated to more complex organisms such as plants and mammals.

In yeast, nucleosome-deprived promoter regions offer transcription factor binding sites that lead to the recruitment of RNA polymerase II (Kaplan et al. 2009). The boundaries of nucleosome-free regions are demarcated by a single nucleosome containing the histone variant H2A.Z that also signals the direction of transcription. Bidirectional promoters contain an H2A.Z variant at either end, whereas largely directional promoters contain one H2A.Z variant close to the transcription start site (Bagchi et al. 2020). Both features, nucleosome depletion and the presence of histone H2A.Z, were assessed genome-wide and correlated with transcriptome data, giving a detailed account of expressed antisense (and sense) transcripts (Gu et al. 2015). Accordingly, 20% of protein-coding genes contain an H2A.Z peak in the 3′ region, indicative of antisense transcription, compared to 49% in the 5′ region.

DNA-binding proteins, including transcription factors in promoter regions, pose obstacles to RNA Pol II and slow down or halt its progression (Hao et al. 2023). Moreover, active transcription leads to modification in DNA packaging and histone content (Kulaeva et al. 2007). The chromatin remodeling capacity of active RNA Pol II and the concept of roadblocks may imply a model of how antisense transcription interferes with the expression of the sense gene. Accordingly, RNA PolII interacts with histone methyl transferases that methylate H3K4 and H3K36, which in turn promote histone deacetylation and more condensed chromatin, in effect sealing the

DNA for spurious transcription initiation. Upon progression into a nucleosome-depleted sense promoter, a competition unfolds between the nucleosome remodeling activity of RNA Pol II and transcription termination promoted by roadblocks of transcription factors and the transcription preinitiation complex (Novačić et al. 2022). The outcome of the conflict will primarily depend on the number of transcription factors and how tightly they are bound to the promoter region. Accordingly, lowly expressed genes are particularly affected by antisense transcription, whereas the level of highly expressed genes remains unchanged (Nevers et al. 2018). In yeast, this applies to genes that respond to environmental changes, such as different carbon sources or stress, that preferentially express antisense transcripts and would benefit from an on-off regulation (Xu et al. 2011). More generally, NATs may provide a barrier to suppress the spurious transcription of protein-coding genes, enhanced by bidirectional promoters and the small genome size.

The extent to which the scenario established in yeast applies to more complex organisms with larger genomes is unclear, but evidence gathered with mammalian cell lines suggests that at least basic mechanistic principles are conserved. Accordingly, the knock-out of several lncRNAs in antisense orientation affected the size and frequency of bursts associated with the transcription of the sense gene. Remarkably, both increased and decreased signals were observed, in line with earlier findings that NATs can stimulate as well as suppress sense transcription (He et al. 2008). These changes are also reflected by altered chromatin signatures. Furthermore, catalytically dead CRISPR-Cas (dCas) protein is being used to position artificial roadblocks and has proven to be an efficient inhibitor of transcription (Hao et al. 2023). However, comparative studies into the mechanism of transcriptional interference between sense and antisense transcripts with a focus on homologous, bidirectionally transcribed loci in yeast and humans (or metazoans in general), have not been done.

Both genome-wide investigations and experiments focusing on a specific bidirectionally transcribed gene have associated antisense transcription with the recruitment of chromatin modifiers, predominantly DNA methyltransferases and histone modifiers that eventually affect transcriptional activity at the particular locus. This is reflected by either accessible chromatin with reduced DNA methylation and/or modifications at histone H3, predominantly at lysines 4, 9, and 27. Promoters of actively transcribed genes display trimethylation of H3K4 (H3K4me3) and acetylation of H3K9 and H3K27 (H3K9Ac, H3K27Ac), whereas repressed promoters feature trimethylation of H3K9 and H3K27 (H3K9me3, H3K27me3). Enzymes that are recruited either directly or indirectly by NATs include the histone acetyltransferase EP300 (Liu et al. 2022; Wang et al. 2019), the histone demethylase UTX-1 (Dong et al. 2019), or the histone methyltransferases G9a and EZH2 (Li et al. 2018). Likewise, DNA methyltransferases have been shown to interact with NATs and become either confined to the promoter region of the sense gene or sequestered elsewhere, leading to inhibitory or stimulatory effects, respectively (Heilmann et al. 2017; Tufarelli et al. 2003; Zinad et al. 2022).

A number of recent studies have also scrutinized structural features associated with antisense transcription that affect the transcriptional activity of the locus, including double-stranded RNA formed between sense and antisense strands, R-loops (RNA/DNA hybrids), or triple helix formation between the NAT and DNA. The formation of these distinct structures precedes chromatin changes and helps recruit the chromatin modifiers (Postepska-Igielska et al. 2015; Ryczek et al. 2025). The timing would support a model where antisense-induced RNA and DNA structures create binding sites for DNA and histone-modifying enzymes that eventually shape the epigenetic signature of the promoter (Fig. 14.2). However, RNA-RNA and DNA-RNA hybrids are potentially toxic and need to be resolved quickly (Camino et al. 2023).

6 Condensate Formation

An additional intriguing possibility of NATs interfering with the transcription of the sense gene has emerged recently, related to the role of noncoding RNAs in promoting phase separation to form a condensate with Pol II and protein factors to stimulate transcriptional activity. Because of the close spatial relation between sense- and antisense transcripts, this mechanism appears to be a particularly relevant contribution of NATs to the regulation of the sense gene.

Biomolecular condensates are membrane-less compartments that accumulate protein and RNA components in a thermodynamically driven process. Whereas the nucleolus, P-bodies, stress granules, or paraspeckles, for example, are well-known examples of condensates in the nucleus and the cytoplasm of eukaryotic cells, it is less well acknowledged that phase separation may also tune transcriptional activity in bacteria and eukaryotic cells alike (Ladouceur et al. 2020; Pei et al. 2025). Intrinsically disordered regions (IDRs) of proteins and the flexible backbone of RNA offer a wealth of low-energy interactions that allow phase separation and the formation of Ribo-protein condensates. The process is not only driven by the concentration of the relevant RNA and protein molecules but also depends on RNA/protein modifications, ion composition, and temperature.

Because of their relative size, RNA molecules can act as scaffolds, whereas bound proteins are known as clients that are attracted through interactions with the scaffold (Somasundaram et al. 2022). Intriguingly, the impact of RNA concentration on condensate formation shows a bimodal distribution, with low concentrations favoring phase separation, whereas high concentrations disrupt the condensate (Natarajan et al. 2023). In the context of NATs, a scenario is conceivable where low levels of antisense transcription favor condensate formation around the sense promoter, which then leads to efficient sense transcription. Indeed, increased levels of NATs preceding enhanced sense expression have been reported in mammalian cells (Zinad et al. 2022).

7 RNA Maturation

The presence of a NAT offers various possibilities to interact with the sense mRNA and affect its maturation. In fact, regulatory interference of NATs on the processing of the primary sense transcript has been demonstrated at the level of splicing, polyadenylation, and RNA modification (Werner et al. 2024) and potentially involves the formation of sense/antisense RNA duplexes. Such RNA hybrids can be experimentally detected by RNase A protection or with the help of a dsRNA-specific antibody (Ling et al. 2011; Portal et al. 2015; Michael et al. 2011; Li et al. 2022a). DsRNA is a potent stimulator of innate immunity, especially when present in the cytoplasm. Hence, sense-antisense hybrids are expected to occur in the nucleus or only at very low levels in the cytoplasm.

The first step where an antisense transcript can modulate the processing of the sense transcript is splicing. Steric hindrance of splicing factor binding leads to alternative splicing of the sense RNA in response to altered expression of the NAT. Accordingly, the synthesis of specific alternative splice isoforms is directly affected by the expression level of the related antisense transcripts (Morrissy et al. 2011). The alternative splice events are more likely to occur in regions of sense/antisense complementarity, supporting a mechanism that involves sense-antisense interaction. The study was done in lymphoblastoid cell lines, though further comparative analyses have established a correlation between antisense transcription and splicing in other metazoan species (Morrissy et al. 2011).

These bioinformatic findings are supported by investigations into specific bidirectionally transcribed genes where the NAT influences the relative expression of sense mRNA splice forms (Chawla and Lazar 1993; Huang et al. 2017; Ma et al. 2021; Niehus et al. 2019). The effect of NATs on alternative polyadenylation is less well documented. Nevertheless, Shen and coworkers found a significant correlation in datasets for NATs and alternative polyadenylation in human tissues when comparing the two features. They went on to investigate a specific sense-antisense transcript pair, *RNASEH2C-KAT5*. The over-expression of *KAT5* provoked the use of a distal polyadenylation signal, the expression of an extended *RNASEH2C* mRNA, and reduced protein levels. The mechanism was only observed in human cells and did not apply to mice (Shen et al. 2018).

8 RNA Masking

RNA masking describes a mechanism where the NAT hybridizes to the sense transcript, most likely in the 3′ untranslated region, and interferes with the binding of miRNAs or regulatory proteins. Since double-stranded RNA formation elicits A-to-I editing by nuclear and cytoplasmic forms of ADAR (adenosine deaminase acting on RNA), co-expressed sense and antisense transcripts are expected to be edited. Genome-wide identification of RNA-edited sites, however, has revealed that

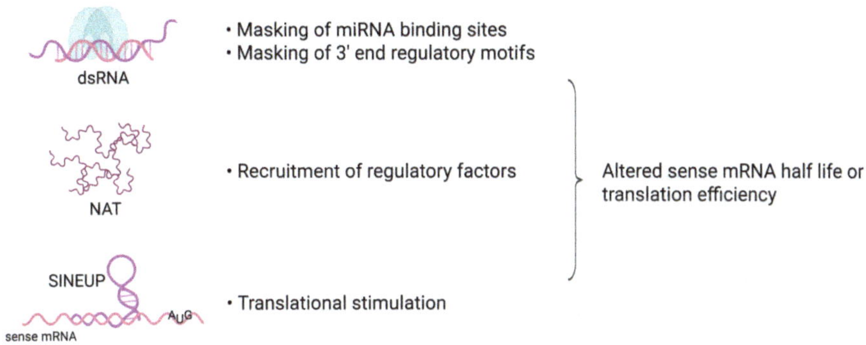

Fig. 14.3 Cytoplasmic actions of NATs. In the cytoplasm, NATs can form RNA hybrids with the cognate sense transcript and mask miRNA binding sites or AU-rich motifs that affect mRNA stability or translation efficiency (top panel). Alternatively, the NAT can recruit RNA modifiers (middle panel). SINEUPs contain a targeting sequence complementary to the 5′ end of the sense mRNA and a hairpin loop that mimics a ribosome entry site. As a result of SINEUP expression, translation of the sense mRNA is moderately upregulated (lower panel)

hairpin loops formed by inverted Alu repeats are the primary target of nuclear ADAR1. Since Alu elements are found in 12.4% of human mRNAs, A-to-I editing is widespread among these transcripts, as demonstrated after targeted deep sequencing (Bazak et al. 2014). Editing was identified in 207 human genes, though without reference to the antisense transcription (Li et al. 2009). In contrast to humans and mice, other species such as ctenophores, sponges, C. *elegans,* or sea squirts show significant enrichment of RNA editing in complementary sense-antisense regions (Zhang et al. 2023).

Genome-wide mapping of RNA editing sites has shown a clear association with immune and autoimmune diseases. Remarkably, the edited loci are not only highly enriched in inverted Alu repeats, well-known sources of dsRNA, but also in sense-antisense transcript pairs (Li et al. 2022b). These findings suggest that a fine balance between the formation and the detection/elimination of RNA-RNA hybrids is maintained. Accordingly, both over-expression of the RNAs and altered sensitivity of the dsRNA sensor MDA5 may lead to unwanted activation of innate immunity (Sadeq et al. 2021).

These findings also validate the many reported examples of NATs masking regulatory motifs on the sense mRNA, such as (de)stabilizing motifs (Lu et al. 2016) or miRNA binding sites (Fig. 14.3). The interference with miRNA binding is arguably the best documented mode of action of NATs in this category. For example, the antisense transcript *BACE1-AS* is complementary to a miRNA binding site on *BACE1* mRNA that encodes β-secretase, which plays an important role in the development of Alzheimer's disease (Faghihi et al. 2008, 2010).

The mechanism of RNA masking raises questions because of the generally unequal expression levels of sense- and antisense transcripts, as well as the immunogenic nature of dsRNA. It has been suggested that the antisense transcript only binds transiently to the sense RNA to elicit its function, i.e., obstructing the binding of a miRNA, an mRNA decay factor, or a stabilization protein. The so-called "hit and run" action would allow a few copies of a NAT to affect a much larger number of sense transcripts (Mahmoudi et al. 2009; Jadaliha et al. 2018). Likewise, the aforementioned association between RNA editing sites in areas of sense-antisense transcript complementarity and innate immunity suggests that such dsRNA structures do occur but may be compartmentalized or below a critical level to trigger the dsRNA sensors (MDA5 and/or PKR) (Chen and Hur 2022).

9 Translational Regulation

Natural antisense transcripts can affect the translation efficiency of the sense transcript. Hereby, the NAT displays a modular structure with a binding domain complementary to the 5′ end of the sense transcript and a fold that mimics an internal ribosomal entry site (IRES). The RNAs are denoted SINEUPs (SINE, small interspersed nuclear element; UP, upregulation of translation) and can have a stimulatory effect by enhancing polysome loading, or, conversely, compete with generic ribosome entry and reduce translation efficiency. The stem-loop fold that generates the IRES derives from inverted Alu/SINE repeats (Fig. 14.3). The structural requirements for SINEUPS (5′ complementarity and inverted SINE) have been identified at over 120 locations in the human genome (Schein et al. 2016) and 30 times in the mouse genome (Espinoza et al. 2021). The expression of SINEUPs has a moderate 1.5–3 times effect on the expression of the related protein. Conversely, another type of NAT, denoted *MAPT-AS*, that reduces translation of the corresponding sense mRNA, *MAPT*, has recently been characterized. The transcript has a similar building plan as SINEUPs and is complementary to the 5′ end of the human gene *MAPT* (encoding TAU protein) and contains a MIR element (Mammalian-wide Interspersed Repeat), i.e., a nonautonomous tRNA-derived retrotransposon. The MIR element forms a secondary structure that competes with an IRES in the *MAPT* mRNA. As a consequence, translation is reduced, leading to decreased expression of the tau protein. Again, the structural features of *MAPT-AS* (the term "SINEDOWNs" has not been coined) informed a genome-wide search for similar transcripts, and 460 genes were identified (Simone et al. 2021). However, the potential of *MAPT-AS* to reduce TAU has recently been questioned (Policarpo et al. 2025). The modular character of SINEUPS and MAPT-AS, as well as the moderate effect on translation of the target mRNA, enables the construction of targeted translational enhancers or repressors.

10 NATs and Disease

In the early days, the discovery of NATs was serendipitous, often linked to "negative" controls with sense-strand probes on northern blots or in situ hybridization experiments that yielded a distinct positive signal (Sun et al. 2017). With the surge in transcriptomic sequencing, differential gene expression analyses identified many noncoding RNAs and NATs. In addition, the comparison of patient samples with healthy controls identified many NATs that show significantly altered expression in pathological conditions. In fact, many of the lncRNAs that helped to investigate the biology of long non-protein coding RNAs and identify their mode of action can be classified as NATs, including *HOTAIR*, *TSIX*, *FENDRR*, *ANRIL*, or *AIRN*, to name a few (Mattick et al. 2023). Even the highly abundant lncRNAs *MALAT1* (metastasis-associated lung adenocarcinoma transcript 1) and *NEAT1* (nuclear-enriched abundant transcript 1) that are associated with nuclear speckles and paraspeckles derive from bidirectionally transcribed loci. In various screens, all of the aforementioned lncRNAs have been associated with cancer development and progression, for example, *ANRIL* (Pasmant et al. 2007), *HOTAIR* (Gupta et al. 2010), *TSIX* (Salama et al. 2019), *FENDRR* (Xu et al. 2014), *AIRN* (Liu et al. 2017), *MALAT1* (Gutschner et al. 2013), or *NEAT1* (Yu et al. 2017). An atlas of lncRNAs (Cancer LncRNA Census; CLC), which drive tumorigenesis, was recently published by Joana Carlevaro-Fita and included 122 transcripts associated with cancer phenotypes (Carlevaro-Fita et al. 2020). It has to be noted that not all identified lncRNAs are NATs, though a large proportion have also been found in a screen for bidirectional promoters (head-to-head NATs) (Lepoivre et al. 2013). In agreement with NATs playing a pivotal role in cancer, bidirectional promoters are significantly enriched in genes associated with breast and ovarian cancer genes (Yang et al. 2007). The role of NATs in developing disease has been extensively studied in the context of cancer; remarkably, the changes observed for NATs and lncRNAs are not mere consequences of altered expression programs driven by (protein-coding) oncogenes, but often, mutations in the regulatory RNAs drive the disease (Huarte 2015).

NATs have been less well investigated in the context of other diseases, though they may well play similarly important roles as in cancer. Accordingly, NATs are essentially involved in cardiovascular disease, Alzheimer's, psoriasis, and kidney disease (Cai et al. 2024; Celik et al. 2025; Shen et al. 2018; Rakhshan et al. 2020), though this list is by no means comprehensive and rapidly growing.

11 NATs in Evolution

The spatial proximity of NATs with their potential target, the cognate sense transcript, offers opportunities for regulatory interventions without off-target effects. With the increasing genome sizes of complex organisms, the co-localization of regulator and target could well provide essential precision in gene expression.

Accordingly, the impact of NATs and lncRNAs on development has been demonstrated in various model systems where interference with the expression levels of the RNA regulator leads to morphological defects (Grote et al. 2013; Piatek et al. 2017; Sauvageau et al. 2013). Changes in behavior caused by altered expression levels of NATs are more difficult to establish, particularly in model systems where the assessment of behavioral traits is not standardized. However, there are a number of NATs (*MAALIN*, *FEDORA*, or *Gm2694*) characterized in mice and humans that are linked to anxiety, aggression, or depression, for example. This list of lncRNAs affecting behavioral traits is rapidly growing, including transcripts that qualify as NATs (Dayal et al. 2024).

NATs also play a vital role in male germ cell development. Testes display the most complex transcriptome of all organs, including protein-coding and noncoding RNAs. NATs seem to play a particular role during germ cell development for two reasons: First, NATs show a distinct expression pattern with peak levels in pachytene spermatocytes (Werner et al. 2007). Second, in germ cells, sense-antisense transcript pairs form RNA hybrids that are potentially processed into endogenous siRNAs (Song et al. 2011). The association between bidirectional genes, dsRNA, and endo-siRNAs has only been observed in mouse testes and not in a comparable dataset from the mouse liver (Werner et al. 2021). A mechanism that involves a highly immunogenic dsRNA intermediate may depend on the immune-privileged environment of germ cells (Zhao et al. 2014) and would not be tolerated in somatic tissues (liver).

The complex transcriptome informed two hypotheses to explain the widespread expression of NATs in spermatocytes. The first one predicts that the tail-to-tail-oriented NATs stabilize mRNAs via RNA hybrid formation and prevent premature translation. During male germ cell development, transcription occurs in two waves. The first wave unfolds in spermatogonia as a consequence of programmed DNA demethylation. Hypomethylation enables genome-wide transcription and also leads to transposon mobilization. The second wave of transcription occurs toward the end of meiosis and coincides with the detection of NATs and dsRNA. During the later stages of development (spermiogenesis), chromatin is compacted, and transcription ceases (Zhu et al. 2015). At this point, the sense-antisense hybrids may be resolved to maintain protein synthesis required for germ cell maturation (Lin et al. 2016; Gan et al. 2013). The second hypothesis proposes a mechanism where antisense transcription is used to monitor the expression levels of the sense transcript. Accordingly, the endo-siRNAs of bidirectionally transcribed genes will preferentially interact with the more abundant transcript. Comparably low levels of sense transcript—because of inefficient transcription, processing, or mRNA stability—would leave an excess of endo-siRNA-protein complexes with complementarity to the sense transcript. These targeted complexes could then trigger a pathway to eliminate the particular germ cell with the underperforming, potentially mutated sense gene (Werner et al. 2015).

12 NATs and Evolution in Yeast

As highlighted in previous paragraphs, the synthesis of NATs has potentially detrimental consequences on cellular homeostasis. Transcriptional interference and dsRNA formation may constitute evolutionary pressure that cells need to mitigate to thrive. The following two examples from yeast and *C. elegans* provide circumstantial evidence to underpin NATs-driven adaptive changes in transcription termination and genome organization, respectively.

Bidirectional promoters in densely packed yeast genomes carry the risk of producing antisense transcripts that interfere with the expression of adjacent genes. Accordingly, antisense transcription leads to reduced levels of the related sense transcripts (Nevers et al. 2018; Wery et al. 2018). To avoid unwanted interference, efficient transcription termination is required. Two mechanisms characterized in *S. cerevisiae* are the CPA pathway (mRNA cleavage and polyadenylation) and the NNS (Nrd1-Nab3-Sen1) complex. The CPA pathway involves the cleavage of the nascent transcript after the polyadenylation motif, followed by the displacement of the Pol II complex (West et al. 2004). It predominantly applies to the termination of mRNAs and some noncoding (antisense) transcripts. For the latter, the NNS complex is more important, together with specific transcription factors (Reb1, Abf1, and Rap) acting as a roadblock. The stalled RNA Pol II is then displaced by the NNS complex, and transcription is terminated (Candelli et al. 2018; Roy et al. 2016). DNA supercoiling that destabilizes or removes DNA-binding proteins (roadblocks) negatively affects termination and increases read-through transcription (Patel et al. 2023). NNS-mediated transcription termination is not conserved in fission yeast and metazoans, where both coding and noncoding transcripts involve the CPA complex for 3′ end processing (Heo et al. 2021).

In larger genomes, transcription termination is more relaxed. Even in closely spaced tandem genes (in the human genome), the intergenic regions are associated with stalled RNA Pol II as well as RNA interference effectors AGO1 and AGO2 (Nissani and Ulitsky 2022) rather than with markers for transcription termination. An intriguing hypothesis suggests that RNA interference, which is extant in *S. pombe* and higher eukaryotes but lost in *S. cerevisiae,* may provide additional protection against pervasive antisense transcripts (Soudet and Stutz 2019).

An alternative genome-wide role for NATs in yeast has recently been put forward by Coban et al., based on their observation that sense-antisense transcripts in *S cerevisiae* form dsRNA structures that are enriched in the cytoplasm. They found that the RNA hybrids are preferentially exported compared to the sense mRNA alone, leading to moderately enhanced translation of the sense mRNA (Coban et al. 2024). The mechanism coined as RNA-mediated gene boosting may be intriguing in yeast and provides an explanation for the concordant expression of sense and antisense transcript pairs. However, extrapolating the mechanism to humans fails to account for the predominant expression of NATs in the testis, and the immunogenic nature of dsRNA is likely incompatible with the formation of sense-antisense hybrids at a significant scale.

13 NATs and Evolution in *C. elegans*

The *C. elegans* model organism has been at the forefront of noncoding RNA research since the discovery of RNA interference in 1998, though most prominently in the fields of small RNAs and RNA-induced gene silencing (Fire et al. 1998). *C. elegans* has a comparably small genome of just over 100 Mb, encoding 20,000 protein-coding genes; to compare, the human genome encodes the same number of proteins but is 30 times larger. Remarkably, only a limited number of lncRNAs have been identified in *C. elegans*. Depending on the approach, 3392 unspliced and 143 spliced non-protein-coding transcripts (Akay et al. 2019) were reported, or 170 lncRNAs and 58 antisense transcripts in an earlier study (Nam and Bartel 2012). The number of 58 NATs might be even lower, as about half of the corresponding sense genes are hypothetical without confirmed expression. Furthermore, the levels of the NATs appear not to affect the expression level of the sense genes. The unusually low number of NATs in *C. elegans* compared to other model systems is well documented (Sun et al. 2006) and raises the question of what specific traits may have selected against the transcription of complementary RNAs.

C. elegans has a particularly diverse RNA interference machinery that acts at both the transcriptional and posttranscriptional levels. A comprehensive description of *C. elegans* RNAi is beyond the scope of this perspective; however, two aspects are relevant in the context of antisense transcription, namely, the signal amplification by RNA-dependent RNA polymerase (RdRP) and transgenerational small RNA-mediated gene silencing. Transgenerational silencing depends on RdRP-mediated amplification of an RNA signal and specific members of the Argonaute protein family, HRDE-1/WAGO-9, in the germline. Gene silencing can be triggered by the PIWI pathway (Shirayama et al. 2012), intronless transcripts (Makeyeva et al. 2021), and dsRNA-induced RNA interference. The fact that dsRNA signals trigger strong and lasting gene silencing implies that sense-antisense hybrids are potentially prime targets for transgenerational silencing. To mitigate the risk of inactivating protein-coding genes, a genomic architecture that avoids the formation of dsRNA may have evolved in *C. elegans*. Of note, the worm genome is not devoid of bidirectionally transcribed loci; about a third of all genes produce antisense transcripts, but all potentially complementary regions are spliced out (Ohhata et al. 2022). The splicing complex inhibits endo-siRNA-primed argonaute surveillance, thus providing further protection against ectopic silencing. In contrast, an argonaute complex loaded with endo-siRNAs from an intronless transcript or exogenous siRNAs is competent to silence a target sequence in a multi-exon gene (Fire et al. 1998; Makeyeva et al. 2021). As already seen in yeast, the expression of the target gene crucially determines silencing efficiency also in *C. elegans*. Efficient RNA splicing and high levels of mRNA provide efficient protection against uncontrolled transgenerational silencing in the *C. elegans* germline. Moreover, the risk may be reduced even more by a genomic architecture that minimizes sense-antisense RNA hybrids.

14 Conclusions

NATs constitute a particular and versatile family of noncoding RNAs. The fact that they originate from the same genetic locus as their potential target ensures localized actions and reduces off-target effects. NATs can also form distinct structures, such as R-loops, triple helices, hairpins, and dsRNA with the sense transcript, that offer a myriad of interfaces with enzymes and transcription factors that can control the expression level of the sense gene and establish the epigenetic fingerprint of the bidirectionally transcribed locus. Moreover, the hybridization between sense/antisense transcripts in the cytoplasm and the formation of distinct hairpin structures in the fully processed NAT (SINEUPs) offer additional regulatory possibilities in the cytoplasm at the level of RNA stability or translational efficiency.

It comes, therefore, as a little surprise that altered levels of NATs are associated with developmental defects and disease, most prominently in cancer. The regulatory power of NATs is also documented by significant changes in genome organization or cellular biology (efficient transcriptional termination in organisms with small genomes, suppression of complementary sense/antisense transcripts, or compartmentalization to avoid dsRNA immune signaling) observed in certain organisms. The hypothesis predicts that these features evolved to mitigate the deleterious consequences of inadequate NAT expression. Hence, the dual consequences of bidirectional transcription (NATs as a regulatory tool vs. cytotoxic dsRNA formation) suggest that NATs are essential evolutionary drivers, a role that should be acknowledged in an integrative theory of evolution.

References

Akay A, Jordan D, Navarro IC, Wrzesinski T, Ponting CP, Miska EA, Haerty W (2019) Identification of functional long non-coding RNAs in C. elegans. BMC Biol 17:14

Bagchi DN, Battenhouse AM, Park D, Iyer VR (2020) The histone variant H2A.Z in yeast is almost exclusively incorporated into the +1 nucleosome in the direction of transcription. Nucleic Acids Res 48:157–170

Bazak L, Haviv A, Barak M, Jacob-Hirsch J, Deng P, Zhang R, Isaacs FJ, Rechavi G, Li JB, Eisenberg E, Levanon EY (2014) A-to-I RNA editing occurs at over a hundred million genomic sites, located in a majority of human genes. Genome Res 24:365–376

Cai J, Ni YQ, Liu YS (2024) Exploring the frontier: antisense long non-coding RNAs as key regulators in Alzheimer's disease. Aging Dis

Camino LP, Dutta A, Barroso S, Pérez-Calero C, Katz JN, García-Rubio M, Sung P, Gómez-González B, Aguilera A (2023) DICER ribonuclease removes harmful R-loops. Mol Cell 83: 3707–3719.e5

Candelli T, Challal D, Briand JB, Boulay J, Porrua O, Colin J, Libri D (2018) High resolution transcription maps reveal the widespread impact of roadblock termination in yeast. EMBO J 37

Carlevaro-Fita J, Lanzós A, Feuerbach L, Hong C, Mas-Ponte D, Pedersen JS, Johnson R (2020) Cancer LncRNA Census reveals evidence for deep functional conservation of long noncoding RNAs in tumorigenesis. Commun Biol 3:56

Cawley S, Bekiranov S, Ng HH, Kapranov P, Sekinger EA, Kampa D, Piccolboni A, Sementchenko V, Cheng J, Williams AJ, Wheeler R, Wong B, Drenkow J, Yamanaka M, Patel S, Brubaker S, Tammana H, Helt G, Struhl K, Gingeras TR (2004) Unbiased mapping of transcription factor binding sites along human chromosomes 21 and 22 points to widespread regulation of noncoding RNAs. Cell 116:499–509

Celik S, Hyrefelt L, Czuba T, Li Y, Assis J, Martinez J, Johansson M, André O, Synnergren J, Sandstedt J, Nordenfelt P, Vukusic K, Smith JG, Gidlöf O (2025) Antisense-mediated regulation of exon usage in the elastic spring region of Titin modulates sarcomere function. Cardiovasc Res 121:629–642

Chawla A, Lazar MA (1993) Induction of Rev-ErbA alpha, an orphan receptor encoded on the opposite strand of the alpha-thyroid hormone receptor gene, during adipocyte differentiation. J Biol Chem 268:16265–16269

Chen YG, Hur S (2022) Cellular origins of dsRNA, their recognition and consequences. Nat Rev Mol Cell Biol 23:286–301

Coban I, Lamping JP, Hirsch AG, Wasilewski S, Shomroni O, Giesbrecht O, Salinas G, Krebber H (2024) dsRNA formation leads to preferential nuclear export and gene expression. Nature 631: 432–438

Dayal S, Chaubey D, Joshi DC, Ranmale S, Pillai B (2024) Noncoding RNAs: emerging regulators of behavioral complexity. Wiley Interdiscip Rev: RNA 15:e1847

Dong Z, Li S, Wu X, Niu Y, Liang X, Yang L, Guo Y, Shen S, Liang J, Guo W (2019) Aberrant hypermethylation-mediated downregulation of antisense lncRNA ZNF667-AS1 and its sense gene ZNF667 correlate with progression and prognosis of esophageal squamous cell carcinoma. Cell Death Dis 10:930

Espinoza S, Bon C, Valentini P, Pierattini B, Matey AT, Damiani D, Pulcrano S, Sanges R, Persichetti F, Takahashi H, Carninci P, Santoro C, Cotella D, Gustincich S (2021) SINEUPs: a novel toolbox for RNA therapeutics. Essays Biochem 65:775–789

Faghihi MA, Modarresi F, Khalil AM, Wood DE, Sahagan BG, Morgan TE, Finch CE, St Laurent G 3rd, Kenny PJ, Wahlestedt C (2008) Expression of a noncoding RNA is elevated in Alzheimer's disease and drives rapid feed-forward regulation of beta-secretase. Nat Med 14: 723–730

Faghihi MA, Wahlestedt C (2009) Regulatory roles of natural antisense transcripts. Nat Rev Mol Cell Biol 10:637–643

Faghihi MA, Zhang M, Huang J, Modarresi F, Van Der Brug MP, Nalls MA, Cookson MR, St-Laurent G 3rd, Wahlestedt C (2010) Evidence for natural antisense transcript-mediated inhibition of microRNA function. Genome Biol 11:R56

Fire A, Xu S, Montgomery MK, Kostas SA, Driver SE, Mello CC (1998) Potent and specific genetic interference by double-stranded RNA in Caenorhabditis elegans. Nature 391:806–811

Gan H, Cai T, Lin X, Wu Y, Wang X, Yang F, Han C (2013) Integrative proteomic and transcriptomic analyses reveal multiple post-transcriptional regulatory mechanisms of mouse spermatogenesis. Mol Cell Proteomics 12:1144–1157

Georg J, Hess WR (2011) Cis-antisense RNA, another level of gene regulation in bacteria. Microbiol Mol Biol Rev 75:286–300

Grote P, Wittler L, Hendrix D, Koch F, Währisch S, Beisaw A, Macura K, Bläss G, Kellis M, Werber M, Herrmann BG (2013) The tissue-specific lncRNA Fendrr is an essential regulator of heart and body wall development in the mouse. Dev Cell 24:206–214

Gu M, Naiyachit Y, Wood TJ, Millar CB (2015) H2A.Z marks antisense promoters and has positive effects on antisense transcript levels in budding yeast. BMC Genomics 16:99

Gupta RA, Shah N, Wang KC, Kim J, Horlings HM, Wong DJ, Tsai MC, Hung T, Argani P, Rinn JL, Wang Y, Brzoska P, Kong B, Li R, West RB, Van De Vijver MJ, Sukumar S, Chang HY (2010) Long non-coding RNA HOTAIR reprograms chromatin state to promote cancer metastasis. Nature 464:1071–1076

Gutschner T, Hämmerle M, Diederichs S (2013) MALAT1 -- a paradigm for long noncoding RNA function in cancer. J Mol Med (Berl) 91:791–801

Hao N, Donnelly AJ, Dodd IB, Shearwin KE (2023) When push comes to shove - RNA polymerase and DNA-bound protein roadblocks. Biophys Rev 15:355–366
He Y, Vogelstein B, Velculescu VE, Papadopoulos N, Kinzler KW (2008) The antisense transcriptomes of human cells. Science 322:1855–1857
Heilmann K, Toth R, Bossmann C, Klimo K, Plass C, Gerhauser C (2017) Genome-wide screen for differentially methylated long noncoding RNAs identifies Esrp2 and lncRNA Esrp2-as regulated by enhancer DNA methylation with prognostic relevance for human breast cancer. Oncogene 36:6446–6461
Heo DH, Kuś K, Grzechnik P, Tan-Wong SM, Birot A, Kecman T, Nielsen S, Zenkin N, Vasiljeva L (2021) Transcription and chromatin-based surveillance mechanism controls suppression of cryptic antisense transcription. Cell Rep 36:109671
Hezroni H, Koppstein D, Schwartz MG, Avrutin A, Bartel DP, Ulitsky I (2015) Principles of long noncoding RNA evolution derived from direct comparison of transcriptomes in 17 species. Cell Rep 11:1110–1122
Huang GW, Zhang YL, Liao LD, Li EM, Xu LY (2017) Natural antisense transcript TPM1-AS regulates the alternative splicing of tropomyosin I through an interaction with RNA-binding motif protein 4. Int J Biochem Cell Biol 90:59–67
Huarte M (2015) The emerging role of lncRNAs in cancer. Nat Med 21:1253–1261
Inouye M (1988) Antisense RNA: its functions and applications in gene regulation--a review. Gene 72:25–34
Jadaliha M, Gholamalamdari O, Tang W, Zhang Y, Petracovici A, Hao Q, Tariq A, Kim TG, Holton SE, Singh DK, Li XL, Freier SM, Ambs S, Bhargava R, Lal A, Prasanth SG, Ma J, Prasanth KV (2018) A natural antisense lncRNA controls breast cancer progression by promoting tumor suppressor gene mRNA stability. PLoS Genet 14:e1007802
Kaplan N, Moore IK, Fondufe-Mittendorf Y, Gossett AJ, Tillo D, Field Y, Leproust EM, Hughes TR, Lieb JD, Widom J, Segal E (2009) The DNA-encoded nucleosome organization of a eukaryotic genome. Nature 458:362–366
Katayama S, Tomaru Y, Kasukawa T, Waki K, Nakanishi M, Nakamura M, Nishida H, Yap CC, Suzuki M, Kawai J, Suzuki H, Carninci P, Hayashizaki Y, Wells C, Frith M, Ravasi T, Pang KC, Hallinan J, Mattick J, Hume DA, Lipovich L, Batalov S, Engström PG, Mizuno Y, Faghihi MA, Sandelin A, Chalk AM, Mottagui-Tabar S, Liang Z, Lenhard B, Wahlestedt C (2005) Antisense transcription in the mammalian transcriptome. Science 309:1564–1566
Kiyosawa H, Yamanaka I, Osato N, Kondo S, Hayashizaki Y (2003) Antisense transcripts with FANTOM2 clone set and their implications for gene regulation. Genome Res 13:1324–1334
Kulaeva OI, Gaykalova DA, Studitsky VM (2007) Transcription through chromatin by RNA polymerase II: histone displacement and exchange. Mutat Res 618:116–129
Ladouceur AM, Parmar BS, Biedzinski S, Wall J, Tope SG, Cohn D, Kim A, Soubry N, Reyes-Lamothe R, Weber SC (2020) Clusters of bacterial RNA polymerase are biomolecular condensates that assemble through liquid-liquid phase separation. Proc Natl Acad Sci USA 117:18540–18549
Lepoivre C, Belhocine M, Bergon A, Griffon A, Yammine M, Vanhille L, Zacarias-Cabeza J, Garibal MA, Koch F, Maqbool MA, Fenouil R, Loriod B, Holota H, Gut M, Gut I, Imbert J, Andrau JC, Puthier D, Spicuglia S (2013) Divergent transcription is associated with promoters of transcriptional regulators. BMC Genomics 14:914
Li D, Shen L, Zhang X, Chen Z, Huang P, Huang C, Qin S (2022a) LncRNA ELF3-AS1 inhibits gastric cancer by forming a negative feedback loop with SNAI2 and regulates ELF3 mRNA stability via interacting with ILF2/ILF3 complex. J Exp Clin Cancer Res 41:332
Li JB, Levanon EY, Yoon JK, Aach J, Xie B, Leproust E, Zhang K, Gao Y, Church GM (2009) Genome-wide identification of human RNA editing sites by parallel DNA capturing and sequencing. Science 324:1210–1213
Li Q, Dong C, Cui J, Wang Y, Hong X (2018) Over-expressed lncRNA HOTAIRM1 promotes tumor growth and invasion through up-regulating HOXA1 and sequestering G9a/EZH2/Dnmts away from the HOXA1 gene in glioblastoma multiforme. J Exp Clin Cancer Res 37:265

Li Q, Gloudemans MJ, Geisinger JM, Fan B, Aguet F, Sun T, Ramaswami G, Li YI, Ma JB, Pritchard JK, Montgomery SB, Li JB (2022b) RNA editing underlies genetic risk of common inflammatory diseases. Nature 608:569–577

Lin X, Han M, Cheng L, Chen J, Zhang Z, Shen T, Wang M, Wen B, Ni T, Han C (2016) Expression dynamics, relationships, and transcriptional regulations of diverse transcripts in mouse spermatogenic cells. RNA Biol 13:1011–1024

Ling KH, Hewitt CA, Beissbarth T, Hyde L, Cheah PS, Smyth GK, Tan SS, Hahn CN, Thomas T, Thomas PQ, Scott HS (2011) Spatiotemporal regulation of multiple overlapping sense and novel natural antisense transcripts at the Nrgn and Camk2n1 gene loci during mouse cerebral corticogenesis. Cereb Cortex 21:683–697

Liu L, Yu D, Shi H, Li J, Meng L (2017) Reduced lncRNA Aim enhances the malignant invasion of triple-negative breast cancer cells mainly by activating Wnt/β-catenin/mTOR/PI3K signaling. Pharmazie 72:599–603

Liu Y, Wang X, Zhu Y, Cao Y, Wang L, Li F, Zhang Y, Li Y, Zhang Z, Luo J, Deng X, Peng C, Wei G, Chen H, Shen B (2022) The CTCF/LncRNA-PACERR complex recruits E1A binding protein p300 to induce pro-tumour macrophages in pancreatic ductal adenocarcinoma via directly regulating PTGS2 expression. Clin Transl Med 12:e654

Lu Z, Zhang QC, Lee B, Flynn RA, Smith MA, Robinson JT, Davidovich C, Gooding AR, Goodrich KJ, Mattick JS, Mesirov JP, Cech TR, Chang HY (2016) RNA duplex map in living cells reveals higher-order transcriptome structure. Cell 165:1267–1279

Ma J, Du WW, Zeng K, Wu N, Fang L, Lyu J, Yee AJ, Yang BB (2021) An antisense circular RNA circSCRIB enhances cancer progression by suppressing parental gene splicing and translation. Mol Ther 29:2754–2768

Mahmoudi S, Henriksson S, Corcoran M, Méndez-Vidal C, Wiman KG, Farnebo M (2009) Wrap53, a natural p53 antisense transcript required for p53 induction upon DNA damage. Mol Cell 33:462–471

Makeyeva YV, Shirayama M, Mello CC (2021) Cues from mRNA splicing prevent default Argonaute silencing in C. elegans. Dev Cell 56:2636–2648.e4

Mattick JS, Amaral PP, Carninci P, Carpenter S, Chang HY, Chen LL, Chen R, Dean C, Dinger ME, Fitzgerald KA, Gingeras TR, Guttman M, Hirose T, Huarte M, Johnson R, Kanduri C, Kapranov P, Lawrence JB, Lee JT, Mendell JT, Mercer TR, Moore KJ, Nakagawa S, Rinn JL, Spector DL, Ulitsky I, Wan Y, Wilusz JE, Wu M (2023) Long non-coding RNAs: definitions, functions, challenges and recommendations. Nat Rev Mol Cell Biol 24:430–447

Michael DR, Phillips AO, Krupa A, Martin J, Redman JE, Altaher A, Neville RD, Webber J, Kim MY, Bowen T (2011) The human hyaluronan synthase 2 (HAS2) gene and its natural antisense RNA exhibit coordinated expression in the renal proximal tubular epithelial cell. J Biol Chem 286:19523–19532

Millar JA, Raghavan R (2020) Modulation of bacterial fitness and virulence through antisense RNAs. Front Cell Infect Microbiol 10:596277

Morrissy AS, Griffith M, Marra MA (2011) Extensive relationship between antisense transcription and alternative splicing in the human genome. Genome Res 21:1203–1212

Nam JW, Bartel DP (2012) Long noncoding RNAs in C. elegans. Genome Res 22:2529–2540

Natarajan P, Shrinivas K, Chakraborty AK (2023) A model for cis-regulation of transcriptional condensates and gene expression by proximal lncRNAs. Biophys J 122:2757–2772

Nevers A, Doyen A, Malabat C, Néron B, Kergrohen T, Jacquier A, Badis G (2018) Antisense transcriptional interference mediates condition-specific gene repression in budding yeast. Nucleic Acids Res 46:6009–6025

Ni T, Tu K, Wang Z, Song S, Wu H, Xie B, Scott KC, Grewal SI, Gao Y, Zhu J (2010) The prevalence and regulation of antisense transcripts in Schizosaccharomyces pombe. PLoS One 5: e15271

Niehus SE, Allister AB, Hoffmann A, Wiehlmann L, Tamura T, Tran DDH (2019) Myc/Max dependent intronic long antisense noncoding RNA, EVA1A-AS, suppresses the expression of Myc/Max dependent anti-proliferating gene EVA1A in a U2 dependent manner. Sci Rep 9: 17319

Nissani N, Ulitsky I (2022) Unique features of transcription termination and initiation at closely spaced tandem human genes. Mol Syst Biol 18:e10682

Novačić A, Menéndez D, Ljubas J, Barbarić S, Stutz F, Soudet J, Stuparević I (2022) Antisense non-coding transcription represses the PHO5 model gene at the level of promoter chromatin structure. PLoS Genet 18:e1010432

Ohhata T, Suzuki M, Sakai S, Ota K, Yokota H, Uchida C, Niida H, Kitagawa M (2022) CCIVR facilitates comprehensive identification of cis-natural antisense transcripts with their structural characteristics and expression profiles. Sci Rep 12:15525

Pasmant E, Laurendeau I, Héron D, Vidaud M, Vidaud D, Bièche I (2007) Characterization of a germ-line deletion, including the entire INK4/ARF locus, in a melanoma-neural system tumor family: identification of ANRIL, an antisense noncoding RNA whose expression coclusters with ARF. Cancer Res 67:3963–3969

Patel HP, Coppola S, Pomp W, Aiello U, Brouwer I, Libri D, Lenstra TL (2023) DNA supercoiling restricts the transcriptional bursting of neighboring eukaryotic genes. Mol Cell 83:1573–1587.e8

Pei G, Lyons H, Li P, Sabari BR (2025) Transcription regulation by biomolecular condensates. Nat Rev Mol Cell Biol 26:213–236

Piatek MJ, Henderson V, Fearn A, Chaudhry B, Werner A (2017) Ectopically expressed Slc34a2a sense-antisense transcripts cause a cerebellar phenotype in zebrafish embryos depending on RNA complementarity and Dicer. PLoS One 12:e0178219

Policarpo R, Wolfs L, Martínez-Montero S, Vandermeulen L, Royaux I, Van Peer G, Mestdagh P, De Strooper B, Sierksma A, D'Ydewalle C (2025) The MIR-NAT MAPT-AS1 does not regulate Tau expression in human neurons. PLoS One 20:e0314973

Portal MM, Pavet V, Erb C, Gronemeyer H (2015) Human cells contain natural double-stranded RNAs with potential regulatory functions. Nat Struct Mol Biol 22:89–97

Postepska-Igielska A, Giwojna A, Gasri-Plotnitsky L, Schmitt N, Dold A, Ginsberg D, Grummt I (2015) LncRNA Khps1 regulates expression of the proto-oncogene SPHK1 via triplex-mediated changes in chromatin structure. Mol Cell 60:626–636

Rakhshan A, Zarrinpour N, Moradi A, Ahadi M, Omrani MD, Ghafouri-Fard S, Taheri M (2020) A single nucleotide polymorphism within HOX Transcript Antisense RNA (HOTAIR) is associated with risk of psoriasis. Int J Immunogenet 47:430–434

Roy K, Gabunilas J, Gillespie A, Ngo D, Chanfreau GF (2016) Common genomic elements promote transcriptional and DNA replication roadblocks. Genome Res 26:1363–1375

Ruiz-Orera J, Albà MM (2019) Conserved regions in long non-coding RNAs contain abundant translation and protein-RNA interaction signatures. NAR Genom Bioinform 1:e2

Ryczek N, Łyś A, Wanowska E, Kozłowska-Masłoń J, Makałowska I (2025) Mechanism of expression regulation of head-to-head overlapping protein-coding genes INO80E and HIRIP3. Commun Biol 8:391

Sadeq S, Al-Hashimi S, Cusack CM, Werner A (2021) Endogenous Double-Stranded RNA. Noncoding RNA 7

Salama EA, Adbeltawab RE, El Tayebi HM (2019) XIST and TSIX: novel cancer immune biomarkers in PD-L1-overexpressing breast cancer patients. Front Oncol 9:1459

Sauvageau M, Goff LA, Lodato S, Bonev B, Groff AF, Gerhardinger C, Sanchez-Gomez DB, Hacisuleyman E, Li E, Spence M, Liapis SC, Mallard W, Morse M, Swerdel MR, D'Ecclessis MF, Moore JC, Lai V, Gong G, Yancopoulos GD, Frendewey D, Kellis M, Hart RP, Valenzuela DM, Arlotta P, Rinn JL (2013) Multiple knockout mouse models reveal lincRNAs are required for life and brain development. elife 2:e01749

Schein A, Zucchelli S, Kauppinen S, Gustincich S, Carninci P (2016) Identification of antisense long noncoding RNAs that function as SINEUPs in human cells. Sci Rep 6:33605

Shen T, Li H, Song Y, Yao J, Han M, Yu M, Wei G, Ni T (2018) Antisense transcription regulates the expression of sense gene via alternative polyadenylation. Protein Cell 9:540–552

Shendure J, Church GM (2002) Computational discovery of sense-antisense transcription in the human and mouse genomes. Genome Biol 3:Research0044

Shirayama M, Seth M, Lee HC, Gu W, Ishidate T, Conte D Jr, Mello CC (2012) piRNAs initiate an epigenetic memory of nonself RNA in the C. elegans germline. Cell 150:65–77

Simone R, Javad F, Emmett W, Wilkins OG, Almeida FL, Barahona-Torres N, Zareba-Paslawska J, Ehteramyan M, Zuccotti P, Modelska A, Siva K, Virdi GS, Mitchell JS, Harley J, Kay VA, Hondhamuni G, Trabzuni D, Ryten M, Wray S, Preza E, Kia DA, Pittman A, Ferrari R, Manzoni C, Lees A, Hardy JA, Denti MA, Quattrone A, Patani R, Svenningsson P, Warner TT, Plagnol V, Ule J, De Silva R (2021) MIR-NATs repress MAPT translation and aid proteostasis in neurodegeneration. Nature 594:117–123

Somasundaram K, Gupta B, Jain N, Jana S (2022) LncRNAs divide and rule: the master regulators of phase separation. Front Genet 13:930792

Song R, Hennig GW, Wu Q, Jose C, Zheng H, Yan W (2011) Male germ cells express abundant endogenous siRNAs. Proc Natl Acad Sci USA 108:13159–13164

Soudet J, Stutz F (2019) Regulation of gene expression and replication initiation by non-coding transcription: a model based on reshaping nucleosome-depleted regions: influence of pervasive transcription on chromatin structure. BioEssays 41:e1900043

Statello L, Guo CJ, Chen LL, Huarte M (2021) Gene regulation by long non-coding RNAs and its biological functions. Nat Rev Mol Cell Biol 22:96–118

Sun M, Hurst LD, Carmichael GG, Chen J (2006) Evidence for variation in abundance of antisense transcripts between multicellular animals but no relationship between antisense transcriptionand organismic complexity. Genome Res 16:922–933

Sun Y, Li D, Zhang R, Peng S, Zhang G, Yang T, Qian A (2017) Strategies to identify natural antisense transcripts. Biochimie 132:131–151

Timmons JA, Good L (2006) Does everything now make (anti)sense? Biochem Soc Trans 34: 1148–1150

Tufarelli C, Stanley JA, Garrick D, Sharpe JA, Ayyub H, Wood WG, Higgs DR (2003) Transcription of antisense RNA leading to gene silencing and methylation as a novel cause of human genetic disease. Nat Genet 34:157–165

Veeramachaneni V, Makałowski W, Galdzicki M, Sood R, Makałowska I (2004) Mammalian overlapping genes: the comparative perspective. Genome Res 14:280–286

Wang YQ, Jiang DM, Hu SS, Zhao L, Wang L, Yang MH, Ai ML, Jiang HJ, Han Y, Ding YQ, Wang S (2019) SATB2-AS1 suppresses colorectal carcinoma aggressiveness by inhibiting SATB2-dependent snail transcription and epithelial-mesenchymal transition. Cancer Res 79: 3542–3556

Werner A, Clark JE, Samaranayake C, Casement J, Zinad HS, Sadeq S, Al-Hashimi S, Smith M, Kotaja N, Mattick JS (2021) Widespread formation of double-stranded RNAs in testis. Genome Res 31:1174–1186

Werner A, Kanhere A, Wahlestedt C, Mattick JS (2024) Natural antisense transcripts as versatile regulators of gene expression. Nat Rev Genet 25:730–744

Werner A, Piatek MJ, Mattick JS (2015) Transpositional shuffling and quality control in male germ cells to enhance evolution of complex organisms. Ann N Y Acad Sci 1341:156–163

Werner A, Schmutzler G, Carlile M, Miles CG, Peters H (2007) Expression profiling of antisense transcripts on DNA arrays. Physiol Genomics 28:294–300

Wery M, Gautier C, Descrimes M, Yoda M, Vennin-Rendos H, Migeot V, Gautheret D, Hermand D, Morillon A (2018) Native elongating transcript sequencing reveals global anti-correlation between sense and antisense nascent transcription in fission yeast. RNA 24:196–208

West S, Gromak N, Proudfoot NJ (2004) Human 5' --> 3' exonuclease Xrn2 promotes transcription termination at co-transcriptional cleavage sites. Nature 432:522–525

Wood EJ, Chin-Inmanu K, Jia H, Lipovich L (2013) Sense-antisense gene pairs: sequence, transcription, and structure are not conserved between human and mouse. Front Genet 4:183

Xu TP, Huang MD, Xia R, Liu XX, Sun M, Yin L, Chen WM, Han L, Zhang EB, Kong R, De W, Shu YQ (2014) Decreased expression of the long non-coding RNA FENDRR is associated with poor prognosis in gastric cancer and FENDRR regulates gastric cancer cell metastasis by affecting fibronectin1 expression. J Hematol Oncol 7:63

Xu Z, Wei W, Gagneur J, Clauder-Münster S, Smolik M, Huber W, Steinmetz LM (2011) Antisense expression increases gene expression variability and locus interdependency. Mol Syst Biol 7:468

Yang MQ, Koehly LM, Elnitski LL (2007) Comprehensive annotation of bidirectional promoters identifies co-regulation among breast and ovarian cancer genes. PLoS Comput Biol 3:e72

Yassour M, Pfiffner J, Levin JZ, Adiconis X, Gnirke A, Nusbaum C, Thompson DA, Friedman N, Regev A (2010) Strand-specific RNA sequencing reveals extensive regulated long antisense transcripts that are conserved across yeast species. Genome Biol 11:R87

Yu X, Li Z, Zheng H, Chan MT, Wu WK (2017) NEAT1: A novel cancer-related long non-coding RNA. Cell Prolif 50

Zhang P, Zhu Y, Guo Q, Li J, Zhan X, Yu H, Xie N, Tan H, Lundholm N, Garcia-Cuetos L, Martin MD, Subirats MA, Su YH, Ruiz-Trillo I, Martindale MQ, Yu JK, Gilbert MTP, Zhang G, Li Q (2023) On the origin and evolution of RNA editing in metazoans. Cell Rep 42:112112

Zhao S, Zhu W, Xue S, Han D (2014) Testicular defense systems: immune privilege and innate immunity. Cell Mol Immunol 11:428–437

Zhu ZJ, Yang S, Li Z (2015) Transcriptome research on spermatogenic molecular drive in mammals. Asian J Androl 17:961–971

Zinad HS, Sae-Lee C, Ariza-Mateos MA, Adamson G, Khazeem MM, Knox A, Chung G, Mann J, Werner A (2022) Interdependent transcription of a natural sense/antisense transcripts pair (SLC34A1/PFN3). Noncoding RNA 8

Chapter 15
System-Evolutionary View of RNA Presence in Living Cells

Margareta Segerståhl and Matti Jalasvuori

Abstract RNA predates cells, but today it "lives" inside them. Most current RNA is short-lived, folded, and often edited from its chromosomal origins, but it governs its own phenotypes. Our guiding question for this article was: How did RNA adjust its life when the RNA world era came to an end? RNA has its practical execution side in the whole-cell systems setup, serving critical functions in metabolism, compartmentalization, and in the RNA–DNA–PROTEIN network, coupling information handling and different events of macromolecule production into a self-sustaining loop. Aligning with current views of cooperation, we find that RNA life is distributed across a population of information-bearing entities persisting by compatibility, with DNA serving as a long-term memory and proteins offering the catalytic reach. We suggest that the reproduction–survival (R/S) axis, relevant to higher organisms, is also manifested by RNA in many forms, including RNA primers, mRNAs, tRNAs, and rRNAs. Each of these has contrasting, highly adapted molecular phenotypes. Ultimately, we wish to emphasize the evolutionary narrative where "RNA did not lose" in exchange for DNA genes or otherwise. Ultimately, RNA just changed strategy, and in its various forms, it has an active and critically important presence in all cells at all times.

M. Segerståhl (✉)
Department of Biological and Environmental Science, Nanoscience Center, University of Jyväskylä, Jyväskylä, Finland

Present Address: Loviisa, Finland

M. Jalasvuori
Department of Biological and Environmental Science, Nanoscience Center, University of Jyväskylä, Jyväskylä, Finland

PrecisionPhage Oy, Jyväskylä, Finland

G. Witzany (ed.), *Integrative Theory of Evolution - Aspects and Insights*,
https://doi.org/10.1007/978-3-032-24295-2_15

1 Background and Motivation

RNA is a key substance of biological life that dates back to ancient biochemical systems, predating the evolution of the last universal common ancestor (LUCA), or a community of common ancestors, from which extant cellular life descends (Koonin and Martin 2005). The initial understanding of cellular RNA molecules was mainly based on the significant universal contribution of messenger RNA (mRNA), transfer RNA (tRNA), and ribosomal RNA (rRNA) molecules to the mechanisms of protein-coding gene expression. What was realized is that functional RNA sequence is, in general, synonymous with (functional) life. This, on the other hand, stems the question of how much of life is linked to RNA. Here, we argue that ultimately RNA just changed its original strategy, and life, as we know it, is derived from this strategic adaptation.

The past few decades of experimental work have led to the remarkable realization that RNA presence in cells is much more diverse and complex than previously thought. In addition to tRNA, mRNA, and rRNA, there are also many additional types of RNAs with diverse functional roles, including regulators of gene expression (reviewed, e.g., by Morris and Mattick 2014). The mechanisms of action vary. Not only can noncoding (nc) RNAs modify other RNA transcripts and regulate the dynamics of many different steps of protein-coding gene expression, but they can also influence what genes can be active in the first place by introducing epigenetic modifications, such as methylations or adenylations, to DNA nucleotides, that can change expression patterns without changing the actual genomic sequence of DNA strands. Many ncRNAs can regulate cellular development and organismal differentiation and act as causal agents of phenotypic change that may even be inherited by progeny, meaning that ncRNAs can introduce evolutionary novelty to the genetic information content of cells (reviewed, e.g., in Fitz-James and Cavalli 2022; Sánchez-Romero and Casadesús 2020) and, together with other distinct mechanisms, such as horizontal gene transfer (HGT), transposable elements, and gene duplication (e.g., Schaack et al. 2010; Soucy et al. 2015), give rise to new phenotypic lineages influencing evolutionary trajectories.

The realization that RNA strands are significant functional components of cell biology, regulating and generating phenotypic diversity, has had an impact on evolutionary theory. It has even been argued that in its current state, evolutionary theory is unable to accommodate and adequately explain the adaptive significance behind the observation that natural living cells have many kinds of nonrandom cell biological mechanisms for generating phenotypic diversity and that some revision or update of the theory is required (see, e.g., Shan 2024; Witzany 2021, for general accounts). In practical observational terms, various studies have, of course, shown phenotypic plasticity in organisms arising from various environmental or other factors. Nevertheless, this issue is a complex problem that touches many areas of biological study where phenotype ultimately links to RNA in one way or another and it poses a *challenge for the long-term dogma that all generation of diversity (*i.e., *material for natural selection) comes down to the emergence of new hereditary* *phenotypes by random mutation* of *DNA nucleotides*. Resolving it may well require making major changes in some previously preferred paradigmatic viewpoints, and it might be especially important to find new methods for holistic management and integration of information between different fields, especially between molecular cell biology and evolutionary biology (Shan 2024). Various perspectives, such as the organismal or genetic level, can be taken to better understand the reality behind all this. Here, we, in a sense, interpolate viewpoints and provide an account from the perspective of RNA.

RNA is a key part of cellular life. Not only does it have the above-mentioned functions, but with increasing knowledge of cellular long noncoding RNAs (lncRNAs), the research is already progressing toward potential applications even in medical studies (Mattick et al. 2023). Having a sound theoretical understanding of the evolutionary foundation of the numerous different kinds of noncoding RNAs is not merely an academic issue, but one that connects us with the very basis of our ability to understand the cell and its behavior—as an individual unit and as a part of the whole. In an attempt to better understand the many roles that RNA has in cellular life, we focus in this article on the topic of combining evolutionary and cell biological viewpoints in a new way, using a novel system-level model of cell biology as a framework (Segerståhl and Segerståhl 2022). This model was initially developed to describe the main material components and production events for realizing the system-level MIC properties (i.e., metabolism, hereditary information, and compartmentalization) of biological cellular organization. These properties are well recognized in the theoretical study of life (reviewed in Cornish-Bowden and Cárdenas 2020), major evolutionary transitions (Smith and Szathmáry 1997), and the minimal chemical foundations of life (Buddingh' and Van Hest 2017; Fellermann et al. 2007; Gánti 1975, 2003).

We will use the new MIC model of biomolecular cellular organization for holistic system-evolutionary contextualization of RNA presence in cell biology. At the same time, we acknowledge the RNA world hypothesis, which posits that cellular life was preceded by an era of molecular-level RNA "life" where RNA molecules were both chemical catalysts and carriers of genetic information (Fine and Pearlman 2023; Gilbert 1986; Müller et al. 2022). This era came to an end when cells evolved and the information storage function was transferred to DNA and the catalytic function to protein enzymes, and RNA molecules became strongly connected to the mechanisms of cellular protein production. We will be using the RNA world hypothesis as an additional systemic reference for examining a new practical viewpoint for the characterization of RNA molecules and their presence in cells. And as synthetic biology already studies adaptive evolutionary capabilities of molecular-level entities (Arnold 2017), we will go one step further and suggest that RNA molecules of cellular life can even be seen as bona fide molecular-level counterparts of organismal entities with genetically determined individual-level phenotypes contributing to their chance to (re)produce and survive in living cells. This is not a mere play with perspectives, but instead, it can help us build a view where we can understand the real actors of the biomolecular realm, especially RNA. The extended view that we propose provides more comprehensive ground for the translation of the resulting, more integrated theoretical understanding to meet the practice.

2 Metabolism, Hereditary Information, and Compartmentalization: A Conceptual System-Level Outline of Cellular Biomolecular Organization

The whole-cell model that we use as a framework for considering RNA presence in cells is an original model by Segerståhl and Segerståhl (2022) describing a core set of components and processes that are universally needed by all cells to maintain a material foundation for realizing the MIC properties of biological cellular organization (Fig. 15.1). Unlike many minimal models[1] of living cells, the MIC model specifically remains connected to the general current knowledge of molecular cell biology, while at the same time, it is connected to the theoretical study of life and living system formalisms (Cornish-Bowden and Cárdenas 2020; Gánti 1975, 2003; Varela et al. 1974). Instead of repeating biochemical detail, the MIC network model provides an abstracted view of higher-level modular organizational aspects of molecular cell biology. This required taking terminology to a more general level than what is typical when describing molecular networks of cell biology. The terminology that is used provides a link between the viewpoints of biomolecular research and the more general theoretical study of life and organismal entities. This intermediate level of abstraction describes how molecular cell biology realizes an organization for the cell as a living entity.

An interesting feature of the minimal MIC network model is that it visualizes self-production both for *material closure,* which results from the cell producing the components from which it consists, and for *catalytic closure,* based on the structural properties of the network. It can be defined as an *autocatalytic set,* meaning that the model describes biologically relevant systemic organization of components that collectively are capable of maintaining and reproducing themselves and the network of catalytic interactions that they form (Kauffman 1993), and even as a so-called *reflexively autocatalytic food-dependent (RAF) set*, which means that the network constructs its own catalysts using only simple raw materials that it acquires as "food" (Hordijk and Steel 2015, 2018). In other words, the oval nodes in Fig. 15.1 represent material products that the cell synthesizes and uses for realizing its MIC properties, and *at the same time*, these nodes refer to the components that provide the collective catalytic contribution for the synthesis of these same substances that establish the biological cell as a particular kind of distinctly identifiable organizational arrangement of energy and matter in the natural world.

Cellular system-level compartmentalization is universally realized by being surrounded by a cell membrane. It mediates the flow of energy and matter between the cell and the environment, including the acquisition of new raw materials for metabolic reactions. Protein enzymes are featured in the model as the material foundation on which cellular metabolism and the production of complex

[1] Focused on describing a minimal set of necessary components, aiming for models that contain as few details as possible.

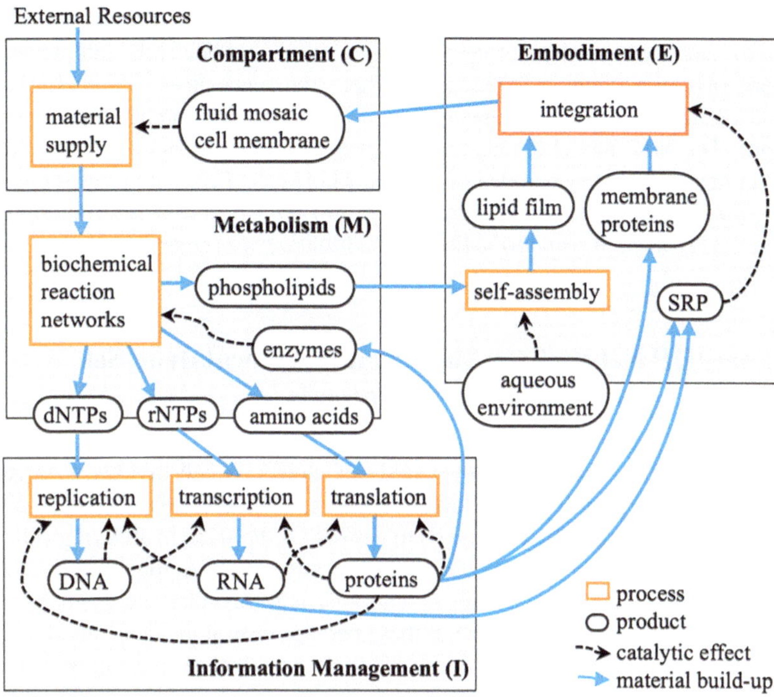

Fig. 15.1 MIC network model of the biological cell. Minimal system view of the universal organizational aspect of molecular cell biology, that metabolism, information management, and compartmentalization (i.e., MIC features) are the main system-level properties, which together with the encapsulation events produce embodiment and closure for an autocatalytic set of components and processes that constitute the cell as a holistic autopoietic living system entity. dNTPs deoxyribonucleoside triphosphates, rNTPs ribonucleoside triphosphates, SRP signal recognition particle. (Modified from Segerståhl and Segerståhl 2022)

biomaterials are based, including the synthesis of cell membrane lipids, nucleotide monomers of RNA and DNA strand synthesis, and amino acids of protein strand synthesis. The cell membrane is the most complex product type, as it consists of chemical components that, even individually, are complex macromolecules (various kinds of lipids and proteins that can be modified or present as multimeric complexes). This complex level of material organization is generated by mechanisms that include signal recognition particle (SRP)-mediated insertion of membrane proteins into the lipid phase and self-organizing behavior of lipids in aqueous environments.

A notable feature of MIC network modularity is the conceptual separation of RNA, DNA, and protein production from the rest of the cell metabolism. There are two main reasons for this. The synthesis of these three types of substances through template-based polymerization of their constituent monomer units—ribonucleotides, deoxyribonucleotides, and amino acids, respectively—into long linear information-bearing strands already makes these events and products unique among all other polymerization events and products of cell biology (e.g., long carbohydrates or fatty acids). In addition to this, while the mechanisms of RNA, DNA, and protein synthesis are operationally autonomous, they form a collectively interdependent network of components and processes where the mechanisms of each process depend on having some components that are produced by another process of the network. The MIC model shows that this network exists as another, subsystemic RAF set inside the larger one that covers the whole cell. *This network will be at the center of our attention as we proceed with this article, and we will be calling it the RDP (RNA, DNA, and protein) network of cell biology for easy reference.*

3 The RDP Network: A Subsystemic Autocatalytic Set of RNA, DNA, and Protein Synthesis

A main incentive of the system-level study of molecular cell biology is to gain a more holistic understanding at the level of functional modules (Hartwell et al. 1999). The main outline of the subsystemic RDP network is the same in all living cells and would have been present in some basic form already in LUCA, together with a lipid membrane, a viral defense mechanism, and a basis for cellular energy metabolism (Moody et al. 2024), although some of these perhaps only at the later stages of early cellular evolution (Jalasvuori and Bamford 2008). Nonetheless, we will be using the RDP network as a distinct reference module against which RNA-related aspects of molecular cell biology can be reflected in an orderly manner. Figure 15.2 provides a more focused view of it.

The process events of the network are transcriptional RNA synthesis, polymerization of new DNA by the mechanisms of semiconservative DNA replication, and ribosomal synthesis of protein strands. The autocatalytic (RAF) network structure arises from the following connections: (1) DNA strands serve as templates both for the synthesis of new DNA and RNA strands; (2) the complex machinery of ribosomal protein synthesis requires cooperative contribution from mRNA, rRNA, and tRNA molecules; and (3) proteins contribute significantly to the production of RNA, DNA, and protein strands. Additionally, the basic structure of the RDP network acknowledges the contribution of RNA primers to the mechanisms of semiconservative DNA replication—all cells need them to initiate DNA replication both on the leading strand and on every Okazaki fragment of the lagging strand. In all this, we turn our attention specifically to the constituent RNAs of this network. It is typical to use the functional perspective to describe what the RNAs "deliver," as mentioned before, but to widen the scope, we provide an additional schematic survey from an individual-level viewpoint of molecular phenotypes. This will be done by analyzing them against a simple formalism of general RNA shapes. This is motivated by the general notion that organismal phenotypes usually contain both reproduction- and survival-related characteristics and justified by molecular cell biology being a realm where chemistry and biology overlap and structural characteristics of biomolecules can be approached from either perspective (Arnold 2017). This will enable making comparisons from a holistic, system-evolutionary perspective of RDP network

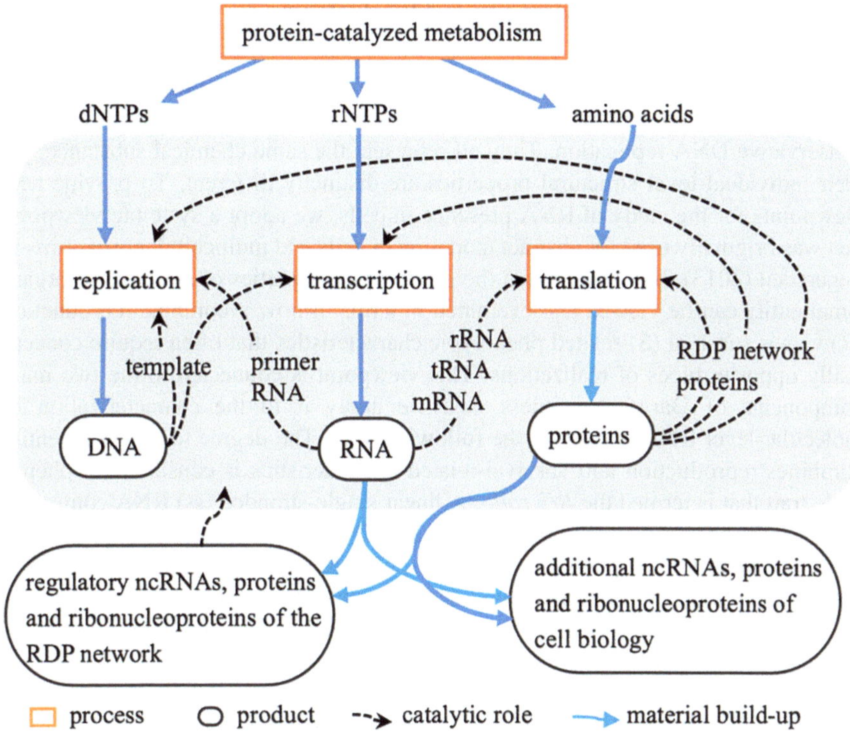

Fig. 15.2 The RDP network of molecular cell biology. A schematic representation of the autocatalytic (RAF) set arrangement of components and processes constituting the biosynthesis network for the production of new RNA, DNA, and protein (i.e., RDP) strands in living cells (gray background). The monomer units of RNA, DNA, and protein strands—ribonucleoside triphosphates (rNTPs), deoxyribonucleoside triphosphates (dNTPs), and amino acids—are considered to represent raw materials ("food")

organization and dynamics between *the current existence of the RDP network RNAs that are locked in the defining core of the RDP network interaction dynamics* and the viewpoints of the *RNA world hypothesis,* regarding the events, selection pressures, and consequences of RNA becoming connected to the DNA–protein synthesis axis of protein-coding gene expression of cellular life.

4 Holistic System-Level View of RNA Presence in Living Cells

Although focusing on a high level of abstraction, the RDP network describes a specific circuit module of molecular cell biology and remains connected to the conventional description paradigms of cell biology and biochemistry. Various types of relationships that exist in cells, both within and between the organizational levels of biopolymers, the RDP network, and the whole cell, can be examined from an RNA-centered holistic systemic perspective.

4.1 Organismal Phenotypes of RNA Entities

The four constituent RNA types of the RDP network are the mRNA, tRNA, and rRNA molecules of protein-coding gene expression and the primer RNA of semi-conservative DNA replication. They all represent the same chemical substance, yet their individual-level structural properties are distinctly different. To provide new viewpoints for the study of RNA presence in cells, we adopt a systemic viewpoint that was originally used for abstract modeling of cells and multicellular organisms in Segerståhl (2013). It considers that the phenotypic properties of any kind of organismal entity can be viewed and examined in terms of how it combine reproduction (R) versus survival (S)-related phenotypic characteristics that often require conceptually opposite types of realizations. This viewpoint is connected to the two main components of Darwinian fitness, and we apply it to the characterization of molecular-level RNA entities in the following way: The degree to which an entity combines reproduction and survival-related characteristics is considered a phenotypic trait that is termed the *R/S ratio*. A linear single-stranded (ss) RNA conformation with the bases fully exposed and available for chemical interactions corresponds to a conceptually open phenotype and has a high R/S ratio. A conceptually closed double-stranded (ds) RNA phenotype where complementary base pairing provides physical protection for sequence identity corresponds to having a low R/S ratio. However, the chemical structure of RNA strands is such that they tend to fold into conceptually intermediate conformations showing different kinds of combinations of ds-stalks and ss-loops. These structurally more complex phenotypes are considered to be conceptually intermediate in the light of the R/S formalism, providing a more balanced contribution to the two components of the R/S ratio. Based on this, a theoretical system axis as visualized in Fig. 15.3 can be formed and used as a reference standard against which R/S phenotypes of real biological RNAs can be compared in an orderly manner.

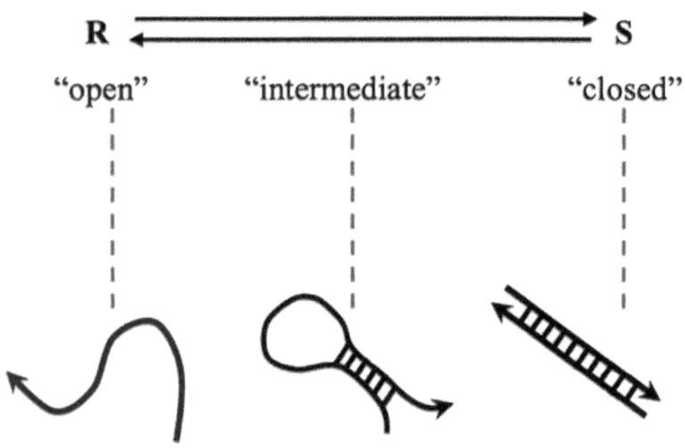

Fig. 15.3 Schematic reference framework of organismal RNA phenotypes based on the ratio of reproduction (R) versus survival (S)-related features in the expressed molecular-level folding pattern and how much it protects the sequence ("open" vs. "closed")

4.2 RNA Molecules of the RDP Network

Based on what is generally known of the four classes of RDP network RNAs, we can observe that they perform their functional roles using very different kinds of naturally evolved, highly specialized R/S phenotypes. The RNA primers of semi-conservative DNA replication are short-lived structures that do not exhibit any individual-level folding according to their sequence to perform their part. Instead, their presence creates functionally important RNA–DNA hybrid locations on chromosomal DNA, and the sequence patterns of primer RNA only need to mirror that of the DNA templates accurately enough to remain in place until removed by cellular machinery. The physiologically expressed phenotype of primer RNA strands could be deemed as being linear from the individual perspective (high R/S ratio) yet, at the same time, being conceptually closed and protected because while bound to the DNA, the RNA bases are chemically inaccessible and thus unavailable for chemical interactions.

In stark contrast to primer RNA, the other three RNA types of the RDP network express specialized individual-level phenotypes, making use of diverse possibilities of RNA folding. The individual-level identity of mRNAs is, above all, based on the long single-stranded sequence patterns of their coding regions. However, real biological mRNA also contains some sequence-dependent secondary structures in the untranslated regions, sometimes even in the coding region (Lewis et al. 2017). These are important for the ability of mRNA to function properly with the rest of the components of the protein-synthesizing genetic translation machinery. Intron splicing is an additional dynamic aspect of conformational mRNA complexity. And while the coding region is eventually only truly functioning by exposing its bases (while being managed by the ribosomal subunits), it can be protected by mRNA-binding proteins from the moment it is transcribed until the moment it is being translated, during which time it effectively exists in a protected and nonoperational, conceptually closed hybrid (ribonucleoprotein) state. Considering all these higher-level organizational aspects, we conclude that real biological mRNA phenotypes are structurally plastic and dynamic. Based on the linear single-strand sequence of the coding regions being the most significant organizational level of the phenotype structure, they can be defined as having, on average, intermediate but R-biased R/S ratios.

tRNAs constitute the phenotypically most uniform class of RDP network RNAs. The prototypic sequence-based tRNA phenotype is a characteristic combination of conceptually open ss-loops and closed ds-stalks, and it is often used as a model example to demonstrate how RNA strands typically fold. Their folding happens efficiently via the formation of a 2D clover-leaf shape before the characteristic 3D L-shaped phenotype is adopted (e.g., Zhang 2024). In addition to the ones of the anticodon loop, there are also other key nucleotides in exact locations of the 1D ss-strand that together define the functional precision of tRNA-ribosome interactions. Numerous posttranscriptional nucleotide modifications add further detail to the shape and the surface charge distribution patterns of functional tRNA molecules and their exact chemical interaction abilities.

And finally, the ribosomal RNA strands. These long, well-structured RNAs express highly complex and precise individual-level 3D phenotypes while containing the oldest functional RNA motifs of cellular life (Bokov and Steinberg 2009). rRNAs have a sequence-based, higher-than-average tendency to form double-stranded segments, but otherwise their grouping together is contextual (based on belonging to the ribosome) rather than based on structural or sequence similarities. Their ability to contribute to cell biology as core components of the RDP network stems from the sequence-based structural properties that enable them to assemble and act in concert with the ribosomal proteins and each other. The rRNA phenotypes are complex, higher-level organizational realizations of RNA folding, with ds-stalks and ss-loops as lower-scale motifs that can interact over longer distances, forming even higher-level structural arrangements while being kept together by the single-strand stretches that extend between the motifs in the 3D ribosomal subunit space. Thus, with the additional support of nucleotide modifications and protein assistance, the pattern formation events of rRNA folding reach into far higher scales of physical organizational complexity than what a conceptual treatment of rRNA sequence patterns merely in the light of the R/S formalism or molecular-level physical chemistry might suggest, and we are, for now, left to wonder how exactly these impressive molecules should be fitted against the observation framework of R/S phenotypes.

Together, these four different kinds of RNAs of the RDP network form the core foundation of RNA presence in living cells. Many aspects of these real biological RNA phenotypes, such as the effective linearity of mRNA or the high-dimensional folding of rRNAs in association with ribosomal proteins, are not revealed directly by the respective linear sequence information that is preserved in DNA genes and knowledge of the folding dynamics of isolated RNA strands (minimizing free energy for thermodynamic stability) (Herschlag et al. 2018; Schroeder et al. 2004). The resulting biologically relevant individual-level properties of these RNA molecules—size, shape, and the distribution of reactive sites on the resulting molecular surface topology—are the key to the type of physical and chemical interactions and organizational behavior that they can have in cells. All this complexity is most relevant not only directly to the concrete material production of new DNA and proteins in living cells but also more profoundly to the system-level maintenance of RDP network integrity that all cells depend on for their holistic existence.

4.3 Noncoding RNAs and the RDP Network

It is currently known that the RNA products of the RDP network's biosynthetic activities include numerous different kinds of ncRNAs with varying lengths, 3D structures, and interaction properties. Especially the discovery of a multitude of lncRNAs with diverse dynamic roles in eukaryotic cell biology has been paradigm-shifting (Mattick et al. 2023; Patil et al. 2013). (l)ncRNAs can regulate and control ongoing gene expression events; cause metabolic, behavioral, or phenotypic change;

control multicellular development or life cycle progression of unicellular organisms; provide protection against viral infections (e.g., RNA interference and CRISPR–Cas systems); and serve as the template for telomere elongation in the eukaryotic telomerase ribonucleoprotein complex, only to mention some examples (see, e.g., Guil and Esteller 2015; Ma et al. 2022, for a wider view).

The RDP network conceptualization provides a new kind of possibility to *formally differentiate between RNAs that belong to the system-defining core components of the RDP network and other types of cellular RNAs*. It enables us to make a novel kind of statement using new terminology that, notably, most (l)ncRNAs perform their functions specifically by interacting with the components and processes of the RDP network, while not belonging to the system-defining core components of the network (see Fig. 15.2). But they can regulate and modulate the synthesis of new RNA, DNA, and protein strands in many concrete ways using different kinds of mechanisms and be involved, for instance, in the insertion (or removal) of epigenetic markings or nonrandom genetic change to the chromosomal DNA, influencing thereby gene expression patterns.

The RDP network also produces all proteins of the cell—the constituent core proteins of the RDP network, metabolic enzymes, structural proteins, and functional proteins, including those that are involved in signal transduction and information processing at the whole-cell level. Many protein-instantiated signals that originate in different parts of the cell and travel inside the cell space, for instance, in response to extracellular stimuli, in a sense "return" to the level of the RDP network that initially synthesized the respective proteins, to modulate and modify RDP-network dynamics, with the final steps being cooperatively completed with additional (l)ncRNA and protein regulators of the RDP network core components. The resulting effects on gene expression can be very precise and effective and may even significantly change the state of the whole cell (Lu and Lionnet 2021; Paget 2015). From a wider system-level perspective, (l)ncRNAs thus contribute alongside proteins to the holistic ability of biological cells to lead adaptive evolutionary lives in various kinds of environments.

4.4 Unique Opportunity for RNA Presented by the Membrane Dimension

The cell membrane is a complex organizational substance dimension of macromolecular arrangements providing significant stability for cellular evolution, with current life being based on only three main types of solutions: *the eubacterial, archaeal, and eukaryotic cell membranes* (Koga 2011; Nicolson 2014; Zhukov and Popov 2023). But within each type, the compositional and structural details are very malleable, allowing for functional adaptive variability and phenotypic plasticity. The exact selection of proteins and lipids that are present at a particular instance, and their spatial arrangement within the membrane, are key parameters that cells adjust to maintain a properly functioning cell membrane as the system-level interface that connects them to the wider environment.

The membrane is a unique dimension for RNA in the cell system. It generates an intracellular niche-like space for localized coexistence and proximity of chemical substances and the chemical reaction networks of enzyme-based biochemistry (Buddingh' and Van Hest 2017; Torday 2016). The shape, size, and stability of early (proto)cellular lipid membrane assemblies probably varied (Deamer 2016; Lombard et al. 2012), and permanent compartmentalization may have been a later adaptation coupled to the evolution of membrane-bending proteins (Laiterä and Lehto 2009; Mulkidjanian et al. 2009). Stabilization of the cell–environment interface was a significant enabling factor for long-term adaptive coevolution of cell membrane compositions, cellular biochemical networks, and RDP network dynamics, as natural environments are seldom non-changing. An adaptive, plastic, well-functioning gene expression–cell membrane axis is a significant system-level feature by which cellular life and the RNAs in it claim living space in nature, and the cell signaling and control aspects featured in Sect. 4.3 are part of this holistic information-based big picture of molecular cell biology.

The RDP network controls the membrane dimension also rather directly by synthesizing the metabolic enzymes of membrane lipid synthesis as well as the proteins of functional cell membranes. In addition to this, the predominant mechanism that all cells use to insert proteins into the lipid phase is managed by the ancient SPR ribonucleoprotein complex that includes an RNA type that has been present in cells since they first emerged (Harris and Goldman 2021; Rosenblad et al. 2009). We might even extrapolate that from a general system-evolutionary perspective, there is no particular reason why the RDP network could not use its ability to control the cell membrane dimension to manage whole cells so that they become constituent units of even higher-level biological assemblies of complex organismal existence (e.g., colony formation and multicellular development). Maybe the RDP network could be seen as some kind of built-in enabling factor behind the ability of cells to find ever-new ways to exist in nature, if necessary, by evolving into *increasingly more complex types of realizations of MIC system organization.* This type of integrating systemic view of the large-scale evolutionary dynamics of molecular cell biology could perhaps be taken under closer examination in the future using, for instance, the kind of practical abstractions that we are presenting in this article as an attempt to deal with the organizational evolutionary complexity of RNA presence in living cells.

The membrane aspect might be especially connected to the origin and evolution of the eukaryotic cell, which, even in its simplest form, includes distinct additional membrane-based compartments. The nuclear envelope, endomembrane system, mitochondria, chloroplasts and other plastids, endosymbionts, and multicellularity are true examples of membrane-based higher-level modularity in the physical realization of cells, and it is a defining cell-level characteristic of eukaryotic life (Dacks and Field 2007). Why such higher-level assemblies would stabilize (as has occurred multiple times during the natural evolution history of life) is another question and part of the ongoing issue of how evolutionary theory can explain why the lower-level units do not disrupt higher-level assemblies (Bourrat et al. 2022; Smith and Szathmáry 1997; West et al. 2015). What we do know, however, is that it is precisely in the eukaryotes that epigenetics and (l)ncRNAs have been found to have an especially significant and powerful presence as regulatory and control elements of the RDP network dynamics and gene expression (Duclos et al. 2019; Mattick et al. 2023).

5 Integrating Viewpoints

In the course of this article, we have been integrating multiple different viewpoints in search of a new kind of holistic system-level view of the type of relationship that exists between RNA and cell biology. This integration has, however, been rather implicit, as it is more about the methodology than a main thing to notice, but now, we provide a breakdown of the four main perspectives that we have used.

The first viewpoint has been an organismal molecular-level view of individual RNA entities that belong to the constituent core components of the highly evolved and organized RDP network. They have been described as R/S entities that rely on their individual-level evolvable phenotypes for their collective ability to contribute to protein synthesis in living cells and, thereby, to their own cell-bound existence in nature.

The second one is the conventional viewpoint of the central dogma (Crick 1970), which focuses on the transfer of genetic hereditary sequence information between RNA, DNA, and protein molecules. This can be seen, for example, in the terminology that is used in Figs. 15.1 and 15.2, where the process nodes of RNA and protein synthesis are termed "transcription" and "translation," respectively, although knowing very well that the correspondence of these terms to all that the nodes represent in these system models might not be an exact match. Using these terms is, however, a practical way to maintain a connection to the conventional prevailing description paradigm of molecular cell biology while searching for novel conceptual ways to describe further abstract organizational aspects of cell biology.

The third viewpoint is a system-organizational viewpoint that focuses on the RDP network being organized as an autocatalytic (RAF) set. Autocatalytic sets have been used to study the chemical origins of life and real biochemical networks of living cells (Hordijk and Steel 2015, 2018; Sousa et al. 2015; Xavier et al. 2020). However, the MIC model is, to our knowledge, the first model of real biological cells that clearly identifies the RDP network as a distinct organizational module of cell biological network circuitry by showing it is a RAF set. The fourth viewpoint is that of the holistic MIC model approach, where the RDP network is one part in a larger network abstraction that focuses on the system-level analysis of higher-level organizational aspects of cell biology and the whole cell. For instance, in the MIC model, all RNA contributions that support the "translation" node events are grouped together and represented by one arrow to avoid any unnecessary detail in what is intended to be a minimal model of the whole cell.

When studying complex system phenomena, the choice of viewpoints that are going to be used is part of the larger issue of how to find the right kind of questions to be asked, what kind of methods are selected, and how the results are planned to be reported, and ideally, the choice of viewpoints should be based on active contemplation of different alternatives and traceable, informed decisions (e.g., Keating et al. 2001; Levins 2006). This could be something to focus on more carefully when studying complex issues of biological organization.

6 System-Evolutionary Contemplations

We have provided an RNA-centered, system-level view of the main types of RNAs that are fundamental to all cellular life. This view uses new kinds of holistic viewpoints that have been provided especially by the MIC network model of biological cellular organization. This has been the first instance of applying the abstract viewpoints of the MIC model in a practical way to better understand the evolutionary and functional relationships that connect RNAs to the rest of molecular cell biology.

6.1 *Distributed System for the Usage and Storage of RNA Information*

From a holistic system-evolutionary perspective, the RDP network presents itself as a distributed system for the usage and storage of RNA information, providing a stable foundation for the adaptive evolutionary existence of RNA entities in living cells. The RNA, DNA, and protein components of the network (and of the whole cell) provide three qualitatively different kinds of material interfaces for presenting structural patterns of the cell's genetic hereditary information to the physical environment. While proteins connect the network to the realm of enzyme-catalyzed chemical reactions and energy metabolism, alpha-helical dsDNA is a superior substance for the physical preservation of sequence information (in comparison to dsRNA), and it is normally never degraded by active cellular mechanisms. With the many kinds of additional replication-based DNA repair mechanisms, the genetic information of chromosomal DNA forms a permanent record of the set of RNA molecules that inhabit the cell. From an RNA perspective, chromosomal DNA has effectively been assigned to take on the task of protecting RNA sequence information (sensu R/S formalism, see Sect. 4.1), and every time that the protein components of the gene expression machinery access chromosomal DNA, they (re)produce RNA using the DNA record of RNA information as template.

With the "parental" role allocated to the much more stable chromosomal DNA that contains the template strands of cellular RNA molecules, the latter can be seen as a "sterile" class of evolvable entities that are reproduced by proteins and stored by DNA. Individual RNAs can be temporarily lost without losing their defining information from the system, allowing RNA molecules to have, from an organismal perspective, finite life spans. RNA, DNA, and protein strands also use their sequence-based phenotypes to interact with each other in concrete physical ways, and their structures can coevolve based on having their defining information held together by chromosomal DNA. But as the RDP network is a chemical system, this requires compartmentalization, as otherwise the molecular components would diffuse away and the RDP network would cease to exist.

6.2 *Did RNA Suffer or Benefit from the Transition to Cellular Life?*

Evolutionary study has been treating "independence" for a long time as a key criterion of evolutionary "success," and the transition of RNA "life" from an era of "independent" RNA world existence to the type of presence that it has in currently living cells with DNA and proteins is typically referred to in a negative sense. The standard view that has remained implicitly present in the way in which the RNA world hypothesis discusses the end of the RNA world era is, that RNA suffered some kind of loss or defeat when it became part of cell biology (Gilbert 1986). However, given the new kinds of viewpoints that we have developed in this article, we are willing to challenge this view.

RNA could have been all the time leading an active and adaptive coevolutionary molecular-level "life" in living cells together with DNA, proteins, and lipid membranes. Cooperation and altruism are currently widely recognized as potentially highly significant supporting factors of adaptive organismal evolution, based on mounting experimental and empirical evidence from natural examples (e.g., Pennisi 2009). In this scenario, RNA is taking full advantage of the physical properties of multiple additional substance dimensions that can express its information-related properties, each in its own respective way, resulting in a far more complex and evolutionarily efficient biomolecular living system organization than what RNA alone could have accomplished. In this scenario, RNA exists ultimately by managing the life that is led by biological cells. If viewed from this perspective, the stabilization of the RDP network as a fundamental early aspect of cellular biomolecular organization would have been a major advantageous evolutionary event for (populations of) RNA "life," releasing RNA from the double-duty constraint of having to be both the catalysts as well as the keepers of their own system-defining hereditary information at all times.

6.3 On Stability and Variability

The main structure of the RDP network and the MIC organization of the cell have remained the same for billions of years in connection to remarkable evolutionary emergence, stabilization, and diversification of numerous different kinds of additional variable features that have enabled cells to evolve and adapt and eventually to occupy nearly all kinds of Earth environments imaginable. Cellular life has tolerated, and in some cases perhaps even been looking for possible new ways to diversify genomes in support of dynamic, explorative molecular cell biology that can extend life to ever-new ecological realms and spatiotemporal resolution scales of biological organization. The total complexity of RNA presence in currently living cells includes the everlasting coexistence of cellular life with viruses and the nowadays well-recognized consequences of HGT and how it may shape the divergence of genetic codes (Jalasvuori 2020). In the light of all this, there is a distinct possibility that rather than maximizing the preservation of the cellular genome as unchanged and "intact" as possible, the ultimate system-level evolutionary guideline of molecular cell biology might be *the preservation of RDP network integrity and functional integrity of the cell as a holistic MIC system entity*. Endosymbionts, plasmids, organelles, and HGT would fall relatively effortlessly into this bigger picture. These genomes might be small, but when active, they manifest *yet another way in which RNA information can be connected to biological cellular life*. There are organelles that have extremely reduced genomes (e.g., the mitochondrion) and might not use semiconservative replication to replicate their remaining chromosomal DNA, but they hold on to having an RDP network in some form, not just as a reminder of their past as cell-level MIC system entities but it keeps them still partly organismal (Kelly 2021; Roger et al. 2017) as opposed to just being complex macromolecular assemblies.

The large-scale evolution of molecular cell biology has also been very dynamic from an energetic perspective, especially in connection to historical macroevolutionary transitions when entirely new kinds of cells and organisms have emerged (Judson 2017; Okie et al. 2016). The predominant forms of energy that cells preferentially use and harvest from the environment have changed and diversified many times in the course of the natural history of biological evolution, yet the basic structure of the RDP network and of the MIC system organization of molecular cell biology have remained the same, as if being biological invariant organizational constants that define cells as living system entities. All this could be seen as having provided further support for active evolutionary RNA presence in living cells. From an RNA perspective, the amount of ongoing gene expression is basically the quantity that determines how active and reproductive an RNA presence is on Earth. The cutoff for what type and amount of novelty is good and what is bad, and what level of foreign elements or alterations can be tolerated, ultimately comes down to the reproduction and survival ability (the R/S ratio) of cells, whether existing alone or as constituent components of higher-level organizational assemblies and entities. This ratio must be a highly complex variable, as we must not forget that all viral

RNA is also produced in cells, and even virus evolution exploits the full range of strategic alternatives in terms of how intensely viral genes are expressed and how much stress infections impose on the cells and organisms (Jalasvuori and Koonin 2015).

6.4 Ribosomal RNA and a Possible General Selection Pressure

We finish this article with a brief discussion of the ribosome. It can be approached as an ancient organizing center of RDP network cell biology (e.g., Fer et al. 2025). Ribosomal proteins and rRNAs are closely coupled both from the evolutionary perspective and from the cell biological viewpoint of realizing the RDP network events in cells (Reuveni et al. 2017; Root-Bernstein and Root-Bernstein 2019). The early evolution of ribosomal assemblies was probably strongly connected to the aspect of protein synthesis, as the rRNA ribozyme that catalyzes peptide bond formation in all cells has one of the most ancient and preserved RNA patterns of cellular life (Bokov and Steinberg 2009). But in turn, ribosomal proteins are needed alongside various RNA-binding proteins to assist in the high-dimensional folding process of long rRNAs, which begins as soon as the strands begin to form (Rodgers and Woodson 2021), and ribosomes possibly still exhibit complex organizational features that evolved initially to optimize their own autocatalytic self-assembly as ribonucleoproteins (Reuveni et al. 2017).

In the bigger picture of organizational RDP network evolution and in search of holistic new perspectives and system-level viewpoints, we are left to wonder if the tight ribosomal RNA–protein couplings could be evolutionarily connected to avoiding the formation of dsRNA. This is because cells and viruses have most likely existed side by side in nature as long as cells have existed (Jalasvuori and Bamford 2008, 2009), as even LUCA most likely already had a viral defense mechanism (Moody et al. 2024), and a distinct unifying aspect of cellular viral defense mechanisms is to use dsRNA to detect infections. Thus, it may have been advantageous for early RNAs of (proto)cellular life to be associated with proteins, thereby avoiding inherent formation of long double-stranded regions that could result from folding, or binding between complementary strands. Removing the risk that dsRNA is "domestic" would make it easier for cells to protect RDP network integrity and thereby themselves against genetic invaders. Somewhat interestingly, using dsDNA as the storage medium (possibly in early evolutionary stages in the form of RNA–DNA ds-hybrids), as opposed to relying on dsRNA (the formal alternative in an RNA-only scenario, see Sect. 4.1), could also be seen from the viewpoint of this same potential selection pressure of allocating the use of dsRNA to viral defense mechanisms. Perhaps even the distinct absence of RNA-dependent RNA synthesis from the repertoire of system-defining cell biological mechanisms could be linked to this same potential selection pressure, as it would produce dsRNA if the old and new strands were not separated. We know that the reason for this absence is not that such mechanisms could not exist, as, for example, RNA phages use RNA-dependent mechanisms of RNA synthesis (Jalasvuori and Koonin 2015).

7 Conclusion

In this article, we have shown new ways to apply novel abstract models of biomolecular cellular organization to better understand the evolutionary complexity of real molecular cell biology, and how new kinds of system-level viewpoints can be connected to existing ones in new ways. This is part of a wider search for concrete, practical ways to reach some kind of new, more holistic theoretical understanding of the biomolecular foundations of complexity evolution of cells, and the multitude of biological mechanisms that can give rise to phenotypic plasticity and generate diversity at different organizational levels of biological organismal life and evolution. There is certainly still a long way to go. However, this system-oriented re-examination of existing knowledge of molecular cell biology and RNA presence in living cells that we have now conducted, might shed some light on the question of why RNA presence in living cells would be a much more complex and diverse phenomenon than what was initially believed. Furthermore, it no longer seems entirely confusing to us, that in the course of the long adaptive evolutionary history of biological life many kinds of nonrandom mechanisms have evolved for causing, as well as for dealing with, epigenetic changes of gene expression and nonrandom genetic alterations of chromosomal DNA sequences, generating phenotypic diversity effectively in an open-ended way, being thereby a major force of biological evolutionary life behind the existence of all kinds of cells and organisms that ever have inhabited the Earth.

References

Arnold FH (2017) Directed evolution: bringing new chemistry to life. Angew Chem Int Ed Engl 57(16):4143. https://doi.org/10.1002/anie.201708408

Bokov K, Steinberg S (2009) A hierarchical model for evolution of 23S ribosomal RNA. Nature 457:977–980. https://doi.org/10.1038/nature07749

Bourrat P, Doulcier G, Rose CJ et al (2022) Tradeoff breaking as a model of evolutionary transitions in individuality and limits of the fitness-decoupling metaphor. elife 11:e73715. https://doi.org/10.7554/eLife.73715

Buddingh' BC, Van Hest JC (2017) Artificial cells: synthetic compartments with life-like functionality and adaptivity. Acc Chem Res 50(4):769–777. https://doi.org/10.1021/acs.accounts.6b00512

Cornish-Bowden A, Cárdenas ML (2020) Contrasting theories of life: historical context, current theories. In search of an ideal theory. Biosystems 88:104063. https://doi.org/10.1016/j.biosystems.2019.104063

Crick F (1970) Central dogma of molecular biology. Nature 227(5258):561–563. https://doi.org/10.1038/227561a0
Dacks JB, Field MC (2007) Evolution of the eukaryotic membrane-trafficking system: origin, tempo and mode. J Cell Sci 120(17):2977–2985. https://doi.org/10.1242/jcs.013250
Deamer D (2016) Membranes and the origin of life: a century of conjecture. J Mol Evol 83(5): 159–168. https://doi.org/10.1007/s00239-016-9770-8
Duclos KK, Hendrikse JL, Jamniczky HA (2019) Investigating the evolution and development of biological complexity under the framework of epigenetics. Evol Dev 21(5):276–293. https://doi.org/10.1111/ede.12301
Fellermann H, Rasmussen S, Ziock HJ et al (2007) Life cycle of a minimal protocell—a dissipative particle dynamics study. Artif Life 13(4):319–345. https://doi.org/10.1162/artl.2007.13.4.319
Fer E, Yao T, McGrath KM et al (2025) The origins and evolution of translation factors. Trends Genet 41(7):590–600. https://doi.org/10.1016/j.tig.2025.02.004
Fine JL, Pearlman RE (2023) On the origin of life: an RNA-focused synthesis and narrative. RNA 29(8):1085–1098. https://doi.org/10.1261/rna.079598.123
Fitz-James MH, Cavalli G (2022) Molecular mechanisms of transgenerational epigenetic inheritance. Nat Rev Genet 23:325–341. https://doi.org/10.1038/s41576-021-00438-5
Gánti T (1975) Organization of chemical reactions into dividing and metabolizing units: the chemotons. Biosystems 7(1):15–21. https://doi.org/10.1016/0303-2647(75)90038-6
Gánti T (2003) The principles of life. Oxford University Press, New York
Gilbert W (1986) The RNA world. Nature 319:618. https://doi.org/10.1038/319618a0
Guil S, Esteller M (2015) RNA–RNA interactions in gene regulation: the coding and noncoding players. Trends Biochem Sci 40(5):248–256. https://doi.org/10.1016/j.tibs.2015.03.001
Harris A, Goldman AD (2021) The very early evolution of protein translocation across membranes. PLoS Comput Biol 17(3):e1008623. https://doi.org/10.1371/journal.pcbi.1008623
Hartwell LH, Hopfield JJ, Leibler S et al (1999) From molecular to modular cell biology. Nature 402(Suppl 6761):C47–C52. https://doi.org/10.1038/35011540
Herschlag D, Bonilla S, Bisaria N (2018) The story of RNA folding, as told in epochs. Cold Spring Harb Perspect Biol 10(10):a032433. https://doi.org/10.1101/cshperspect.a032433
Hordijk W, Steel M (2015) Autocatalytic sets and boundaries. J Syst Chem 6(1):1. https://doi.org/10.1186/s13322-014-0006-2
Hordijk W, Steel M (2018) Autocatalytic networks at the basis of life's origin and organization. Life (Basel) 8(4):62. https://doi.org/10.3390/life8040062
Jalasvuori M (2020) Silent rain: does the atmosphere-mediated connectivity between microbiomes influence bacterial evolutionary rates? FEMS Microbiol Ecol 96(7):fiaa096. https://doi.org/10.1093/femsec/fiaa096
Jalasvuori M, Bamford JK (2008) Structural co-evolution of viruses and cells in the primordial world. Orig Life Evol Biosph 38(2):165–181. https://doi.org/10.1007/s11084-008-9121-x
Jalasvuori M, Bamford JK (2009) Did the ancient crenarchaeal viruses from the dawn of life survive exceptionally well the eons of meteorite bombardment? Astrobiology 9(1):131–137. https://doi.org/10.1089/ast.2007.0189
Jalasvuori M, Koonin EV (2015) Classification of prokaryotic genetic replicators: between selfishness and altruism. Ann N Y Acad Sci 1341(1):96–105. https://doi.org/10.1111/nyas.12696
Judson OP (2017) The energy expansions of evolution. Nat Ecol Evol 1(6):0138. https://doi.org/10.1038/s41559-017-0138
Kauffman SA (1993) The origins of order: Self-organization and selection in evolution. Oxford University Press, New York
Keating CB, Kauffmann P, Dryer D (2001) A framework for systemic analysis of complex issues. J Manag Dev 20(9):772–784. https://doi.org/10.1108/02621710110405479
Kelly S (2021) The economics of organellar gene loss and endosymbiotic gene transfer. Genome Biol 22:345. https://doi.org/10.1186/s13059-021-02567-w
Koga Y (2011) Early evolution of membrane lipids: how did the lipid divide occur? J Mol Evol 72: 274–282. https://doi.org/10.1007/s00239-011-9428-5

Koonin EV, Martin W (2005) On the origin of genomes and cells within inorganic compartments. Trends Genet 21(12):647–654. https://doi.org/10.1016/j.tig.2005.09.006

Laiterä T, Lehto K (2009) Protein-mediated selective enclosure of early replicators inside of membranous vesicles: first step towards cell membranes. Orig Life Evol Biosph 39(6): 545–558. https://doi.org/10.1007/s11084-009-9171-8

Levins R (2006) Strategies of abstraction. Biol Philos 21(5):741–755. https://doi.org/10.1007/s10539-006-9052-8

Lewis C, Pan T, Kalsotra A (2017) RNA modifications and structures cooperate to guide RNA–protein interactions. Nat Rev Mol Cell Biol 18:202–210. https://doi.org/10.1038/nrm.2016.163

Lombard J, López-García P, Moreira D (2012) The early evolution of lipid membranes and the three domains of life. Nat Rev Microbiol 10:507–515. https://doi.org/10.1038/nrmicro2815

Lu F, Lionnet T (2021) Transcription factor dynamics. Cold Spring Harb Perspect Biol 13(11): a040949. https://doi.org/10.1101/cshperspect.a040949

Ma X, Zhao F, Zhou B (2022) The characters of non-coding RNAs and their biological roles in plant development and abiotic stress response. Int J Mol Sci 23(8):4124. https://doi.org/10.3390/ijms23084124

Mattick JS, Amaral PP, Carninci P et al (2023) Long non-coding RNAs: definitions, functions, challenges and recommendations. Nat Rev Mol Cell Biol 24:430–447. https://doi.org/10.1038/s41580-022-00566-8

Moody ER, Álvarez-Carretero S, Mahendrarajah TA et al (2024) The nature of the last universal common ancestor and its impact on the early Earth system. Nat Ecol Evol 8(9):1654–1666. https://doi.org/10.1038/s41559-024-02461-1

Morris K, Mattick JS (2014) The rise of regulatory RNA. Nat Rev Genet 15(6):423–437. https://doi.org/10.1038/nrg3722

Mulkidjanian AY, Galperin MY, Koonin EV (2009) Co-evolution of primordial membranes and membrane proteins. Trends Biochem Sci 34(4):206–215. https://doi.org/10.1016/j.tibs.2009.01.005

Müller F, Escobar L, Xu F et al (2022) A prebiotically plausible scenario of an RNA–peptide world. Nature 605:279–284. https://doi.org/10.1038/s41586-022-04676-3

Nicolson GL (2014) The fluid-mosaic model of membrane structure: still relevant to understanding the structure, function and dynamics of biological membranes after more than 40 years. Biochim Biophys Acta 1838(6):1451–1466. https://doi.org/10.1016/j.bbamem.2013.10.019

Okie JG, Smith VH, Martin-Cereceda M (2016) Major evolutionary transitions of life, metabolic scaling and the number and size of mitochondria and chloroplasts. Proc Biol Sci 283(1831): 20160611. https://doi.org/10.1098/rspb.2016.0611

Paget MS (2015) Bacterial sigma factors and anti-sigma factors: structure, function and distribution. Biomolecules 5(3):1245–1265. https://doi.org/10.3390/biom5031245

Patil VS, Zhou R, Rana TM (2013) Gene regulation by non-coding RNAs. Crit Rev Biochem Mol Biol 49(1):16–32. https://doi.org/10.3109/10409238.2013.844092

Pennisi E (2009) On the origin of cooperation. Science 325:1196–1199. https://doi.org/10.1126/science.325_1196

Reuveni S, Ehrenberg M, Paulsson J (2017) Ribosomes are optimized for autocatalytic production. Nature 547:293–297. https://doi.org/10.1038/nature22998

Rodgers ML, Woodson SA (2021) A roadmap for rRNA folding and assembly during transcription. Trends Biochem Sci 46(11):889–901. https://doi.org/10.1016/j.tibs.2021.05.009

Roger AJ, Muñoz-Gómez SA, Kamikawa R (2017) The origin and diversification of mitochondria. Curr Biol 27(21):R1177–R1192. https://doi.org/10.1016/j.cub.2017.09.015

Root-Bernstein R, Root-Bernstein M (2019) The Ribosome as a missing link in prebiotic evolution III: over-representation of tRNA- and rRNA-like sequences and plieofunctionality of ribosome-related molecules argues for the evolution of primitive genomes from ribosomal RNA modules. Int J Mol Sci 20(1):140. https://doi.org/10.3390/ijms20010140

Rosenblad MA, Larsen N, Samuelsson T et al (2009) Kinship in the SRP RNA family. RNA Biol 6(5):508–516. https://doi.org/10.4161/rna.6.5.9753

Sánchez-Romero MA, Casadesús J (2020) The bacterial epigenome. Nat Rev Microbiol 18:7–20. https://doi.org/10.1038/s41579-019-0286-2
Schaack S, Gilbert C, Feschotte C (2010) Promiscuous DNA: horizontal transfer of transposable elements and why it matters for eukaryotic evolution. Trends Ecol Evol 25(9):537–546. https://doi.org/10.1016/j.tree.2010.06.001
Schroeder R, Barta A, Semrad K (2004) Strategies for RNA folding and assembly. Nat Rev Mol Cell Biol 5:908–919. https://doi.org/10.1038/nrm1497
Segerståhl M (2013) Formal model of living systems. In: ECAL 2013: The twelfth European conference on artificial life, Taormina, September 2013. Artificial Life Conference Proceedings. One Rogers Street, Cambridge, MA 02142-1209, USA journals-info@ mit. edu. MIT Press, pp 745–749. https://doi.org/10.1162/978-0-262-31709-2-ch106
Segerståhl M, Segerståhl B (2022) Modeling the cell as a network of parallel processes—a new approach. In: ALIFE 2022: The 2022 conference on artificial life, July 2022. Artificial life conference proceedings. One Rogers Street, Cambridge, MA 02142-1209, USA journals-info@ mit. edu. MIT Press, p 72. https://doi.org/10.1162/isal_a_00506
Shan Y (2024) The extended evolutionary synthesis: an integrated historical and philosophical examination. Philos Compass 19(6):e13002. https://doi.org/10.1111/phc3.13002
Smith JM, Szathmáry E (1997) The major transitions in evolution. Oxford University Press, Oxford
Soucy S, Huang J, Gogarten J (2015) Horizontal gene transfer: building the web of life. Nat Rev Genet 16:472–482. https://doi.org/10.1038/nrg3962
Sousa FL, Hordijk W, Steel M et al (2015) Autocatalytic sets in *E. coli* metabolism. J Syst Chem 6: 4. https://doi.org/10.1186/s13322-015-0009-7
Torday JS (2016) The cell as the first niche construction. Biology 5(2):1. https://doi.org/10.3390/biology5020019
Varela FG, Maturana HR, Uribe R (1974) Autopoiesis: the organization of living systems, its characterization and a model. Biosystems 5(4):187–196. https://doi.org/10.1016/0303-2647(74)90031-8
West SA, Fisher RM, Gardner A et al (2015) Major evolutionary transitions in individuality. PNAS 112(33):10112–10119. https://doi.org/10.1073/pnas.1421402112
Witzany G (2021) Preliminary remarks to an integrative theory of evolution. Theor Biol Forum 114(2):41–59
Xavier JC, Hordijk W, Kauffman S et al (2020) Autocatalytic chemical networks at the origin of metabolism. Proc Biol Sci 287(1922):20192377. https://doi.org/10.1098/rspb.2019.2377
Zhang J (2024) Recognition of the tRNA structure: everything everywhere but not all at once. Cell Chem Biol 31(1):36–52. https://doi.org/10.1016/j.chembiol.2023.12.008
Zhukov A, Popov V (2023) Eukaryotic cell membranes: structure, composition, research methods and computational modelling. Int J Mol Sci 24(13):11226. https://doi.org/10.3390/ijms241311226

Chapter 16
Archaeology as a Toolkit for mRNA: History and Meaning in Use

Ascensión Ariza-Mateos, M. Díaz-Zorita Bonilla, and Jordi Gómez

Abstract Using archaeological concepts, we treat mRNAs as bodies whose historical and meaningful dimensions emerge through experimental practice, focusing on structurally instantiated constraints rather than nucleotide sequences approached primarily as symbolic, text-like objects in phylogenetic analysis. Ancient, structure-dependent probes—such as RNase P, RNase III, UV-activated ribozymes, and the 40S ribosomal subunit—reveal that organizational principles of early RNA evolution, including recognition compatibilities and interaction rules, are presently operating. We propose that RNA viruses may complement the study as historically derived perturbing agents, potentially activating interactions or gene expression patterns normally silent. Our results support a model in which proto-mRNAs were assemblies of RNA structural modules capable of sensing, responding to, and interacting with their molecular and environmental context, operating communally. Structural motifs of present-day mRNAs may reflect these early principles, forming a context-dependent, socially communicative molecular "language" arising from interactions among coding and structural elements as well as external factors, situating mRNA study within a biocommunicative theory.

A. Ariza-Mateos
Laboratory of RNA Archaeology, Instituto de Parasitología y Biomedicina 'López-Neyra' (CSIC), Granada, Spain

Department of Biochemistry and Molecular Pharmacology, New York University Grossman School of Medicine, New York, NY, USA

Ronald O. Perelman Department of Dermatology, New York University Grossman School of Medicine, New York, USA

M. D.-Z. Bonilla
Departmento de Prehistoria y Arqueología, Universidad de Granada, Granada, Spain
e-mail: mdzb@ugr.es

J. Gómez (✉)
Laboratory of RNA Archaeology, Instituto de Parasitología y Biomedicina 'López-Neyra' (CSIC), Granada, Spain
e-mail: jgomez@ipb.csic.es

G. Witzany (ed.), *Integrative Theory of Evolution - Aspects and Insights*,
https://doi.org/10.1007/978-3-032-24295-2_16

1 Introduction

1.1 Archaeological Approach for Molecular Biology Studies of mRNA

> Archaeology is not a return to the innermost secret of the origin; it is the systematic description of a discourse-object. (Michael Foucault 2013)

Archaeological metaphors have long been used in molecular biology. Terms such as "Rosetta Stone", "relics", "living fossils", "molecular imprints", or "fossil signatures" reflect the intuition that evolution is discontinuous and that biological information may persist in material configurations (Ariza-Mateos et al. 2019a, b) beyond nucleotide sequences approached mainly as textual objects of comparison (Zuckerkandl and Pauling 1965a). This perspective is supported by both theoretical and empirical work. The RNA world hypothesis emphasized the catalytic and informational roles of RNA as primordial relics (Gilbert 1986). In addition, aspects of the modern metabolism (Benner et al. 1989), the virus-like elements (Krupovic et al. 2020), and the continued presence of nucleotide-derived coenzymes and other RNA-based regulatory elements in modern cells (Breaker 2012; Nelson and Breaker 2017) have been interpreted as palimpsests of ancient RNA-based systems. Strong evidence for this view comes from crystallographic (Schluenzen et al. 2000; Ban et al. 2000; Wimberly et al. 2000) and phylogenetic analyses of the ribosome (Caetano-Anollés 2002a; Petrov et al. 2014), which reveal it as a deeply RNA-based machine that preserves early modes of interaction between nucleic acids and peptides.

All this evidence (theoretical and structural) suggests an idea proposed by Woese. He argues that, at first, evolution was communal, in an RNA-dominated world, and genetic information was widely shared among organisms or communities. Over time, a key transition occurred: the genetic code became established and stabilized. Once this change was crossed—what Woese called the "Darwinian threshold"—evolution became primarily Darwinian, meaning it relied more on the differentiation of evolutionary lineages and less on widespread genetic exchange (Woese 2002). Following this transition, nucleic acid acquired a representational character, their information being realized in an amino acid "language", and proteins assumed dominant functional roles. However, earlier RNA functions and structural principles were not completely eliminated. Rather than being fully integrated into the coding regime, they persisted as remnants of prior organizational modes. Although the evolutionary significance of mRNAs has also been addressed in the context of different historical layers of genetic code formation (Vetsigian et al. 2006) (Caetano-Anollés, Book Chap. 17), their material, structural, and relational dimensions as archaeological sites remain largely underexplored. Unlike many noncoding RNAs with well-defined secondary structures, mRNAs exhibit dynamic and context-dependent structural configurations (Wan et al. 2011; Yu et al. 2021), forming alternative folds and interaction surfaces that mediate regulation, translation, and recognition by diverse cellular partners (Bevilacqua et al. 2016). Because these structural configurations are conditional on molecular context, such as binding factors, ribosome engagement, stress conditions, etc., they may reveal material traces of past organizational constraints that can be accessed through specific experiments (Bevilacqua et al. 2016; Kertesz et al. 2010; Spitale et al. 2015). From this perspective, mRNAs are not merely carriers of coded information for proteins but material repositories in which multiple historical layers of RNA-based organization (Woese 2002) may remain accessible to archaeological investigation.

At the same time, adopting an archaeological perspective requires abandoning several intuitive assumptions about how knowledge of the past is obtained. Archaeology does not treat material traces as transparent windows onto earlier organizational structures, nor does it assume that historical knowledge arises from isolated objects or from the simple accumulation of facts. Material traces of the past rarely speak without mediation, only becoming evidence through interpretive scaffolding provided by theory and practice (Chapman and Wylie 2014). It does not aim to restore a record of what once existed, nor to reconstitute past states through representation—understood representation here as the assumption that material entities directly reflect an underlying reality, as if their meaning were fully contained within them (Foucault 1978). Instead, archaeology treats material remains as active components of knowledge production. Objects, structures, and residues are epistemic elements through which questions, hypotheses, and interpretations may be formulated (Chapman and Wylie 2014; Huvila et al. 2023; Lucas 2012, 2018). To operationalize this perspective, we begin with three core notions: documents, the archaeological record, and the archive that form the toolkit we use throughout the chapter.

Unlike history, which primarily engages written records and narrative accounts, archaeology reconstructs past human practices through the contextual analysis of material remains, including artifacts, architecture, organic materials, and bioarchaeological evidence that records biological responses to past environments and activities. It examines how their use, persistence, transformation, and spatial organization condition what can be inferred about earlier forms of organization and practice. Knowledge is produced through contextual analysis, in which material remains are related to one another, their depositional environments, and interpretive frameworks, emphasizing an integrated ecological perspective—including zooarchaeology, geoarchaeology, archaeobotany, and related disciplines—that foregrounds human–environment interactions and multiple perspectives on context (Butzer 1980). In this way, archaeology complements but remains distinct from history, emphasizing material evidence and the operational conditions under which it becomes intelligible. To make the analogy with the study of nucleic acids, molecular phylogeny, which relies on sequence analysis, is akin to a historical analysis based on reading texts, whereas the approach we describe aligns more closely with the reasoning and methods of archaeology.

1.2 Archaeological Concepts: Documents, the Archaeological Record, and the Archive

From a material perspective, Archaeology redefines the notion of the document. Documents are physical entities that, beyond carrying symbolic information, are material substrates—parchment, stone, paper, or reused architectural elements—that shape what can be inscribed, preserved, erased, or overwritten. A document cannot be reduced to the information it contains, as its physical support actively influences what can be recorded and how it endures (Piquette and Whitehouse 2013). Archaeological documents often exhibit layered histories: reused materials, inscriptions over earlier texts, or objects repurposed under new functions. These palimpsest-like configurations condense multiple temporalities within a single material form (Lucas 2004; Bailey 2007).

The archaeological record differs from individual documents. While documents are entities, the record emerges from relationships established among material remains—documents, or other objects—when they are removed from their original context and analyzed using disciplined procedures. These procedures determine how objects are grouped, which features are considered relevant, and how traces are compared (Lucas 2012). A document enters the record only when treated as evidence within a defined analytical framework. The record is therefore not merely "what remains", but the product of operations that render material features detectable, comparable, and interpretable. Its significance depends on the interpretive frameworks connecting remains, not on intrinsic properties of the objects themselves (Chapman and Wylie 2014). To illustrate this process, consider the case of the "Lovers of Modena" (Fig. 16.1), a pair of human skeletons buried hand in hand from the early medieval period. Initially interpreted as a male–female couple in accordance with prevailing assumptions about gender, the same later yielded a different historical meaning when those gender assumptions were questioned and peptide-based analyses revealed that both individuals were male (Lugli et al. 2019). The skeletal remains themselves did not change, but the archaeological record—how these remains functioned as evidence—was reconfigured. Changes in interpretive frameworks altered the conditions under which the remains should be analyzed, yielding a different historical account from the same material substrate.

Similarly, the archaeological archive is the ensemble of practices, instruments, spatial arrangements, classification schemes, and routines that determine what can be observed, measured, and acted upon (Lucas 2012). For instance, we could also understand bodies, and, therefore, skeletons are biological archives (Sofaer 2006). Archives are not neutral repositories of past information but active systems that govern what can be recorded and rendered meaningful. They are inherently incomplete, not only due to accidental loss but also because they are produced within specific practical and conceptual contexts (Lucas 2010). Archives both enable and constrain knowledge, preserving certain aspects of the past while excluding others. To illustrate, we take the hospital as an example of an institution functioning as an archival system. Medieval hospitals, for instance, were not designed for systematic

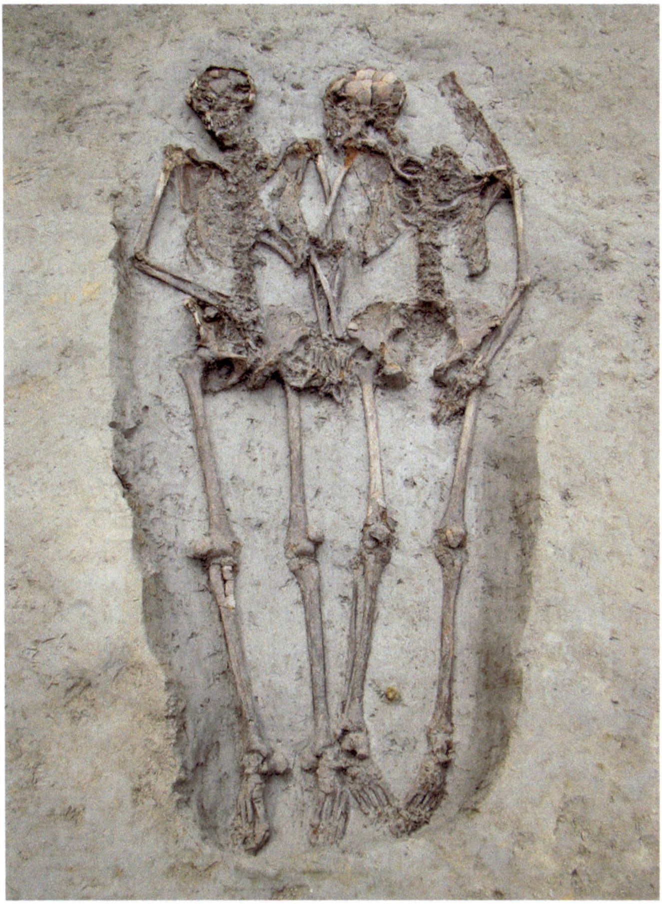

Fig. 16.1 The Lovers of Modena. Medieval burial initially interpreted as a heterosexual couple; later osteological analysis revised this reading, indicating two male individuals. This shift reflects not only new evidence but also a change in interpretive context: as gender assumptions became explicit, the burial was re-examined under a different analytical framework, illustrating how archaeological meaning emerges through situated practice (Source: Lugli et al. 2019)

observation of disease; instead, they functioned primarily as spaces for care, shelter, social regulation, and religious service, and consequently, they did not produce consistent medical knowledge. Disease became a standardized, recordable phenomenon only once systems for systematic observation and intervention were established, enabling consistent recording, classification, comparison, and statistics in modern clinical practice (Foucault 2002). Thus, hospitals as archives produce medical knowledge only when operating through clinical practices that render disease intelligible.

Taken together, the archaeological document, record, and archive differ in ontological status and relational character. The document is an individual material entity;

the archaeological record emerges from relationships among traces through comparison, interpretation, and contextualization; and the archive defines the material, technical, and conceptual conditions under which traces can be recognized and interpreted. Material traces enter the record only when engaged within these operational systems, and archives acquire significance only through active use (Lucas 2012, 2018; Gouda et al. 2009). They do not mirror a preexisting past but define the conditions under which traces appear, persist, and become intelligible (Foucault 2013; Lucas 2012). Archives determine what counts as a trace, what can be recognized as relevant, and what remains invisible or unintelligible. Meaning arises not from the document itself but through its use within specific archival regimes (Foucault 2013; Lucas 2012).

In this presentation, we want to emphasize that empirical objects do not present themselves independently of the conditions, practices, and discourses through which they are recognized—an insight long established in BioArchaeology but still insufficiently acknowledged in contemporary molecular biology that should be reframed (Gowland 2025). As Karel Kleisner has noted (Kleisner 2015), "*Many biologists still believe in a sort of post-Cartesian foundation of reality wherein objects are independent of the subjects which cognize them*". Taking this shift seriously invites a reconsideration of how the past is approached in biological inquiry. From an archaeological perspective, inquiry is directed toward identifying the material and operational conditions that allow traces of earlier organizational regimes to persist and to become intelligible in the present.

1.3 mRNAs as Material Documents

An archaeological perspective applied to mRNAs presupposes that a range of historical events from their emergence may be at least partially embedded within their molecular structure. Viewed archaeologically, mRNAs resemble material documents whose informational value is not limited to what they explicitly encode but also resides in how they are materially organized. Just as inscriptions, parchments, or architectural fragments can preserve traces of earlier regimes of use beneath later modifications, mRNAs may retain structural residues (forms) of ancient RNA organizations beneath subsequent coding functions (Ariza-Mateos et al. 2019a, b). However, as with material remains, the most recent functional layer—in this case, protein coding—tends to dominate interpretation unless earlier organizational features are explicitly sought (Harris 2014; Lucas 2002; Mattick 2009; Eddy 2014).

Molecular evolution has traditionally relied on comparisons of linear nucleotide sequences to reconstruct historical relationships. This approach is valid as long as sequence variation accumulates in a way that preserves genealogical signal. However, this assumption becomes problematic when considering the pre-coding ancestors of mRNA molecules. Unlike rRNAs or tRNAs (Woese and Fox 1977; Olsen and Woese 1993; Sun and Caetano-Anollés 2008), cellular mRNAs have not conserved ancestral sequences, as they have primarily evolved to encode a diverse array of proteins. Once RNA molecules became committed to coding functions, their sequences were progressively reshaped by constraints related to amino acid specification and protein fitness. Evolutionary processes such as exon shuffling, recombination, insertion of mobile genetic elements (Francisco Sanchez-Luque, Book Chap. 9), codon usage biases, viral infections, and compatibility with host or parasitic elements have further contributed to mRNA innovation and diversification. Nevertheless, they have also erased much of the sequence continuity linking modern mRNAs to their precursors (Forterre 2005; Sela et al. 2010; Morris and Mattick 2014). As a result, phylogenetic analyses have provided insights into only a limited number of gene families, notably those encoding ribosomal proteins and chaperones, for which sequence conservation remains sufficient to infer deep evolutionary relationships (Jeffares et al. 1998; Poole et al. 1998). For the vast majority of mRNAs, however, sequence-based phylogenetic reconstruction fails to reach their earliest evolutionary stages. Consequently, the origin of mRNA remains hypothetical. Existing hypotheses are consistent with experimental observations but have not been directly tested through experiments specifically designed to address mRNA origins. Different lines of research propose that mRNAs may have emerged from ancient introns that separated active ribozymes, from proto-tRNAs that guided consecutive peptidyl–tRNA interactions, or from fragments derived from different rRNA segments (Di Giulio 2015; Root-Bernstein and Root-Bernstein 2019).

However, this does not mean that all traces of evolutionary history are lost. Rather, structural continuity may persist even in the absence of recognizable sequence homology. Elements such as internal ribosome entry sites (IRES), riboswitches, and replication mechanisms such as rolling-circle replication in viroids and the hepatitis delta agent could represent remnants of earlier RNA-based organizational regimes inherited from the RNA world (Branch and Robertson 1984; Breaker 2012). This means that ancestral mRNA material properties may persist and be repeatedly repurposed through the superimposition of new functional constraints, in a manner analogous to architectural overbuilding in archaeology. The weak correlation between sequence space and structure space allows structural motifs to persist despite extensive sequence changes (Schuster et al. 1994). Indeed, a variety of structural motifs are recurrently observed across viral RNA genera, persisting despite high mutation rates and extensive sequence turnover (Brown et al. 1992; Thurner et al. 2004). Treating mRNAs as material documents therefore shifts the analytical focus from linear sequences of symbols to the material persistence of mRNA forms and from traditional phylogenetic reconstruction to the experimental identification of conserved structural motifs, patterns, and constraints.

1.4 The Archaeological Record of mRNA

If mRNAs can be approached as material documents, the next question concerns what, within them, can be considered part of the archaeological record. In archaeology, the record is composed of traces that have fallen out of their original contexts of use and have entered a state of persistence, availability, and potential reactivation through analytical practice (Patrik 1985; Lucas 2012). Translating this notion to RNA biology suggests that not all structural features of an mRNA should be considered part of that record. Instead, it comprises structural elements that have potentially persisted from the pre-coding era through their redeployment within cellular contexts different from those in which they originally arose (Lucas 2012). In this sense, one might think of them as analogous to a Roman temple column reused in a Renaissance palace. Accordingly, in both cellular and viral mRNAs, numerous structural motifs have been described. Within the present framework, however, only those motifs that preserve properties plausibly established in the pre-coding era can be regarded as remnants—forms that persist even after being detached from their original contextual functions. Importantly, they become detectable only within specific observational and analytical settings. In practice, candidate remnants may first be suggested by in silico secondary-structure modeling (e.g., *mfold*) or by comparative evidence for conserved structural organization (for instance, conserved base-pairing patterns across related sequences). At this stage, however, these signals remain hypotheses. Just as archaeological records do not preexist excavation, putative structural remnants within mRNAs enter the record only when they are made legible as comparable and interpretable features through specific analytical experiments (Lucas 2012). In our work, this engagement has been achieved through in vitro enzymatic assays or factor-binding experiments that expose mRNAs to structure-dependent activities.

In operational terms, the archaeological record of mRNAs is not the sum of all RNA structures found in mRNAs. It comprises only those structural motifs that meet two conditions: (i) they remain within the recognition range of evolutionarily ancient, structure-dependent RNA-processing activities, and (ii) they persist as reproducible, materially instantiated features in present-day mRNAs. This framework may also contribute to the reconstruction of ancestral molecules. However, its central aim is to delineate historically conserved constraints—enduring organizational and recognition properties that were once operative and remain available for reactivation. Accordingly, these residues are not evolutionary "fossils" in a representational sense: they do not depict past states. Rather, they are material survivals defined by continued recognizability, that is, by their capacity to engage interaction modes inherited from early stages of molecular evolution.

Enzymes as Archaeological Archives: Operational Memory and Recognition

Enzymes have memory and operational activity (Monod 1965; Kull 2009). In the case of evolutionarily ancient RNA structure-dependent enzymes (Fig. 16.2), they "remember" the past by matching present RNA forms to past structures, and they define which material configurations can be detected and interpreted as archaeological traces. These enzymatic whose specificities were established at very early stages of biological evolution, reflect ancient regimes of RNA metabolism, preceding the

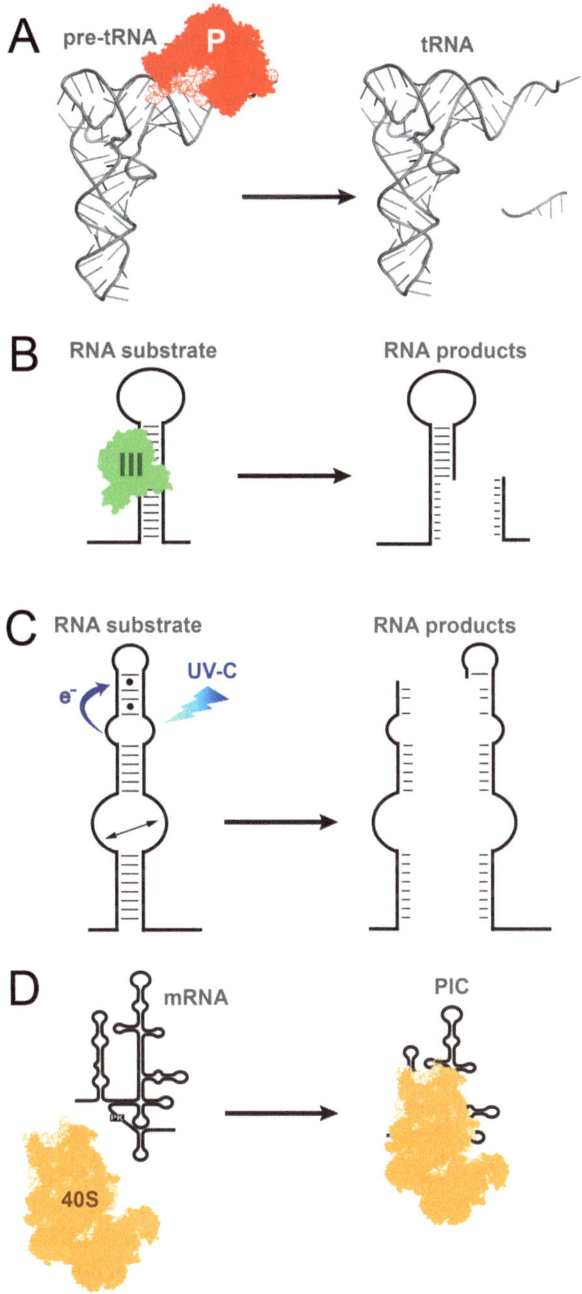

Fig. 16.2 Structure-dependent RNA recognition and processing by ancient RNA-directed activities. Schematic representation of four structure-dependent reactions defined as experimental archives of mRNA organization. (**a**) RNase P (red "P") cleaves a pre-tRNA substrate to generate a mature tRNA and a short leader fragment, illustrating recognition of tRNA-like architecture. (**b**) *Escherichia coli* RNase III (green "III") cleaves an imperfect double-stranded RNA (dsRNA) stem within a hairpin-like substrate, producing two RNA fragments. (**c**) UV-C irradiation can induce structure-dependent RNA reactivity, including site-specific self-cleavage (represented in the figure) and/or crosslinking (not represented) within particular RNA folds (represented here as a structured RNA motif), yielding discrete RNA products. (**d**) The eukaryotic 40S ribosomal subunit (orange "40S") binds directly to a structured mRNA region (including pseudoknot-like features, "PK"), promoting formation of a pre-initiation complex (PIC) and highlighting direct ribosome–RNA compatibility beyond canonical cap-dependent initiation

emergence of modern protein coding mRNAs or at their initial phases. These enzymes do not function as representational repositories; in other words, they do not accumulate signals that stand in for past RNA events. Instead, they impose recognition criteria shaped by ancient biological constraints and operate as systems that determine which material configurations enter the archaeological record. This approach relies on enzymes and ribozymes that recognize RNA primarily based on structure rather than linear sequence. Although limited sequence requirements may exist (Chelladurai et al. 1991; Chamberlain et al. 2024; Ariza-Mateos et al. 2012), their defining feature is the ability to identify conserved folds, long RNA duplexes, or higher-order configurations independently of coding function. When we deliberately deploy these enzymes in this interpretive way, we effectively turn them into historical archives.

RNase P provides a paradigmatic example because its cleavage activity is performed by its RNA moiety, making it a ribozyme (Guerrier-Takada et al. 1983). Originally characterized as the enzyme responsible for processing precursor tRNAs (Robertson et al. 1972), RNase P represents one of the most ancient and conserved RNA-based activities known (Walker and Engelke 2006). Remarkably, bacterial and human RNase P enzymes are able to recognize and correctly cleave each other's pre-tRNA substrates, underscoring the deep evolutionary conservation of the structural criteria that govern substrate recognition (Ariza-Mateos et al. 2019a, b). This conservation indicates that RNase P specificity was established prior to major phylogenetic divergence and has been maintained despite extensive molecular evolution (Ariza-Mateos et al. 2019a, b; Phan et al. 2021; Schencking et al. 2020). When applied experimentally to contemporary mRNAs, RNase P can detect tRNA-like folds, some of which have cloverleaf structure, that are no longer used as processing substrates in vivo nor donors of Aa in protein synthesis (Nadal et al. 2002; Guerrier-Takada et al. 1988; Díaz-Toledano and Gómez 2015). These structures are embedded within mRNAs, not as functional tRNAs, but as material residues compatible with an ancient recognition system (Ariza-Mateos et al. 2019a, b).

RNase III is another archive. The *Escherichia coli* RNase III enzyme recognizes extended regions of double-stranded RNA and cleaves them at defined positions (Robertson and Dunn 1975; Robertson 1990). Its activity is not restricted to perfectly paired duplexes, but also extends to RNA regions containing 10–15 base-pair helices with mismatches and imperfections (Robertson et al. 1977; Dunn 1976). RNase III plays a central role in RNA processing and regulation across all domains of life, and its broad phylogenetic distribution suggests an early origin linked to primordial RNA-based regulatory mechanisms (Lejars et al. 2021; Szczepanska et al. 2025). When E. coli RNase III is confronted with cellular or viral mRNAs in vitro, it reveals extended dsRNA segments that are compatible with RNase III–type recognition. In vivo, these segments may remain unprocessed (Nicholson 2014; Beguiristain et al. 2005) either because they are protected within their native context or because the corresponding RNase III activity is absent or operates under different substrate constraints. Importantly, what becomes legible as a trace depends on the archive deployed. Here, E. coli RNase III functions as a structure-dependent probe that discloses a particular recognition space rather than reproducing the exact processing landscape of any given cell. These structures can therefore be interpreted as archaeological traces.

UV-C light can induce structure-dependent ribozymic activity. UV-C irradiation can induce RNA–RNA inter-strand crosslinks (Branch et al. 1985; Lyons et al. 2001; Wurtmann and Wolin 2009; Wollenzien 1988) and activate site-specific self-cleavage at defined local structures (Ariza-Mateos et al. 2012), such as Loop E motifs (Leontis and Westhof 1998) and other, still uncharacterized configurations (Ariza-Mateos et al. 2012). The relevance of this archive lies in its hypothesized prebiotic or proto-biological origin. Before the formation of an ozone layer, UV-C radiation would have reached the Earth's surface at intensities sufficient to drive photochemical reactions involving the RNA (Ariza-Mateos et al. 2012). The presence of UV-C–reactive conformations in modern RNAs points to conserved structural constraints that connect present-day RNA molecules to early RNA modes of RNA organization (Ariza-Mateos et al. 2019a, b), even though the basis for their persistence remains unknown. UV-C exposure makes these constraints experimentally accessible by reactivating structure-dependent chemical reactivity.

Finally, the eukaryotic 40S ribosomal subunit is the smaller component of the eukaryotic ribosome, playing a central role in decoding and initiation of protein synthesis. Comparative phylogenetic analyses further suggest that components of the small subunit evolved earlier than the catalytic core responsible for peptide bond formation, implying that elements of the decoding apparatus may predate the full emergence of the protein synthesis (Harish and Caetano-Anolles 2012). Despite its role in the highly complex machinery of cellular mRNA translation, it has been shown to bind directly to structured regions within eukaryotic viral mRNAs, including complex folds such as pseudoknots (Lytle et al. 2002; Pestova et al. 1998; Kieft et al. 1999; Pestova et al. 2004). These interactions do not merely reflect its participation in translation initiation but reveal a compatibility between contemporary mRNA structures and ribosomal recognition surfaces (Spahn et al. 2001) shaped early in the evolution of gene expression. In particular, the hepatitis C virus (HCV) IRES establishes direct contacts with the 40S subunit that position the start codon in the P site and remodel the ribosome through interactions involving the E-site region, highlighting how viral RNA can engage conserved ribosomal architecture that likely emerged during the earliest stages of protein synthesis (Harish and Caetano-Anolles 2012).

It is worth noting that RNase H is absent from this list of archives based on structure-specific RNases, despite sharing with RNase P and RNase III a mechanistic mode of RNA cleavage that distinguishes it from most degradative RNases. In principle, RNase H could therefore constitute an additional archaeological archive. However, at present, we have not identified any RNA substrate that is efficiently cleaved by RNase H in the absence of complementary DNA (Moelling et al. 2017), limiting its applicability within the present framework.

Taken together, RNase P, RNase III, UV-C–activated ribozymes, and 40S ribosomal subunits function as enzymatic archives in an archaeological sense. They do not directly encode or retrieve historical information. Rather, part of the historical "signal" resides in RNA material form and part in the recognition constraints enacted by these factors, which determine which configurations can enter the record. In this way, they enact historically established modes of recognition and embody constraints fixed in early RNA-based systems, delimiting what can be recognized and processed as an RNA form. These constraints were established at, or even before, the origins of coding-based gene expression and have remained operational ever since. Their specificity reflects early metabolic requirements centered on structural compatibility, and these inherited recognition criteria continue to delimit their activity even when they are incubated with contemporary mRNAs with fully developed coding functions. In this sense, enzymatic archives operationalize historical continuity by selectively responding to molecular configurations whose forms remain compatible with their ancestral substrates. Following Lucas, the archive is not a passive repository but an operational system: it establishes the conditions under which traces become recognizable and can be activated as evidence through practice (Lucas 2012). In Foucauldian terms, it functions as a field of constraints that governs what can appear as a record in the first place (Foucault 2013).

Viruses as Archaeological Archives: Operational Memory and Recognition

In this section, we suggest that RNA viruses can serve as archaeological archives of cellular RNA organization. This builds on our previous proposal of viruses as archaeological tools for uncovering ancient molecular relationships (Ariza-Mateos et al. 2023). Here, we extend that view by reframing viral infection itself as an archive-like system. Infection can reactivate and render legible residual molecular interactions that are not expressed in the absence of viral entry. It can also expose gene expression patterns that remain silent under standard physiological conditions. This perspective does not describe an established field of research. Rather, it outlines a conceptual and experimental framework for future investigation, supported by some theoretical and a few empirical observations.

We have previously defined viruses as signals for host cells (Gómez et al. 2015). In this view, the viral genome functions as a signal interpreted by the host cell: it acts as a sign vehicle, engages specific host cellular configurations (its semiotic "object"), and elicits host molecular responses (the interpretant) (Chandler 2022; Nöth 2001; Nöth et al. 2020; Ariza-Mateos et al. 2024a). We propose that, during a host jump, step-wise adaptation to an entirely new set of molecular interactions would be difficult, since productive infection would require many of these interactions to be available simultaneously. Rather, infection may succeed because the host already provides accessible interaction networks—often latent, repressed, or normally unused—that can be mobilized upon viral entry so that the virus only needs to engage a limited number of key activation points. The resulting cellular reorganization would thus reflect preexisting, likely ancient, rules of sign-mediated interactions. In Wittgenstein's terms, rule-following is a social practice rather than a one-off event (Wittgenstein 1953). Productive infection is therefore more consistent with the activation of preexisting, sign-mediated, rule-governed (Witzany 2000, 2010; Ariza-Mateos et al. 2024a) responses than with the de novo invention of entirely new ones. In this sense, viruses can be understood less as mere agents of perturbation and more as experimental probes that operate as archaeological archives, by at least partially reactivating latent or potential molecular relationships rooted in the cell's evolutionary past. In doing so, they render these relationships experimentally legible as historical traces, expose operational constraints inherited from earlier cellular regimes, and provide experimental access to historically embedded cellular potentials.

From this perspective, each virus can be viewed as a specific form of cellular alterity. Upon entering the cell, viral RNAs introduce structural motifs, some of which correspond to configurations recognizable by ancient enzymatic activities and give rise to modes of molecular interaction distinct from those normally operating in healthy cells. These motifs thus become legible as archaeological traces when probed by ancient structure-dependent factors. For example, the IRES elements of HCV (Lytle et al. 2002) or cricket paralysis virus (CrPV) (Pestova et al. 2004) (sensitive to human RNase P (Nadal et al. 2002; Lyons and Robertson 2003)) interact with regions of the 40S ribosomal subunit that are not contacted by cellular mRNAs during canonical translation. Similarly, in several plant RNA viruses, the specific aminoacylation of their 3′ genomic end by aminoacyl-tRNA synthetases is essential for infection, despite the fact that this reaction does not occur on cellular mRNAs and that the aminoacylated tRNA-like motif does not act as an amino acid donor in the translation reaction (Matsuda and Dreher 2004, 2007). Although such structures are often interpreted as the outcome of evolutionary strategies such as viral mimicry (Florentz and Giegé 1994; Giegé et al. 1998; Dreher 2009; Ariza-Mateos and Gómez 2017)—typically assumed to serve the sequestration of cellular functions—we propose a different explanation (Ariza-Mateos et al. 2023). Consistent with Wiener and Maizels (Weiner and Maizels 1987; Maizels and Weiner 1994), we suggest that at least some of these RNA motifs and their recognition principles reflect deeply conserved molecular relationships, plausibly rooted in early stages of life, when tRNA-like structures may have functioned as replication-linked tags before being fully integrated into modern translation. In this sense, viral infection can reactivate latent molecular functionalities that are no longer engaged under healthy cellular conditions.

In addition, viral infection reorganizes the intracellular landscape of expressed mRNAs and their molecular interactions, often leading to the concurrent activation of partially incompatible cellular programs (Ariza-Mateos et al. 2019a, b, 2024b). In this view, infection does not simply pit "virus" against "host" but precipitates an intracellular conflict between alternative regulatory states that the cell can enact. It mobilizes antiviral sensors, RNA degradation pathways, and restriction factors that limit viral propagation, including activities that mutagenize viral genomes, such as those mediated by ADAR-1 or APOBEC enzymes (Samuel 2011). At the same time, other cellular pathways can become engaged that amplify viral signals or stabilize persistent infection, which we treat here as a distinct cellular alternative. An illustrative example is the interferon response (Schoggins et al. 2014). Although it is triggered by viral infection and plays a central role in the antiviral defense response, nearly one-third of interferon-stimulated genes, when considered individually, facilitate the replication of certain viruses. In line with our interpretation of a virus as a signal (Gómez et al. 2015), we propose that both sides of virus-generated conflict—antiviral defense and the subversion of cellular homeostasis—result from the actualization of preexisting cellular capacities in response to viral signals.

The virus-specific redistribution of molecular activities and the selective expression of particular mRNA programs—i.e., which pathways become activated or repressed during infection—can reveal functional molecular associations that remain undetectable in the absence of infection. Although a global consensus on the origin of viruses—whether RNA or DNA—remains lacking (Domingo 2020), according to one of the most widely accepted hypotheses, RNA viruses and, later, DNA viruses were already present at the origin of life and emerged concurrently with early cellular systems (Domingo 2020).

Taken together, these examples position RNA viruses as active archaeological operators. From this perspective, viral RNAs do not transmit a preserved molecular memory, nor do they reproduce ancestral states. Rather, they can partially reinstate interaction modes constrained by earlier organizational programs. In this way, infection foregrounds persistent material constraints of recognition and compatibility. Unlike phylogenetic analyses, which infer history from sequence continuity and descent, this approach identifies historically conserved constraints through their continued operational effects in contemporary cells. It thereby helps determine which viral RNA structures remain compatible with ancient recognition rules and which host responses can still engage older layers of RNA-based organization.

Archaeological Limits in mRNA Evolution

A challenging aspect of the archaeological method applied to mRNA is that it does not produce linear evolutionary timelines. Instead, it offers a stratified view of molecular history (Ariza-Mateos et al. 2019a, b), in which evolutionary ordering is conditional on sampling strategies (i.e., analyzing mRNAs coding for ribosomal proteins or including histone coding proteins) and on the convergence or divergence of multiple experimental proxies (Ariza-Mateos et al. 2019a, b). This is not a limitation of the approach, which identifies persistent structural and recognition properties in contemporary mRNAs, but reflects the scope of what can be inferred: enduring properties whose temporal origin is distinct from their material persistence. Relative chronology, therefore, requires complementary strategies, such as RNA-structure-based phylogenetic analyses, which can reconstruct relationships where sequence-based approaches fail (Caetano-Anollés 2002a, b). These methods have been applied to model RNAs, including archaeal 5S RNA, the RNA Y1 component of the Ro ribonucleoprotein complex, and tRNAs (Sun and Caetano-Anollés 2008, 2009; Caetano-Anollés 2002b)—molecules comparable in length to the RNase P RNA substrates. By combining archaeological identification of operational constraints with structural phylogenies, it may be possible to situate ancient recognition modes within evolutionary history.

2 Experimental Identification of Ancient Recognition Constraints

RNase P detects structure-dependent signals within cellular mRNAs and within viral mRNAs from several genera, including HCV, CrPV, classical swine fever virus (CSFV), bovine viral diarrhea virus (BVDV), and foot-and-mouth disease virus (FMDV) (Ariza-Mateos et al. 2019a, b; Nadal et al. 2002; Sabariegos et al. 2004; Piron et al. 2005; Serrano et al. 2007; Lyons and Robertson 2003). RNase III similarly reveals structure-dependent dsRNA signals within interferon-induced mRNAs and within the mRNAs of HCV and related pestiviruses (CSFV and BVDV) (Beguiristain et al. 2005; Diaz-Toledano et al. 2009; Ariza-Mateos et al. 2019a, b). UV-C irradiation can induce inter-strand crosslinking and activate self-cleavage in noncoding regions of RNA pathogens, including HCV, CSFV, HDV, and viroids (Branch et al. 1985; Lyons et al. 2001; Branch et al. 1989; Ariza-Mateos et al. 2012). Likewise, the 40S ribosomal subunit shows direct compatibility with highly structured viral RNAs, including the 5′ regions of HCV, CSFV, and BVDV (Pestova et al. 1998; Pestova and Hellen 1999), as well as the intercistronic region of CrPV (Pestova et al. 2004; Jan and Sarnow 2002). Taken together, these observations support the existence of persistent RNA structural residues that become experimentally detectable through ancient structure-dependent activities, despite being largely silent under standard physiological conditions (Beguiristain et al. 2005; Diaz-Toledano et al. 2009; Ariza-Mateos et al. 2019a, b), (Piron et al. 2005).

Unlike conventional molecular biology, where different techniques are expected to converge on the same molecular properties, e.g., chemical and enzymatic RNA structure mapping (Díaz-Toledano and Gómez 2015) or sequence- and structure-based phylogeny (Sun and Caetano-Anollés 2009), the archaeological approach deliberately exploits non-convergence. RNase P, RNase III, UV-C irradiation, and 40S ribosomal subunits tested in virus-infected or healthy cell mRNAs do not provide alternative views of a single preexisting feature; rather, each engages distinct recognition specificities inherited from early RNA metabolism. Divergence between experimental readouts is therefore not treated as noise or methodological failure but as evidence that multiple material traces of past organizational constraints coexist within the same mRNA, becoming accessible only through the specific experiments that identified them. The process can be further expanded by probing identified individual mRNA motifs with other RNA structure-dependent enzymes of distinct recognition specificities and historical origins, such as 3′ RNA-modifying enzymes—including aminoacyl-tRNA synthetases, the 3' CCA-adding enzyme, and RNase Z—and by systematically testing RNase P, RNase III, and small ribosomal subunits from diverse branches of archaea, bacteria, and eukarya. These approaches would expand the historical archive accessible through structure-dependent probing and refine the conserved recognition constraints across evolutionary lineages, clarifying how they continue to manifest in contemporary RNA molecules.

Taken together, the patterns revealed by RNase P, RNase III, UV-C irradiation, and the 40S (Fig. 16.3) show a recurrent convergence of an RNase P signal positioned adjacent to an RNase III signal in several viral mRNA genera and in the cellular IFNA5 mRNA (Ariza-Mateos et al. 2019a, b). From an archaeological perspective, the primary significance of this convergence points to the persistence of shared recognition constraints that remain operative across evolutionary transitions. These constraints include, first, the tolerance of RNA structures compatible with ancient structure-dependent enzymatic activities, such as RNase P and RNase III; and second, the recurrent association, which implies physical association of distinct RNA motifs—such as tRNA-like folds and extended duplex regions—within the same mRNA molecule. This recurrence is compatible with the possibility that such configurations constituted functional elements in proto-mRNAs, a hypothesis that would ultimately require validation through comparative structural phylogeny. The coexistence of these motifs indicates that their present-day activities are functionally related and that in early RNA-based systems, distinct activities may also have been functionally associated, without implying that these ancestral activities were the same as those observed today. Accordingly, although the dsRNA/tRNA-like configuration may initially appear as a conserved remnant, its archaeological relevance lies in the convergence of multiple constraints that together form a meaningful assemblage, rather than in the preservation of a discrete molecular ancestor. This interpretation aligns with Woese's critique of classical Darwinian thinking, in which evolution is understood not merely as a "procession of forms" but as the persistence of underlying organizational principles (Woese 2002). The archaeological signal, therefore, resides in the enduring combinations of structural compatibilities that continue to shape RNA organization across evolutionary transitions.

3 From the Genetic Code Correspondences to Its Contextual Use: An Archaeological Approach to mRNA Meaning

Classical molecular biology has typically treated nucleic acid strands as carriers of digital information, assuming that evolutionary information is primarily encoded in the sequence itself (Crick 1958; Zuckerkandl and Pauling 1965a). Within this view, historical relationships—such as temporal divergence or genealogical proximity—are inferred by comparing linear nucleotide sequences or, in some cases, conserved structural features (Zuckerkandl and Pauling 1965a, b; Woese 1987). This approach has proven extraordinarily fruitful, forming the basis of molecular phylogenetics and comparative genomics. More generally, the genetic code has been regarded as analogous to a formal language, in which discrete symbols (nucleotides and amino acids) are combined according to evolutionarily fixed rules of correspondence (Eigen et al. 1981). Within this view, each informational unit—whether a codon or a structural motif in an mRNA—is assumed to possess an intrinsic meaning that can

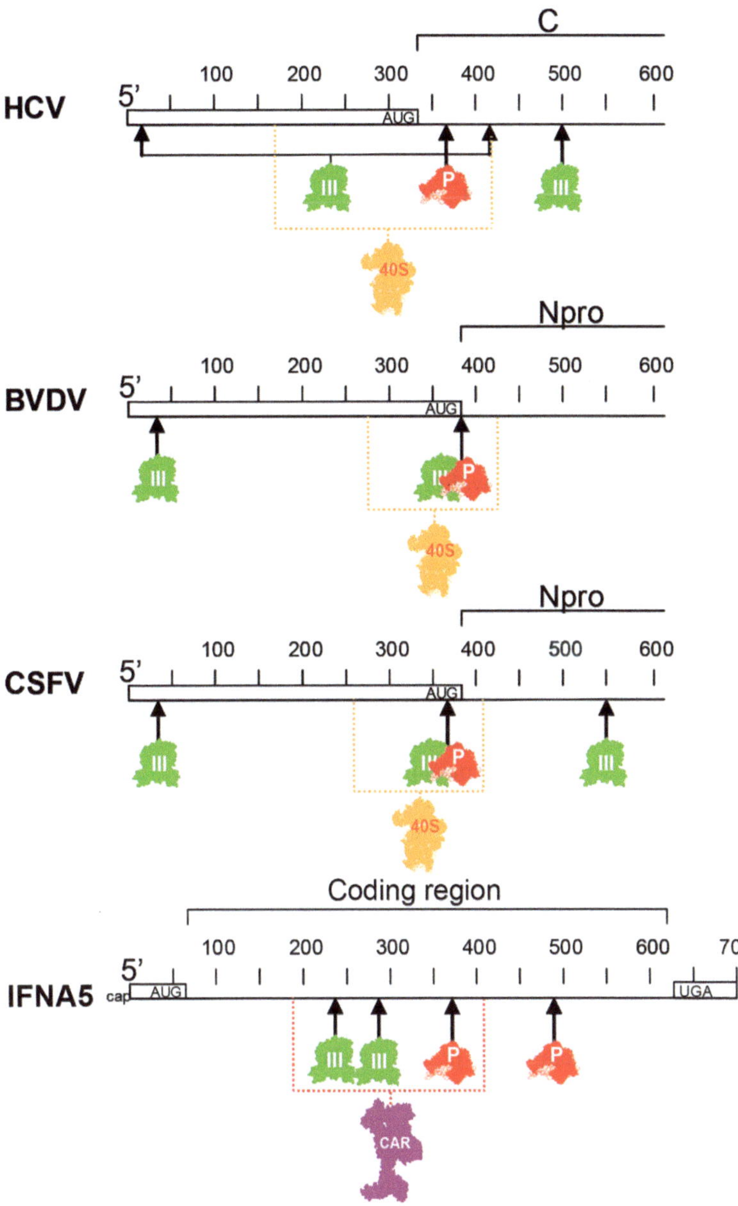

Fig. 16.3 Conserved co-occurrence of RNase P and RNase III cleavage sites in viral and cellular mRNAs. Schematic representation of structure-dependent cleavage sites identified in the 5′ regions of several viral RNAs—hepatitis C virus (HCV), bovine viral diarrhea virus (BVDV), and classical swine fever virus (CSFV)—and in the cellular interferon alpha 5 mRNA (IFNA5). Colored symbols indicate the positions of experimentally detected cleavage sites: RNase P (red "P"), RNase III (green "III"), and the 40S ribosomal subunit binding site (RBS) (orange "40S"). The dashed boxes highlight regions where RNase P and RNase III sites cluster and frequently co-occur. In several mRNAs, a conserved RNase P-sensitive site is observed around ~370 nt, commonly accompanied by one or more adjacent RNase III-sensitive sites. Structural mapping of these regions revealed the

be decoded independently of context, functioning as a record of a molecular transaction. For example, the sequence AUGGGC is interpreted unambiguously as methionine–glycine, just as each tRNA identity determinant specifies a particular aminoacylation reaction. This "intrinsic meaning" view has been explicitly challenged by biocommunicative accounts, which argue that biological signs acquire meaning through context-dependent rule-governed use rather than through sequence alone (Witzany 2006, 2010, 2013, 2014a; Witzany and Baluška 2012).

Building on this biocommunicative critique, our archaeological approach interprets mRNAs not merely as linear sequences encoding information but as assemblies of preexisting RNA structures—proto-mRNAs—upon which encoded information was later superimposed. As Wittgenstein proposed, "the meaning of a word (or, in our case, a sign or code) is its use." (Wittgenstein 1953) These motifs acquire meaning through interactions with molecular partners, structural context, and environmental conditions. Components of proto-mRNAs may have included ancient aptamers, riboswitches, ribozymes, or tRNA fragments, functioning cooperatively, parasitically, or relationally within early RNA-based systems. Their activity depended on inter-motif and inter-molecular interactions, reflecting a form of "social" behavior among RNA elements, with meaning emerging communally rather than residing solely within the individual motif. The original function of each motif was therefore inseparable from its molecular partners, spatial constraints, and environmental conditions, prefiguring the context-dependent usage rules observed in modern mRNAs. Accordingly, the archaeological method gains explanatory power by relating identified motifs to others within the same mRNA and across different RNA populations, types, and interacting protein factors, in which the same structure may acquire radically different meanings. As in archaeology, an object's significance increases when contextualized with surrounding artifacts and conditions of discovery (Hodder and Hutson 2003; Schalk and Binford 1977). Here, meaning does not reside in the RNA structure itself, but emerges from context—at multiple levels of organization, including *cis* and *trans* molecular interactions, cellular state, organismal physiology, ecological interactions, and the external environment.

Several well-documented examples from viral mRNAs illustrate our approach. First, genetic code usage itself can be contextually regulated through RNA structure. In the hepatitis A virus, rare codons within the capsid coding region slow translation elongation, thereby modulating protein folding in a manner essential for viral stability during fecal–oral transmission (Pintó et al. 2018). Second, ancient RNA structures can acquire new functions in modern contexts. tRNA-like structures at the 3′ ends of certain plant RNA viruses may originally have acted as sorting signals distinguishing RNAs destined for replication from those for translation (Weiner and

Fig. 16.3 (continued) juxtaposition of a tRNA-like fold (with or without pseudoknots) and a neighboring extended, imperfect dsRNA-like element, despite limited or absent sequence similarity. In multiple cases, this region overlaps with the 40S RBS, suggesting a conserved structural context associated with ribosome recruitment. The cytoplasmic accumulation region interacting with the TREX complex in violet

Maizels 1987). Today, these same structures function as regulatory elements in viral replication, suggesting that rules based on structure remain operative through modified uses (Dreher 2010). Third, individual mRNAs can act as regulators of global gene expression. The 5′ double-stranded RNA motif of hepatitis C virus mRNA interacts with miR-122, stabilizing the viral genome and enhancing translation while simultaneously sequestering the microRNA from its cellular targets (Jopling et al. 2005; Shimakami et al. 2012; Díaz-toledano et al. 2009; García-Sacristán et al. 2015). This single structural interaction reshapes cellular regulatory networks and contributes to pathological outcomes such as liver cancer (Luna et al. 2015). This proposal is consistent with broader biocommunicative and virocentric perspectives that have pioneered the view that important forms of biological novelty can emerge from collective interaction rather than isolated informational units (Domingo and Witzany 2024; Villarreal and Witzany 2013, 2015, 2021, 2023; Witzany 2014b).

Together, these examples support the view that proto-RNAs were composed of functional structural modules adapted to ancestral environments and that the material constraints underlying these functions persist today, shaping context-dependent patterns of genetically coded information usage. Within the archaeological framework, mRNA molecules are not treated as functionally communicative agents in the cell but as material forms embedded in historically established systems of identification. But the results from the archaeological experimentation can serve as the basis for new cellular studies in communication. For example, the specific processing of a liver-expressed interferon-alpha mRNA by host RNase P in vitro indicates that an mRNA carrying a processing signal faces the risk of cleavage and inactivation. Notably, the motif identified coincides with the most conserved structural regions among interferon-alpha mRNAs, suggesting evolutionary conservation of a site that is potentially vulnerable. This observation motivates the hypothesis that the processing signal must be protected from endogenous RNase P activity. Intriguingly, the RNase P–recognized cloverleaf-like motif overlaps the minimum binding region of the TREX transport factor complex, which is involved in recognizing, compacting, and protecting mRNAs during export (Pacheco-Fiallos et al. 2023). From this overlap, one can hypothesize a testable module of addiction or toxin–antitoxin type, in which protection or masking by the TREX complex shields the mRNA from destructive endogenous RNase P cleavage. These ideas remain experimentally approachable and could illustrate how ancient structural constraints and modern recognition systems, acting as context-dependent rules of use, jointly shape the knowledge we can obtain about mRNA functions. By extending the archaeological perspective from conceptual framing to concrete cellular experiments, we can translate these hypotheses into testable biochemical models. In this sense, the archaeological approach provides a direct experimental scaffold for a pragmatic view of biological communication, in which the significance of RNA motifs does not lie in fixed codes or intrinsic meanings but in their context-dependent use—defined by concrete patterns of recognition, tolerance, masking, and processing.

4 Conclusion

Taken together, the archaeological approach developed in this chapter reframes mRNAs as material objects whose experimentally accessible properties depend on the specific systems that recognize them, rather than as fixed entities awaiting description. In this perspective, structure-dependent probes such as RNase P, RNase III, UV-C irradiation, and the 40S ribosomal subunit do not converge on a single underlying "true" feature of an mRNA. Instead, each engages distinct recognition specificities inherited from early RNA metabolism, and the divergence between readouts becomes informative rather than problematic.

Importantly, this framework is not only conceptual. Across multiple viral mRNAs and in the cellular IFNA5 mRNA, we observe a recurrent co-occurrence of a tRNA-like element positioned adjacent to a dsRNA-like motif, as independently revealed by RNase P and RNase III. Rather than treating this configuration as a direct molecular fossil, we interpret it as evidence for persistent constraints of compatibility and recognition that remain operative across evolutionary transitions. From an archaeological standpoint, the "signal" therefore lies less in the preservation of ancestral structures than in the persistence of meaningful assemblages of structural tolerances and motif associations that continue to shape RNA organization today.

This perspective resonates with Lucas's view of the archive as an operational system that governs conditions of recognizability and determines how traces become activated as evidence through practice. It also aligns with Foucault's archaeological stance, in which archaeology is not a return to an origin but the systematic description of a discourse-object constituted by historically situated procedures of identification. In our case, the "discourse-object" is the mRNA as experimentally disclosed through structure-dependent recognition regimes.

Finally, this approach offers a complementary way of thinking about biological meaning. Rather than treating meaning as intrinsically contained in an mRNA sequence, we emphasize that mRNA motifs acquire functional significance through their use: through interactions with molecular partners, structural context, and cellular conditions that enable or suppress particular modes of recognition. By treating mRNAs as archaeological documents and ancient enzymes and viral infections as operational archives, this approach provides an experimental scaffold to rethink molecular history beyond sequence-centered reconstruction and to explore how historically embedded recognition rules continue to operate within contemporary cells.

Bibliography

Ariza-Mateos A, Gómez J (2017) Viral tRNA mimicry from a biocommunicative perspective. Front Microbiol 8. https://doi.org/10.3389/fmicb.2017.02395

Ariza-Mateos A, Prieto-Vega S, Díaz-Toledano R et al (2012) RNA self-cleavage activated by ultraviolet light-induced oxidation. Nucleic Acids Res 40. https://doi.org/10.1093/nar/gkr822

Ariza-Mateos A, Briones C, Perales C et al (2019a) The archaeology of coding RNA. Ann N Y Acad Sci 1447. https://doi.org/10.1111/nyas.14173
Ariza-Mateos A, Briones C, Perales C et al (2019b) The archaeology of coding RNA. Ann N Y Acad Sci 1447:119–134
Ariza-Mateos A, Briones C, Perales C et al (2023) Viruses as archaeological tools for uncovering ancient molecular relationships. Ann N Y Acad Sci 1529(1):3–13
Ariza-Mateos A, Briones C, Perales C et al (2024a) Natural languages and RNA virus evolution. J Physiol 602(11):2565–2580
Ariza-Mateos A, Briones C, Perales C et al (2024b) Archaeological approaches to RNA virus evolution. J Physiol 602(11):2469–2478
Bailey G (2007) Time perspectives, palimpsests and the archaeology of time. J Anthropol Archaeol 26:198–223
Ban N, Nissen P, Hansen J et al (2000) The complete atomic structure of the large ribosomal subunit at 2.4 Å resolution. Science 289:905–920
Beguiristain N, Robertson HD, Gómez J (2005) RNase III cleavage demonstrates a long range RNA: RNA duplex element flanking the hepatitis C virus internal ribosome entry site. Nucleic Acids Res 33. https://doi.org/10.1093/nar/gki822
Benner SA, Ellington AD, Tauer A (1989) Modern metabolism as a palimpsest of the RNA world. Proc Natl Acad Sci 86:7054–7058
Bevilacqua PC, Ritchey LE, Su Z, Assmann SM (2016) Genome-wide analysis of RNA secondary structure. Annu Rev Genet 50:235–266
Branch AD, Robertson HD (1984) A replication cycle for viroids and other small infectious RNA's. Science 223:450–455
Branch AD, Benenfeld BJ, Robertson HD (1985) Ultraviolet light-induced crosslinking reveals a unique region of local tertiary structure in potato spindle tuber viroid and HeLa 5S RNA. Proc Natl Acad Sci 82. https://doi.org/10.1073/pnas.82.19.6590
Branch AD, Benenfeld BJ, Baroudy BM et al (1989) An ultraviolet-sensitive RNA structural element in a viroid-like domain of the hepatitis delta virus. Science 243:649–652
Breaker RR (2012) Riboswitches and the RNA world. Cold Spring Harb Perspect Biol 4:a003566
Brown EA, Zhang H, Ping L-H, Lemon SM (1992) Secondary structure of the 5′ nontranslated regions of hepatitis C virus and pestivirus genomic RNAs. Nucleic Acids Res 20:5041–5045. https://doi.org/10.1093/nar/20.19.5041
Butzer KW (1980) Context in archaeology: an alternative perspective. J F Archaeol 7:417–422
Caetano-Anollés G (2002a) Tracing the evolution of RNA structure in ribosomes. Nucleic Acids Res 30:2575–2587. https://doi.org/10.1093/nar/30.11.2575
Caetano-Anollés G (2002b) Evolved RNA secondary structure and the rooting of the universal tree of life. J Mol Evol 54:333–345
Chamberlain AR, Huynh L, Huang W et al (2024) The specificity landscape of bacterial ribonuclease P. J Biol Chem 300
Chandler D (2022) Semiotics: the basics, 4th edn. Routledge, New York, 10158
Chapman R, Wylie A (2014) Material evidence: learning from archaeological practice. Routledge, New York and London, pp 1–20
Chelladurai BS, Li H, Nicholson AW (1991) A conserved sequence element in ribonuclease III processing signals is not required for accurate in vitro enzymatic cleavage. Nucleic Acids Res 19:1759–1766
Crick FH (1958) On protein synthesis. In: Symp soc exp biol, p 8. Cambridge University Press. Cambridge UK
Di Giulio M (2015) A model for the origin of the first mRNAs. J Mol Evol 81:10–17
Díaz-Toledano R, Gómez J (2015) Messenger RNAs bearing tRNA-like features exemplified by interferon alfa 5 mRNA. Cell Mol Life Sci 72. https://doi.org/10.1007/s00018-015-1908-0
Diaz-Toledano R, Ariza-Mateos A, Birk A et al (2009) In vitro characterization of a miR-122-sensitive double-helical switch element in the 5′ region of hepatitis C virus RNA. Nucleic Acids Res 37:5498–5510

Díaz-toledano R, Ariza-mateos A, Birk A et al (2009) In vitro characterization of a miR-122-sensitive double-helical switch element in the 5′ region of hepatitis C virus RNA. Nucleic Acids Res 37. https://doi.org/10.1093/nar/gkp553
Domingo E (2020) Introduction to virus origins and their role in biological evolution. Virus Popul:1–33
Domingo E, Witzany G (2024) Quasispecies productivity. Naturwissenschaften 111:11
Dreher TW (2009) Role of tRNA-like structures in controlling plant virus replication. Virus Res 139:217–229
Dreher TW (2010) Viral tRNAs and tRNA-like structures. Wiley Interdiscip Rev RNA 1:402–414
Dunn JJ (1976) RNase III cleavage of single-stranded RNA. Effect of ionic strength on the fideltiy of cleavage. J Biol Chem 251:3807–3814
Eddy SR (2014) Computational analysis of conserved RNA secondary structure in transcriptomes and genomes. Annu Rev Biophys 43:433–456
Eigen M, Gardiner W, Schuster P, Winkler-Oswatitsch R (1981) The origin of genetic information. Sci Am 244:88–119
Florentz C, Giegé R (1994) tRNA-like structures in plant viral RNAst+. tRNA Struct Biosynthesis, Funct 10:141–163
Forterre P (2005) The two ages of the RNA world, and the transition to the DNA world: a story of viruses and cells. Biochimie 87:793–803
Foucault M (1978) Nietzsche, genealogy, history
Foucault M (2002) The birth of the clinic: an archaeology of medical perception. London: Routledge, UK
Foucault M (2013) Archaeology of knowledge. Routledge classics. London and New York. UK, USA
García-Sacristán A, Moreno M, Ariza-Mateos A et al (2015) A magnesium-induced RNA conformational switch at the internal ribosome entry site of hepatitis C virus genome visualized by atomic force microscopy. Nucleic Acids Res 43. https://doi.org/10.1093/nar/gku1299
Giegé R, Frugier M, Rudinger J (1998) tRNA mimics. Curr Opin Struct Biol 8:286–293. https://doi.org/10.1016/S0959-440X(98)80060-2
Gilbert W (1986) Origin of life: the RNA world. Nature 319:618
Gómez J, Ariza-Mateos A, Cacho I (2015) Virus is a signal for the host cell. Biosemiotics 8. https://doi.org/10.1007/s12304-015-9245-0
Gouda F, Raben R, Nordholt HS, Stoler AL (2009) Ann Laura Stoler, along the archival grain; epistemic anxieties and colonial common sense. Bijdr tot taal-, land-en volkenkunde/Journal Humanit Soc Sci Southeast Asia 165:551–567
Gowland RL (2025) Skeleton matter (s): reframing current approaches in bioarchaeology towards a more inclusive future. Int J Paleopathol 51:73–81
Guerrier-Takada C, Gardiner K, Marsh T et al (1983) The RNA moiety of ribonuclease P is the catalytic subunit of the enzyme. Cell 35:849–857
Guerrier-Takada C, van Belkum A, Pleii CWA, Altman S (1988) Novel reactions of RNAase P with a tRNA-like structure in turnip yellow mosaic virus RNA. Cell 53:267–272
Harish A, Caetano-Anolles G (2012) Ribosomal history reveals origins of modern protein synthesis. PLoS One 7:e32776
Harris EC (2014) Principles of archaeological stratigraphy. Academic Press, London, UK
Hodder I, Hutson S (2003) Reading the past: current approaches to interpretation in archaeology. Cambridge University Press. Cambridge UK
Huvila I, Sköld O, Andersson L (2023) Knowing-in-practice, its traces and ingredients. In: The posthumanist epistemology of practice theory: re-imagining method in organization studies and beyond. Springer, pp 37–69. Cham (Switzerland): Palgrave Macmillan
Jan E, Sarnow P (2002) Factorless ribosome assembly on the internal ribosome entry site of cricket paralysis virus. J Mol Biol 324:889–902. https://doi.org/10.1016/S0022-2836(02)01099-9
Jeffares DC, Poole AM, Penny D (1998) Relics from the RNA world. J Mol Evol 46:18–36

Jopling CL, Yi M, Lancaster AM et al (2005) Modulation of hepatitis C virus RNA abundance by a liver-specific MicroRNA. Science 80-) 309:1577–1581

Kertesz M, Wan Y, Mazor E et al (2010) Genome-wide measurement of RNA secondary structure in yeast. Nature 467:103–107

Kieft JS, Zhou K, Jubin R et al (1999) The hepatitis C virus internal ribosome entry site adopts an ion-dependent tertiary fold. J Mol Biol 292:513–529

Kleisner K (2015) Semantic organs: the concept and its theoretical ramifications. Biosemiotics 8. https://doi.org/10.1007/s12304-015-9246-z

Krupovic M, Dolja VV, Koonin EV (2020) The LUCA and its complex virome. Nat Rev Microbiol 18:661–670

Kull K (2009) Umwelt and modelling. In: Cobley P (ed) The Routledge companion to semiotics. Routledge, pp 43–56. Routledge and Taylor & Francis. London and New York

Lejars M, Kobayashi A, Hajnsdorf E (2021) RNase III, ribosome biogenesis and beyond. Microorganisms 9:2608

Leontis NB, Westhof E (1998) A common motif organizes the structure of multi-helix loops in 16 S and 23 S ribosomal RNAs. J Mol Biol 283:571–583

Lucas G (2002) Critical approaches to fieldwork: contemporary and historical archaeological practice. London, UK

Lucas G (2004) The archaeology of time. Routledge

Lucas G (2010) Time and the archaeological archive. Rethink Hist 14:343–359. https://doi.org/10.1080/13642529.2010.482789

Lucas G (2012) Understanding the archaeological record. Cambridge UK

Lucas G (2018) Writing the past: knowledge and literary production in archaeology. London, UK

Lugli F, Di Rocco G, Vazzana A et al (2019) Enamel peptides reveal the sex of the Late Antique 'Lovers of Modena'. Sci Rep 9:13130

Luna JM, Scheel TKH, Danino T et al (2015) Hepatitis C virus RNA functionally sequesters miR-122. Cell 160:1099–1110

Lyons AJ, Robertson HD (2003) Detection of tRNA-like structure through RNase P cleavage of viral internal ribosome entry site RNAs near the AUG start triplet. J Biol Chem 278:26844–26850

Lyons AJ, Lytle JR, Gomez J, Robertson HD (2001) Hepatitis C virus internal ribosome entry site RNA contains a tertiary structural element in a functional domain of stem-loop II. Nucleic Acids Res 29. https://doi.org/10.1093/nar/29.12.2535

Lytle JR, Wu L, Robertson HD (2002) Domains on the hepatitis C virus internal ribosome entry site for 40s subunit binding. RNA 8:1045–1055

Maizels N, Weiner AM (1994) Phylogeny from function: evidence from the molecular fossil record that tRNA originated in replication, not translation. Proc Natl Acad Sci 91:6729–6734

Matsuda D, Dreher TW (2004) The tRNA-like structure of turnip yellow mosaic virus RNA is a 3′-translational enhancer. Virology 321:36–46

Matsuda D, Dreher TW (2007) Cap- and initiator tRNA-dependent initiation of TYMV polyprotein synthesis by ribosomes: evaluation of the trojan horse model for TYMV RNA translation. RNA 13. https://doi.org/10.1261/rna.244407

Mattick JS (2009) The genetic signatures of noncoding RNAs. PLoS Genet 5:e1000459

Moelling K, Broecker F, Russo G, Sunagawa S (2017) RNase H as gene modifier, driver of evolution and antiviral defense. Front Microbiol 8:1745

Monod J (1965) On the nature of allosteric transitions: a plausible model. J Mol Biol 12:88–118

Morris KV, Mattick JS (2014) The rise of regulatory RNA. Nat Rev Genet 15:423–437

Nadal A, Martell M, Robin Lytle J et al (2002) Specific cleavage of hepatitis C virus RNA genome by human RNase P. J Biol Chem 277. https://doi.org/10.1074/jbc.M203595200

Nelson JW, Breaker RR (2017) The lost language of the RNA World. Sci Signal 10:eaam8812

Nicholson AW (2014) Ribonuclease III mechanisms of double-stranded RNA cleavage. Wiley Interdiscip Rev RNA 5:31–48

Nöth W (2001) Protosemiotics and physicosemiosis. Sign Syst Stud 29:13–25

Nöth W, Kull K, de Camargo CEP (2020) Virus semiosis: uma entrevista com Kalevi Kull. TECCOGS Rev Digit Tecnol Cogn

Olsen GJ, Woese CR (1993) Ribosomal RNA: a key to phylogeny. FASEB J 7:113–123

Pacheco-Fiallos B, Vorländer MK, Riabov-Bassat D et al (2023) mRNA recognition and packaging by the human transcription–export complex. Nature 616:828–835

Patrik LE (1985) Is there an archaeological record? In: Advances in archaeological method and theory. Academic Press, New York, pp 27–62

Pestova TV, Hellen CUT (1999) Internal initiation of translation of bovine viral diarrhea virus RNA. Virology 258:249–256

Pestova TV, Shatsky IN, Fletcher SP et al (1998) A prokaryotic-like mode of cytoplasmic eukaryotic ribosome binding to the initiation codon during internal translation initation of hepatitis C and classical swine fever virus RNAs. Genes Dev 12:67–83. https://doi.org/10.1101/gad.12.1.67

Pestova TV, Lomakin IB, Hellen CUT (2004) Position of the CrPV IRES on the 40S subunit and factor dependence of IRES/80S ribosome assembly. EMBO Rep 5:906–913

Petrov AS, Bernier CR, Hsiao C et al (2014) Evolution of the ribosome at atomic resolution. Proc Natl Acad Sci 111:10251–10256

Phan H-D, Lai LB, Zahurancik WJ, Gopalan V (2021) The many faces of RNA-based RNase P, an RNA-world relic. Trends Biochem Sci 46:976–991

Pintó RM, Pérez-Rodríguez F-J, D'Andrea L et al (2018) Hepatitis A virus codon usage: implications for translation kinetics and capsid folding. Cold Spring Harb Perspect Med 8:a031781

Piquette KE, Whitehouse RD (2013) Introduction: developing an approach to writing as material practice. In: Writing as material practice. Ubiquity Press, London (UK)

Piron M, Beguiristain N, Nadal A et al (2005) Characterizing the function and structural organization of the 5′ tRNA-like motif within the hepatitis C virus quasispecies. Nucleic Acids Res 33. https://doi.org/10.1093/nar/gki290

Poole AM, Jeffares DC, Penny D (1998) The path from the RNA world. J Mol Evol 46:1–17

Robertson HD (1990) Escherichia coli ribonuclease III. In: Methods in enzymology. Elsevier, New York (Estados Unidos), pp 189–202

Robertson HD, Dunn JJ (1975) Ribonucleic acid processing activity of Escherichia coli ribonuclease III. J Biol Chem 250:3050–3056

Robertson HD, Altman S, Smith JD (1972) Purification and properties of a specific Escherichia coli ribonuclease which cleaves a tyrosine transfer ribonucleic acid precursor. J Biol Chem 247. https://doi.org/10.1016/S0021-9258(19)44963-6

Robertson HD, Dickson E, Dunn JJ (1977) A nucleotide sequence from a ribonuclease III processing site in bacteriophage T7 RNA. Proc Natl Acad Sci 74:822–826

Root-Bernstein R, Root-Bernstein M (2019) The ribosome as a missing link in prebiotic evolution III: over-representation of tRNA-and rRNA-like sequences and plieofunctionality of ribosome-related molecules argues for the evolution of primitive genomes from ribosomal RNA modules. Int J Mol Sci 20:140

Sabariegos R, Nadal A, Beguiristain N et al (2004) Catalytic RNase P RNA from Synechocystis sp. cleaves the hepatitis C virus RNA near the AUG start codon. FEBS Lett:577. https://doi.org/10.1016/j.febslet.2004.10.059

Samuel CE (2011) Adenosine deaminases acting on RNA (ADARs) are both antiviral and proviral. Virology 411:180–193

Schalk RF, Binford LR (1977) For theory building in archaeology. The structure of an anadromous fish resource, p 207–249

Schencking I, Rossmanith W, Hartmann RK (2020) Diversity and evolution of RNase P. In: Evolutionary biology—a transdisciplinary approach. Springer, Cham (Switzerland): Palgrave Macmillan, pp 255–299

Schluenzen F, Tocilj A, Zarivach R et al (2000) Structure of functionally activated small ribosomal subunit at 3.3 Å resolution. Cell 102:615–623

Schoggins JW, MacDuff DA, Imanaka N et al (2014) Pan-viral specificity of IFN-induced genes reveals new roles for cGAS in innate immunity. Nature 505:691–695

Schuster P, Fontana W, Stadler PF, Hofacker IL (1994) From sequences to shapes and back: a case study in RNA secondary structures. Proc Biol Sci 255:279–284

Sela N, Kim E, Ast G (2010) The role of transposable elements in the evolution of non-mammalian vertebrates and invertebrates. Genome Biol 11:R59

Serrano P, Gomez J, Martínez-Salas E (2007) Characterization of a cyanobacterial RNase P ribozyme recognition motif in the IRES of foot-and-mouth disease virus reveals a unique structural element. RNA 13. https://doi.org/10.1261/rna.506607

Shimakami T, Yamane D, Jangra RK et al (2012) Stabilization of hepatitis C virus RNA by an Ago2–miR-122 complex. Proc Natl Acad Sci 109:941–946

Sofaer JR (2006) The body as material culture: a theoretical osteoarchaeology. Cambridge University Press, Cambridge (UK)

Spahn CMT, Kieft JS, Grassucci RA et al (2001) Hepatitis C virus IRES RNA-induced changes in the conformation of the 40*S* ribosomal subunit. Science 291:1959–1962. https://doi.org/10.1126/science.1058409

Spitale RC, Flynn RA, Zhang QC et al (2015) Structural imprints in vivo decode RNA regulatory mechanisms. Nature 519:486–490

Sun F-J, Caetano-Anollés G (2008) The origin and evolution of tRNA inferred from phylogenetic analysis of structure. J Mol Evol 66:21–35

Sun F-J, Caetano-Anollés G (2009) The evolutionary history of the structure of 5S ribosomal RNA. J Mol Evol 69:430–443

Szczepanska A, Wojcik K, Kurzynska-Kokorniak A (2025) RNase III enzymes and their accessory domains. In: Molecular evolution of RNA regulatory enzymes and their systems. Springer, Cham (Switzerland), pp 161–201

Thurner C, Witwer C, Hofacker IL, Stadler PF (2004) Conserved RNA secondary structures in Flaviviridae genomes. J Gen Virol 85:1113–1124

Vetsigian K, Woese C, Goldenfeld N (2006) Collective evolution and the genetic code. Proc Natl Acad Sci 103:10696–10701

Villarreal L, Witzany G (2013) Genomics insights: the DNA habitat and its RNA inhabitants: at the Dawn of RNA sociology. Genomics Insights 6. https://doi.org/10.4137/GEI.S11490

Villarreal LP, Witzany G (2015) When competing viruses unify: evolution, conservation, and plasticity of genetic identities. J Mol Evol 80:305–318

Villarreal L, Witzany G (2021) Social networking of quasi-species consortia drive virolution via persistence. AIMS Microbiol 7. https://doi.org/10.3934/microbiol.2021010

Villarreal L, Witzany G (2023) Self-empowerment of life through RNA networks, cells and viruses. F1000Res 12:138

Walker SC, Engelke DR (2006) Ribonuclease P: the evolution of an ancient RNA enzyme. Crit Rev Biochem Mol Biol 41:77–102

Wan Y, Kertesz M, Spitale RC et al (2011) Understanding the transcriptome through RNA structure. Nat Rev Genet 12:641–655

Weiner AM, Maizels N (1987) tRNA-like structures tag the 3'ends of genomic RNA molecules for replication: implications for the origin of protein synthesis. Proc Natl Acad Sci 84:7383–7387

Wimberly BT, Brodersen DE, Clemons WM Jr et al (2000) Structure of the 30S ribosomal subunit. Nature 407:327–339

Wittgenstein L (1953) Philosophical investigations. Basil Blackwell Ltd, Oxford

Witzany G (2000) Language and communication - an approximation. In: Life: the communicative structure a new philosophy of biology. A new philosophy of biology. Libri BooksonDemand, Norderstedt, pp 29–45

Witzany G (2006) The logos of the bios. 1. Contributions to the foundation of a three-leveled biosemiotics. Umweb Publications, Helsinki (Finland)

Witzany G (2010) Biocommunication and natural genome editing. World J Biol Chem 1. https://doi.org/10.4331/wjbc.v1.i11.348

Witzany G (2013) Why biocommunication of animals? In: Biocommunication of animals. Springer, Dordrecht. Netherlands, pp 1–6

Witzany G (2014a) Pragmatic turn in biology: from biological molecules to genetic content operators. World J Biol Chem 5. https://doi.org/10.4331/wjbc.v5.i3.279

Witzany G (2014b) RNA sociology: group behavioral motifs of RNA consortia. Life (Basel) 4:800–818

Witzany G, Baluška F (2012) Life's code script does not code itself. EMBO Rep 13. https://doi.org/10.1038/embor.2012.166

Woese CR (1987) Bacterial evolution. Microbiol Rev 51:221–271

Woese CR (2002) On the evolution of cells. Proc Natl Acad Sci 99:8742–8747

Woese CR, Fox GE (1977) Phylogenetic structure of the prokaryotic domain: the primary kingdoms. Proc Natl Acad Sci 74:5088–5090

Wollenzien PL (1988) [21] Isolation and identification of RNA cross-links. In: Methods in enzymology. Elsevier, New York (USA), pp 319–329

Wurtmann EJ, Wolin SL (2009) RNA under attack: cellular handling of RNA damage. Crit Rev Biochem Mol Biol 44:34–49

Yu G, Zhu H, Chen X, Yang J-R (2021) Specificity of mRNA folding and its association with evolutionarily adaptive mRNA secondary structures. Genomics Proteomics Bioinformatics 19: 882–900

Zuckerkandl E, Pauling L (1965a) Molecules as documents of evolutionary history. J Theor Biol 8. https://doi.org/10.1016/0022-5193(65)90083-4

Zuckerkandl E, Pauling L (1965b) Evolutionary divergence and convergence in proteins. In: Evolving genes and proteins. Elsevier, New York (USA) pp 97–166

Chapter 17
The Proteomic and Coevolutionary Roots of the Genetic Code

Gustavo Caetano-Anollés

Abstract The genetic code is primarily implemented by aminoacyl-tRNA synthetases (aaRSs), cognate tRNA adapters, the ribosomal decoding site, and the atomic structure of proteins. Its origin and evolution are here elucidated by integrating evidence from phylogenies of tRNA, evolutionary timelines of protein structural domains, and the deep history of dipeptide sequences in proteomes within a biocommunication and coevolutionary framework. Phylogenetic analyses of tRNA sequence and structure reveal that aminoacylation specificities for Sec, Tyr, Ser, and Leu are ancient, whereas those for Asn, Met, and Arg are evolutionarily derived. These timelines highlight early reliance on the second and then first codon bases, the antiquity of Ala and Pro codons, and major evolutionary takeovers associated with structural innovations such as the loss of the long variable arm of tRNA. Complementary chronologies of the structural domains of aaRSs show that an operational RNA code embedded in the tRNA acceptor arm emerged well before the standard anticodon-based code and that early aaRSs with architectures homologous to modern TyrRS and SerRS likely catalyzed both aminoacylation and peptide bond formation. These interactions exemplify early molecular biocommunication, in which primitive peptides and tRNA cofactors evolved by fostering catalytic efficiency, editing accuracy, specificity, and structural recruitment to improve folding and protein flexibility. Independent support for this early operational code comes from a global phylogeny of 400 canonical dipeptides reconstructed from billions of sequences across thousands of proteomes. Dipeptides containing Leu, Ser, and Tyr emerged first, followed by Val, Ile, Met, Lys, Pro, and Ala, reflecting overlapping timelines of amino acid entry into the code. The parallel emergence of dipeptide–antidipeptide pairs further supports an ancient bidirectional (sense-antisense) coding system previously revealed by synthetic biology studies. Together, these findings offer a unified retrodictive view linking early protein structural constraints, coevolutionary signaling processes, and the rise of the genetic code.

G. Caetano-Anollés (✉)
Department of Crop Sciences and Carl R. Woese Institute for Genomic Biology, University of Illinois at Urbana-Champaign, Urbana, IL, USA
e-mail: gca@illinois.edu

G. Witzany (ed.), *Integrative Theory of Evolution - Aspects and Insights*,
https://doi.org/10.1007/978-3-032-24295-2_17

1 Introduction

The genetic code transforms a vocabulary that stores genetic information in nucleic acid molecules into a vocabulary of protein structure responsible for the molecular entities of the cell. Since its experimental discovery (Nirenberg and Matthaei 1961), the origin and evolution of the genetic code has been a vexing problem, squarely positioned between understanding the origin and early evolution of peptides and polypeptides and that of complex RNA structure.

The formation of amino acids is a prerequisite for peptide synthesis. Early Earth's capacity to generate proteinogenic and non-proteinogenic amino acids was first demonstrated in the landmark Miller-Urey experiment (Miller 1953). This experiment used an electrical "silent" discharge to mimic lightning in a reducing atmosphere of CH_4, NH_3, H_2, and H_2O, building upon earlier work by Löb (1913) but specifically designed to test prebiotic chemistry hypotheses. Subsequent findings showed similar amino acid synthesis in nonreducing environments of CO_2, N_2, and H_2O (Plankensteiner et al. 2004), aligning with modern views of Earth's early atmosphere. Once present, amino acids themselves likely catalyzed the formation of peptides under prebiotic conditions (Rode et al. 2007), aided by minerals, organic molecules, and gas-phase interfaces known to stabilize bonds against hydrolysis. The generation of peptides requires the formation of peptide bonds by condensation (water removal) reactions, with free energy demands (2–4 kcal/mol) often met by phosphate-activated intermediates like aminoacyl adenylates. In modern biology, these are provided by aminoacyl-tRNA synthetases (aaRSs) (Gomez and Ibba 2020), but prebiotic scenarios propose thioesters as energy sources (Jakubowski 2017), leveraging Earth's ancient redox chemistry. Once formed, peptide bonds exhibit remarkable stability with an average half-life of up to 267 years (Smith and Hansen 1998) but remain vulnerable to hydrolysis, racemization, deamidation, β-elimination, and other processes of degradation (Mahesh et al. 2018). Calculation of the free energies of peptide bond hydrolysis and formation shows that dipeptide synthesis requires eight times more energy than elongating existing peptides (about 1.2 kcal/mol) (Martin 1998). This fact alone suggests that "nucleation" processes that speed up energy dissipation facilitated abiogenic peptide bond formation. Recent studies are particularly revealing. The use of aminoacyl-phosphate esters as analogs of aminoacyl adenylates shows spontaneous self-assembly into supramolecular structures, including fibers and ribbons, that template amide bond formation in water (Sharma et al. 2025). Similarly, water-air interfaces in amino acid-containing water microdroplets serve as "drying surfaces" favoring millisecond peptide synthesis under ambient conditions and without other reagents, catalysts, or radiation (Holden et al. 2022). These mechanisms align with geochemical settings like hot springs, hydrothermal vents, tidal pools, and atmospheric aerosols—all plausible sites for life's origins. Such cooperativity rivals that expected from prebiotic chemistries, which acted as seeds for building blocks such as amino acids, peptides, and ribonucleotide triphosphates. For example, crystallization plays an important role in amino acid polymerization when using crystalline phosphoric acid salts of amino acids and low temperatures (Šponer et al. 2024). Moreover, RNA containing noncanonical bases found in modern tRNA and rRNA molecules can self-decorate with peptides, suggesting efficient prebiotic chemistries operating in a plausible RNA-peptide primordial world (Müller et al. 2022). Remarkably, a forward-synthesis AI platform (Allchemy) that utilizes 614 expertly curated reaction rules grounded in synthetic organic chemistry principles (ensuring experimental validity) generated a comprehensive prebiotic reaction network in only seven iterative synthesis generations, starting with H_2O, N_2, HCN, NH_3, CH_4, and H_2S substrates (Wołos et al. 2020). This network encompassed 82 biotic molecules—including amino acids, nucleobases, peptides, carbohydrates, and proto-metabolic intermediates—and 36,603 abiotic compounds, while uncovering both established and novel routes to biological targets. Prebiotic interactions demonstrated chemical emergence through self-regenerating cycles resembling autocatalytic motifs in modern metabolism, primitive catalytic pathways with feedback loops akin to early biochemical regulation, and compartmentalization drivers such as prebiotic fatty acid and peptide surfactants capable of forming micelles and vesicles. Strikingly, the AI-driven algorithm recapitulates approximately 70% of historically reported prebiotic syntheses while proposing novel, experimentally verifiable pathways and nonlinear interactions between abiotic systems that could seed molecular evolution.

The formation of complex RNA structure is also at the center of genetic code history. The building blocks of nucleic acids, some of which are readily represented in the Allchemy prebiotic networks, have been regenerated through prebiotic chemistries in the lab. Examples include a high-yielding regioselective purine prebiotic pathway mediated by the condensation of prebiotically reasonable formamidopyrimidine precursors (Becker et al. 2016) and the synthesis of pyrimidine nucleosides from small molecules and ribose driven by wet-dry cycles (Becker et al. 2019). The early appearance of RNA precursor molecules on Early Earth is highly likely, guaranteeing the early polymerization of nucleotides into nucleic acid precursors of modern RNA (Guo et al. 2023). Transfer RNA (tRNA) and ribosomal RNA (rRNA) molecules interact with each other and, together with RNase P RNA, are believed to be very ancient. In fact, catalogs of these RNA molecules by the Rfam database show that these families are universally present in cellular life (Hoeppner et al. 2012). tRNA is particularly central. Interactions between cognate tRNA and aaRSs define the specificity of the genetic code by selectively charging amino acids to their corresponding acceptor molecules (Gomez and Ibba 2020). These aminoacyl-tRNAs (aa-tRNAs) interact with translation factors to deliver amino acids to the peptidyl transfer center (PTC) of the large subunit of the ribosome for protein biosynthesis guided by a messenger RNA (mRNA) "tape" necessary for universal Turing machine computation (Caetano-Anollés 2023). The central information role of tRNA is made explicit by both establishing catalytic aaRS charging specificities and guiding amino acid order in polypeptides through anticodon loop-mRNA interactions. These functions are facilitated by the ribosome machinery, which ratchets the tRNA molecules toward the PTC for polypeptide biosynthesis in processes enabled by an atomic constellation that favors GTP hydrolysis and changes the structure of translation factors and the ribosome (Jobe et al. 2019). tRNA and rRNA appear to have common origins, an insight already anticipated by David Bloch and colleagues almost half a century ago (Bloch et al. 1984, 1989). The idea that the ribosome was built piecemeal from primordial tRNA modules ("genes") acting as both acceptor molecules ("adapters") and genetic repositories ("messengers") (Eigen and Winkler-Oswatitsch 1981; Weiner and Maizels 1987; Maizels and Weiner 1994) seems appealing and increasingly supported by evidence of ancient traces left as relics in RNA sequence and structure (Davidovich et al. 2009; Michel 2012; Farias 2013; Farias et al. 2014, 2021; Root-Bernstein and Root-Bernstein 2015; Caetano-Anollés and Caetano-Anollés 2016; Demongeot et al. 2023).

Here, I revisit the evolutionary genomic analysis of proteins and RNA molecules and describe how phylogenomic-based chronological information helps untangle the origin and evolution of the genetic code. I start by placing the genetic code within a framework of biomolecular communication. I then explore the origin and evolution of tRNA, synthetases, and dipeptide sequences. I finally discuss how the origin of the genetic code is linked to the early evolution of proteomes.

2 The Molecular Communication Underlying the Genetic Code

2.1 The Genetic Code Is a Natural Code

Communication is pervasive in living systems (Witzany 2016). It unfolds in complex networks of interactions at widely different timescales. Communication at molecular and cellular levels demands interactions that involve "signs," elements of information and life that communicate meaning, take the form of icons, indices, or symbols, and are causally related to objects through interpretation (Caetano-Anollés 2023; Ariza-Mateos et al. 2024).

Communication in natural languages requires that entities convey meaning through "messages" via a wordbook of signs (lexicon). This wordbook can be used at higher levels of language communication (Morris 1968) by systematically combining signs through "grammatical" rules (syntax), by defining the meaning of signs and their combination in linguistic expressions (semantics), and by establishing context-dependent relationships according to rules governing successful or unsuccessful sign-mediated interactions (pragmatics). I previously showed that there are clear correspondences between the *Lexicon* ⇒ *Syntax* ⇒ *Semantics* ⇒ *Pragmatics* progression of language organization and the *Module* ⇒ *Structure* ⇒ *Function* ⇒ *Fitness* paradigms of biology (Caetano-Anollés 2021). I showed that correspondences in protein organization are constrained by statistical laws that are likely universal, including Zipf's law of scale-free distributions, Heaps' law of sublinear growth, and the Menzerath-Altmann law of decreasing returns. I also showed that the existence of trade-off relationships between the engineering principles of economy, flexibility, and robustness exists in protein language organization. Consequently, there is an expectation that the same patterns, processes, and principles that apply to communication in protein organization would apply to communication underlying the genetic code and its organization. I caution that while the genetic code has been considered a natural language that links genetics to molecular functions (Eigen 1971), drawing analogies between language and biology can be limiting, even if language is a direct product of biology. Instead, I prompt the use of analogies as useful metaphors that can guide understanding of the principles of communication.

For the genetic code to qualify as a natural code—that is, a natural system of signs and rules—three conditions must be met: *combinatorial rules* must govern how code parts are connected, *content rules* must assign meaning to those parts by relating signs to the objects they describe, and *contextual rules* must embed the code within its functional setting. In natural languages, syntax refers to the systematic combination of lexical units (e.g., words) into larger expressions such as clauses, sentences, texts, and narratives. These expressions convey propositions shaped by grammar rules. The genetic code mirrors this principle through its own syntax: code elements are combined into triplets, which are read in nonoverlapping reading frames, with start and stop signals defining the boundaries of code translation. Semantics in natural language involves assigning meaning to linguistic expressions by linking signs to the objects they denote. The genetic code embodies semantics in a lookup table that translates code elements into amino acids according to a largely universal set of rules. Finally, pragmatics in natural language concerns how signs are interpreted within a communicative context. In the genetic code, pragmatics likewise emerges through interactions between code elements, the translation machinery, and the cellular context that ultimately gives meaning and purpose to the coding process.

2.2 Three Communication Elements

The following three elements of communication are necessary for the existence of a genetic code:

- Signs must be embedded in a nucleic acid molecule, typically as combinations of nucleotides (Fig. 17.1a). Their sequences will be largely defined by the energetics of base pair stacking interactions, their ability to base pair with complementary sequences, and their ability to stabilize structure.
- Pairs of consecutive signs must encode a *dipeptide sequence* embedding a *peptide bond* (Fig. 17.1b). Each dipeptide sequence must be constrained by the energetics and conformation of its 3-dimensional atomic structure, typically visualized through degrees of freedom endowed by ϕ (phi) and ψ (psi) torsion angles around backbone Cα atoms in a Ramachandran plot.
- Signs must be interpreted by a molecular system that matches them to amino acids in dipeptide sequences and makes genetic encoding persistent. This system is expected to grow in complexity during evolution and is majorly exemplified by tRNA-aaRS interactions responsible for maintaining the rules of the genetic code (Fig. 17.1c).

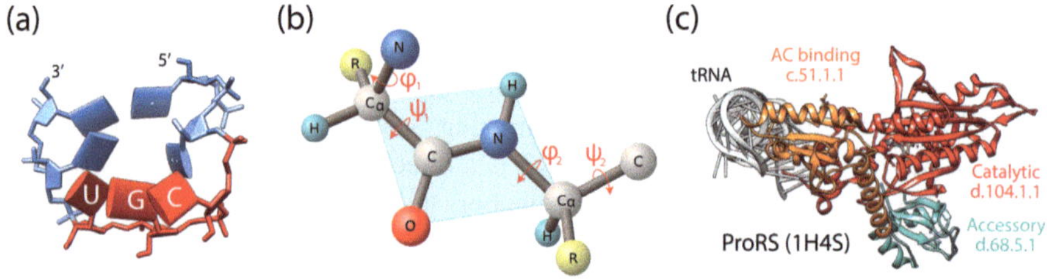

Fig. 17.1 The *raison d'être* of the genetic code. (**a**) The loop structure of the anticodon-binding stem of a tRNA harbors a sign (nucleotides in red) that codes for a specific amino acid (alanine). (**b**) The partial double-bond character of the peptide bond (π orbital-mediated resonance) in a dipeptide molecule establishes an amide plane (light blue) and a relatively large dipole moment ($\mu = 3.5$ D). This makes main chain conformations effectively defined by pairs of ϕ (phi) and ψ (psi) torsion angles centered around backbone Cα atoms that hold amino acid side chains (R). It also provides opportunities to line up dipole moments in structures such as α–helices. (**c**) The tRNA-aaRS enzymatic complex responsible for the specificity of the genetic code is illustrated with a ProRS structure (a Class II enzyme) in complex with tRNA. This complex is at the center of a sign vehicle-object-interpretant triangle describing a communication triad

The canonical lexicon (inventory of signs) of the modern genetic code involves groups of three nucleotides located in the anticodon loop of tRNA (bases 34, 35, and 36; Fig. 17.1a). These are the anticodons. Anticodons recognize mRNA through base pair interactions with complementary codon sequences when the molecule is scanned by the ribosome (codon bases 1, 2, and 3, which pair with anticodon bases 36, 35, and 34, respectively, in the ribosomal decoding site). The rules of correspondence between the 64 possible codons made up of nucleotides G, C, U, and A and the 20 proteinogenic amino acids carried by the 20 canonical types of tRNA (plus start and termination signals) define the lookup table of the genetic code. This code is *degenerate* (redundant) because only two amino acids are coded by a single codon, while the rest are specified by 2, 3, 4, and 6 codons. However, the genetic code is *unambiguous* because every codon codes for only one amino acid and *exhibits regularities* that allow partial repertoires of tRNA in different evolving lineages.

Besides anticodons, other identity elements in tRNA play important roles in structuring the genetic code vocabulary during aminoacylation (Giegé et al. 1998; Giegé and Eriani 2023). They include the "discriminator" nucleotide at position 73 and base pairs 1–72, 2–71, 3–70, and 4–69 of the acceptor arm (Acc), nucleotide 37 and base pairs 32–38 and 31–39 of the anticodon loop of the anticodon arm (AC), and the terminal nucleotides of the variable arm (Fig. 17.2). Identity elements cluster in the terminal region of the Acc arm, in the AC arm, and around the central junction but are scarce in the T-arm (Fig. 17.2b). Notably, in the atypical $tRNA^{Ser}$ and $tRNA^{Leu}$ type II molecules, identity elements shift to the variable arm (V) and are absent in the anticodon loop.

Crucially, the discriminator base identity element (Crothers et al. 1972) splits the code into two groups according to stacking and base pairing energetics, NWN and

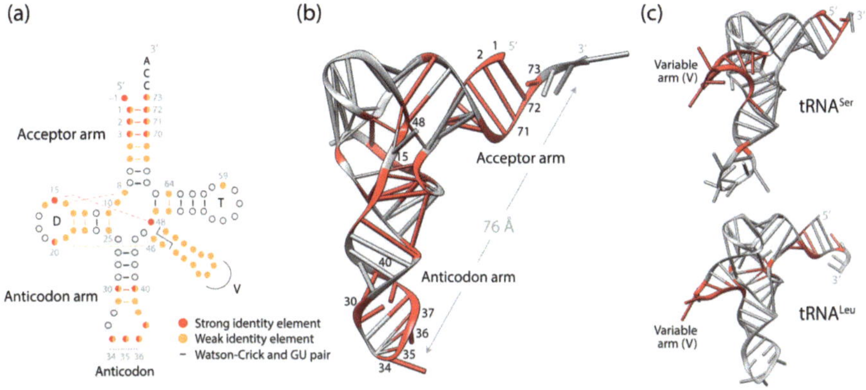

Fig. 17.2 The tRNA molecule and its identity elements. (**a**) The secondary structure of a generic tRNA molecule with strong and weak identity elements is traced along the different arms of the molecule. (**b**) Identity elements for all amino acids (in red) traced along the backbone of $tRNA^{Phe}$ (PDB entry 1EHZ). Identity elements of the variable arm (V) cannot be shown in the structural model. (**c**) The atypical structures of $tRNA^{Ser}$ (8P7D) and $tRNA^{Leu}$ (1WZ2) have identity elements present in their variable arms but absent in their anticodon loops

NSN, where W = A or U and S = G or C, and the 2–71 base further splits them into four groups, NUN, NAN, NGN, and NCN, where N = G, C, U, or A (any nucleotide) (Caetano-Anollés et al. 2013). NUN codes for hydrophobic amino acids, NAN codes for hydrophilic amino acids, and NGN and NCN code for semipolar amino acids. These four groups will then give rise to the condensed *vis-à-vis* degenerate genetic code representation of Rodin and Rodin (2008), which minimizes ambiguity in anticodon-codon recognition. This primordial complementarity manifests in pairs YUN:NAR and RUN:NAY associated with the NUN and NAN split and pairs YGN:NCR and RGN:NCY associated with the NGN and NCN split, where R = G or A and Y = C or U (Caetano-Anollés et al. 2013; Wang et al. 2025). Figure 17.3 presents a circular depiction of the genetic code induced by the decision tree. The depiction places codons in a circular arrangement that uncovers remarkable symmetries. The four quadrants, defined by RNA complementarity and amino acid physicochemical properties, can be further bisected according to minor-major groove recognition of tRNA by aaRS enzymes, which divides the aaRSs into two equally populated groups. This splits the code into two equal sectors, rotated 45° relative to the NWN and NSN division. The center of the circular diagram should be conceptualized as a primordial developing code, an "operational" RNA code structured by identity elements in tRNA and supported by early biochemical experiments (Schimmel et al. 1993; Hou and Schimmel 1988), phylogenomic analyses of tRNA and aaRS coevolution (Caetano-Anollés et al. 2013), and regression modeling of major and minor groove recognition (Carter Jr and Wills 2018). The peripherical code represents the modern "standard" code, in which all three codon positions contribute to thermodynamics and sequence complementarity.

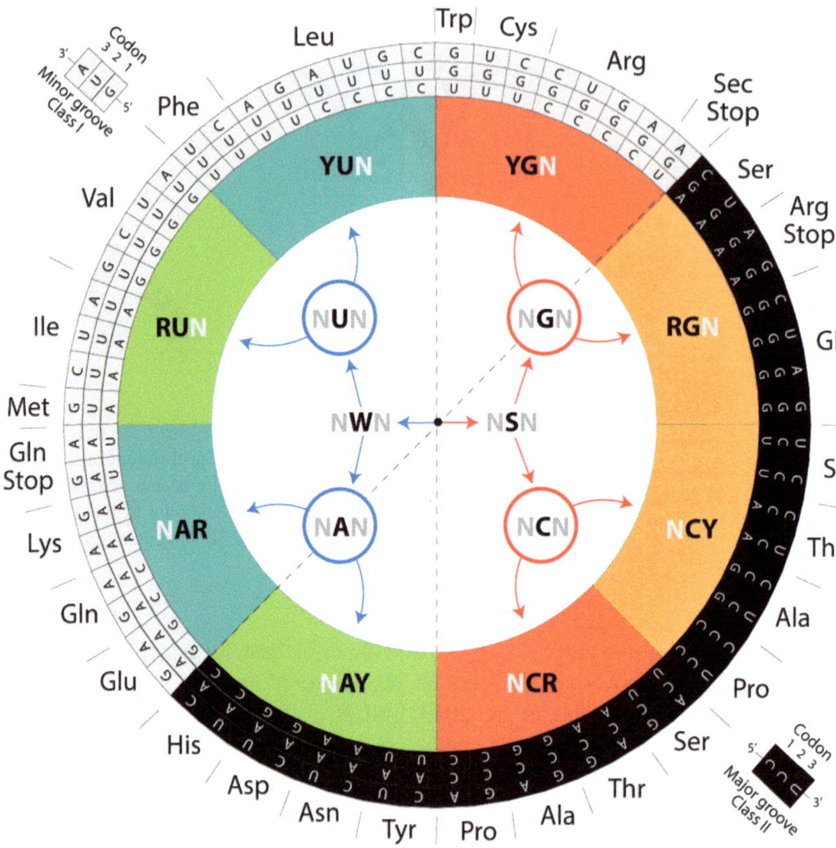

Fig. 17.3 A circular depiction of the genetic code uncovers remarkable symmetries. Codon sequences coding for amino acids labeled in the periphery of the diagram are given in grey boxes for those involving minor groove recognition (Class I) and black boxes for those involving major groove recognition (Class II). The four major quadrants divide the code according to the second codon position: NGN and NCN (in warm colors) code for semipolar amino acids, and NAN and NUN (cool colors) code for hydrophilic and hydrophobic amino acids, respectively. Quadrant subdivisions follow a decision tree induced by identity elements located in the acceptor stem of tRNA (Caetano-Anollés et al. 2013). The center of the circular diagram, therefore, represents evolutionary steps taken during the emergence of an operational code. Peripheral sectors depict the modern standard code. Sectors with the same color have codons that are reverse complementary to each other, showing that this circular depiction follows the *vis-á-vis* sense-antisense complementarity representation of Rodin and Rodin (2008)

Signs in nucleic acid molecules encode information that affects the atomic structure of proteins. Different amino acid pairs favor or disfavor certain regions in the Ramachandran space, altering local secondary structure propensities that are responsible for the formation of α-helices, β-sheets, turns, and other structural components (Hollingsworth and Karplus 2010). For example, dihedral angle motifs of the type X-Y-X-Y-X, where X and Y represent amino acid residues in any of the three major regions of the Ramachandran plot, are highly represented in known protein structures and foster helix-loop-helix and β-hairpin structures that facilitate the return of the polypeptide chain (Riziotis and Glykos 2019). These local preferences scale up to affect the folding and energetics of entire protein molecules. Consequently, dicodon sequences (6 contiguous nucleotides in mRNA) encoding dipeptide sequences are expected to embed a more complex genetic lexicon (see below). Remarkably, an exhaustive global study of the distribution of 544 physicochemical or biological properties of amino acids drawn from the *amino acid index* database (*aaindex1;* Kawashima and Kanehisa 2000) along the rows and columns of the genetic code table (Caldararo and Di Giulio 2022) revealed that the Miyazawa-Jernigan (MJ) partition energies (Miyazawa and Jernigan 1999) were the most significantly autocorrelated properties. These properties were close to a global optimum, suggesting they arose from natural selection likely imposed by coevolutionary forces acting on genetic code-related enzyme catalysis. Note that the MJ partition energies are knowledge-based potentials derived from statistical analysis of residue contacts in protein structures that are foundational for knowledge-based protein modeling, particularly for hydrophobicity-driven folding. MJ energies are energy differences between protein residues in contact (folded state) versus those in a reference state (e.g., fully solvated or extended conformation) that are calculated using the Bethe approximation, which relates observed contact frequencies in proteins to equilibrium probabilities. The origin and organization of the genetic code are therefore driven by the encoded amino acids rather than the nucleic acid signs, challenging the influential stereochemical theory that poses a complementarity between codons and amino acids (a chemical necessity) drove the origin of the genetic code (Gamow 1954) and the error-minimization theory that poses the driver was the minimization of the biological effects of translation errors (Woese 1965).

Finally, sign interpretation depends on the context of the entire translation system that is responsible for the synthesis of proteins, including the specificities imposed by the interpreting agents of the genetic code, including the aaRS genetic code guardians, the tRNAs, the tight nucleic acid interactions that occur in the decoding site of the small subunit of the ribosome, the vestibular cotranslational mechanisms that help fold nascent polypeptides as they come out of the ribosomal exit tunnel, and the functions of the translated proteins that ultimately modulate the entire translation process. For instance, a meta-analysis of aaRS structures indicates that decoding genetic information is governed by the chemical makeup of amino acid binding pockets, the spectrum of noncovalent interactions, editing functions, and steric constraints (Kaiser et al. 2020). Likewise, studies of peptide libraries suggest that the foldability of early proteins was a major determinant of the canonical amino acid set (Makarov et al. 2023). Note that sign interpretation is an agent-based social endeavor (Witzany and Baluška 2012). These agents cannot work in isolation. Instead, they generate codes or rules within a social and communicative syntactic, semantic, and pragmatic environment.

3 Building Blocks, Evolutionary Genomics and the Search for Origins

The principle of continuity demands the existence of *prior molecular states*, intermediate forms, or building blocks that mediate the evolution of the central information carriers of the cell, proteins and nucleic acid molecules (Caetano-Anollés et al. 2024). These evolutionary precursors are characterized by dynamic stability, evolutionary conservation, and modularity, properties that facilitate their reuse across biological systems. The existence of building blocks, a concept already intimated by Eck and Dayhoff (1966) in their seminal search for origins of the ferredoxin molecule, explains modern structural and functional complexity. They typically exist in nested hierarchies. In proteins, structural domains are evolutionary, structural, and functional units that often combine with each other to form multidomain proteins (Aziz and Caetano-Anollés 2021). Similarly, loops are structurally diverse polypeptide segments that brace secondary structures, contribute to protein flexibility and function, and combine in evolution to form structural domains (Berezovsky et al. 2000; Aziz et al. 2016, 2023). At a more granular level, *k*-mer motifs are short amino acid sequences within loops that carry historical signatures and distinguish genomic lineages (Pe'er et al. 2004; Choi and Kim 2020). They include dipeptide sequences, foundational 2-mer units influencing protein folding and adaptability (Amangeldina et al. 2024). In nucleic acids, RNA molecules exhibit hierarchical structural organization akin to protein loops and domains. Their folding begins with local hydrogen bonding between base pairs, forming double helices separated by single-stranded segments (Schuster 2010). This process is marked by energetic frustration due to competing stabilizing/destabilizing interactions. Advanced techniques such as fluorescence resonance energy transfer, atomic force microscopy, and optical tweezers reveal folding pathways (Li et al. 2008) dominated by a "plastic repertoire" of conformations, constrained evolutionarily and guided by free energy landscapes that funnel toward functionally viable tertiary structures (Fontana 2002). Initial helical stems act as structural nuclei, enabling long-distance tertiary contacts and molecular collapse stabilized by counterions. The formation of multibranched loops known as junctions is stabilized with non-Watson-Crick base pairs and coaxial stacking interactions (Leontis and Westhof 2001; Lescoute and Westhof 2006), mirroring the modularity of protein loops and domains. RNA junctions form evolutionarily conserved scaffolds that balance flexibility with constrained dynamics, fostering functional complexity through reusable helical arrays (Besseová et al. 2010). This hierarchy—from base-paired stems to modular junctions—parallels the nested structural units of proteins, underscoring a universal principle: biological complexity arises through the combinatorial reuse of conserved, dynamically stable building blocks.

Molecular evolution studies have, for the most part, embraced the use of the most elementary building block, the amino acid or nucleotide component of nucleic acid or protein sequences. The study of these components requires alignment-dependent methodologies, which are subject to mutational saturation of unconstrained regions and the erasure of information and history (Sober and Steel 2014). However, building *phylogenies* (ideographic statements of history) with more complex building block structures untangles deeper evolutionary processes and history (Caetano-Anollés 2021). The historical memory of modules such as *k*-mers, loops, or domains is enhanced by their ability to remain persistent due to their resistance to mutational change (increased conservation) and to horizontal transfers and recruitments that foster their spread. These processes increase their abundance and decrease their return to basal levels by loss in the absence of strong reductive evolutionary tendencies. Therefore, abundance can be coded as ordered multistate phylogenetic characters for phylogeny reconstruction, unfolding the unique power of "serial homologies" that is typical of morphological characters but rarely used in molecular studies.

Phylogenomic approaches that take advantage of large "omic" datasets (e.g., structures, functions) extracted from the genomes of thousands of reference organisms use serial homologies to build chronologies describing the first evolutionary appearance of building blocks (their time of origin), including structural domains or substructures that make up RNA (Caetano-Anollés et al. 2021). The general phylogenomic strategy dissects the evolution of structural parts and wholes by building and transposing (flipping) data matrices for phylogeny reconstruction (Fig. 17.4a). Once optimal trees are identified by traveling through tree solution space via branch-swapping operations, they are rooted a posteriori using Weston's generality criterion and Lundberg to establish an evolutionary order of branching (reviewed in Caetano-Anollés et al. 2018).

At the proteome level, Caetano-Anollés and Caetano-Anollés (2003) reconstructed global phylogenies of structural domains and their folds, establishing chronologies for their emergence. Numerous updates and applications have since been published (Caetano-Anollés et al. 2021). For example, Fig. 17.4b shows a tree of proteomes (wholes) depicting the evolutionary relationships of 420 free-living organisms spanning Archaea, Bacteria, and Eukarya (Kim and Caetano-Anollés 2012). This tree was built from structural domains defined at the fold family level of SCOP protein classification (Murzin et al. 1995). By transposing the data matrix, a corresponding tree of structural domains and its chronology were obtained (Fig. 17.4c). A molecular clock of folds, calibrated with biomarkers and geomarkers (Wang et al. 2011), placed domain families along a timescale of billions of years (Gy). Domains of aaRS enzymes were indexed along this chronology, revealing the sequential rise of molecular recognition elements and the later emergence of anticodon-binding elements essential for genetics (Caetano-Anollés et al. 2013).

At the nucleic acid level, Caetano-Anollés (2002a, b) was the first to reconstruct phylogenetic trees of RNA molecules or their substructures from structural (geometrical) or statistical features of the molecules. This approach made it possible to generate chronologies describing the history and stepwise development (accretion) of macromolecules and complexes directly from trees of substructures. Such chronologies were derived for tRNA (Sun and Caetano-Anollés 2008a), 5S rRNA (Sun and Caetano-Anollés 2009), the large and small ribosomal subunits (Caetano-Anollés 2002a, b; Harish and Caetano-Anollés 2012), RNase P RNA (Sun and

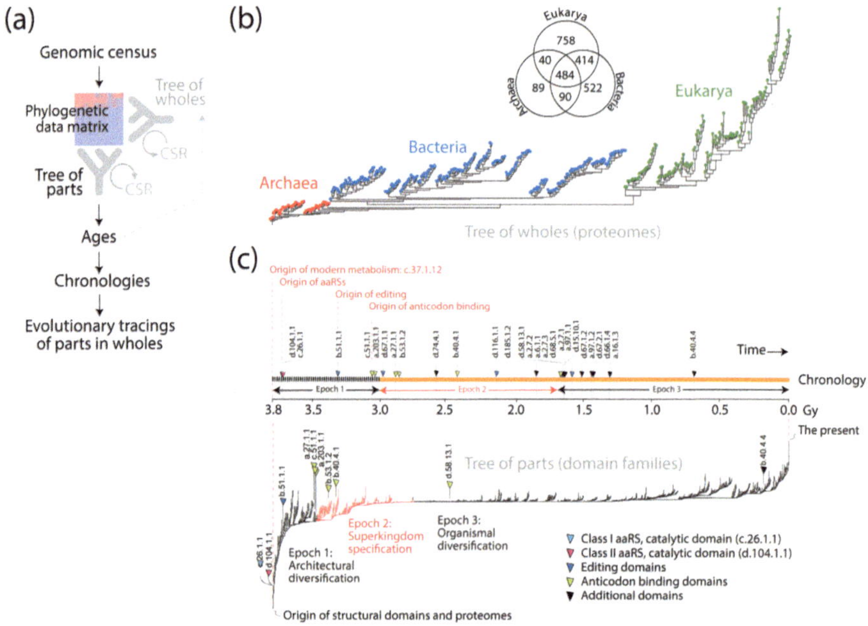

Fig. 17.4 A general phylogenomic strategy to study the evolution of parts and wholes, illustrated with the evolution of structural domains in proteomes. (**a**) A flow diagram describes the reconstruction of trees of parts and trees of wholes from genomic data and the production of chronologies and evolutionary tracings embedding the history of systems. CSR, character state reconstruction. (**b**) A tree of proteomes (wholes) describes the evolution of 48 proteomes from Archaea, 239 from Bacteria, and 133 from Eukarya. The most parsimonious phylogenetic tree (length = 128,371 steps; CI = 0.103; RI = 0.760; $g_1 = 0.241$) was reconstructed from genomic abundances of 2397 SCOP domain families, 2.262 of which were parsimony informative. The Venn diagram describes the distributions of families among the three cellular superkingdoms of life. (Data from Kim and Caetano-Anollés (2012)). (**c**) A tree of domain families (parts) describes the evolution of 2397 SCOP families. The most parsimonious phylogenetic tree (length = 177,864 steps; CI = 0.030; RI = 0.749; $g_1 = -0.070$) was reconstructed from a genomic census of 2397 families in the 420 proteomes described above. The chronology above the tree orders families according to their time of origin in billions of years (Gy) using a molecular clock of folds delimited by biomarkers and geo-markers. Domain families of aminoacyl-tRNA synthetases (aaRSs) are indexed to show the rise of molecular recognition components of translation. The chronological line in black describes the evolution of the protein world prior to modern genetics (anticodon-binding recognition elements). (Data from Kim and Caetano-Anollés (2012) and Caetano-Anollés et al. (2013))

Caetano-Anollés 2010a), and SINE RNA (Sun et al. 2007). In the presence of complex evolutionary signatures, chronologies were also inferred by forcing tRNA molecules into predefined monophyletic groups and testing competing and noncompeting evolutionary hypotheses (Sun and Caetano-Anollés 2008a, b, c). This involved reconstructing suboptimal trees of tRNAs under constrained groupings of taxa. Competing hypotheses were evaluated quantitatively by the number of additional steps (S) required relative to the optimal tree: hypotheses that were closer to parsimony were retained. This allowed, for example, the comparison of alternative scenarios of genetic code history. In contrast, noncompeting hypotheses were ranked according to their *S* values and used to infer timelines for the emergence of amino acid specificities and codon assignments. Hypotheses with smaller *S* values—reflecting greater parsimony—were considered less influenced by lineage-specific recruitment and therefore more likely to represent ancient evolutionary events. Put differently, lineages supported by these hypotheses tended to coalesce more readily when traced backward in time. This analytical strategy, known as "constraint analysis," is widely applied in phylogenetics to evaluate hypotheses of monophyly (Doyle 2006) and in cybernetics to deconstruct complex systems (Ashby 1956).

Together, protein and RNA chronologies provide fundamental insights into the origin and evolution of tRNA and the genetic code, as I will now describe.

4 Origin and Evolution of tRNA

tRNA is central to the translation system. As an adapter, it channels amino acids to the ribosomal PTC through mRNA recognition at the ribosomal decoding site. As a recognition determinant, it carries identity elements essential for maintaining the fidelity of the genetic code. As a ligand, it interacts with aaRSs to enable aminoacylation. As a sign-object-interpretant, it mediates communication (translation) at higher levels of complexity. In fact, the centrality of tRNA makes it a primordial demiurge—a modular entity capable of giving origin to the first genes (mRNA), the nucleic acid scaffolds of the ribosome (rRNA), and the genetic code itself (Farias and José 2020). Early spectral covariance studies by David Block and colleagues detected puzzling periodicities in ancestral homologies between tRNA and rRNA (Nazarea et al. 1985). They termed these periodicities "archeomodular," describing modular relics of spectral amplitude centered at a repeat length of nine nucleotides, three times that of the repeating pattern found in tRNA (Eigen and Winkler-Oswatitsch 1981). Implicit in this study was the existence of a primordial nucleic acid building block of sufficient size to form a short RNA hairpin, combinable in evolution to form more complex molecules. Previously, Jacques Demongeot proposed a 22-nucleotide RNA ring (a cyclic code) that represented a nondegenerate summary of the genetic code, including start and stop codons, and embodied a hairpin configuration of a primordial tRNA (Demongeot 1978). Following several iterations, this initial ring constructed on an economy principle has become an "archetypal loop" (AL) acting as a unique barycenter to another 18 out of 29,520 optimal ring solutions (Demongeot 2025).

The idea of tRNA arising from a simple stem loop structure had already been anticipated by Carl R. Woese decades earlier, when he suggested that *"primitive nucleic acid duplex structures"* were embedded in rRNA and other functional RNA molecules (Woese 1969). He predicted these structures would take the form of tRNA-like arms with helical stems about five base pairs long and loops of 3–10 nucleotides in length. Longer stem loop precursors were later proposed that were homologous to the acceptor arm (Hopfield 1978; Eigen and Winkler-Oswatitsch

1981). Alternative hypotheses and evolutionary pathways involving one or multiple RNA duplications have also been suggested. For example, Szathmáry (1999) considered the AC arm acting as a cofactor-delivering "handle" expanded to form mature tRNA; Di Giulio (1992) argued that the cloverleaf structure arose through duplication or ligation of two minihelix structures; Lei and Burton (2023) proposed that three 32-nucleotide-long minihelices were ligated to form a tRNA precursor; and Demongeot and Norris (2019) proposed a minimalistic RNA ring sequence that in a stable hairpin structure form, would combine in groups of four to build tRNA. Other scenarios have since been advanced, some anecdotal, which I will not detail here.

Instead of relying on structural, statistical, mathematical, or algorithmic arguments, we produced phylogenies that trace the history of tRNA from the structural or statistical features of the molecules themselves (Sun and Caetano-Anollés 2008a, b, c, d; Caetano-Anollés and Sun 2014). We analyzed 571 cloverleaf-structured tRNAs with information on base modifications from part II of the Bayreuth tRNA Database by scoring 42 structural phylogenetic characters and reconstructing phylogenies of tRNAs and their substructures (Sun and Caetano-Anollés 2008a). Phylogenetic analyses of 56 data matrices, organized by taxonomy, amino acid specificity, and biological source, revealed little support for the monophyly of tRNAs across superkingdoms or amino acid specificity. This likely reflects phylogenetic signal loss or the possibility that tRNA diversification preceded organismal diversification. Instead, substructure phylogenies based on structural and statistical characters were congruent, consistently placing the Acc arm as the most ancient feature, followed by the T and AC arms, and then the D and variable arms. Figure 17.5 illustrates our findings with an analysis of the 571 tRNAs with four

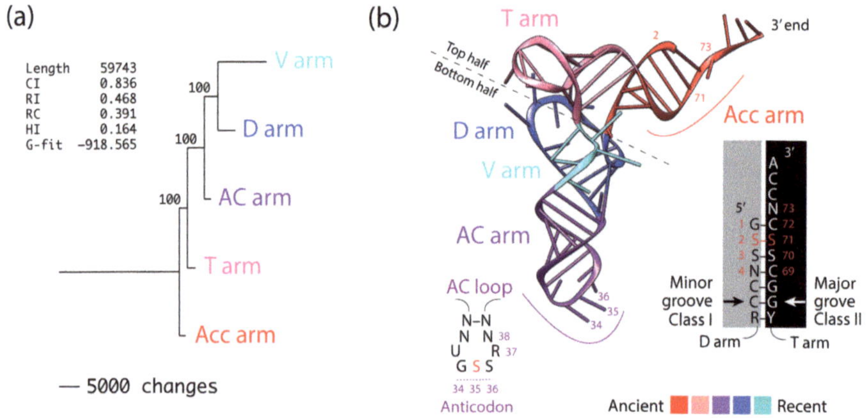

Fig. 17.5 Evolution of the tRNA molecule. (**a**) Phylogeny describing the evolution of the substructures of 571 tRNA structures generated from an analysis of 2284 statistical phylogenetic characters. Bootstrap support values were assessed with 2000 branch-and-bound replicates. Data from Sun and Caetano-Anollés (2005). (**b**) The times of origin of individual substructures were traced along the backbone of the tRNA molecule

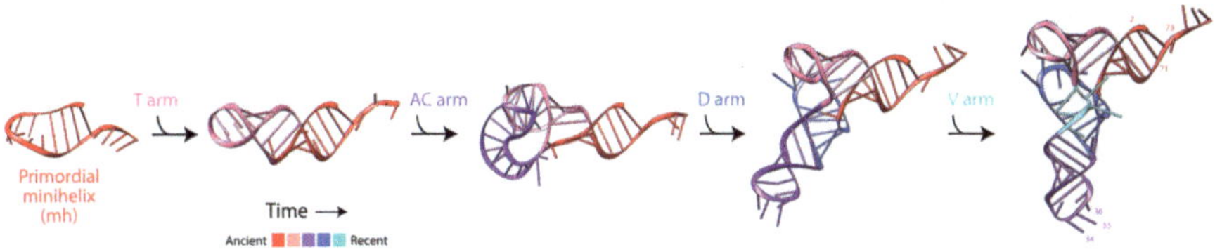

Fig. 17.6 Evolution of the tRNA molecule. The atomic structures of evolving tRNA structures are modeled using deep learning trRosettaRNA from an ancestor sequence reconstructed from a tRNA dataset of 571 molecules using maximum likelihood methods

statistical characters defining a complete structural morphospace: (i) the Frobenius norm of the probability matrix (F), which evaluates the Euclidean spread of base pairing interactions; (ii) the Shannon entropy of the probability matrix (Q), which captures conflicting intermolecular and intramolecular interactions; (iii) the base pair propensity (P); and (iv) the mean length of helical stems (S), both describing the degree of folding and ramification in RNA (Sun and Caetano-Anollés 2026). The basal placement of the Acc arm in the rooted phylogeny indicates an origin of the molecule in a relatively short stem loop structure, which expanded in evolution by gradual accretion of helical segments (Fig. 17.5a). To model this process, we reconstructed the sequence of the ancestor of the set of the 571 tRNAs recovered using maximum likelihood methods of character state reconstruction (Harish and Caetano-Anollés 2012), partitioned the ancestral sequence into regions corresponding to the five tRNA arms, and used these sequences to simulate the structural growth of the molecule based on the substructure phylogeny (Fig. 17.6). Three-dimensional structural predictions of evolutionary intermediates were then generated at the atomic level using the deep learning trRosettaRNA prediction suite (Wang et al. 2023), with most per-residue confidence levels comparable to those of experimentally determined crystallographic structures. All such reconstructions consistently support a model of tRNA evolution in which the cloverleaf structure arose by gradual enlargement leading to the stepwise addition of helical segments, with the variable arm added last (Sun and Caetano-Anollés 2008a, d). This model highlights the ancestral role of the Acc and T arms, the top-half of the molecule, consistent with the genomic tag hypothesis that tRNAs originated as ancient telomeres (Weiner and Maizels 1987; Maizels and Weiner 1994).

Note that phylogenetic constraint analyses, akin to those previously used to test competing models of 5S rRNA growth (Sun and Caetano-Anollés 2010b), showed that a single duplication of the type proposed by Di Giulio (1992) was less parsimonious than the helical accretion model supported by the phylogenetic reconstruction of Fig. 17.5 (Caetano-Anollés, ms. submitted). As we have shown for 5S rRNA (Sun and Caetano-Anollés 2010a, b), the evidence indicates that tRNA evolution cannot be explained by a single large duplication, challenging the validity of recent parsimony arguments against multiple duplication events (Di Giulio 2020). In fact, a more extensive exploration uncovering the existence of Menzerath-Altmann's law of syntax in RNA accretion history showed that change occurred gradually in RNA evolution (Sun and Caetano-Anollés 2021). Character state reconstruction along the branches of phylogenetic trees of molecules and substructures of tRNA, but also RNase P RNA, 5S rRNA, and small and large subunits of rRNA, revealed a principle of diminishing returns in which smaller helical segments tended to become smaller and larger segments tended to become larger in evolution. This contrasted with unpaired regions, where gains were consistently favored over losses of average lengths, with frequencies decreasing with nucleotide length. This differential evolutionary behavior reflects the well-known frustrated energetics and folding landscape of RNA, in which helical and unpaired segments stabilize or destabilize the structure, respectively (Schuster 2010). It also reveals pushes toward an economy of size, which manifests in significant negative correlations and strong associations between the length of helical stems and their time of origin. These correlations are probably counterbalanced in smaller molecules such as tRNA by "canalization" mechanisms, including the use of modified bases, wobble base pairing, and pseudo-knot lock-in configurations.

To evaluate the frustrated energetics that drive the folding landscape of the primordial stem loop structure with a statistical mechanics approach, Fig. 17.7 explores the folding kinetics of the reconstructed tRNA minihelix ancestor (mh) of Fig. 17.6 (Sun and Caetano-Anollés 2026) and the "cyclosome" AL structure that acts as the barycenter of theoretical primordial RNA rings (Demongeot 2025). Energy and folding landscapes were represented as barrier trees, hierarchical structures that capture alternative folding pathways, in which RNA conformers traverse saddle points to overcome the highest energy barriers in a quest to achieve local energy minima (Flamm et al. 2002). Remarkably, both reconstructed and theoretical tRNA ancestors displayed similar *funnel-like* energy landscapes, in sharp contrast to the *flat* and rugged profiles of randomized derivatives. Unlike the randomized counterparts, ancestral tRNA adapters avoided landscapes prone to deep, frequent, and often irreversible misfolds. Instead, their folding funnels efficiently guided conformations toward the native state, consistent with active evolutionary canalization (Fontana 2003). Furthermore, the minimum free energies (MFE) of the optimal secondary structures in both the reconstructed and ring models were comparable and encompassed a substantial fraction of their respective thermodynamic ensembles (72.6 and 50.3%). These results reinforce the principles of plastogenetic congruence and canalization that drive RNA evolution (Fontana 2003).

5 Distinct Origins and Evolution of Amino Acid Charging and Codon Identity Functions

Phylogenomic reconstruction of trees of molecules showed that tRNAs with a variable arm, a structural hallmark of type II tRNAs, appeared at the base of the phylogenies (Sun and Caetano-Anollés 2008b, c, d). These tRNAs, which code for Sec, Ser, Tyr, and Leu, were also ancestral compared to a minority of type I $tRNA^{Sec}$,

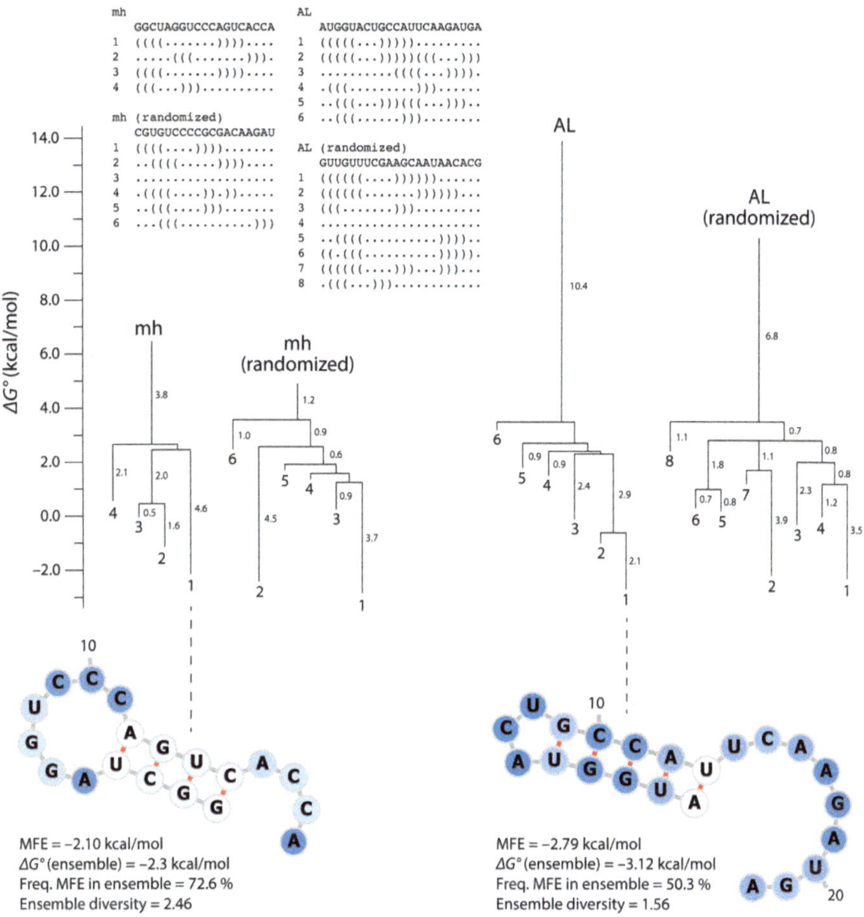

Fig. 17.7 Barrier trees describing the folding landscapes of reconstructed mini-hairpin (mh) and archetypal loop (AL) structures and randomized derivatives. Barrier trees were built using the *barriers* web server (Flamm et al. 2002) of the Vienna RNA web package (Gruber et al. 2008). A total of 23, 23, 128, and 172 substructures covered an energy range of 40 kcal/mol above the minimum free energy (MFE) for mh, mh (randomized), AL, and AL (randomized), respectively. The alternative secondary structures of the leaves of the barrier trees are defined using bracket notation. Nucleotides of secondary structures are colored according to the probabilities of being paired or unpaired, with deeper shades representing higher probabilities

tRNA^{Ser}, tRNA^{Tyr}, and tRNA^{Leu} isoacceptors lacking the variable arm. Figure 17.8 shows example trees reconstructed from data matrices partitioned according to organismal identity, which always place type II tRNAs at their base. Results suggest that the variable arm is an ancient development and that Sec, Ser, Tyr, and Leu were among the first amino acids to gain functional specificity, with UGA—the Sec codon—possibly representing the oldest modern stop codon (Sun and Caetano-Anollés 2008d). The phylogeny also suggests that amino acid charging by modern

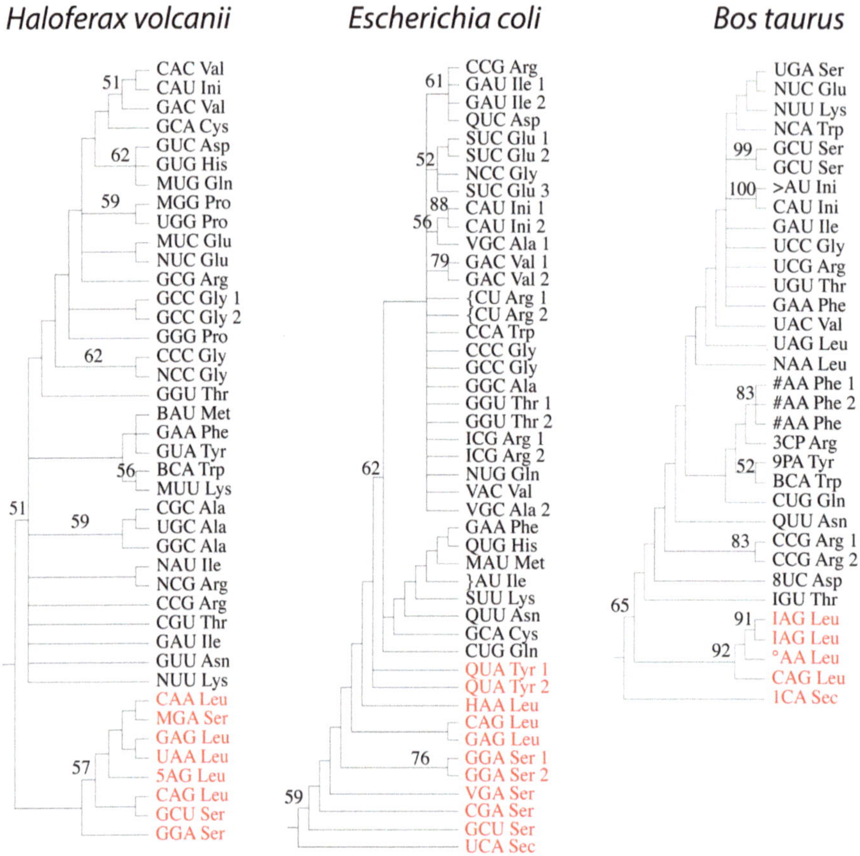

Fig. 17.8 Phylogenetic trees of tRNA from representative organisms of the three superkingdoms of life present in part II of the Bayreuth database. Taxa are indexed with anticodons followed by amino acid specificities and accession numbers. Type II tRNAs are labeled in red. The numbers above the branches are bootstrap values. (Data from Sun and Caetano-Anollés (2008b))

aaRSs, with all its potentialities, likely emerged only after the canonical cloverleaf was fully established but before diversification of tRNA specificity and superkingdom lineages (Sun and Caetano-Anollés 2008a, b, c). This aligns with sequence matching, and randomization experiments support the cloverleaf structure emerging prior to the divergence of tRNA specificities and the three domains of life (Widmann et al. 2005).

Bernhardt (2016) noted a potential contradiction: although the variable arm represented the last substructural addition to the expanding tRNA structure, type II tRNAs containing this feature evolved earlier than type I tRNAs. The contradiction is resolved by recognizing that phylogenies of substructures trace the evolution of sequence and structure, not function. Functional capacities must instead be addressed through phylogenies of entire tRNA molecules, which encompass full complements of amino acid charging and anticodon recognition properties. Amino acid charging relies on the development of identity elements in tRNA, which are weak in seryl-tRNA and leucyl-tRNA synthetases and emerged gradually—and perhaps loosely—during the early stages of the genetic code (the operational RNA code). Many of these features arose before the establishment of interactions between the anticodon loop and mRNA, the ribosomal decoding site, and anticodon-binding domains of aaRSs, all of which became essential for specifying codon identity (Caetano-Anollés et al. 2013).

We investigated the origin and evolution of codon and amino acid identities using constraint analyses to circumvent the lack of monophyletic groupings of tRNAs with shared charging or codon identities. Hypotheses yielding smaller *S* values, being more parsimonious, were interpreted as less affected by lineage-specific recruitment and therefore more representative of ancient evolutionary events. Supporting this "polarization" assumption, we demonstrated that lower *S* values correlated with shorter phylogenetic distances from a hypothetical ancestral tRNA (Sun and Caetano-Anollés 2008b).

Analysis of anticodon positions revealed a temporal ordering in the history of codon assignments (Sun and Caetano-Anollés 2008b). As shown in Table 17.1, constraint analyses that forced tRNAs with shared bases at the first, second, or third codon positions into monophyly during phylogeny reconstruction produced average *S* values of 368.50, 310.25, and 659.75, respectively. A similar pattern was observed when constraints were applied across all four bases of each codon position, yielding *S* values of 708 (first), 649 (second), and 1597 (third). These results indicate that the second codon position was recruited earliest for functional differentiation, followed by the first, with the third "wobble" position diversifying most recently. This

Table 17.1 Number of additional steps (*S*) needed to force tRNAs sharing the first, second, or third codon positions into monophyly during phylogeny reconstruction

Codon position	Constraint	Ancestry (*S*)
First	**A**NN	372
	CNN	466
	GNN	247
	UNN	389
	XNN	708
Second	N**A**N	345
	N**C**N	267
	N**G**N	330
	N**U**N	299
	N**X**N	649
Third	NN**A**	393
	NN**C**	817
	NN**G**	533
	NN**U**	896
	NN**X**	1597

Data from Sun and Caetano-Anollés (2008b)

N stands for a degenerate position that accepts any of the four bases, and X stands for constraints on all four bases on a codon position

ordering aligns with extensive prior evidence, particularly biochemical and energetic constraints on base pairing and the requirement for robust early anticodon-codon recognition (Woese et al. 1966; Haig and Hurst 1991; Grosjean and Westhof 2016). The findings also provide independent support for the decision tree (Caetano-Anollés et al. 2013) that organizes the circular genetic code of Fig. 17.3, in which the four sectors (NGN, NCN, NAN, and NUN) are arranged by the second codon position. To accommodate insights from tRNA phylogenies, Fig. 17.9 modifies this circular code by subdividing each of the four initial sectors into subsectors according to the first codon position. This generates eight octants, YGN, RGN, SCN, WCN, WAN, SAN, RUN, and YUN, partitioning the NAN sector of Fig. 17.3 and

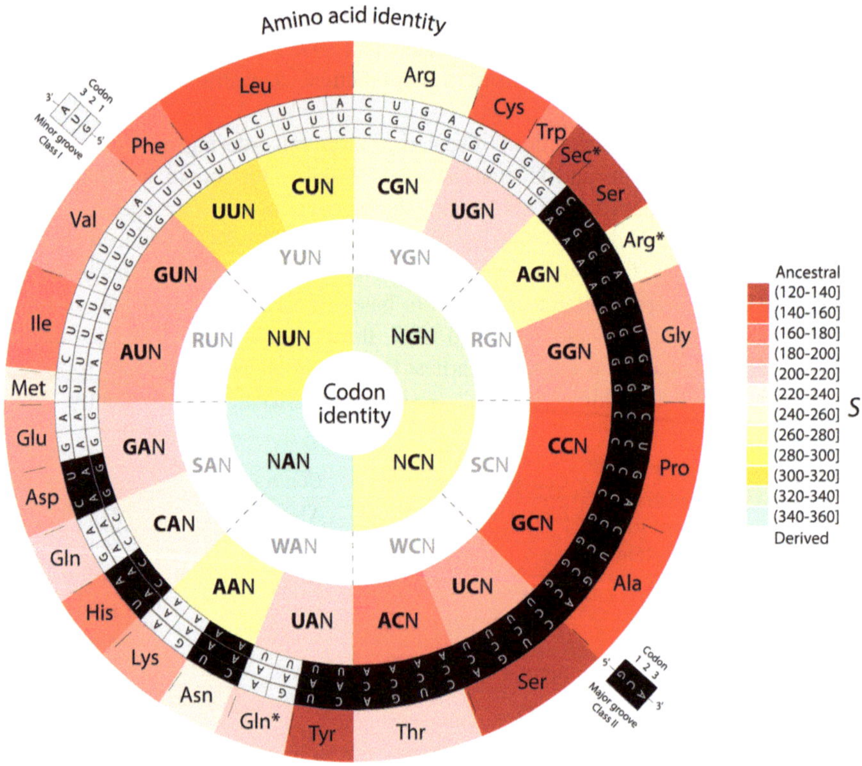

Fig. 17.9 Phylogenetic constraint analyses of codon (center) and amino acid (periphery) identities in tRNA sequence and structure mapped onto a circular depiction of the genetic code. This depiction splits the code into two halves, NWN and NSN, and then organizes codons according to the second and then the first codon position. This divides the code first into four quadrants, NGN, NCN, NAN, and NUN, and then into eight octants, YGN, RGN, SCN, WCN, WAN, SAN, RUN, and YUN. Note that such an arrangement ungroups the Class I major groove recognition mode of codons (black cells) in the NAN quadrant of the circular depiction of the genetic code described in Fig. 17.3. Ancestries (*S*) were derived by constraining sets of tRNA molecules into monophyletic groups, which were then painted with different shades onto boxes according to a scale of values. Ancestries in concentric circles represent noncompeting hypotheses. Asterisks represent stop codons

ungrouping codons in the Class I major groove recognition mode. In doing so, the revised circular code disrupts the *vis-á-vis* sense-antisense complementarity patterns of the genetic code described by Rodin and Rodin (2008) but makes explicit the contribution of the first codon position to genetic code history.

A timeline of codon identity functions generated by constraining tRNAs with shared bases at both the second and first codon positions was also traced onto the modified circular code (Fig. 17.9). With one exception (CGN coding for Arg), codons with strong/moderate base interactions with anticodons (5–6 hydrogen bonds for the first two positions) were more ancestral than those with weaker base interactions, which were generally associated with larger *S* values. The oldest codons belonged to the NCN quadrant, followed closely by those of the NUN and then the NGN quadrants. This pattern matches the data of Table 17.1. Codons coding for Ala and Pro emerge as particularly ancient, suggesting these codons were stabilized early in genetic code history. They were followed by codons coding for Thr, Ser, Gly, and Val.

As with codon identities, analysis of amino acid charging revealed a temporal ordering of tRNA-associated amino acid identities (Sun and Caetano-Anollés 2008b), which were also traced onto the circular code (Fig. 17.9). Timelines showed that Tyr, Sec, Ser, and Leu were among the earliest charged amino acids, whereas Asn, Met, and Arg appeared as the most derived group. The timeline mirrored, but did not completely coincide with, selected hypotheses of amino acid entry into the genetic code. Constraint analyses revealed that the consensus chronology of Trifonov (2004), the coevolutionary sequence of Wong (1981) that dissected prebiotically synthesized amino acids (Miller 1987) from those synthesized in primordial metabolic pathways, and chronologies derived from amino acid usage rates that were coupled to Bayesian reasoning (Brooks et al. 2002), a focus on ribosomal proteins and ATPase subunits (Fournier and Gogarten 2007), or 3-way alignments of orthologous proteins (Jordan et al. 2005), in that order, were increasingly suboptimal and unlikely. Discrepancies point to a dynamic recruitment history rather than a linear, strictly chronological codon-amino acid mapping that often describes distinct evolutionary stages or organizational levels (e.g., prebiotic synthesis and abundance relationships), making direct comparison and congruence assessments difficult.

Inspection of the phylogenetic tracings of Fig. 17.9 shows mismatches in the evolutionary trajectories of tRNA codon recognition (anticodon identity) and aminoacylation (amino acid charging identity). Phylogenetic analyses indicate that these two functional domains of tRNA followed largely independent evolutionary paths. This suggests that the genetic code did not arise from a single coevolutionary process linking codons to amino acids. Instead, tRNAs defined by specific charging identities may have been subsequently reassigned to different anticodons—and vice versa—through processes such as takeover, co-option, or horizontal transfer. The lack of strict congruence challenges classical "frozen accident" or simple coevolution models of the genetic code.

In conclusion, tRNA phylogenies support the existence of an ancient, versatile tRNA world in which molecules fulfilled both informational and functional roles of the type seen in modern codon–anticodon interactions and aminoacylation enzymatic functions. Over evolutionary time, these dual functions became increasingly specialized and decoupled through repeated co-option events. Consequently, the genetic code did not originate from a single tightly coupled process but rather from a layered evolutionary history shaped by molecular takeovers. Such a scenario must by necessity involve a coevolutionary process.

6 tRNA-aaRS Coevolution

To explore the coevolution of tRNA and cognate aaRS enzymes, times of origin (ancestries or *nd*) of proofreading and anticodon-binding activities—required for amino acid and tRNA discrimination—were mapped onto charging and codon identities in tRNA, respectively (Caetano-Anollés et al. 2013). We assumed the ancestry of a domain family reflects the ancestry of its associated function. As shown in Fig. 17.4c, catalytic cores of Class I and II aaRSs appeared among the oldest protein folds, preceding both editing and anticodon-binding domains, in that order. This ordering already supports a primitive aminoacylation system that operated through structural features in the acceptor stem—particularly the discriminator base and nearby identity elements—long before codon–anticodon interactions were established. When coupled with tRNA history, however, the phylogenomic timeline of structural domains fine-tunes the reconstruction of coevolutionary histories of charging and coding functions.

Amino acids are small molecules with relatively similar physicochemical properties. Consequently, discrimination against noncognate amino acids requires exclusion by size, charge, or metal binding; editing mechanisms in catalytic and editing domains; and the activity of freestanding factors (Gomez and Ibba 2020). Ten aaRSs harbor editing functions that remove incorrect amino acids before or after the transfer of amino acids to tRNAs. Class I MetRS and Class II SerRS and LysRS harbor pre-transfer editing activities, often at the active site of the catalytic domain. Class I IleRS, LeuRS, and ValRS and Class II AlaRS, ThrRS, and PheRS harbor post-transfer editing domains that hydrolyze misacylated amino acids; Class II ProRS harbors a similar activity but is located in an insertion (INS) of the catalytic domain. When plotting evolutionary ancestries, we uncovered statistically significant temporal correlations between these pre-transfer and post-transfer editing functions and charging identities in tRNA ($F = 9.10$; $P < 0.012$), which unfolded from 3.7 to 3 Gy ago (Caetano-Anollés et al. 2013). The regression established an early evolutionary sequence of charged amino acids associated with an “operational” RNA code: Tyr, Ser, and Leu appeared first, followed by Val, Ile, Met, Lys, and Pro, and then Thr and Phe. This sequence strongly suggests the earliest phases of translation were governed primarily by the recognition of tRNA acceptor stem elements rather than anticodon triplets. The analysis further defined three groups of aaRSs based on the evolutionary relationships of their catalytic and editing domains: *Group I* recognizing Tyr, Ser, and Leu; *Group II* recognizing Val, Ile, Met, Lys, Pro, Thr, and Phe; and *Group III* encompassing the remaining amino acids.

While editing functions proofread the activation and transfer of amino acids to the acceptor arm of tRNAs, anticodon-binding domains of aaRSs selectively interact with identity elements in the anticodon loops of cognate tRNAs, which are 76 Å away from the 5′ and 3′ ends of the acceptor stem (Fig. 17.2b). All aaRSs have anticodon-binding or tRNA-binding domains, except for SerRSs, which have a coiled-coil N-terminal tRNA-binding structure. Again, plotting ancestries uncovered temporal correlations between aaRS anticodon-binding and codon identities in tRNA ($F = 20.8$; $P < 0.001$), which unfolded from 3 to 2.7 Gy ago. The regression established an evolutionary sequence of codon recognition associated with tRNA-binding and the standard genetic code: Pro, Ala, Gly, Thr, and His (Group A), followed by Val, Met, Ile, Cys, Gln, Arg, and Leu (Group B), and then Asp, Asn, Lys, Ser, Glu, Tyr, Lys, Phe, and Trp (Group C), in that order. Note that all N-terminal anticodon-binding domain structures (Asn, Asp, Lys, Ser, and Phe) appear in Group C, making their N-terminal placement an evolutionarily derived feature. Importantly, the evolutionary sequence defines a chronology of expansion of the standard genetic code and its anticodon alphabet, which can be visualized within the condensed *vis-à-vis* representation of the code (Fig. 17.10). Notably, proline—the founding amino acid—already engages the second and first codon positions (C pairing identity elements G35 and G36), while the third codon position (pairing G34) is first utilized by threonine and subsequently by histidine (the final two recruitments of domain c.51.1.1) as the code expands into its triplex form. Finally, the Yin–Yang complementarity pattern reaches completion with the recruitment of a.27.1.1, coinciding with the establishment of the modern tetraplex code. This chronology highlights a central transition between major- and minor groove modes of tRNA recognition that accounts for both the most parsimonious number of anticodon positions and the key identity elements of the anticodon loop.

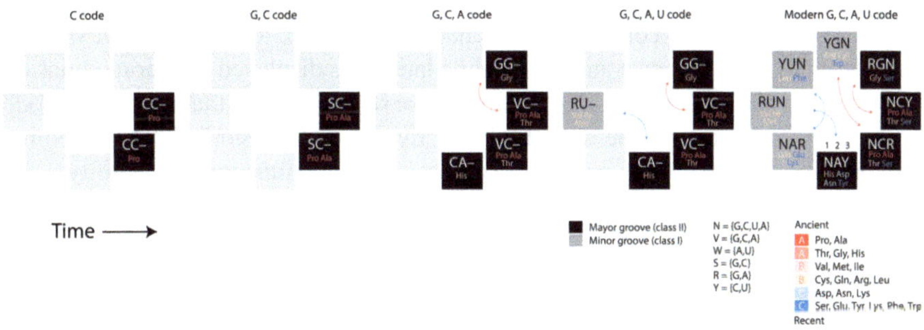

Fig. 17.10 Progression of the codon alphabet during the evolution of the "standard" genetic code. Ancestries of codon identities and anticodon-binding define a timeline of code expansion, which is traced in the octants of the circular depiction of the genetic code described in Fig. 17.3. Arrows indicate codons that are reverse complementary to each other, following the *vis-á-vis* sense-antisense complementarity representation of Rodin and Rodin (2008)

To summarize, tRNA and aaRS structures coevolved to expand and refine translational specificity. Editing functions and new tRNA features (D-arm, variable loop, and anticodon arm) emerged sequentially, giving rise to a dual recognition system that gradually linked amino acid identity to nucleotide triplets. The eventual addition of anticodon-binding domains completed this transition, coupling operational aminoacylation rules to an iconic codon-based genetic code. This staged progression—from an enzymatically grounded operational code to a modern informational code—illustrates how translation fidelity could increase in concert with the diversification of amino acid biosynthesis and ribosomal architecture. The genetic code can therefore be portrayed as a historical layering of molecular innovations, reflecting a long coevolutionary interplay between protein catalysts and their tRNA substrates.

7 The Biosynthetic Origins of the Genetic Code and Its Link to Protein Dynamics

Understanding the proteomic origins of the genetic code requires linking the emergence of coding specificity to the evolutionary history of the structure and dynamics of proteins. Structural phylogenomics offers such a framework by treating modern proteomes as molecular fossils that record the diversification of proteins and their functions through time. A global phylogeny of protein structural domains at the domain family level of SCOP fold classification across hundreds of proteomes revealed that amino acid recruitment followed clear evolutionary trends shaped by biosynthetic and structural innovation (Caetano-Anollés et al. 2013). We found that amino acid residues that are biosynthetically simple and prebiotically plausible were broadly represented throughout the timeline, whereas amino acids derived from more complex biosynthetic pathways appeared progressively as metabolism diversified. This progressive diversification paralleled the emergence of new metabolic capabilities and increasingly sophisticated protein structures. As I will discuss, amino acid recruitment correlated with changes in protein topology and flexibility: early proteins were compact and rigid, while later ones displayed greater modularity, disorder, and conformational adaptability. These results suggest that the genetic code evolved in concert with the structural and dynamic expansion of the proteome, driven by both biosynthetic innovation and selection for functional versatility.

7.1 Catalytic Relics of Pre-ribosomal Peptide Synthesis

Enzyme homologs of the most ancient catalytic domains of aaRSs support the early rise of precursor (core) structures capable of interacting with primordial tRNA molecules to jumpstart an early operational RNA code. Structures homologous to the most ancestral *Group 1* catalytic domains display atypical biosynthetic functions (Caetano-Anollés et al. 2013). For instance, amino acid–[acyl-carrier-protein]–ligases (aaACPLs), which are homologous to seryl-tRNA synthetases (SerRSs), aminoacylate 4′-phosphopantetheinyl and other prosthetic groups of carrier proteins—activities characteristic of non-ribosomal peptide synthetase (NRPS) modules (Mocibob et al. 2010). Similarly, cyclodipeptide synthases (CDPSs), homologous to tyrosyl-tRNA synthetases (TyrRSs) and involved in numerous secondary metabolism biosynthetic pathways, catalyze the formation of cyclodipeptides from charged tRNAs specified by *Groups 1* and *2* domains (Gondry et al. 2009; Sauguet et al. 2011). Structural alignments of aaACPLs and CDPSs with their close aaRS structural neighbors reveal highly conserved structural cores and sequence motifs at ligand-contacting sites, suggesting that these enzymes are relics and possible founders of archaic biosynthetic activities (Caetano-Anollés et al. 2013). CDPSs are particularly relevant. Structural alignments of a CDPS representative (Snou-CDPS, formerly AlbC from *Streptomyces noursei*; Sauguet et al. 2011) against a PDB structural set using DALI conservation mapping, a powerful remote homology detection tool (Holm et al. 2023), showed that the closest structural neighbors of CDPSs harbored catalytic cores of Class Ic aaRS enzymes, TyrRSs, and TrpRSs (Fig. 17.11a). Their highly conserved catalytic cores (Fig. 17.11b) suggest CDPSs are relics and founders of archaic biosynthetic activities.

CDPSs employ a ping-pong mechanism to form two successive aminoacyl-enzyme intermediates using two aminoacyl-tRNA (aa-tRNA) molecules and

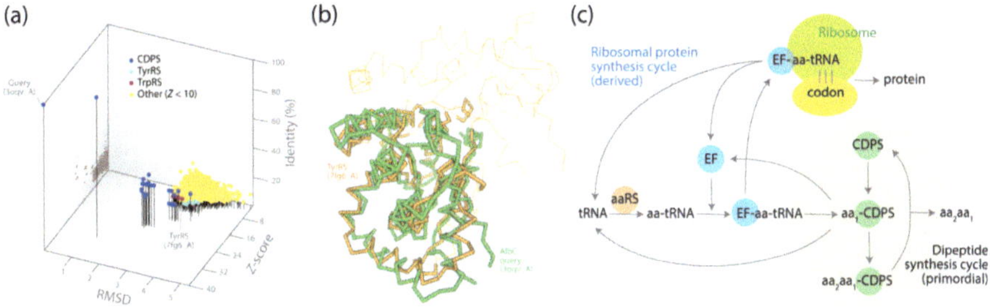

Fig. 17.11 Remote homologies support the primordial biosynthesis of dipeptides prior to ribosomal protein synthesis. (**a**) Structural alignment of a cyclodipeptide synthase from *Streptomyces noursei* (*Snou*-CDPS) (AlbC; PDB entry 3oqv) to structural neighbors using DALI conservation mapping. The three-dimensional plot describes the RMSD, *Z*-score, and sequence identity (%) of 1858 structural neighbors of the CDPS query with *Z*-scores above 2 in an analysis of 678,865 PDB chains conducted on Nov 10, 2025. The closest neighbors of CDPSs (blue symbols) were TyrRS structures. (**b**) Structural alignment of the AlbC query to one of the closest TyrRS neighbors (PDB entry 7 fg6 from *Nanoarchaeum equitans*; RMSD = 3.6, *Z*-score = 11.1, identity = 15) shows that structural cores are highly conserved, suggesting CDPS functions are relics and founders of archaic *Group 1* aaRS biosynthetic activities. Structures are visualized as C-alpha traces following ligand removal. (**c**) Coupled biosynthetic cycles. CDPSs divert aminoacylated tRNAs (aa-tRNA) produced by aaRSs from their canonical role in ribosomal protein synthesis to produce dipeptides (aa_2aa_1) with the help of elongation factors (EFs) (Bourgeois et al. 2020). We propose that dipeptide synthesis is ancestral compared to ribosomal protein synthesis. (Caetano-Anollés et al. 2013)

involving two deep solvent-accessible pockets, P1 and P2 (Sauguet et al. 2011; Moutiez et al. 2014; Bourgeois et al. 2020). In the XYP subfamily of CDPSs, the β2 and β7 strands of the Rossmann fold typical of Class I aaRSs interact with the major groove of the acceptor stem of the first aa-tRNA, bending the aminoacyl moiety through a channel formed by two catalytic loops (CL1 and CL2) and transferring it to a conserved Ser residue in P1 to form the aminoacyl-enzyme intermediate (Bourgeois et al. 2020). The aminoacyl moiety of the second aa-tRNA in P2 then forms the dipeptidyl-enzyme intermediate. The final step involves the nucleophilic attack of the enzyme on the ester bond, facilitated by a Tyr residue, leading to the release of the cyclodipeptide. Figure 17.12 describes a sequence and structural alignment of 20 structural neighbors of Fig. 17.11a, spanning CDPSs, TyrRSs, TrpRSs, and other structures, indexed with residues lining the P1 and P2 pockets, catalytic loops, functional important residues of the AlbC CDPS structure, and active sites (AC) listed in PDBsum (Laskowski 2022). The alignment defines the common structural core of CDPSs and associated aaRSs.

Since CDPSs depend on charged tRNAs to function, it is plausible that primordial aaRS-like enzymes capable of both aminoacylation and peptide bond formation acted cooperatively to perform early peptide synthesis prior to the emergence of the ribosome (Fig. 17.11c). This hypothesis, first elaborated by Caetano-Anollés et al. (2012), situates early aaRS-like structures at the heart of a pre-ribosomal biosynthetic system where aminoacylation, peptide bond formation, and operational RNA code evolution were mechanistically intertwined. Phylogenomic analysis of

Fig. 17.12 Structural alignment without gaps of 20 representative neighbors of *Snou*-CDPS defined by DALI conservation mapping, indexed with structural and functional properties. Residues of the P1 (33, 35, 65, 67, 119, 185, 186, and 200) and P2 (152, 155, 156, 159, 204, 206, and 207) pockets, PDBsum active sites, and residues of central catalytic activities of Snou-CDPS (PDB entry 3oqv) are indexed above the sequence alignment. DSSP secondary structures of the structural alignment describing helix (H/h), strand (E/e), and coil (L/l) regions (uppercase letters represent structurally equivalent positions) are indexed below the structural alignment, together with catalytic loops (CL1 and CL2) that channel the aminoacyl moiety of tRNA to the active site in P1. A topology diagram of the *Snou*-CDPS query generated by the *topology* program of PDBsum is displayed on the right; α-helices are shown as cylinders and β-strands as arrows (named according to Sauguet et al. 2011). Note that the highly conserved residue S37 of PDB entry 4q24_A was modified (S37C) by site-directed mutagenesis. (Moutiez et al. 2014)

domain families revealed that catalytic domain structures of aaRSs and CDPSs appeared before ribosomal protein structures, supporting an early non-ribosomal origin of translation. Several structural and functional properties of CDPSs support the model. CDPSs and elongation factors (EFs) interact with opposite sides of the acceptor arm of tRNA, facilitating the rerouting of aa-tRNA away from ribosomal protein synthesis (Bourgeois et al. 2020). Moreover, CDPSs are promiscuous, producing a main dipeptide and various other products, with specificities of aminoacyl moieties in P1 being favored over those in P2 (Gondry et al. 2018; Harding et al. 2021). CDPSs produce 75 known cyclodipeptides (Gondry et al. 2018), many of which contain amino acid residues belonging to *Groups 1* and *2*, which are linked to the origin of the operational code (Wang et al. 2025). Analysis of consensus motifs defining the residues lining pockets P1 and P2 showed they were particularly enriched with the oldest *Groups 1* and *2* residues, particularly those lining P1, suggesting P1 may have preceded P2 in evolution. Remarkably, CDPSs synthesize dipeptides from small activated amino esters, demonstrating their biosynthetic versatility and RNA-independent catalytic abilities (Harding et al. 2021). Such promiscuity is a hallmark of early biochemical systems (Ycas 1974; Kacsey and Beeby 1984), a property that supports the early rise of dipeptide synthases in protein evolution (Caetano-Anollés et al. 2013).

7.2 Dipeptide Composition and Ancestral Proteomic Signatures

One prediction of our hypothesis is that primordial dipeptides influenced the amino acid composition of emerging peptides and proteins. To test this conjecture, we analyzed the expected impact of early dipeptide synthesis on the amino acid and dipeptide composition of domain structures in proteomes (Caetano-Anollés et al. 2013). The sequentially overlapped dipeptide sequences embodying about 400 possible canonical peptide bond "types" provide unique structural signatures of protein folds, with predictive power comparable to that of domain-level classifications. To minimize artifacts arising from modern domain recruitment and rearrangement, we focused on a dataset of 2384 single-domain proteins with 1475 domain family assignments drawn from over 200,000 structural entries. This allowed accurate tracing of dipeptide patterns in high-quality structural models.

Heat map representations of amino acid and dipeptide compositions across the evolutionary timeline revealed distinct compositional "fingerprints" for each domain family (Fig. 17.13). These signatures uncovered strong and systematic biases in amino acid and dipeptide usage. The distributions were bimodal: both the most ancient and the most recent domains showed lower abundance but higher diversity of amino acid and dipeptide sequences, suggesting evolutionary constraints and diversification at opposite ends of proteomic history. Among individual residues, Ala was the most abundant throughout, followed by Asp, His, Leu, and Val. The

Kyte-Doolittle hydropathy indices of the most frequent half of amino acids (−0.27) were higher than those of the less frequent half (−1.27), highlighting the early importance of hydrophobic interactions in stabilizing protein structures. The most frequent dipeptides were Leu-Leu, followed by 13 dipeptides, including Ala-Leu, Ala-Ala, Leu-Ala, and Glu-Leu, all approximately 25% more frequent than other combinations. Their Kyte-Doolittle hydropathy indices were elevated (0.93) compared to the rest, again supporting the central role of amino acid hydrophobicity.

Statistical analyses revealed significant evolutionary trends in the relative frequencies of amino acids associated with *Groups 1, 2*, and *3* aaRS domains. *Group 1* amino acids were consistently most abundant across the timeline, supporting their ancestral character (ANOVA, $p < 0.0001$). Dipeptide analyses reinforced this pattern: amino acids linked to *Group 1* and *2* editing domains were overrepresented in dipeptides, while those associated with *Group 3* domains were underrepresented (χ^2 test, $p < 0.0001$). These patterns persisted across the evolutionary timeline, indicating that dipeptide composition preserves ancestral aaRS domain signatures.

We identified 33 dipeptides significantly enriched (hypergeometric test, $p < 0.01$) in the earliest domains that appeared 3.0 Gy ago prior to anticodon-binding domains and modern genetics (time of origin nd = 0.2) during epoch 1 (Fig. 17.4c) (Caetano-Anollés et al. 2013). Nearly all involved *Groups 1* and *2* amino acids. However, Ser was notably absent, implying that early SerRSs may have functioned primarily as

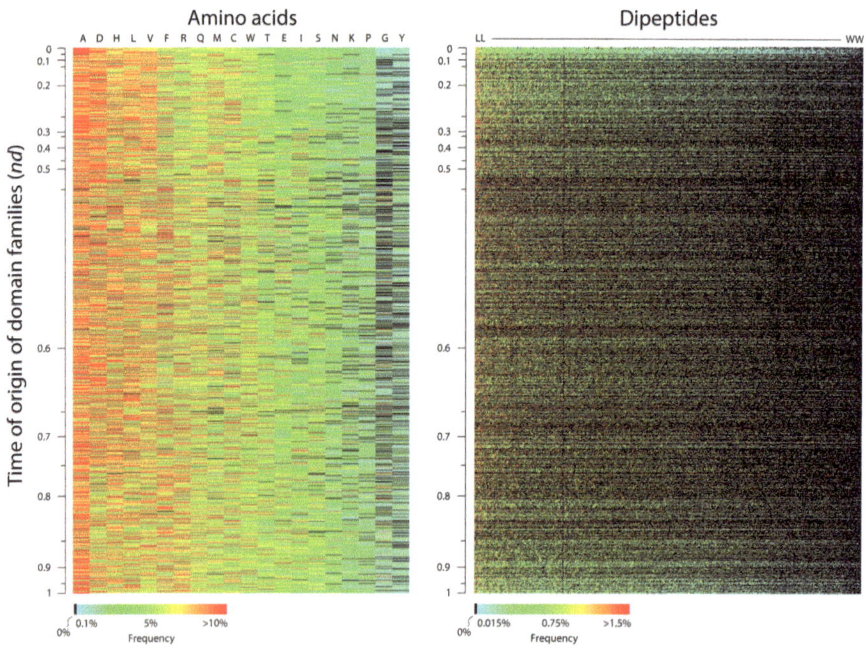

Fig. 17.13 Evolutionary heat maps describing the amino acid and dipeptide sequence makeup of 1475 domain families along a chronology of domains. The 20 standard amino acids and 400 dipeptides are ordered from left to right according to frequency in domains

aaACPL-like ligases rather than canonical aminoacylating agents. Among the enriched residues, Ala, Val, Ile, Leu, Gly, Glu, and Asp—amino acids known to be prebiotically abundant—dominated the dipeptide makeup, consistent with experimental prebiotic chemistry. Structural mappings showed that these ancient, enriched dipeptides were statistically disfavored in loops and loop-regular structure boundaries but favored at boundaries with turns, structural elements believed to be ancient (Sobolevsky et al. 2013). Loops themselves are compact, non-regular (lacking fixed internal hydrogen bonding patterns), highly flexible structures that often host molecular functions (Leszczynski and Rose 1986). These patterns suggest that the emergence of the standard genetic code introduced flexible, partially unstructured regions into ancient proteins, enhancing catalysis and allosteric regulation through diverse topologies. Thus, a cryptic but fundamental link emerges between dipeptide makeup and code specificity: the genetic code arose as an "exacting process" coupling protein structure and dynamics to protein–nucleic acid interactions, driving early peptide biosynthesis. This framework provides clear evolutionary drivers for code evolution—improved folding stability and accessibility, along with greater conformational flexibility and molecular specificity essential for cellular persistence (Caetano-Anollés et al. 2013). Two contemporary studies reinforce this view. First, phylogenomic analysis of nonlocal contacts (contact order) shows an early trend toward faster folding (Debès et al. 2013), which correlates with increased flexibility. Second, in vitro evolution of a protein–RNA ligase from a non-catalytic scaffold led to loss of helices and the acquisition of a long, dynamic unstructured loop (Chao et al. 2013), echoing ancient links between flexibility, function, and the rise of coding. Remarkably, a study of the evolution of intrinsic disorder in the structural domains of viral and cellular proteomes, a flexibility property known to be

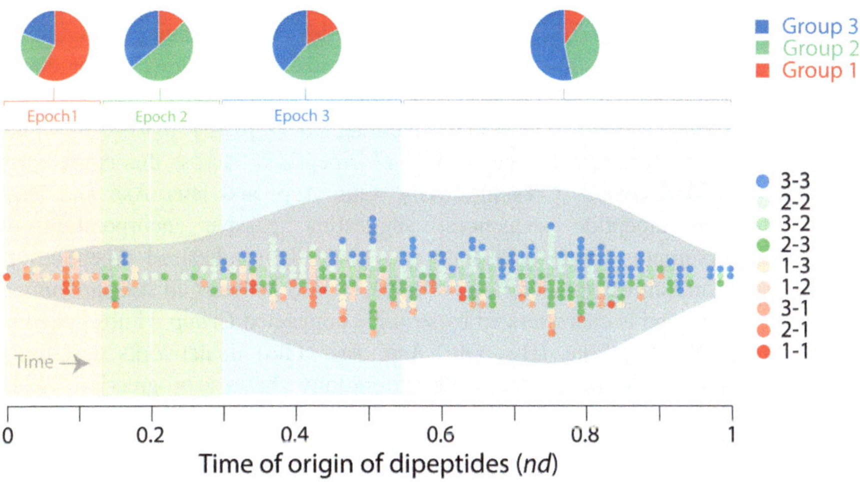

Fig. 17.14 Chronology of dipeptide sequences visualized as a bee-swarm scatterplot, indexed with categories specified by *Groups 1, 2,* and *3* of amino acids. Epochs are driven by the abundance of those amino acids in dipeptide makeup. (Data is from Wang et al. (2025))

associated with loop regions, showed that ancient domains were ordered and that disorder evolved as a benefit acquired later in evolution (Mughal and Caetano-Anollés 2025).

8 The Proteomic Origins of the Genetic Code

Research described so far has shown that dipeptide sequences are nonrandom foundational units that arose from the demands of protein folding. Since these elemental units played a key role in the emergence of the genetic code, we reconstructed a phylogeny of dipeptide sequences from an analysis of about 4.3 billion dipeptide sequences present in over 1500 proteomes (Wang et al. 2025). The phylogenetic matrix for tree building has rows representing the closed set of 400 dipeptides (taxa) and columns representing proteomes (characters), with cell values describing the abundance of dipeptide sequences in proteomes. The rooted phylogeny allowed us to build a dipeptide chronology describing the time or origin (*nd*) of each one of the 400 dipeptide sequences. This chronology established a robust order in which dipeptides entered the evolutionary repertoire, with the earliest dipeptides—those closest to the root—reflecting primordial peptide chemistry. Figure 17.14 displays a bee-swarm scatter plot of the reconstructed chronology, organized into nine dipeptide categories defined by amino acids in *Groups 1*, *2,* and *3*. Each category, denoted X-Y, comprises dipeptides whose N-terminal residue belongs to *Group X* and whose C-terminal residue belongs to *Group Y* (e.g., category 1–2 contains dipeptides made from *Group 1* and *Group 2* amino acids, in that order).

The scatter plot reveals clear, sequential shifts in amino acid usage during dipeptide evolution. The earliest sequences, spanning $nd = [0, 0.137]$, consist exclusively of 21 Leu- and Ser-containing dipeptides (with LL being the earliest), indicating strong enrichment of *Group 1* amino acids. This *Group 1*-driven epoch gives way to increasing representation of Tyr and *Group 2* amino acids during the subsequent *Group 2*-dominated epoch, spanning $nd = [0.137–0.301)$, in which *Group 2* residues occurred in roughly 80% of dipeptides. Across this entire early interval, only five *Group 3* residues (Gly, Glu, Asp, and later Asn and Arg) participated in dipeptide sequences, suggesting passive incorporation of prebiotically common amino acids into enzymatically synthesized dipeptides at this stage. A third epoch spanning $nd = [0.301, 0.548)$ includes all standard amino acids except Cys and is characterized by strongly connected *Group 3* hubs linked to anticodon-binding functions (His, Gln, Asp, Asn, Glu) in networks describing dipeptide makeup. The remainder of the chronology shows a progressive rise in *Group 3* participation. *Group 2* hubs (e.g., Met, Pro, Lys) dominate initially, but they are soon joined or overtaken by *Group 3* hubs (e.g., Gln, Trp, Cys). A final epoch ($nd \geq 0.794$) is marked by the late rise of Cys-containing dipeptides.

Stepwise patterns in amino acid usage are further supported by tracing the nine dipeptide categories along the bee-swarm scatter plot (Fig. 17.14) and by tracking dipeptide evolution across the 20 overlapping timelines of dipeptides holding specific amino acids (Wang et al. 2025). These overlapping timelines highlight the temporal order in which amino acids entered the operational and standard codes. First occurrences of dipeptides increased systematically along amino acid inception order, and most initial appearances fell below a diagonal marking the earliest-to-latest dipeptide range, a pattern consistent with coevolution of dipeptide and genetic code chronologies. Exceptions (e.g., Thr, Phe, Gln, Asp) represent transitional phases. Boxplots of times of origin distributions showed increasing median ages for *Group 1* and early *Group 2* amino acids, matching their inferred roles in establishing the earliest operational code. Remarkably, the overlapping temporal emergence of dipeptides containing Leu, Ser, and Tyr, followed by those containing Val, Ile, Met, Lys, Pro, and Ala, is consistent with the amino acid inception orders inferred from phylogenomic analyses of tRNA and aaRS evolution associated with the operational RNA code.

Proteomes use all 400 possible dipeptides, yet many dipeptides occur at different frequencies than their reversed counterparts (antidipeptides). Previous work by Carugo (2013) documented strong abundance asymmetries among the 190 dipeptide–antidipeptide pairs but found no underlying mechanism. Using the dipeptide chronology, we examined the evolutionary origins of these pairs by comparing their times of origin (Δnd). Nearly half of all pairs originated simultaneously ($\Delta nd = 0$), despite their unequal abundances in modern proteomes. Out of the rest, nearly all originated within $\Delta nd < 0.1$, while the rest were evolutionarily derived. This striking and widespread evolutionary symmetry suggests a deeper molecular duality that is not directly reflected in present-day proteomic composition.

To test whether simultaneous emergence might reflect the sense-antisense codon relationships described by Zull and Smith (1990) and illustrated in Fig. 17.3, we assessed which pairs could arise from strict codon exchanges. Only 26 out of the jointly arising pairs satisfied these strict exchanges. They appeared throughout the timeline and were distributed among the three sense-antisense amino acid exchange groups associated with α-helices, turn motifs, and β-strands. While strict sense-antisense coding explains only a fraction of the observed symmetry, the duality could also be explained by symmetrical signatures of catalytic specificity or promiscuity in the dual active sites of primordial dipeptide synthases. The joint evolutionary appearance of dipeptide and antidipeptide pairs now necessitates a deeper understanding of the molecular determinants of this emerging phenomenon.

Together, proteomic evidence positions dipeptides as powerful phylogenomic markers that encode traces of ancient chemical and biological organization. Their ordered appearance integrates peptide chemistry, protein folding, and tRNA-based aminoacylation into a unified historical framework, in which proteins and the genetic code co-emerged through strong reciprocal influence. Within this framework, Fig. 17.15 illustrates the evolution of the genetic code as a series of episodes of recruitment. By reconstructing this early chronology, the study provides one of the clearest molecular narratives yet for how the genetic code took shape, how early proteins achieved functional specificity, and how life's most fundamental molecular systems transitioned from simple peptides to the sophisticated translation machinery observed today.

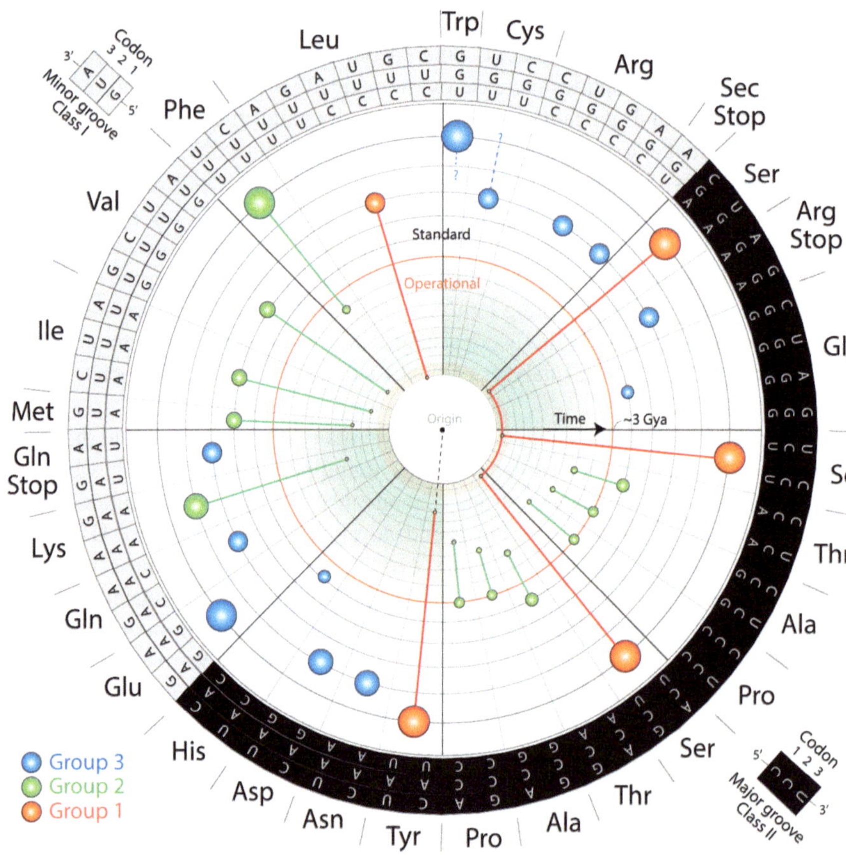

Fig. 17.15 Evolution of the genetic code. The circular diagram follows the format introduced in Fig. 17.3. Its four quadrants reflect the condensed vis-à-vis complementarity scheme, partitioning the code according to the nucleotide in the second codon position. Each quadrant is further subdivided using a decision tree based on tRNA identity elements (Caetano-Anollés et al. 2013). Spheres marking the points at which amino acids entered the operational and standard codes are scaled to their inferred time of inception and colored according to aaRS structure-defined *Groups 1*, *2,* and *3* amino acids. Lines connecting spheres depict recruitment pathways. Temporal progression is radial, extending from the center outward. The innermost red circle marks the approximate emergence of the standard genetic code (~3 Gya). (Diagram modified from Wang et al. (2025))

9 Support from Synthetic Biology

Advances in synthetic biology have enabled the engineering of minimal catalytic systems that recapitulate key features of early aaRSs (Carter Jr et al. 2025). These systems provide experimental access to testing hypotheses describing the primordial stages of translation. Their design is grounded in protein engineering and evolutionary reasoning: rather than attempting to resurrect precise ancestral sequences, they strip modern aaRSs down to the smallest conserved catalytic motifs capable of supporting core biochemical functions.

Protozymes represent the most minimal of these engineering constructs. These short peptides, typically fewer than 50 residues, incorporate ATP-binding elements that are conserved across aaRS superfamilies and can accelerate amino acid activation by several orders of magnitude, demonstrating that rudimentary peptides can support essential steps in premodern translation (Martinez-Rodriguez et al. 2015; Onodera et al. 2021). Urzymes, in contrast, are somewhat larger polypeptides of 80–130 residues (Pham et al. 2007; Tang et al. 2024). They retain conserved catalytic cores sufficient to aminoacylate RNA adapters, albeit with reduced catalytic efficiency and broad substrate tolerance. One of these studies is particularly relevant. Engineered leucyl-tRNA urzyme variants aminoacylated an acceptor stem minihelix with roughly tenfold greater efficiency than full-length cognate tRNAs (Tang et al. 2024). This finding reinforces the idea supported by phylogenomic and biochemical studies that the top half of the tRNA molecule played a central functional and coevolutionary role early in translation, carrying an operational code within the acceptor stem. In addition, the stronger activity toward Ileu over Leu observed in these experiments highlights the broad specificity and catalytic promiscuity that likely characterized the earliest coding systems. This molecular promiscuity suggests that ancestral synthetases functioned as dynamic "quasispecies," capable of activating multiple amino acids prior to the evolution of stringent substrate discrimination (Carter Jr et al. 2025). The promiscuous behavior of early catalysts may have conferred adaptive advantages to early organisms living in fluctuating chemical environments but was eventually replaced by higher specificity as mutation and natural selection favored faster growth (Kacsey and Beeby 1984). Remarkably, modern organisms appear to retain traces of the primordial flexibility: environmentally responsive codon reassignment, such as the decoding of the CUG codon as Ser rather than Leu in certain fungi (Gomes et al. 2007), demonstrates that misacylation can be selectively tolerated and can be occasionally advantageous. Such historical traces also support the connection between Leu and Ser at the top of the decision tree induced by the emergence and diversification of tRNA identity elements. The tree organizes the circular code of Fig. 17.3 by gradually reducing ambiguity in the signs that are embedded in the growing nucleic acid molecules, by increasing sign multiplicity (degeneration), and by tightening aaRS specificities through coevolving recognition events. Phylogenomic data, therefore, aligns with promiscuities unfolded in synthetic biology experimentation.

Building on the observation that sense and antisense codons exhibit structured exchange relationships (Zull and Smith 1990) and the proposal that the earliest Class I and Class II aaRSs were encoded on opposite strands of a single primitive gene (Rodin and Ohno 1995), synthetic biology experimentation has been directed to a primordial molecular duality (Carter Jr and Wills 2021). Notably, Martinez-Rodriguez et al. (2015) provided direct evidence that designed 46-residue peptides containing ATP-binding sites of functional Class I tryptophanyl-tRNA and Class II histidyl-tRNA protozymes encoded on opposite strands of a gene could catalyze amino acid activation. Peptides demonstrated saturable kinetics, sensitivity to active-site mutations, and catalytic efficiencies ($k_{(cat)}/K_m$) on the order of 3–7 $\times$ 10^{-3} $M^{-1}s^{-1}$, which corresponds to about 10^6-fold rate acceleration over the uncatalyzed reaction. These findings lend strong experimental support to bidirectional (sense-antisense) coding, a property that ensures *economy* through coordinated evolution of both classes of aaRSs, *reflexivity* by aaRSs carrying and enforcing the genetic code, and *symmetry* matching genetic and functional complementarity.

Beyond catalysis, the structural duality between Class I and Class II enzymes is also mirrored in how they recognize tRNA. Regression models show that groove discrimination in tRNA depends on specific bases (discriminator base 73, base pairs 1–72, base 2) in patterns that differ between classes in a way that seems ancestral (Carter Jr and Wills 2018). This suggests that the operational RNA code may have coevolved with ancestral synthetases while being shaped by the emergent base pairing rules of nucleic acids. The duality is evident in the mirror modes of tRNA and aaRS recognition and genetic complementarity (Rodin and Ohno 1995; Rodin and Rodin 2008), the phylogenomic-inspired decision tree (Caetano-Anollés et al. 2013) that defines the circular code of Fig. 17.3, and the evolutionary rise of dipeptide and anti-dipeptide pairs in near synchrony along the entire chronology of dipeptide sequences (Wang et al. 2025). A duality in dipeptide sequences of proteomes probably reflects historically deeper biochemical or structural symmetries in early peptide synthesis, possibly rooted in the dual active-site architectures of ancestral dipeptide synthases and their promiscuous interactions with RNA and non-RNA ligands.

Finally, the duality is also embodied in modern regulatory activities of the cell. For example, *natural antisense transcripts* (NATs) are part of a major class of long noncoding RNAs (lncRNAs) important for cell and developmental biology (Werner et al. 2024). They are widely expressed from the opposite strand of protein-coding and noncoding genes, participating in regulatory pathways that control expression, splicing stability, or translation of their cognate sense genes. Similarly, *target-directed microRNA degradation* (TDMD) is a specialized regulatory mechanism that induces the degradation of microRNAs (miRNAs) (Buhagiar and Kleaveland 2024). miRNAs remodel the transcriptome by accelerating the degradation of specific "target" mRNAs. TDMD reverses the process through highly complementary "trigger" RNAs that induce the degradation of miRNAs. Even Chargaff's second parity rule, which states that a nucleic acid strand contains roughly equal numbers of a sequence and its reverse complement, can be explained by physical constraints as sequences tend toward maximum randomness, with biological selection acting as a "soft" driver (Fariselli et al. 2021). These statistical and physical symmetries, which give rise to a genomic duality, stem from constraints inherent to the DNA double helix. As a result, genomes exhibit scale-free *k*-mer frequency distributions and characteristic inverse-symmetry decay (Warr and Hatton 2025).

10 Conclusions

Phylogenomic, structural, and synthetic biology reconstructions collectively reveal that evolutionary trends in amino acid usage and dipeptide composition encode a molecular memory of ancient enzymatic systems and biosynthetic innovation. The correspondence between aaRS domain histories, dipeptide phylogenies, and fold structures indicates that proteomic diversification both recorded and shaped the emergence of coding specificity. As metabolic pathways expanded to generate new amino acids, the structural evolution of the proteome accommodated and selected for these residues, linking metabolic innovation, peptide assembly, and coding evolution into a self-reinforcing and self-empowering process.

These insights converge on a coherent narrative of early translation in which primitive peptides were highly promiscuous yet catalytically competent, capable of sustaining essential chemistry despite minimal structural elaboration. Evidence from genetic duality models shows that complementary strands of ancestral genes could encode distinct catalytic modules, embedding functional duality into the earliest protein–RNA systems. Such minimalist catalysts—ancestral aaRS-like enzymes—likely acted both as aminoacylating agents and primitive dipeptide synthases, forging the biochemical link between tRNA charging, peptide bond formation, and the emergence of coding itself.

Together, these lines of evidence support the view that the origin and refinement of the genetic code were deeply proteomic in nature. The code's expansion mirrored the progressive enrichment of early proteomes in chemically diverse residues, enabling increased structural flexibility, catalytic range, and eventual specificity through interactions with evolving tRNAs. By integrating molecular symmetries, ancestral catalytic promiscuity, and the structural evolution of peptide motifs, this framework positions proteins not merely as interpreters of the code but as central architects in its evolutionary construction. Understanding the interplay between these minimalist systems and the broader pattern of proteomic evolution offers vital insight into the molecular logic that shaped life's first coded polymers.

Acknowledgments I thank Guenther Witzany and Jordi Gomez for their valuable comments. Research was supported by grants from the National Science Foundation (MCB-0749836 and OISE-1132791) and the United States Department of Agriculture (ILLU-802-909 and ILLU-483-625) and by Blue Waters supercomputer allocations from the National Center for Supercomputing Applications (NCSA) to GCA.

References

Amangeldina A, Tan ZW, Berezovsky I (2024) Living in trinity of extremes: genomic and proteomic signatures of halophilic, thermophilic and pH adaptation. Curr Res Struct Biol 7: 100129

Ariza-Mateos A, Briones C, Perales C, Sobrino F, Domingo E, Gómez J (2024) Natural languages and RNA virus evolution. J Physiol 602(11):2565–2580

Ashby WR (1956) An introduction to cybernetics. Chapman and Hall, London, p 295

Aziz MF, Caetano-Anollés G (2021) Evolution of networks of protein domain organization. Sci Rep 11:12075

Aziz MF, Caetano-Anollés K, Caetano-Anollés G (2016) The early history and emergence of molecular functions and modular scale- free network behavior. Sci Rep 6(1):25058

Aziz MF, Mughal F, Caetano-Anollés G (2023) Tracing the birth of structural domains from loops during protein evolution. Sci Rep 13(1):14688

Becker S, Thoma I, Deutsch A, Gehrke T, Mayer P, Zipse H, Carell T (2016) A high-yielding, strictly regioselective prebiotic purine nucleoside formation pathway. Science 352:833–836
Becker S, Feldmann J, Wiedemann S, Okamura H, Carell T (2019) Unified prebiotically plausible synthesis of pyrimidine and purine RNA ribonucleotides. Science 366:76–82
Berezovsky IN, Grosberg AY, Trifonov EN (2000) Closed loops of nearly standard size: common basic element of protein structure. FEBS Lett 466:283–286
Bernhardt HS (2016) Clues to tRNA evolution from the distribution of class II tRNAs and serine codons in the genetic code. Life (Basel) 6:10
Besseová I, Réblová K, Leontis NB, Sponer J (2010) Molecular dynamics simulations suggest that RNA three-way junctions can act as flexible RNA structural elements in the ribosome. Nucleic Acids Res 38:6247–6264
Bloch D, McArthur B, Widdowson R, Spector D, Guimarães RC, Smith J (1984) tRNA-rRNA sequence homologies: a model for the origin of a common ancestral molecule, and prospects for its reconstruction. Orig Life 14:571–578
Bloch DP, McArthur B, Guimarães RC, Smith J, Staves MP (1989) tRNA–rRNA sequence matches from inter-and intraspecies comparisons suggest common origins for the two RNAs. Braz J Med Biol Res 22:931–944
Bourgeois G, Seguin J, Babin M, Gondry M, Mechulam Y, Schmitt E (2020) Structural basis of the interaction between cyclodipeptide synthases and aminoacylated tRNA substrates. RNA 26: 1589–1602
Brooks DJ, Fresco JR, Lesk AM, Singh M (2002) Evolution of amino acid frequencies in proteins over deep time: inferred order of introduction of amino acids into the genetic code. Mol Biol Evol 19:1645–1655
Buhagiar AF, Kleaveland B (2024) To kill a microRNA: emerging concepts in target-directed microRNA degradation. Nucleic Acids Res 52:1558–1574
Caetano-Anollés G (2002a) Evolved RNA secondary structure and the rooting of the universal tree of life. J Mol Evol 54(3):333–345
Caetano-Anollés G (2002b) Tracing the evolution of RNA structure in ribosomes. Nucleic Acids Res 30(11):2575–2587
Caetano-Anollés G (2021) The compressed vocabulary of microbial life. Front Microbiol 12: 655990
Caetano-Anollés G (2023) Agency in evolution of biomolecular communication. Ann N Y Acad Sci 1525:88–103
Caetano-Anollés D, Caetano-Anollés G (2016) Piecemeal buildup of the genetic code, ribosomes, and genomes from primordial tRNA building blocks. Life (Basel) 6:43
Caetano-Anollés G, Caetano-Anollés D (2003) An evolutionarily structured universe of protein architecture. Genome Res 13:1563–1571
Caetano-Anollés G, Sun F-J (2014) The natural history of transfer RNA and its interactions with the ribosome. Front Genet 5:127
Caetano-Anollés G, Wang M, Caetano-Anollés D (2013) Structural phylogenomics retrodicts the origin of the genetic code and uncovers the evolutionary impact of protein flexibility. PLoS One 8(8):e72225
Caetano-Anollés G, Nasir A, Kim KM, Caetano-Anollés D (2018) Rooting phylogenies and the tree of life while minimizing ad hoc and auxilliary assumptions. Evol Bioinforma 14: 1176934318805101
Caetano-Anollés G, Aziz MF, Mughal F, Caetano-Anollés D (2021) Tracing protein and proteome history with chronologies and networks: folding recapitulates evolution. Expert Rev Proteomics 18(10):863–880
Caetano-Anollés K, Aziz MF, Mughal F, Caetano-Anollés G (2024) On protein loops, prior molecular states and common ancestors of life. J Mol Evol 92:624–646
Caldararo F, Di Giulio M (2022) The genetic code is very close to a global optimum in a model of its origin taking into account both the partition energy of amino acids and their biosynthetic relationships. Biosystems 215:104613

Carter CW Jr, Wills PR (2018) Hierarchical groove discrimination by class I and II aminoacyl-tRNA synthetases reveals a palimpsest of the operational RNA code in the tRNA acceptor-stem bases. Nucleic Acids Res 46:9667–9683
Carter CW Jr, Wills PR (2021) The roots of genetic coding in aminoacyl-tRNA synthetase duality. Annu Rev Biochem 90:349–373
Carter CW Jr, Tang GQ, Patra SK, Dieckhaus H, Kuhlman B, Douglass J, Wills PR, Bouckaert R, Popovic M, Ditzler MA (2025) Structural enzymology, phylogenetics, differentiation, and symbolic reflexivity at the dawn of biology. Genome Biol Evol 17:evaf095
Carugo O (2013) Frequency of dipeptides and antidipeptides. Comput Struct Biotechnol J 11: e201308001
Chao F-A, Morelli A, Haughner JC III, Chuchfield L, Hagmann LN et al (2013) Structure and dynamics of a primordial catalytic fold generated by in vitro evolution. Nat Chem Biol 9:81–83
Choi J, Kim S-H (2020) Whole-proteome tree of life suggests a deep burst of organism diversity. Proc Natl Acad Sci USA 117(7):3678–3686
Crothers DM, Seno T, Söll DG (1972) Is there a discriminator site in transfer RNA? Proc Natl Acad Sci USA 69:3063–3067
Davidovich C, Belousoff M, Bashan A, Yonath A (2009) The evolving ribosome: from non-coded peptide bond formation to sophisticated translation machinery. Res Microbiol 160:487–492
Debès C, Wang M, Caetano-Anollés G, Gräter F (2013) Evolutionary optimization of protein folding. PLoS Comput Biol 9:e1002861
Demongeot J (1978) Sur la possibilité de considérer le code génétique comme un code à enchaînement. Revue Biomaths 62:61–66
Demongeot J (2025) Traces in current genomes of a primitive RNA ring. Biology 14:538
Demongeot J, Norris V (2019) Emergence of a "cyclosome" in a primitive network capable of building "infinite" proteins. Life (Basel) 9:51
Demongeot J, Waku J, Cohen O (2023) Combinatorial and frequency properties of the ribosome ancestors. Math Biosci Eng 21:884–902
Di Giulio M (1992) On the origin of the transfer RNA molecule. J Theor Biol 159:199–214
Di Giulio M (2020) An RNA ring was not the progenitor of the tRNA molecule. J Mol Evol 88: 228–233
Doyle JA (2006) Seed ferns and the origin of angiosperms. J Torrey Bot Soc 133:169–209
Eck RV, Dayhoff MO (1966) Evolution of the structure of ferredoxin based on living relics of primitive amino acid sequence. Science 152:363–366
Eigen M (1971) Self organization of matter and the evolution of biological macromolecules. Naturwissenschaften 58:465–523
Eigen M, Winkler-Oswatitsch R (1981) Transfer-RNA, an early gene? Naturwissenschaften 68: 282–292
Farias ST (2013) Suggested phylogeny of tRNAs based on the construction of ancestral sequences. J Theor Biol 335:245–248
Farias ST, Rêgo TG, José MV (2014) Origin and evolution of the peptidyl transferase center from proto-tRNAs. FEBS Open Bio 4:175–178
Farias ST, Rêgo TG, José MV (2021) Origin of the 16S ribosomal molecule from ancestor tRNAs. J Mol Evol 89:249–256
Farias ST, José MV (2020) Transfer RNA: the molecular demiurge in the origin of biological systems. Prog Biophys Mol Biol 153:28–34
Fariselli P, Taccioli C, Pagani L, Maritan A (2021) DNA sequence symmetries from randomness: the origin of the Chargaff's second parity rule. Brief Bioinform 22:2172–2181
Fournier GP, Gogarten JP (2007) Signature of a primitive genetic code in ancient protein lineages. J Mol Evol 65:425–436
Flamm C, Hofacker IL, Stadler PF, Wolfinger MT (2002) Barrier trees of degenerate landscapes. Z Phys Chem 216:1–19
Fontana W (2002) Modelling 'evo-devo' with RNA. BioEssays 24:1164–1177

Gamow G (1954) Possible relation between deoxyribonucleic acid and protein structures. Nature 173:318
Giegé R, Eriani G (2023) The tRNA identity landscape for aminoacylation and beyond. Nucleic Acids Res 51:1528–1570
Giegé R, Sissler M, Florentz C (1998) Universal rules and idiosyncratic features in tRNA identity. Nucleic Acids Res 26:5017–5035
Gomes AC, Miranda I, Silva RM, Moura G, Thomas B, Akoulitchev A, Santos MAS (2007) A genetic code alteration generates a proteome of high diversity in the human pathogen *Candida albicans*. Genome Biol 8:R206
Gomez MAR, Ibba M (2020) Aminoacyl-tRNA synthetases. RNA 26:910–936
Gondry M, Sauguet L, Belin P, Thai R, Amouroux R et al (2009) Cyclodipeptide synthetases are a family of tRNA-dependent peptide-bond-forming enzymes. Nat Chem Biol 5:414–420
Gondry M, Jacques IB, Thai R, Babin M, Canu N, Seguin J, Belin P, Pernodet J-L, Moutiez M (2018) A comprehensive overview of the cyclodipeptide synthase family enriched with the characterization of 32 new enzymes. Front Microbiol 9:46
Grosjean H, Westhof E (2016) An integrated, structure- and energy-based view of the genetic code. Nucleic Acids Res 44:8020–8040
Gruber AR, Lorenz R, Bernhart SH, Neuböck R, Hofacker IL (2008) The Vienna RNA websuite. Nucleic Acids Res 36:W70–W74
Guo X, Fu S, Ying J, Zhao Y (2023) Prebiotic chemistry: a review of nucleoside phophorylation and polymerization. Open Biol 13:220234
Haig D, Hurst LD (1991) A quantitative measure of error minimization in the genetic code. J Mol Evol 33:412–417
Harding CJ, Sutherland E, Hanna JG, Houston DR, Czekster CM (2021) Bypassing the requirement for aminoacyl-tRNA by a cyclodipeptide synthase enzyme. RSC Chem Biol 2:230–240
Harish A, Caetano-Anollés G (2012) Ribosomal history reveals origins of modern protein synthesis. PLoS One 7:e32776
Hoeppner MP, Gardner PP, Poole AM (2012) Comparative analysis of RNA families reveals distinct repertoires for each domain of life. PLoS Comput Biol 8:e1002752
Holden DT, Morato NC, Cooks RG (2022) Aqueous microdroplets enable abiotic synthesis and chain extension of unique peptide isomers from free amino acids. Proc Natl Acad Sci USA 19: e2212642119
Hollingsworth SA, Karplus PA (2010) A fresh look at the Ramachandran plot and the occurrence of standard structures in proteins. Biomol Concepts 1:271–283
Holm L, Laiho A, Törönen P, Salgado M (2023) DALI shines a light on remote homologs: one hundred discoveries. Protein Sci 32:e4519
Hopfield JJ (1978) Origin of the genetic code: a testable hypothesis based on tRNA structure, sequence, and kinetic proofreading. Proc Natl Acad Sci USA 75:4334–4338
Hou YM, Schimmel P (1988) A simple structural feature is a major determinant of the identity of a transfer RNA. Nature 333:140–145
Jakubowski H (2017) Homocysteine editing, thioester chemistry, coenzyme a, and the origin of coded peptide synthesis. Life (Basel) 7(1):6
Jobe A, Liu Z, Gutierrez-Vargas C, Frank J (2019) New insights into ribosome structure and function. Cold Spring Harb Perspect Biol 11(1):a032615
Jordan IK, Kondrashov FA, Adzhubei IA, Wolf YI, Koonin EV et al (2005) A universal trend of amino acid gain and loss in protein evolution. Nature 433:633–638
Kacsey H, Beeby R (1984) On the origin of enzyme species by means of natural selection. J Mol Evol 20:38–51
Kaiser F, Krautwurst S, Salentin S, Haupt VJ, Leberecht C, Bittrich S, Labudde D, Schroeder MS (2020) The structural basis of the genetic code: amino acid recognition by aminoacyl-tRNA synthetases. Sci Rep 10:12647
Kawashima S, Kanehisa M (2000) AAindex: amino acid index database. Nucleic Acids Res 28:374

Kim KM, Caetano-Anollés G (2012) The evolutionary history of protein fold families and proteomes confirms that the archaeal ancestor is more ancient than the ancestors of the other superkingdoms. BMC Evol Biol 12:13

Laskowski RA (2022) PDBsum*1*: a standalone program for generating PDBsum analyses. Protein Sci 31:e4473

Lei L, Burton Z (2023) The 3 32 nucleotide minihelix tRNA evolution theorem and the origin of life. Life (Basel) 12:2224

Leontis NB, Westhof E (2001) Geometric nomenclature and classification of RNA base pairs. RNA 7:499–512

Lescoute A, Westhof E (2006) Topology of three-way junctions in folded RNAs. RNA 12:83–93

Leszczynski JF, Rose GD (1986) Loops in globular proteins: a novel category of secondary structure. Science 234:849–855

Li PTX, Vieregg J, Tinoco I (2008) How RNA unfolds and refolds. Annu Rev Biochem 77:77–100

Löb W (1913) Über das Verhalten des Formamids unter der Wirkung der stillen Entladung Ein Beitrag zur Frage der Stickstoff-Assimilation. Ber Dtsch Chem Ges 46:684–697

Mahesh S, Tang K-C, Raj M (2018) Amide bond activation of biological molecules. Molecules 23: 2615

Maizels N, Weiner AM (1994) Phylogeny from function: evidence from the molecular fossil record that tRNA originated in replication, not translation. Proc Natl Acad Sci USA 91:6729–6734

Makarov M, Sanchez Rocha AC, Krystufek R, Cherepashuk I, Dzmitruk V, Charnavetz T, Faustino AM, Lebl M, Fujishima K, Fried SD, Hlouchova K (2023) Early selection of the amino acid alphabet was adaptively shaped by biophysical constraints on foldability. J Am Chem Soc 145: 5320–5329

Martin RB (1998) Free energies and equilibria of peptide bond hydrolysis and formation. Biopolymers 45:351–353

Martinez-Rodriguez L, Erdogan O, Jimenez-Rodriguez M, Gonzalez-Rivera K, Williams T, Li L, Weinreb V, Collier M, Chandrasekaran SN, Ambroggio X, Kuhlman B, Carter CW Jr (2015) Functional class I and II amino acid activating enzymes can be coded by opposite strands of the same gene. J Biol Chem 290:19710–19725

Michel CJ (2012) Circular code motifs in transfer and 16S ribosomal RNAs: a possible translation code in genes. Comput Biol Chem 37:24–37

Miller SL (1953) Production of amino acids under possible primitive Earth conditions. Science 117: 528–529

Miller SL (1987) Which organic compounds could have occurred on the prebiotic earth? Cold Spring Harb Symp Quant Biol 52:17–27

Miyazawa S, Jernigan RL (1999) Self-consistent estimation of inter-residue protein contact energies based on an equilibrium mixture approximation of residues. Proteins 34:49–68

Mocibob M, Ivic N, Bilokapic S, Maier T, Luic M et al (2010) Homologs of aminoacyl-tRNA synthetases acylate carrier proteins and provide a link between ribosomal and nonribosomal peptide synthesis. Proc Natl Acad Sci USA 107:14585–14590

Morris C (1968) Signification and significance. MIT Press, Cambridge MA

Moutiez M, Schmitt E, Sequin J, Thai R, Favry E, Belin P, Mechulam Y, Gondry M (2014) Unravelling the mechanism of non-ribosomal peptide synthesis by cyclodipeptide synthases. Nat Commun 5:5141

Mughal F, Caetano-Anollés G (2025) Evolution of intrinsic disorder in the structural domains of viral and cellular proteomes. Sci Rep 15:2878

Müller F, Escobar L, Xu F, Wegrzyn E, Nanyte M, Amatov T, Chan G-Y, Pichler A, Carell T (2022) A prebiotically plausible scenario of an RNA-peptide world. Nature 605:279–284

Murzin AG, Brenner SE, Hubbard T, Chothia C (1995) SCOP: a structural classification of proteins database for the investigation of sequences and structures. J Mol Biol 247:536–540

Nazarea AD, Bloch DP, Semrau AC (1985) Detection of a fundamental modular format common to transfer and ribosomal RNA: second order spectral analysis. Proc Natl Acad Sci USA 82:5337–5341

Nirenberg MW, Matthaei JH (1961) The dependence of cell-free protein synthesis in E. coli upon naturally occurring or synthetic polyribonucleotides. Proc Natl Acad Sci USA 47(10): 1588–1602
Onodera K, Suganuma N, Takano H, Sugita Y, Shoji T, Minobe A, Yamaki N, Otsuka R, Mutsuro-Aoki H, Umehara T, Tamura K (2021) Amino acid activation analysis of primitive aminoacyl-tRNA synthetases encoded by both strands of a single gene using the malachite green assay. Biosystems 208:104481
Pe'er I, Felder CE, Man O, Silman I, Sussman JI, Beckmann JS (2004) Proteomic signatures: amino acid and oligopeptide compositions differentiate among taxa. Proteins 54:20–40
Pham Y, Li L, Kim A, Erdogan O, Weinreb V, Butterfoss G, Kuhlman B, Carter CW Jr (2007) A minimal TrpRS catalytic domain supports sense/antisense ancestry of class I and II aminoacyl-tRNA synthetases. Mol Cell 25:851–862
Plankensteiner K, Reiner H, Schranz B, Rode BM (2004) Prebiotic formation of amino acids in a neutral atmosphere by electric discharge. Angew Chem Int Ed 43:1886–1888
Riziotis IG, Glykos NM (2019) On the presence of short-range periodicities in protein structures that are not related to established secondary structure elements. Proteins 87:966–978
Rode BM, Fitz D, Jakschitz T (2007) The first steps of chemical evolution towards the origin of life. Chem Biodivers 4:2674–2702
Rodin SN, Ohno S (1995) Two types of aminoacyl-tRNA synthetases could be originally encoded by complementary strands of the same nucleic acid. Orig Life Evol Biosph 25:565–589
Rodin SN, Rodin AS (2008) On the origin of the genetic code: signatures of its primordial complementarity in tRNAs and aminoacyl-tRNA synthetases. Heredity 100:341–355
Root-Bernstein M, Root-Bernstein R (2015) The ribosome as a missing link in the evolution of life. J Theor Biol 367:130–158
Sauguet L, Moutiez M, Li Y, Belin P, Seguin J, Le Du M-H, Thai R, Masson C, Fonvielle M, Pernodet J-L, Charbonnier J-B, Gondry M (2011) Cyclodipeptide synthases, a family of class-I aminoayl-tRNA synthetase-like enzymes involved in non-ribosomal peptide synthesis. Nucleic Acids Res 39:4475–4489
Schimmel P, Giegé R, Moras D, Yokohama S (1993) An operational RNA code for amino acids and possible relationship to genetic code. Proc Natl Acad Sci USA 90:8763–8768
Schuster P (2010) Genotypes and phenotypes in the evolution of molecules. In: Caetano-Anollés G (ed) Evolutionary bioinformatics and systems biology. Wiley-Blackwell, Hoboken, pp 123–152
Sharma A, Dai K, Pol MD, Thomann R, Thomann R, Roy SK, Pappas CG (2025) Selective peptide bond formation via side chain reactivity and self-assembly of abiotic phosphates. Nat Commun 16:1306
Smith RM, Hansen DE (1998) The pH-rate profile for the hydrolysis of a peptide bond. J Am Chem Soc 120:8910–8913
Sober E, Steel M (2014) Time and knowability in evolutionary processes. Phys Sci 81:537–557
Sobolevsky Y, Guimaraes RC, Trifonov EN (2013) Towards functional repertoire of the earliest proteins. J Biomol Struct Dyn 31:1293–1300
Šponer JE, Coulon R, Otyepka M, Šponer J, Siegle AF, Trapp O, Ślepokura K, Zdráhal Z, Šedo O (2024) Phosphoric acid salts of amino acids as a source of oligopeptides on the early Earth. Commun Chem 7:185
Sun F-J, Caetano-Anollés G (2008a) The origin and evolution of tRNA inferred from phylogenetic analysis of structure. J Mol Evol 66:21–35
Sun F-J, Caetano-Anollés G (2008b) Evolutionary patterns in the sequence and structure of transfer RNA: a window into early translation and the genetic code. PLoS One 3:e2799
Sun F-J, Caetano-Anollés G (2008c) Evolutionary patterns in the sequence and structure of transfer RNA: early origins of Archaea and viruses. PLoS Comput Biol 4:e1000018
Sun F-J, Caetano-Anollés G (2008d) The evolutionary significance of the long variable arm in transfer RNA. Complexity 14:26–39
Sun F-J, Caetano-Anollés G (2009) The evolutionary history of the structure of 5S ribosomal RNA. J Mol Evol 69:430–443
Sun F-J, Caetano-Anollés G (2010a) The ancient history of the structure of ribonuclease P and the early origins of Archaea. BMC Bioinform 11:153

Sun F-J, Caetano-Anollés G (2010b) The origin of modern 5S rRNA: a case relating models of structural history to phylogenetic data. J Mol Evol 71:3–5
Sun F-J, Caetano-Anollés G (2021) Menzerath-Altmann's law of syntax in RNA accretion history. Life (Basel) 11:489
Sun F-J, Caetano-Anollés G (2026) Tracing the origin and evolution of the secondary structure of RNA. Meth Mol Biol 2981:323–345
Sun F-J, Fleurdépine S, Bousquet-Antonelli C, Caetano-Anollés G, Deragon J-M (2007) Common evolutionary trends for SINE RNA structures. Trends Genet 23:26–33
Szathmáry E (1999) The origin of the genetic code: amino acids as cofactors in an RNA world. Trends Genet 15:223–229
Tang GQ, Hu H, Douglas J, Carter CW Jr (2024) Primordial aminoacyl-tRNA synthetases preferred tRNA minihelix substrates over full-length tRNA. Nucleic Acids Res 52:7096–7111
Trifonov EN (2004) The triplet code from first principles. J Biomol Struct Dyn 22:1–11
Wang M, Jiang YY, Kim KM, Wu G, Ji H-F, Mittenthal JE, Zhang H-Y, Caetano-Anollés G (2011) A universal molecular clock of protein folds and its power in tracing the early history of aerobic metabolism and planet oxygenation. Mol Biol Evol 28:567–582
Wang M, Aziz MF, Caetano-Anollés G (2025) Tracing the origin of the genetic code and thermostability to dipeptide sequences in proteomes. J Mol Biol. 438(4):169396
Wang W, Feng C, Han R, Wang Z, Ye L, Du Z, Wei H, Zhang F, Peng Z, Yang J (2023) trRosettaRNA: automated prediction of RNA 3D structure with transformer network. Nat Commun 14:7266
Warr G, Hatton L (2025) The architecture of the genome integrates scale independence with inverse symmetry. Acad Mol Biol Genom 2(2). https://doi.org/10.20935/AcadMolBioGen7650
Weiner AM, Maizels N (1987) tRNA-like structures tag the 3' ends of genomic RNA molecules for replication: implications for the origin of protein synthesis. Proc Natl Acad Sci USA 84:7383–7387
Werner A, Kanhere A, Wahlestedt C, Mattick JS (2024) Natural antisense transcripts as versatile regulators of gene expression. Nat Rev Genet 25:730–744
Widmann J, Di Giulio M, Yarus M, Knight R (2005) tRNA creation by hairpin duplication. J Mol Evol 61:524–530
Witzany G (2016) The biocommunication method: on the road to an integrative biology. Commun Integr Biol 9:e1164374
Witzany G, Baluška F (2012) Life's code script does not code itself. EMBO Rep 13(12):1054–1056
Woese CR (1965) Order in the genetic code. Proc Natl Acad Sci USA 54:71–75
Woese CR (1969) The biological significance of the genetic code. In: Agranoff BW, Davies J, Hahn FE, Mandel G, Scott NS, Smillie RM, Wose CR (eds) Progress in molecular and subcellular biology, vol 1. Springer-Verlag, Berlin Heidelberg, pp 5–46
Woese CR, Dugre DH, Dugre SA, Kondo M, Saxinger WC (1966) On the fundamental nature and evolution of the genetic code. Cold Spring Harb Symp Quant Biol 31:723–736
Wong JT (1981) Coevolution of the genetic code and amino acids biosynthesis. Trends Biochem Sci 6:33–36
Wołos A, Roszak R, Zadło-Dobrowolska A et al (2020) Synthetic connectivity, emergence, and self-regeneration in the network of prebiotic chemistry. Science 369:eaaw1955
Ycas M (1974) On earlier states of the biochemical system. J Theor Biol 44:145–160
Zull JE, Smith SK (1990) Is genetic code redundancy related to retention of structural information in both DNA strands? Trends Biochem Sci 15:257–261

Chapter 18
Inteins: Emerging Regulatory Roles and Their Exaptation for Hedgehog Signaling

Christopher W. Lennon, Brian P. Callahan, and Marlene Belfort

Abstract Inteins, the polycatalytic mobile genetic elements of the microbial world, are co-translated as intervening regions within essential host proteins. Protein splicing activity of the intein is necessary to bring the host protein to a fully active state. Due to their additional mobility at the DNA level, inteins have long been characterized as a class of selfish genetic sequences. In some cases, however, inteins appear capable of functioning adaptively as environmentally responsive posttranslational regulatory elements. In this chapter, we discuss some recent breakthroughs in the developing field of conditional protein splicing (CPS), whereby the rate and accuracy of intein activity are functionally linked to intra- and extracellular conditions. We highlight examples of inteins acting both as "pause buttons" and "rheostats" to provide an underappreciated means of posttranslational regulation. Lastly, we discuss evidence for an evolutionary link between CPS and the autoprocessing activity of eukaryotic hedgehog family proteins. Hedgehog autoprocessing yields a vital extracellular signaling factor with potent morphogenic activity. The close mechanistic parallels and structural similarities between intein splicing and hedgehog autoprocessing support speculation that an intein was exapted to catalyze a vital transformation for animal development.

1 Introduction

Inteins, or intervening proteins, are translated within host proteins and excised by a self-mediated protein splicing reaction. In this process, the flanking sequences of the host protein, termed N- and C-exteins, are ligated with a peptide bond following

C. W. Lennon (✉)
Department of Biological Sciences, Murray State University, Murray, KY, USA
e-mail: clennon1@murraystate.edu

B. P. Callahan
Chemistry Department, Binghamton University, Binghamton, NY, USA

M. Belfort
Department of Biological Sciences and RNA institute, University at Albany, Albany, NY, USA

G. Witzany (ed.), *Integrative Theory of Evolution - Aspects and Insights*,
https://doi.org/10.1007/978-3-032-24295-2_18

intein excision (Fig. 18.1a; Mills et al. 2014; Wood et al. 2023). The unique ability of inteins to precisely make and break peptide bonds in complex proteins, and do so in a traceless manner, has enabled many biotechnological applications (Wood and Camarero 2014; Sarmiento and Camarero 2019), with new uses frequently emerging (Baloban et al. 2025; Hua et al. 2025; Janjuha et al. 2025). Given their powerful chemistry, a major focus of research into inteins has centered on their use as protein engineering tools.

The biological roles of inteins have remained enigmatic. Many inteins house homing endonuclease domains that catalyze mobility at the DNA level, promoting intein spread within and between populations. Due to their role as mobile genetic elements, inteins have historically been considered "selfish DNA" or "molecular parasites" that provide no benefit to the host organisms they invade (Gogarten and Hilario 2006). Like other mobile genetic elements, however, the view of inteins is expanding, with a growing appreciation for the complex evolutionary interplay between mobile element and host (Benler and Koonin 2022; Lennon et al. 2022; Fueyo et al. 2022). In this chapter, we discuss the possible mutualistic function of inteins as adaptive, posttranslational regulatory elements, as well as the evidence supporting the hypothesis that hedgehog signaling—a vital cell/cell communication mechanism in animals—is derived from the evolutionary repurposing of an intein.

2 Inteins as Emergent Posttranslational Regulatory Elements

Inteins have only been detected in the genomes of single-celled organisms and viruses, with a strong tendency toward retention in archaeal (~47%) and bacterial (~24%) genomes compared to eukaryotic (~1%) genomes (Novikova et al. 2016). The distribution of inteins within archaea and bacteria is idiosyncratic—abundant in some genera, absent in others. It seems that certain organisms favor the retention of inteins. For example, the hyperthermophilic archaeon *Thermococcus kodakarensis* houses 15 inteins, 11 of which are within proteins involved in DNA replication, recombination, and repair (RRR proteins) (Liman et al. 2024). In total, inteins account for about 1% of an otherwise tightly packed *T. kodakarensis* genome (2.09 Mbp).

In archaea and bacteria, the distribution of inteins is heavily biased toward RRR proteins, with nearly two-thirds found within these often essential proteins (Novikova et al. 2016). Inteins also tend to map to critical functional domains, with about 70% in ATPases (Novikova et al. 2016). In many of these RRR proteins, inteins have been retained independently in non-orthologous genes of homologous function. For example, the replicative DNA helicases—MCM in archaea and DnaB in bacteria—are the most common intein-containing proteins in their respective domains (Novikova et al. 2016). While both proteins function as replicative helicases, they evolved their DNA unwinding abilities independently (Leipe et al. 2000; Novikova et al. 2016).

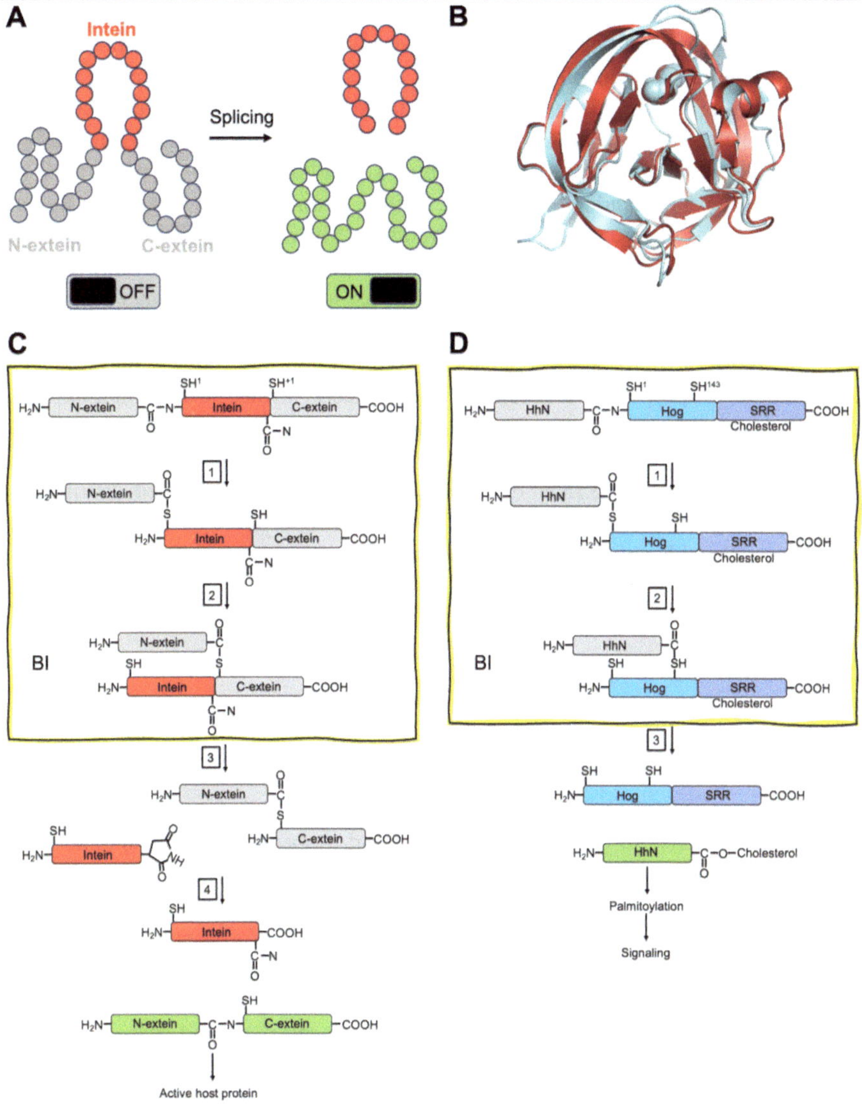

Fig. 18.1 Conditional protein splicing and the structural and mechanistic similarities between inteins and hedgehog. (**a**) Conditional Protein Splicing (CPS). The rate of intein excision and the accuracy of extein ligation are responsive to environmental and cellular conditions to the benefit of the intein-containing organism. In the unprocessed precursor, host protein function is disrupted by the intein (red). Exteins are gray (prior to splicing); Intein-catalyzed excision and ligation of the flanking exteins fully activate host protein function (exteins are green post splicing). (**b**) Structural conservation between intein (red; PDB: 2IN0) and hedgehog (Hh) C-terminal autoprocessing domain (blue; PDB: 1AT0). The N-terminal substrate residue for both the intein and Hh domains are shown as spheres. Images generated by Pymol. (**c**, **d**) The mechanisms of class 1 protein splicing (**c**) and cholesterol-dependent Hh autoprocessing (**d**). Green color in panels **c** and **d** indicates a fully functional host protein following intein splicing (**c**) or autoprocessing of the Hh signaling domain (**d**). In panels **c** and **d**, yellow boxes highlight shared mechanistic steps between protein splicing and Hh autoprocessing. Details of these transformations are included in the main text. BI indicates branched intermediate formation

It would stand to reason that if all inteins were merely parasitic elements, they would tend toward minimizing their negative impact on the host organism. As the function of an intein-containing protein is almost certainly compromised prior to intein excision (Fig. 18.1a), splicing should therefore occur either co-translationally or immediately following translation. Robust intein activity should be maintained over a wide range of environmental conditions. Contrary to these expectations, the rate of intein excision and the fidelity of extein ligation are variable between inteins, as well as under different solution conditions in vitro (Belfort 2017; Lennon and Belfort 2017; Wood et al. 2023). Inteins from the same organism can exhibit rates of splicing that vary by orders of magnitude under identical conditions (Ellilä et al. 2011). It is also reasonable to hypothesize that because inteins from, for example, hyperthermophiles have adapted to their environment, they now require elevated temperature to properly fold and splice. Indeed, while several inteins from hyperthermophiles require high temperatures to splice, some hyperthermophilic inteins splice rapidly at low temperatures (Topilina et al., 2015b; Lennon et al. 2018). Could the variable rates of protein splicing, as well as biased distribution, indicate that selected inteins have been exapted—or repurposed by evolution—from selfish parasite to beneficial regulatory element?

3 Conditional Protein Splicing

In contrast to the traditional perspective of inteins as exclusively parasitic, numerous examples of conditional protein splicing (CPS) have been described where the rate of splicing is sensitive to environmental cues. These include, but are not limited to, temperature, reactive oxygen, nitrogen, and chlorine species, metals, salt, and DNA damage (Belfort 2017; Callahan et al. 2011; Ciragan et al. 2016; Green et al. 2019; Kelley et al. 2018; Lennon et al. 2016, 2018, 2021; Lennon and Belfort 2017; Mills and Paulus 2001; Reitter et al. 2016; Wood et al. 2023; Woods et al. 2020; Yalala et al. 2022). Inteins have even been trained to splice in response to small molecules through directed evolution (Buskirk et al. 2004; Peck et al. 2011; Skretas and Wood 2005), which begs the question—if protein engineers can find a use for CPS, couldn't nature as well?

We posit that some inteins are beneficial to host organisms through at least two regulatory strategies. First, inteins can act as "pause buttons" to temporarily block splicing of any newly synthesized or yet unspliced protein under stress. Then, once that stress has been alleviated, splicing can resume, and the intein-containing protein can be reconstituted. Second, inteins can function as "rheostats," where the unspliced intein-containing protein requires a regulatory signal to trigger activity, in turn leading to immediate dose-dependent splicing. In this way, the splicing of the intein-containing precursor can be dialed up to match the concentration of a specific environmental cue. Later, we explore some of the CPS examples that highlight inteins as pause buttons and rheostats. This is not a complete list of all conditions that promote or prohibit protein splicing but rather focuses on a handful of well-understood examples.

3.1 Inteins as Pause Buttons: Metal Ion and Redox Stress

Over two decades ago, it was demonstrated that divalent cations, especially zinc and copper, can block protein splicing, an observation proposed to have usefulness in the preparation of unmodified protein splicing precursors for structural studies (Mills and Paulus 2001). Although structural analysis of unspliced precursors has remained elusive, the biomedical relevance of splicing's metal inhibition has emerged for inteins of the human pathogens *Mycobacterium tuberculosis* and *M. leprae* (Wall et al. 2021). The nonpathogenic model organism *M. smegmatis* also contains inteins and was used to assess the effect of zinc on splicing of an intein from the replicative helicase DnaB. Using in vivo and in vitro splicing reporters, zinc was found to block protein splicing (Woods et al. 2020). Zinc is bound by this intein in a pocket that includes a nucleophilic cysteine required to initiate the protein splicing (Woods et al. 2020). Prp8, an essential component of the spliceosome complex for intron removal from pre-mRNA, is a common site for intein insertion in unicellular eukaryotes (Green et al. 2019). One such unicellular eukaryote, the fungal pathogen *Cryptococcus neoformans*, contains an intein in Prp8 that can be reversibly inhibited by copper and zinc (Green et al. 2019). Presumably, when copper and zinc levels are high in the environment, splicing is temporarily inhibited in order to conserve resources. The human immune response utilizes these very metals to fight pathogens (Neyrolles et al. 2015), suggesting another molecular front in host/pathogen battles.

As inteins are found within essential proteins of devastating human pathogens, most notably *M. tuberculosis* and *C. neoformans*, and are absent in humans, inteins represent novel antimicrobial targets. As such, several studies have sought to develop novel antimicrobial strategies based on the inhibition of protein splicing, some using platinum-based compounds (Chan et al. 2016; Li et al. 2021; Wall et al. 2021; Tharappel et al. 2022). Metal inhibition of protein splicing underscores how understanding intein-mediated stress responses has practical implications beyond deepening our appreciation of molecular regulation.

Inteins employ cysteine-based chemistry to mediate protein splicing (Mills et al. 2014; Wood et al. 2023). As cysteines are particularly sensitive to oxidative stress (Lennon et al. 2021), various oxidizing agents have been shown to reversibly block intein activity, providing a convenient pathway to redox-switchable splicing (Callahan et al. 2011). Unlike established strategies to impart CPS through mutagenesis of the intein, this effort focused on engineering residues of the flanking N-extein. Taking the redox-switchable thioredoxin protein as inspiration, the last three N-extein residues of a DnaE intein were substituted with Cys(-3)Pro(-2)Gly(-1) with the goal of disulfide/dithiol interchange with the intein's Cys(1) residue. Operationally, oxidizing conditions favored Cys(-3)-to-Cys(1) disulfide bonding, which effectively handcuffed the intein in an unreactive off state, whereas reduction of the Cys(-3)-to-Cys(1) disulfide (via addition of TCEP) reversed the engineered disulfide to the free thiols, liberating the intein's Cys(1) sidechain to initiate splicing. The disulfide-bonded, off state of this engineered N-extein/intein precursor proved sufficiently stable for crystallization, offering new structural insight into splicing's molecular mechanism and inspiring novel intein biotechnology (Callahan et al. 2013). The versatility of a redox-switchable intein was not lost on nature, as multiple intein precursor proteins with N-extein Cys(-3) have been identified in the sequence databases. At least one, an oxygen-sensitive protein from *Pyrococcus abyssi* (MoaA), has been shown to mirror DnaE's N-extein-mediated redox splicing behavior.

The same DnaB intein from *M. smegmatis* mentioned above can also be reversibly inhibited by the reactive oxygen species hydrogen peroxide, as well as the reactive chlorine species hypochlorous acid found in household bleach (Kelley et al. 2018; Lennon et al. 2021). In the presence of hydrogen peroxide, a cysteine required for splicing forms a reversible disulfide bond with an adjacent cysteine to block protein splicing (Kelley et al. 2018). While DNA replication requires a large number of proteins, DnaB is essential to the process, and inhibition of this one intein—which is sensitive to a variety of relevant environmental stressors (Kelley et al. 2018; Woods et al. 2020; Lennon et al. 2021)—would be enough to block the process as a whole.

3.2 Inteins as Rheostats: Temperature and DNA Damage

Several inteins from hyperthermophilic archaea have been shown to require elevated temperatures similar to their natural environment to promote protein splicing (Mills et al. 2005). In some cases, like the PolD (also refered to as PolII) intein from *Pyrococcus abyssi*, the requirement of high temperature for splicing occurs within non-native extein splicing reporters, meaning this property is inherent to the intein. In some cases, however, the relationship between temperature and splicing is more complex.

The RadA intein from *Pyrococcus horikoshii* also requires high temperatures (>65 °C) to splice, but this feature depends on the extein context. The *P. horikoshii* RadA intein was first identified as an exceptionally active intein within several non-native extein splicing reporters at relatively low temperatures (~20–37 °C) (Ellilä et al. 2011; Oeemig et al. 2012; Topilina et al. 2015b). Unexpectedly, when this intein was purified within its native exteins, splicing was inhibited at temperatures below 65 °C (Topilina et al., 2015b). Through a combination of modeling and mutagenesis, it was shown that interactions between the intein and the C-extein block splicing at low temperatures and that these inhibitory interactions are broken above 65 °C (Topilina et al., 2015b).

When comparing the splicing of the *P. horikoshii* RadA intein to a homologous intein from *Thermococcus sibiricus*, a similar intein-C-extein interaction was observed (Lennon et al. 2018). While both are hyperthermophilic, *P. horikoshii* grows optimally at about 20 °C higher temperatures than *T. sibiricus*, and the temperature required to stimulate splicing was approximately 15 °C lower for *T. sibiricus* compared to *P. horikoshii*, correlating with growth temperature (Lennon et al. 2018).

The most specific example of the rheostat model of posttranslational regulation by inteins described to date also comes from the *P. horikoshii* RadA intein (Lennon et al. 2016, 2019). While temperature was first shown to induce protein splicing in the native extein context, single-stranded DNA (ssDNA) resulted in a 50-fold increase in the rate of splicing compared to temperature alone. Archaeal RadA is the homolog of bacterial RecA and eukaryotic RadA51, and serves as the primary homologous recombinase used to repair DNA in response to damage (Lennon et al. 2016). Additionally, ssDNA has long been known to stimulate the ATPase activity of RecA in vitro (Shibata et al. 1979). Double-stranded DNA of the same sequence as the ssDNA had no effect, nor did RNA or free deoxynucleotides (Lennon et al. 2016). Mechanistically, ssDNA interacts with RadA via the C-extein, presumably breaking intein-C-extein interactions (Lennon et al. 2016). Biochemical analysis demonstrated that ssDNA binding by the RadA C-extein compresses the intein active site to promote splicing (Lennon et al. 2019). The model of ssDNA-induced RadA splicing seems logical, as RadA activity could be detrimental to the cell in the absence of DNA damage by wasting ATP and stimulating aberrant recombination events. By using the substrate of RadA as the inducer of protein splicing, the presence of the intein provides an elegant means of CPS.

3.3 Conditional Accuracy of Protein Splicing

Inteins can undergo off-pathway side reactions where either the N- or C-extein is ejected from the intein prior to ligation to the partner extein. Therefore, just like rates of splicing, accuracy can be highly variable as well, which ultimately dictates how much of the active intein-containing protein is available to the cell. Two cases of increased off-pathway cleavage under stress come from inteins within the human pathogen *M. tuberculosis*. In one case, an intein found within the iron-sulfur cluster assembly protein SufB was shown to be inhibited by oxidative and nitrosative stressors (Topilina et al. 2015a). When treated with hydrogen peroxide, the extent of N-terminal cleavage increased markedly (Topilina et al. 2015a). In another example, the *M. tuberculosis* RecA intein was shown to undergo N-terminal cleavage within *M. tuberculosis* cells (Schneider et al. 2024). Intriguingly, expression of the *M. tuberculosis* RecA N-extein was sufficient to activate LexA and trigger the SOS response in *E. coli*, the first experimental result to suggest that off-pathway products may have roles within the cell. Finally, in the abovementioned example of ssDNA-accelerated splicing of the RadA intein from *P. horikoshii*, ssDNA drastically improved the accuracy of splicing (Lennon et al. 2016). In the absence of ssDNA, however, RadA splicing is inefficient, with most precursors forming off-pathway products (Lennon et al. 2016).

3.4 The Benefits of CPS in the Native Host

The probability of inteins as emergent regulatory elements is strengthened by the finding that evolution can tailor the rate and accuracy of protein splicing to a wide variety of environmental cues (e.g., oxidative stress) or to signals specific to the intein-containing protein (e.g., DNA damage). In this way, intein-containing organisms can develop customized responses adapted for their specific niches. As inteins are particularly abundant in hyperthermophilic archaea (Novikova et al. 2016), consider temperature as an environmental condition where intein-based regulation may be useful. If a particular intein-containing ATPase is only required at high temperature, it would be uneconomical for this protein to retain any activity at low temperature. This would be akin to running an air conditioner on a cold winter's day. But when the environmental conditions warm, much like switching on the air conditioning on a hot day, the intein can now splice out, and the ATPase can become active. Importantly, it is not our argument that all inteins have been exapted to function as regulatory elements. Indeed, many inteins likely exert a significant fitness cost without any known benefit (Naor et al. 2016). One might argue that these are *en route* to expulsion or exaptation. Regardless, the evidence gathered over the past decade that some inteins are likely beneficial regulatory elements is compelling.

While an intein-based fitness advantage has yet to be described, the power of these elements to control host physiology has been recently shown in vivo in an organism that naturally contains inteins (Liman et al. 2024). Specifically, it was demonstrated that protein splicing can control one of the most fundamental cellular processes, DNA replication. Like most organisms, the archaeon *T. kodakarensis* can perform traditional origin-based DNA replication. However, *T. kodakarensis* is naturally polyploid and prefers to use recombination-based DNA replication under fast growth conditions (Gehring et al. 2017). Recombination-based replication relies on the intein-containing recombinase RadA, which is an essential gene in *T. kodakarensis* (Gehring et al. 2017). Using isogenic strains of *T. kodakarensis* with chromosomal modifications to the RadA intein, as well as suboptimal temperature, it was shown that inhibition of RadA splicing not only resulted in a growth defect and increased sensitivity to ultraviolet radiation, but it also mediated a switch from recombination-based to origin-based replication (Liman et al. 2024). Interestingly, RadA splicing was only inhibited by ~50%, meaning that a relatively modest inhibition of splicing can have an enormous influence on the intein-containing organism. These results clearly demonstrate the power of an intein to control a fundamental cellular process.

4 Hedgehog Signaling Is Dependent on a CPS-Like Autoprocessing Event

Hedgehog (Hh) family proteins are conserved throughout animals, acting as crucial extracellular signaling factors during embryonic development and promoting stem cell homeostasis in adults (Zhang and Beachy 2023).

The existence of an intein-like autoprocessing event in the biosynthesis of mature Hh protein was established through biochemical studies on *Drosophila melanogaster* Hh (Porter et al. 1996), representing the founding member of this protein family (Nüsslein-Volhard and Wieschaus 1980). The molecular transformation, described below, might represent the clearest example of conditional protein splicing, except that Hh autoprocessing does not result in protein-protein ligation but rather protein-cholesterol esterification. Cholesterol serves as both substrate and inducer for the Hh reaction, producing a cleaved and C-terminally cholesterylated Hh signaling factor. All enzymatic activity for this single-turnover cleavage/lipidation resides in the Hh precursor protein. Before extracellular secretion, a second, mechanistically independent lipidation with palmitate takes place at the Hh signaling factor's N-terminus (Chamoun et al. 2001). Hh's two lipids serve, at least in part, as membrane anchors that modulate extracellular dispersal (Siebold and Rohatgi 2023). Under conditions of acute cholesterol depletion or when autoprocessing is blocked by mutation, the unprocessed Hh precursor is retained in the secretory pathway and quickly degraded by the proteasome, thereby eliminating downstream biological Hh signaling (Guy 2000; Chen et al. 2011; Huang et al. 2013).

More distant and largely uncharacterized Hh-related proteins (Bürglin 2008) have been identified in the genomes of the freshwater protist *Monosiga ovata* (Snell et al. 2006) and the mycorrhizal fungus *Glomus mosseae* (Requena et al. 2002). Fungi and protists are noteworthy in harboring both inteins and Hh-related proteins. Whether these more evolutionarily distant Hh-related proteins bring about intein-adjacent chemistry and whether that putative activity is conditional on cholesterol or some other small molecule remains unknown.

5 Evidence of the Evolutionary Link Between Inteins and Hedgehog Proteins

It has long been hypothesized based on structural homology and mechanistic similarities that an intein was exapted to catalyze the biosynthesis of Hh signaling proteins (Hall et al. 1997; Adamska et al. 2007). The first piece of evidence of shared ancestry between inteins and Hh is in the atomic structure of these elements (Fig. 18.1b). The hedgehog precursor protein is composed of two domains—an N-terminal signaling domain (HhN), a C-terminal domain (HhC) composed of an enzymatically active "Hog" subdomain followed by the sterol recognition region (SRR). The Hog domain of Hh proteins and inteins shares a common "HINT" (hedgehog/INTein) fold, which resembles a β-sheet- and loop-rich horseshoe (Fig. 18.1b). When aligning an intein and Hog domain, the structures are remarkably similar, with an RMSD of only 1.36 Angstroms (Hall et al.1997; Kandel and Wang 2022; Van Roey et al. 2007). While inteins are poorly conserved in sequence outside of splicing blocks, their fold is strikingly well conserved, even between different intein classes (Kelley et al. 2018). The structural similarities of inteins and the Hh Hog domain provide strong support for common ancestry.

Mechanistic similarities in the self-mediated chemistry performed by each element offer additional compelling evidence of shared ancestry (Fig. 18.1c, d). The canonical class 1 mechanism of protein splicing is a four-step process (Mills et al. 2014; Wood et al. 2023). In the first step of splicing, a cysteine or serine at the N-terminus of the intein ("1 position") performs a nucleophilic attack on the preceding peptide bond that separates the N-extein from the intein, resulting in an amide-to-(thio)ester rearrangement (Fig. 18.1c). In the second step of protein splicing, a transesterification occurs where N-extein is transferred to the side chain sulfur (Cys) or oxygen (Ser or Thr) of the first residue of the C-extein ("+1 residue"), resulting in the formation of a branched intermediate (BI) (Fig. 18.1c). Following BI formation, in the third step, the last residue of the intein, generally an Asp, cyclizes to release the intein. In the fourth and final step, the (thio)ester separating the N- and C-exteins spontaneously rearranges to a standard backbone peptide bond (Fig. 18.1c).

The first two steps of protein splicing are closely mimicked in Hh autoprocessing (Fig. 18.1c, d). In the first step, a conserved cysteine at the N-terminus of the Hog domain mediates an amide-to-thioester shift analogous to the first step of protein splicing (Wall et al. 2021; Faris et al. 2025). Until recently, it was thought that a cholesterol molecule bound to the Hh SSR directly leads to cleavage of this thioester bond and subsequent formation of cholesterylated HnN. Recent results have suggested that the second step of Hh autoprocessing involves the formation of a BI thioester via another conserved Hh Hog domain cysteine, mirroring the second step of protein splicing (Faris et al. 2025). Specifically, this work showed that Hh residue Cys143 serves an equivalent role to the +1 residue after an intein to drive BI formation. Taken together, the chemistry of the first two catalytic steps of protein splicing and Hh autoprocessing are essentially identical (Fig. 18.1c, d).

6 How Might an Intein Have Been Exapted to Form the Hedgehog Hog Domain?

Most inteins house a nested homing endonuclease (HEN) domain that promotes horizontal gene transfer of the intein between genomes. HENs are highly specific, cleaving intein-lacking alleles of the same gene. For example, if a helicase intein with an active HEN domain encounters the same helicase allele without an intein, it will induce a double-strand DNA break that can be repaired through homologous recombination in the presence of the intein-containing helicase allele. Mini-inteins are those that have lost their HEN domains but retain all necessary features to mediate protein splicing. When considering the structure of the HINT domain, this refers specifically to the portion of the intein required for splicing.

When comparing mini-inteins and the Hh Hog domain, in addition to the structural and mechanistic features detailed above, both can be thought of as single-turnover enzymes that modify surrounding sequences. In the case of inteins, a ligation occurs, and for the Hh Hog domain, a cleavage/lipidation occurs. One could imagine an ancient mini-intein gene duplication event, with one copy retaining protein splicing activity and the other copy evolving to perform the new cleavage function linked to cholesterylation for signaling. Over time, this new signaling function was retained, while the mini-intein sequence was lost.

Due to the lack of sequence conservation, as well as the ancient nature of both inteins and Hh, it is impossible to say with certainty that one gave rise to the other. It is possible that the HINT fold shared between inteins and the Hh Hog domain may be a result of convergent evolution. Additionally, the striking mechanistic similarities might too have arisen independently. While the structural and mechanistic commonalities are compelling, we concede that both, indeed, may be molecular red herrings. However, if the hypothesis that an intein was the progenitor of the Hh Hog domain is correct, a plausible sequence of events could be the following (Fig. 18.2). First, an intein invades an essential fungal or protist cell-surface protein in a critical hinge region that separates the protein's N-terminal ectodomain from the C-terminal cholesterol-binding membrane domain. The essential function of the host protein is unaffected owing to the intein's splicing activity. Second, a duplication event occurs, creating a second copy of the host/intein gene. One version continues to behave as before, with host protein function dependent on successful intein splicing. Third, mutations accumulate in the duplicated copy, largely deleterious, but over an extended period, selected substitutions lead fortuitously to a novel transformation. In this novel activity, the cholesterol molecule bound by the host protein's C-terminal membrane domain, originally as a means of membrane attachment, now begins to displace the intein thioester, diverting intein splicing toward an N-terminal cleavage/cholesterylation event. Finally, as biosynthesis of an essential signaling factor is now conditional on the concentration of an essential lipid, this confers an advantage to the host as a means of nutrient sensing. Over additional time, the original intein/gene copy becomes superfluous to the host and is lost.

7 Concluding Remarks

While once considered exclusively parasitic sequences, mobile genetic elements and specifically inteins are uniquely positioned to function as drivers of evolutionary adaptation, and their exaptation is emerging as a common theme in nature (Benler

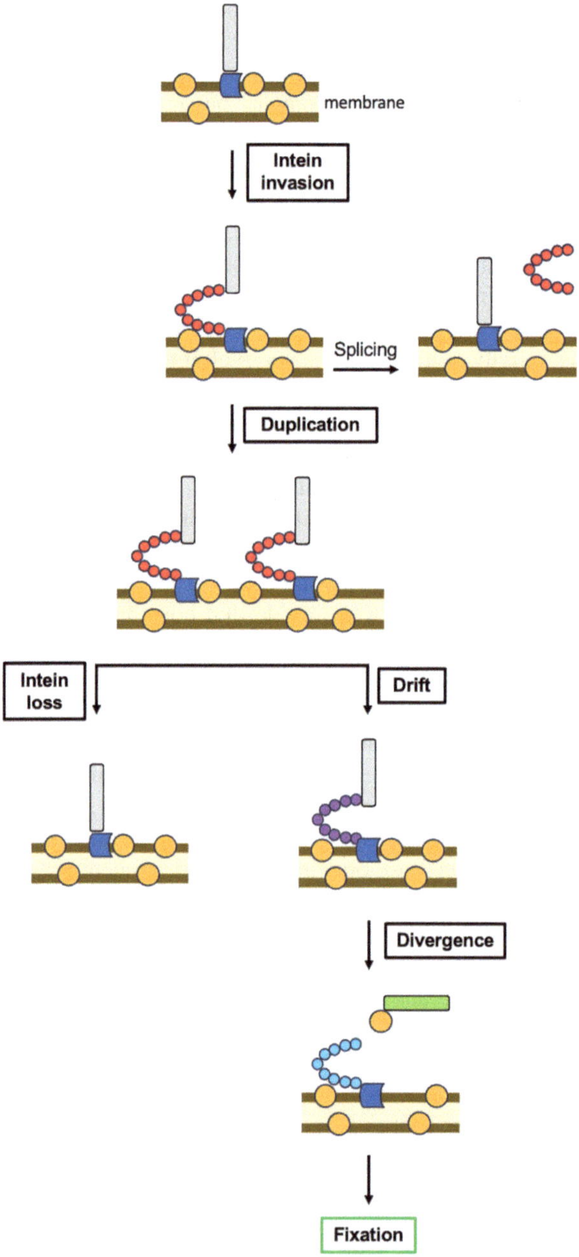

Fig. 18.2 Cholesterol-induced autoprocessing from autonomous splicing intein. Ancestral invasion and splicing. *Ancestral intein invasion:* an intein (red) invades the hinge region of an essential cell-surface protein, separating the N-terminal ectodomain (gray) from the C-terminal transmembrane domain (slate; cholesterol is represented as yellow circles). Host protein function is preserved via canonical protein splicing. *Duplication and drift:* a gene duplication event creates two copies. One copy maintains the original splicing function while the second accumulates mutations

and Koonin 2022; Lennon et al. 2022; Fueyo et al. 2022). The evidence of the repurposing of inteins from selfish to beneficial elements by some organisms grows stronger every year, as does their connection to hedgehog protein autoprocessing. Moving forward, we therefore must reconsider and reimagine the diverse and potentially beneficial roles of mobile DNA.

Acknowledgments Research reported in this publication was supported by the National Institute of General Medical Sciences of the National Institutes of Health under Award Number R15GM143662 to CWL. The content is solely the responsibility of the authors and does not necessarily represent the official views of the National Institutes of Health.

References

Adamska M, Matus DQ, Adamski M et al (2007) The evolutionary origin of hedgehog proteins. Curr Biol 17:R836–R837
Baloban M, Manoilov KY, Karasev MM et al (2025) Photoswitchable intein for light control of covalent protein binding and cleavage. Nat Commun 16:8263
Belfort M (2017) Mobile self-splicing introns and inteins as environmental sensors. Curr Opin Microbiol 38:51–58
Benler S, Koonin EV (2022) Recruitment of mobile genetic elements for diverse cellular functions in prokaryotes. Front Mol Biosci 9:821197
Bürglin TR (2008) The Hedgehog protein family. Genome Biol 9:241
Buskirk AR, Ong Y-C, Gartner ZJ et al (2004) Directed evolution of ligand dependence: small-molecule-activated protein splicing. Proc Natl Acad Sci USA 101:10505–10510
Callahan BP, Topilina NI, Stanger MJ et al (2011) Structure of catalytically competent intein caught in a redox trap with functional and evolutionary implications. Nat Struct Mol Biol 18:630–633
Callahan BP, Stanger M, Belfort M (2013) A redox trap to augment the intein toolbox. Biotechnol Bioeng 110:1565–1573
Chamoun Z, Mann RK, Nellen D et al (2001) Skinny hedgehog, an acyltransferase required for palmitoylation and activity of the hedgehog signal. Science 293:2080–2084
Chan H, Pearson CS, Green CM et al (2016) Exploring intein inhibition by platinum compounds as an antimicrobial strategy. J Biol Chem 291:22661–22670
Chen X, Tukachinsky H, Huang C-H et al (2011) Processing and turnover of the Hedgehog protein in the endoplasmic reticulum. J Cell Biol 192:825–838
Ciragan A, Aranko AS, Tascon I et al (2016) Salt-inducible protein splicing in cis and trans by inteins from extremely halophilic archaea as a novel protein-engineering tool. J Mol Biol 428: 4573–4588

Fig. 18.2 (continued) (ancestral intein now purple). *Divergence and fixation:* in the mutated copy, fortuitous substitutions allow positioning cholesterol associated with the C-terminal membrane domain to function as a nucleophile, displacing the intein-mediated thioester intermediate and diverting the reaction from splicing to N-terminal cleavage and cholesterylation (ancestral intein now light blue). The N-terminal ectodomain is now green following cholesterlyation. This new conditional activity couples the production of the essential signaling factor (N-terminal domain) to cholesterol availability, creating a nutrient-sensing mechanism. The original gene copy becomes redundant and is lost, fixing cholesterol-dependent autoprocessing

Ellilä S, Jurvansuu JM, Iwaï H (2011) Evaluation and comparison of protein splicing by exogenous inteins with foreign exteins in Escherichia coli. FEBS Lett 585:3471–3477
Faris S, Xia K, Wagner AG et al (2025) Conserved C143 forms a branched intermediate in Hedgehog autoprocessing: a cancer drug discovery target against Hedgehog signaling. Proc Natl Acad Sci USA 122:e2415144122
Fueyo R, Judd J, Feschotte C et al (2022) Roles of transposable elements in the regulation of mammalian transcription. Nat Rev Mol Cell Biol 23:481–497
Gehring AM, Astling DP, Matsumi R et al (2017) Genome replication in Thermococcus kodakarensis independent of Cdc6 and an origin of replication. Front Microbiol 8:2084
Gogarten JP, Hilario E (2006) Inteins, introns, and homing endonucleases: recent revelations about the life cycle of parasitic genetic elements. BMC Evol Biol 6:94
Green CM, Li Z, Smith AD et al (2019) Spliceosomal Prp8 intein at the crossroads of protein and RNA splicing. PLoS Biol 17:e3000104
Guy RK (2000) Inhibition of sonic hedgehog autoprocessing in cultured mammalian cells by sterol deprivation. Proc Natl Acad Sci USA 97:7307–7312
Hall TM, Porter JA, Young KE et al (1997) Crystal structure of a Hedgehog autoprocessing domain: homology between Hedgehog and self-splicing proteins. Cell 91:85–97
Hua Y, Tay NES, Ye X et al (2025) Protein editing using a coordinated transposition reaction. Science 388:68–74
Huang C-H, Hsiao H-T, Chu Y-R et al (2013) Derlin2 protein facilitates HRD1-mediated retro-translocation of sonic hedgehog at the endoplasmic reticulum. J Biol Chem 288:25330–25339
Janjuha S, Haenggi T, Chamberlain TC et al (2025) Spatial profiling of gene editing by in situ sequencing in mice and macaques. Nat Biomed Eng. https://doi.org/10.1038/s41551-025-01512-7
Kandel N, Wang C (2022) Hedgehog autoprocessing: from structural mechanisms to drug discovery. Front Mol Biosci 9:900560
Kelley DS, Lennon CW, Li Z et al (2018) Mycobacterial DnaB helicase intein as oxidative stress sensor. Nat Commun 9:4363
Leipe DD, Aravind L, Grishin NV et al (2000) The bacterial replicative helicase DnaB evolved from a RecA duplication. Genome Res 10:5–16
Lennon CW, Belfort M (2017) Inteins. Curr Biol 27:R204–R206
Lennon CW, Stanger M, Belfort M (2016) Protein splicing of a recombinase intein induced by ssDNA and DNA damage. Genes Dev 30:2663–2668
Lennon CW, Stanger M, Banavali NK et al (2018) Conditional protein splicing switch in hyperthermophiles through an intein-extein partnership. MBio 9:e02304–e02317
Lennon CW, Stanger MJ, Belfort M (2019) Mechanism of single-stranded DNA activation of recombinase intein splicing. Biochemistry 58:3335–3339
Lennon CW, Wahl D, Goetz JR et al (2021) Reactive chlorine species reversibly inhibit DnaB protein splicing in mycobacteria. Microbiol Spectr 9:e0030121
Lennon CW, Callahan BP, Cousineau B et al (2022) Editorial: Genetically mobile elements repurposed by nature and biotechnologists. Front Mol Biosci 9:992664
Li Z, Tharappel AM, Xu J et al (2021) Small-molecule inhibitors for the Prp8 intein as antifungal agents. Proc Natl Acad Sci USA 118:e2008815118
Liman GLS, Lennon CW, Mandley JL et al (2024) Intein splicing efficiency and RadA levels can control the mode of archaeal DNA replication. Sci Adv 10:eadp4995
Mills KV, Paulus H (2001) Reversible inhibition of protein splicing by zinc ion. J Biol Chem 276:10832–10838
Mills KV, Dorval DM, Lewandowski KT (2005) Kinetic analysis of the individual steps of protein splicing for the Pyrococcus abyssi PolII intein. J Biol Chem 280:2714–2720
Mills KV, Johnson MA, Perler FB (2014) Protein splicing: how inteins escape from precursor proteins. J Biol Chem 289:14498–14505

Naor A, Altman-Price N, Soucy SM, Green AG, Mitiagin Y, Turgeman-Grott I, Davidovich N, Gogarten JP, Gophna U (2016) Impact of a homing intein on recombination frequency and organismal fitness. Proc Natl Acad Sci USA 113(32):E4654–E4661. https://doi.org/10.1073/pnas.1606416113. Accessed 26 Jul 2016. PMID: 27462108; PMCID: PMC4987785
Neyrolles O, Wolschendorf F, Mitra A et al (2015) Mycobacteria, metals, and the macrophage. Immunol Rev 264:249–263
Novikova O, Jayachandran P, Kelley DS et al (2016) Intein clustering suggests functional importance in different domains of life. Mol Biol Evol 33:783–799
Nüsslein-Volhard C, Wieschaus E (1980) Mutations affecting segment number and polarity in drosophila. Nature 287:795–801
Oeemig JS, Zhou D, Kajander T et al (2012) NMR and crystal structures of the Pyrococcus horikoshii RadA intein guide a strategy for engineering a highly efficient and promiscuous intein. J Mol Biol 421:85–99
Peck SH, Chen I, Liu DR (2011) Directed evolution of a small-molecule-triggered intein with improved splicing properties in mammalian cells. Chem Biol 18:619–630
Porter JA, Young KE, Beachy PA (1996) Cholesterol modification of hedgehog signaling proteins in animal development. Science 274:255–259
Reitter JN, Cousin CE, Nicastri MC et al (2016) Salt-dependent conditional protein splicing of an intein from Halobacterium salinarum. Biochemistry 55:1279–1282
Requena N, Mann P, Hampp R et al (2002) Early developmentally regulated genes in the arbuscular mycorrhizal fungus Glomus mosseae: identification of GmGIN1, a novel gene with homology to the C-terminus of metazoan hedgehog proteins. Plant Soil 244:129–139
Sarmiento C, Camarero JA (2019) Biotechnological applications of protein splicing. CPPS 20:408–424
Schneider RF, Hallstrom K, DeMott C et al (2024) Conditional protein splicing of the Mycobacterium tuberculosis RecA intein in its native host. Sci Rep 14:20664
Shibata T, DasGupta C, Cunningham RP et al (1979) Purified Escherichia coli recA protein catalyzes homologous pairing of superhelical DNA and single-stranded fragments. Proc Natl Acad Sci USA 76:1638–1642
Siebold C, Rohatgi R (2023) The inseparable relationship between cholesterol and Hedgehog signaling. Annu Rev Biochem 92:273–298
Skretas G, Wood DW (2005) Regulation of protein activity with small-molecule-controlled inteins. Protein Sci 14:523–532
Snell EA, Brooke NM, Taylor WR et al (2006) An unusual choanoflagellate protein released by Hedgehog autocatalytic processing. Proc Biol Sci 273:401–407
Tharappel AM, Li Z, Li H (2022) Inteins as drug targets and therapeutic tools. Front Mol Biosci 9:821146
Topilina NI, Green CM, Jayachandran P et al (2015a) SufB intein of Mycobacterium tuberculosis as a sensor for oxidative and nitrosative stresses. Proc Natl Acad Sci USA 112:10348–10353
Topilina NI, Novikova O, Stanger M et al (2015b) Post-translational environmental switch of RadA activity by extein-intein interactions in protein splicing. Nucleic Acids Res 43:6631–6648
Van Roey P, Pereira B, Li Z et al (2007) Crystallographic and mutational studies of Mycobacterium tuberculosis recA mini-inteins suggest a pivotal role for a highly conserved aspartate residue. J Mol Biol 367:162–173
Wall DA, Tarrant SP, Wang C et al (2021) Intein inhibitors as novel antimicrobials: protein splicing in human pathogens, screening methods, and off-target considerations. Front Mol Biosci 8:752824
Wood DW, Belfort M, Lennon CW (2023) Inteins-mechanism of protein splicing, emerging regulatory roles, and applications in protein engineering. Front Microbiol 14:1305848
Wood DW, Camarero JA (2014) Intein applications: from protein purification and labeling to metabolic control methods. J Biol Chem 289(21):14512–14519. https://doi.org/10.1074/jbc.R114.552653. Accessed 2 Apr 2014. PMID: 24700459; PMCID: PMC4031509

Woods D, Vangaveti S, Egbanum I et al (2020) Conditional DnaB protein splicing is reversibly inhibited by zinc in mycobacteria. MBio 11:e01403–e01420
Yalala VR, Lynch AK, Mills KV (2022) Conditional alternative protein splicing promoted by inteins from haloquadratum walsbyi. Biochem 61(4):294–302. https://doi.org/10.1021/acs.biochem.1c00788. Accessed 24 Jan 2022. PMID: 35073064; PMCID: PMC8847336
Zhang Y, Beachy PA (2023) Cellular and molecular mechanisms of Hedgehog signalling. Nat Rev Mol Cell Biol 24:668–687

Chapter 19
Opinion: On RNA, Virus, and Groups

Luis P. Villarreal

How RNA group interactions underlie and integrate the evolution of life seems a counterintuitive question, as we have long accepted the central role of DNA in evolution. Furthermore, the fact that most such RNA interactions stem from viruses and parasitic ancestral agents (junk) also seems illogical from a gene-centric perspective. And finally, that such RNA interactions behave and evolve in interacting groups does not appear to adhere to Darwinian mechanisms, which operate on the fittest type and incorporate error-based events as the basis of adaptation. Yet these are the main premises being considered in this book, which are well justified.

The RNA group interactions being considered involve both positive (complementary) and negative (interference) interactions that result in coherent group behaviors. But historically, group behaviors are not primarily the interactional entities that underlie thinking in evolutionary biology. RNA group concepts (i.e., quasi-species) were initially developed mathematically from the premise of the fittest individual type replication with errors to create related populations. But later experimental measurements showed viral RNA quasi-species display various group behaviors, including the capacity to identify and compete with nonmembers. Such observations relate directly to what coherent behaviors emerge, including memory and group identity. All social entities (including human cognition and culture) must have some rules and memory for membership. These rules are found in the foundations of RNA group interactions.

The abundant RNA entities so prominent in complex life are mostly derived from diffuse colonizing viral/parasitic RNA ancestors; they have the inherent capacity for de novo network production via scattered but RNA-coordinated colonization events that retain interactional motifs and can produce new interacting RNA groups. Relevant empirical knowledge of RNAs has exploded and includes many "RNA-

L. P. Villarreal (✉)
Department of Molecular Biology and Biochemistry, University of California, Irvine, CA, USA
e-mail: lpvillar@uci.edu

G. Witzany (ed.), *Integrative Theory of Evolution - Aspects and Insights*,
https://doi.org/10.1007/978-3-032-24295-2_19

omics" type studies with massive data content. The term "multi-omics" has been applied to such diverse data. This is in addition to many other types of diverse and massive data sets. It now seems clear to me that no individual researcher remains able to stay on top of so much data. We need "groups" of scholars and researchers working cooperatively (and competitively), attempting to integrate it all and draw meaning from it.

My own opinion is that an individual researcher, such as I was for many years, can no longer fully comprehend and integrate all the relevant and needed knowledge and concepts, let alone perform all the needed studies. We must become a group entity in order to encompass all this and be dedicated to integrating this knowledge. However, I do not know if supporting such groups is even possible in our current competitive and consensus-based paradigm of science. But this inherent conflict is just like a living entity or quasi-species itself, which includes opposing and seemingly contradictory elements that, combined, produce coherence. The whole is much more than the sum of seemingly incoherent parts. We scholars all resemble individual "trees" in a diverse forest, never able to see the whole, but each of us is able to bear fruit and alter the course of forest growth and also introduce and "seed" new and potential and transforming growth. An excellent example of an individual scholar that provided fruitful (infectious, branching) ideas and thinking is my long-time colleague, Guenther Witzany. By adopting the premise that nucleic acids provide a living language of life but, like all languages, must have agents that convey meaning within a proper context, we arrive at a transforming new perspective on the central role of communication for life and that the agents of life provide meaning itself.

I can only give an outline of my own experience as a way of noting how my "tree" initiated the development of my thinking along this RNA group-based path. I started conducting bench experiments in the early 1970s with the late Dr. John J. Holland in San Diego to study the molecular and biochemical basis of RNA virus (VSV) replication and transcription. The existence of virus-encoded and bound RNA polymerase had recently been discovered. While growing and purifying the virus, I frequently encountered the presence of defective or truncated versions of the virus. As I was then characterizing in vitro RNA synthesis, I also examined the defective virus. Much to my surprise, I observed that the defectives did not synthesize the usual monocistronic mRNAs or full genome RNA of the virus but instead made a large quantity of small and heterogeneous RNAs derived mostly from the regulatory ends of the genome. Later, it was shown that small regulatory RNAs are used by essentially all viruses for various regulatory purposes, but this particular population of RNAs seemed able to actively interfere with the replication of the full-length infectious virus. These were then defective interfering viruses (a defective viral parasite). The defective "mutant" competed with the fittest type of full-length virus to replicate itself, but also its small RNAs acted in trans to interfere with full-length virus replication. The defective one thus outcompeted the "fit" version of itself. By itself, however, the defective could not propagate; it needed to interact with the full virus. Clearly, these were due to RNAs interacting at a distance (trans). Such behavior had been proposed to allow persistent infections to be established in infected host cells by controlling virus replication, which I was able to show later.

Yet I was struck by how this easily observed behavior seemed to be at odds with the idea that only an individual type could be the fittest type. Clearly, a group interaction was present and prevailed. Six decades later, we start to see that such RNA-based interactions are ubiquitous and may indeed hold the clues to integrating our understanding of life and viruses. But for this path to grow, we need research groups to function interactively and collectively, as none of us seems fit to do this alone. I hope that this book could provide such a beginning.

Chapter 20
The Viral Turn in Biology

Robert Prinz and Günther Witzany

Abstract Important conclusions about viruses and their role in life drawn from a post-millennial, i.e., post-genomic, perspective are presented. Transcending *Neo-Darwinism* means not only following the path of the *New Synthesis*, or walking *The Third Way* of evolution, but also paving the way toward a much more revolutionary route of thinking in biology. Viruses can no longer be regarded as parasites or infectious agents alone. They are agents of change, sculptors of genomes, engineers of immune systems, and creators of *real novelty*. Their ability to convey new properties to hosts affects not only a few individuals but also all cellular life, their communication, and social life. The origin of *life itself*, the evolution of (meta)-genomes, and multicellularity are closely connected to viruses. The genetic code and its use, i.e., codon usage bias, are also influenced by viral activity. Furthermore, *Code Biology* showed that many biological codes have their roots in viral processes, rendering viruses actual codemakers, programmers of life, or original biohackers, if you will. A holistic view of viral processes leads to new insights and a better understanding of how living systems adapt, change, and create new realities. From our perspective, viruses and their relatives are at the bottom of any genetic novelty and in all processes of natural genome editing.

Keywords Viruses · Origin of life · Evolution · Agency · Novelty · Code biology

R. Prinz (✉)
Rechenkraft.net e.V., Marburg, Germany
e mail: rp@rechenkraft.net

G. Witzany
Telos-Philosophische Praxis, Buermoos, Austria
e-mail: witzany@sbg.at

G. Witzany (ed.), *Integrative Theory of Evolution - Aspects and Insights*,
https://doi.org/10.1007/978-3-032-24295-2_20

1 Introduction

"Where there is life, there are viruses" (Fermin 2018). This bold and contentious claim should illustrate the prevalence of past and present viral activity, either as acute or persistent infection or as ancient traces of viruses left in all genomes. Viruses are special in many ways: not only do they defy common attempts to classify them as living or nonliving (Gortner 1938; Forterre 2010; Koonin and Starokadomskyy 2016; Caetano-Anollés 2024), but they also exceed all other organisms by magnitudes in their number (Anthony et al. 2013; Mushegian 2020). Another hallmark is their ability to horizontally and vertically spread (Stewart et al. 2005; García-Ordóñez and Pagán 2024). They come as naked DNA or RNA or dressed with proteins or membranes as intricate, functional clothes enabling them to enter cells, camouflage from immune responses, and integrate into their host genomes (Gorbalenya et al. 2020; Louten 2023).

Viruses, ever since their discovery, have had a negative connotation; e.g., they were considered *enemies* when invading the host organism, penetrating cell walls, etc. Similar to bacterial infections, the immune system fights viral invaders, yet once viruses have successfully overcome the first line of immune defense and established a "niche," they are considered *parasites*. The arms race between viruses and their targeted organisms is an endless war, with individual infections being single combats and pandemics being battle grounds of global reach (Forterre and Prangishvili 2009). This pre-millennial, Darwinian-grounded terminology continued (Li and Hayward 2011; Hwang et al. 2016; Maxmen 2017; Barreat and Katzourakis 2023) but was stepwise complemented by neo-Darwinian worldviews of the twentieth century, when viruses still played no role on the main stage of evolution. The gene-centric view of the time led to the perception of *selfish elements*—genes and genome sequences that exist exclusively for their reproductive benefit by using the organism as a vehicle (Dawkins 2016). Viruses and mobile elements perfectly match such an egoistic narrative, and therefore, the attribution of hostile terms sustains.

Militaristic terminology has a long tradition in the context of the immune system and (viral) infections (Benzi and Novarese 2022). Contemporary bioweapon research further fuels the spread of this kind of language, as viruses are actively investigated or even engineered for bioterrorism (Morse and Meyer 2017). With the COVID pandemic, the war metaphor experienced another peak; however, the impact of the terminology also depends on cultural background (Schnepf and Christmann 2022). In other words, positive connotations can rarely be found and mainly belong to their beneficial use as biotechnological tools (Varanda et al. 2021).

The dawn of the new millennium demarks fundamental changes in biology, most prominently due to the advent of high-throughput genome sequencing. Exploration of microbial genomes revealed the extent of the virosphere and the role viruses must have played during evolution (Koonin and Wolf 2012; Wallace et al. 2025). The deciphering of the human genome and the sequencing of many other genomes, including metagenomes from diverse ecological communities, brought to light that viruses are present in almost all possible environments and leave their ubiquitous "signatures" in (meta)genomes (Sommers et al. 2021; Schulz et al. 2022). This astonishing discovery can either fuel the existing war metaphors, meaning there is a true *world war* taking place everywhere on earth, or raise the question if the distribution and activity of viruses can be seen in a different light. The *full-spectrum dominance* viruses exhibit in terms of various roles they take in (re)engineering molecular pathways, immune responses, editing genomes, and determining the composition of entire populations opens the way to a very different interpretation. Viruses are essential drivers of evolution and are symbiotic rather than parasitic in nature (Ryan 2009, Villarreal and Witzany 2010). Beyond that, viruses shape host and group identity and are key determinants of social life (Villarreal 2009a; Villarreal and Witzany 2019, 2021).

The viral turn in biology is the recognition that (a) viral destruction and construction are two sides of the same coin, (b) overall, viral activity was and still is a key driver of evolution, and (c) the dynamics of life can only be understood when the contribution of viruses is accounted for.

2 The Planetary Virosphere

Viruses are the most abundant biological entities on earth. One milliliter of seawater contains one million bacteria but ten times more viral particles. It is hard to imagine, but *life as defined* until recently, i.e., as exclusively cellular, represents rare islands in a sea of viruses (Villarreal 2005; Rohwer et al. 2014). The cellular life of prokaryotes and eukaryotic domains is constantly affected by infectious genetic parasites (viruses and their relatives) since the beginning of life. The dominant viral lifestyle is, however, host persistence, either as a dormant virus or as a defective virus, which means serving as effective RNA tools for all kinds of cellular gene regulation (Villarreal 2009b; Hughes and Coffin 2016).

The DNA of cellular life is the preferred habitat for noncoding RNA tool-like elements derived from viral infection events (Villarreal and Witzany 2013). In the DNA, be it in the form of bacterial plasmids or eukaryotic chromosomes, the tools find a proper place for short-term storage or rest or for long-term dormancy or hibernation—to draw analogies to plant and animal behavior. Cells themselves are the perfect habitat for reproduction and (co)evolution of fully fledged viruses, i.e., nucleic acids and capsid structures.

The biosphere of cells, as a result of continued infection events that provide all kinds of RNA networks as adaptive and, therefore, multipurpose tools, represents the ideal resource for viral spread and reproduction. There is not only infectious pressure on cellular life with its core operative motifs of reproduction, transcription, translation, repair, immunity, and epigenetic imprinting, but also constant cooperation between viral groups to access cellular targets. Viruses may recombine from single-stranded RNA to double-stranded DNA viruses and cooperate in infection strategies in various ways (Segredo-Otero and Sanjuan 2022). RNA viruses may generate unique sequences de novo "that never existed before and will never exist again" (Holland et al. 1992). This means *de novo*, virus-generated sequences are not the result of replication errors in previously existing genomes but literally are real novelties. Therefore, the traditional term "error replication" used to explain increasing quasispecies mutant spectra could be replaced by "Quasispecies Productivity," which better fits empirical data of RNA variant production, integrating also the role of RNA minority genomes (Domingo and Witzany 2024).

2.1 Viral (Street) Gangs

The ligation of random RNA oligomers is a key driver of functional complexity since the beginning of the RNA world (Hayden and Lehman 2006; Briones et al. 2009; Gwiazda et al. 2012; Vaidya et al. 2012; Broecker 2021). Also, today, RNA networks can be found in all regulatory tools of noncoding RNAs of cellular life. They can loosely assemble through RNA stem loops, thus building a unique group identity that may change quickly from competitive to cooperative behavior depending on circumstances such as binding to foreign RNAs, DNA, and proteins (Manrubia and Briones 2007, Mercer and Mattick 2013). Such acquisition of foreign RNA stem loop structures may lead to context-dependent rejection in the next moment. What is of current need determines what RNA components are recruited or deleted. It functions similarly to street "gangs" in which members are recruited or refuted, depending on the case of need (see Fig. 20.1 and Villarreal 2015). It has been shown that single RNA stem loops react physically and chemically exclusively. Interestingly, at a certain density of RNA stem loop structures, biological selection starts (Hayden and Lehman 2006; Higgs and Lehman 2015). Additionally, it was shown that cooperative RNA stem loop structures outcompete selfish ones (Vaidya et al. 2013). Evolutionarily conserved RNA network group building is well documented in the assembly history of the subunits of the ribosome or the evolution of tRNAs (Petrov et al. 2015; Root-Bernstein 2016; de Farias et al. 2016).

The RNA stem loop gangs can be considered precursors of viral RNA toolkits, and viruses still show many characteristics of this "gang" behavior (Mattick and Gagen 2001; Vignuzzi et al. 2006).

2.2 Viral Addict(s)

In most cases, competing viral clouds together with host immune systems reach an equilibrium that is inheritable and dramatically changes host genetic identity such that the host acquires and uses defective viral sequence parts to better adapt to environmental conditions. The resulting addiction module now serves as an immune function against related viral gangs to prohibit their access to the host genome, like a genetic doorman (Villarreal 2012, 2016a, b). The parts encoded by such addiction modules are stuck to one another, e.g., a stable toxin and an unstable antitoxin.

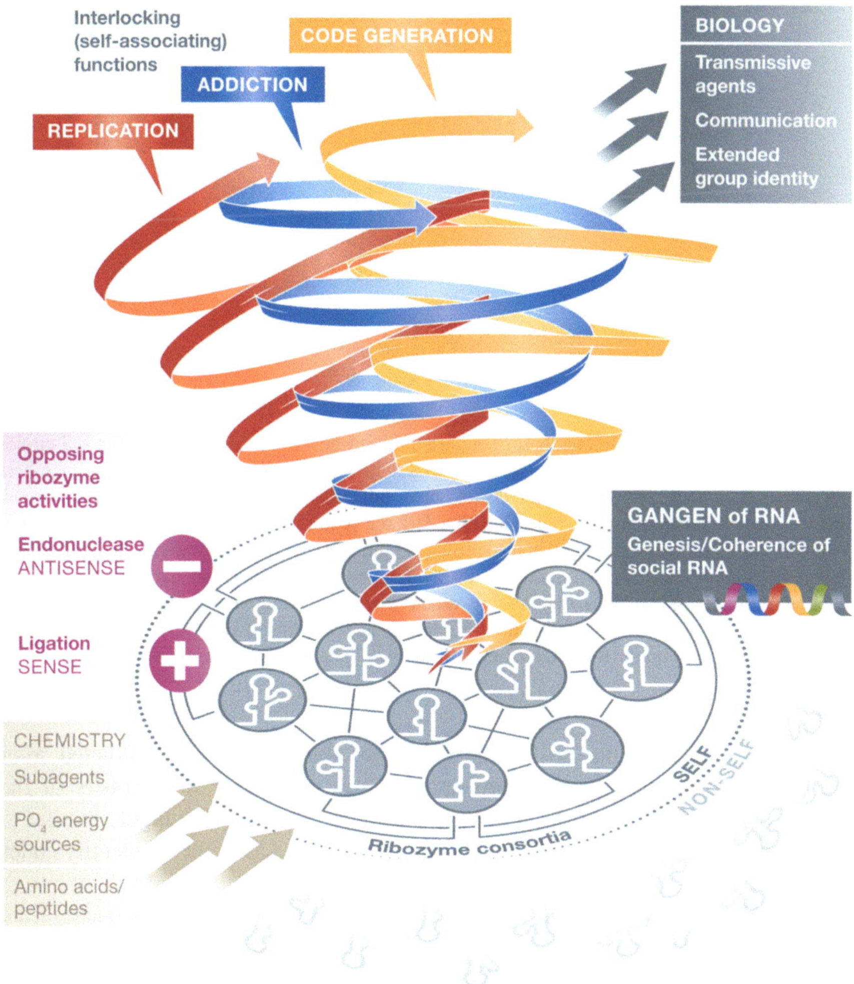

Fig. 20.1 The RNA "gang" hypothesis: RNA stem loop groups assemble like street gangs. According to Villarreal, group identity and cooperativity of an RNA collective require opposing functions, which are essential for the genesis of life (social behavior of agents). Interestingly, at the very beginning of the RNA world, ribozyme consortia with group identities initiated the differentiation competence of "self" from "nonself," like "gangs," according to dynamically varying contextual requirements. The figure compiles all key topics: abiotic–biotic interactions, consortial life and identity building, inclusion–preclusion, opposing ribozyme activities (endonuclease–ligase), basics of cell biology, and code emergence out of interactions. The red arrow flow represents all the RNA-mediated regulations necessary in replication steps; the blue arrow represents all RNA-mediated regulations integrating two competing genctic parasites into the host genome ("addiction module"); and the yellow arrow designates all RNA-mediated interactions leading to the emergence of the genetic code. (From Villarreal 2015; with permission)

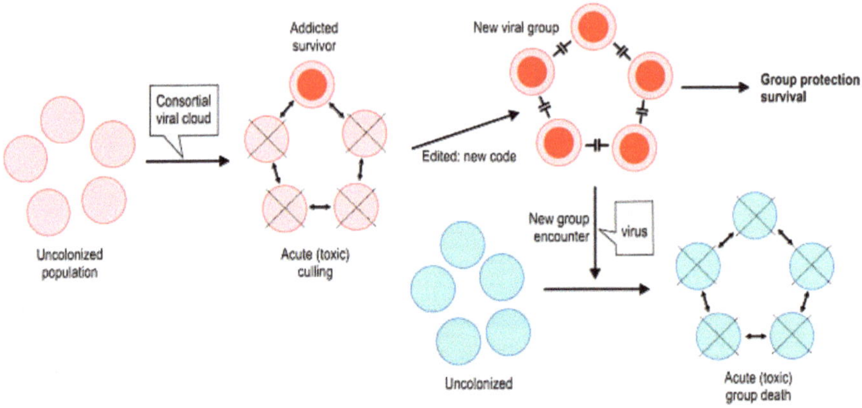

Fig. 20.2 Schematic of virus effects on population-based host survival. The five light red circles represent a host population free of the infectious virus in question. When exposed, many members will succumb to the toxic (acute) effects of virus infection (crossed light red circles). Some, however, may be stably colonized (shown with a dark red center). This host population has acquired a new virus-derived instruction set that also provides immunity to the same (and often other) viruses (shown by broken lines between cells). If this population retains some capacity to produce infectious virus, or if the virus remains prevalent, when it encounters another naive population (blue circles), the uncolonized population will crash due to virus toxicity. The virus-colonized population will be favored. (Reproduced from reference Villarreal 2012, with permission)

Without such a pair, the host cannot survive; thus, it becomes addicted after infection. On the other hand, it protects the host, compared to members of the same (host) population that do not possess this module. They are highly similar to restriction-modification systems and have been suggested as primeval identity codes (Prinz 2023).

Best known examples of addiction modules that can be found in bacteria are toxin/antitoxin and restriction/modification modules, yet there are also other insertion/deletion modules (Fig. 20.2) (Mruk and Kobayashi 2014). Even many other counter-regulated pathways may derive from such competitive viral groups with host equilibrium status, originally evolved for immune functions against related genetic parasites. This would explain the insertion and multiplication of introns (with their repetitive sequence grammar) into cellular genomes. Interestingly, in all known cases, the viral infection does not destroy or delete well-established protein-coding regions but almost exclusively inserts into regions with (repetitive) noncoding DNA (Brady et al. 2009). The repetitive sequence grammar of noncoding DNA is highly species-specific and more suitable for the determination of species than the coding sequences (Villarreal 2005). This means the repetitive sequence structure is more binding-prone to nonself repetitive RNA stem loops and additionally is more relevant to identify genetic content identity by foreign RNA stem loops for insertion. “Binding-prone” means single RNA sequences generally tend to bind to foreign single RNA sequences via complementary base pairing. For example, to find a complementary RNA sequence under foreign repetitive RNA sequences for TTTAAGGG is much more probable than to find one for TACGATGC.

The addiction module can become unstable again through different circumstances in which host organisms are involved, e.g., different types of stress. This may disturb the balance between toxin and antitoxin, leading to virulence that can prevail for some time, as demonstrated by various forms of symptoms of persistent human herpes virus, in which they change the (epigenetic) regulation of their persistent status by rewiring host chromosomes (Gonzalez-Almela et al. 2025). Addiction modules ensure the integration of new infection-derived properties into a host genome in a mostly stable persistence, which can become unstable again in certain life circumstances of the host (e.g., stress)

3 Viral (Genome) Editors

All empirical data show that viruses share natural genome editing competencies, acting as agents of genetic change and innovation. They share the capability to integrate, replicate, transcribe, translate, silence, and even repair genetic sequences. For those processes, they dock on or make use of the existing codes of those fundamental processes (see Prinz 2022, 2023). Viruses may insert, delete, and recombine genetic sequences within host organisms. Striking examples are the abundance of mobile genetic elements (transposons and retrotransposons) and the variety of "defectives" of former viral infection events that now serve as "effective" tools in host gene regulation and innovation, e.g., SINEs, LINEs, Alu's, LTRs, and non-LTRs (Stoddard and Belfort 2010; Jangam et al. 2017). Most importantly, they may generate RNA/DNA sequences de novo, i.e., they act as persistent agents in genetic innovation for host organisms. Many examples support this view, including viral involvement in the origins or modulation of: replicase, polymerase, integrase, DNA repair, restriction/modification, DNA methylation, the eukaryotic nucleus (bilayer nuclear envelope, division of transcription and translation, nuclear pores, linear chromosomes, and a tubulin system that enables separation of duplicated chromosomes), chitin and calcification, cartilage and bones, skin, dermal glands for poison, mucus and milk, larvae, eggs, placenta, neuronal arc proteins, and flowering plants, as well as innate and adaptive immune systems (Villarreal 2005; Bell 2020; Takemura 2020; Barreat and Katzourakis 2023). Especially the infection events of endogenous retroviruses that led to the integration of the syncytin genes and, therefore, to the evolution of the placenta and viviparous mammals were of major public interest because they led to an understanding of the lethal consequences if syncytin genes are not transcribed during pregnancy (Perot et al. 2012; Lavialle et al. 2013; Villarreal 2016a, b). But also, the signal-mediated interaction (communication) of neuronal synapses in animals, which evolved by the integration of arc proteins by retroviral infection events, is an impressive example of how natural genome editing led to innovation (Hantak et al. 2021; Sullivan et al. 2025). More recently, it has been found that giant viruses also integrate persistently into unicellular eukaryotes (Erazo-Garcia et al. 2025).

3.1 Natural Biological Codes: Regulations, Counter-Regulations, Instructions

The diversity and functional complexity of cellular life are possible through an abundant number of conserved regulations and instructions that ensure all cellular processes. We term this a "code" similar to a language. A natural code consists of adaptors that connect different domains of the living realm. Often, adaptors can be compared to signs that introduce directionality to processes in a way that entities from one domain convey their presence to the other domain through the adaptor. Due to this directionality, one can speak of an input and output domain. There are complex composite inputs, like spatiotemporal changes of solid, gaseous, or fluid patterns, mediated through sensory systems (e.g., tactile, electric) that are actual signals. Signs and signals in living coding systems emerge as instructions due to the arbitrariness of their interaction rules. We may find structural and regulatory codes (Prinz 2023). Structural codes are mainly involved in fundamental processes of life, like protein synthesis, transport, and metabolism—all processes that produce biomacromolecules, nutrients, or secondary metabolites—and also in cell signaling, body patterning, and development. Structural codes determine that a process happens. Regulatory codes are mainly involved in finetuning basic processes enabled by structural codes. They determine the when and how of those processes, i.e., the context sensitivity. The use of a natural code has a clear goal. When the goal is reached, feedback ensures a halt to the used code. The coherent line-up of these sub-regulations and their chronological sequence is conserved. These natural codes, which are essential to all cellular processes, appear in specific biological contexts within the organism. It is also important to note that many of them interact and communicate with each other, often sharing certain biological components. Overlapping communication pathways are shared by a number of real-life code users (cells, tissues, organs, organisms), such as signaling molecules. Depending on context, codes can be activated or not. The evolution of such natural codes is the history of many steps of genetic innovation, their biological selection, the conserved function (meaning), and their possible co-option and exaptation. This means several codes or parts of them may have evolved for another original reason, which later on got a new function within a new context. Their meaning may change through evolution depending on the context of cellular needs. As we have learned that cellular regulations, together with their counter-regulations, are most often derived from persistent viral infection events, the abundance of natural biological codes represents a glimpse of the role of viruses in the evolution of organismal complexity (Witzany 2017). Deciphering the viral impacts on the evolution of the diversity of natural biological codes is not part of this article but should be kept in evidence for further investigations.

3.2 Vintage or Retro? Old Elements for New Identities

Retroelements are mobile genetic elements found in various groups of eukaryotes that significantly shaped animal genomes (Nefedova and Kim 2022). Phylogenetic analysis indicates that the reverse transcriptase is derived from a very ancient retroelement. Retroviruses and their relatives, such as LTR retrotransposons, non-LTR retrotransposons, caulimoviruses, and hepadnaviruses, are called retroelements (Arkhipova and Yushenova 2023). They use tRNAs to initiate reverse transcription. Obviously, reverse transcriptase played a key role in the evolution of DNA out of an early RNA world because it is the only known competent agent that transfers single-stranded RNA into double-stranded DNA. Many derivatives from LTR retrotransposons play important roles in host gene regulation, DNA repair, and immunity, such as group II introns, LINEs, SINEs, and Ty1 (Chuong et al. 2016; González-Delgado et al. 2021; Hoyt et al. 2022; Carotti et al. 2023; Gebrie 2023). Retroelements are known to be effective mobile genetic elements that can move and replace themselves within host genomes and, therefore, are seen as important genetic content operators or, more recently, as natural genome editors (Edgell et al. 2011; Cordaux and Batzer 2009; Witzany 2016). They may change the genetic identity of the host in the long run, as documented in primate linages (Chen et al. 2025). A characteristic feature is the repetitive sequence structure, which indicates, as all repetitive sequence structures do, a descent from the early RNA world (Jurka et al. 2007). Repetitive sequence structures in host genomes are obviously all derived from infection events by genetic parasites (viruses and their relatives) and dominate genetic content order in all domains of life (Shapiro and Sternberg 2005; de Koning et al. 2011; Lambowitz and Zimmerly 2011; Villarreal and Witzany 2018). As demonstrated, the retroviral integration sites in the human genome of, e.g., only three different kinds of retroviruses are inconceivably numerous (Fig. 20.3).

3.3 Reversing the Central Dogma

With the detection of reverse transcriptase, the *central dogma* of molecular biology collapsed (Mattick 2009a; Shapiro 2009). Additionally, the self-splicing retrotransposons (group II introns) are one of the key drivers to generate multiple protein forms out of an identical gene mRNA by implementing the rules of the splicing code (Manet et al. 1989; Barash et al. 2010; Baralle and Baralle 2018; Bao et al. 2019; Capitanchik et al. 2025). The splicing process is able to generate a diversity of mature mRNAs out of the pre-mRNA, combining various exons. This was the end of the decades-long *one-gene one-protein* narrative.

Another key factor in generating multiple protein forms out of its original mRNA sequence is the editosome, which is a ribonucleoprotein complex (rnp) that edits RNA sequences by insertion/deletion of nucleotides in the mitochondrial mRNA. Interestingly, the editosome shares some similarities with proteins important in viral

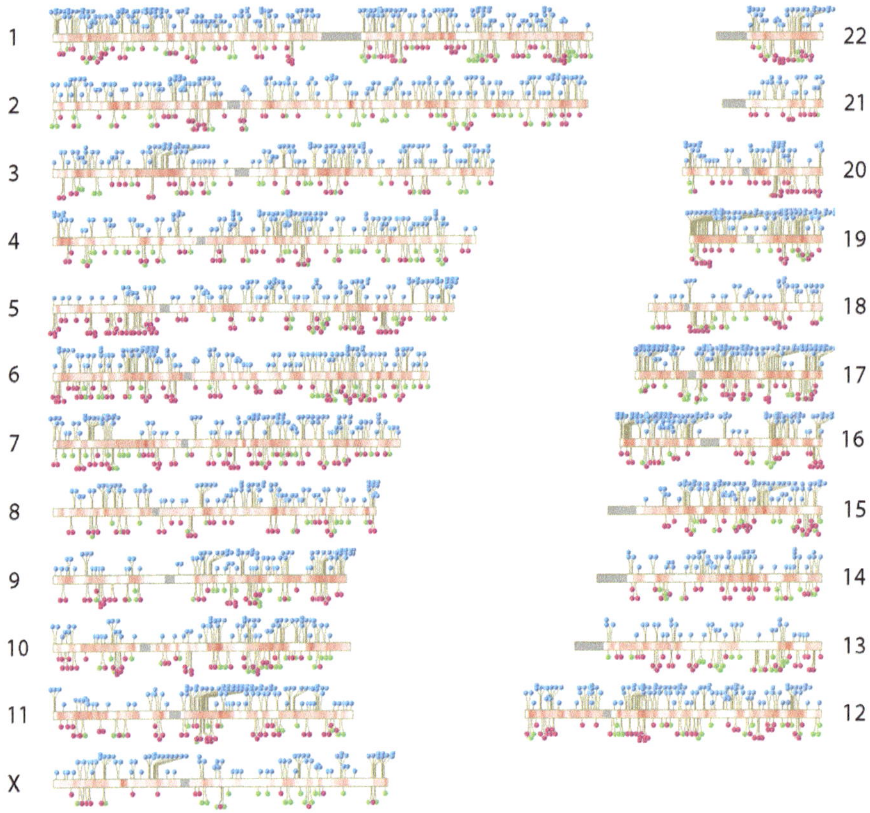

Fig. 20.3 Relationship between integration sites and transcriptional intensity in the human genome. The human chromosomes are shown numbered. HIV (human immunodeficiency virus) integration sites from all datasets in Table 1 are shown as blue "lollipops." MLV (murine leukemia virus) integration sites are shown in lavender, and ASLV (avian sarcoma-leukosis virus) integration sites are shown in green. Transcriptional activity is shown by the red shading on each of the chromosomes. Centromeres, which are mostly unsequenced, are shown as gray rectangles. (Mitchell et al. 2004; with permission)

replication and RNA binding (Witzany 2014a). This leads to the speculation that all these features imply a dramatic increase in the complexity of natural genome editing as a result of an escape strategy of reverse transcriptase, such as group II introns, out of the highly competitive early RNA world because storage of genetic information into DNA provides immunity against nearly all invasive RNA networks.

The coevolution of these retroelements started with their invasion (evolution) of eubacteria and archaea bacteria and diversification as splice-only elements in mitochondria and chloroplasts (Zimmerly and Semper 2015). The reverse transcriptase derivatives or fragments play multiple roles in the evolution of prokaryotic and eukaryotic domains (Tang et al. 2024). They significantly shaped animal genomes (Nefedova and Kim 2022). Dysregulation of mobile elements may cause disease, e.g., cancer. Therefore, organisms must organize an efficient immune response that regulates and confines the migration of mobile genetic elements. This is achieved by DNA methylation, which originally evolved to mark bacteriophage sequences to be differentiated from the bacterial host genome—a simple self vs. nonself discrimination—and later on was exapted for epigenetic imprinting. Retroelements finally ended as persistent relics of infection events that nowadays serve as important resources of noncoding RNAs being exapted and coopted by host organisms for RNA-mediated gene regulation in all cellular functions of transcription, translation, immunity, repair, and epigenetic (re)programming (Witzany 2009; Matylla-Kulinska et al. 2014; Majorek et al. 2014; Wilusz 2015; Moelling and Broecker 2015; Broecker and Moelling 2019; Sadeq et al. 2021, Handa et al. 2025).

4 The Whole Is More than the Sum of Its Parts

The natural genome editing competencies of viruses are also demonstrated in their capability of multiplicity reactivation (Bernstein et al. 2018; Sanjuán 2021). For example, Phycodnaviruses and Cyanophages are able to repair UV-light-induced damage to their genomes and can reproduce even if the host organisms are exposed to lethal doses of radiation. This feature depends on the ability to express functional subsets of genes that are retained even when each individual viral genome has been lethally damaged and on multiple independent viral infections of the same type. Once functional complementation of expressed genes is accomplished, viral replication and assembly can be initiated based on the necessary combinations of intact viral genome sequences. Therefore, viruses are the only living entities capable of meaningfully recombining text fragments of a damaged genome into a fully functional viral genome capable of self-replication (Villarreal 2005). This process is crucial to understanding the dynamics of common flu (Farrell et al. 2023) or the impact of viral infections on crops (Sánchez-Tovar et al. 2025). In other words, viruses appear as distributed systems, or roaming gangs, and the multiplicity of infection by different viruses increases this complex scenario even further and is also crucial during phage therapy (Pradhan et al. 2022; Nang et al. 2023).

5 Reading the Past, Writing the Future

Regulatory pathways being differentially active during development, pattern formation (morphology), cell fate and identity determination, (stress) physiology, genome regulation, immunity, and memory/learning depend on epigenetic coding (Gayathiri et al. 2025; Wagner 2025, Xiang et al. 2026). This means that certain patterns of protein modifications are written by distinct enzymes on DNA and/or histones and are read by adaptors that mediate certain effectors (Jenuwein and Allis 2001; Harbison et al. 2004; Turner 2000, 2007). Genetic sequences of all organisms in all domains of life can be marked according to their context-dependent (environmental and social) experiences (Shapiro 2014). RNA networks ensure such regulatory instructions and the (re)programming of these instructions (Mattick and Amaral 2022). With the capability of epigenetic memory, organisms can annotate historical and context-dependent experiences (Shapiro 2014). Evolution learned to learn (Mattick 2009b). Learning in this context means organisms can memorize (recall) certain, now standardized, molecular responses elicited by proximate surrogates for distant causes, like symptoms of stress (e.g., temperature, mechanical forces). The readers of such epigenetic memories contain domains (adaptors) that connect to certain actions or orchestrate multiple actions, sometimes referred to as (re)-programming. The arbitrariness in linking markings (memory) to context is flexible, and over multiple generations, new outcomes, new "interpretations" of old (epigenetic) annotations, can emerge, indicating adaptation.

The epigenetic code (Jeffers 2024; Marunde et al. 2024; Ueberheide et al. 2024) is one of the most complex codes known in molecular biology, and its origins may range back to the ancestral virosphere (Frías-Lasserre 2012; Li et al. 2014; Erwin and Richards 2024). Epigenetic regulation emerges as a fine-tuned genome-wide network that can rapidly remodel and reprogram genetic content. Interestingly, epigenetic switching outcompetes genetic mutations (error replications) during adaptation to changing lifeworlds (Stajic et al. 2022). Epigenetic markings can have both short-term and long-term functional effects, such as soma-to-germline inheritance (Korolenko and Skinner 2024). However, inheritance of acquired characteristics is only one of the many examples of the explanatory power of epigenetics. Behavioral epigenetics demonstrates the way in which environmental and social experiences produce individual differences in behavior, cognition, personality, and mental health (Moore 2017).

From the epigenetic perspective, phenotypic variation occurs due to the encoding of environmental parameters as epigenetic marks that can cover a range of values, i.e., individual differences (Gaggi et al. 2025). Phenotypic variants may better map old or quicker map new contexts and situations to appropriate behavior and avoid the conflicts or disadvantages of their ancestors' encoded behavioral programs and transfer this into the next generations. The adaptive benefit from this kind of environment-to-genome *interaction* has long been refuted but is now a well-accepted empirical fact (Skinner 2015). The epigenetic toolkit for such molecular marks dates back to viral infection events early in the history of life: DNA methylation evolved out of an immune function (restriction/modification) in akaryotes to mark the (nonself) identity of phages, which later on was exapted in eukaryotic life but was originally derived from addiction modules (Waldern et al. 2021). Interestingly, the histone modifications that early evolved in archaea most probably are exapted from viruses that encode histone-like proteins (Liu et al. 2021). Consequently, evolutionary impact on host genome evolution derives from natural genome editing by so-called genetic parasites on the one side and, on the other side, from the mapping of environmental influences and circumstances to gene expression patterns. Gene expression is regulated by complex transcription factor codes (Torcal Garcia and Graf 2021; Zrimec et al. 2021) that tightly interplay with epigenetic and genome regulatory codes (Rister and Desplan 2010; Basu et al. 2022). The (repetitive) sequences are derived from viruses, e.g., mobile genetic elements, and influence genomic plasticity and effective adaptation of the cellular host.

6 Completing the Puzzle

The role of viruses as literally creative evolutionary driving forces has only emerged in the post-genomic era, when their formative impact on genomes became clear (Koonin et al. 2006; Villarreal 2009a, 2011; Rasmussen and Katze 2016; Koonin and Makarova 2017; Wolf et al. 2018 Koonin and Krupovic 2022). While it was indirectly evident from their detrimental or destructive effects in terms of (lethal) diseases that shaped populations, they were never really considered architects or builders of genomes, engineers of immune systems, or similar. Whether virus-first, cell-first, or coevolutionary views on life's origin, evidence overwhelmingly points toward the crucial role viruses played.

The prevailing lifestyle of viruses is persistence in host organisms (lysogeny). Their traces left, or the sown seeds, comprise non-transcribed genomic DNA and transcribed noncoding RNAs that have been exapted by cells for regulatory needs in all steps and sub-steps of coordination of transcription, translation, repair, and immunity and ultimately germinated in epigenetic (re)programming. Importantly, viral systems can also significantly engage in the horizontal transfer of genetic information created by viruses (Lin et al. 2016; Athanasouli and Rödelsperger 2022; Muller et al. 2025). The various involved mobile elements, often referred to as the mobilome, are especially important for microbial communities to share and spread resistance genes (Saranya et al. 2025).

Cellular life is in constant exchange with competing viral groups, often called "clouds." In a joint effort with the host immune system, this can lead to an unstable equilibrium and integration of multiple competing sequence structures into the host within a single infection event, with serious consequences to host genetic identity. The competence of cellular life to adapt is ensured by both (i) natural genome editing competencies of persistent viral settlers and their defectives (noncoding RNA networks) and (ii) epigenetic imprinting, which serves as a crucial integration tool between environmental circumstances and cellular phenotypes. The underlying genomic sequences of viral origin were exapted for cellular needs, leading to the conclusion that there is no life (as we know it) without infection. Life is social and fueled by infectious events through the actions of viral consortia, organized "gangs" that cooperate and complement their competencies based on which exaptation enables cells to explore new niches and environments. Multicellularity, the cornerstone of eukaryotic social life, cannot be imagined without life's viral faculty.

The new millennium is also the age of the human (gut) microbiome (Kurilovich and Geva-Zatorsky 2025; Rubas et al. 2025). Its importance for health and disease is undeniable, and only recently has the human (gut) virome been acknowledged as an equally important and integrative part of the human holobiont (Liang and Bushman

2021; Bhagchandani et al. 2023; Ma et al. 2024). The many communications and (genetic) exchange processes happening in the various symbioses are the deep oceans of unexplored biological codes, and only the tip of the iceberg in this regard has been described, e.g., the BAFF code (Mackay and Schneider 2009), which may significantly interact with virus infections and, e.g., may influence transplantation outcome (Xu et al. 2014). The immune codes (Kim et al. 2018; Chi et al. 2020; Gray et al. 2025) are central to understanding self vs. nonself discrimination, not only in the context of infections or transplantations (Xu et al. 2014) but also in oncology and virus-induced cancer (Milavetz and Balakrishnan 2015; MacLennan and Marra 2023; Alsaadawe et al. 2025). The cancer-virus axis clearly shows that viruses play much more intricate roles in health and disease than the common flu or more aggressive viral infections.

7 New Metaphors for Old Viruses

Viruses were and are seen as warriors, or collectively as strong-armed forces, with cells and organisms as their enemies, respectively. If this were true, then viruses were winners (Cobián Güemes et al. 2016). However, this reflects an incomplete perspective. The term “genetic parasite” is also not appropriate, as it does not reflect the benefits most hosts gain from their presence, at least in the long run. They can also not be considered “selfish” anymore, as they convey important tools for the host. So, symbiosis may be the most appropriate among the commonly used terms (Villarreal 2016a, b).

The multiple sequence insertions into genomes are the seed for future functionalities that may be beneficial to both the viruses and their hosts (Koonin et al. 2006). In a way, like *farmers* seed crops, viruses seed (gene) sequences that germinate in later generations of tools—nucleic acids or proteins (Jiang et al. 2025). The case of multiplicity of infection (Bhat et al. 2022; Budzyńska et al. 2022) is comparable to a gang of viruses occupying a house (cell) wherein they chose different chromosomes (rooms) as their place to be. They become actual *squatters* that, over time, convert to a *flat-sharing community* and are indistinguishable from their host as they take over (or are brought in) essential maintenance tasks. They turned into *facility managers*, if you will. Originally, this might have appeared as an act of hostile takeover or break-ins, but today it has grown into a mutual interdependency. Many viruses are “polite” cellular guests, and unlike some human thieves who leave a house by climbing out of the window, most viruses follow cellular exit codes by coordinately leaving a cell through well-defined pathways (Altan-Bonnet and Panigrahi 2025; Bołoz et al. 2025; de Aquino et al. 2025). One may also think of viruses as *visitors* that leave or decide to stay and become *inhabitants* (Ottmann 2022; Villarreal and Witzany 2013). Clearly, the observable behavior and group dynamics of virus-host systems resemble *social* life (Dolgin 2019; Villarreal and Witzany 2021).

Viruses call for comparisons with computer viruses. Whereas computer viruses need skilled programmers, biological viruses are actual *programmers* themselves.

This is literally true, as they can create (write) new codes and (re)program cellular processes (Dupré and Guttinger 2016). Whereas computer programs, including viruses and malware, follow standardized and formal programming language, viral *gangs(ters)* develop their own slang or language (Simón et al. 2021; Ventoso 2025). That "slang" offers means to track the identity of viral vs. host proteins and to follow evolutionary trajectories (Ofer and Linial 2025). In a way, viruses are a faculty of life, as language is a faculty of the mind (Pinker and Jackendoff 2005; Pietroski and Crain 2012). So, whereas computer viruses are a kind of text (instruction) based on a formal language, viruses are *authors* posting nonformal texts on social media (cells).

Paleovirology has uncarved viruses captured in time capsules, i.e., conserved dead bodies, and helped decipher the history of viral infections (Barreat and Katzourakis 2022; Emerman and Malik 2010; Patel et al. 2011). At the same time, such actual excavations, complemented by extensive sequence analysis, prove the long, yet dynamic, history of viral strains with their hosts. More importantly, it turns out that host genomes are actual time capsules in their own right, as they contain the aforementioned viral sequences. Drawing on such findings, viruses have been proposed as *archaeological tools* and shortcuts to detect pathways and origins of cellular function (Ariza-Mateos et al. 2023; Ariza-Mateos et al. 2024).

We like to conclude that it's time to change our opinion about viruses and sharpen our perception regarding (meta)genomics, epigenomics, and the true origin of biological codes and novelty. It's a conserved change right in front of us, and it's not only a metaphor!

8 Outlook

Turning up and down scientific paradigms has a long tradition with often decades- or even centuries-lasting consequences (Kuhn 1962). Although Ernst Mayr tried to refute the fact that such paradigmatic changes happen also in biology, today there is no doubt about such fundamental changes in biology (Mayr 1982; Gilbert 1991). Some of these "turns" are explicitly described in the history of science by Thomas Kuhn. For biology, such turns have not been explicitly described, so we claim here biology's first turn: the viral turn.

The viral turn is of comparable importance for biology, as are the linguistic turn and the pragmatic turn for philosophy of science discourse between 1920 and 1980 and the central question: What is an exact natural science, and what language is the appropriate one to describe observations and construct theories about how to interpret what has been described? This was a crucial step out of the confusion derived from the former conviction; scientific sentences could depict reality in a 1:1 manner to the fact that social groups of researchers formulate commonly shared convictions of how to interpret empirical data according to the present consensus of the research community (Witzany 2014b). The viral turn highlights the importance of viral toolkits in actively shaping the evolutionary trajectory of life on earth and, in addition to their destructive side, acknowledging the constructive side. Viruses are entities that transcend gradualistic, i.e., neo-Darwinian, views on evolution, as they are able first to introduce large-scale genomic shuffling and, second, independently create de novo sequences. Those large-scale rearrangements and confined sequence randomness generate the seed for real novelties, both in terms of DNA and/or RNA structures, as well as regarding proteins, like reverse transcriptase, or epigenetic writers and readers, and the codes associated with those systems (Villarreal 2005; Nasir et al. 2012).

The question of who wrote the book of life is inextricably linked with the question of who reads the book of life. The epigenetic code, as far as derived from viral activity, clearly answers these questions (Lieberman 2016; Balakrishnan and Milavetz 2017; Tsai and Cullen 2020; Rehman et al. 2023). In light of the epigenetic code and its terminology of readers and writers, we conclude that the continuum of viruses, from small naked single-stranded RNA to large capsulated double-stranded DNA viruses, comprises the readers and writers of the book of life. At the same time, they produce the paper (DNA and RNA) on which these epigenetic texts are written.

The new millennium begins to fully embrace viruses as what they (also) are: viruses are "more friends than foes," as Karin Mölling points out (Moelling 2020), and are beneficial in abundant respects for their host (Roossinck 2011). Mostly unnoticed, viruses and virus-derived systems are used for (epi)genome engineering in various biotechnological applications (Tang and Sternberg 2023), making use of their very original capabilities. This really highlights the positive power of harnessing viral creativity. An *Integrative Theory of Evolution* has to integrate that.

References

Alsaadawe M, Radman BA, Hu L et al (2025) From viral infection to malignancy: the dual threat of EBV and COVID-19 in cancer development. Viruses 17:1195

Altan-Bonnet N, Panigrahi M (2025) Nonlytic egress and transmission in the virus world. Annu Rev Biochem 94:531–560

Anthony SJ, Epstein JH, Murray KA et al (2013) A strategy to estimate unknown viral diversity in mammals. mBio 4(5)

Ariza-Mateos A, Briones C, Perales C et al (2023) Viruses as archaeological tools for uncovering ancient molecular relationships. Ann N Y Acad Sci 1529:3–13

Ariza-Mateos A, Briones C, Perales C et al (2024) Archaeological approaches to RNA virus evolution. J Physiol 602(602):2469–2478

Arkhipova IR, Yushenova IA (2023) To be mobile or not: the variety of reverse transcriptases and their recruitment by host genomes. Biochemistry (Moscow) 88:1754–1762

Athanasouli M, Rödelsperger C (2022) Analysis of repeat elements in the Pristionchus pacificus genome reveals an ancient invasion by horizontally transferred transposons. BMC Genomics 23:523

Balakrishnan L, Milavetz B (2017) Epigenetic regulation of viral biological processes. Viruses 9: 346

Bao S, Moakley DF, Zhang C (2019) The splicing code goes deep. Cell 176:414–416

Baralle M, Baralle FE (2018) The splicing code. Biosystems 164:39–48

Barash Y, Calarco JA, Gao W et al (2010) Deciphering the splicing code. Nature 465:53–59

Barreat JGN, Katzourakis A (2022) Paleovirology of the DNA viruses of eukaryotes. Trends Microbiol 30:281–292
Barreat JGN, Katzourakis A (2023) A billion years arms-race between viruses, virophages, and eukaryotes. Elife 12:RP86617
Basu A, Bobrovnikov DG, Cieza B et al (2022) Deciphering the mechanical code of the genome and epigenome. Nat Struct Mol Biol 29:1178–1187
Bell PJL (2020) Evidence supporting a viral origin of the eukaryotic nucleus. Virus Res 289:198168
Benzi M, Novarese M (2022) Metaphors we Lie by: our 'War' against COVID-19. Hist Philos Life Sci 44:18
Bernstein H, Bernstein C, Michod RE (2018) Sex in microbial pathogens. Infect, Genetics Evol 57: 8–25
Bhagchandani T, Nikita VA et al (2023) Exploring the human virome: composition, dynamics, and implications for health and disease. Curr Microbiol 81:16
Bhat T, Cao A, Yin J (2022) Virus-like particles: measures and biological functions. Viruses 14:383
Bołoz A, Lannoy V, Olszak T, Drulis-Kawa Z, Augustyniak D (2025) Interplay between bacterial extracellular vesicles and phages: receptors, mechanisms, and implications. Viruses 17:1180
Brady T, Lee YN, Ronen K, Malani N, Berry CC, Bieniasz PD, Bushman FD (2009) Integration target site selection by a resurrected human endogenous retrovirus. Genes Dev 23:633–642
Briones C, Stich M, Manrubia SC (2009) The dawn of the RNA World: toward functional complexity through ligation of random RNA oligomers. RNA 15:743–749
Broecker F (2021) Genome evolution from random ligation of RNAs of autocatalytic sets. Int J Mol Sci 22:13526
Broecker F, Moelling K (2019) Evolution of immune systems from viruses and transposable elements. Front Microbiol 10:51
Budzyńska D, Zwart MP, Hasiów-Jaroszewska B (2022) Defective RNA particles of plant viruses-origin, structure and role in pathogenesis. Viruses 14:2814
Caetano-Anollés G (2024) Are viruses taxonomic units? A protein domain and loop-centric phylogenomic assessment. Viruses 16:1061
Capitanchik C, Wilkins OG, Wagner N et al (2025) From computational models of the splicing code to regulatory mechanisms and therapeutic implications. Nat Rev Genet 26:171–190
Carotti E, Carducci F, Barucca M et al (2023) Transposable elements: epigenetic silencing mechanisms or modulating tools for vertebrate adaptations? Two sides of the same coin. Int J Mol Sci 24:11591
Chen X, Zhang Z, Yan Y et al (2025) A phylogenetic approach uncovers cryptic endogenous retrovirus subfamilies in the primate lineage. Sci Adv. 11(29):eads9164
Chi X, Li Y, Qiu X (2020) V(D)J recombination, somatic hypermutation and class switch recombination of immunoglobulins: mechanism and regulation. Immunology 160:233–247
Chuong EB, Elde NC, Feschotte C (2016) Regulatory evolution of innate immunity through co-option of endogenous retroviruses. Science 351:1083–1087
Cobián Güemes AG, Youle M, Cantú VA et al (2016) Viruses as winners in the game of life. Ann Rev Virol 3:197–214
Cordaux R, Batzer MA (2009) The impact of retrotransposons on human genome evolution. Nat Rev Genet 10:691–703
Dawkins R (2016) The selfish gene, 4th edn. Oxford University Press, Oxford
de Koning AP, Gu W, Castoe TA et al (2011) Repetitive elements may comprise over two-thirds of the human genome. PLoS Genet 7(12):e1002384
de Aquino ILM, Azevedo BL, Arias NEC et al (2025) The final cut: how giant viruses of protists are released from their hosts' cells. Arch Virol 170:77
de Farias ST, Rêgo TG, José MV (2016) tRNA core hypothesis for the transition from the RNA world to the ribonucleoprotein world. Life (Basel) 6:15
Dolgin E (2019) The secret social lives of viruses [News Feature]. Nature 570:290–292
Domingo E, Witzany G (2024) Quasispecies productivity. Naturwissenschaften. 111:11

Dupré J, Guttinger S (2016) Viruses as living processes. Stud Hist Phil Biol Biomed Sci 59:109–116

Edgell DR, Chalamcharla VR, Belfort M (2011) Learning to live together: mutualism between self-splicing introns and their hosts. BMC Biol 9:22

Emerman M, Malik HS (2010) Paleovirology – modern consequences of ancient viruses. PLOS Biol 8(2):e1000301

Erazo-Garcia MP, Sheyn U, Barth ZK et al (2025) Cryptic infection of a giant virus in a unicellular green alga. Science 388(6748):eads6303

Erwin NAT, Richards TA (2024) Self-assembling viral histones are evolutionary intermediates between archaeal and eukaryotic nucleosomes. Nat Microbiol 9:1713–1724

Farrell A, Phan T, Brooke CB, Koelle K, Ke R (2023) Semi-infectious particles contribute substantially to influenza virus within-host dynamics when infection is dominated by spatial structure. Virus Evol 9(1):vead020

Fermin G (2018) Host range, host–virus interactions, and virus transmission. In: Tennant P, Fermin G, Foster JE (eds) Viruses. Academic, pp 101–134

Forterre P (2010) Defining life: the virus viewpoint. Orig Life Evol Biosph 40:151–160

Forterre P, Prangishvili D (2009) The great billion-year war between ribosome- and capsid-encoding organisms (cells and viruses) as the major source of evolutionary novelties. Ann N Y Acad Sci 1178:65–77

Frías-Lasserre D (2012) Non coding RNAs and viruses in the framework of the phylogeny of the genes, epigenesis and heredity. Int J Mol Sci 13:477–490

Gaggi G, Hausman C, Cho S et al (2025) LncRNAs ride the storm of epigenetic marks. Genes (Basel) 16:313

García-Ordóñez L, Pagán I (2024) Vertical and horizontal transmission of plant viruses: two extremes of a continuum? Viruses 2:18

Gayathiri E, Viswanathan D, Prakash P et al (2025) A critical review on epigenetics and epigenomics in plant development and stress resilience. Braz J Bot 48:51

Gebrie A (2023) Transposable elements as essential elements in the control of gene expression. Mob DNA 14:9

Gilbert W (1991) Towards a paradigm shift in biology. Nature. 349(6305):99

González-Delgado A, Mestre MR, Martínez-Abarca F, Toro N (2021) Prokaryotic reverse transcriptases: from retroelements to specialized defense systems. FEMS Microbiol Rev 45(6):fuab025

González-Almela E, Castells-Garcia A, Le Dily F et al (2025) Herpes simplex virus type 1 reshapes host chromatin architecture via transcription machinery hijacking. Nat Commun 16:5313

Gorbalenya AE, Krupovic M, Mushegian A et al (2020) The new scope of virus taxonomy: partitioning the virosphere into 15 hierarchical ranks. Nat Microbiol 5:668–674

Gortner RA (1938) Viruses—living or non-living? Science 87:529–530

Gray GI, Chukwuma PC, Eldaly B et al (2025) The evolving T cell receptor recognition code: the rules are more like guidelines. Immunol Rev 329(1):e13439

Gwiazda S, Salomon K, Appel B et al (2012) RNA self-ligation: from oligonucleotides to full length ribozymes. Biochimie 94:1457–1463

Handa S, Biswas T, Chakraborty J et al (2025) RNA control of reverse transcription in a diversity-generating retroelement. Nature. 638(8052):1122–1129

Hantak MP, Einstein J, Kearns RB et al (2021) Intercellular communication in the nervous system goes viral. Trends Neurosci 44:248–259

Harbison CT, Gordon DB, Lee TI et al (2004) Transcriptional regulatory code of a eukaryotic genome. Nature 431:99–104

Hayden EJ, Lehman N (2006) Self-assembly of a group I intron from inactive oligonucleotide fragments. Chem Biol 13:909–918

Higgs PG, Lehman N (2015) The RNA World: molecular cooperation at the origins of life. Nat Rev Genet 16:7–17

Holland JJ, de La Torre JC, Steinhauer DA (1992) RNA virus populations as quasispecies. Curr Top Microbiol Immunol 176:1–20
Hoyt SJ, Storer JM, Hartley GA et al (2022) From telomere to telomere: the transcriptional and epigenetic state of human repeat elements. Science 376:eabk3112
Hughes SH, Coffin JM (2016) What integration sites tell us about HIV persistence. Cell Host Microbe 19:588–598
Hwang J, Jurado KA, Fikrig E (2016) Genetics of war and truce between mosquitos and emerging viruses. Cell Host Microbe 19:583–587
Jangam D, Feschotte C, Betrán E (2017) Transposable element domestication as an adaptation to evolutionary conflicts. Trends Genet 33:817–831
Jeffers V (2024) Histone code: a common language and multiple dialects to meet the different developmental requirements of apicomplexan parasites. Curr Opin Microbiol 79:102472
Jenuwein T, Allis CD (2001) Translating the histone code. Science 293(5532):1074–1080
Jiang R, Zhou J, Liu Y et al (2025) Endogenous retroviruses in host-virus coevolution: from genomic domestication to functional innovation. Genes (Basel) 16:964
Jurka J, Kapitonov VV, Kohany O et al (2007) Repetitive sequences in complex genomes: structure and evolution. Annu Rev Genomics Hum Genet 8:241–259
Kim M-S, Chuenchor W, Chen X et al (2018) Cracking the DNA code for V(D)J recombination. Molecular Cell 70(2):358–370.e354
Koonin EV, Krupovic M (2022) A life LINE for large viruses. Elife 11:e83488
Koonin EV, Makarova KS (2017) Mobile genetic elements and evolution of CRISPR-Cas systems: all the way there and back. Genome Biol Evol 9:2812–2825
Koonin EV, Senkevich TG, Dolja VV (2006) The ancient Virus World and evolution of cells. Biol Direct 1:29
Koonin EV, Starokadomskyy P (2016) Are viruses alive? The replicator paradigm sheds decisive light on an old but misguided question. Stud Hist Philos Biol Biomed Sci 59:125–134
Koonin EV, Wolf YI (2012) Evolution of microbes and viruses: a paradigm shift in evolutionary biology? Front Cell Infect Microbiol 2:119
Korolenko A, Skinner MK (2024) Generational stability of epigenetic transgenerational inheritance facilitates adaptation and evolution. Epigenetics. 19(1):2380929
Kuhn TS (1962) The structure of scientific revolutions. University of Chicago Press, Chicago
Kurilovich E, Geva-Zatorsky N (2025) Effects of bacteriophages on gut microbiome functionality. Gut Microbes 17:2481178
Lambowitz AM, Zimmerly S (2011) Group II introns: mobile ribozymes that invade DNA. Cold Spring Harb Perspect Biol 3(8):a003616
Lavialle C, Cornelis G, Dupressoir A et al (2013) Paleovirology of 'syncytins', retroviral env genes exapted for a role in placentation. Philos Trans R Soc Lond B Biol Sci 1626:20120507
Li R, Hayward SD (2011) The Ying-Yang of the virus-host interaction: control of the DNA damage response. Future Microbiol 6:379–383
Li S, Kong L, Yu X, Zheng Y (2014) Host–virus interactions: from the perspectives of epigenetics. Rev Med Virol 24:223–241
Liang G, Bushman FD (2021) The human virome: assembly, composition and host interactions. Nat Re Microbiol 19:514–527
Lieberman PM (2016) Epigenetics and genetics of viral latency. Cell Host Microbe 19:619–628
Lin X, Faridi N, Casola C (2016) An ancient transkingdom horizontal transfer of penelope-like retroelements from arthropods to conifers. Genome Biol Evol 8:1252–1266
Liu Y, Bisio H, Toner CM et al (2021) Virus-encoded histone doublets are essential and form nucleosome-like structures. Cell 184:4237–4250.e19
Louten J (2023) Virus structure and classification. In: Louten J (ed) Essential human virology, 2nd edn. Academic, pp 19–31
Ma Z, Zuo T, Frey N et al (2024) A systematic framework for understanding the microbiome in human health and disease: from basic principles to clinical translation. Sign Transd Targ Ther 9: 237

Mackay F, Schneider P (2009) Cracking the BAFF code. Nat Rev Immunol 9:491–502
MacLennan SA, Marra MA (2023) Oncogenic viruses and the epigenome: how viruses hijack epigenetic mechanisms to drive cancer. Int J Mol Sci 24:9543
Majorek KA, Dunin-Horkawicz S, Steczkiewicz K et al (2014) The RNase H-like superfamily: new members, comparative structural analysis and evolutionary classification. Nucleic Acids Res 42: 4160–4179
Manet E, Gruffat H, Trescol-Biemont MC et al (1989) Epstein-Barr virus bicistronic mRNAs generated by facultative splicing code for two transcriptional trans-activators. EMBO J 8:1819–1826
Manrubia SC, Briones C (2007) Modular evolution and increase of functional complexity in replicating RNA molecules. RNA 13:97–107
Marunde MR, Fuchs HA, Burg JM et al (2024) Nucleosome conformation dictates the histone code. Elife 13:e78866
Mattick JS (2009a) Deconstructing the dogma: a new view of the evolution and genetic programming of complex organisms. Ann N Y Acad Sci 1178:29–46
Mattick JS (2009b) Has evolution learnt how to learn? EMBO Rep. 10(7):665
Mattick JS, Gagen MJ (2001) The evolution of controlled multitasked gene networks: the role of introns and other noncoding RNAs in the development of complex organisms. Mol Biol Evol 18:1611–1630
Mattick JS, Amaral P (2022) RNA. The epicenter of genetic information. CRC Press, Boka Raton
Matylla-Kulinska K, Tafer H, Weiss A et al (2014) Functional repeat-derived RNAs often originate from retrotransposon-propagated ncRNAs. Wiley Interdiscip Rev RNA 5:591–600
Maxmen A (2017) A billion-year arms race against viruses shaped our evolution. Nature. https://doi.org/10.1038/nature.2017.22191
Mayr E (1982) The growth of biological thought. Diversity, evolution, and inheritance. The Belknap Press of Harvard University, Cambridge, MA/London
Mercer TR, Mattick JS (2013) Structure and function of long noncoding RNAs in epigenetic regulation. Nat Struct Mol Biol 20:300–307
Milavetz BI, Balakrishnan L (2015) Viral epigenetics. In: Verma M (ed) Cancer epigenetics: risk assessment, diagnosis, treatment, and prognosis. Springer, New York, pp 569–596
Mitchell RS, Beitzel BF, Schroder AR et al (2004) Retroviral DNA integration: ASLV, HIV, and MLV show distinct target site preferences. PLoS Biol. 2(8):E234
Moelling K (2020) Viruses more friends than foes. Electroanalysis 32:669–673
Moelling K, Broecker F (2015) The reverse transcriptase-RNase H: from viruses to antiviral defense. Ann N Y Acad Sci 1341:126–135
Moore DS (2017) Behavioral epigenetics. Wiley Interdiscip Rev Syst Biol Med. 9(1)
Morse SA, Meyer RF (2017) Viruses and bioterrorism. Reference module in life sciences. 2017: B978-0-12-809633-8.11007-6
Mruk I, Kobayashi I (2014) To be or not to be: regulation of restriction-modification systems and other toxin-antitoxin systems. Nucleic Acids Res 42:70–86
Muller H, Savisaar R, Peccoud J et al (2025) Phylogenetic relatedness rather than aquatic habitat fosters horizontal transfer of transposable elements in animals. Genome Res 35:2011–2022
Mushegian AR (2020) Are there 10(31) virus particles on Earth, or more, or fewer? J Bacteriol 202(9)
Nang SC, Lin YW, Petrovic Fabijan A et al (2023) Pharmacokinetics/pharmacodynamics of phage therapy: a major hurdle to clinical translation. Clin Microbiol Infect 29:702–709
Nasir A, Kim KM, Caetano-Anollés G (2012) Viral evolution: primordial cellular origins and late adaptation to parasitism. Mob Genet Elements 2:247–252
Nefedova LN, Kim AI (2022) The role of retroelements in the evolution of animal genomes. Biol Bull Rev 12:29–40
Ofer D, Linial M (2025) Protein language models expose viral immune mimicry. Viruses 17:1199
Ottmann M (2022) These viruses that inhabit and visit us: the human virome. Med Sci (Paris) 38: 1028–1038

Patel MR, Emerman M, Malik HS (2011) Paleovirology—ghosts and gifts of viruses past. Curr Opin Virol 1:304–309
Pérot P, Bolze PA, Mallet F (2012) From viruses to genes: syncytins. In: Witzany G (ed) Viruses: essential agents of life. Springer, Dordrecht, pp 325–351
Petrov AS, Gulen B, Norris AM et al (2015) History of the ribosome and the origin of translation. Proc Natl Acad Sci U S A 112:15396–15401
Pietroski PM, Crain S (2012) The language faculty. In: Margolis E, Samuels R, Stich SP (eds) The Oxford handbook of philosophy of cognitive science. Oxford University Press
Pinker S, Jackendoff R (2005) The faculty of language: what's special about it? Cognition 95:201–236
Pradhan S, Varsani A, Leff C et al (2022) Viral aggregation: the knowns and unknowns. Viruses 14: 438
Prinz R (2022) Code biology database – a list of biological codes. International Society of Code Biology. Retrieved 16 November from http://www.codebiology.org/database.pdf
Prinz R (2023) Biological codes: a field guide for code hunters. Biol Theory 19:120–136. https://doi.org/10.1007/s13752-023-00444-2
Rasmussen AL, Katze MG (2016) Genomic signatures of emerging viruses: a new era of systems epidemiology. Cell Host Microbe 19:611–618
Rehman UU, Ghafoor D, Ullah A et al (2023) Epigenetics regulation during virus-host interaction and their effects on the virus and host cell. Microb Pathogen 182:106271
Rohwer F, Youle M, Maughan H et al (2014) Life in our phage world. A centennial field guide to the Earth's most diverse inhabitants. Wholon, San Diego
Rister J, Desplan C (2010) Deciphering the genome's regulatory code: the many languages of DNA. Bioessays 32:381–384
Root-Bernstein R, Root-Bernstein M (2016) The ribosome as a missing link in prebiotic evolution II: ribosomes encode ribosomal proteins that bind to common regions of their own mRNAs and rRNAs. J Theor Biol 397:115–127
Roossinck MJ (2011) The good viruses: viral mutualistic symbioses. Nat Rev Microbiol 9:99–108
Rubas NC, Torres A, Maunakea AK (2025) The gut microbiome and epigenomic reprogramming: mechanisms, interactions, and implications for human health and disease. Int J Mol Sci 26:8658
Ryan F (2009) Virolution – the most important evolutionary book since Dawkins' Selfish Gene. Collins, London
Sadeq S, Al-Hashimi S, Cusack CM et al (2021) Endogenous double-stranded RNA. noncoding. RNA. 7(1):15
Sánchez-Tovar MR, Rivera-Bustamante RF, Saavedra-Trejo DL, Guevara-González RG, Torres-Pacheco I (2025) Mixed plant viral infections: complementation, interference and their effects, a review. Agronomy 15:620
Sanjuán R (2021) The social life of viruses. Ann Rev Virol 8:183–199
Saranya SV, Prathiviraj R, Chellapandi P (2025) Mobilome-mediated speciation: genomic insights into horizontal gene transfer in methanosarcina. J Basic Microbiol 65(7):e70013
Schnepf J, Christmann U (2022) "It's a war! It's a battle! It's a fight!": do militaristic metaphors increase people's threat perceptions and support for COVID-19 policies? Int J Psychol 57:107–126
Schulz F, Abergel C, Woyke T (2022) Giant virus biology and diversity in the era of genome-resolved metagenomics. Nat Rev Microbiol 20:721–736
Segredo-Otero E, Sanjuán R (2022) Cooperative virus-virus interactions: an evolutionary perspective. Biodes Res 9819272
Shapiro JA (2009) Revisiting the central dogma in the 21st century. Ann N Y Acad Sci 1178:6–28
Shapiro JA, von Sternberg R (2005) Why repetitive DNA is essential to genome function. Biol Rev Camb Philos Soc 80:227–250
Shapiro JA (2014) Epigenetic control of mobile DNA as an interface between experience and genome change. Front Genet 5:87

Simón D, Cristina J, Musto H (2021) Nucleotide composition and codon usage across viruses and their respective hosts [Original Research]. Front Microbiol 12:646300

Skinner MK (2015) Environmental epigenetics and a unified theory of the molecular aspects of evolution: a Neo-Lamarckian concept that facilitates Neo-Darwinian evolution. Genome Biol Evol 7:1296–1302

Sommers P, Chatterjee A, Varsani A et al (2021) Integrating viral metagenomics into an ecological framework. Ann Rev Virol 8:133–158

Stajic D, Bank C, Gordo I (2022) Adaptive potential of epigenetic switching during adaptation to fluctuating environments. Genome Biol Evol 14(5):evac065

Stewart AD, Logsdon JM, Kelley SE (2005) An empirical study of the evolution of virulence under both horizontal and vertical transmission. Evolution 59:730–739

Stoddard B, Belfort M (2010) Social networking between mobile introns and their host genes. Mol Microbiol 78:1–4

Sullivan KR, Ravens A, Walker AC, Shepherd JD (2025) Arc – a viral vector of memory and synaptic plasticity. Curr Opin Neurobiol 91:102979

Takemura M (2020) Medusavirus ancestor in a proto-eukaryotic cell: updating the hypothesis for the viral origin of the nucleus. Front Microbiol 11:571831

Tang S, Sternberg SH (2023) Genome editing with retroelements. Science 382:370–371

Tang S, Conte V, Zhang DJ et al (2024) De novo gene synthesis by an antiviral reverse transcriptase. Science. 386(6717):eadq0876

Torcal Garcia G, Graf T (2021) The transcription factor code: a beacon for histone methyltransferase docking. Trends Cell Biol 31:792–800

Tsai K, Cullen BR (2020) Epigenetic and epitranscriptomic regulation of viral replication. Nat Rev Microbiol 18:559–570

Turner BM (2000) Histone acetylation and an epigenetic code. Bioessays 22:836–845

Turner BM (2007) Defining an epigenetic code. Nat Cell Biol 9:2–6

Ueberheide BM, Mollah S, Garcia BA (2024) On the hunt for the histone code. Mol Cell Proteom 23(12)

Varanda C, Félix MDR, Campos MD et al (2021) An overview of the application of viruses to biotechnology. Viruses 13(10)

Vaidya N, Manapat M, Chen I et al (2012) Spontaneous network formation among cooperative RNA replicators. Nature 491:72–77

Vaidya N, Walker SI, Lehman N (2013) Recycling of informational units leads to selection of replicators in a prebiotic soup. Chem Biol 20:241–252

Ventoso I (2025) Codon usage bias in human RNA viruses and its impact on viral translation, fitness, and evolution. Viruses 17:1218

Vignuzzi M, Stone JK, Arnold JJ et al (2006) Quasi-species diversity determines pathogenesis through cooperative interactions in a viral population. Nature 439:344–348

Villarreal LP (2005) Viruses and the evolution of life. ASM Press, Washington

Villarreal LP (2009a) Persistence pays: how viruses promote host group survival. Curr Opin Microbiol 12:467–472

Villarreal LP (2009b) The source of self: genetic parasites and the origin of adaptive immunity. Ann N Y Acad Sci 1178:194–232

Villarreal LP (2011) Viral ancestors of antiviral systems. Viruses 3:1933–1958

Villarreal LP (2012) The addiction module as a social force. In: Witzany G (ed) Viruses: essential agents of life. Springer, Dordrecht, pp 107–145

Villarreal LP (2015) Force for ancient and recent life: viral and stem-loop RNA consortia promote life. Ann N Y Acad Sci 1341:25–34

Villarreal LP (2016a) Viruses and the placenta: the essential virus first view. APMIS 124:20–30

Villarreal LP (2016b) Persistent virus and addiction modules: an engine of symbiosis. Curr Opin Microbiol 31:70–79

Villarreal LP, Witzany G (2010) Viruses are essential agents within the roots and stem of the tree of life. J Theor Biol 262:698–710

Villarreal LP, Witzany G (2013) The DNA habitat and its RNA inhabitants: at the dawn of RNA sociology. Genomics Insights 6:1–12
Villarreal LP, Witzany G (2018) Editorial: genome invading RNA networks. Front Microbiol 9:581
Villarreal LP, Witzany G (2019) That is life: communicating RNA networks from viruses and cells in continuous interaction. Ann N Y Acad Sci 1447:5–20
Villarreal LP, Witzany G (2021) Social networking of quasi-species consortia drive virolution via persistence. AIMS Microbiol 7:138–162
Wagner W (2025) Epigenetic networks coordinate DNA methylation across the genome. Mol Ther 33(10)
Waldern JM, Smith D, Piazza CL et al (2021) Methylation of rRNA as a host defense against rampant group II intron retrotransposition. Mob DNA 12:9
Wallace MA, Wille M, Geoghegan J et al (2025) Making sense of the virome in light of evolution and ecology. Proc Biol Sci 292(2044):20250389
Wilusz JE (2015) Repetitive elements regulate circular RNA biogenesis. Mob Genet Elements 5:1–7
Witzany G (2009) Noncoding RNAs: persistent viral agents as modular tools for cellular needs. Ann N Y Acad Sci 1178:244–267
Witzany G (2014a) RNA sociology: group behavioral motifs of RNA consortia. Life (Basel) 4:800–818
Witzany G (2014b) Pragmatic turn in biology: from biological molecules to genetic content operators. World J Biol Chem 5:279–285
Witzany G (2016) Crucial steps to life: from chemical reactions to code using agents. Biosystems. 140:49–57
Witzany G (2017) Two genetic codes: repetitive syntax for active non-coding RNAs; non-repetitive syntax for the DNA archives. Commun Integr Biol 10(2):e1297352
Wolf YI, Kazlauskas D, Iranzo J et al (2018) Origins and evolution of the global RNA virome. mBio 9(6)
Xiang Y, Qian Y, Li Z et al (2026) Endogenous retroviruses synthesize heterologous chimeric RNAs to reinforce human early embryo development. Science. 391(6783):eadv5257
Xu H, Dong P, He X et al (2014) B-cell-activating factor code and human cytomegalovirus infection in renal transplant recipients. Microbiol Immunol 58:439–448
Zrimec J, Buric F, Kokina M et al (2021) Learning the regulatory code of gene expression. Front Mol Biosci 8:673363
Zimmerly S, Semper C (2015) Evolution of group II introns. Mobile DNA 6:7

Chapter 21
Viral Quasispecies as Models of Biological Complexity: New Findings with SARS-CoV-2

Esteban Domingo, Brenda Martínez-González, María Eugenia Soria, Ana I. de Ávila, and Celia Perales

Abstract Evolution has several narratives and formulations, and their integration is a pending challenge. RNA viruses, the most abundant subcellular replicative entities in the biosphere, are part of a widespread class of RNA genetic elements that evolve at rates that exceed by almost a million-fold the rates exhibited by their host organisms. Since they use the same tools for genome variation as cells, viruses have been highly informative of molecular and population dynamics routes for adaptation and persistence. Here, we review basic features of viral RNA evolution, including new observations obtained by application of ultra-deep sequencing to the severe acute respiratory syndrome coronavirus 2 (SARS-CoV-2) populations. Mutual interactions between complex arrays of variant genomes and host cells underlie a number of implications for viral pathogenesis and the control of viral disease. Genome variations, which are largely presided by stochastic events, can give rise to emergent phenotypic traits in what has been termed "quasispecies productivity." It relates to tinkering in evolutionary constructions, as expressed by F. Jacob 50 years ago. By virtue of their accelerated evolution, RNA viruses can help in the understanding of biological complexity as an input for evolutionary processes.

Keywords Viral dynamics · Mutation · Recombination · Genome segment reassortment · Mutant spectrum · Stochastic event · Accelerated evolution · Trait emergence · Viral disease

---

E. Domingo (✉) · A. I. de Ávila
Microbes in Health and Welfare Program, Centro de Biología Molecular Severo Ochoa (CSIC-UAM), Campus de Cantoblanco, Madrid, Spain
e-mail: edomingo@cbm.csic.es

B. Martínez-González · M. E. Soria · C. Perales (✉)
Microbes in Health and Welfare Program, Centro de Biología Molecular Severo Ochoa (CSIC-UAM), Campus de Cantoblanco, Madrid, Spain

Department of Clinical Microbiology, Instituto de Investigación Sanitaria-Fundación Jiménez Díaz University Hospital, Universidad Autónoma de Madrid (IIS-FJD, UAM), Madrid, Spain
e-mail: cperales@cbm.csic.es

G. Witzany (ed.), *Integrative Theory of Evolution - Aspects and Insights*,
https://doi.org/10.1007/978-3-032-24295-2_21

1 Introduction: Scope of the Quasispecies Concept and Its Scientific Impact

The study of the evolution of RNA viruses (those that have RNA as genetic material) has found support in the conceptual framework contributed by quasispecies theory, formulated by M Eigen and P Schuster 50 years ago. The term "quasispecies" refers to assemblies of closely related but nonidentical entities that replicate and transmit information. Assemblies can exhibit collective behavior, achieved through interactions among their component individuals. Quasispecies was initially developed as a theory of self-organization and adaptability of simple replicons in the origin of life (Eigen and Schuster 1979). The initial mathematical, deterministic formulation, which referred to mutant distributions in equilibrium, was extended to cover more realistic scenarios of genome replication. Relevant developments—that rendered the theory suited to RNA virus dynamics—have consisted in the inclusion of recombination and gene transfers as variation mechanisms, complementation, stochastic effects, and consideration of finite populations in complex fitness landscapes, away from equilibrium [as representative conceptual articles and books, see (Boerlijst et al. 1996; Domingo et al. 2001; Eigen 2013; Figlerowicz et al. 2003; Jacobi and Nordahl 2006; Mustonen and Lassig 2009; Nowak 1992; Park and Deem 2007; Saakian 2018; Saakian and Hu 2006, 2016; Sardanyés et al. 2024; Schuster 2016)].

In addition to the understanding of primeval life-generating mechanisms and evolutionary optimization, quasispecies has a continuing influence on other research areas such as replicator networks, RNA-RNA interactions, chromosomal instability, dynamics of cellular collectivities (bacteria, eukaryotic parasites, and tumor cells), viral pathogenesis, antiviral drug resistance, the error threshold for maintenance of genetic information and its application to antiviral designs, or conformation heterogeneity of prions [as reviews of the scope of quasispecies influence, see (Domingo et al. 2021b; Domingo and Schuster 2016; Sardanyés et al. 2024)]. Viral quasispecies fill a small (yet highly relevant) part of a wide conceptual space. The seemingly simplistic approach of viewing RNA virus dynamics through a theory of life's origin has provided an interpretation of the adaptive potential of viruses, which is pertinent to viral pathogenesis, even considering the presence of virulence factors (whose identification by deletion is often difficult to distinguish from fitness-decreasing genome modifications). Reflection on what it implies that viruses are mutant clouds has facilitated the interpretation of antiviral treatment failures and requirements for vaccine efficacy [i.e., inadequacy of monotherapies or peptide vaccines (Domingo, 1989; Domingo et al. 2025)]. It has also stimulated a reconsideration of disease control strategies, including the introduction of lethal mutagenesis (virus extinction by an excess of mutations evoked by mutagenic nucleotide analogues) as an antiviral approach (Sect. 5).

Regarding their biomedical impact, the concepts on population dynamics and adaptability that we discuss for RNA viruses also apply to DNA viruses whose replication is catalyzed by low-fidelity DNA polymerases.

In the present article, we describe some indeterminations in the molecular basis of RNA variations that are the substrate of evolution. We also consider the utility of RNA viruses as experimental models to approach biological complexity. Previous quantifications of population heterogeneity of other RNA viruses are compared with those recently obtained with SARS-CoV-2 for two main reasons: (i) because such results extend quasispecies implications to coronaviruses, despite their large (circa 30 Kb) genome, which encodes a proofreading-repair activity, and (ii) because adherence of quasispecies behavior to genomes with several regulatory regions and the capacity to encode around 27 different proteins reinforces the value of RNA viruses as laboratory models of biological complexity. We underline the quasispecies productivity (Domingo and Witzany 2024), meaning "creativity from errors," to move beyond an error replication narrative and to emphasize emergent properties within viral populations.

2 Multiple RNA Genome Variation Mechanisms: Participation of the Host Cells

RNA viruses exhibit mutation rates that average 10^{-4} mutations introduced per nucleotide copied, values that are 10^5- to 10^6-fold higher than the error rate displayed by cellular DNA during its replication in normal cells. Introduction of point mutations due to the limited copying fidelity of the replicative machinery is a universal mechanism of RNA virus variation. RNA genomes may undergo recombination (that can give rise to mosaic genomes consisting of RNA regions copied from different template molecules, as well as to genomes containing insertions, deletions, duplications, translocations, or inversions). The frequency of intra-molecular and inter-molecular recombination varies among viruses, largely depending on the processivity of the polymerase complex. Processivity is defined as the capacity of the complex to remain copying the same template string without detachment and subsequent binding to another template molecule to continue nucleotide polymerization. Segment reassortment [in viruses with a segmented genome, typically the human influenza viruses (IVs) and several plant viruses] is yet another source of genome variations.

Drastic genome alterations—in particular deletions, abundant in the case of SARS-CoV-2—often lead to defective genomes (DGs), which are defined as those that, in isolation, cannot complete an infectious cycle (meaning the steps that go from virus entry into the cells to the release to the extracellular environment of newly synthesized infectious particles). DGs can be maintained in viral populations by complementation, effected by *trans*-acting proteins expressed and supplied by other swarm members. Because they are not functional per se, DGs can accumulate genome variations at higher rates than their viable counterparts (similarly to cellular pseudo-genes versus functional genes). Therefore, they constitute a reservoir of deviant sequences that can be reintroduced into the viable pool of viral genomes by recombination.

Coexistence in the same evolving population of genomes with different lengths—with molecular mechanisms that render possible transitions among them—has led to the introduction of the *ultracube* concept (Sardanyés et al. 2024). It is an extension of the hypercube representation that serves to visualize movements in the sequence space of a genome with a given length (Swetina and Schuster 1982). Such movements can occur within a hypercube or between hypercubes, in the domain of the *ultracube*. This notion opens a new facet of quasispecies dynamics that embraces length transitions. It is an extension that can be framed within a needed integration of quasispecies with epidemiology and ecology in multilayer networks (Sardanyés et al. 2024).

In addition to variation mechanisms inherent to the replicative machinery (mutation and recombination) or to the fragmented genome structure (segment reassortment), point mutations can be introduced in viral RNA by cellular editing activities. The latter include those displayed by some members of the adenosine deaminases acting on double-stranded RNA (ADAR) (Zhang et al. 2025) or the apolipoprotein B mRNA editing complex (APOBEC) (Xu et al. 2020) protein families. Except when editing is overtly manifested in the form of biased hypermutation at expected sites (that is, the same mutation type is repeated to reach a frequency of around 10^{-2} mutations per nucleotide and affects sites that comply with the expected 3′ and 5′ nucleotide neighbors), the extent of the contribution of editing activities to mutations in viable RNA genomes is a subject of debate. This is particularly true in the interpretation of mutation types observed in the SARS-CoV-2 (Begum et al. 2024; de Ávila et al. 2025; Picardi et al. 2021; Pu et al. 2023; Ringlander et al. 2022). The virus-dependent and cell-dependent mechanisms of genome variation are mutually compatible, and their relative participation in the generation of viable virus variants depends on the virus and the host cells in ways that are not well understood. Other mutagenic factors, such as free radicals (that can result from metabolic oxidative stress or cosmic radiation), are thought to have a limited contribution to the mutational load of RNA genomes.

A critical element that promotes RNA genome variation is that the majority of RNA viral genomes do not encode a proofreading-repair protein, i.e., a 3′ to 5′ exonuclease that can excise mispaired nucleotides in the nascent progeny RNA during template copying. This absence has been evidenced by functional analyses and comparative structural studies with RNA-dependent RNA polymerases and DNA-dependent DNA polymerases (Bernad et al. 1989; Ferrer-Orta et al. 2009; Friedberg et al. 2006; Steinhauer et al. 1992). Despite the presence of repair and correction mechanisms in some RNA viruses, such as the ExoN of SARS-CoV-2, they do not prevent a remarkable mutational input in the corresponding viral genomes [reviewed in (Domingo et al. 2021a) (see Sect. 8)].

Host cells have a direct role in virus variation, and, in turn, variations in viruses may affect cellular functions in a two-way relationship (Castello-Sanjuan et al. 2025; Shaukat and Zhang 2025). Replication complexes often include host proteins as subunits, and editing activities are genuinely cellular. One of the results of variation, competition, and selection acting on viruses (the classic Darwinian principles) is an increase in their fitness, which is defined as the capacity to yield infectious progeny in a given environment. High fitness is a driver of virus survival and resilience, including the surmounting of antiviral interventions [reviewed in (Domingo et al. 2019)]. Interestingly, viral fitness is not a neutral trait regarding cellular epigenetic markers. Viruses can dysregulate cellular epigenetic mechanisms and gene expression profiles in a fitness-dependent manner (Cervera et al. 2018; Dandri 2020; Wang et al. 2019). High-fitness hepatitis C virus (HCV) decreased the levels of several epigenetic markers in human hepatoma cells; especially noteworthy was that the decrease of H4K20m3 (histone 4 tri-methylated at Lys 20)—which is a hallmark of cellular transformation—was significant only when the infecting HCV exhibited high replicative fitness (Garcia-Crespo et al. 2023). Viruses gain fitness upon multiple passages in the same host cell type, and once they have gained fitness, they may modify the expression pattern in those cells. There is still a poorly understood mutual influence between virus replicative fitness and cellular modifications that, in turn, give rise to a modified cellular environment for virus multiplication.

The plastic trait conglomerate displayed by RNA viruses and other RNA genetic elements led to the coining of the term "*RNA genetics*." It recapitulates the contrast between the dynamism of the present-day, subcellular replicative RNA world and that of the cellular DNA world, where the RNA genomes are necessarily immersed (Holland et al. 1982).

3 Hidden Information: Consensus Sequences and Mutant Spectra in the Assessment of Virus Adaptability

Consensus sequences of RNA virus isolates determined experimentally are a weighted average of the individual sequences that compose the virus sample. They are those generally deposited in data banks and serve as the major source of information for evolutionary studies and for preventive and therapeutic designs (particularly for the preparation of pan-genomic antiviral agents and universal vaccines). Current estimates suggest that an individual genome differs in 2–20 mutations from the other genomes of the same population. Only a minority of mutations are repeated in many genomes, and a population usually comprises (at least) thousands of genomes. As a consequence, an individual genomic molecule with a nucleotide sequence identical to that of the consensus may not even exist in the population. Moreover, residue (nucleotide and amino acid) conservation quantified with consensus sequences is far less strict than conservation quantified with mutant spectra (García-Crespo et al. 2020). Organized data banks with sequences from mutant spectra would represent an important impulse for medical virology.

Consensus sequences help in interpreting an environment-promoted selection of a virus variant once the selective episode has already taken place; they do not embody the information to explain the adaptation. Such information lies in the mutant cloud, which contains low-frequency genomes with a sequence that differs from that of the consensus and that rise toward dominance in the adaptation process. Such minority variants, either individually or collectively, may display higher fitness than the parental population when the ensemble meets a modified environment.

Viral populations can generally find multiple molecular pathways for fitness gain, unless population size limitations restrict the number of variants available to respond to selective forces (Domingo and Perales 2012; Escarmís et al. 1999; Jerzak et al. 2005; Moreno et al. 2017; Nijhuis et al. 2009; Novella et al. 1995a, b, 1999). Adaptive pathways expand subsets of minority genomes, replacing those that were previously dominant. Production of new variants does not cease under the blind molecular instruction represented by low-fidelity replication, together with other mechanisms of genome variation (Sect. 2). Not surprisingly, fitness gain pathways are staggering and discontinuous, as suggested by mutational waves identified in sequential samples of an evolving virus (Gallego et al. 2020). All adds to the unpredictable composition of mutant spectra and of the evolutionary trajectories they initiate.

4 Several Sources of Indetermination and Permanently Unfinished Trends

The genome variation mechanisms are based on stochastic events in their origin, and they are part of what has been termed the expanded realms of stochastic thinking in biology, which concerns evolutionary as well as biochemical processes (Domingo et al. 2023; Graham et al. 2017). Some viral genome features (homopolymeric tracts, short tandem repeats, and biased nucleotide abundances in template molecules) and cellular traits (nucleotide pool composition and ionic environment) may favor the occurrence of some mutation types over others. They may also promote polymerase detachment from template molecules that precedes binding to another template molecule (or to a different site of the same template molecule) in the process of recombination. Likewise, editing activities may exhibit preferences for the location of the RNA site that they alter. Despite these biases, overall, randomness plays a (probably prominent) role in variant generation. An example of a source of stochasticity is provided by the quantum-mechanical indeterminations behind the electronic fluctuations that ultimately give rise to point mutations (Fedeles et al. 2021).

Short-term functional competence of the mutant and recombinant molecules that emerge intracellularly (an early ingredient of viral fitness) acts as an immediate sieve for survival. Overcoming the timely sieve means the continuation of the involvement of the genomic molecules in the replication cycle inside the infected cell, as opposed to being excluded and entering degradation pathways. The available evidence suggests that only a few of the many generated point mutants, recombinants, and edited versions of viral RNAs overcome the immediate functional sieve (Alnaji et al.

2022; Bentley et al. 2021; Domingo et al. 2023; Lowry et al. 2014; Sztuba-Solinska et al. 2011), and studies are quoted in these references.

Less obvious is how the alternative packaging models that have been proposed for the eight IV genomic segments [random (Bancroft and Parslow 2002) and selective (Fujii et al. 2003)] may influence the extent of randomness in genome segment reassortment (Du et al. 2023). Single-cell analysis of IV-infected cells has shown nonrandom reassortment with few genotypes present at high frequency and many genotypes at low frequency (Chen et al. 2023). Variants are produced without anticipation of their fate; functionality is the test faced by randomness.

In addition to indeterminations that have a molecular origin, population dynamics itself is a source of uncertainties regarding evolutionary trajectories. Well-documented perturbations are viral population size expansions and reductions, the latter realized in the form of bottlenecks of varying intensity (number of genomes involved). The viral particles that are transported to another location (within the same or a different host individual) originate a new quasispecies, again under the lottery of which variants arise and overcome the early filter of molecular competence.

Each mutant spectrum of a virus isolate is unique, including the mutant spectra of sequential samples of the same evolving population [(Delgado et al. 2021, 2024; Martínez-González et al. 2025), among other studies]. A population equilibrium (defined as a steady-state distribution of mutant genomes) cannot be reached; the pursuit of equilibrium in viral populations should be regarded as a permanently unsatisfied tendency that promotes exploration of new adaptive pathways (García-Crespo et al. 2021). The molecular and population dynamics toolbox for novelty is remarkable, as stated by J. Lederberg at a time when our analytical capacities were more limited than those of today (Lederberg 1993). This incredible productivity and generation of new phenotypes have been distinguished from the simple error replication narrative (Domingo and Witzany 2024). New traits may result from complementation and cooperation among expression products of variant genomes within a population (Shirogane et al. 2016, 2023; Taye et al. 2024; Yousaf et al. 2023). Emergent phenotypic traits may be favored by block transmission of multiple particles within vesicles into susceptible cells (Kerviel et al. 2021). Again, it cannot be predicted how diverse are the individuals that travel within a vesicle.

5 Basis of the Biomedical Impact of Viral Quasispecies

Complex sets of genetically related genomes, not a population with a defined sequence, are the real entities that are subjected to a continuous process of variation, competition, and selection in infected hosts. The cell interactions exerted by a heterogeneous collection of viral particles explain the biomedical relevance of quasispecies dynamics in viruses. Each viral population has a certain probability of including a low-frequency variant type that interacts with the host in a way that differs from the interaction observed with the complete population or with the majority of members of the population (Fig. 21.1).

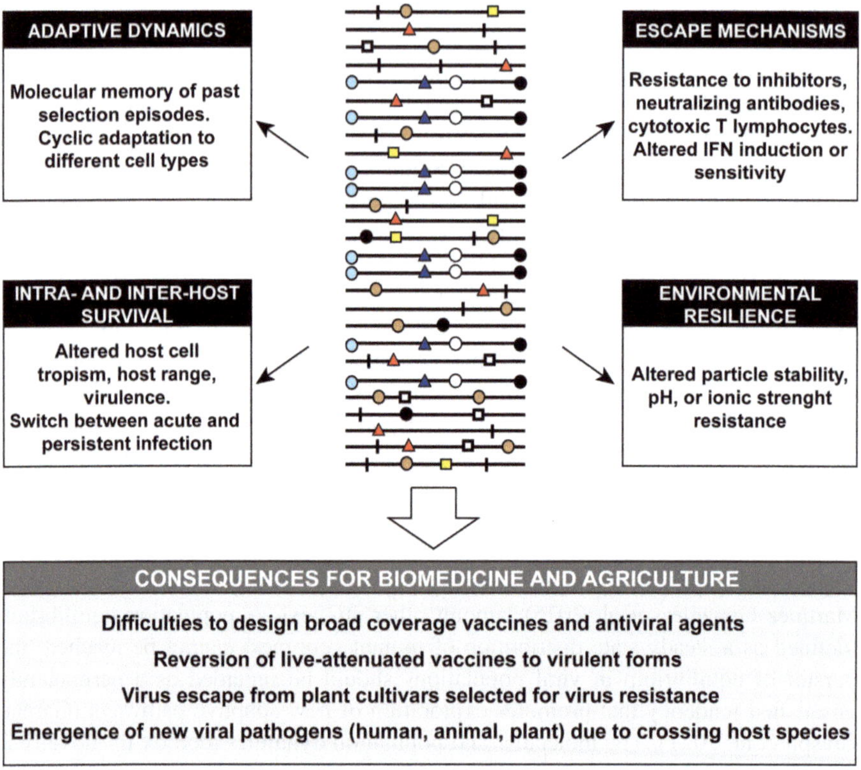

Fig. 21.1 The behavior of a viral quasispecies may differ from the behavior of its individual components when the latter are withdrawn from the ensemble and allowed to grow. The central genome distribution depicts a viral quasispecies frozen at a given time point in a replicative process. Individual genomes are represented as horizontal lines, and genomic lesions (often point mutations) are represented by various symbols. The four lateral boxes summarize phenotypic traits that, in many experimental studies (described in references quoted in the text), have been shown to vary between a complete population and its individual components. The main difference between the drawing and the real virus world is that viral populations often include many thousands of genomic molecules in a single infected cell, and such thousands of genomes are a trait reservoir. The bottom box summarizes some important consequences of viruses existing as mutant clouds or swarms

Examination of the traits that have been identified as hidden in quasispecies reveals that they are pertinent to adaptive dynamics, escape mechanisms, intra- and inter-host survival, or environmental resilience. A significant example is offered by minority variants that differ in cell tropism preferences (Baranowski et al. 2001, 2003). Such variants may favor invasion of specific organs (i.e., neuro-invasion), fuel ongoing systemic infections, or mediate the establishment of viral persistence, in addition to being a factor in the emergence and reemergence of viral disease (Domingo et al. 2025; Gonzalez Aparicio et al. 2022; Holland et al. 1982; Sardanyés et al. 2024). It should be underlined that medical interventions (vaccination, administration of antiviral agents or monoclonal antibodies, organ transplantation, and immunosuppressive treatments) represent environmental changes for the viruses that reside in the treated organism. Viruses respond to such constraints with variable success. Selection of antiviral- or monoclonal antibody-resistant mutants, in addition to viral divergence that affects B-cell and T-cell epitopes in viral proteins, contributes to vaccine or treatment ineffectiveness; an example is the withdrawal by the Food and Drug Administration (FDA) of the USA of *evusheld* due to its inefficacy against many SARS-CoV-2 variants (www.fda.gov/drugs). A single antiviral agent, a monoclonal antibody, or a peptide vaccine administered alone cannot control genetically variable viruses. Complexity cannot be combated with simplicity.

Quasispecies theory has furnished conceptual support to improve control strategies for viral diseases via two main manifestations: (i) the recognition of the need to implement combination antiviral therapies to suppress mutant swarms (a treatment mode that four decades ago was considered bad practice in standard pharmacology); and (ii) the introduction of lethal mutagenesis and synergistic lethal mutagenesis in antiviral therapy, in coherence with the error threshold concept included in quasispecies theory (Domingo 1989; Eigen 2002; García-Crespo et al. 2024; Perales et al. 2019, 2020). Although not universally practiced, the requirement of avoiding monotherapies to combat entities exhibiting quasispecies behavior (particularly RNA viruses and cancer) is increasingly accepted. Evolutionary arguments should influence treatment strategies whenever the medical target is characterized by heterogeneity and diversification potential.

6 Darwinian Principles Meet Complexity: Time-Space Coordinates of Evolutionary Events

Virus behavior in terms of variation, competition, and selection adheres to the Darwinian principles on which quasispecies theory was built (Eigen and Schuster 1979; Schuster and Stadler 2023). Yet, it is an adherence that deviates from a linear view of evolutionary events (sequential replacement of a variant form by another) for two reasons: (i) the continuous perturbation evoked by the incessant input of variations (see Sect. 2) and (ii) random events that impede viruses reaching a population equilibrium (see Sect. 4). New variant clouds enter the game of competition and selection when the existing clouds are in the process of gaining fitness. The latest clouds are, in turn, supervened by others, and so forth and so on. This is what we mean by Darwinian principles meeting complexity. The time frame in which such cloud dynamics take place in relation to the time needed for an environmental change to be sensed by the virus is an unsolved question.

While a valid argument is that the same conceptual quasispecies framework used for RNA viruses actually applies to any biological system, the time-space coordinates in which events develop distinguish viruses from differentiated organisms. For humans to be perceived as dynamic genome clouds over centuries—as we do with viruses during a pandemic or a laboratory experiment—the time frame between

Homo habilis (about 28 million years ago) and us should be compressed by a factor of 10^3. This elongated time interval for human variation (which is true also for many cellular organisms) has to be added to a lower proportion of points of the genomic space accessible by organisms, as compared with viruses [the genomic space is given by 4 (number of different nucleotides) to the power of the genome length]. The time-space coordinates of the replicative events, together with genome size, reinforce the unique implications of quasispecies dynamics for the behavior of simple, subcellular RNA genetic elements (Domingo 2020).

Complexity has at least two meanings in viral genetics. It can refer to the amount of information conveyed by a viral genome, which is estimated by nucleotide length, number of encoded proteins, and number of noncoding regulatory regions [for example, the SARS-CoV-2 genome (29.6 Kb) is more complex than the bacteriophage Qβ genome (4.2 Kb)]. Another meaning of complexity is the extent of intra-population diversity for any given genome length, and the interactions that may be established among individual components. Such interactions may give rise to emergent behavior, that is, a behavior that cannot be readily explained by the sum of actions of individual constituents (Solé and Goodwin 2000), as we have explained in Sect. 4. This second meaning of complexity is the one we discuss in the present article.

Complexity of a viral population can be quantified by several diversity indices, some classically used in virology and others adopted from population ecology (Fuhrmann et al. 2021; Gregori et al. 2016). They offer complementary information on the number of residues that differ between any two genomes following sequence alignment, or the probability of finding two identical genomes when they are taken at random from a population. Comparison of diversity indices among different populations of the same virus is providing new insights into the molecular basis of lethal mutagenesis, as well as adaptive and pathogenic mechanisms of highly variable viruses (see Sect. 8).

7 Viruses as Experimental Models for Biological Complexity

Viruses are suitable experimental models to address some questions of population dynamics because their genome variations can be analyzed more easily than those undergone by cellular collectivities. Acute or persistent viral infections can be established and monitored in cell or organoid cultures, and the virus may respond in days (on occasion, even hours) to environmental modifications. Relevant information on virus behavior has been obtained by experimental designs in which both the variant composition of the initial virus and the environmental conditions were controlled [as an example of group selection of antigenic variants, see Martín and Domingo (2008). Although still in their infancy, related designs have the potential to inquire into the behavior of complex biological systems (Solé and Goodwin 2000).

Viral populations reconstructed with variants that may evoke (or be subjected to) complementation or cooperation among cloud subsets should inform us of the factors (critical variant frequency and environmental parameters) that either promote or inhibit the emergence of new traits.

Remarkable progress in viral genomics has been driven by the massive medical and public health research response to the emergence of SARS-CoV-2. Worthy of being mentioned are the millions of SARS-CoV-2 genomic sequences deposited in data banks [14,843,380 sequences deposited in GISAID until October, 2025 (https://gisaid.org)] and the high-resolution penetration into the composition of individual populations by application of ultra-deep sequencing technology (Andrés et al. 2020; Delgado et al. 2024; Martínez-González et al. 2022a, b, 2024, 2025). Results from the analysis of mutant spectra question the view that minority mutations should be regarded as merely unavoidable and of marginal significance.

8 Ultra-Deep Sequencing of SARS-CoV-2: Unveiling Limits of Viral Complexity?

Decades before the emergence of SARS-CoV-2, successive procedures (T1 oligonucleotide fingerprinting, molecular cloning and Sanger sequencing, pyrosequencing, and first deep sequencing protocols) contributed to establishing the quasispecies (extremely heterogeneous) nature of RNA viruses and their biological relevance, despite initial incredulity and skepticism [reviewed in (Domingo et al. 2021b)]. The results of ultra-deep sequencing have amply confirmed and quantitatively refined the conclusion advanced by previous methods that RNA virus populations exist as mutant clouds. A new challenge brought about by methodological advances is to reliably identify minority haplotypes, rank them according to frequency, and trace relationships among them. Although we are still halfway through this endeavor, several bioinformatics tools, artificial intelligence (AI) designs, and statistical methods have been described to quantify and organize minority haplotypes from viral samples [among many studies, see (Baccam et al. 2001; Cai et al. 2022; Chen et al. 2018; Delgado et al. 2015, 2021, 2024; Freire et al. 2021; Fuhrmann et al. 2024; Glebova et al. 2017; Grubaugh et al. 2019; Jayasundara et al. 2015; Perales et al. 2026; Skums et al. 2018; Sung and Baaijens 2025)]. We have adapted an Illumina ultra-deep sequencing procedure and a bioinformatics protocol to reliably identify point mutations and deletions in short SARS-CoV-2 amplicons (circa 400 nucleotides) with a cut-off frequency of 01% (Delgado et al. 2024; Martínez-González et al. 2022a, 2025).

Experimental and bioinformatics controls have established that the great majority of individual mutations or deletions found with a detection limit of 01% frequency are real (meaning that they are present in the template RNA molecules of the sample). The controls that have established such reliability are the following: (i) The average clean read coverage is around 100,000. (ii) A quality score

(Q) higher than 30 (calculated according to Illumina specifications) is attained for around 90% of the sequenced residues. (iii) The distribution of mutations among the three residues of triplets in the coding regions is the same when setting the frequency cut-off either at 05% or at 01%. (iv) There is agreement in the mutation and deletion repertoires obtained with separate sequencing runs of the same sample (for example, duplicate sequencing runs from the same patient sample yielded a robustly similar number of point mutations and their frequencies) ($P < 0.0001$ in all cases; Pearson correlation test) (Delgado et al. 2024). (v) As many as 97% of the amino acid substitutions predicted from the composition of minority haplotypes are represented in the outbreakinfo database (https://outbreak.info/) (Gangavarapu et al. 2023); such representation is the same with the frequency cut-off set at 05% or 01%. (vi) The average acceptability of amino acid substitutions according to the PAM250 score, or their functional effect according to the SNAP2 predictor, is the same with the substitution repertoire identified with a 05% and with a 01% frequency cut-off. (vii) In improving the resolution from 05% to 01% frequency detection, the total number of mutations identified increased 55-fold for the nsp12 (polymerase)-coding region and 97-fold for the S (spike)-coding region. (viii) Significant differences in mutant spectrum complexity were calculated among isolates from patients that suffered from mild, moderate, or severe disease, between uncloned SARS-CoV-2 populations and biological clones retrieved from them, and among isolates from the first and sixth COVID-19 waves; in these comparative analyses, the starting amounts of template nucleic acids for amplification and sequencing were the same. The observations listed in (i)–(viii) are incompatible with sequencing mistakes accounting for a significant number of mutations scored with a frequency cut-off of 0.1%.

Prevention of cross-contamination among samples involved: (i) Ensuring absence of contacts among samples at the time of their collection in hospital settings and their transport to the BSL-3 facility. (ii) Handling of different samples at separate times in the high containment cabins, with disinfection and UV irradiation following work with each sample. (iii) Preparation of all reagents used for amplification in a separate cabin that never contained viral RNA samples. (iv) Inclusion of negative control amplifications (with all reagents except template RNA) performed in all experiments. As additional proof, no viral RNA could be amplified from the water contents of the tube left open during a sample treatment process in the high containment cabin.

Further details on the mutant detection reliability, cross-contamination controls, and statistical procedures used in our studies have been published by (de Ávila et al. 2025; Delgado et al. 2024; Martínez-González et al. 2024, 2025; Perales et al. 2026). They have been outlined here because they are critical for the assessment of SARS-CoV-2 complexity and dynamics that we summarize next, and that are pivotal to our view of evolution.

There is extensive evidence of SARS-CoV-2 intra-host population heterogeneity and quasispecies dynamics obtained with different COVID-19 patient cohorts (Al Khatib et al. 2020; Andrés et al. 2020; Armero et al. 2021; Bader et al. 2022; Capobianchi et al. 2020; Dudouet et al. 2022; Gaiarsa et al. 2022; Ghorbani et al. 2020; Jary et al. 2020; Karamitros et al. 2020; Kimbrel et al. 2022; Lau et al. 2021; Li et al. 2022; Martínez-González et al. 2022a; Messali et al. 2023; Rueca et al. 2020; Sun et al. 2021). This new evidence follows the one that was previously reported for other pathogenic coronaviruses (Borucki et al. 2016; Liu et al. 2005; Park et al. 2016; Tang et al. 2006; Xu et al. 2004). We emphasize this point here because when a new viral pathogen emerges, the ingrained tendency of the scientific community is to underestimate its genetic variability, often with incomplete evidence. This has happened with several viral pathogens and once again with SARS-CoV-2, perhaps reinforced by expectations from its large genome size that encodes an ExoN proofreading-repair activity. Medical implications summarized in Sect. 5 are very likely to be valid for SARS-CoV-2, as suggested by multiple reports on antiviral resistance of SARS-CoV-2 and decreased efficacy of anti-COVID-19 vaccines in protecting against emerging viral lineages. As more isolates are analyzed, the more evident diversity is. With the support of advanced computational methods, it is realistic to continue defining variants, clades, lineages, and sub-lineages as a useful epidemiological survey. It may serve as a significant alert, however, that in the 1980s, serological subtyping of foot-and-mouth disease virus (FMDV)—a well-established diagnosis method at that time—was discontinued when it became apparent that virtually any new isolate of the virus could be classified as a new subtype.

A quantitative difference between SARS-CoV-2 and other RNA viruses was unveiled by comparison of the genomic consensus sequences of clonal SARS-CoV-2 populations produced in cell culture (de Ávila et al. 2025). An average number of 2.8 variations (mutations and deletions) per viable genome was counted relative to the consensus sequence of the parental population used to obtain the biological clones. This value is about 6.5-fold lower than estimates for clonal populations of bacteriophage Qβ, FMDV, or HCV in similar experimental designs. These results hint at a limited tolerance of viable SARS-CoV-2 to accept mutations, a trait that is also suggested by the haplotype composition of SARS-CoV-2 mutant spectra (see Sect. 8.3).

The analysis of SARS-CoV-2 quasispecies has unveiled some interesting features (Table 21.1). Noteworthy among them are the different mutation preferences found in mutant spectra as compared with consensus sequences, the abundance of low-frequency mutations in mutant spectra, and the presence of minority haplotypes that correspond to sequences that served to define other clades of the virus. Next, we discuss the relevance of some of these observations. We put the focus on the distinction between the information embodied in consensus sequences versus mutant spectra, the fitness effect of mutations, and limitations in occupation of sequence space that may be faced by a complex RNA virus (complex in the sense of genome size and functional codification).

8.1 The Puzzle of Mutation Type Preferences

In alignments of SARS-CoV-2 consensus sequences from many isolates of a COVID-19 patient cohort from Madrid (Soria et al. 2021), the most frequent mutation types, counted relative to some ancestral reference sequence, were

Table 21.1 Observations deduced from high-resolution, ultra-deep sequencing of SARS-CoV-2 populations[a]

Mutant spectrum complexity is an epidemiologically evolvable trait. It may vary according to the severity of COVID-19
SARS-CoV-2 mutant spectra include an overwhelming majority of low frequency mutations
There is a deficit of intermediate frequency haplotypes as compared with other RNA viruses such as hepatitis C virus
There is a significant difference in mutation type preferences within mutant spectra and among consensus sequences
Minority haplotypes in SARS-CoV-2 isolates may be identical to dominant haplotypes that served to define other clades of the virus
Incipient minority amino acid substitutions in virus from infected patients altered essential viral functions

[a]These statements are based on the following references Martínez-González et al. (2022a, b, 2024, 2025) and Delgado et al. (2024)

C → U followed by G → U. In mutant spectra of the nsp12 (polymerase) and S (spike) of the isolates, the most frequent mutation types were U → C and A → G, counted relative to the consensus sequence of the corresponding isolate. The origin of the difference in mutation type preference between consensus sequences and mutant spectra—which has been noted also with isolates from other patient cohorts—is not clear. We have suggested two possibilities: (i) that on average U → C mutations inflict a larger fitness cost than C → U mutations on the genomes that acquire them. The consequence would be that genomes with acquired U → C mutations would be less frequent in the population and, therefore, have less weight in defining the consensus sequence. In fact, there is evidence that A → G and U → C mutations are more likely to elicit modifications of RNA and protein structure than other mutation types (Mendez et al. 2010). The fitness-decreasing effect that many mutations are likely to have may be accentuated in the large coronavirus genome. (ii) That part of the mutations recorded in consensus sequences was due to cellular editing activities, which may target the most abundant genomes during intracellular viral replication. Further arguments in support or against these two nonexclusive possibilities have been previously discussed (de Ávila et al. 2025). The results emphasize the different information on evolutionary events and virus adaptability that is gathered from consensus sequences versus mutant spectra (Bader et al. 2022; Domingo and Perales 2019; Martínez-González et al. 2025).

8.2 Dominance of Minorities Within Minorities

Mutant spectra of SARS-CoV-2 exhibit a very large dominance (on average 95% of the total) of low-frequency mutations, those in the range of 0.10–0.49%, and a remarkable deficit of mutations present at intermediate frequencies. A significant comparison with another virus was provided by the number of predicted amino acid substitutions from the sequences of several genomic regions of SARS-CoV-2 and HCV, detected with a 1% frequency cut-off, using comparable bioinformatics pipelines. The percentage of substitutions that fell in the 10–90% frequency level was 10.5% for HCV and 1.6% for SARS-CoV-2; when the comparison was restricted to the corresponding viral polymerases, the values were 8.1% and 0%, respectively. The differences were statistically significant (Martínez-González et al. 2022a). We outlined a number of possible mechanisms to explain this interesting difference, whose interpretation remains an open question; no solid evidence supports that the detrimental fitness effect of mutations may solely be attributable to the large coronavirus genome size [discussed in (Martínez-González et al. 2022a)].

8.3 Clade Discordances: A Sign of Adaptability Limitations?

A remarkable finding with SARS-CoV-2 regarding the minority haplotype composition in clinical isolates, laboratory populations, and biological clones isolated in cell culture has been the presence of clade-discordant residues. They are defined as mutations or deletions present in the mutant spectrum of a virus of a given clade, but that are typical of other clades; in fact, they served to define the other clades where the same mutation and deletion clusters were dominant and typical. This finding concerned not only individual mutations and amino acid substitutions (Martínez-González et al. 2022b; Sun et al. 2021), but also clusters of two or more genome modifications (mutations, deletions, or insertions) present in the same or different amplicons and sometimes in the same haplotype within an amplicon (de Ávila et al. 2025; Martínez-González et al. 2024).

Experimental results and statistical calculations suggest that clade-discordant residues are not the result of sequencing mistakes, cross-contaminations, or chance occurrences of the relevant mutation and deletion clusters. Rather, their presence probably reflects limitations of SARS-CoV-2 genomes for the occupation of many regions of sequence space that, in theory, could be accessible to them. This limitation operates at least within the context of the currently dominant nucleotide sequences that surround the nsp12 and S regions analyzed. Indeed, clade discordances have been identified in mutant spectra of the two genomic regions, and it is not known whether they will also be found in other genomic regions. We have suggested that they may have a predictive value of SARS-CoV-2 sequences that will be dominant at later stages of the evolution of the virus (de Ávila et al. 2025; Martínez-González et al. 2024).

8.4 Mutant Spectra as Evolvable and Instructive Ensembles

The complexity of SARS-CoV-2 mutant spectra of nasopharyngeal isolates decreased as the successive COVID-19 waves reached Madrid (Spain) from 2020 until 2022. Thus, mutant spectrum complexity can be an epidemiologically evolvable trait (Martínez-González et al. 2025). This finding has two possible main interpretations: that the replicative machinery of the virus that circulated during late COVID-19 waves produced a lower number of mutations or that pathogenesis-epidemiological influences were responsible for the mutant spectrum shrinkage. The second possibility is more likely because when populations (and biological clones retrieved from them) were multiplied in cell culture, no difference in mutant spectrum complexity was noted between viruses from early and late COVID-19 waves. In addition, no dominant amino acid substitutions that might affect polymerase fidelity were identified when comparing the viruses under study (Martínez-González et al. 2025). In support of viral pathogenic factors being a major influence in the decrease of complexity of the nasopharyngeal viral population of late COVID-19 waves is the evidence in a feline infection model that SARS-CoV-2 Omicron BA.11 (which emerged late in 2021) displayed a more restricted tissue distribution than previous variants (Martins et al. 2022). The nasopharyngeal environment gathers viruses from several other body sites, thus enriching its viral diversity from different sources. Mutant spectra may be informative not only of the error rate of the virus and of the intervening number of replication rounds since a clonal origin but also of the prior diversification of the virus that reached the isolation site, pushed by environmental heterogeneity (de Ávila et al. 2025; Martínez-González et al. 2025).

In summary, SARS-CoV-2 population complexity has reinforced the value of quasispecies dynamics even for remarkably complex RNA genomes. Its study is opening new questions for research whose approach may help to understand in greater detail the forces that drive viral dynamics.

9 Concluding Comments: Creativity from Errors and Tinkering

As ultra-deep sequencing of cellular and subcellular biological collectivities (pan-genomic analyses) improves, a paradox becomes apparent: there is a dramatic extinction of many biological species attributable to human intervention [environmental alterations, climate change, and global warming (Steffen et al. 2015)], and a parallel recognition that the intra-species diversity of the surviving species is larger than anticipated a few decades ago. Not that such newly discovered complexity can compensate for the irreversible consequences of biological species loss, certainly not. The novelty is that complexity, with all its connotations and implications (Solé and Goodwin 2000), permeates the biological world. Dynamic diversity is the norm, not an exception. How does such diversity originate, and how is it maintained? Is it regulated? How do useful traits originate? Are there unpredictable events that go unnoticed and that influence discrepancies in experimental results?

In the scenario learned from the accelerated version of evolution displayed by RNA viruses, summarized in the present chapter, we come close to a view of evolution that fits an old concept that was delineated by F. Jacob half a century ago: evolution and tinkering (Jacob 1977). Viruses (as well as cells, as explained in the original article) take advantage of what is available at any time in their survival as a powerful selection force. The proposal is analogous to language construction that uses words (incorporated or modified) from previous languages; borrowed terms lead to complex etymological trails, as also does tinkering for viral phylogenies. The alphavirus Western equine encephalitis virus was probably the outcome of a recombination event between viruses similar to present-day Sindbis virus and Eastern equine encephalitis virus; the coronavirus mouse hepatitis virus most likely captured its hemagglutinin-esterase gene from an IV type C; cytopathic variants of bovine viral diarrhea virus capture cellular tRNA fragments in their 5′ termini; insertions of rRNA sequences in hemagglutinin RNA were reported for some pathogenic IVs [additional examples of the likely role of recombination in the origin of new viruses—or of functional domains transported from a virus group to another—are included in (Domingo 2020)]. The tinkering concept fits the modular theory of evolution (Botstein 1980; Koonin and Dolja 2014; Vinicius 2010; Zimmern 1988). The usually large viral population numbers, randomness in the molecular instructions to generate variation, and selection for transient aptness mark evolutionary trajectories, with many viruses still to be identified (Lipkin and Anthony 2015).

It is truly remarkable how events that are random in their inception can produce the array of biological solutions for viruses in the form of resilient survival in a diverse cellular environment. It may pay to look carefully at how the accelerated versions of biological evolution provided by viruses succeed in producing meaningful biological organizations. Despite being short-lived in their molecular solutions, such organizations are long-lasting in producing recognizable lineages, which express key traits that extend in time.

Acknowledgments This work was supported by the Spanish Ministry of Science and Innovation, grants PID2020-113888RB-I00/AEI/1013039/501100011033 and 202220I116; by the Ministerio de Ciencia, Innovación y Universidades (MICIU), grant PID2023-146622OB-I00 financed by MICIU/AEI/1013039/501100011033; by the Fondo Europeo de Desarrollo Regional (FEDER), UE, through the "Severo Ochoa" Program for Centers of Excellence in Research and Development (CEX2021-001154-S); and by the European Commission-Next Generation EU (regulation EU 2020/2024) through the Consejo Superior de Investigaciones Científicas (CSIC's) Global Health Platform [Plataforma Temática Interdisciplinar (PTI) Salud Global]. The work was also funded by grants PI21/00139 from Instituto de Salud Carlos III, CSIC-COV19-014 from CSIC, and project 525/C/2021 from Fundació La Marató de TV3; grants 202136-30 and 202136-31. We also acknowledge the projects TEC-2024/BIO-66 (SALAINDEC-CM from Comunidad de Madrid/FEDER) and S2018/BAA-4370 (PLATESA2 from Comunidad de Madrid/FEDER). Institutional grants from the Fundación Ramón Areces and Banco Santander to the Centro de Biología Molecular (CBM) are also acknowledged. The team at Centro de Biología Molecular (CBM) belongs to the Global Virus Network and to the Biomedicine Unit of Universidad de Castilla-La Mancha, associated with CSIC. BM G is supported by a predoctoral contract PFIS FI19/00119 from the Instituto de Salud Carlos III, cofinanced by Fondo Social Europeo (FSE), "El FSE invierte en tu futuro."

References

Al Khatib HA, Benslimane FM, Elbashir IE, Coyle PV, Al Maslamani MA, Al-Khal A, Al Thani AA, Yassine HM (2020) Within-host diversity of SARS-CoV-2 in COVID-19 patients with variable disease severities. Front Cell Infect Microbiol 10:575613

Alnaji FG, Bentley K, Pearson A, Woodman A, Moore J, Fox H, Macadam AJ, Evan DJ (2022) Generated randomly and selected functionally? The nature of enterovirus recombination. Viruses 14(5):916

Andrés C, García-Cehic D, Gregori J, Pinana M, Rodríguez-Frías F, Guerrero-Murillo M, Esperalba J, Rando A, Goterris L, Codina MG, Quer S, Martín MC, Campins M, Ferrer R, Almirante B, Esteban JI, Pumarola T, Antón A, Quer J (2020) Naturally occurring SARS-CoV-2 gene deletions close to the spike S1/S2 cleavage site in the viral quasispecies of COVID19 patients. Emerg Microbes Infect 9:1900–1911

Armero A, Berthet N, Avarre JC (2021) Intra-host diversity of SARS-Cov-2 should not be neglected: case of the State of Victoria, Australia. Viruses 13(1):133

Baccam P, Thompson RJ, Fedrigo O, Carpenter S, Cornette JL (2001) PAQ: partition analysis of quasispecies. Bioinformatics 17:16–22

Bader W, Delerce J, Aherfi S, La Scola B, Colson P (2022) Quasispecies analysis of SARS-CoV-2 of 15 different lineages during the first year of the pandemic prompts scratching under the surface of consensus genome sequences. Int J Mol Sci 23(24):15658

Bancroft CT, Parslow TG (2002) Evidence for segment-nonspecific packaging of the influenza a virus genome. J Virol 76:7133–7139

Baranowski E, Ruiz-Jarabo CM, Domingo E (2001) Evolution of cell recognition by viruses. Science 292:1102–1105

Baranowski E, Ruíz-Jarabo CM, Pariente N, Verdaguer N, Domingo E (2003) Evolution of cell recognition by viruses: a source of biological novelty with medical implications. Adv Virus Res 62:19–111

Begum MM, Bokani A, Rajib SA, Soleimanpour M, Maeda Y, Yoshimura K, Satou Y, Ebrahimi D, Ikeda T (2024) Potential role of APOBEC3 family proteins in SARS-CoV-2 replication. Viruses 16(7):1141

Bentley K, Alnaji FG, Woodford L, Jones S, Woodman A, Evans DJ (2021) Imprecise recombinant viruses evolve via a fitness-driven, iterative process of polymerase template-switching events. PLoS Pathog 17(8):e1009676

Bernad A, Blanco L, Lazaro JM, Martin G, Salas M (1989) A conserved 3′–5′ exonuclease active site in prokaryotic and eukaryotic DNA polymerases. Cell 59:219–228

Boerlijst MC, Boenhoefer S, Nowak MA (1996) Viral quasispecies and recombination. Proc R Soc Lond B 263:1577–1584

Borucki MK, Lao V, Hwang M, Gardner S, Adney D, Munster V, Bowen R, Allen JE (2016) Middle East respiratory syndrome coronavirus intra-host populations are characterized by numerous high frequency variants. PLoS One 11:e0146251

Botstein D (1980) A theory of modular evolution for bacteriophages. Ann N Y Acad Sci 354:484–491

Cai D, Shang J, Sun Y (2022) HaploDMF: viral haplotype reconstruction from long reads via deep matrix factorization. Bioinformatics 38:5360–5367

Capobianchi MR, Rueca M, Messina F, Giombini E, Carletti F, Colavita F, Castilletti C, Lalle E, Bordi L, Vairo F, Nicastri E, Ippolito G, Gruber CEM, Bartolini B (2020) Molecular characterization of SARS-CoV-2 from the first case of COVID-19 in Italy. Clin Microbiol Infect 26: 954–956

Castello-Sanjuan M, González R, Romoli O, Blanc H, Nigg JC, Saleh MC (2025) Persistent viral infections impact key biological traits in Drosophila melanogaster. PLoS Biol 23:e3003437

Cervera H, Ambros S, Bernet GP, Rodrigo G, Elena SF (2018) Viral fitness correlates with the magnitude and direction of the perturbation induced in the host's transcriptome: the tobacco etch potyvirus-tobacco case study. Mol Biol Evol 35:1599–1615

Chen J, Zhao Y, Sun Y (2018) De novo haplotype reconstruction in viral quasispecies using paired-end read guided path finding. Bioinformatics 34:2927–2935

Chen KY, Karuppusamy J, O'Neill MB, Opuu V, Bahin M, Foulon S, Ibanez P, Quintana-Murci L, Ozawa T, van der Werf S, Nghe P, Naffakh N, Griffiths A, Isel C (2023) High-throughput droplet-based analysis of influenza A virus genetic reassortment by single-virus RNA sequencing. Proc Natl Acad Sci USA 120:e2211098120

Dandri M (2020) Epigenetic modulation in chronic hepatitis B virus infection. Semin Immunopathol 42:173–185

De Ávila AI, Soria ME, Martínez-González B, Somovilla P, Mínguez P, Salar-Vidal L, Esteban-Muñoz M, Martín-García M, Zuñiga S, Sola I, Enjuanes L, Gadea I, Perales C, Domingo E (2025) SARS-CoV-2 biological clones are genetically heterogeneous and include clade-discordant residues. J Virol 99(5):e0225024

Delgado S, Morán F, Mora A, Merelo JJ, Briones C (2015) A novel representation of genomic sequences for taxonomic clustering and visualization by means of self-organizing maps. Bioinformatics 31:736–744

Delgado S, Perales C, García-Crespo C, Soria ME, Gallego I, de Ávila AI, Martínez-González B, Vázquez-Sirvent L, López-Galíndez C, Morán F, Domingo E (2021) A two-level, intramutant spectrum haplotype profile of hepatitis C virus revealed by self-organized maps. Microbiol Spectrum 9(3):e0145921

Delgado S, Somovilla P, Ferrer-Orta C, Martínez-González B, Vázquez-Monteagudo S, Muñoz-Flores J, Soria ME, García-Crespo C, de Ávila AI, Durán-Pastor A, Gadea I, López-Galíndez C, Morán F, Lorenzo-Redondo R, Verdaguer N, Perales C, Domingo E (2024) Incipient functional SARS-CoV-2 diversification identified through neural network haplotype maps. Proc Natl Acad Sci USA 121(10):e2317851121

Domingo E (1989) RNA virus evolution and the control of viral disease. Prog Drug Res 33:93–133

Domingo E (2020) Virus as populations, 2nd edn. Academic/Elsevier, Amsterdam

Domingo E, Perales C (2012) From quasispecies theory to viral quasispecies: how complexity has permeated virology. Math Model Nat Phenom 7:32–49

Domingo E, Perales C (2019) Viral quasispecies. PLoS Genet 15(10):e1008271

Domingo E, Schuster P (2016) Quasispecies: from theory to experimental systems. In: Current topics in microbiology and immunology, vol 392. Springer

Domingo E, Witzany G (2024) Quasispecies productivity. Naturwissenschaften 111(2):11

Domingo E, Biebricher C, Eigen M, Holland JJ (2001) Quasispecies and RNA virus evolution: principles and consequences. Landes Bioscience, Austin

Domingo E, de Ávila AI, Gallego I, Sheldon J, Perales C (2019) Viral fitness: history and relevance for viral pathogenesis and antiviral interventions. Pathog Dis 77(2):ftz021

Domingo E, Garcia-Crespo C, Lobo-Vega R, Perales C (2021a) Mutation rates, mutation frequencies, and proofreading-repair activities in RNA virus genetics. Viruses 13(9):1882

Domingo E, García-Crespo C, Perales C (2021b) Historical perspective on the discovery of the quasispecies concept. Annu Rev Virol 8(1):51–72

Domingo E, Martínez-González B, García-Crespo C, Somovilla P, de Ávila AI, Soria ME, Durán-Pastor A, Perales C (2023) Puzzles, challenges, and information reservoir of SARS-CoV-2 quasispecies. J Virol 97(12):e0151123

Domingo E, Martínez-González B, Somovilla P, García-Crespo C, Soria ME, de Ávila AI, Gadea I, Perales C (2025) A general and biomedical perspective of viral quasispecies. RNA 31(3): 429–443

Du R, Cui Q, Chen Z, Zhao X, Lin X, Rong L (2023) Revisiting influenza A virus life cycle from a perspective of genome balance. Virol Sin 38(1):1–8

Dudouet P, Colson P, Aherfi S, Levasseur A, Beye M, Delerce J, Burel E, Lavrard P, Bader W, Lagier JC, Fournier PE, La Scola B, Raoult D (2022) SARS-CoV-2 quasi-species analysis from patients with persistent nasopharyngeal shedding. Sci Rep 12(1):18721

Eigen M (2002) Error catastrophe and antiviral strategy. Proc Natl Acad Sci USA 99:13374–13376

Eigen M (2013) From strange simplicity to complex familiarity. Oxford University Press, Oxford

Eigen M, Schuster P (1979) The hypercycle: a principle of natural self-organization. Springer, Berlin

Escarmís C, Dávila M, Domingo E (1999) Multiple molecular pathways for fitness recovery of an RNA virus debilitated by operation of Muller's ratchet. J Mol Biol 285:495–505

Fedeles BI, Li D, Singh V (2021) Structural insights into tautomeric dynamics in nucleic acids and in antiviral nucleoside analogs. Front Mol Biosci 8:823253

Ferrer-Orta C, Agudo R, Domingo E, Verdaguer N (2009) Structural insights into replication initiation and elongation processes by the FMDV RNA-dependent RNA polymerase. Curr Opin Struct Biol 19:752–758

Figlerowicz M, Alejska M, Kurzynska-Kokorniak A, Figlerowicz M (2003) Genetic variability: the key problem in the prevention and therapy of RNA-based virus infections. Med Res Rev 23: 488–518

Freire B, Ladra S, Parama JR, Salmela L (2021) Inference of viral quasispecies with a paired de Bruijn graph. Bioinformatics 37:473–481

Friedberg EC, Walker GC, Siede W, Wood RD, Schultz RA, Ellenberger T (2006) DNA repair and mutagenesis. American Society for Microbiology, Washington, DC

Fuhrmann L, Jablonski KP, Beerenwinkel N (2021) Quantitative measures of within-host viral genetic diversity. Curr Opin Virol 49:157–163

Fuhrmann L, Langer B, Topolsky I, Beerenwinkel N (2024) VILOCA: sequencing quality-aware viral haplotype reconstruction and mutation calling for short-read and long-read data. NAR Genom Bioinform 6(4):lqae152

Fujii Y, Goto H, Watanabe T, Yoshida T, Kawaoka Y (2003) Selective incorporation of influenza virus RNA segments into virions. Proc Natl Acad Sci USA 100(4):2002–2007

Gaiarsa S, Giardina F, Batisti Biffignandi G, Ferrari G, Piazza A, Tallarita M, Novazzi F, Bandi C, Paolucci S, Rovida F, Campanini G, Piralla A, Baldanti F (2022) Comparative analysis of SARS-CoV-2 quasispecies in the upper and lower respiratory tract shows an ongoing evolution in the spike cleavage site. Virus Res 315:198786

Gallego I, Soria ME, Garcia-Crespo C, Chen Q, Martínez-Barragan P, Khalfaoui S, Martínez-González B, Sánchez-Martín I, Palacios-Blanco I, de Ávila AI, García-Cehic D, Esteban JI, Gómez J, Briones C, Gregori J, Quer J, Perales C, Domingo E (2020) Broad and dynamic diversification of infectious hepatitis C virus in a cell culture environment. J Virol 94(6): e01856–e01819

Gangavarapu K, Latif AA, Mullen JL, Alkuzweny M, Hufbauer E, Tsueng G, Haag E, Zeller M, Aceves CM, Zaiets K, Cano M, Zhou X, Qian Z, Sattler R, Matteson NL, Levy JI, Lee RTC, Freitas L, Maurer-Stroh S, Core G, Curation T, Suchard MA, Wu C, Su AI, Andersen KG, Hughes LD (2023) Outbreakinfo genomic reports: scalable and dynamic surveillance of SARS-CoV-2 variants and mutations. Nat Methods 20(4):512–522

García-Crespo C, Soria ME, Gallego I, de Ávila AI, Martínez-González B, Vázquez-Sirvent L, Gómez J, Briones C, Gregori J, Quer J, Perales C, Domingo E (2020) Dissimilar conservation pattern in hepatitis C virus mutant spectra, consensus sequences, and data banks. J Clin Med 9(11):3450

García-Crespo C, Gallego I, Soria ME, De Ávila AI, Martínez-González B, Vázquez-Sirvent L, Lobo-Vega R, Moreno E, Gómez J, Briones C, Gregori J, Quer J, Domingo E, Perale C (2021) Population disequilibrium as promoter of adaptive explora-tions in hepatitis C virus. Viruses 13(4):616

García-Crespo C, Francisco-Recuero I, Gallego I, Camblor-Murube M, Soria ME, López-López A, de Ávila AI, Madejón A, García-Samaniego J, Domingo E, Sánchez-Pacheco A, Perales C (2023) Hepatitis C virus fitness can influence the extent of infection-mediated epigenetic modifications in the host cells. Front Cell Infect Microbiol 13:1057082

García-Crespo C, de Ávila AI, Gallego I, Soria ME, Durán-Pastor A, Somovilla P, Martínez-González B, Muñoz-Flores J, Mínguez P, Salar-Vidal L, Esteban-Muñoz M, Canar-Camacho E, Ferrer-Orta C, Zuñiga S, Sola I, Enjuanes L, Esteban J, Fernández-Roblas R, Gadea I, Gómez J, Verdaguer N, Domingo E, Perales C (2024) Synergism between remdesivir and ribavirin leads to SARS-CoV-2 extinction in cell culture. Br J Pharmacol 181(15):2636–2654

Ghorbani A, Samarfard S, Ramezani A, Izadpanah K, Afsharifar A, Eskandari MH, Karbanowicz TP, Peters JR (2020) Quasi-species nature and differential gene expression of severe acute respiratory syndrome coronavirus 2 and phylogenetic analysis of a novel Iranian strain. Infect Genet Evol 85:104556

Glebova O, Knyazev S, Melnyk A, Artyomenko A, Khudyakov Y, Zelikovsky A, Skums P (2017) Inference of genetic relatedness between viral quasispecies from sequencing data. BMC Genomics 18:918

González Aparicio LJ, López CB, Felt SA (2022) A virus is a community: diversity within negative-sense RNA virus populations. Microbiol Mol Biol Rev 86:e0008621

Graham JE, Marians KJ, Kowalczykowski SC (2017) Independent and stochastic action of DNA polymerases in the replisome. Cell 169(7):1201–1213.e17

Gregori J, Perales C, Rodríguez-Frías F, Esteban JI, Quer J, Domingo E (2016) Viral quasispecies complexity measures. Virology 493:227–237

Grubaugh ND, Gangavarapu K, Quick J, Matteson NL, De Jesus JG, Main BJ, Tan AL, Paul LM, Brackney DE, Grewal S, Gurfield N, Van Rompay KKA, Isern S, Michael SF, Coffey LL, Loman NJ, Andersen KG (2019) An amplicon-based sequencing framework for accurately measuring intrahost virus diversity using PrimalSeq and iVar. Genome Biol 20(1):8

Holland JJ, Spindler K, Horodyski F, Grabau E, Nichol S, VandePol S (1982) Rapid evolution of RNA genomes. Science 215(4540):1577–1585

Jacob F (1977) Evolution and tinkering. Science 196(4295):1161–1166

Jacobi MN, Nordahl M (2006) Quasispecies and recombination. Theor Popul Biol 70(4):479–485

Jary A, Leducq V, Malet I, Marot S, Klement-Frutos E, Teyssou E, Soulie C, Abdi B, Wirden M, Pourcher V, Caumes E, Calvez V, Burrel S, Marcelin AG, Boutolleau D (2020) Evolution of viral quasispecies during SARS-CoV-2 infection. Clin Microbiol Infect 26(11):1560.e1–1560.e4

Jayasundara D, Saeed I, Maheswararajah S, Chang BC, Tang SL, Halgamuge SK (2015) ViQuaS: an improved reconstruction pipeline for viral quasispecies spectra generated by next-generation sequencing. Bioinformatics 31(6):886–896

Jerzak G, Bernard KA, Kramer LD, Ebel GD (2005) Genetic variation in West Nile virus from naturally infected mosquitoes and birds suggests quasispecies structure and strong purifying selection. J Gen Virol 86:2175–2183

Karamitros T, Papadopoulou G, Bousali M, Mexias A, Tsiodras S, Mentis A (2020) SARS-CoV-2 exhibits intra-host genomic plasticity and low-frequency polymorphic quasispecies. J Clin Virol 131:104585

Kerviel A, Zhang M, Altan-Bonnet N (2021) A new infectious unit: extracellular vesicles carrying virus populations. Annu Rev Cell Dev Biol 37:171–197

Kimbrel J, Moon J, Avila-Herrera A, Marti JM, Thissen J, Mulakken N, Sandholtz SH, Ferrell T, Daum C, Hall S, Segelke B, Arrildt KT, Messenger S, Wadford DA, Jaing C, Allen JE, Borucki MK (2022) Multiple mutations associated with emergent variants can be detected as low-frequency mutations in early SARS-CoV-2 pandemic clinical samples. Viruses 14(12): 2775

Koonin EV, Dolja VV (2014) Virus world as an evolutionary network of viruses and capsidless selfish elements. Microbiol Mol Biol Rev 78:278–303

Lau BT, Pavlichin D, Hooker AC, Almeda A, Shin G, Chen J, Sahoo MK, Huang CH, Pinsky BA, Lee HJ, Ji HP (2021) Profiling SARS-CoV-2 mutation fingerprints that range from the viral pangenome to individual infection quasispecies. Genome Med 13(1):62

Lederberg J (1993) Viruses and humankind: intracellular symbiosis and evolutionary competition. In: Morse SS (ed) Emerging viruses. Oxford University Press, Oxford, pp 3–9

Li J, Du P, Yang L, Zhang J, Song C, Chen D, Song Y, Ding N, Hua M, Han K, Song R, Xie W, Chen Z, Wang X, Liu J, Xu Y, Gao G, Wang Q, Pu L, Di L, Li J, Yue J, Han J, Zhao X, Yan Y, Yu F, Wu AR, Zhang F, Gao YQ, Huang Y, Wang J, Zeng H, Chen C (2022) Two-step fitness selection for intra-host variations in SARS-CoV-2. Cell Rep 38(2):110205

Lipkin WI, Anthony SJ (2015) Virus hunting. Virology 479-480:194–199

Liu J, Lim SL, Ruan Y, Ling AE, Ng LF, Drosten C, Liu ET, Stanton LW, Hibberd ML (2005) SARS transmission pattern in Singapore reassessed by viral sequence variation analysis. PLoS Med 2(2):e43

Lowry K, Woodman A, Cook J, Evans DJ (2014) Recombination in enteroviruses is a biphasic replicative process involving the generation of greater-than genome length 'imprecise' intermediates. PLoS Pathog 10(6):e1004191

Martín V, Domingo E (2008) Influence of the mutant spectrum in viral evolution: focused selection of antigenic variants in a reconstructed viral quasispecies. Mol Biol Evol 25:1544–1554

Martínez-González B, Soria ME, Vázquez-Sirvent L, Ferrer-Orta C, Lobo-Vega R, Mínguez P, de la Fuente L, Llorens C, Soriano B, Ramos-Ruíz R, Corton M, López-Rodríguez R, García-Crespo C, Somovilla P, Durán-Pastor A, Gallego I, de Ávila AI, Delgado S, Morán F, López-Galíndez C, Gómez J, Enjuanes L, Salar-Vidal L, Esteban-Muñoz M, Esteban J, Fernández-Roblas R, Gadea I, Ayuso C, Ruíz-Hornillos J, Verdaguer N, Domingo E, Perales C (2022a) SARS-CoV-2 mutant spectra at different depth levels reveal an overwhelming abundance of low frequency mutations. Pathogens 11:662

Martínez-González B, Vázquez-Sirvent L, Soria ME, Mínguez P, Salar-Vidal L, García-Crespo C, Gallego I, Ávila A, Llorens C, Soriano B, Ramos-Ruíz R, Esteban J, Fernández-Roblas R, Gadea I, Ayuso C, Ruíz-Hornillos J, Pérez-Jorge C, Domingo E, Perales C (2022b) Vaccine-breakthrough infections with SARS-CoV-2 Alpha mirror mutations in Delta Plus, Iota and Omicron. J Clin Invest 132:e157700

Martínez-González B, Soria ME, Mínguez P, Lorenzo-Redondo R, Salar-Vidal L, López-García A, Esteban-Muñoz M, Durán-Pastor A, Somovilla P, García-Crespo C, de Ávila AI, Gómez J, Esteban J, Fernández-Roblas R, Gadea I, Domingo E, Perales C (2024) SARS-CoV-2 mutant spectra as variant of concern nurseries: endless variation? Front Microbiol 15:1358258

Martínez-González B, Soria ME, de Ávila AI, Somovilla P, Aguilar-Sabido C, Mínguez P, Ferrer-Orta C, Salar-Vidal L, Lorenzo-Redondo R, Delgado S, Morán F, Verdaguer N, Gadea I, Domingo E, Perales C (2025) SARS-CoV-2 mutant spectrum complexity is an epidemiologically evolvable trait. Proc Natl Acad Sci USA 122(39):e2515706122

Martins M, do Nascimento GM, Nooruzzaman M, Yuan F, Chen C, Caserta LC, Miller AD, Whittaker GR, Fang Y, Diel DG (2022) The Omicron variant BA11 presents a lower pathogenicity than B1 D614G and Delta variants in a feline model of SARS-CoV-2 infection. J Virol 96: e0096122

Mendez R, Fritsche M, Porto M, Bastolla U (2010) Mutation bias favors protein folding stability in the evolution of small populations. PLoS Comput Biol 6:e1000767

Messali S, Rondina A, Giovanetti M, Bonfanti C, Ciccozzi M, Caruso A, Caccuri F (2023) Traceability of SARS-CoV-2 transmission through quasispecies analysis. J Med Virol 95: e28848

Moreno E, Gallego I, Gregori J, Lucia-Sanz A, Soria ME, Castro V, Beach NM, Manrubia S, Quer J, Esteban JI, Rice CM, Gomez J, Gastaminza P, Domingo E, Perales C (2017) Internal disequilibria and phenotypic diversification during replication of Hepatitis C virus in a noncoevolving cellular environment. J Virol 91:e02505–e02516

Mustonen V, Lassig M (2009) From fitness landscapes to seascapes: non-equilibrium dynamics of selection and adaptation. Trends Genet 25:111–119

Nijhuis M, van Maarseveen NM, Boucher CA (2009) Antiviral resistance and impact on viral replication capacity: evolution of viruses under antiviral pressure occurs in three phases. Handb Exp Pharmacol 189:299–320

Novella IS, Clarke DK, Quer J, Duarte EA, Lee CH, Weaver SC, Elena SF, Moya A, Domingo E, Holland JJ (1995a) Extreme fitness differences in mammalian and insect hosts after continuous replication of vesicular stomatitis virus in sandfly cells. J Virol 69:6805–6809

Novella IS, Duarte EA, Elena SF, Moya A, Domingo E, Holland JJ (1995b) Exponential increases of RNA virus fitness during large population transmissions. Proc Natl Acad Sci USA 92:5841–5844

Novella IS, Quer J, Domingo E, Holland JJ (1999) Exponential fitness gains of RNA virus populations are limited by bottleneck effects. J Virol 73:1668–1671
Nowak MA (1992) What is a quasispecies? Trends Ecol Evol 4:118–121
Park JM, Deem MW (2007) Phase diagrams of quasispecies theory with recombination and horizontal gene transfer. Phys Rev Lett 98:058101
Park D, Huh HJ, Kim YJ, Son DS, Jeon HJ, Im EH, Kim JW, Lee NY, Kang ES, Kang CI, Chung DR, Ahn JH, Peck KR, Choi SS, Kim YJ, Ki CS, Park WY (2016) Analysis of intrapatient heterogeneity uncovers the microevolution of Middle East respiratory syndrome coronavirus. Cold Spring Harb Mol Case Stud 2:a001214
Perales C, Gallego I, de Ávila AI, Soria ME, Gregori J, Quer J, Domingo E (2019) The increasing impact of lethal mutagenesis of viruses. Future Med Chem 11:1645–1657
Perales C, Ortega-Prieto AM, Beach NM, Sheldon J, Menéndez-Arias L, Domingo E (2020) Quasispecies and drug resistance. In: Handbook of antimicrobial resistance. Springer, New York
Perales C, Soria ME, De Ávila AI, Martínez-González B, Gallego I, García-Crespo C, Somovilla P, Mínguez P, Gadea I, Domingo E (2026) Methods to characterize mutant spectra of RNA viruses the case of SARS-CoV-2. Methods Mol Biol (Evolutionary Genomics), in press
Picardi E, Mansi L, Pesole G (2021) Detection of A-to-I RNA editing in SARS-CoV-2. Genes 13(1):41
Pu X, Xu Q, Wang J, Liu B (2023) The continuing discovery on the evidence for RNA editing in SARS-CoV-2. RNA Biol 20:219–222
Ringlander J, Fingal J, Kann H, Prakash K, Rydell G, Andersson M, Martner A, Lindh M, Horal P, Hellstrand K, Kann M (2022) Impact of ADAR-induced editing of minor viral RNA populations on replication and transmission of SARS-CoV-2. Proc Natl Acad Sci USA 119(6):e2112663119
Rueca M, Bartolini B, Gruber CEM, Piralla A, Baldanti F, Giombini E, Messina F, Marchioni L, Ippolito G, Di Caro A, Capobianchi MR (2020) Compartmentalized replication of SARS-CoV-2 in upper vs lower respiratory tract assessed by whole genome quasispecies analysis. Microorganisms 8(9):1302
Saakian DB (2018) Looking for the optimal rate of recombination for evolutionary dynamics. Phys Rev E 97(1–1):012409
Saakian DB, Hu CK (2006) Exact solution of the Eigen model with general fitness functions and degradation rates. Proc Natl Acad Sci USA 103:4935–4939
Saakian DB, Hu CK (2016) Mathematical models of quasi-species theory and exact results for the dynamics. Curr Top Microbiol Immunol 392:121–139
Sardanyés J, Perales C, Domingo E, Elena SF (2024) Quasispecies theory and emerging viruses: challenges and applications. NPJ Viruses 2:54
Schuster P (2016) Quasispecies on fitness landscapes. In: Domingo E, Schuster P (eds) Quasispecies: from theory to experimental systems. Current topics in microbiology and immunology, vol 392, pp 61–120
Schuster P, Stadler PF (2023) Virus evolution on fitness landscapes. Curr Top Microbiol Immunol 439:1–94
Shaukat AN, Zhang J (2025) Conduits, doppelgangers, and chimeras: RNA structures at the virus-host Interface. Annu Rev Virol 12(1):277–298
Shirogane Y, Watanabe S, Yanagi Y (2016) Cooperative interaction within RNA virus mutant spectra. Curr Top Microbiol Immunol 392:219–229
Shirogane Y, Harada H, Hirai Y, Takemoto R, Suzuki T, Hashiguchi T, Yanagi Y (2023) Collective fusion activity determines neurotropism of an en bloc transmitted enveloped virus. Sci Adv 9(4): eadf3731
Skums P, Zelikovsky A, Singh R, Gussler W, Dimitrova Z, Knyazev S, Mandric I, Ramachandran S, Campo D, Jh D, Bunimovich L, Costenbader E, Sexton C, O'Connor S, Xia GL, Khudyakov Y (2018) QUENTIN: reconstruction of disease transmissions from viral quasispecies genomic data. Bioinformatics 34:163–170
Solé R, Goodwin B (2000) Signs of life how complexity pervades biology. Basic Books, New York

Soria ME, Cortón M, Martínez-González B, Lobo-Vega R, Vázquez-Sirvent L, López-Rodríguez R, Almoguera B, Mahíllo I, Mínguez P, Herrero A, Taracido JC, Macias-Valcayo A, Esteban J, Fernández-Roblas R, Gadea I, Ruíz-Hornillos J, Ayuso C, Perales C (2021) High SARS-CoV-2 viral load is associated with a worse clinical outcome of COVID-19 disease. Access Microbiol 3:000259

Steffen W, Broadgate W, Deutsch L, Gaffney O, Ludwig C (2015) The trajectory of the Anthropocene: the great acceleration. Anthropocene Rev 2:81–98

Steinhauer DA, Domingo E, Holland JJ (1992) Lack of evidence for proofreading mechanisms associated with an RNA virus polymerase. Gene 122:281–288

Sun F, Wang X, Tan S, Dan Y, Lu Y, Zhang J, Xu J, Tan Z, Xiang X, Zhou Y, He W, Wan X, Zhang W, Chen Y, Tan W, Deng G (2021) SARS-CoV-2 quasispecies provides an advantage mutation pool for the epidemic variants. Microbiol Spectrum 9:e0026121

Sung WJ, Baaijens JA (2025) Algorithms for short-read viral haplotype reconstruction: challenges, solutions, and perspectives. Methods Mol Biol 2955:89–109

Swetina J, Schuster P (1982) Self-replication with errors a model for polynucleotide replication. Biophys Chem 16:329–345

Sztuba-Solinska J, Urbanowicz A, Figlerowicz M, Bujarski JJ (2011) RNA-RNA recombination in plant virus replication and evolution. Annu Rev Phytopathol 49:415–443

Tang JW, Cheung JL, Chu IM, Sung JJ, Peiris M, Chan PK (2006) The large 386-nt deletion in SARS-associated coronavirus: evidence for quasispecies? J Infect Dis 194:808–813

Taye B, Yousaf I, Navaratnarajah CK, Schroeder DC, Pfaller CK, Cattaneo R (2024) A measles virus collective infectious unit that caused lethal human brain disease includes many locally restricted and few widespread copy-back defective genomes. J Virol 98(11):e0123224

Vinicius L (2010) Modular evolution how natural selection produces biological complexity. Cambridge University Press, Cambridge

Wang C, Wang C, Zou J, Yang Y, Li Z, Zhu S (2019) Epigenetics in the plant-virus interaction. Plant Cell Rep 38(9):1031–1038

Xu D, Zhang Z, Wang FS (2004) SARS-associated coronavirus quasispecies in individual patients. N Engl J Med 350:1366–1367

Xu WK, Byun H, Dudley JP (2020) The role of APOBECs in viral replication. Microorganisms 8(12):1899

Yousaf I, Hannon WW, Donohue RC, Pfaller CK, Yadav K, Dikdan RJ, Tyagi S, Schroeder DC, Shieh WJ, Rota PA, Feder AF, Cattaneo R (2023) Brain tropism acquisition: the spatial dynamics and evolution of a measles virus collective infectious unit that drove lethal subacute sclerosing panencephalitis. PLoS Pathog 19:e1011817

Zhang Y, Bryant J, Herron L, Mali P (2025) ADARs: pleiotropy in function, versatility in application. Nucleic Acids Res 53(13):gkaf672

Zimmern D (1988) Evolution of RNA viruses. In: Domingo E, Holland JJ, Ahlquist P (eds) RNA genetics. CRC Press Inc, Florida, pp 211–240

Chapter 22
Wholobionts and Viruses

Mark Little and Forest Rohwer

Abstract Host-associated communities are increasingly studied through the *holobiont* concept, which has been valuable for integrating hosts with their microbial partners across ecology and evolution. Here, we clarify its application by separating an evolutionary scale, where animals protect the germline and inherit a genome-encoded platform for managing microbial exchange through hypervariable elements and fixed genetic modules, from a within-and-across lifetime scale, where host interfaces assemble open communities from environmental pools through combinatorial elements. We restore and formalize the term "*wholobiont*" to name this state-dependent ecological arena of viruses, microbes, and hosts, preserving the original coral framing of the host as habitat while avoiding claims that the entire association functions as a unit of selection or a collectively inherited genome. Using corals as a focal system, we synthesize evidence for compartmentalized and context-dependent viromes and microbiomes, limited consortium-level inheritance, and stress-linked transitions to dysbiosis. We propose hypervariability and selection as unifying mechanisms: hosts deploy combinatorial elements of surface chemistry (mucins, antimicrobial peptides, diverse glycans) that bias attachment, persistence, and invasion, while viruses concentrate at boundary layers and drive transient gene fluxes that generate interaction diversity, regulate microbial mortality, mediate gene flow, and shape immune ecology at mucus and tissue interfaces. By distinguishing heritable host interface capacity from within-lifetime community assembly, the *wholobiont* framework explains how host-associated systems remain coherent through hypervariability and selection, with viruses acting as central regulators of stability and stress-driven transitions.

M. Little (✉)
Scripps Institute of Oceanography, San Diego, CA, USA

San Diego State University, San Diego, CA, USA

F. Rohwer
San Diego State University, San Diego, CA, USA

G. Witzany (ed.), *Integrative Theory of Evolution - Aspects and Insights*,
https://doi.org/10.1007/978-3-032-24295-2_22

1 Introduction

Host-associated communities shape development, physiology, and resilience across the tree of life, and the term "*holobiont*" has become a cornerstone for thinking about these integrated systems, capturing the simple idea that hosts and their microbial partners function together. Yet its meaning still varies across studies, because terminology and theoretical framing have not converged on a single definition, and multiple frameworks are often applied to the same biological phenomena. Much of the ambiguity arises because "*holobiont*" is used to describe two different biological scales: (i) what is inherited across generations as part of the host lineage, and (ii) what is assembled and maintained within a lifetime at host interfaces once the organism is exposed to the environment.

At the evolutionary scale, animals can be understood as lineages that protect the germline from environmental organisms and typically begin life without microbes in the germline, while still allowing that some systems include obligate, vertically transmitted symbionts. In this view, the heritable substrate is not a fixed community membership list but the host's encoded capacity to manage exchange with the microbial world. Hosts deploy diverse, heritable interface phenotypes through combinatorial chemistry, including mucins and glycan architectures, antimicrobial peptides, oligo- and polysaccharides, and regulatory and epigenetic programs. Some of this capacity is encoded in hypervariable elements (for example, expanded immune gene families), whereas much of the expressed diversity is generated within a lifetime by combinatorial elements (for example, dynamic glycosylation patterns and context-dependent antimicrobial chemistry). Together, these hypervariable, genome-encoded traits shape which microbes can colonize, which are excluded, and how rapidly the interface can be reconfigured as conditions change.

At the ecological scale, once an organism is born or settled, it must exchange with its environment, and that exchange creates both opportunity and risk: symbiosis and functional complementation on one hand and invasion, predation, and opportunism on the other. Here, persistence and "health" often reflect interaction dynamics at host interfaces, including niche exclusion by established symbionts, virome and microbiome turnover under changing conditions, and viral-mediated controls through bacterial predation and transient gene fluxes that can rapidly reshape microbial traits and competitive outcomes. In particular, bacteriophage adherence to mucus (BAM) represents a form of animal–phage symbiosis that can position phages at the surface where colonization and invasion occur (Barr et al. 2013), while broader virome dynamics can contribute to exclusion or suppression of pathogens, as illustrated by well-known vertebrate examples such as EBV-like viruses (Barton et al. 2007).

In coral reef biology, Nancy Knowlton and one of us introduced the coral *holobiont* in the early 2000s to emphasize that coral colonies consist not only of animal tissue but also of viruses, bacteria, archaea, fungi, and algal symbionts and that coral performance depends on interactions across this entire community (Rohwer et al. 2002; Knowlton and Rohwer 2003). This framing was explicitly ecological rather than evolutionary: the host was treated as habitat, with multiple micro-niches that could support distinct symbiont assemblages and interaction networks (Knowlton and Rohwer 2003). The original manuscript proposed the term "*wholobiont*," but editorial intervention led to *holobiont* instead. Over time, this ecological usage was further complicated by conflation with Lynn Margulis's earlier use of *holobiont* to describe evolutionary individuals, such as the eukaryotic cell and its mitochondria (Margulis and Fester 1991), contributing to the term's dual use as both an ecological descriptor and a proposed unit of selection. We use macrobe here to denote the host organism itself, while reserving *holobiont* for evolutionary framings that require strong partner fidelity.

Subsequent work reinforced the ecological interpretation in corals, documenting that coral-associated viruses and microbes participate in nutrient cycling, disease dynamics, and stress responses and that reef decline often involves shifts in microbial communities rather than simple loss of a single partner (Bourne et al. 2009; Ainsworth et al. 2010; Thompson et al. 2014; van Oppen et al. 2009; Bettarel et al. 2015; Vega Thurber et al. 2017). The *holobiont* concept was also extended into the hologenome theory of evolution, which proposed that selection can act on the combined genomes of the host and microbiome and that microbial turnover could contribute heritable variation at the level of the association (Zilber-Rosenberg and Rosenberg 2008; Rosenberg et al. 2009; Rosenberg and Zilber-Rosenberg 2018; Roughgarden 2023). This extension has been most informative when it motivates explicit tests of fidelity, transmission, and levels of selection, particularly in systems where viral and microbial membership is dynamic, compartmentalized, and environmentally responsive (Theis et al. 2016; Webster and Reusch 2017; Voolstra et al. 2024) Box 11.1.

Box 11.1: Defines Key Terms Used throughout to Separate Heritable Host-Associated Features from within-Lifetime Community Assembly and Gene Flow

Inheritance Associated with Macrobe

Holobionts are evolutionary assemblages whose combined genomes form a hologenome, and natural selection will act on the host plus the obligate-associated viruses (e.g., endogenous retroviruses) and microbes (e.g., Buchnera, some Wolbachia), as well as organelles like mitochondria and plastids. This evolutionary framing is plausible in systems with absolute/very strong vertical transmission and is most likely Margulis' original endosymbiotic work.

Hypervariable elements refer to loci or gene families that harbor many different alleles encoded in the genome and transmitted across generations. Hypervariable systems are fodder for evolutionary adaptation. Examples are gene family expansions like the diversification of TNF receptor families and other innate immune genes in corals.

(continued)

Box 11.1 (continued)

Combinatorial elements are created by heritable genetic elements that produce functional diversity within a lifetime, BUT the resulting diversity is not heritable. Examples include somatic rearrangement of T-cell receptors and antibodies in vertebrates, dynamic glycosylation patterns of mucins that alter microbial and phage binding landscapes, and context-dependent modulation of antimicrobial chemistry or epigenetic programs.

Fixed genetic modules are horizontally acquired DNA module elements stably integrated and inherited across generations of the macrobe. These modules often drive punctuated evolutionary innovation and long-term host–symbiont adaptation. Examples include endogenized viral elements, CRISPR-Cas acquisition, and stable modular recombination events.

Inheritance Associated with Virome and Microbiome

Wholobionts are the macrobial host with opportunistically associated viruses, microbes, protists, et cetera. Wholobionts are not a unit of evolutionary selection. Functional coherence emerges through coevolutionary dynamics, often at host interfaces, where hypervariable loci and combinatorial elements interact with diversity-generating elements. This is most likely Rohwer et al. (2002) coral *holobiont* concept and is the most common type of macrobial-microbial symbiosis.

Diversity-Generating Elements refer to molecular mechanisms that actively create new genetic variants that are inherited. These mechanisms work through targeted mutation and recombination. These processes are especially prominent in viruses and microbes and include diversity-generating retroelements and gene amplification (Andersson et al. 1998), often coupled with error-prone DNA polymerases.

Transient gene fluxes include strain-level variation in symbionts like proviruses, *Endozoicomonas* in corals, and viral modular recombination, and prophage-mediated gene flux can rapidly supply novel variants that reshape acclimatization.

In corals, multiple empirical studies demonstrate that microbial communities are spatially structured across compartments, highly variable among individuals, and responsive to environmental context, with limited evidence for stable vertical inheritance of specific microbial taxa as a rule (Sweet et al. 2011; Hester et al. 2016; Boilard et al. 2020; Little et al. 2021; Voolstra et al. 2024). This pattern suggests that much of the relevant organization occurs through selection acting on hypervariable host interfaces and on equally dynamic viral and microbial traits. Here, we therefore restore and formalize the concept of the *wholobiont*, defined as the ecological assemblage of viruses, microbes, and host (Figs. 22.1 and 22.2). *Wholobionts* are

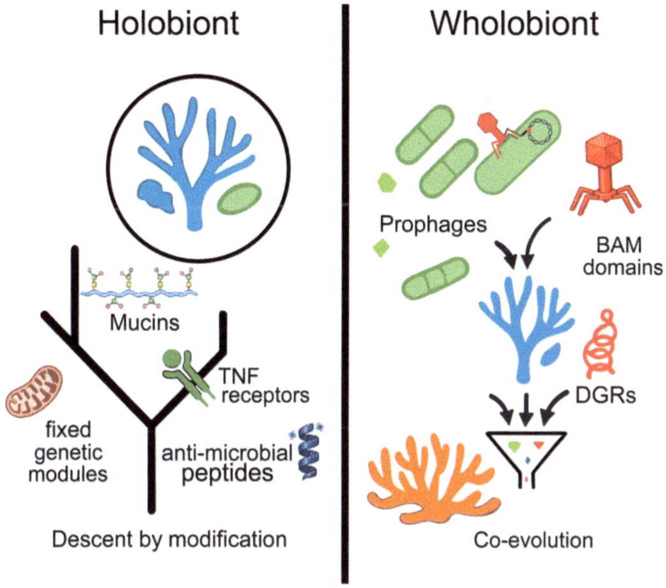

Fig. 22.1 *Holobiont* and *wholobiont* frameworks across evolutionary and ecological scales. Left—*Holobiont*: The *holobiont* is shown as the host lineage and its heritable, genome-encoded capacity to manage exchange with microbes. In most animals, the germline is protected from environmental organisms, so what is reliably inherited is not a fixed community membership list but host traits that bias colonization and exclusion across generations. Examples of heritable interface modules are shown, including mucins and glycan architectures, antimicrobial peptides, and diversified innate immune receptor families such as TNF receptors. Canonical fixed genetic modules (for example, mitochondria and other obligate, vertically transmitted partners where they exist) represent the subset of associations that can contribute to high-fidelity inheritance. This side emphasizes descent by modification through host-controlled interface phenotypes. Right (*Wholobiont*): The *wholobiont* is defined as the open, state-dependent ecological assemblage of host, viruses, and microbes at host interfaces. Its functional coherence does not require group-level inheritance or a collective hologenome. Instead, coherence emerges from continual hypervariability across partners and real-time ecological filtering at interfaces. Viral processes are highlighted as major generators and regulators of this variability, including prophages and their cargo genes, diversity-generating retroelements (DGRs), and phage capsid domains that enable bacteriophage adherence to mucus (BAM). Arrows depict continuous variant generation and interaction, and the funnel represents selection through interface conditions, density-dependent dynamics, and niche exclusion that shape which variants persist. The outcome is a repeatable but dynamic community state that can stabilize function or shift under changing conditions, linking within-host ecology to longer-term coevolutionary trajectories

not units of selection and are not encoded in a collective hologenome. Instead, their functional coherence emerges from continual hypervariability across partners and selection through interactions at host interfaces (Figs. 22.1 and 22.2). This reframing preserves the ecological insight of the concept while avoiding the requirement for multilevel selection.

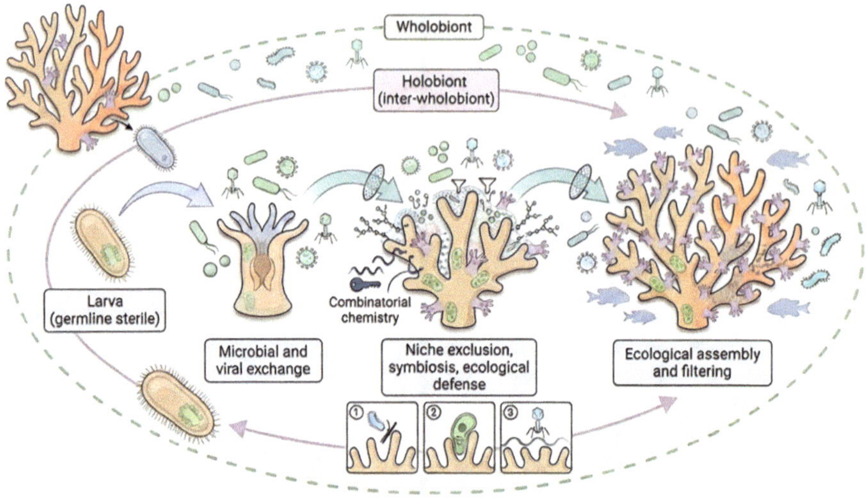

Fig. 22.2 *Holobiont* and *wholobiont* as complementary scales of host-associated biology. A coral is shown from larval settlement through colony growth to reproduction to illustrate how host-associated systems operate across generations and within a lifetime. The *holobiont* (inter-*wholobiont*) label refers to the evolutionary, generation-to-generation arena. Here, the host lineage begins with a larva that is germline sterile and must manage contact with environmental organisms while protecting sensitive host lineages. Rather than inheriting a fixed membership list of microbes, the heritable substrate is the host's encoded capacity to regulate exchange using hypervariable interface traits produced by combinatorial chemistry, including mucus and mucin glycan architectures, antimicrobial peptides, and other surface and immune-associated molecules that bias colonization, persistence, and exclusion. The dashed green envelope labeled *wholobiont* denotes the within-lifetime ecological arena in which the host, viruses, and microbes form an open assemblage that is continually reshaped by environmental exposure. As the colony develops, viral and microbial exchange with surrounding habitats supplies potential colonists and viral predators, while host interfaces and resident communities impose selection. This selection is expressed as niche exclusion, symbiosis, and ecological defense, including microbiome turnover under changing conditions and virus-mediated controls on bacterial competitors. The rightmost panel emphasizes that mature colonies maintain coherence primarily through ecological assembly and selection at interfaces, not through strict inheritance of an entire consortium. Small inset cartoons (1–3) summarize representative interface mechanisms: host-structured microhabitats, symbiont-mediated niche exclusion, and virus-mediated suppression of incoming bacteria at the surface. Arrows indicate the continuous coupling between host development, exchange with the environment, and interface-driven selection that together generate repeatable outcomes despite dynamic membership

2 Wholobionts as Arenas of Hypervariability

The *wholobiont* framework explains how host-associated systems can be coherent despite taxonomic turnover and uneven intergenerational fidelity by shifting emphasis from membership to process. Viruses, microbes, and hosts evolve independently, yet their traits interact within a shared ecological arena. Across generations, heritable change is carried primarily by the host genome (and, in some systems, by obligate symbionts with high partner fidelity), including fixed genetic modules and hypervariable elements that shape interfaces such as mucus composition, glycan presentation, and antimicrobial repertoires. Within a single lifetime, those inherited interface capacities interact with locally available microbial and viral pools, so selection is expressed most strongly through interaction traits and functional outcomes at boundaries, with additional diversity produced by combinatorial elements, rather than requiring the whole assemblage to behave as a single heritable unit (Knowlton and Rohwer 2003). In practice, stable function can emerge even when membership turns over, because hosts can repeatedly select and reassemble communities through compartmentalized microhabitats and chemically and physically structured interfaces (Table 22.1).

This framing is consistent with a broader emphasis on separating the descriptive utility of host plus microbiome language from stronger claims about levels of selection. It also fits with evidence that microbes can influence host evolution and diversification through reciprocal selection on interacting traits, even when partner identities are not faithfully transmitted as a package (Theis et al. 2016; Stencel and Wloch-Salamon 2018; O'Brien et al. 2019).

The *wholobiont* framework provides a mechanistic basis for why host-associated systems can remain functionally coherent even when membership turns over. Coherence arises from two coupled processes: continuous hypervariability across hosts, viruses, and microbes and persistent selection at host interfaces. At these interfaces, niche exclusion among resident microbes and spatially structured viral dynamics, including BAM, can suppress invasion and rapidly reshape competitive outcomes (Barr et al. 2013; Silveira and Rohwer 2016). Figure 22.4 summarizes this logic as a recurring cycle in which variants are generated across partners, assembled into local communities, selected through ecological interactions, and continually regenerated through ongoing diversification (Vega Thurber et al. 2017; Wallace et al. 2025).

3 Hypervariability across Hosts, Microbes, and Viruses

Across all partners of the *wholobiont*, conserved molecular strategies generate extreme diversity at specific loci. In hosts, these include mucins and innate immune receptor families such as tumor necrosis factor (TNF) receptors, which show diversification and expansion in corals (Shinzato et al. 2014; Quistad et al. 2014; Daniels et al. 2015). These expansions are consistent with the idea that coral immunity relies on families of related receptors and signaling modules with variable binding and regulatory properties rather than a small set of fixed recognition elements (Quistad et al. 2014; Vega Thurber et al. 2017).

In bacteria, hypervariability arises through strain-level differences in surface adhesins, secretion systems, mobile genetic elements, and prophage integration. Culture-based work underscores this point: Li et al. (2022) cultured a large collection of bacteria from *Pocillopora* and documented extensive functional diversity across isolates, demonstrating that strain-level variation can supply multiple ecological

Table 22.1 Axioms of the *wholobiont* framework and their empirical support

Axiom	Core claim	Mechanistic basis	Empirical evidence	Representative references
A1. *Wholobionts* are ecological, not evolutionary, units	Host-associated communities function as ecological assemblages, not units of selection	Independent evolution of partners; lack of heritable covariance	No consistent vertical inheritance of microbiomes; conceptual critiques of group selection	Rohwer et al. (2002), Knowlton and Rohwer (2003), Theis et al. (2016), Stencel and Wloch-Salamon (2018), and Rosenberg and Zilber-Rosenberg (2018) (contrast)
A2. Stable function without stable membership	Functional coherence persists despite taxonomic turnover	Ecological assembly and redundancy	No ubiquitous microbial taxa across hosts; stability emerges with sampling depth	Hester et al. (2016), Sweet et al. (2011), and Bernasconi et al. (2019)
A3. Host compartmentalization structures communities	Hosts create distinct microbial niches	Mucus, tissue, skeleton as different selections	Strong compartment-specific microbiomes and immune regulation	Sweet et al. (2011), Kvennefors et al. (2012), Little et al. (2021), and Pogoreutz et al. (2022)
A4. Hypervariability is a shared design principle	Hosts, viruses, and microbes generate diversity at specific loci	Gene family expansion, mobile elements, viral recombination	Diversified immune receptors, strain-level bacterial diversity, viral genomic plasticity	Shinzato et al. (2014), Quistad et al. (2014), Li et al. (2022), and Weynberg et al. (2017)
A5. Viruses are central regulators, not peripheral agents	Viruses structure communities via predation, gene flow, and immunity	Lytic control, horizontal gene transfer, immune modulation	High viral abundance in mucus; recurrent coral viromes	Marhaver et al. (2008), Bettarel et al. (2015), Vega Thurber et al. 2017, and Zhou et al. (2024)
A6. Viral strategies are context-dependent	Lysis and lysogeny shift with ecological conditions	Piggyback-the-winner; metabolic control of switching	Increased lysogeny at high host density; metabolism-regulated lysis	Silveira et al. (2023), Kelman et al. (2025), and Chuang et al. (2025)

(continued)

Table 22.1 (continued)

Axiom	Core claim	Mechanistic basis	Empirical evidence	Representative references
A7. Ecological immunity emerges at interface selection	Immunity is an emergent ecological property, not host-only	Mucus scaffolding, BAM, apoptotic selection	BAM; viral immune-like genes; TNF diversification	Barr et al. (2013), Sweet and Bythell (2017), Quistad et al. (2017), and Correa et al. (2016)
A8. Dysbiosis reflects state shifts, not single causes	Bleaching and disease reflect reconfiguration	Stress alters selection surfaces	Microbial and viral restructuring under heat stress	Bourne et al. (2009), Vega Thurber et al. (2009), Boilard et al. (2020), and Howe-Kerr et al. (2023)
A9. Pathobiomes arise from altered states	Parasites and opportunists expand under specific states	Conditional pathogenicity	Aquarickettsia linked to susceptibility; vibrio blooms under heat stress	Klinges et al. (2020), Baker et al. (2022), and Martin-Cuadrado et al. (2024)
A10. *Wholobionts* are arenas of coevolution	Coevolution occurs through interacting traits, not inheritance	Reciprocal selection on hypervariable systems	Coevolutionary signals without group selection	O'Brien et al. (2019), Thompson et al. (2014), and Webster and Reusch (2017)

This table summarizes the core axioms underlying the *wholobiont* framework and links each axiom to its mechanistic basis, empirical evidence, and representative literature. Here, the term "axiom" is used to denote foundational principles that are constrained by existing empirical data, rather than assumptions adopted without evidence. Each axiom synthesizes results from multiple independent studies across coral, microbial, and viral ecology and reflects recurring patterns observed across systems

roles even within a single host species. In dominant coral-associated lineages such as *Endozoicomonas*, genomic plasticity and viral integration provide additional routes for diversification: bacteriophages associated with *Endozoicomonas* exhibit temperate lifestyles and strain-specific genomic features, embedding viral variability directly into a major coral-associated bacterial clade (Pogoreutz et al. 2022; Chuang et al. 2025).

In viruses, diversity is generated through mutation, recombination, modular genome architectures, and, in some phage lineages, through diversity-generating retroelements and hypervariable binding domains (Weynberg et al. 2017; Vega Thurber et al. 2017; Wallace et al. 2024). Importantly, genome-resolved coral viromics increasingly show that these viral repertoires include widely distributed viral genomes and gene inventories consistent with structured ecological roles and with multi-partner molecular interactions (Cárdenas et al. 2020; Wallace et al. 2024). Rather than representing noise, hypervariability supplies a portfolio of possible interactions on which ecological selection can act (Vega Thurber et al. 2017; Varona et al. 2024).

4 Viruses as Central Regulators of Wholobiont Ecology

Viruses are abundant in coral-associated habitats and are strongly enriched at host boundary layers. Viral particle concentrations in coral mucus frequently exceed those in surrounding seawater and can rival or exceed bacterial abundance in that layer (Marhaver et al. 2008; Bettarel et al. 2015; Ambalavanan et al. 2021; Zhou et al. 2024). At these densities, viruses exert strong control over microbial mortality, horizontal gene transfer, and competitive dynamics on ecological time scales (Rohwer and Thurber 2009; van Oppen et al. 2009; Vega Thurber et al. 2017; Varona et al. 2024).

Viromic surveys consistently identify a recurring set of viral groups associated with corals, including tailed bacteriophages and diverse eukaryotic viral lineages (Bettarel et al. 2015; Cárdenas et al. 2020; Vega Thurber et al. 2017; Sweet and Bythell 2017; Ambalavanan et al. 2021). A key recent shift is that coral viral ecology is increasingly supported by genome-resolved approaches. Wallace et al. (2024) assembled a large coral-associated viral sequence resource and showed that many viral genomes are globally distributed, while their gene content is consistent with structured ecological roles and with multi-partner molecular interactions, including viral genes plausibly involved in direct interactions with the eukaryotic host (Weynberg et al. 2017; Wallace et al. 2024). This strengthens a central *wholobiont* claim: viruses are not peripheral because their genomes encode mechanisms that can reshape bacterial function, regulate microbial competitors, and interact with host processes (Vega Thurber et al. 2017; Wallace et al. 2024).

In the *wholobiont* framework, viruses are not merely pathogens. They are regulators that generate diversity, prune microbial populations, redistribute genes, and modulate immune landscapes. This regulatory role is conceptualized in Fig. 22.3 as coupled lytic predation in outer mucus layers and lysogenic stabilization in deeper microhabitats (Silveira and Rohwer 2016; Silveira et al. 2023; Wallace et al. 2025).

5 Viral Dynamics in Mucus and Ecological Immunity

Viruses are typically the most abundant biological entities in microbial ecosystems and can strongly influence trophic structure, population turnover, and gene flow. In host-associated systems, this abundance means that *wholobionts* can also function as persistent viral reservoirs that are continually exchanging viruses with surrounding habitats and with neighboring hosts. A key question, especially in benthic

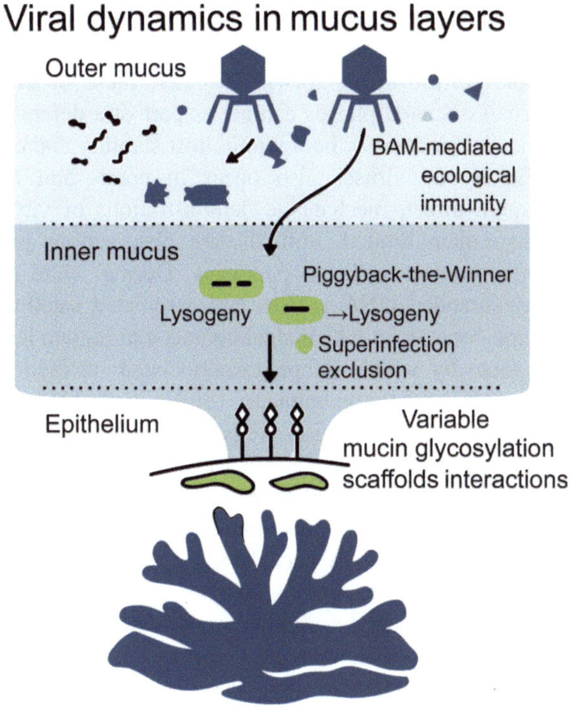

Fig. 22.3 Viral dynamics in mucus layers. The mucus layer above epithelia provides a structured environment where phage-host interactions drive *wholobiont* stability. In outer mucus, lytic phages adhering to mucins attack incoming bacteria, providing BAM-mediated ecological immunity. In deeper mucus, commensal bacteria carrying prophages benefit from lysogeny and superinfection exclusion, as described by the Piggyback-the-Winner model. The host maintains these dynamics by secreting mucins with variable glycosylation patterns that scaffold viral and microbial interactions

competition, is how host interfaces and community structure create conditions that retain viruses and, in doing so, shape infection risk and cross-host spillover dynamics.

In coral-dominated reef states, most viruses detected at host interfaces are bacteriophages (phages) that infect bacterial members of the microbiome (Silveira et al. 2020). Many phages act as predators via lytic infection, amplifying within bacterial cells and releasing progeny through lysis, while others persist as temperate viruses that can switch between lysogeny and lysis depending on ecological context (Barr et al. 2013; Silveira and Rohwer 2016). This creates a plausible route for "nonhost encoded" defense at animal surfaces: phages enriched at mucus interfaces can suppress incoming bacterial colonists and reduce the probability that potential pathogens establish footholds (Barr et al. 2013; Silveira and Rohwer 2016). In coral–algal competitive settings, comparative work has reported higher concentrations of viral-like particles and lower bacterial densities near corals relative to adjacent algal competitors, yielding elevated virus-to-microbe ratios at the coral interface (Roach et al. 2020). Functional specificity is also possible within this framework. For example, phages detected in coral-associated settings have included cyanophages that can infect cyanobacteria that are common components of turf-associated microbial assemblages, providing a mechanistic route by which one *wholobiont* can harbor viruses with consequences for neighboring competitors (Little et al. 2021). More directly, coral mucus has been reported to contain phages infecting the coral pathogen *Vibrio mediterranei*, consistent with in situ host–phage associations that could contribute to pathogen suppression at the mucus boundary (Rubio-Portillo et al. 2024). Together, these observations support the idea that mucus-associated phages can act as part of a defensive ecological layer, with outcomes that matter for both within host stability and cross-host interactions.

Eukaryotic viruses also occur in coral- and algae-associated assemblages, although direct, mechanistic demonstrations of virus transfer between coral and algae remain limited. Still, disease observations provide relevant precedent for benthic contact-mediated pathways. During white plague disease, Rep-encoding single-stranded DNA viruses and associated satellites (SCSDVs) were linked to disease contexts in which algal contact can initiate lesions, highlighting one plausible route by which competitor-associated interactions could intersect with viral dynamics at the tissue boundary (Soffer et al. 2014).

Beyond defense, viruses are potent agents of ecological and evolutionary change through gene movement in microbial hosts. Under conditions favoring lysogeny, temperate phages can integrate and persist, and prophage cargo genes can alter host phenotypes, including traits linked to antimicrobial resistance, virulence, and metabolism (Bondy-Denomy and Davidson 2014; McKerral et al. 2023). In coral–algal comparisons, metagenomic patterns consistent with higher prophage-associated gene content in algal-associated microbial communities, coupled with lower apparent free-virus signal, have been interpreted as a shift toward temperate strategies in benthic algae relative to corals (Roach et al. 2020). This aligns with Piggyback-the-Winner dynamics, in which high microbial abundance and reduced lytic pressure favor lysogeny and prophage enrichment (Knowles et al. 2016; Coutinho et al. 2017). Importantly, prophage enrichment has been linked in multiple systems to increased community pathogenic potential through prophage-encoded virulence factors and shifts in bacterial competitive traits (Brüssow et al. 2004; Munro et al. 2003; Brüssow 2007; Weynberg et al. 2015; Little et al. 2020). At the reef scale, these viral and microbial shifts are consistent with broader trajectories described as reef microbialization, where ecosystem state change amplifies microbial biomass and alters host–microbe interaction regimes (Haas et al. 2016). In that context, viral strategy and viral gene flow are central drivers that can determine whether benthic assemblages remain functionally stable or instead shift toward dysbiosis and disease-prone states.

The mucus layer overlying coral epithelia is a structured ecological matrix that concentrates viruses and microbes and imposes steep physical and chemical gradients (Sweet et al. 2011; Sweet and Bythell 2017). Within this environment, bacteriophages can adhere to mucins via capsid immunoglobulin-like domains, as formalized in the BAM model (Barr et al. 2013). Phage Ig-like domains are structurally conserved yet sequence-diverse, enabling adjustment to changing mucin glycosylation patterns. Coral mucins themselves exhibit extensive glycan heterogeneity across individuals and physiological states, creating a shifting receptor landscape. The interaction between these two hypervariable systems produces a moving selective interface that enriches phages in mucus and increases encounter rates with incoming bacteria (Nguyen-Kim et al. 2015; Sweet and Bythell 2017;

Silveira and Rohwer 2016). This process constitutes a form of ecological immunity, in which protection emerges from continual matching between viral and host variability rather than from fixed immune specificity. Figure 22.3 illustrates this mechanism as a spatially stratified interface in which viral processes shape microbial colonization and invasion dynamics (Barr et al. 2013; Sweet and Bythell 2017; Vega Thurber et al. 2017; Silveira and Rohwer 2016).

6 Viral Innovations as Engines of Variability

Viral life-history strategies often track ecological context because the payoffs of lysis versus temperate integration depend on host density, resource supply, and the stability of local microbial populations. Under conditions of high microbial abundance and comparatively steady growth, temperate strategies can become advantageous, and the Piggyback-the-Winner framework predicts increased lysogeny with prophages contributing superinfection exclusion, accessory functions, and altered competitive outcomes (Knowles et al. 2016; Silveira and Rohwer 2016; Silveira et al. 2023). In host-associated systems, this expectation becomes especially relevant because host-created microhabitats can maintain persistently high microbial densities and steep chemical gradients over small spatial scales.

Building on this, Silveira and Rohwer (2016) proposed that mucus-associated ecosystems can support layered viral regimes, where lytic and temperate dynamics partition across surface-proximal versus deeper mucus microenvironments. In this view, the same host-associated habitat can simultaneously support stabilization through lysogeny (via immunity and accessory gene carriage in resident bacteria) and exclusion through lytic predation (via suppression of invaders or competitively weaker strains), with the balance shaped by microbial density, nutrient and oxygen gradients, and host-controlled mucus turnover and chemistry (Silveira and Rohwer 2016; Vega Thurber et al. 2017). By tying microenvironmental structure to lytic versus temperate partitioning, this framework yields clear expectations for when viral processes stabilize residents versus exclude invaders at the host boundary (Fig. 22.3).

Mechanistic work also supports the plausibility of context-sensitive switching by demonstrating that viral decisions can be coupled to bacterial physiology. Kelman et al. (2025) showed that bacterial central carbon metabolism can regulate lytic switching, directly tying metabolic state to induction potential. In coral mucus and tissue microhabitats, where bacteria can experience high nutrient availability, intense competition, and fluctuating redox conditions, these physiological controls provide a concrete route by which local interface conditions could bias communities toward lysogeny, induction, or prophage-mediated trait exchange (Sweet and Bythell 2017; Silveira and Rohwer 2016).

Genome-resolved studies of coral-associated prophages add direct empirical support for this mechanism. Wallace et al. (2025) found that in *Madracis mirabilis*, successful colonies were associated with higher viral and microbial richness and

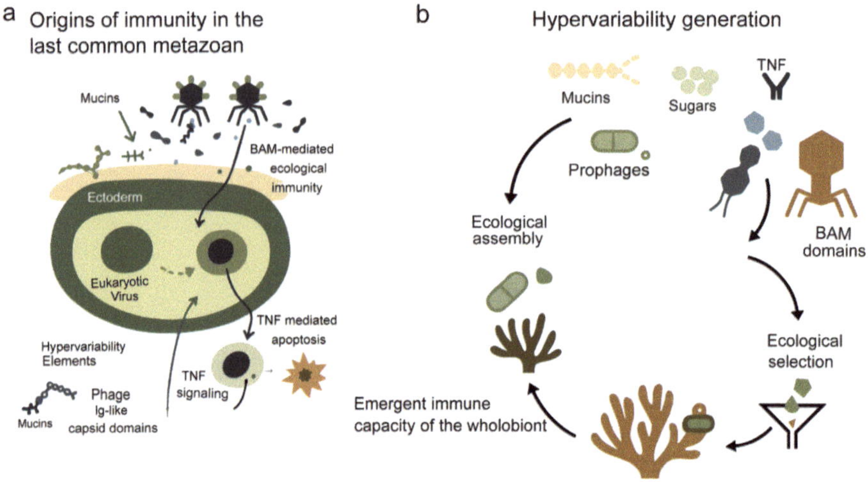

Fig. 22.4 Hypervariability as the origin and engine of *wholobiont* immunity. (**a**) Origins of immunity in the last common metazoan ancestor (LCMA). The LCMA is depicted as a simple, two-layer organism with a mucus-secreting ectoderm that creates a structured interface for viral and microbial interactions. At the external surface, mucus-associated phages adhering to mucins provide a first line of ecological immunity by increasing encounters with invading bacteria and helping exclude potential pathogens (BAM-mediated ecological immunity). Within tissues, intracellular stressors, including eukaryotic viral infection, can be tolerated when low-cost or controlled; when damaging or pathogenic, they can be countered by conserved TNF receptor ligand signaling that triggers apoptosis to eliminate compromised cells. In both compartments, immune function depends on hypervariability: diversification of host mucin glycans and phage Ig-like capsid domains at the surface, and diversification of TNF receptor families and downstream signaling internally. (**b**) Coevolutionary cycle of hypervariability and ecological assembly. Across viruses, microbes, and hosts, new variants are continually generated through mucin diversification and mucus sugars, BAM domains, prophages, diversity-generating retroelements (DGRs), and immune receptor families. These variants assemble into dynamic communities at host interfaces, where ecological selection favors functional combinations and removes maladaptive ones. Repeated cycles of generate, assemble, and select produce the emergent immune capacity of the *wholobiont*, maintaining functional coherence without requiring that the *wholobiont* act as an evolutionary unit of selection

functional redundancy, alongside enrichment of proviruses and prophage-linked genes. Reported prophage gene content included functions plausibly relevant to bacterial performance under stress and competition, including antibiotic resistance, central metabolism, and oxidative stress tolerance. Rather than implying community-level inheritance, these patterns are consistent with prophages acting as a flexible reservoir of accessory functions that can reshape bacterial ecological strategies while remaining constrained by host selection and local environmental conditions (Vega Thurber et al. 2017; Wallace et al. 2025) (Figs. 22.3 and 22.4).

Endozoicomonas provides a complementary illustration of how an abundant bacterial associate can be repeatedly maintained at a host interface without requiring fixation of a single strain. The lineage is often enriched in coral tissues yet exhibits substantial strain-level diversity across colonies and regions, consistent with strong host selection of interaction traits and functional potential rather than strict taxonomic identity (Sweet et al. 2011; Hester et al. 2016; Pogoreutz et al. 2022). Coral-derived cues can prime *Endozoicomonas* for symbiosis and shift transcriptional programs linked to host interaction, immune-relevant processes, and metabolite exchange (Pogoreutz et al. 2022). Genomic plasticity, including mobile elements and temperate phage interactions, then provides a route for continual diversification within that host-selected niche, with *Endozoicomonas*-associated phages showing temperate strategies and strain-specific genomic features that embed viral variability within bacterial genomes (Chuang et al. 2025). Together, these data support a mechanism in which host-controlled microhabitats recurrently favor *Endozoicomonas*-like traits, while prophages and other mobile elements generate variants that are differentially retained under shifting nutrient, immune, and stress regimes, stabilizing the association at the level of functions and interaction traits rather than fixed strain identity (Pogoreutz et al. 2022; Chuang et al. 2025; Wallace et al. 2025).

7 Stress-Linked Pathobiome Shifts: Intracellular Parasites, Vibrios, and Virus-Mediated Competitive Cascades

The *wholobiont* framework accommodates mutualists, commensals, parasites, and opportunists within the same host-associated arena. Therefore, the virome and microbiome exist as a state-dependent assemblage shaped by host control, resource gradients, and biotic interactions at host interfaces. Under relatively stable conditions, these selections can constrain opportunists and limit the proliferation of harmful associates. Under heat stress, bleaching, or disease, however, the physicochemical environment of the *wholobiont* can shift rapidly, for example, through changes in oxygen availability, dissolved organic matter, and immune regulation. These shifts alter the selective landscape experienced by viruses and microbes and can open pathways for pathobiont expansion, parasite proliferation, and community-wide restructuring (Bourne et al. 2009; Vega Thurber et al. 2009; Littman et al. 2011; Boilard et al. 2020; Sweet and Bythell 2017; Nunn et al. 2025) (Fig. 22.4).

One increasingly well-documented class of coral-associated microbes that fits this "state shift" framing is intracellular Rickettsiales bacteria, including *Candidatus Aquarickettsia rohweri*. *Aquarickettsia* is best interpreted as a host-dependent intracellular associate with parasitic tendencies, and its abundance can rise to become a major component of the coral-associated bacterial signal under host and environmental contexts. In *Acropora cervicornis*, higher *Aquarickettsia* abundance has been associated with disease susceptibility, while disease-resistant genotypes tend to show lower *Aquarickettsia* loads and more even, diverse microbiomes (Klinges et al. 2019, 2020). The key inference is not that *Aquarickettsia* enhances *holobiont* adaptability but that host genotype and physiological condition can bias the accessibility of pathobiome states without implying inheritance of a stable, beneficial microbiome (Boilard et al. 2020; Voolstra et al. 2024).

Bleaching-linked dysbiosis provides a broader community-scale example of the same principle. Boilard et al. (2020) frame bleaching as staged dysbiosis, ranging from reversible community shifts to maladaptive states characterized by opportunistic invasion and altered microbial functions. Among the taxa repeatedly implicated as conditional pathogens are *Vibrio* spp., which can persist at low abundance in apparently healthy corals but increase under heat stress alongside restructuring of microbial and viral assemblages (Bourne et al. 2009; Vega Thurber et al. 2009; Sweet and Bythell 2017; Martin-Cuadrado et al. 2024; Voolstra et al. 2024). Viral and multi-omic evidence further supports that bleaching-associated state shifts can include coordinated changes in viral repertoires and host-associated metabolism, rather than only bacterial replacement (Marhaver et al. 2008; Vega Thurber et al. 2017; Roach et al. 2021; Nunn et al. 2025; Zhang et al. 2024).

Viruses add an additional layer because they can reshape competitive outcomes among bacteria and shift quickly with community context. Under stress, prophage induction and shifts between lysogeny and lysis can convert latent viral reservoirs into active ecological forces, changing which bacterial strategies succeed and which collapse (Silveira et al. 2023; Wallace et al. 2025; Vega Thurber et al. 2017). In this framing, disease emergence is often most interpretable as an interaction-network transition, in which host physiology, opportunistic bacteria, and viral dynamics jointly reconfigure the community, rather than as a single agent acting in isolation (van Oppen et al. 2009; Vega Thurber et al. 2017; Wallace et al. 2024) (Fig. 22.4).

8 Inheritance

A central motivation of *holobiont* models is to explain how host-associated microbial communities might contribute to evolutionary change by adding a heritable component of variation to host phenotype. In the hologenome theory of evolution, selection is proposed to act on the combined genomes of host and microbiome, and changes in community composition or function are framed as potentially contributing to heritable variation at the level of the association (Zilber-Rosenberg and Rosenberg 2008; Rosenberg et al. 2009; Rosenberg and Zilber-Rosenberg 2018). This framing does not require strict vertical transmission of every microbial member, but it does require that some aspects of the association (specific partners, functional capacities, or consistently recruited guilds) show sufficient fidelity across generations to generate parent-to-offspring covariance in fitness-relevant traits (Theis et al. 2016; Webster and Reusch 2017). In corals, the question is therefore not whether microbes matter, but which forms of fidelity exist and at what biological scales they are strong enough to support inheritance-like dynamics (Bourne et al. 2009; Thompson et al. 2014; Voolstra et al. 2024).

To clarify what "fidelity" can realistically mean in corals, it is useful to separate two coupled arenas of *wholobiont* biology (Ferretti et al. 2025). In an inter-*wholobiont* (*holobiont*) arena, any high-fidelity partnerships must be routed through reproduction and early development while microbial access to sensitive host lineages is restricted. Here, the heritable substrate is less a taxonomic "package" of microbes than the host's evolved, high-dimensional interface traits that bias which symbionts can persist or re-establish each generation, via combinatorial chemistry such as mucus-associated glycans and mucin-like matrices, antimicrobial compounds, and other surface and immune-associated molecules that tune colonization success (Nyholm and McFall-Ngai 2003; Sweet et al. 2011; Kvennefors et al. 2012; Sweet and Bythell 2017). In an intra-*wholobiont* arena, within- and across-lifetime stability and invasion resistance can emerge through symbiosis, niche exclusion, and density-dependent viral and microbial interactions even when membership turns over, because the stabilizing force is selection acting at host interfaces rather than strict consortium inheritance (Bourne et al. 2009; Sweet and Bythell 2017; Vega Thurber et al. 2017). This separation aligns with broader cautions that host specificity or intimacy does not, by itself, demonstrate coevolution or *holobiont* multilevel selection without direct evidence for the required inheritance structure (Moran and Sloan 2015).

Empirical work in corals highlights constraints on consortium-level inheritance. Microbial communities are strongly compartmentalized within colonies, with distinct assemblages across mucus, tissue, and skeleton, consistent with corals functioning as mosaics of microhabitats that differ in turnover, resource supply, and exposure (Sweet et al. 2011; Little et al. 2021). This implies that any inheritance-like fidelity should be evaluated at the level of compartments and interfaces rather than treated as a single uniform "microbiome." Corals can also regulate bacterial communities through immune-like processing, including antimicrobial activity that shapes persistence and limits overgrowth or invasion (Bourne et al. 2009; Kvennefors et al. 2012). Disturbance further complicates inference: bleaching and disease are often characterized as transitions between eubiotic and dysbiotic states involving community restructuring, and metagenomic studies report shifts in viral and microbial assemblages and functional gene profiles under stress, highlighting rapid environmental reconfiguration of membership and function (Vega Thurber et al. 2009; Littman et al. 2011; Boilard et al. 2020; Vega Thurber et al. 2017; Sweet and Bythell 2017; Voolstra et al. 2024; Nunn et al. 2025).

Community-scale analyses also show that apparent stability can arise without strict taxonomic fixation. The Abundance–Ubiquity framework identified a high-diversity component that becomes statistically stable with sufficient sampling alongside a sporadic component that varies across hosts and sites, emphasizing that stability can reflect consistent ecological structure rather than the universal presence of particular taxa (Hester et al. 2016). From a *holobiont* perspective, this suggests inheritance-like effects, if present, may involve conserved functional roles or reliably enriched groups rather than transmission of a fixed membership list (Webster and Reusch 2017; Voolstra et al. 2024). Consistent with this, network analyses of algal symbiont–bacteria co-occurrence show strong context dependence and host or basin specificity alongside a smaller shared component, again supporting structured assembly over fixed taxonomic inheritance (Bernasconi et al. 2019).

Transmission mode provides a complementary lens. Vertical transmission can generate parent-to-offspring covariance, whereas horizontal acquisition shifts systems toward repeated assembly each generation (Roughgarden et al. 2018). Rosenberg and Zilber-Rosenberg (2018) emphasize vertical transmission as a route to higher-fidelity hologenomic inheritance while noting that horizontal acquisition can expand access to novel microbial variation. In corals, Symbiodiniaceae transmission varies with reproductive mode: vertical transmission is common in brooding corals, whereas many broadcast spawners acquire symbionts from the environment, with mixed modes also occurring (Thompson et al. 2014; Boilard et al. 2020; Stévenne et al. 2021; Gantt et al. 2025). For bacterial partners, evidence supports multiple routes, including associations with gamete bundles or larval stages and parent-mediated enrichment that biases early recruitment, rather than strict transmission of an entire microbiome as a fixed package (Thompson et al. 2014; Boilard et al. 2020; Voolstra et al. 2024). Together, these patterns map onto the inter vs. intra distinction: inter-*wholobiont* inheritance is most plausible for partner-specific, high-fidelity symbioses routed through constrained transmission opportunities and selection, whereas intra-*wholobiont* stability and defense are maintained by ecological dynamics within compartments and at interfaces.

Overall, decades of work support a consistent picture: viral and microbial partners can strongly influence coral phenotype, while the persistence of specific partnerships varies in fidelity and mechanism. The dominant empirical signal is that coral-associated communities are structured by microhabitat, selection, and environmental context, with substantial context dependence in membership and function (Sweet et al. 2011; Kvennefors et al. 2012; Bourne et al. 2009; Vega Thurber et al. 2009; Littman et al. 2011; Boilard et al. 2020; Voolstra et al. 2024). This does not negate hologenome arguments, but it suggests that the strongest and most testable predictions in corals are likely to apply to specific high-fidelity host–symbiont pairings or reproducible functional modules, rather than to the entire virome and microbiome as a single inherited unit (Theis et al. 2016; Webster and Reusch 2017; Zilber-Rosenberg and Rosenberg 2008; Rosenberg et al. 2009; Rosenberg and Zilber-Rosenberg 2018). These considerations motivate the conceptual distinction in Fig. 22.1: *wholobionts* can be coherent and biologically consequential through structured variability and selection at host interfaces even when consortium-level inheritance is limited, uneven, or partner-specific.

9 Microbiome-Based Interventions as Experiments in Wholobiont Selection

A practical implication of the *wholobiont* framework is that it generates testable expectations for when microbiome manipulation should work. If host-associated communities are shaped by continual variant generation and selection at interfaces, then interventions should be most effective when they either (i) introduce traits that can pass through existing host selection or (ii) shift the selections themselves by altering resource landscapes, antagonistic interactions, or local redox conditions

(Webster and Reusch 2017; Voolstra et al. 2021, 2024). Recent syntheses of the coral microbiome explicitly position microbiome-based interventions both as experimental tools for identifying mechanisms that underpin coral health and as potential strategies for promoting resilience under climate change (Voolstra et al. 2024).

A key concern in open marine systems is whether introduced strains can measurably restructure host-associated communities without producing off-target effects in surrounding habitats. Delgadillo-Ordoñez et al. (2024) provide a direct, field-based test by repeatedly inoculating healthy *Pocillopora verrucosa* colonies in situ in the Red Sea with beneficial microorganisms for corals (BMCs) over three months. The inoculated genera (including *Halomonas*, *Pseudoalteromonas*, and *Bacillus*) were enriched in treated corals; treatment was associated with increases in other groups often framed as beneficial and with decreases in potential pathogens such as *Vibrio*, and surrounding seawater and sediment communities did not show detectable changes at the studied scale. Although coral health outcomes could not be evaluated because both treated and control colonies remained healthy, the study still supports a clear inference: even comparatively stable microbiomes can be shifted in the field, but sustained inoculation may be needed to overcome the inertia of an established, host-selected community (Voolstra et al. 2024).

Phototrophic bacteria motivate a complementary intervention logic. Osman et al. (2025) synthesize evidence that phototrophic bacteria could support coral nutrition and reduce oxidative stress during bleaching through photosynthesis and antioxidant production, while emphasizing that their functional contribution to *holobiont* energy budgets and heat stress resilience remains underexplored (Roach et al. 2020; Nunn et al. 2025). In the *wholobiont* framework, these microbes are therefore interesting not simply as "added members," but as potential modifiers of energetic and redox conditions at the host interface, with downstream consequences for which viral and microbial strategies are favored during bleaching-associated transitions.

Together, these intervention studies reinforce the central *wholobiont* claim that coherence is maintained by ecological processes acting on variant repertoires. Microbiome manipulation can therefore be interpreted as a controlled perturbation of ecological assembly and trait distributions, and its successes and failures can be used to identify which interfaces, traits, and viral strategies most strongly stabilize function (Fig. 22.4) (Voolstra et al. 2021, 2024; Vega Thurber et al. 2017).

10 Viral–Host Interactions and the Origins of Immunity

The *wholobiont* framework reframes the origins of metazoan immunity by centering mucus interfaces and viral pressure as persistent, reciprocal selection forces rather than treating immunity as a purely host-encoded trait. The last common metazoan ancestor likely possessed a mucus-secreting ectoderm overlying simple internal tissues, creating a boundary layer where microbial encounter rates were high and where immune function could emerge through control of adhesion, exposure, and local mortality (Fig. 22.4). In such a system, external microbial selection could be mediated in part by phages adhering to mucins, as formalized in the BAM model, which concentrates viruses at the precise interface where bacterial colonization and invasion occur (Barr et al. 2013). Because mucus is continuously produced and shed, and because mucins present a chemically heterogeneous glycan landscape, this boundary layer should be understood as a dynamic ecological matrix rather than a static barrier, with immune function expressed as a moving selection that shapes which microbes can persist at the surface (Sweet et al. 2011; Kvennefors et al. 2012; Sweet and Bythell 2017; Silveira and Rohwer 2016).

A key implication is that early immunity could be partly ecological and spatial. By enriching phages in mucus, the host effectively positions a lytic and gene-transfer-capable agent at its outer surface, increasing the likelihood that incoming bacteria encounter phages before they establish deeper associations. This is not equivalent to adaptive immune specificity. It is closer to an interface defense that depends on densities, encounter rates, and trait matching, and that can remain effective even when community membership turns over, because selection is acting on interaction traits rather than on a fixed list of symbionts (Barr et al. 2013; van Oppen et al. 2009; Sweet and Bythell 2017; Vega Thurber et al. 2017). In the *wholobiont* framing, the crucial point is that both partners can continuously generate variation relevant to this interface: hosts vary mucus composition and glycan presentation, while phages vary binding and host-range traits, allowing selection at the surface to remain responsive (Barr et al. 2013; Nguyen-Kim et al. 2015; Vega Thurber et al. 2017; Varona et al. 2024).

Internal tissues add a second, complementary layer of selection shaped strongly by eukaryotic viruses and by host pathways that control infected or damaged cells. Coral genomics supports the plausibility of this architecture by showing that cnidarians possess expanded and diversified innate immune repertoires, including immune signaling components and receptor families that can mediate apoptosis and inflammatory-like responses (Shinzato et al. 2014; Daniels et al. 2015). Quistad et al. (2014) specifically highlighted diversification of tumor necrosis factor (TNF) receptor repertoires in corals, consistent with an immune strategy built around families of related receptors and signaling modules rather than a small set of fixed recognition elements. Within the *wholobiont* framework, this diversification is the molecular substrate that allows internal immune control to track a shifting landscape of viral challenges and to manage the tradeoffs between tolerance and elimination through regulated cell death and cytokine-like signaling (Shinzato et al. 2014; Quistad et al. 2014).

Importantly, viral communities associated with corals appear positioned to interact with these immune processes rather than simply co-occurring. Quistad et al. (2017) used coral viromes to predict and contextualize novel immune-related proteins and domains and report signatures consistent with viral interaction with core innate immune architectures, including TNF/TNFR-linked signaling context and additional innate immune pathway components (for example, TLR and NLR contexts) alongside apoptosis-associated functions (Quistad et al. 2017; Vega Thurber et al. 2017). Separately, algal symbiont and coral-associated eukaryotic viruses include genes and domains that plausibly interface with host immunity, including ankyrin-repeat proteins reported in Symbiodiniaceae viral contexts under heat stress (Brüwer et al. 2017; Weynberg et al. 2017; Vega Thurber et al. 2017). Evidence for eukaryotic viral infections within the coral symbiosis further strengthens the internal-pressure component of Fig. 22.4. Correa et al. (2016) demonstrated that eukaryotic viral infections in Symbiodiniaceae are detectable and relevant to bleaching contexts, and additional work shows that viruses associated with Symbiodiniaceae can be widespread and responsive to thermal stress (Howe-Kerr et al. 2023; Brüwer et al. 2017). Taken together, these findings support interpreting immunity as an emergent outcome of interacting hypervariable systems rather than as a host-only trait: a surface layer shaped by mucus chemistry and phage enrichment and an internal layer shaped by diversified immune receptor families and apoptosis pathways under persistent viral selection (Barr et al. 2013; Shinzato et al. 2014; Quistad et al. 2014, 2017; Correa et al. 2016; Brüwer et al. 2017) (Fig. 22.4).

In this view, immune capacity is continually renewed because each partner generates variant repertoires on ecological time scales, while selection at interfaces stabilizes function without requiring inheritance of a fixed community. This is the mechanistic bridge between the immunity schematic (Fig. 22.4) and the broader *wholobiont* cycle (Fig. 22.4): immunity arises from repeated generation of variation in mucus traits, viral binding and infection traits, and host immune receptor and apoptosis pathways, with selection acting through real-time interactions at host boundary layers and within host tissues rather than through the inheritance of a single, stable "immune microbiome" (Barr et al. 2013; Nguyen-Kim et al. 2015; Quistad et al. 2014, 2017; Vega Thurber et al. 2017).

11 Summary

Macrobes begin life with a heritable, genome-encoded platform that can generate hypervariability and constrain microbial colonization. This platform includes protecting sensitive host lineages during reproduction and deploying combinatorial surface chemistry, such as mucins, antimicrobial peptides, and diverse glycans, that shape which microbes can attach, persist, or invade. In that sense, evolutionary change primarily acts on host traits and on any high-fidelity symbioses that are consistently routed across generations.

After birth, however, most day-to-day performance is governed by the *wholobiont*: an open, state-dependent ecological arena in which viral, microbial, and host traits interact, compete, and are continually filtered at host interfaces. Within this arena, viruses are central regulators rather than background components. They generate interaction diversity, control microbial mortality, redistribute genes, and shape immune ecology at mucus and tissue boundaries.

This framing reduces confusion without requiring a single paradigm to fit all systems. It explains how host-associated assemblages can remain coherent and repeatable through structured variability and selection, even when consortium-level inheritance is limited and why dysbiosis is best interpreted as a shift in interface conditions and interaction networks rather than the loss of a single correct community. It also clarifies why viral life histories and viral gene flow are expected to be major forces governing stability versus transitions to disease-prone states. By separating inheritance of fixed genetic modules and hypervariable elements from within-lifetime dynamics driven by combinatorial elements, diversity-generating elements, and transient gene fluxes, we show how macrobes can be evolutionarily tuned for microbial control while remaining ecologically open to continual viral and microbial exchange.

Acknowledgments This work was supported by the National Science Foundation PRFB award No. 2209377 to Mark Little.

References

Ainsworth TD, Thurber RV, Gates RD (2010) The future of coral reefs: a microbial perspective. Trends Ecol Evol 25(4):233–240

Ambalavanan L, Iehata S, Fletcher R, Stevens EH, Zainathan SC (2021) A review of marine viruses in coral ecosystem. J Mar Sci Eng 9(7):711

Andersson JO, Slechta ES, Roth JR (1998) Directed evolution of bacterial surface proteins by a diversity-generating retroelement. Science 281:548–550

Barr JJ, Auro R, Furlan M, Whiteson KL, Erb ML, Pogliano J et al (2013) Bacteriophage adhering to mucus provide a non–host-derived immunity. Proc Natl Acad Sci USA 110(26): 10771–10776

Barton ES, White DW, Cathelyn JS, Brett-McClellan KA, Engle M, Diamond MS et al (2007) Herpesvirus latency confers symbiotic protection from bacterial infection. Nature 447(7142): 326–329

Baker LJ, Reich HG, Kitchen SA, Klinges JG, Koch HR, Baums IB et al (2022) The coral symbiont Candidatus Aquarickettsia is variably abundant in threatened Caribbean acroporids and transmitted horizontally. ISME J 16(2):400–411

Bernasconi R, Stat M, Koenders A, Huggett MJ (2019) Global networks of Symbiodinium-bacteria within the coral holobiont. Microb Ecol 77(3):794–807

Bettarel Y, Bouvier T, Nguyen HK, Thu PT (2015) The versatile nature of coral-associated viruses: viruses and the coral holobiont. Environ Microbiol 17(10):3433–3439

Boilard A, Dubé CE, Gruet C, Mercière A, Hernandez-Agreda A, Derome N (2020) Defining coral bleaching as a microbial dysbiosis within the coral holobiont. Microorganisms 8(11):1682

Bondy-Denomy J, Davidson AR (2014) When a virus is not a parasite: the beneficial effects of prophages on bacterial fitness. J Microbiol 52(3):235–242

Bourne DG, Garren M, Work TM, Rosenberg E, Smith GW, Harvell CD (2009) Microbial disease and the coral holobiont. Trends Microbiol 17(12):554–562

Brüssow H (2007) Bacteria between protists and phages: from antipredation strategies to the evolution of pathogenicity. Mol Microbiol. Available from: https://onlinelibrary.wiley.com/doi/abs/10.1111/j.1365-2958.2007.05826.x

Brüssow H, Canchaya C, Hardt W-D (2004) Phages and the evolution of bacterial pathogens: from genomic rearrangements to lysogenic conversion. Microbiol Mol Biol Rev 68(3):560–602

Brüwer JD, Agrawal S, Liew YJ, Aranda M, Voolstra CR (2017) Association of coral algal symbionts with a diverse viral community responsive to heat shock. BMC Microbiol 17(1):174

Cárdenas A, Ye J, Ziegler M, Payet JP, McMinds R, Vega Thurber R et al (2020) Coral-associated viral assemblages from the Central Red Sea align with host species and contribute to holobiont genetic diversity. Front Microbiol 11:572534

Chuang PS, Chen M, Chiang P-W, Chen H-J, Wang F-Q, Chen Y-H, et al (2025) Bacteriophages of the predominant coral symbiont Endozoicomonas: novel models for coral holobiont interactions. bioRxiv. Available from: https://doi.org/10.1101/2025.10.29.685362

Correa AMS, Ainsworth TD, Rosales SM, Thurber AR, Butler CR, Vega Thurber RL (2016) Viral outbreak in corals associated with an in situ bleaching event: atypical herpes-like viruses and a new Megavirus infecting Symbiodinium. Front Microbiol 7:127

Coutinho FH, Silveira CB, Gregoracci GB, Thompson CC, Edwards RA, Brussaard CPD et al (2017) Marine viruses discovered via metagenomics shed light on viral strategies throughout the oceans. Nat Commun 8:15955

Daniels CA, Baumgarten S, Yum LK, Michell CT, Bayer T, Arif C et al (2015) Metatranscriptome analysis of the reef-building coral Orbicella faveolata indicates holobiont response to coral disease. Front Mar Sci 2. Available from: https://doi.org/10.3389/fmars.2015.00062

Delgadillo-Ordoñez N, Garcias-Bonet N, Raimundo I, García FC, Villela H, Osman EO et al (2024) Probiotics reshape the coral microbiome in situ without detectable off-target effects in the surrounding environment. Commun Biol 7(1):434

Ferretti P, Martignoni MM, McManus LC, Sakal T, Liaghat A, Stevens B et al (2025) Theory of host-microbe symbioses: challenges and opportunities. Cell Host Microbe 33(7):1052–1056

Gantt SE, Keister EF, Jerry SE, Ferguson Tindal R, Manfroy AA, Merck DE et al (2025) Physiological and symbiotic flexibility of reef-building corals to new habitats: insights from clonal colony transplants. J Appl Ecol 62(2):207–219

Haas AF, Fairoz MFM, Kelly LW, Nelson CE, Dinsdale EA, Edwards RA et al (2016) Global microbialization of coral reefs. Nat Microbiol 1(6):16042

Hester ER, Barott KL, Nulton J, Vermeij MJ, Rohwer FL (2016) Stable and sporadic symbiotic communities of coral and algal holobionts. ISME J 10(5):1157–1169

Howe-Kerr LI, Grupstra CGB, Rabbitt KM, Conetta D, Coy SR, Klinges JG et al (2023) Viruses of a key coral symbiont exhibit temperature-driven productivity across a reefscape. ISME Commun 3(1):27

Kelman J, Khan M, Umunna C, Brainard R, Donohue G, Edwards R, et al (2025) Central carbon metabolism switching in lytic versus temperate coral reef viral communities. bioRxivorg. Available from: https://doi.org/10.1101/2025.09.25.676695

Klinges JG, Rosales SM, McMinds R, Shaver EC, Shantz AA, Peters EC et al (2019) Phylogenetic, genomic, and biogeographic characterization of a novel and ubiquitous marine invertebrate-associated Rickettsiales parasite, Candidatus Aquarickettsia rohweri, gen nov, sp nov. ISME J 13(12):2938–2953

Klinges G, Maher RL, Vega Thurber RL, Muller EM (2020) Parasitic 'Candidatus Aquarickettsia rohweri' is a marker of disease susceptibility in Acropora cervicornis but is lost during thermal stress. Environ Microbiol 22(12):5341–5355

Knowlton N, Rohwer F (2003) Multispecies microbial mutualisms on coral reefs: the host as a habitat. Am Nat 162(4 Suppl):S51–S62

Knowles B, Silveira CB, Bailey BA, Barott K, Cantu VA, Cobián-Güemes AG et al (2016) Lytic to temperate switching of viral communities. Nature 531(7595):466–470

Kvennefors ECE, Sampayo E, Kerr C, Vieira G, Roff G, Barnes AC (2012) Regulation of bacterial communities through antimicrobial activity by the coral holobiont. Microb Ecol 63(3):605–618

Li J, Zou Y, Yang J, Li Q, Bourne DG, Sweet M et al (2022) Cultured bacteria provide insight into the functional potential of the coral-associated microbiome. mSystems 7(4):e0032722

Little M, Rojas MI, Rohwer F (2020) Bacteriophage can drive virulence in marine pathogens. In: Behringer DC, Silliman BR, Lafferty KD (eds) Marine disease ecology. Oxford University Press, Oxford, pp 73–82. https://doi.org/10.1093/oso/9780198821632.003.0004

Little M, George EE, Arts MGI, Shivak J, Benler S, Huckeba J et al (2021) Three-dimensional molecular cartography of the Caribbean reef-building coral Orbicella faveolata. Front Mar Sci 8. Available from: https://www.frontiersin.org/articles/10.3389/fmars.2021.627724

Littman R, Willis BL, Bourne DG (2011) Metagenomic analysis of the coral holobiont during a natural bleaching event on the great barrier reef: metagenomics analysis of bleached coral. Environ Microbiol Rep 3(6):651–660
Margulis L, Fester R (eds) (1991) Symbiosis as a source of evolutionary innovation: speciation and morphogenesis. MIT Press, Cambridge
Marhaver KL, Edwards RA, Rohwer F (2008) Viral communities associated with healthy and bleaching corals. Environ Microbiol 10(9):2277–2286
Martin-Cuadrado A-B, Rubio-Portillo E, Rosselló F, Antón J (2024) The coral Oculina patagonica holobiont and its response to confinement, temperature, and Vibrio infections. Microbiome 12(1):222
McKerral JC, Papudeshi B, Inglis LK, Roach MJ, Decewicz P, McNair K, et al (2023) The promise and pitfalls of prophages. bioRxiv. Available from: https://doi.org/10.1101/2023.04.20.537752
Moran NA, Sloan DB (2015) The hologenome concept: helpful or hollow? PLoS Biol 13(12): e1002311
Munro J, Oakey J, Bromage E, Owens L (2003) Experimental bacteriophage-mediated virulence in strains of Vibrio harveyi. Dis Aquat Org 54(3):187–194
Nguyen-Kim H, Bouvier T, Bouvier C, Bui V, Le-Lan H, Bettarel Y (2015) Viral and bacterial epibionts in thermally-stressed corals. J Mar Sci Eng 3(4):1272–1286
Nunn BL, Brown T, Timmins-Schiffman E, Mudge MC, Riffle M, Axworthy JB et al (2025) Protein signatures predict coral resilience and survival to thermal bleaching events. Commun Earth Environ 6(1)
Nyholm SV, McFall-Ngai MJ (2003) Dominance of Vibrio fischeri in secreted mucus outside the light organ of Euprymna scolopes: the first site of symbiont specificity. Appl Environ Microbiol 69(7):3932–3937
O'Brien PA, Webster NS, Miller DJ, Bourne DG (2019) Host-microbe coevolution: applying evidence from model systems to complex marine invertebrate holobionts. MBio 10(1)
Osman EO, Garcias-Bonet N, Cardoso PM, Rosado PM, García FC, Ferrier-Pagès C et al (2025) Phototrophic bacteria as potential probiotics for corals. NPJ Biodivers 4(1):16
Pogoreutz C, Oakley CA, Rädecker N, Cárdenas A, Perna G, Xiang N et al (2022) Coral holobiont cues prime Endozoicomonas for a symbiotic lifestyle. ISME J 16(8):1883–1895
Quistad SD, Stotland A, Barott KL, Smurthwaite CA, Hilton BJ, Grasis JA et al (2014) Evolution of TNF-induced apoptosis reveals 550 my of functional conservation. Proc Natl Acad Sci USA 111(26):9567–9572
Quistad SD, Grasis JA, Barr JJ, Rohwer FL (2017) Viruses and the origin of microbiome selection and immunity. ISME J 11(4):835–840
Roach TNF, Dilworth J, Christian Martin H, Jones AD, Quinn RA, Drury C (2021) Metabolomic signatures of coral bleaching history. Nat Ecol Evol 5(4):495–503
Roach TNF, Little M, Arts MGI, Huckeba J, Haas AF, George EE et al (2020) A multiomic analysis of in situ coral–turf algal interactions. Proc Natl Acad Sci USA 117(24):13588–13595
Rohwer F, Thurber RV (2009) Viruses manipulate the marine environment. Nature 459(7244): 207–212
Rohwer F, Seguritan V, Azam F, Knowlton N (2002) Diversity and distribution of coral-associated bacteria. Mar Ecol Prog Ser 243:1–10
Rosenberg E, Zilber-Rosenberg I (2018) The hologenome concept of evolution after 10 years. Microbiome 6(1):78
Rosenberg E, Sharon G, Zilber-Rosenberg I (2009) The hologenome theory of evolution contains Lamarckian aspects within a Darwinian framework. Environ Microbiol 11(12):2959–2962
Roughgarden J (2023) Holobiont evolution: population theory for the hologenome. Am Nat 201(6): 763–778
Roughgarden J, Gilbert SF, Rosenberg E, Zilber-Rosenberg I, Lloyd EA (2018) Holobionts as units of selection and a model of their population dynamics and evolution. Biol Theory 13(1):44–65
Rubio-Portillo E, Robertson S, Antón J (2024) Coral mucus as a reservoir of bacteriophages targeting Vibrio pathogens. ISME J 18:wrae017

Shinzato C, Inoue M, Kusakabe M (2014) A snapshot of a coral 'holobiont': a transcriptome assembly of the scleractinian coral, porites, captures a wide variety of genes from both the host and symbiotic zooxanthellae. PLoS One 9(1):e85182
Silveira CB, Rohwer FL (2016) Piggyback-the-Winner in host-associated microbial communities. NPJ Biofilms Microbiomes 2:16010
Silveira CB, Coutinho FH, Cavalcanti GS, Benler S, Doane MP, Dinsdale EA et al (2020) Genomic and ecological attributes of marine bacteriophages encoding bacterial virulence genes. BMC Genomics 21(1):126
Silveira CB, Luque A, Haas AF, Roach TNF, George EE, Knowles B et al (2023) Viral predation pressure on coral reefs. BMC Biol 21(1):77
Soffer N, Brandt ME, Correa AMS, Smith TB, Thurber RV (2014) Potential role of viruses in white plague coral disease. ISME J 8(2):271–283
Stencel A, Wloch-Salamon DM (2018) Some theoretical insights into the hologenome theory of evolution and the role of microbes in speciation. Theory Biosci 137(2):197–206
Stévenne C, Micha M, Plumier J-C, Roberty S (2021) Corals and sponges under the light of the holobiont concept: how microbiomes underpin our understanding of marine ecosystems. Front Mar Sci 8. Available from: https://doi.org/10.3389/fmars.2021.698853
Sweet M, Bythell J (2017) The role of viruses in coral health and disease. J Invertebr Pathol 147: 136–144
Sweet MJ, Croquer A, Bythell JC (2011) Bacterial assemblages differ between compartments within the coral holobiont. Coral Reefs 30(1):39–52
Theis KR, Dheilly NM, Klassen JL, Brucker RM, Baines JF, Bosch TCG et al (2016) Getting the hologenome concept right: an eco-evolutionary framework for hosts and their microbiomes. mSystems 1(2)
Thompson JR, Rivera HE, Closek CJ, Medina M (2014) Microbes in the coral holobiont: partners through evolution, development, and ecological interactions. Front Cell Infect Microbiol 4:176
van Oppen MJH, Leong J-A, Gates RD (2009) Coral-virus interactions: a double-edged sword? Symbiosis 47(1):1–8
Varona NS, Hesketh-Best PJ, Coutinho FH, Stiffler AK, Wallace BA, Garcia SL et al (2024) Host-specific viral predation network on coral reefs. ISME J 18(1):wrae240
Vega Thurber R, Payet JP, Thurber AR, Correa AMS (2017) Virus-host interactions and their roles in coral reef health and disease. Nat Rev Microbiol 15(4):205–216
Vega Thurber R, Willner-Hall D, Rodriguez-Mueller B, Desnues C, Edwards RA, Angly F et al (2009) Metagenomic analysis of stressed coral holobionts. Environ Microbiol 11(8):2148–2163
Voolstra CR, Suggett DJ, Peixoto RS, Parkinson JE, Quigley KM, Silveira CB et al (2021) Extending the natural adaptive capacity of coral holobionts. Nat Rev Earth Environ 2(11): 747–762
Voolstra CR, Raina J-B, Dörr M, Cárdenas A, Pogoreutz C, Silveira CB et al (2024) The coral microbiome in sickness, in health and in a changing world. Nat Rev Microbiol 22(8):460–475
Wallace BA, Varona NS, Hesketh-Best PJ, Stiffler AK, Silveira CB (2024) Globally distributed bacteriophage genomes reveal mechanisms of tripartite phage-bacteria-coral interactions. ISME J 18(1). Available from: https://doi.org/10.1093/ismejo/wrae132
Wallace BA, Varona NS, Stiffler AK, Vermeij MJA, Silveira C (2025) High microbial diversity, functional redundancy, and prophage enrichment support the success of the yellow pencil coral, Madracis mirabilis, in Curaçao's coral reefs. mSystems 10(11):e0120825
Webster NS, Reusch TBH (2017) Microbial contributions to the persistence of coral reefs. ISME J 11(10):2167–2174
Weynberg KD, Voolstra CR, Neave MJ, Buerger P, van Oppen MJH (2015) From cholera to corals: viruses as drivers of virulence in a major coral bacterial pathogen. Sci Rep 5:17889
Weynberg KD, Laffy PW, Wood-Charlson EM, Turaev D, Rattei T, Webster NS et al (2017) Coral-associated viral communities show high levels of diversity and host auxiliary functions. PeerJ 5: e4054

Zhang Z, Tong M, Ding W, Liu S, Jong M-C, Radwan AA et al (2024) Changes in the diversity and functionality of viruses that can bleach healthy coral. mSphere 9(12):e0081624

Zhou K, Zhang T, Chen X-W, Xu Y, Zhang R, Qian P-Y (2024) Viruses in marine invertebrate holobionts: complex interactions between phages and bacterial symbionts. Annu Rev Mar Sci 16:467–485

Zilber-Rosenberg I, Rosenberg E (2008) Role of microorganisms in the evolution of animals and plants: the hologenome theory of evolution. FEMS Microbiol Rev 32(5):723–735

Chapter 23
A Hypercycle for Ribozymes or Viroids as Origin of Life

Karin Moelling

Abstract Viruses can cause diseases, but their more important role is as drivers of evolution. They may have contributed to the origin of life as noncoding RNA, which we know today from viroids and ribozymes. Viroids/ribozymes are the most ancient entities that can replicate and evolve. They may have been autonomous, helping build up larger structures, protocells, and cells. An often underestimated trend in evolution is loss of function (LoF) in contrast to gain of function (GoF). For the simplest virus, phage Qbeta, gene loss allows an increased speed of replication in vitro (Spiegelman's Monster). The model of how genes may have evolved has been designed by M. Eigen with his hypercycle, which can be applied to RNA only and reduces the almost infinite possibilities of sequences to realistic numbers. Retroviruses with and without deletions and, to a lesser extent, other viruses, accumulated in the human and other genomes, where they contributed to new properties such as, allowing embryonic growth inside the uterus. Viruses are active in horizontal gene transfer; they can transport gene regulators and recombine with growth-promoting genes such as oncogenes. Many of them are the strongest cancer genes we know of. Transposable elements are important for gene modifications and can be defined as endogenous rudimentary viruses lacking coats. RNA viral diseases are models for evolution in time-lapse. Viruses introduced immunity into their host cells. Many properties of viruses resemble social traits that are reminiscent of the social behavior of all living species.

K. Moclling (✉)
Institute of Medical Microbiology, University Zürich, Zürich, Switzerland

Max Planck-Institute Molecular Genetics, Berlin, Berlin, Germany
e-mail: moelling@molgen.mpg.de

G. Witzany (ed.), *Integrative Theory of Evolution - Aspects and Insights*,
https://doi.org/10.1007/978-3-032-24295-2_23

1 Introduction

Viruses can cause diseases, but their more important role is as drivers of evolution. They contributed to evolution all the way back to the beginning of life. The early world was an RNA world and started with noncoding RNA, which we know today from viroids and ribozymes. Viroids/ribozymes are the most ancient entities that can replicate, undergo mutations, and evolve. They may have been autonomous in the early world, helping build up larger structures, protocells, and cells. Later on, during evolution, gene reduction may have made them parasites as we know them today.

A model of how life may have evolved has been designed by Manfred Eigen by demonstrating chemical evolution with prebiotic molecules. This occurred before the biological evolution of living matter. Furthermore, M. Eigen designed a hypercycle, an abstract model of an autocatalytic cycle, where self-organization occurs, self-replicating members cooperate, and are linked in a positive feedback loop. It can be applied to primordial RNA molecules such as ribozymes, which are essential throughout the evolution of life until today. The hypercycle model allows the reduction of the almost infinite possibilities of RNA sequences to realistic numbers and the generation of information from random starts. Self-organization appears to be a law of nature. What can one learn from early terrestrial life for exoplanetary life?

2 Early Earth

Earth is the only case in which we know about life, microbial life, and "intelligent" life. The latter definition is often applied to the life of higher organisms, including mankind. The origin of life on Earth and the living conditions known in our planetary system may serve as a lead to look for similar conditions on other systems. The recent progress in cosmology by telescopes such as the Hubble Telescope or the James Webb Space Telescope allows new insights into the origin of the Universe and stars, their composition, development, and the detection of biomarkers as indirect measures of life.

The number of stars in a galaxy is 100 bio, and the same number holds for galaxies, amounting in total to 10^{22} stars, all of which may even be exoplanets, consisting of planetary systems with moons. This gigantic number suggests that life may also exist in extraterrestrial space. A team of scientists has evaluated the Earth-like conditions and how often these may occur in the Universe, coming up with 3×10^8. Other estimates go up to 20% of the stars of our Milky Way; this would be 5×10^9. We simply do not know.

Life on Earth may help as an initial search for possible extraterrestrial life. The knowledge about the evolution of the Universe, the environmental conditions on other planetary systems, as well as exoplanets, may guide researchers to detect life, even if only one approach is based on Earth-like conditions.

The elements on Earth, the basis for life, organized in the periodic system, arose in the Universe in three different ways. Hydrogen and some helium originate from the Universe's very beginning, around the time of the Big Bang. Hydrogen is throughout the Universe by far most abundant element. The next heavier elements are produced within the stars under nuclear fusion conditions. This occurs with the emission of energy up to the size of iron, Fe52. Nuclear fusion of small elements is how stars produce light, which can be seen after about 400 million years after the Big Bang. Heavier molecules than iron require energy input, which is supplied by the explosion of supernovae, the end of big enough stars. They produce shock waves that spread the heavier elements throughout the Universe. They are then taken up by newly forming stars.

What is needed for the formation of life? The Universe is full of hydrogen and some helium. The basic reaction within stars is the nuclear fusion of hydrogen to helium. All elements we know of on Earth are available throughout the Universe. This supports the argument that life can form at any location since the elements to form the building blocks for life are everywhere available. Yet this is not enough; many Earth-like conditions are also required.

3 Chemicals for Life: C, H, N, O, P, S

Two chemists, Miller and Urey, showed that biomolecules, the building blocks for life on Earth, can be produced in a test tube. In their famous classic experiment from 1952, they described the formation of organic compounds from simple organic precursors such as water, methane, ammonia, and hydrogen. In a hypothetical primordial Earth-like atmosphere, they exposed their compounds to electric sparks to simulate lightning as a source of energy. They found, among others, three amino acids, glycine, alanine, and perhaps aspartic acid, and many other organic compounds. This supports the idea of a chemical origin of life. They did not find any larger compounds, nucleic acids, or their precursors, though, nor RNA or DNA (Miller and Urey 1959).

An important recent experiment is almost a follow-up of the Urey-Miller experiment performed in the laboratory by *Sutherland and colleagues, the "one pot synthesis."* They used the most prevalent elements C, H, N, O, P, and S and succeeded in the simultaneous production of these three most relevant life-promoting building blocks. In a single reaction vessel, they synthesized: 1. Lipids, leading to micelles, which could be considered as droplets or compartments as an early version of a cell, and two. The nucleobases, as part of the ribonucleic acids, are the carriers of our genes, and three. Many amino acids, which lead to proteins, such as enzymes, are important for life and metabolism. 20 amino acids are enough; they can polymerize to polypeptide chains or bigger proteins. These are the essential ingredients that life requires (Patel et al. 2015).

This is a surprising result. Out of 118 elements listed in the periodic system, only a few are life-essential—only six elements. Many others serve additional or luxury functions. Furthermore, they are the most abundant ones present in the Universe.

In the Universe, apparently, some of the monomers can be formed, but no polymers. This is known from meteorites that hit the Earth and reveal what kind of molecules have been formed on other planets or in supernovae, even before the beginning of our planetary system. A meteorite that fell in Australia, the Murchison Meteorite, contained 80–90 different amino acids. This is in contrast to the fact that life on Earth only uses 20 of them. Also, di-amino acids were found, and fullerenes, which are composed of 60 carbons, consisting of sixfold and fivefold structures, like the classical football—a platonic body, which we find as a structure in many viruses and in almost all phages (Meierhenrich et al. 2004, 2008). This raises the question of whether on other stars, more or different amino acids may have generated some different forms of life. Perhaps one can use AI to design life with such ingredients—possibly with new types of proteins and different types of metabolism.

The black smokers in the deep sea are locations where biomolecules could have formed. There, water, temperature gradients, biomolecules, niches reminiscent of Darwin's little warm pond for the formation of life, various pressures, and metal elements are available. All this may have promoted life. Also, shallow water with tides or wet-and-dry cycles is being discussed as a putative location for the origin of life.

Our solar system began 4.567 bio years ago, a long period after the Big Bang, which is dated back to 13.8 bio years by red-shift studies.

The *six* elements are so ubiquitous that they are most likely not the limiting factor for the formation of life outside of Earth. Yet the development of stars all depends on different conditions due to their environment, other stars and therefore undergoes different histories. Among the many stars that exist, there are probably nonidentical ones; each one has its own starting conditions and its history depending on the influence from neighboring stars.

Self-organization is an important mechanism for the formation of early life. Similarly, self-organization is the basis for the formation of stars, planetary systems, and galaxies.

Thus, self-organization appears to be a principle of nature—a law of physics.

Erwin Schrödinger and Max Delbrück suspected that life is characterized by a new, different, or unknown type of physics. This is not confirmed, as far as life, metabolism, replication, and other biological processes are concerned. Even the rather new transgenerational inheritance is approaching molecular understanding based on known physical laws (Moelling 2024). Yet we do not understand the human brain, such as consciousness. Also, unknown physics still exists in the Universe, e.g., in the form of dark matter or dark energy, which are not yet understood.

4 Prebiotic Molecules

The concept of an "Early RNA World" was already proposed in 1965/1968 by C. Woese, F. Crick, and L. Orgel (Woese 1965; Crick 1968; Orgel 1968). The molecules formed are ribozymes/viroids and tRNAs. The formation of such structures occurs by self-organization.

We know that protein synthesis today is performed by ribosomes. Ribosomes are machines that link amino acids to longer polypeptide chains. The synthesis of polypeptide chains is based on the enzymatic activity of a molecule, the *ribozyme*. The word is derived from "RNA enzyme," or ribonucleotide enzyme. Ribozymes are related to viroids, virus-like elements. Even today, ribozymes are the catalytic center within about 100 ribosomal proteins, which serve as scaffold proteins, without being involved in direct catalytic activity for protein synthesis. This brings the timeline in the tree of life as far back as possible, back to ribozymes, catalytic RNA preceding DNA and proteins. Tom Cech defined this by stating, "ribosomes are ribozymes" (Cech 2000). The supply of amino acids for peptide bond formation can happen spontaneously, whereby transfer RNAs, the tRNAs, speed up the process. Today, a tRNA is only meaningful if it contains the amino acid triplet code and thereby directs the tRNA to deliver the required amino acids to a growing polypeptide chain. This may have evolved from much simpler primitive mechanisms, even without a genetic code, which speed up the synthesis of polypeptide chains required for metabolism and the biological world. The protein enzymes arose later. It all depends to this day on ribozymes for peptide bond formation. Ribozymes fulfill this function to this day, indicating that they are extremely robust and fitter than any other molecule, because no protein enzyme displaced them.

tRNAs also originate from ribozymes (Kowiatek 2022). They have been shown to bind amino acids, which, later in evolution, in the protein world, were performed by aminoacyl synthetases. Halves of tRNA molecules or even smaller fragments play important roles in gene regulation in all species.

tRNAs are extremely stable and contain pseudo-uridines, which are present there to this very day and are not detected elsewhere. They play an important role in today's mRNA vaccines against coronaviruses, because the modifications make the RNA nonimmunogenic. This allowed the production of mRNA as a vaccine without toxicity.

Ribozymes are characterized by partial double-stranded regions and loops, described as hairpins, in a closed circular form. They are normally 75 to a few 100 ribonucleotides long. Longer structures are at risk for accumulation of mutations or errors, which, at too high a frequency, can destroy their function.

The energy required for RNA synthesis could have been derived from chemical energy on the early Earth, because deep down in the ocean below 200 m, there is darkness. Ribozymes or viroids may have started to form at hydrothermal vents or in ice, where RNA is surprisingly stable and reactive because it gets concentrated in liquid water niches between the ice crystals (Moelling and Broecker 2019a, b).

Base pairing is an essential mechanism to stabilize RNA structures. This is required for codon-anti-codon recognition and is typical of ribozyme structures, which are composed of numerous short stems, loops, and bulges within closed circular hairpin-loop structures.

Ribozymes are supposedly the only and earliest entities with catalytic activities and the ability to replicate as well as to mutate. Ribozymes can cleave and join each other and can add on nucleotides, allowing growth of polypeptide chains in the absence of ribosomes and also in the absence of a template. They can polymerize new RNAs without protein enzymes (Szostak et al. 2001). "RNA can do it all," was the statement by Sir Francis Crick (1968). Thus, one might place RNA viroids/ribozymes at the bottom of the tree of life.

The *ribozymes/viroids* were reproduced in a test tube by Jack W. Szostak and colleagues. They put RNA of about 100 ribonucleotides in length into a vessel and waited. The sequence space of the RNA molecules was enormous, and functional sequences were rare but detected within an estimated 10^{15} RNA species. Among many products, ribozymes as well as tRNAs arose. They are characteristic of the prebiotic chemical world (Wilson and Szostak 1999).

G. Joyce described ribozymes with self-replicating activity in vitro, which can produce progeny by cleaving and joining bonds (Joyce 2002; Lincoln and Joyce 2009).

By now, eight different kinds of ribozymes are known (Winkler et al. 2004). Ribozymes/viroids do not code for triplets; they harbor noncoding sequences but exhibit structural features. Structure is the basic information of molecules. They are naked molecules without proteins, so they lack environmental protection, but on the other hand, RNases did not yet exist in the pre-protein world. Some ribozymes have lost their enzymatic activity during evolution. The ribozyme-related viroids indeed have often maintained enzymatic activities, but many of them have lost such activity and rely on host cellular support in today's world. tRNAs are derived from pre-RNAs by processing and cleavage through an RNase P, which is not an enzyme but a ribozyme. It cleaves the 5` leader sequence of precursor tRNAs as described by Altman. S. Altman and T. Cech received the Nobel Prize in 1989 together for their discoveries of ribozymes.

RNA came before DNA, and DNA was then a major improvement for stabilization, but at the expense of innovation, which is the advantage of single-stranded RNA molecules. DNA may have come from RNA by deoxygenation or later through protein synthesis via a reverse transcriptase (Moelling and Broecker 2015, 2017).

5 Viruses

Ribozymes are closely related to virus-like elements, the viroids; they consist of naked but structured RNA only, noncoding RNA without the triplet code. Because of the lack of a protein coat, which was then considered an essential component of a virus when they were discovered, they were designated as viroids, viroid means virus-like (Gross et al. 1978; Diener 2016).

Stem-loop-structured regions in noncoding RNA are still present in all RNA viruses, not only viroids, today, where, especially at their termini, the RNAs also regulate processes such as viral replication, expression, and protein synthesis. At the termini of today's RNA viruses, ribozyme structures are conserved, redirecting the host cell and forcing it to allow replication and translation of the virus, not the cell. Especially at the untranslated termini, tRNAs are important elements in RNA viruses. tRNAs are known today for the transfer of amino acids to the ribosome, which gave them their name. Subsequently, the ribozyme, the catalytic activity of the ribosome, synthesizes the polypeptide chains from the amino acids transferred by the tRNAs. The tRNAs are among the first structured RNA molecules in the early RNA world, produced by two ribozymes or even smaller RNA fragments. tRNAs were involved in a wide variety of biological processes, including cell survival, cell proliferation, cell differentiation, metabolism of amino acids, cell signaling, a stress response program, and HIV-1 priming.

Thus, early tRNAs were multifunctional.

The interaction of the ends of RNA viruses can form by fold-back mechanisms, which can even be as complicated as pseudoknots. They are composed of two ribozymes, where one loop forms hydrogen bonds with the stem of a second one. These are also located at the ends of human chromosomes as telomeres and can have regulatory functions. They are present in the simplest RNA viruses we know of, the tRNA-like viruses (Matsuda and Dreher 2004; Dreher 2008). But even the protein-coding regions of viral RNAs of today's viruses are arranged like a series of ribozymes and tRNAs, and some viruses harbor tRNA structures at their untranslated termini or at the end of sub-genomic RNAs. Among these elements are polio viruses, phages, bacteria, plant viruses such as turnip yellow mosaic viruses (TYMV), Hepatitis B, C, and D viruses (HBV, HCV, hepatitis delta virus (HDV)) (Taylor 2009; Piron et al. 2005). Two tRNAs were also detected within the HCV genome, not only at the termini. TYMV, a plant virus, exhibits a tRNA at a sub-genomic message and a pseudoknot, which is also found in phage Q beta RNA (Mans et al. 1991) and hepatitis delta virus (HDV). HDV is the smallest human RNA virus with an RNA of 1680 nucleotides in a rod-like closed structure. It is a ribozyme forming up to 70% double-stranded regions. It codes for a single protein, the Hepatitis Delta protein, probably derived from recombination with host information, and uses the coat protein of HBV for export. Autocatalytic self-cleaving ribozymes are present in the genome and its complement, the antigenome, which are related but not identical. The ribozyme activity requires a complex double pseudoknot RNA structure (Doudna and Cech 2002) (Fig. 23.1).

A variation of the use of tRNAs is a melted tRNA primer used by retroviruses as primers for a DNA copy of the genomic RNA by reverse transcription. Retroviruses adopt special tRNAs from the cellular pool to bind to the viral Primer Binding Site, PBS, with their 5′ acceptor arm, up to about 100 nucleotides in length. These tRNAs initiate replication. Later on, this region is also used for translation. The start codons for replication and translation are superimposed (Moelling and Broecker 2015). Ribozymes developed from simple structures into complex hyper-structured introns and further into retrovirus genomes (Moelling et al. 2017) (Fig. 23.1).

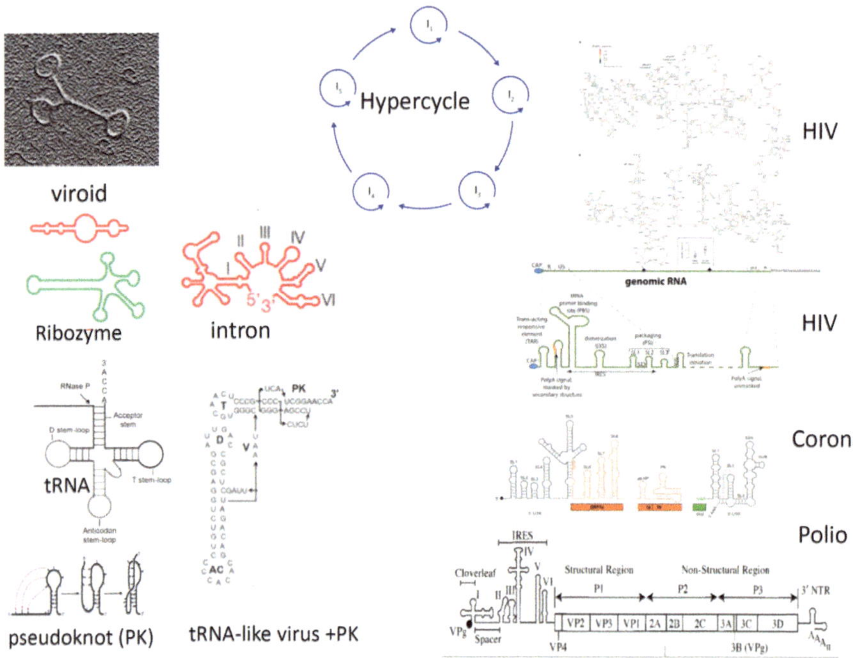

Fig. 23.1 *Left:* An electron micrograph shows a viroid. Models of a viroid, a ribozyme, an intron, a tRNA, pseudoknots, and the RNA of a tRNA-like virus with a pseudoknot (PK) are shown as described in the text, including the references. *Right:* Furthermore, complex viral hyperstructures with ribozymes and tRNAs are shown for the whole genome of HIV; the termini of HIV; the termini of coronavirus, including ribozyme structures in the open reading frames (ORF); and poliovirus with tRNA structures at both termini. *On top,* the simplified hypercycle is shown. Several replicating entities, e.g., ribozymes, are indicated as small cycles, which feed information into the next cycle. The cycles form a hypercycle with feedback to the origin. (Modified from Szostak et al. 2016)

The prominent role of RNA is further illustrated by the fact that RNA can perform de novo polymerization in contrast to DNA, which needs an RNA primer, as in the case of retroviruses. DNA requires, in general, RNA as a starting point.

There are some primordial noncoding RNA "viruses" with tRNA-like structures (TLS), which bind to single individual amino acids. Three types of 3′-terminal tRNA-like structures exist, which bind each of the three amino acids valine, histidine, and tyrosine as in TYMV, Tobacco Mosaic virus (TMV), and Brome Mosaic virus, respectively. Valine, furthermore, binds preferentially to the TYMV genome, and tyrosine to the Barley Stripe Mosaic virus genome. They often differ from regular tRNAs by the presence of a pseudoknot structure at their 3′ ends. They also occur in HDV. One could speculate that the viral tRNA-like structures reflect archaic intermediates of evolution manifested by viruses toward the development of early steps in protein synthesis and conserved to this day. They are regulators of plant virus replication, while normally, canonical tRNAs play a role in translation. Thus, mutations toward tRNA-like structures lead to another function, replication instead of translation (Dreher 2008). tRNA-like structures have been demonstrated in many plant viral genomes. They play a role in the life cycle of these viruses; they act as 3′-translational enhancers and regulators of minus-strand synthesis.

It came as a surprise that protein functions were replaced by RNA, as stated by Tom Cech: "RNA finds a simpler way" (Cech 2004). *He refers to riboswitches*, a bacterial regulation system with a "twist," an on–off cleavage element with enzymatic activity. It is 75 nucleotides long—similar to the size of tRNA. The riboswitches form a pocket where metabolites or other small molecules, proteins, nucleic acids or aptamers can bind. The riboswitches induce gene regulation and decide between two states, on–off, for further activities, a property that was attributed to proteins only before (Winkler et al. 2004; Ishida et al. 2020).

All RNA viruses contain proteins that bind to their RNA as chaperones or melting proteins for protection. They are mostly positively charged and help with translation and replication by disentangling the hyper-structured single-stranded RNAs. Polypeptides and protein fragments can also stimulate ribozyme activity significantly in vitro (Müller et al. 1994). Perhaps in the early RNA world, where single amino acids but no polymers existed, they might have bound to ribozymes as a series of monomers to increase their activity. This can be up to 1000-fold in vitro (Müller et al. 1994). Ribozymes have been addressed as drugs against cancer for cleavage of the bcr-abl oncogene in Chronic Myelogenous Leukemia, CML (Moelling et al. 1994; Pachuk et al. 1994).

In summary, ribozymes exist in every living cell on our planet as regulators and especially as prerequisites of protein synthesis, where they are hidden among many protein scaffold molecules today. They are multifunctional, and up to hundreds of ribozyme derivatives with variations exist in today's world. This affects the whole tree of life. RNAs, as ribozymes or viroids, are involved in the regulation of gene silencing as siRNAs, which can regulate other regulators as chief regulators in eukaryotic cells, named circRNA from circular RNA (Hansen et al. 2013). Their behavior has been described by the authors as sponge-like.

A recent discovery is the presence of small tRNA fragments bound to the RNase H-like enzyme PIWI in the sperm of mammals as piRNA, where it plays a role in transgenerational inheritance (Moelling 2024). PAZ-PIWI is related to the reverse transcriptase-RNase H complex known for replication of retroviruses. In the case of regulatory silencing by siRNAs, one can detect one-to-one similarities between the components of the silencing machinery and components of retroviruses (Moelling et al. 2006). Who learned from whom?

Accumulation of RNA fragments in male sperm counteracts the hyperactivity of detrimental transposons. Furthermore, if the RNase H-like enzyme PIWI is mutated and not capable of silencing transposon activity, the sperm are not fertile (Moelling 2024).

Ribozymes and viroids may have been autonomous elements in the early RNA world. They existed in niches, replacing cells. Viruses may have been independent and may have been the builders of early pre-cells. They lost their autonomy and freedom by gene loss for the gain of specialization of functions as an energy-saving intracellular lifestyle. Indeed, the viroids have become host-dependent as we know them today. This reduces their energy consumption since it is supplied by the host. A major part of ribozymes even lost their enzymatic activity, still surviving in nature not by catalysis but by replication through cellular mechanisms called the rolling cycle (Moelling and Broecker 2019b, 2021).

All viruses today are host-dependent. They found the most efficient, successful survival and replication strategy on Earth, where they are the most abundant, the fittest species of about 10^{33} particles, surpassing bacteria by about 100-fold, which the bacterial viruses, the phages, use for replication. Viruses may have generated cellular precursors and may have given up an independent, autonomous lifestyle to become "lazy" as cellular parasites, whereby laziness is better characterized as gene reduction for energy saving. This may be attributable to energy-saving replication and translation and requires that the viruses carry the necessary instructions on their genomes for the cell to serve the virus first and not the cell. Gene loss is known in many processes described as endosymbiosis; the best examples are mitochondria in mammalian cells.

The gene loss accompanied by accelerated replication has been demonstrated in a test tube by an extremely important experiment by Sol Spiegelman, then at Columbia University in New York. He performed serial passages of aliquots from one reaction vessel to the next. Initially, the vessel contained RNA of a phage, Q beta, a replicase, and building blocks for new RNA. The initial size of the RNA comprised 3300 nucleotides and was reduced within 74 transfers to 470 noncoding nucleotides with the advantage of much higher replication speed. This RNA was called Spiegelman's Monster because of its new life-like properties and small size (Spiegelman et al. 1965).

Another example of optimization and self-organization comes from the world of viruses containing both RNA and proteins, a step further in evolution from naked RNA. Tobacco mosaic virus, or TMV, is an RNA virus that infects plants and is composed of RNA and many identical building blocks.

One can disrupt the complex virus into RNA and protein subunits. When putting them into one vessel, they will lead to the reformation of the rod-like structure, and the virus will be infectious again. That means it can replicate and infect tobacco plants (Butler and Klug 1971).

6 Chemical Evolution

The importance of RNA for the origin of life is beyond question today.

Another important question is not answered as yet, which refers to the mechanism of how information, the genetic code, is generated and how the almost infinite number of random sequences can be reduced. A bacterial cell, such as *Escherichia coli*, is composed of about four Mio nucleotides coding for about 2000 to 3000 different proteins. Starting with RNA as the first biomolecule, one has to consider that four nucleotides allow $10^{2.4\text{millon}}$ alternative sequences (Küppers 1971, 1981, 1985). However, most of these sequences are useless and irrelevant for meaningful information. How nonrandom information may have been created has been described by Manfred Eigen, Bernd-Olaf Küppers, and other colleagues. If one wanted to generate the *E. coli* genome of four Mio nucleotides from random sequences, the sequence alternatives would amount to $10^{2.4\text{million}}$ possibilities. This is extremely unlikely and comparable to the chance of generating an encyclopedia! A selective mechanism for creating something informative, to generate information, and a reduction of the almost infinite sequence space are necessary.

One could start with four Mio nucleotides in a random order and allow them to freely mutate. Information with respect to the goal sequence would be fixed (if it is known), while others would be discarded. This would reduce the number of possibilities by 4, and the synthesis of the whole bacterial genome would require 16 Mio trials. Such a selective approach is much superior to random trials (Küppers 1971, 1990). Behind this procedure of optimization is a principle applied, which M. Eigen described as chemical evolution (Eigen 1971).

The unique properties of chemical evolution are the principles of selection. In biological Darwinian evolution, it is the surrounding, the environmental conditions, the finches differed in their beaks to pick up food specific for their island, which lead biological parameters to evolve and to the survival of the fittest.

In the case of a *chemical evolution*, inherent properties of the molecule decide upon its fitness. The parameters are chemical or physical, not biological in nature. They comprise properties such as the charge of a molecule, bonds, structure, concentration, stability, etc. These are molecular or chemical parameters. As an example, one might think of an RNA molecule, such as a ribozyme or viroid, with a hairpin-loop structure, which can bind other molecules, whereby mistakes and mutations arise during replication and synthesis of slightly different or new molecules to replicate and amplify the building blocks. Mutations could, among others, lead to new properties, such as self-replicating ribozymes. These would be most successful and predestined to win and overgrow the others.

Many of them may arise with minor differences and form a quasispecies. This term was also introduced by M. Eigen, characterizing virus populations with many related sequences, too diverse for a single species. A population of viruses within a host consists of a quasispecies of similar RNA molecules. These molecules may compete with each other for better replication and survival of the fittest. Virus infections such as HIV, as well as other RNA viruses, in the human body show this type of behavior.

Proof for the de novo synthesis and self-organization of RNA has been provided in an experiment. It was shown that RNA can arise by *de novo* synthesis from nucleotides, the building blocks of RNA, in a reaction vessel. The reaction solution contained the nucleotides and a replicase. After some time, RNA molecules had formed. Thus, the result showed that the appearance of RNA without a template, just from the building blocks, is possible. This is proof that the *de novo* generation of RNA by intrinsic properties of single molecules can happen and that self-organization occurs. At the beginning, the sequences are most likely based on chance. Yet self-reproducing species finally competed for the limited number of replicase enzymes and went through a selection process.

Furthermore, when the authors added a toxin, it was shown that the RNA could be produced even in the presence of the toxin. When, afterwards, the toxin was removed, the RNA could not replicate anymore; it had adapted to the toxin; it became addicted. This proves that RNA is able to undergo replication and evolution in vitro. The RNA was able to replicate and mutate, which are the hallmarks of life. However, this is chemical evolution. This is an extension of the Darwinian principle of biological evolution to chemical evolution, which exists in prebiotic molecules. The Darwinian principle also occurs at the prebiotic level and is the basis for the origin of life (Eigen 1971; Sumper and Luce 1975).

7 Hypercycle

M. Eigen and colleagues noticed that a successful molecule cannot become very large because of accumulating errors during replication, whereby the error rate increases with size. Many mutations lead to rapid progress, but too many mutations resemble mistakes and could result in breakdown of information and the accumulation of simply background noise. This can lead to the breakdown and complete loss of information in the system. To cope with this problem, called the "error catastrophe," large components had to be composed of a series of smaller ones. The small components need to cooperate to make larger molecules. A series of cycles may finally end up at the initial cycle, and this closes a ring with a positive feedback mechanism back to the origin. This is what has been defined as a hypercycle (Eigen and Schuster 1977). Each element of the hypercycle is a cycle itself and interacts with the next one. The various molecules need to cooperate and coevolve. The model includes both cooperation and selfishness of the members. The individual replication cycles are thus interdependent and finally form a hypercycle by interacting with the beginning.

Eigen even considered the error rate as a way to destroy genetic information on purpose. He designed an antiviral therapy exploiting such an error catastrophe to inhibit viral replication, such as HIV.

Eigen himself leaves it open what kind of chemical molecule could have started and evolved in the hypercycle; DNA plus protein was his suggestion, but just RNA is sufficient, a ribozyme, because of its dual functions, it can replicate and mutate. Eigen and coworkers designed the hypercycle with replicating DNA, leading to mRNA production and protein synthesis. The mRNA then feeds into an enzyme, a polymerase, for protein synthesis, or a replicase, to generate more mRNAs.

One can simplify the model just based on RNA, which exhibits both hardware and software, corresponding to enzymatic molecules (ribozymes) and information (genetic code) or structural and genetic information. Replicated RNA, produced by the ribozyme activity, can feed into the next cycle, including mutagenesis, and becomes replicated there by another ribozyme activity (Eigen 1971; Eigen and Schuster 1977). Eigen prefers to think of a principle, a universal model, based on entities with self-organization and self-replication.

A simplified hypercycle is shown as published by Natalia Szostak (Szostak et al. 2016) (Fig. 23.1).

Self-learning is the keyword today for AI-driven methods and how ChatGPT learns to improve information. The procedures resemble the mechanisms of learning and optimization as properties of nature, of evolution. The principle of fitness as the best basis for survival holds even true for generating information by ChatGPT (Hopfield 1982).

Even the formation of stars, planetary systems, and galaxies in the Universe seems to follow this rule of self-organization, just like RNA, viruses, living species, and AI machines.

It goes back not only all the way to the origin of life but also to the origin of the Universe.

Thus, autocatalysis and self-organization appear to be principles of nature, laws of physics.

What can we learn from this for life on extraterrestrial exoplanets?

Acknowledgment Not applicable.

Conflict of Interest None.

Ethics Not applicable.

References

Butler PJ, Klug A (1971) Assembly of the particle of tobacco mosaic virus from RNA and disks of protein. Nat New Biol 13:47–50. https://doi.org/10.1038/newbio229047a0

Cech TR (2000) Structural biology. The ribosome is a ribozyme. Science 289:878–879

Cech TR (2004) RNA finds a a simpler Way. Nature 428:263–264

Crick FH (1968) The origin of the genetic code. J Mol Biol 38:367–379

Diener TO (2016) Viroids: "living fossils" of primordial RNAs? Biol Direct 11:15–18. https://doi.org/10.1186/s13062-016-0116-7

Doudna JA, Cech TR (2002) The chemical repertoire of natural ribozymes. Nature 418:222–228

Dreher TW (2008) Role of tRNA-like structures in controlling plant virus replication. Virus Res 139:217–229. https://doi.org/10.1016/j.virusres.2008.06.010

Eigen M (1971) Self organization of matter and the evolution of biological macromolecules. Naturwissenschaften 58:465–523. Springer, Berlin/Heidelberg, ISSN 0028-1042

Eigen M, Schuster P (1977) The Hypercycle, A principle of natural Self-organization. Naturwissenschaften 64(11). https://doi.org/10.1007/BF00623322

Gross HJ, DomdeyH LC, Jank P, Raba M, Alberty H, Sänger HL (1978) Nucleotide sequence and secondary structure of potato spindle tuber viroid. Nature 273:203–208

Hansen TB, Jensen TI, Clausen BH, Bramsen JB, Finsen B, Damgaard CK, Kjems J (2013) Natural RNA circles function as efficient microRNA sponges. Nature 495(7441):384–388

Hopfield JJ (1982) Neural networks and physical systems with emergent collective computational abilities. Proc Natl Acad Sci 79:2554–2558. https://doi.org/10.1073/pnas.79.8.2554

Ishida S, Terasaka N, Katoh T, Suga H (2020) An aminoacylation ribozyme eveolved fram a natural tRNA-sensing T-box riboswitch. Nat Chem Biol 16:702.709

Joyce GF (2002) The antiquity of RNA-based evolution. Nature 418:214–221

Kowiatek BK (2022) Did tRNA evolve from a ribozyme? An in silico study. Int J Front Sci Technol Res 02:024–030
Küppers BO (1971) Molekulare Evolution und der Ursprung des Lebens. Bertelsmann Lexikothek Verlag, pp 348–351
Küppers BO (1981) Evolution im Reagenzglas. In: von Dithfurth H (ed) Mannheimer Forum 80/81
Küppers BO (1985) Molecular theory of evolution. Springer, Berlin/Heidelberg
Küppers BO (1990) Information and the origin of life. MIT Press, Cambridge, MA
Lincoln TA, Joyce GF (2009) Self-sustained replication of an RNA enzyme. Science 323:1229–1232
Mans RM, Pleij CW, Bosch L (1991) tRNA-like structures. Structure, function and evolutionary significance. Eur J Biochem 201:303–324. https://doi.org/10.1111/j.1432-1033.1991.tb16288.x
Matsuda D, Dreher TW (2004) The tRNA-like structure of Turnip yellow mosaic virus RNA is a 3'-translational enhancer. Virology 321:36–46. https://doi.org/10.1016/j.virol.2003.10.023
Meierhenrich UJ (2008) Amino acids and the asymmetry of life. Springer, Berlin Heidelberg
Meierhenrich UJ, Muñoz Caro GM, Bredehöft JH, Thiemann WHP (2004) Identification of diamino acids in the Murchison meteorite. PNAS 101:9182–9186. https://doi.org/10.1073/pnas.0403043101
Miller SL, Urey HC (1959) Organic compound synthesis on the primitive earth. Science 130:245–251
Moelling K (2024) Epigenetics and transgenerational inheritance. J Physiol 602(11):2537–2545
Moelling K, Broecker F (2015) The reverse transcriptase-RNase H: from viruses to antiviral defense. Ann N Y Acad Sci 1341:126–135. https://doi.org/10.1111/nyas.12668
Moelling K, Broecker F (2019a) Viroids-first—a model for life on Earth, Mars and exoplanets. Geosciences 9:241–259
Moelling K, Broecker F (2019b) Viruses and evolution - viruses first? Front Microbiol 10:523–528
Moelling K, Broecker F (2021) Viroids and the origin of life. Int J Mol Sci 22:3476–3490. https://doi.org/10.3390/ijms22073476
Moelling K, Mueller G, Dannull J, Reuss C, Beimling P, Bartz C, Wiedenmann B, Yoon K, Surovoy A, Jung G (1994) Stimulation of Ki-ras ribozyme activity by RNA binding protein, NCp7, *in vitro,* and in pancreatic tumor cell line, Capan1. New York Academy Sciences, Wiley, p 17496632
Moelling K, Matskevich A, Jung JS (2006) Relationship between retroviral replication and RNA interference machineries. Cold Spring Harb Symp Quant Biol 71(0):365–368
Moelling K, Broecker F, Russo G, Sunagawa S (2017) RNase H as gene modifier, driver of evolution and antiviral defense. Front Microbiol 8:1745. https://doi.org/10.3389/fmicb.2017.01745
Müller G, Strack B, Dannull J, Sproat BS, Surovoy A, Jung G, Moelling K (1994) Aminoacid requirements of the nucleocapsid protein of HIV for increasing catalytic activity of a Ki-ras Ribozyne in vitro. J Mol Biol 242:422–429
Orgel LE (1968) Evolution of the genetic apparatus. J Mol Biol 38:381–393
Pachuk CJ, Yoon K, Moelling K, Coney LR (1994) Selective cleavage of bcr-abl chimeric RNAs by a ribozyme targeted to non-contiguous sequences. Nucleic Acids Res. 22:301–307
Patel BH, Percivalle C, Ritson DJ, Duffy CD, Sutherland JD (2015) Common origins of RNA, protein and lipid precursors in a cyanosulfidic protometabolism. Nat Chem 7:301–307. https://doi.org/10.1038/nchem.2202
Piron M, Beguiristain N, Martinez-Salas E, Gomez J (2005) Characterizing the function and structural organization of the 5` tRNA-like motif within the hepatitis C virus quasispecies. Nucleic Acids Res 33:1487–1502. https://doi.org/10.1093/nar/gki290
Spiegelman S, Haruna I, Holland IB, Beaudreau G, Mills D (1965) The synthesis of a self-propagating and infectious nucleic acid with a purified enzyme. Proc Natl Acad Sci USA 54:919–927. https://doi.org/10.1073/pnas.54.3.919
Sumper M, Luce R (1975) Evidence for de novo production of self-replicating and environmentally adapted RNA structures by bacteriophage Qbeta replicase. Proc Natl Acad Sci USA 72:162–166

Szostak JW, Bartel DP, Luisi PL (2001) Synthesizing life. Nature 409:387–390
Szostak N, Wasik S, Blazewicz J (2016) Hypercycle. PLoS Comput Biol 12:e1004853. https://doi.org/10.1371/journal.pcbi.1004853
Taylor JM (2009) Replication of the hepatitis delta virus RNA genome. Adv Virus Res 74:103–121. https://doi.org/10.1016/S0065-3527(09)74003-5
Wilson DS, Szostak JW (1999) In vitro selection of functional nucleic acids. Annu Rev Biochem 68:611–647
Winkler WC, Nahvi A, Roth A, Collins JA, Breaker RR (2004) Control of gene expression by a natural metabolite-responsive ribozyme. Nature 428(6980):281–286. ISSN 1476-4687. https://doi.org/10.1038/nature02362, PMID 15029187
Woese CR (1965) On the evolution of the genetic code. Proc Natl Acad Sci USA 54:1546–1552

Chapter 24
Minimal RNA Replicons in the Origin and Evolution of Life

Olga Rueda and Marcos de la Peña

Abstract Minimal replicons of circular RNA, such as viroids or deltaviruses, occupy the conceptual and evolutionary boundary between living and nonliving matter. They exemplify genomes in which RNA alone integrates heredity, self-cleaving catalysis, and parasitic adaptation to the environment, thereby providing plausible modern analogues of primordial RNA World replicators. Advances in high-throughput sequencing and planetary-scale data mining have transformed these agents from rare curiosities into a vast ecosystem of "insignificant" RNA agents such as zetaviruses, circular RNA viruses, and obelisks, associated with all major environments and host types. This expanding catalog of minimal replicators challenges traditional host-range assumptions, blurs the virus–viroid distinction, and suggests that minimal RNA-based molecular parasites have persisted and diversified throughout Earth's history. By linking molecular fossils such as ribozymes and circular genomes to contemporary microbial ecosystems, these discoveries support the view that the RNA World is not a vanished prebiotic stage but a continuing dimension of life, and that studying minimal RNA replicons can illuminate both the origins and ongoing evolution of biological complexity.

1 Introduction: Life at the Border of Life

The question "What is life?" has puzzled scientists and philosophers for the last millennium ("Life (Stanford Encyclopedia of Philosophy/Fall 2025 Edition)," n.d.). During this time, the boundary between living and nonliving seemed clear: entities that multiply, grow. and adapt. However, this "comfortable" certainty collapses when examined closely, particularly when confronting the simplest replicons in the biosphere, which challenge our most fundamental assumptions about what it means to be alive. The natural phenomenon of life, which we generally contrast with

O. Rueda · M. de la Peña (✉)
Instituto de Biologia Molecular y Celular de Plantas, CSIC-Universidad Politecnica de Valencia, Valencia, Spain
e-mail: rivero@ibmcp.upv.es

G. Witzany (ed.), *Integrative Theory of Evolution - Aspects and Insights*,
https://doi.org/10.1007/978-3-032-24295-2_24

everything inert and inanimate, is fundamentally difficult to define satisfactorily. The etymological origins of “life” itself—from the Latin *vita*, meaning simply to exist—reflect this conceptual challenge. Similar patterns emerge across languages: in English, “life,” or in German, “Leben” (from Indo-European *Leip*, to remain); in Mandarin Chinese, 生 (*sheng*, to exist); and in Hindi, जीविषा (*jijivisha*, to persist). Evidently, not everything that merely exists can be considered alive, and our definitions inevitably remain tied to human experience, employing partly arbitrary criteria to distinguish the living from the nonliving.

Numerous natural and artificial systems simulate or share many aspects of biological life: meteorological phenomena like storms self-organize, exchange energy, and dissipate into their environment; crystals grow, replicate patterns, and compete for resources; even artificial intelligence and computer viruses replicate, mutate, and evolve within digital environments. To distinguish biological life from these simulation systems, scientists have identified exclusive characteristics of biological phenomena: (i) the capacity to transform matter and energy into ordered growth; (ii) the existence of a functional program or purpose encoded in molecular structure that mechanically directs survival and reproduction—known as “internal teleology”; and (iii) the capacity for Darwinian evolution, wherein information is inherited with variation and subjected to selection across generations.

However, life might be understood as a far broader phenomenon if we relax some of these requirements, a necessary step for comprehending two interrelated boundary phenomena in biology: the origin of life (abiogenesis) and biological entities with minimal genomes (viruses and other minimal replicators), where the frontier between the inert and the living becomes remarkably porous.

To reconstruct the first steps of life’s origin, we must acknowledge that the earliest replicators must have been extraordinarily simple entities, far removed from the complexity of the simplest known “canonical” living organism today—typically a bacterium (Moran and Bennett 2014) or even an archaeon (Harada et al. 2025) with a few hundred genes in a genome comprising hundreds of thousands of base pairs. By evolutionary logic and following the principles of maximum parsimony, those primordial replicating genomes must have been small molecules, perhaps only tens or hundreds of nucleotides at most, capable of arising and organizing randomly (Fine and Pearlman 2023). Only when these proto-entities acquired minimal self-copying capacity could the gradual race to increase complexity begin through replication errors and Darwinian selection—a competition that continues to this day. After at least four billion years of evolution since that hypothetical origin, the current tree of life displays astonishing complexity, from enormous multicellular organisms to minimal biological entities so extremely simple that they might potentially emulate the first ancestral replicators.

At the most basic stages of this evolutionary ladder, we find what we generally call “RNA replicons”—a group of viral- and subviral-like entities with RNA genomes that place them at the very boundary of the definition of life (De la Peña et al. 2020) (Fig. 24.1). Extreme examples of such simplicity are plant viroids, discovered roughly 50 years ago (Diener 1971) (Fig. 24.2), consisting of mere naked circular noncoding RNA genomes of only about 300 nucleotides. Since the discovery of plant viroids in potato plants in 1971, up to 44 different species of plant

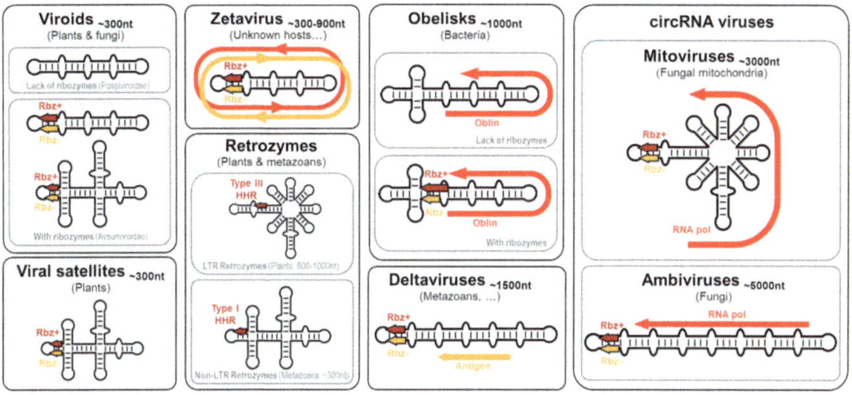

Fig. 24.1 Schematic representation of the principal circRNA entities described to date, arranged by genome size. From left to right and top to bottom: viroids, plant virus RNA satellites, Zetavirus, Plant and metazoan Retrozymes, Obelisks, Deltavirus, and fungal circRNA viruses. Rbz ribozyme, RNA pol RNA polymerase

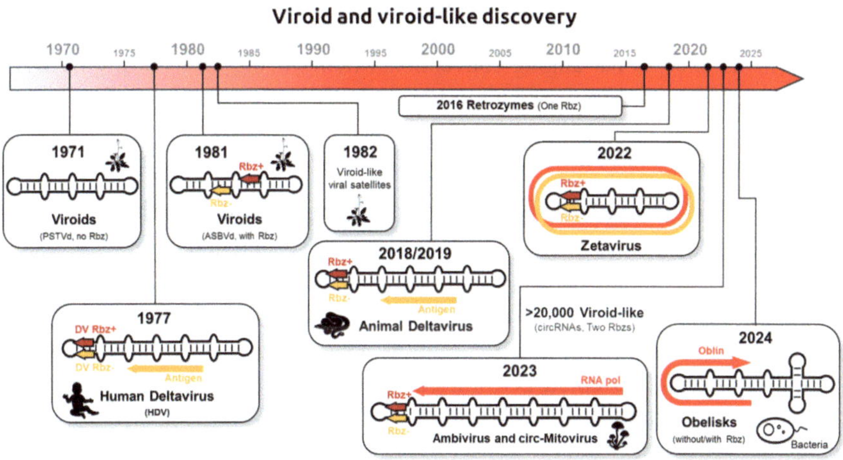

Fig. 24.2 Discovery of the main circular RNA replicons in chronological order. During the decades of the 1970s and 1980s, the first plant viroids, human deltavirus, and plant-viroid–like RNA satellites were described. For 40 years, only new species of plant-viroid RNAs were reported. In 2016, the discovery of retrozymes in plants, followed by the discovery of new animal deltaviruses in 2018, confirmed that circRNAs with ribozymes are much more frequent. Subsequently, zetaviruses with unknown hosts (2022), circular mycoviruses (2023), and bacterial Obelisks (2024) have been reported

viroids have been reported. Most of them (39 species) belong to the *Pospiviroidae* family, which is characterized by a rod-like RNA structure containing a central conserved region (CCR). These viroids are known to replicate in the plant cell nucleus through an asymmetric rolling circle replication mechanism, in which, all the required factors are provided by the host cell, such as RNA polymerase II for replication, RNase III-like enzymes for the processing of oligomeric RNA intermediates, and DNA ligase one for the circularization of monomeric linear forms (Zhang et al. 2025). There is a second family of viroids (5 recognized species), the *Avsunviroidae* family, first characterized in 1981 with the sequencing of avocado sunblotch viroid (ASBVd) (Fig. 24.2) (Symons 1981), and featured by the presence of small RNA self-cleaving hammerhead ribozymes in both RNA polarities (Hutchins et al. 1986). These motifs allow the replication of Avsunviroidae members through symmetric rolling circle mechanism that takes place in the chloroplast of the cell through a nuclear-encoded phage-like RNA polymerase (NEP) (Navarro et al. 2000). Another paradigmatic example of a minimal replicator is the human hepatitis D virus or human deltavirus (HDV) (Rizzetto et al. 1977), which has been historically considered the simplest known molecular parasite in humans (Fig. 24.1). This is a circular rod-like RNA similar to viroids but substantially larger (~1600 nucleotides) and capable of encoding a small polypeptide required for its replication in the cell nucleus through a rolling circle replication catalyzed by RNA polymerase II (Flores et al. 2011). All these minimal replicons completely depend on either plant or human living cells that they parasitize for reproduction. A remarkable characteristic of these microorganisms is that they employ RNA as their genetic material, which usually does not encode any protein but can itself catalyze certain chemical reactions. This is exceptional among conventional living beings, which rely on DNA genomes and biochemistry based almost exclusively on proteins.

2 The RNA World: A Key to Life's Origins on Earth

The discovery of RNAs with catalytic activity, or ribozymes, during the 1980s (Guerrier-Takada et al. 1983; Kruger et al. 1982) revolutionized this narrative. The knowledge that RNA could function both as genetic material (in many viral and subviral replicons) and could perform catalytic functions strongly supported the hypothesis of a primordial RNA World (Gilbert 1986). According to this concept, a prebiotic scenario existed in which self-replicating RNA molecules simultaneously fulfilled genomic and biochemical functions that could gradually evolve into increasingly complex organisms, ultimately yielding DNA- and protein-based life forms. Supporting evidence for this hypothetical world includes the modern ribosome—a ribozyme that decodes genetic information and catalyzes protein synthesis in all living organisms (viruses included)—which represents a crucial molecular fossil from that period (Cech 2000). Additional evidence supporting a prebiotic RNA era includes the minimal RNA genomes present today, which, through their extreme simplicity and frequent presence of ribozymes, emulate the primordial replicating agents (Flores et al. 2014).

However, for half a century, a paradox persisted regarding these simple replicons: why did only very few examples of viroid-like agents exist (approximately 50 species discovered over nearly 50 years), and why did all of them replicate exclusively in the most complex and recently evolved organisms of the biosphere—plants and animals? This observation strongly suggested that these rare circular RNA genomes might not be so closely related to primordial replicating entities but could have more recent origins, or perhaps both ancient and recent entities have coexisted simultaneously (Diener 2016) (Fig. 24.2).

3 Small Ribozymes and circRNAs Are Not Oddities

The surprising solution to this paradox arrived progressively over the past 15 years, driven by unprecedented advances in computational biology and the acquisition of massive genomic data through high-throughput sequencing technologies. The first clues emerged from systematic bioinformatic surveys revealing that small ribozyme motifs—particularly the hammerhead (hhr) and deltavirus (dvr) ribozymes— are widespread among DNA genomes along the tree of life, from bacteria to humans (De la Peña and Garcia-Robles 2010; Hammann et al. 2012; Webb et al. 2009). These discoveries indicated that the catalytic signatures of rare circular RNA replicators were ancient molecular fossils that were extremely common in the biosphere and not merely exotic curiosities confined to the particular subviral field. Moreover, the following discoveries of novel retrotransposons from plants and animals, the retrozymes, exploit small viroid-like circRNAs with ribozymes to spread among eukaryotic genomes (Cervera et al. 2016 Genome Biol; Cervera 2020), revealing that not only the small ribozymes but circRNA entities are far more frequent molecules than expected (Fig. 24.2).

The watershed moment came with the explosion of metatranscriptomic data coupled with increasingly powerful computational analysis, enabling the discovery of novel RNA microorganisms at the viral and subviral scales. A paradigmatic case was the detection of two novel deltaviruses related to the unique human HDV in ducks (Wille et al. 2018) and snakes (Hetzel et al. 2019). These results were followed by the discovery of deltaviruses in many other animals (Chang et al. 2019), including examples of novel variants with typical plant-viroid hhr ribozymes (De la Peña et al. 2021).

A breakthrough began with a critical technological innovation: the development of petabase-scale sequence alignment approaches, which fundamentally changed our capacity to explore viral diversity in planetary-scale datasets. This leap permitted researchers to interrogate billions of genomic and metagenomic sequence libraries for specific molecular signatures such as the presence of RNA virus polymerases, allowing them to identify well over 100,000 novel RNA viruses and expand the known species diversity by approximately tenfold (Edgar et al. 2022). The study also identified previously unknown circular RNA agents related to deltaviruses. Through sequence searching specifically for delta antigen proteins and characteristic ribozymes (either typical dvr or plant-viroid hhr), the researchers identified 14 novel deltaviruses and 39 epsilonviruses (hhr-containing deltaviruses). Whereas a few of them were found in metazoan hosts, most of these new deltavirus-like circRNAs occurred in environmental samples enriched in microorganisms such as unicellular eukaryotes, archaea, or bacteria as the most likely hosts. Moreover, the authors also detected in most of these environmental samples the existence of 311 circRNAs of small size (300–900 nt) with deltavirus-like genomic organization, subsequently termed zetaviruses (Edgar et al. 2022). These enigmatic genomes, found widespread in samples from soils, lakes, and oceans, displayed astonishing characteristics beginning with their mathematically precise size: always a multiple of three nucleotides. They show the capacity to encode in each of their two polarities one ribozyme but also one protein, which lacks stop codons. Given that zetavirus genomes are circular and maintain perfect size multiples of 3, translation theoretically would never terminate, potentially producing "infinite" polypeptides consisting of endless repetitions of the sequence encoded by the zetavirus (Edgar et al. 2022). The actual hosts and biological significance of these characteristics remain mysterious, and may represent entities employing a novel genetic language potentially specialized for parasitizing ribosomes within host cells. These discoveries definitively unveiled that minimal circular RNA replicators can be more widespread than the scientific community had realized, and their evolutionary histories indicate a reasonable common ancestry, potentially solving a long-standing question about the origins of circular RNA subviral agents in higher eukaryotes.

4 The Big circRNA Explosion Through Data Mining

If the study of Edgar and collaborators showed the vastness of RNA virus diversity globally, two complementary mining approaches in 2023 already have confirmed the extraordinary hidden diversity of viroid-like agents (Forgia et al. 2023; Lee et al. 2023). Computational pipelines developed to identify viroid-like entities (i.e., circRNAs with known ribozymes) were applied to thousands of public transcriptomes and metatranscriptomes from diverse organisms and environments. The searches yielded up to 22,000 species-level clusters, representing more than a two-orders-of-magnitude increase compared to previously identified viroid-like elements combined. But particularly remarkable was the discovery of hundreds of large (3000–5000 nt) circRNAs with paired ribozymes that encoded ambivirus- or mitovirus-like RNA-dependent RNA polymerase sequences. Until this study, fungal ambiviruses (Sutela et al. 2020) and mitoviruses (Polashock and Hillman 1994) were reported as linear RNA viruses, and the identification of circular RNA virus genomes represented a novel category of viral agent. Many of these sequences contained matched or mismatched combinations of most known self-cleaving ribozymes, such as hammerhead, delta, hairpin, or even twister ribozymes, suggesting evolutionary connections between typical RNA viruses and minimal viroid-like replicators—a hybrid category previously unknown. The discovery of entirely new and distinct categories of circular RNA agents, each with distinctive characteristics, blurred the boundary between virus and viroid-like entities. But even more intriguingly, the evidence from CRISPR spacer matches in some of these minimal circRNAs also opened a very likely possibility: some of these minimal replicons may replicate in prokaryotic hosts (Lee et al. 2023).

5 The First CircRNA Replicons Discovered in the Bacterial Kingdom

Perhaps most unexpectedly, the 2024 characterization of the family of obelisks—a novel category of circular RNA agent—dramatically extended the host range of viroid-like replicators beyond any previously recognized boundaries (Zheludev et al. 2024). These agents were initially detected in RNA sequencing data from human fecal samples, where approximately 7% of sampled individuals harbored curious circular RNAs of approximately 1000 nucleotides. These RNAs adopted a quasi-rod structure similar to viroids, delta viruses, and ambiviruses, earning the distinct name "Obelisks". Through comprehensive bioinformatic approaches, researchers detected approximately 30,000 distinct obelisk species distributed across virtually all types of environments and ecological niches, with particular prominence in microbiomes of animals throughout the phylogenetic tree (Zheludev et al. 2024). Obelisks encode at least one protein of approximately 200 amino acids—termed oblins—that display no sequence homology to any known protein in scientific databases, rendering their biological function completely mysterious and potentially indicating entirely novel protein functions unknown to biochemistry. Remarkably, certain subfamilies of obelisks encode in both polarities type III hammerhead ribozymes, the characteristic catalytic motifs of plant viroids and many others of the newly discovered viroid-like entities, further reinforcing evolutionary connections between all these minimal circular RNA agents (Zheludev et al. 2024). Most significantly, at least one obelisk species was recovered from pure bacterial cultures of *Streptococcus sanguinis* isolated from the human oral cavity, definitively establishing the first example of a viroid-like circular RNA agent replicating within a prokaryotic organism. This finding somehow closes a circle that began 50 years earlier with the discovery of viroids in plants: now we can confirm that these minimal RNA plant pathogens, historically considered biological anomalies, were merely tiny representatives of an entire new world of circular RNAs inhabiting virtually every ecological niche on this planet, with origins likely extending back billions of years.

6 Implications for the Primordial RNA World Hypothesis

Planetary mining studies revealed that viroid-like agents are distributed geographically worldwide across all ecosystem types, both environmental and animal-associated. Soil, sediment, and water samples proved the richest sources of viroid-like elements, where virtually no plant or animal sequences were present. This distribution pattern contradicts the historical assumption of narrow host ranges for viroid-like agents. These discoveries overturn the notion that viroid-like agents require complex eukaryotic hosts and call for reappraisal of their evolutionary status. Altogether, these revelations carry profound implications for understanding life's origins. The extreme simplicity of viroid-like genomes, combined with their catalytic ribozymes and the evident ubiquity across all domains of life, reinforces the long-standing question about these replicators being modern descendants of the primordial RNA World (Chela-Flores 1994; Diener 1989; Flores et al. 2014). If simple RNA replicons could self-replicate and catalyze reactions billions of years ago, and if organisms as simple as bacteria could maintain association with viroid-like agents, then the evolutionary path from simple RNA molecules to modern cellular life becomes more comprehensible. Moreover, the fact that "minimal RNA replicators" are not rare biological anomalies but rather abundant planetary agents somehow alters the scientific narrative about early life. Rather than being evolutionary dead-ends, viroid-like replicators may represent a vast ecological realm of obligate molecular parasites that have persisted and adapted across the entire evolutionary history of life. The presence of diverse ribozymes, novel protein-coding sequences, and apparent evolutionary connections between distinct circular RNA families suggests that these agents undergo constant evolutionary innovation, indicating that the prebiotic RNA World may not be merely a prebiotic step but rather an ongoing dimension of global biological diversity.

7 Conclusion: A New Paradigm in Virology and Molecular Evolution

The decade of the 2020s represents a watershed moment in virology and computational biology. Through the transformative technologies of petabase-scale data mining, our understanding of global viral and subviral diversity has expanded by orders of magnitude. What was previously understood as a marginal biological phenomenon (a handful of plant viroids and a single human deltavirus) has been revealed as an entire planetary ecosystem of minimal RNA replicators. These discoveries carry implications extending far beyond taxonomy and epidemiology. They strongly suggest that the earliest replicators on Earth, the molecular ancestors of all life, may not be extinct relics preserved only in the modern ribosome, but rather persist as living, evolving components of microbiomes and other ecosystems throughout the biosphere. The boundary between virus and viroid blurs with the discovery of hybrid forms of viroid-like viruses. The exclusive occurrence of subviral agents in higher eukaryotes expands with the identification of viroid-like agents in unicellular eukaryotes and bacterial cells. Our understanding of life and nonlife becomes far more permeable. Years of research still lie ahead to characterize the hosts, replication mechanisms, and evolutionary histories of these thousands of newly discovered circular RNA agents. Yet the fundamental paradigm shift has already occurred: we now understand that we inhabit not just a world of DNA-based life and protein-based biochemistry, but that we coexist with a vast, ancient, and still largely mysterious kingdom of RNA-based minimal replicators, a world that may ultimately illuminate the deepest origins of life itself.

References

Cech TR (2000) The ribosome is a ribozyme. Science (1979) 289:878–879. https://doi.org/10.1126/science.289.5481.878

Cervera A, Urbina D, de la Peña M (2016) Retrozymes are a unique family of non-autonomous retrotransposons with hammerhead ribozymes that propagate in plants through circular RNAs. Genome Biol 17(1):135. https://doi.org/10.1186/s13059-016-1002-4

Cervera A, de la Peña M (2020) Small circRNAs with self-cleaving ribozymes are highly expressed in diverse metazoan transcriptomes. Nucleic Acids Res 48(9):5054–5064. https://doi.org/10.1093/nar/gkaa187

Chang W-S, H-O Pettersson J, Le Lay C, Shi M, Lo N, Wille M, Eden J-S, Holmes EC, Pettersson JH-O, Le Lay C, Shi M, Lo N, Wille M, Eden J-S, Holmes EC (2019) Novel hepatitis D-like agents in vertebrates and invertebrates. Virus Evol 5:vez021. https://doi.org/10.1093/ve/vez021

Chela-Flores J (1994) Are viroids molecular fossils of the RNA world? J Theor Biol 166:163–166

De la Peña M, Garcia-Robles I (2010) Intronic hammerhead ribozymes are ultraconserved in the human genome. EMBO Rep 11:711–716. https://doi.org/10.1038/embor.2010.100

De la Peña M, García-Robles I (2010) Ubiquitous presence of the hammerhead ribozyme motif along the tree of life. RNA 16:1943–1950. https://doi.org/10.1261/rna.2130310

De la Peña M, Ceprián R, Cervera A (2020) A singular and widespread group of mobile genetic elements: RNA circles with autocatalytic ribozymes. Cells 9:2555. https://doi.org/10.3390/cells9122555

De la Peña M, Ceprian R, Casey JL, Cervera A (2021) Hepatitis delta virus-like circular RNAs from diverse metazoans encode conserved hammerhead ribozymes. Virus Evol 7. https://doi.org/10.1093/VE/VEAB016

Diener TO (1971) Potato spindle tuber "virus". IV. A replicating, low molecular weight RNA. Virology 45:411–428

Diener TO (1989) Circular RNAs: relics of precellular evolution? Proc Natl Acad Sci USA 86:9370–9374

Diener TO (2016) Viroids: "living fossils" of primordial RNAs? Biol Direct 11:15. https://doi.org/10.1186/s13062-016-0116-7

Edgar RC, Taylor J, Lin V, Altman T, Barbera P, Meleshko D, Lohr D, Novakovsky G, Buchfink B, Al-Shayeb B, Banfield JF, de la Peña M, Korobeynikov A, Chikhi R, Babaian A (2022) Petabase-scale sequence alignment catalyses viral discovery. Nature 602:1–6. https://doi.org/10.1101/2020.08.07.241729

Fine JL, Pearlman RE (2023) On the origin of life: an RNA-focused synthesis and narrative. RNA 29:1085. https://doi.org/10.1261/RNA.079598.123

Flores R, Grubb D, Elleuch A, Nohales MÁ, Delgado S, Gago S (2011) Rolling-circle replication of viroids, viroid-like satellite RNAs and hepatitis delta virus: variations on a theme. RNA Biol 8:200–206. https://doi.org/10.4161/rna.8.2.14238

Flores R, Gago-Zachert S, Serra P, Sanjuan R, Elena SF (2014) Viroids: survivors from the RNA world? Ann Rev Microbiol 68:395–414. https://doi.org/10.1146/annurev-micro-091313-103416

Forgia M, Navarro B, Daghino S, Cervera A, Gisel A, Perotto S, Aghayeva DN, Akinyuwa MF, Gobbi E, Zheludev IN, Edgar RC, Chikhi R, Turina M, Babaian A, Di Serio F, de la Peña M (2023) Hybrids of RNA viruses and viroid-like elements replicate in fungi. Nat Commun 14:1–11. https://doi.org/10.1038/s41467-023-38301-2

Gilbert W (1986) The RNA world. Nature 319:618
Guerrier-Takada C, Gardiner K, Marsh T, Pace N, Altman S (1983) The RNA moiety of ribonuclease P is the catalytic subunit of the enzyme. Cell 35:849–857. https://doi.org/10.1016/0092-8674(83)90117-4. [pii]
Hammann C, Luptak A, Perreault J, de la Peña M (2012) The ubiquitous hammerhead ribozyme. RNA 18:871–885. https://doi.org/10.1261/rna.031401.111
Harada R, Nishimura Y, Nomura M, Yabuki A, Shiba K, Inaba K, Inagaki Y, Nakayama T (2025) A cellular entity retaining only its replicative core: hidden archaeal lineage with an ultra-reduced genome. bioRxiv. https://doi.org/10.1101/2025.05.02.651781
Hetzel U, Szirovicza L, Smura T, Prähauser B, Vapalahti O, Kipar A, Hepojoki J (2019) Identification of a novel deltavirus in boa constrictors. MBio 10:1–8. https://doi.org/10.1128/MBIO.00014-19
Hutchins CJ, Rathjen PD, Forster AC, Symons RH (1986) Self-cleavage of plus and minus RNA transcripts of avocado sunblotch viroid. Nucleic Acids Res 14:3627–3640. https://doi.org/10.1093/nar/14.9.3627
Kruger K, Grabowski PJ, Zaug AJ, Sands J, Gottschling DE, Cech TR (1982) Self-splicing RNA: autoexcision and autocyclization of the ribosomal RNA intervening sequence of Tetrahymena. Cell 31:147–157
Lee BD, Neri U, Roux S, Wolf YI, Camargo AP, Krupovic M, Simmonds P, Kyrpides N, Gophna U, Dolja VV, Koonin EV (2023) Mining metatranscriptomes reveals a vast world of viroid-like circular RNAs. Cell 186:646–661.e4. https://doi.org/10.1016/J.CELL.2022.12.039
Life (Stanford Encyclopedia of Philosophy/Fall 2025 Edition) [WWW Document] (n.d.). URL https://plato.stanford.edu/archives/fall2025/entries/life/ (accessed 1.26.26)
Moran NA, Bennett GM (2014) The tiniest tiny genomes. Ann Rev Microbiol 68:195–215. https://doi.org/10.1146/ANNUREV-MICRO-091213-112901/CITE/REFWORKS
Navarro JA, Vera A, Flores R (2000) A chloroplastic RNA polymerase resistant to tagetitoxin is involved in replication of avocado sunblotch viroid. Virology 268:218–225. https://doi.org/10.1006/VIRO.1999.0161
Polashock JJ, Hillman BI (1994) A small mitochondrial double-stranded (ds) RNA element associated with a hypovirulent strain of the chestnut blight fungus and ancestrally related to yeast cytoplasmic T and W dsRNAs. Proc Natl Acad Sci USA 91:8680–8684. https://doi.org/10.1073/PNAS.91.18.8680
Rizzetto M, Canese MG, Arico S, Crivelli O, Trepo C, Bonino F, Verme G (1977) Immunofluorescence detection of new antigen-antibody system (delta/anti-delta) associated to hepatitis B virus in liver and in serum of HBsAg carriers. Gut 18:997–1003. https://doi.org/10.1136/GUT.18.12.997
Sutela S, Forgia M, Vainio EJ, Chiapello M, Daghino S, Vallino M, Martino E, Girlanda M, Perotto S, Turina M (2020) The virome from a collection of endomycorrhizal fungi reveals new viral taxa with unprecedented genome organization. Virus Evol 6. https://doi.org/10.1093/VE/VEAA076
Symons RH (1981) Avocado sunblotch viroid: primary sequence and proposed secondary structure. Nucleic Acids Res 9:6527–6537. https://doi.org/10.1093/NAR/9.23.6527
Webb CHT, Riccitelli NJ, Ruminski DJ, Lupták A (2009) Widespread occurrence of self-cleaving ribozymes. Science (1979) 326:953. https://doi.org/326/5955/953 [pii]. https://doi.org/10.1126/science.1178084
Wille M, Netter HJ, Littlejohn M, Yuen L, Shi M, Eden JS, Klaassen M, Holmes EC, Hurt AC (2018) A divergent hepatitis D-like agent in birds. Viruses 10. https://doi.org/10.3390/v10120720
Zhang Y, Bisaro DM, Wu J (2025) Recent advances in viroid research. Virology 604:110424. https://doi.org/10.1016/J.VIROL.2025.110424
Zheludev IN, Edgar RC, Lopez-Galiano MJ, de la Peña M, Babaian A, Bhatt AS, Fire AZ (2024) Viroid-like colonists of human microbiomes. bioRxiv. https://doi.org/10.1101/2024.01.20.576352

Chapter 25
The Many Shades of Compartmentalization in Prebiotic Evolution

Louis M. P. Ter-Ovanessian, Thomas Matreux, Ajay Verma, Tommaso Fraccia, and Philippe Nghe

Abstract The emergence of life hinges on a mutual evolution between compartments and their encapsulated content, addressing the paradox where each one seems to require the other to function. This chapter frames compartmentalization as a gradual progression rather than a sudden biological invention. We first review how primitive compartmentalization, including limited diffusion in geological pores, transient coacervates, and permeable vesicles, supports ground evolutionary properties such as reproduction, heredity, and the mitigation of the so-called "Error catastrophe." We examine the experimental state-of-the-art, highlighting how mineral surfaces scaffold chemical networks and how phase-separated coacervates serve as dynamic, membraneless microenvironments for ribozyme activity. Then, we describe the transition from simple fatty acid vesicles to robust phospholipid membranes, based on "missing link" lipids that bridge this evolutionary gap. Finally, we propose a complete nine-stage evolutionary arc, depicting the putative shift from environment-dependent chemistry to autonomous protocells capable of regulated cell cycles.

L. M. P. Ter-Ovanessian · T. Matreux · A. Verma
Laboratoire de Biophysique et Evolution, UMR CNRS-ESPCI 8231 Chimie Biologie Innovation, ESPCI Paris, Université PSL, Paris, France

T. Fraccia
Dipartimento di Scienze Farmacologiche e Biomolecolari "Rodolfo Paoletti", Università degli Studi di Milano, Milano, Italy

P. Nghe (✉)
Laboratoire de Biophysique et Evolution, UMR CNRS-ESPCI 8231 Chimie Biologie Innovation, ESPCI Paris, Université PSL, Paris, France

Laboratoire de Biologie Structurale de la Cellule, BIOC, CNRS, Ecole polytechnique, Institut Polytechnique de Paris, Palaiseau, France

G. Witzany (ed.), *Integrative Theory of Evolution - Aspects and Insights*,
https://doi.org/10.1007/978-3-032-24295-2_25

1 Introduction

A major challenge for the origins of life is to explain how compartmentalization and evolution emerged as they seem necessary to each other. Canonical Darwinian evolution requires lineages of selection units (Smith and Szathmary 1997), but lineages in biological cells involve elaborate processes of membrane synthesis and control, which result from an evolutionary process. In a prebiotic setting, one rather seeks the simplest possible versions of these processes, but there is still a potential paradox of mutual necessity.

Indeed, along the lines of the RNA world hypothesis, one scenario is that RNA self-replicases (RNA-dependent RNA polymerases made of RNA and capable of copying their own template) first appear in the absence of compartments, assuming that these naked RNAs are self-sufficient Darwinian entities (Joyce and Szostak 2018). However, they are prone to the "Error catastrophe", where erroneous non-functional copies (called parasites) overflow functional ribozymes (Eigen 1971; Ninio 2007). Selection at the compartment level alleviates this phenomenon, as initially proposed in the stochastic corrector model (Szathmáry and Demeter 1987). The mechanism is that compartments that do not contain a sufficient amount of functional RNAs, in particular those that are filled with parasites, do not survive, thus avoiding a collapse of functional replicases. In the end, this mechanism brings protocells back as a prior entity.

It was then considered that a spontaneous cell cycle may pre-exist in the absence of ribozymes, where self-assembled vesicles grow by recruiting further amphiphilic units, then divide once they reach a critical size. In this scenario, vesicles must predate nucleic acid polymer reactions and scaffold their evolution (Chen et al. 2005). While vesicular systems were shown to grow and divide (Bachmann et al. 1992; Miele et al. 2020), stable reproduction coupled with RNA reactions has not yet been established, and prebiotically relevant implementations are missing.

Several alternatives have been proposed to couple evolution and compartments. One of them is that micelles, as ancestors of vesicles, demonstrate a form of Darwinian evolution due to the interaction between diversified membrane components, where this diversity encodes for different compositional states (Lancet et al. 2018). Other prebiotic scenarios propose that reactions between small biopolymers (e.g., RNA, peptide) are sufficient to initiate a form of evolution, before template-based replication (Pavlinova et al. 2023; Rubinov et al. 2009). These dynamics involve rudimentary forms of self-reproduction by autocatalysis, requiring cooperative reaction networks (Kauffman 1986). In this case, too, compartments are needed prior to the reactions, as multicomponent reaction systems must be identified as a unit of selection to participate in evolution (Ameta et al. 2021).

In this chapter, we highlight compartmentalization as a gradual notion and the usefulness of partial forms of compartmentalization in helping chemical or biochemical reaction systems to acquire evolutionary properties. Partial compartmentalization refers to any spatial structure imposed by an open environment, or slow diffusion in a crowded environment, separation in self-assembled entities, or encapsulation in vesicles, toward protocells that grow and divide. Our analysis will lead us to consider non-canonically Darwinian forms of evolution, where progressive couplings between partial compartmentalization and evolutionary properties suggest transitions between successive self-sustained stages, a broader pattern that can be seen as an evolutionary process (Jeancolas et al. 2020).

We will first review how evolutionary properties can be coupled with partial forms of compartmentalization and the benefit of the latter for maintaining these evolutionary properties. We will then focus on the experimental state-of-the-art, starting from open geological environments, then molecular self-assemblies forming coacervates, and vesicles. We conclude by proposing a scenario for the coupled emergence of compartmentalization and protocells.

2 Evolution Before the Cell Cycle

Darwinian evolution results from the combination of variation, reproduction with heredity, and selection. In biology, evolution occurs via the template-based replication of the genome, where base-pair complementarity in DNA ensures heredity and mutations (or any other sequence alteration during copying) lead to variations. In cells, these heritable variations in the genome lead to variations in traits, which in turn determine cell fitness. The phylogenetic tree–like representation of selection units throughout time during such an evolutionary process is tightly associated with cell growth and division. While in biology evolution does not always follow the strict pattern described above, as for the origins of life it is an absolute necessity to broaden the scope of evolution, either by extending Darwinian evolution to other types of molecular processes or by considering non-Darwinian forms of evolution. Below, we describe different evolutionary processes at the theoretical level, in relation to spatial structure, from the least constrained to the most (Fig. 25.1).

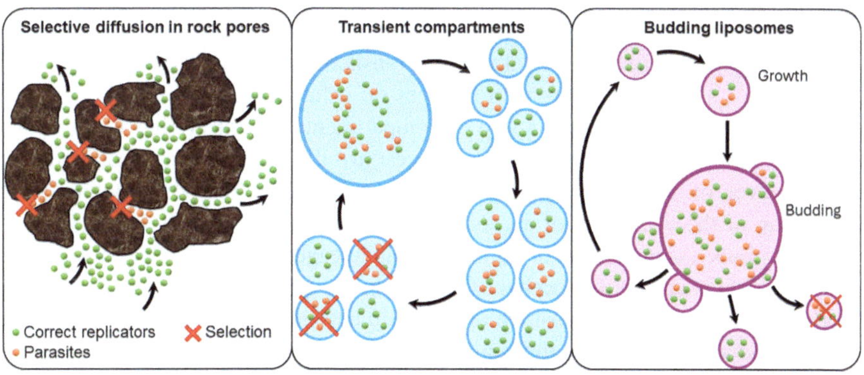

Fig. 25.1 Variety of putative prebiotic compartmentalized evolutionary processes

The most basic form of compartmentalization is limited diffusion (left-hand panel of Fig. 25.1). It may occur when molecules adsorb on surfaces (Wächtershäuser 1988), due to molecular crowding in a concentrated medium, or because the geometry of the reaction vessel is constrained, as in rock pores. Incomplete mixing has important consequences for self-organization in chemical reactions. The most classically known is the formation of patterns, such as static waves as in Turing patterns, or traveling waves as in the Belusov–Zhabotinski reaction, geometric cells, or fractal structures (Maselko et al. 1989). The existence and geometry of these patterns are controlled by the relative timescales of reaction and diffusion. While these phenomena are not directly relevant to molecular evolution, they may help establish spatial structure for later evolutionary processes. For instance, contrasted reaction timescales for different species that induce local crowding can lead to protocellular entities (Kamimura and Kaneko 2010). Now, considering evolutionary properties, limited diffusion on surfaces was proposed as a way to maintain diversity in autocatalytic systems (Plum et al. 2025). This model assumes the existence of multiple autocatalytic regimes that mutually inhibit each other, so that the stochastic triggering of one or the other leads to spatial domains that sustain themselves in time. In the case of more elaborate reproduction by template-based replication, limited diffusion was shown to overcome the "Error catastrophe". Computational simulations demonstrated that when replication parasites appear, they lead to a spatial dead-end, while functional replicators extend along fronts, overall forming spiraling patterns of replication (Takeuchi and Hogeweg 2009). A later analysis showed that there is a threshold phenomenon where a sufficient local accumulation of replicators is needed at the initial stages to nucleate the propagation of replicators (Wu and Higgs 2012). Furthermore, even in the absence of autocatalytic reactions per se, multistable reaction systems can display spatiotemporal dynamics akin to evolution at the population level, due to the propagation of compositional states forming domains that compete for space (Bunin and Rivoire 2025).

Yet another level of spatial structure is transient compartments (central panel of Fig. 25.1). The latter have well-defined boundaries, but their integrity is not maintained in time. While one can conceive a transient version for any type of protocell that can be imagined, transient compartments may also be realized via purely environmental processes, such as condensates forming during drying, droplets forming by weathering or tidal waves. Transient compartments avoid the "Error catastrophe", as compartments with lineages do (Matsumura et al. 2016). The life cycle of replicators comprises a step of compartmentalization, during which replication occurs, followed by selection, where only the compartments containing a certain level of functional molecules can survive, and finally a pooling step, consisting of the disruption of all compartments and the dispersal of their content in a larger bulk. As in the case of the model with lineages, the theory does not specify how selection occurs at the mechanistic level, but it was shown that the mechanism works as soon as there is a positive relationship between the probability of survival and the fraction of functional molecules per compartment (Blokhuis et al. 2018) and that it admits stochasticity in compartment and replicator dynamics (Blokhuis et al. 2020). Here, replicators can be traced through a phylogenetic tree, but compartments cannot. This illustrates a case of non-canonical Darwinian evolution, as selection acts on compartments without parent–offspring relationships, while reproduction and variation occur at the molecular-level following traceable phylogenetic relationships. Later stages may rely on the fractionation into multiple entities or budding of compartments (right-hand panel of Fig. 25.1), before the advent of a well-controlled cell cycle.

The last category we consider here is permeable compartments. As any compartment is permeable to some extent, the question is more precisely that of the control of permeability. Biological cells possess import–export pumps and vesicular systems, which are specific to the exchanged species and can regulate fluxes, a well-known case being energetics management via a proton gradient and ATPase pumps. In a primordial setting, permeability control has not evolved, but any form of semi-permeability (i.e., species-specific differences in permeability or exchange dynamics) might open opportunities for evolutionary properties. For instance, the local sequestration of certain reaction products can led to the growth of a compartment even without a membrane, as a result of liquid–liquid phase separation (Zwicker et al. 2017). Specific exchange properties between three compartments allow autocatalysis to emerge at the compartment level, even using a simple reaction that is not autocatalytic by itself (Blokhuis et al. 2020). Closer to biological cells, it was shown experimentally that when the reaction feedstock exchanges much faster than the autocatalytic products, competition by differential growth spontaneously occurs. The maintenance of composition differences between the inside and the outside is furthermore critical for heredity, which depends on both semi-permeability and the order of the autocatalytic reaction (Sanders et al. 2025).

Overall, we have seen that even mild forms of compartmentalization (limited diffusion, transient compartments, permeable compartments) can support evolutionary properties (reproduction, variation, heredity). We reported here that propositions were mostly theoretical, or were shown in experimental model systems using technologies such as microfluidics to submit replication reactions to compartmentalization cycles. It has not been possible in general to fully integrate chemical reactions capable of evolution together with more natural systems. However, many attempts are on their way, starting to show the potential of various ways of compartmentalization to host and couple with chemical reactions of interest for evolution.

3 Geological Scaffolding

Before the appearance of lipid membranes or protocellular entities such as coacervates, the early Earth itself supplied a framework for organizing chemistry. Geological environments impose constraints and selective enrichments on molecular interactions, effectively scaffolding prebiotic systems long before biology learned to construct its own compartments. This perspective traces back to early conceptual work emphasizing the physical basis of life (Bernal 1949) and has been repeatedly reinforced by later studies highlighting the potentially active role of minerals and planetary processes in shaping chemical evolution (Deamer and Weber 2010; Dass et al. 2016; De Jong et al. 2025). Crucially, geological scaffolds are not limited to hosting reactions; they can actively shape reaction networks by localizing molecules and creating persistent spatial heterogeneity. Compartmentalization, from this viewpoint, is not initially a biological invention but an emergent consequence of geomaterials on the early Earth (Branscomb and Russell 2013; Smith and Morowitz 2016). Mineral surfaces, clay interlayers, and geothermal gradients together reveal geological scaffolding as a hierarchical phenomenon. These forms of organization may have provided continuity between unstructured chemistry and the later emergence of membrane-bound cellular systems.

One of the most immediate mechanisms of geological scaffolding is interaction with mineral surfaces. In bulk aqueous environments, organic molecules are diluted, freely diffusing, and prone to hydrolysis. Adsorption onto mineral surfaces fundamentally alters these constraints (Hazen and Sverjensky 2010; Lambert 2008). Experimental studies have shown that mineral surfaces can promote condensation reactions, stabilize otherwise labile intermediates, and bias reaction outcomes (Cleaves et al. 2012; Ter-Ovanessian et al. 2023). Importantly, adsorption is often reversible, allowing locally elevated concentrations while still permitting molecular exchange under changing pH, temperature, or salinity conditions. Natural mineral surfaces are chemically and structurally heterogeneous, featuring defects, step edges, and compositional variation (Hazen et al. 2008). This heterogeneity creates adjacent microenvironments with distinct catalytic properties, enabling parallel chemical "experiments" to occur side by side.

Layered clay minerals extend this logic by introducing reversible sequestration into geological scaffolding. Their high surface area and ion-exchange capacity allow organic molecules to be selectively taken up into interlayer spaces, where they can be concentrated and partially shielded from degradation. Laboratory experiments have demonstrated adsorption of amino acids, nucleobases, nucleotides, and short oligomers onto clays, in some cases accompanied by enhanced polymerization or selective stabilization (Ferris et al. 1996; Ferris 2005). The importance of clays lies not only in their binding capacity but in the reversibility of that binding. Fluctuations in hydration, pH, temperature, or ionic composition can modulate interlayer spacing, driving cycles of uptake and release that couple molecular persistence to environmental dynamics (Hashizume 2015; Franchi et al. 2003). These cycles allow reaction products to accumulate transiently without becoming evolutionarily isolated. Clay-associated chemistry thus supports a form of dynamic compartmentalization that protects and concentrates molecules while preserving chemical communication with the surrounding environment. From an evolutionary standpoint, such soft compartments may have been especially conducive to prebiotic selection. They allow molecular lineages to persist across cycles while still competing, recombining, and responding to external conditions (Chen and Nowak 2012).

At larger spatial scales, geothermal environments provide geological scaffolding, naturally generating stable thermal and chemical gradients. These gradients partition space into reaction niches characterized by distinct kinetics, equilibria, and molecular lifetimes (Lane and Martin 2012; Sojo et al. 2016). Recent work has significantly deepened our understanding of how gradients function as organizing principles. Heat flows through water-filled rock cracks were shown to drive selective accumulation that is capable of spatially sorting reactants (Matreux et al. 2024) to establish local reaction niches (Matreux et al. 2021). This process, called thermogravitational accumulation, couples thermal convection with thermophoretic molecular transport to produce sustained concentration effects without cellular barriers. Building on this framework, gradient-driven systems were shown to assemble diverse cellular compounds to create a physical non-equilibrium–contained protocell (Floroni et al. 2025).

4 Coacervates

The role of coacervates for the origins of life was first understood by a Russian biochemist, Alexander Oparin, who proposed in 1924 that tiny microspheres arising in colloidal solutions of organic and inorganic compounds, the so-called prebiotic soup, could have served as primitive cells on early Earth (Oparin 1924). Coacervates are associative liquid–liquid phase-separated (LLPS) systems that form when a homogeneous solution separates into two distinct aqueous phases due to various inter- and intramolecular interactions, including electrostatic interactions, hydrogen bonding, π–π stacking, etc. (Lin et al. 2023; Martin 2019; Sing and Perry 2020; van Haren et al. 2020). There are two classes of coacervates: simple and complex coacervates. Simple coacervates form when a single polymer contains multiple interaction sites that promote self-association, causing the polymer to become insoluble and hence phase separate. For example, intrinsically disordered proteins carry both positive and negative charges, which allow them to undergo phase separation on their own (Uversky et al. 2015). In contrast, complex coacervation arises from attractive interactions between different polymers, most commonly electrostatic attraction between oppositely charged polyelectrolytes, resulting in the coexistence of a dense phase—rich in both polymers—and a dilute or supernatant phase—depleted of polymers and rich in counter-ions. Among them, the most interesting for the origins of life investigation are those formed by the complexation of biomolecules, such as anionic peptides or nucleic acids, with cationic peptides, such as polylysine and polyarginine (Aumiller et al. 2016; Vieregg et al. 2018; Jia and Fraccia 2020). Additionally, the word coacervate can be extended to segregative phase separation occurring between mutually repulsive polymers, such as poly (ethylene glycol) and dextran (Albertsson 1970), also named aqueous two-phase systems (Keating 2012).

Coacervates are formed not only by long polymers but also by short ones, including low valence moieties and nucleic acid monomers and oligomers (Cakmak et al. 2020; Lu et al. 2021; Aumiller et al. 2016; Smokers et al. 2022). This is an important aspect to meet the constraints of the expected limited yield of abiotic polymerization processes that could have taken place on early Earth (Caimi et al.

2025; Smokers et al. 2022). Besides still lacking nucleic acids, the formation of microdroplets as a consequence of a polymerization process has been shown for alpha–hydroxy acids subjected to drying at high temperatures (Chen et al. 2025; Jia et al. 2021). The universality of the coacervation process—which extends to nonbiological polymers too—and the robustness and spontaneity of their formation—given the impressive number of combinations of polymers that prefer to phase separate instead of forming homogeneous solutions, even at low concentrations (Walter et al. 1985)—make coacervate microdroplets an intriguing candidate to sustain the origin of primitive compartments in prebiotic scenarios (Ghosh et al. 2021; van Haren et al. 2020). Coacervates' functionality has been probed by testing their capacity to host various chemical and biochemical reactions, which are crucial in prebiotic scenarios such as amide bond formation (Wang et al. 2023), C-C bond formation (Peyraud-Vicré et al. 2025), metabolite conversion (Smokers et al. 2022), and others (Beneyton et al. 2020).

Importantly, coacervates also host proteins and ribozymes without disrupting their secondary structure and, therefore, without impairing the enzymatic activity (Lindhoud et al. 2010). In some cases, the activity is even enhanced due to the concentration effect (Strulson et al. 2012), an advantage that would have helped fight against low reactant concentrations on early Earth (Benner et al. 2012). In this frame, the uptake of guest molecules in coacervates could be selective due to the lack of membranes and to the transport of molecules between the droplets and the supernatant driven by molecular affinity (Nakashima et al. 2019). As a result, coacervates with different scaffolding architectures can provide distinct microenvironments capable of selecting and supporting different biochemical reactions. Also, despite their intrinsic propensity to mix, coacervates can form multiphase compartments, providing distinct spatiotemporal organization, further enabling the separation of biochemical reactions even in close proximity (Lu and Spruijt 2020; Mountain and Keating 2020). Additionally, coacervates can display different, but still fluid, phases from simple liquids like the liquid crystal mesophases (Jia and Fraccia 2020; Fraccia and Zanchetta 2021), which have been shown to enhance DNA and RNA assembly and reactivity (Fraccia and Martin 2023; Todisco et al. 2018; Fraccia et al. 2016, 2015; Smith et al. 2018).

The reported observations suggest that coacervate-based systems can not only act as passive containers of prebiotic molecules but also represent fundamental compartments for sustaining the joint evolution of protocells and their biochemical content (Ghosh et al. 2021; Slootbeek et al. 2022). Sustaining evolutionary dynamics among coacervates requires coacervate droplets to be coupled with a set of chemical/biochemical reaction networks, and combinations of compartments and networks exhibit differential fitness altogether (sketched in Fig. 25.2). In this regard, a study demonstrated that coacervates host self-reproducing reaction networks in a cross-catalytic scenario with enhanced activity, thereby increasing the robustness of self-replication-based systems (Ameta et al. 2023). Furthermore, the stochastic corrector model (Szathmáry and Demeter 1987) correlates the reaction network efficiency with the compartment properties, i.e., selective growth of the compartment and/or differential compartment stability (e.g., coacervates' critical salt concentration, c*)

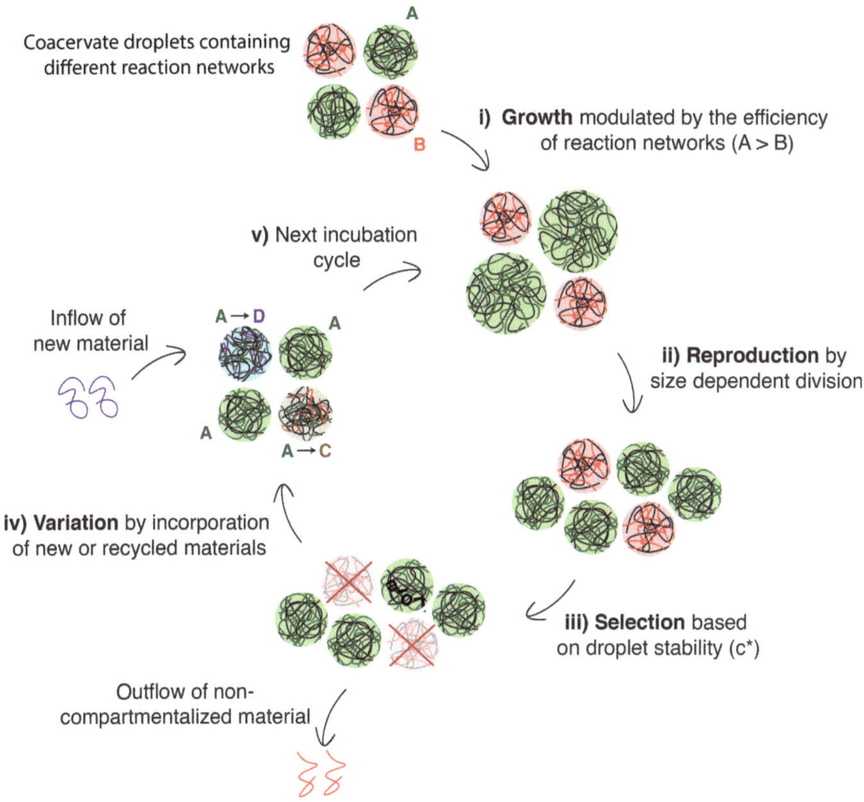

Fig. 25.2 Sketch of the required processes to sustain evolutionary dynamics in a coacervate-based system. The order of the steps (i) to (iii) is interchangeable

influenced by respective compartmentalized reaction networks. Consequently, reproduction can take place by size-dependent coacervate division, i.e., larger droplets reproduce at a larger rate. Applied selection pressure results in the extinction of less stable droplets (based on c*). Depending on the extent of droplets' wetting and content diffusion, droplets can preserve their individual identity to a certain extent or assimilate the components of the dissipated ones or newly coming material, providing variation.

Several studies have tried to probe these steps. For example, one study explored the interaction between prebiotically relevant metabolites and coacervates formed from them (Smokers et al. 2022). The results showed that the reactions in bulk solution involving metabolite conversion induced phase separation, giving rise to coacervates. These newly formed coacervates, in turn, positively influence another reaction by enhancing the metabolite conversion rate, implying that the crosstalk is not restricted to a single catalytic system. Conversely, some metabolite-driven reactions were also found to destabilize coacervation and led to their dissolution, depicting the bidirectional nature of the interaction. Together, these findings demonstrate that reaction networks can regulate compartment stability, opening the possibility to realize compartments with distinct stabilities. By varying the reaction networks, it is possible to achieve differential fitness within a population, a prerequisite for the emergence of Darwinian properties. Environmental pressures could then act as a selection mechanism, driving evolutionary dynamics within these prebiotic compartments.

Coacervates were formed by the combination of azobenzene trimethylammonium bromide (azo-TAB), a photoresponsive amphiphilic azobenzene-based cation, and short anionic double-stranded DNA oligonucleotides (Fraccia and Martin 2023). The interaction between two polyions was altered by UV (favoring azo-TAB cis isomerization) and blue light (favoring azo-TAB trans isomerization), which in turn triggered reversible dissolution and assembly of coacervates, respectively. Furthermore, these light-induced changes also affected the ligation efficiency, leading to the formation of longer dsDNA polymers that increased the coacervate stability over time, establishing a positive feedback loop between the reaction and the compartment stability. In addition, the reaction-driven changes induced a transition from isotropic to liquid crystalline coacervates, promoting further phase separation and the emergence of multiphase organization. Taken together, these findings demonstrate a strong correlation between compartment structure and the reaction occurring within it, with external stimuli such as light regulating the reaction network and giving rise to differentially stable coacervates.

Another study explored similar behavior in ribozyme–peptide–based coacervate systems, which are widely investigated in the origins of life community for their prebiotic relevance under the 'RNA world hypothesis' and their capacity to form autocatalytic reaction networks. In this system, coacervates were formed from the R3C ligase ribozyme and its RNA substrate as the polyanion, together with poly-L-lysine as the polycation, with the rationale that the ribozyme-catalyzed ligation reaction would yield increasingly longer RNA molecules over time, thereby altering the coacervate properties (Le Vay et al. 2023). The results showed that the reaction rate was enhanced due to the concentration effect of coacervation, consistent with observations from other ribozyme systems. Moreover, the longer RNA products modified the coacervate's physical properties by reducing growth, wetting, diffusion, and the transfer of content across droplets, as opposed to the coacervate composed of inactive ribozymes. Nevertheless, the question of how such reaction-driven changes in coacervate properties could be utilized to invoke Darwinian properties at the compartment level remains open and unexplored.

A major limitation of coacervate systems to be suitable to sustain Darwinian evolution is the inherent propensity of membraneless coacervate droplets to fuse through coalescence and Ostwald ripening, leading to the loss of compartment identity (Agrawal et al. 2024). Several strategies have been proposed to mitigate this by stabilizing coacervate solutions, including the formation of hybrid protocells in which coacervates are coated with lipid membranes (Dora Tang et al. 2014; Lu et al. 2023) or the use of triblock copolymers capable of assembling at the phase interface (Mason et al. 2017; Tang et al. 2025). A second challenge is the incorporation of heredity mechanisms, which requires compartment reproduction through division. Although reaction-induced changes in coacervate properties have been demonstrated, preserving out-of-equilibrium states or providing active processes seems to be inescapable to sustain droplet division. Recent studies employed opposite chemical fuels or flows to partially destroy and regenerate coacervates (Donau et al. 2020; Ianeselli et al. 2022), but these approaches resulted in only fission-like events rather than duplication by division. Overall, coacervation offers a promising platform to engineer prebiotically relevant compartments, although several conceptual and experimental gaps remain to be addressed before a purely coacervate-based system could effectively sustain Darwinian evolution (Adamski et al. 2020).

5 Vesicles

Cell membranes were essential for the earliest forms of life (protocells) because of their multiple advantages. They concentrate biomolecules, increase the concentration of intermediates via a permeable barrier, maintain a stable chemical microenvironment, ensuring the sustainability of processes, they protect biomolecules against inhibitors or competitors, and they sequester toxic intermediates (Polka et al. 2016). The formation of enclosed, vesicular structures is therefore a critical intermediate step hypothesized in the chemical evolution toward the emergence of protocells on early Earth (Deamer 2024). This process relies on the unique properties of certain amphiphilic organic molecules.

Amphiphiles, such as the phospholipids that constitute extant cell membranes, are characterized by a molecular structure possessing both a hydrophilic head group and a hydrophobic tail. When suspended in an aqueous environment, these molecules spontaneously minimize the unfavorable contact between the hydrophobic tails and water. This intrinsic drive leads to self-assembly into highly organized, stable structures, most notably the lipid bilayer. The bilayer forms as two molecular monolayers align such that all hydrophilic moieties face outward (contacting the aqueous milieu), while the hydrophobic tails face inward, shielded from the surrounding water. This self-sealing process results in the formation of microscopic, spherical vesicles.

Given that amphiphilic compounds are among the organic molecules hypothesized to have been present on the primitive Earth—whether geochemically synthesized or delivered via extraterrestrial sources—it is highly probable that the spontaneous self-assembly of these molecules constituted a common and fundamental process, offering the first membrane-bound, closed compartment necessary for encapsulating and concentrating the molecular machinery of early life.

Crucially, these early vesicles would have been selectively permeable, allowing small ionic substrates (nutrients) to enter for the polymerization of macromolecules (like RNA), while simultaneously trapping the larger macromolecules inside (Deamer and Barchfeld 1982). This encapsulation would have kept the emerging catalytic and informational molecules localized, enabling the selection of the most effective combinations—a process impossible in solution.

Modern phospholipids, though stable, require complex biosynthetic machinery and are not considered prebiotically plausible, creating a fundamental gap in understanding early cellular evolution. Hence, research must focus on simpler, more plausible amphiphilic structures that could fulfill these essential protocellular functions. Early protocells likely used prebiotically available short-chain fatty acids (C9–C12) for their membranes. While these fatty acid vesicles can grow and divide, they suffer from significant limitations: a narrow range of stability regarding pH, temperature, and especially metal ion concentrations (Mg^{2+}), which are often necessary for RNA function. If the inclusion of other molecules (like fatty alcohols or monoglyceride esters) can help, it doesn't solve the fundamental question of how these primitive systems evolved into the phospholipid-based membranes of extant life.

The recent work of R. Krishnamurthy (Pulletikurti et al. 2024) has shed new light on the role of single-chain cyclic-phospholipids (cPC) and their double-chain derivatives in the chemical evolution of early cell membranes, proposing, as a consequence, a hypothetical pathway for the transition from unstable fatty acid vesicles towards robust phospholipid membranes, closer to the ones in modern cells.

Prior discovery of a one-pot prebiotic reaction using fatty acid, glycerol, and diamidophosphate (DAP) to produce single-chain cyclic-phospholipids (cPCs) set the stage for focusing on phosphoglyceride membrane stability studies (Gibard et al. 2018). The hypothetical pathway suggests that these cPCs and their subsequent transformations could facilitate the crucial evolutionary transition:

1. Incorporation of cPCs: Fatty acid vesicles incorporate the single-chain cPCs (e.g., cPC10), forming chimeric vesicles with enhanced stability and robustness against divalent metal ions and a wider pH range.
2. Ring-opening and esterification: The high-energy cyclic phosphate moiety of the cPCs can undergo hydrolysis and subsequent esterification with a second fatty acid, leading to the formation of double-chain phosphomonoester lipids (glyceryl 2,3-didecanoate 1-phosphate, a.k.a. GDDP1 and glyceryl 1,3-didecanoate 2-phosphate, a.k.a. GDDP2).
3. Future complexity: These GDDPs (diacyl phospholipids) systems represent the next level of complexity, potentially leading to the final stage: double-chain phosphodiester-based lipids (modern phospholipids).

The study experimentally modeled this transition by investigating binary mixtures of cPC10 with decanoic acid (DA) and glyceryl monodecanoate (GMD).

DA: cPC10 Systems: The DA: cPC10 ratio (2:1) formed stable vesicles across a broad pH range (4.0–6.6), with morphology (unilamellar vesicles, multivesicular vesicles, multilamellar vesicles) depending on the pH value, likely due to the varying degrees of protonation on the fatty acid component.

GMD: cPC10 Systems: The 5:1 ratio formed robust vesicles from pH 6.1 to 8.0. Starting with individual components that did not form stable vesicles, the addition of the corresponding component quickly led to the formation of stable, heterogeneous vesicles, suggesting a strong synergistic effect. The critical aggregate concentration (CAC) was significantly lowered in these hybrid systems, consistent with enhanced stabilization, thus confirming the particular combined properties of binary systems.

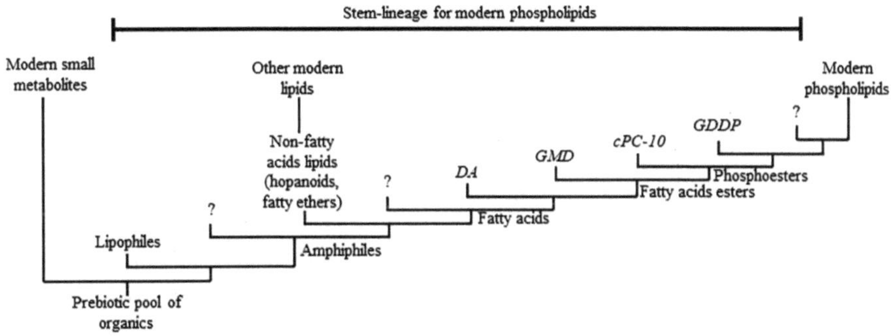

Fig. 25.3 Phylogenetic analogy for transitional evolution of lipids (see also Saha and Chen 2024)

Heating-cooling cycles, generally used to prepare uniform solutions, fully influenced the vesicle morphology, producing giant multilamellar vesicles (MLVs) and oligolamellar vesicles (OLVs) that were stable over days and exhibited dynamic behavior like fusion and division. The cPC inclusion prevented the collapse of DA-based systems into oil droplets upon heating.

Finally, the geochemical plausibility of the entire pathway was supported by simulating a prebiotic reaction using glycerol, fatty acids, DAP, and imidazole, demonstrating that a complex mixture of the necessary lipids could spontaneously form stable vesicles under realistic conditions (wet–dry cycling).

The evolutive reasoning of I. Chen (Saha and Chen 2024) went one step further in the interpretation of the work by considering that the two transitional forms of lipids are truly analogous to biological "missing links" like Archaeopteryx, to bridge an evolutionary gap (see Fig. 25.3). The transition moves from the single-chain fatty acid to the double-chained phospholipid. These missing-link lipids have both characteristics of a single fatty chain (like a fatty acid) and a cyclic glycerol-phosphate head group (resembling a phospholipid). Esterification with a second fatty acid tail and subsequent modification of the head group would yield modern-day phospholipids.

This progression from cPC10-based to GDDP-based vesicles showed a clear, measurable increase in selectable traits, including greater ease of formation and increased robustness to environmental variables like pH, temperature, and metal-ion concentration. This enhanced stability provides the necessary selective advantage—or increased "fitness"—that would drive the evolution of protocells toward more sophisticated membranes through Darwinian selection acting on encapsulated RNA.

Whatever the favored scenario, encapsulation into protocells necessarily required associated mechanisms that enabled efficient exchange between the protocell and its milieu (Dalai and Sahai 2019). Unfortunately, molecular-level understanding regarding the emergence of these cellular barriers remains partly undeciphered (Deamer 2024). However, experimental efforts to model these processes are currently expanding rapidly (Gözen et al. 2022).

6 Conclusion: Selective Advantages of Early Compartments Enabling Their Coevolution with Molecular Processes

We aim to present a putative timeline for compartmentalization, emphasizing the dynamic roles of geological scaffolding, coacervates, and lipid vesicles. In this model, compartments are not merely passive containers; they provide a sequence of strategic metabolic advantages that evolved over time. The emergence of boundaries allowed prebiotic systems to transition from random encounters to regulated organization. These advantages often emerged concomitantly as soon as primitive boundaries formed (see Table 25.1).

The compartments allow for the concentration of biomolecules to escape dilution phenomena. They facilitate an increased concentration of intermediates via a semi-permeable barrier. The compartments maintain a chemical microenvironment essential for ensuring the sustainability of the chemical processes occurring within this envelope. The inner compartments and the external membrane also serve to protect formed biomolecules against potential inhibitors or competitors. Finally, the existence of compartments allows for the sequestration of toxic intermediates (Polka et al. 2016).

Table 25.1 Evolutionary advantages of the emergence of compartments

Advantage	Mechanism	Prebiotic impact
Compartment stability	Self-assembly of amphiphilic molecules into stable bilayers or coacervates	Provides a persistent physical boundary that survives environmental fluctuations
Anti-dilution	Physical containment prevents the entropic dispersal of molecules into the bulk environment	Maintains the critical reactant density required for complex polymers to form
Reaction acceleration	Molecular crowding increases collision frequency and reduces the search space for reactants	Drastically improves the speed and efficiency of primitive metabolic pathways
Selective flux	Semi-permeability controls the passage of molecules via a physical or chemical barrier	Facilitates the filtration of the environment, importing nutrients while retaining catalysts
Primitive homeostasis	Chemical buffering maintains a distinct internal microenvironment (pH, ion concentration)	Creates robust conditions that protect sensitive chemical processes from external fluctuations
Biochemical shielding	Physical barrier blocks the entry of large or harmful molecules	Protects nascent biomolecules from degradation by inhibitors or takeover by competitors
Waste sequestration	Internal partitioning isolates harmful byproducts or intermediates	Prevents the poisoning of the system by sequestering toxic reaction products

These advantages could have concomitantly emerged once compartments had formed sub-compartment. We propose to integrate them into the breakdown of the transition from simple physical chemistry to complex biological organization, moving from purely diffusion-limited interactions (random encounters) to regulated inheritance.

6.1 Step 1: Initial Enhancement of Reactions Via Crowding, Adsorption, and Confinement (Minerals)

Chemical reactivity and selectivity were initially enhanced through the physical modulation of space rather than closed biological boundaries. This process began with molecular crowding, where high local concentrations of oligonucleotides increased collision frequencies and stabilized reaction complexes. This was further supported by surface adsorption on minerals or interfaces, which partially immobilized molecules to improve interaction selectivity and minimize diffusion losses. Finally, physical confinement within natural pores, matrixes, or gels reduced the effective reaction volume, creating a critical kinetic bias that favored productive chemical transformations over the entropy of the open environment.

6.2 Step 2: Local Accumulation of Copies (Mineral Pores)

The second stage marks the transition from simple chemistry to the dawn of spatial heredity, where the local accumulation of copies creates a distinct evolutionary advantage. In this phase, autocatalytic systems generate molecular copies that, due to limited diffusion, remain clustered together rather than dispersing. This proximity leads to the formation of enriched micro-domains populated by identical or closely related sequences. Crucially, this clustering triggers the emergence of a sequence–environment spatial correlation, where a specific location becomes the first medium for heredity. By tying a "genetic" sequence to its immediate physical surroundings, the environment begins to preserve and favor successful molecular lineages, setting the stage for natural selection.

6.3 Step 3: Avoiding the "Error Catastrophe" (Porous Matrixes, Gels)

A third crucial stage involves avoiding the "Error catastrophe," a tipping point where parasitic sequences threaten to overwhelm functional chemistry. In a completely open or well-mixed system, parasitic sequences—which replicate faster but perform no useful function—quickly dominate and degrade the chemical information. However, through limited diffusion and physical confinement, molecular copies remain grouped, effectively "quarantining" parasites and preventing them from dispersing efficiently. This spatial structure enables local selection, which shifts the evolutionary focus away from isolated individuals and toward coherent autocatalytic sets. By favoring groups that work together constructively, the system ensures that functional molecular lineages can persist and evolve without being outcompeted by non-functional replicators.

6.4 Step 4: Emergence of Cooperation Through Proximity (Mineral Pores)

A fourth stage is the emergence of cooperation through proximity, where spatial constraints transform independent molecules into integrated systems. By keeping various molecular species in close physical contact, the environment favors the immediate reuse of reaction products, ensuring that the output of one process becomes the instantaneous input for the next. This proximity facilitates positive feedback loops between distinct molecular activities, where different types of molecules (such as early peptides and RNAs) can mutually stabilize or catalyze each other's formation. Consequently, this leads to the appearance of local functional networks and a rudimentary division of labor, marking the transition from a collection of isolated reactions to a synchronized, protometabolic system.

6.5 Step 5: Reaction Products Modify Compartmentalization (Coacervates)

At this stage, a transformative feedback loop emerges where reaction products begin to modify their own compartmentalization, allowing the system to actively construct its own niche. Rather than being passive subjects of their environment, certain synthesized molecules proactively alter their surroundings by increasing local viscosity or promoting phase separation, which leads to the formation of primitive coacervates. These early systems can then recruit peptides or lipids to the interface, reinforcing the structure and providing new functional properties. This represents a pivotal transition from environment-imposed confinement to a self-constructed boundary, marking the birth of an autonomous entity capable of maintaining its own physical integrity—the essential precursor to a living cell.

6.6 Step 6: Progressive Control of Permeability (Lipid Vesicles)

In this "gatekeeper" stage, the system achieves progressive control over permeability, moving away from simple barriers toward selective interfaces. While these early compartments remain "open" systems, they become increasingly nonhomogeneous, allowing for the preferential entry of small molecules while ensuring the partial retention of larger polymers. This semipermeability creates a powerful selective pressure, where natural selection favors structures that optimize the import of substrates and the retention of essential catalysts. By coupling metabolic efficiency with the physical properties of the boundary, the system ensures that its internal chemistry is both fueled by the environment and protected from the loss of its most complex and hard-won molecular components

6.7 Step 7: Transient Compartments (Coacervates, Liquid Droplets)

Later on, the system passes through a phase dominated by transient compartments, characterized by unstable structures such as coacervates, viscous pockets, or liquid droplets. These precursors lack the permanence of modern cells, exhibiting rapid formation and disappearance as environmental conditions fluctuate. This volatility leads to the frequent mixing of contents, resulting in a state of weak collective heredity where chemical information is often shared and reshuffled. However, this instability serves a critical evolutionary purpose: it acts as a dynamic laboratory for testing diverse chemical architectures and transiently amplifying the most efficient functional networks, allowing only the most robust chemical combinations to persist before more permanent boundaries emerge. This instability is, in practice, the selective filter that permits the emergence of robust systems.

6.8 Step 8: Uncontrolled Growth and Division (Lipid Vesicles)

In the eighth stage, the system enters a phase of uncontrolled growth and division, where physical forces drive the first primitive form of reproduction, a precursor to mitosis. As internal chemical processes flourish, the internal accumulation of products forces a steady increase in the compartment's volume. Eventually, the structure reaches a critical limit where physical instabilities, such as shear stress or surface tension imbalances, trigger a random splitting of the droplet or vesicle. Because this process is purely physical rather than regulated by complex machinery, it results in the imperfect transmission of contents. This creates statistical lineages—descendants that resemble their parents in composition but lack a precise genealogy.

6.9 Step 9: Establishment of a Controlled Cell Cycle (Stable Lipid Bilayers)

The final stage marks the establishment of a controlled cell cycle, where the transition from chaotic physical splitting to regulated biological reproduction is completed. During this phase, evolutionary selection begins to favor systems where growth is tightly coupled to internal production, ensuring that the boundary expands in direct response to metabolic success. Division becomes less stochastic, shifting away from random environmental fragmentation toward a more synchronized process. This represents the final transition of the prebiotic era: the emergence of stable protomembranes with integrated control over permeability, growth, and fission. The result is the birth of the first true protocells, capable of reliable collective heredity and the consistent transmission of life's functional blueprints to future generations.

This nine-stage evolutionary arc (see Table 25.2) follows a journey from environmental dependence to biological autonomy. Although speculative and therefore debatable, this trajectory demonstrates that the cell is not merely a container for life, but the very architectural breakthrough that allowed chemical complexity to transform into a persistent, self-replicating, and evolving lineage. While compartments began to influence reactions, the question of how reactions influence compartments, which is crucial to the concept of natural selection, is still experimentally unresolved. This bidirectional coupling between chemistry and container is a definitive hallmark of the transition from prebiotic matter to living systems.

Table 25.2 The complete evolutionary arc from compartments to protocells

Phase	Transformation	Key outcome
I. Pre-boundaries (1–2)	Physical modulation and spatial heredity	Overcoming dilution and establishing local lineages
II. Proto-networks (3–4)	Error avoidance and cooperation	Protecting functional sets from parasites and building feedback loops
III. Self-organization (5–7)	Niche construction and permeability	The shift to self-made boundaries and selective "gatekeeping"
IV. True cellularity (8–9)	Growth and controlled division	Transitioning from random splitting to a regulated, reliable cell cycle

References

Adamski P, Eleveld M, Sood A, Kun Á, Szilágyi A, Czárán T, Szathmáry E, Otto S (2020) From self-replication to replicator systems en route to de novo life. Nat Rev Chem 4(8):386–403

Agrawal A, Radakovic A, Vonteddu A, Rizvi S, Huynh VN, Douglas JF, Tirrell MV, Karim A, Szostak JW (2024) Did the exposure of coacervate droplets to rain make them the first stable protocells? Sci Adv 10(34):eadn9657

Albertsson P-Å (1970) Partition of cell particles and macromolecules in polymer two-phase systems. Adv Protein Chem 24:309–341

Ameta S, Arsène S, Foulon S, Saudemont B, Clifton BE, Griffiths AD, Nghe P (2021) Darwinian properties and their trade-offs in autocatalytic RNA reaction networks. Nat Commun 12(1):842

Ameta S, Kumar M, Chakraborty N, Matsubara YJ, Prashanth S, Gandavadi D, Thutupalli S (2023) Multispecies autocatalytic RNA reaction networks in coacervates. Commun Chem 6(1):9

Aumiller WM, Pir Cakmak F, Davis BW, Keating CD (2016) RNA-based coacervates as a model for membraneless organelles: formation, properties, and interfacial liposome assembly. Langmuir 32(39):10042–10053

Bachmann PA, Luisi PL, Lang J (1992) Autocatalytic self-replicating micelles as models for prebiotic structures. Nature 357(6373):57–59

Beneyton T, Love C, Girault M, Tang T-YD, Baret J-C (2020) High-throughput synthesis and screening of functional coacervates using microfluidics. ChemSystemsChem 2(6):e2000022

Benner SA, Kim H-J, Carrigan MA (2012) Asphalt, water, and the prebiotic synthesis of ribose, ribonucleosides, and RNA. Acc Chem Res 45(12):2025–2034

Bernal JD (1949) The physical basis of life. Proc Phys Soc Sect A 62(9):537–558

Blokhuis A, Lacoste D, Nghe P, Peliti L (2018) Selection dynamics in transient compartmentalization. Phys Rev Lett 120(15):158101

Blokhuis A, Nghe P, Peliti L, Lacoste D (2020) The generality of transient compartmentalization and its associated error thresholds. J Theor Biol 487:110110

Branscomb E, Russell MJ (2013) Turnstiles and bifurcators: the disequilibrium converting engines that put metabolism on the road. Biochim Biophys Acta (BBA) - Bioenergetics 1827(2):62–78

Bunin G, Rivoire O (2025) Evolutionary features in a minimal physical system: diversity, selection, growth, inheritance, and adaptation. Proc Natl Acad Sci 122(31):e2425753122

Caimi F, Langlais J, Fontana F, Wunnava S, Bellini T, Braun D, Fraccia TP (2025) High-yield prebiotic polymerization of 2′,3′-Cyclic nucleotides under wet–dry cycling. ACS Central Science 11(9):1546–1557

Cakmak FP, Choi S, Meyer MO, Bevilacqua PC, Keating CD (2020) Prebiotically-relevant low polyion multivalency can improve functionality of membraneless compartments. Nat Commun 11(1):5949

Chen IA, Nowak MA (2012) From Prelife to life: how chemical kinetics become evolutionary dynamics. Acc Chem Res 45(12):2088–2096

Chen IA, Salehi-Ashtiani K, Szostak JW (2005) RNA catalysis in model protocell vesicles. J Am Chem Soc 127(38):13213–13219

Chen C, Yi R, Igisu M, Afrin R, Sithamparam M, Chandru K, Ueno Y, Sun L, Laurenzi T, Eberini I, Fraccia TP, Wang A, James Cleaves H, Jia TZ (2025) Primitive homochiral polyester formation driven by tartaric acid and calcium availability. Proc Natl Acad Sci 122(12):e2419554122

Cleaves JH, Scott AM, Hill FC, Leszczynski J, Sahai N, Hazen R (2012) Mineral–organic interfacial processes: potential roles in the origins of life. Chem Soc Rev 41(16):5502

Dalai P, Sahai N (2019) Mineral–lipid interactions in the origins of life. Trends Biochem Sci 44(4): 331–341

Dass AV, Hickman-Lewis K, Brack A, Kee TP, Westall F (2016) Stochastic prebiotic chemistry within realistic geological systems. Chemistryselect 1(15):4906–4926

De Jong TJ, Demertzi AD, Robinson WE, Huck WTS (2025) Environmental history is transferred via minerals altering Formose reaction pathways. Angew Chem 137(23):e202504659

Deamer DW (2024) Perspective: protocells and the path to minimal life. J Mol Evol 92(5):530–538

Deamer DW, Barchfeld GL (1982) Encapsulation of macromolecules by lipid vesicles under simulated prebiotic conditions. J Mol Evol 18:203–206

Deamer DW, Weber AL (2010) Bioenergetics and life's origins. Cold Spring Harb Perspect Biol 2(2):a004929

Donau C, Späth F, Sosson M, Kriebisch BAK, Schnitter F, Tena-Solsona M, Kang H-S, Salibi E, Sattler M, Mutschler H, Boekhoven J (2020) Active coacervate droplets as a model for membraneless organelles and protocells. Nat Commun 11(1):5167

Dora Tang T-Y, Rohaida Che Hak C, Thompson AJ, Kuimova MK, Williams DS, Perriman AW, Mann S (2014) Fatty acid membrane assembly on coacervate microdroplets as a step towards a hybrid protocell model. Nat Chem 6(6):527–533

Eigen M (1971) Selforganization of matter and the evolution of biological macromolecules. Naturwissenschaften 58(10):465–523

Ferris JP (2005) Mineral catalysis and prebiotic synthesis: montmorillonite-catalyzed formation of RNA. Elements 1(3):145–149

Ferris JP, Hill AR, Liu R, Orgel LE (1996) Synthesis of long prebiotic oligomers on mineral surfaces. Nature 381(6577):59–61

Floroni A, Martín NY, Matreux T et al (2025) Membraneless protocell confined by a heat flow. Nat Phys 21(8):1303–1310

Fraccia TP, Martin N (2023) Non-enzymatic oligonucleotide ligation in coacervate protocells sustains compartment-content coupling. Nat Commun 14(1):2606

Fraccia TP, Zanchetta G (2021) Liquid–liquid crystalline phase separation in biomolecular solutions. Curr Opin Colloid Interface Sci 56:101500

Fraccia T, Zanchetta G, Rimoldi V, Clark N, Bellini T (2015) Evidence of liquid crystal–assisted abiotic ligation of nucleic acids. Orig Life 45:51

Fraccia TP, Smith GP, Bethge L, Zanchetta G, Nava G, Klussmann S, Clark NA, Bellini T (2016) Liquid crystal ordering and isotropic gelation in solutions of four-base-long DNA oligomers. ACS Nano 10(9):8508–8516

Franchi M, Ferris JP, Gallori E (2003) Cations as mediators of the adsorption of nucleic acids on clay surfaces in prebiotic environments. Orig Life Evol Biosph 33(1):1–16

Ghosh B, Bose R, Tang T-YD (2021) Can coacervation unify disparate hypotheses in the origin of cellular life? Curr Opin Colloid Interface Sci 52:101415

Gibard C, Bhowmik S, Karki M, Kim EK, Krishnamurthy R (2018) Phosphorylation, oligomerization and self-assembly in water under potential prebiotic conditions. Nat Chem 10(2): 212–217

Gözen I, Köksal ES, Põldsalu I, Xue L, Spustova K, Pedrueza-Villalmanzo E, Ryskulov R, Meng F, Jesorka A (2022) Protocells: milestones and recent advances. Small 18:2106624

van Haren M, Nakashima K, Spruijt E (2020) Coacervate-based protocells: integration of life-like properties in a droplet. J Syst Chem 8:107–120

Hashizume H (2015) Adsorption of nucleic acid bases, ribose, and phosphate by some clay minerals. Life 5(1):637–650

Hazen RM, Sverjensky DA (2010) Mineral surfaces, geochemical complexities, and the origins of life. Cold Spring Harb Perspect Biol 2(5):a002162

Hazen RM, Papineau D, Bleeker W et al (2008) Mineral evolution. Am Mineral 93(11–12): 1693–1720

Ianeselli A, Tetiker D, Stein J, Kühnlein A, Mast CB, Braun D, Dora Tang T-Y (2022) Non-equilibrium conditions inside rock pores drive fission, maintenance and selection of coacervate protocells. Nat Chem 14(1):32–39

Jeancolas C, Malaterre C, Nghe P (2020) Thresholds in origin of life scenarios. iScience 23(11): 101756

Jia TZ, Fraccia TP (2020) Liquid crystal peptide/DNA coacervates in the context of prebiotic molecular evolution. Crystals 10(11):964

Jia TZ, Bapat NV, Verma A, Mamajanov I, Cleaves HJ, Chandru K (2021) Incorporation of basic α-Hydroxy acid residues into primitive polyester microdroplets for RNA segregation. Biomacromolecules 22(4):1484–1493

Joyce GF, Szostak JW (2018) Protocells and RNA self-replication. Cold Spring Harb Perspect Biol 10(9):a034801

Kamimura A, Kaneko K (2010) Reproduction of a protocell by replication of a minority molecule in a catalytic reaction network. Phys Rev Lett 105(26):268103

Kauffman SA (1986) Autocatalytic sets of proteins. J Theor Biol 119(1):1–24

Keating CD (2012) Aqueous phase separation as a possible route to compartmentalization of biological molecules. Acc Chem Res 45(12):2114–2124

Lambert J-F (2008) Adsorption and polymerization of amino acids on mineral surfaces: a review. Orig Life Evol Biosph 38(3):211–242

Lancet D, Zidovetzki R, Markovitch O (2018) Systems protobiology: origin of life in lipid catalytic networks. J R Soc Interface 15(144):20180159

Lane N, Martin WF (2012) The origin of membrane bioenergetics. Cell 151(7):1406–1416

Le Vay KK, Salibi E, Ghosh B, Tang TD, Mutschler H (2023) Ribozyme activity modulates the physical properties of RNA–peptide coacervates. elife 12:e83543

Lin Z, Beneyton T, Baret J, Martin N (2023) Coacervate droplets for synthetic cells. Small Methods 7(12):2300496

Lindhoud S, Norde W, Cohen Stuart MA (2010) Effects of polyelectrolyte complex micelles and their components on the enzymatic activity of lipase. Langmuir 26(12):9802–9808

Lu T, Spruijt E (2020) Multiphase complex coacervate droplets. J Am Chem Soc 142(6): 2905–2914

Lu T, Nakashima KK, Spruijt E (2021) Temperature-responsive peptide–nucleotide coacervates. J Phys Chem B 125(12):3080–3091

Lu T, Javed S, Bonfio C, Spruijt E (2023) Interfacing coacervates with membranes: from artificial organelles and hybrid protocells to intracellular delivery. Small Methods 7(12):2300294

Martin N (2019) Dynamic synthetic cells based on liquid–liquid phase separation. Chembiochem 20(20):2553–2568

Maselko J, Reckley JS, Showalter K (1989) Regular and irregular spatial patterns in an immobilized-catalyst Belousov-Zhabotinskii reaction. J Phys Chem 93(7):2774–2780

Mason AF, Buddingh' BC, Williams DS, Van Hest JCM (2017) Hierarchical self-assembly of a copolymer-stabilized coacervate protocell. J Am Chem Soc 139(48):17309–17312

Matreux T, Le Vay K, Schmid A et al (2021) Heat flows in rock cracks naturally optimize salt compositions for ribozymes. Nat Chem 13(11):11

Matreux T, Aikkila P, Scheu B, Braun D, Mast CB (2024) Heat flows enrich prebiotic building blocks and enhance their reactivity. Nature 628(8006):110–116

Matsumura S, Kun Á, Ryckelynck M, Coldren F, Szilágyi A, Jossinet F, Rick C, Nghe P, Szathmáry E, Griffiths AD (2016) Transient compartmentalization of RNA replicators prevents extinction due to parasites Beating the curse of the parasite. Science 354(6317):1293–1296. https://doi.org/10.1126/science.aag1582

Miele Y, Medveczky Z, Holló G, Tegze B, Derényi I, Hórvölgyi Z et al (2020) Self-division of giant vesicles driven by an internal enzymatic reaction. Chem Sci 11(12):3228–3235

Mountain GA, Keating CD (2020) Formation of multiphase complex coacervates and partitioning of biomolecules within them. Biomacromolecules 21(2):630–640

Nakashima KK, Vibhute MA, Spruijt E (2019) Biomolecular chemistry in liquid phase separated compartments. Front Mol Biosci 6:21

Ninio J (2007) Errors and alternatives in prebiotic replication and catalysis. Chem Biodivers 4:622–632

Oparin A (1924) The origin of life. Moskovskii rabochii

Pavlinova P, Lambert CN, Malaterre C, Nghe P (2023) Abiogenesis through gradual evolution of autocatalysis into template-based replication. FEBS Lett 597(3):344–379

Peyraud-Vicré K, Dechamps C, Martin N, Desvergnes V (2025) Coacervate droplets drive organocatalyzed aqueous C–C bond formation via interfacial activation. J Am Chem Soc 147(41):37337–37346
Plum AM, Kempes CP, Peng Z, Baum DA (2025) Spatial structure supports diversity in prebiotic autocatalytic chemical ecosystems. npj Complexity 2(1):21
Polka JK, Hays SG, Silver PA (2016) Building spatial synthetic biology with compartments, scaffolds, and communities. Cold Spring Harb Perspect Biol 8(8):a024018
Pulletikurti S, Suresh VK, Yadav M, Ashok DA, Krishnamurthy R (2024) Experimentally modeling the emergence of prebiotically plausible phospholipid vesicles. Chem 10(6):1839–1867
Rubinov B, Wagner N, Rapaport H, Ashkenasy G (2009) Self-replicating amphiphilic β-sheet peptides. Angew Chem 121(36):6811–6814
Saha R, Chen IA (2024) Synthetic chemistry recreates transitional forms in prebiotic membrane evolution. Chem 10(6):1631–1633
Sanders QC, Verbeem A, Higgs PG (2025) Maintaining an autocatalytic reaction system in a protocell: nonenzymatic RNA templating and the link between replication and metabolism. Phys Rev E 111(1):014424
Sing CE, Perry SL (2020) Recent progress in the science of complex coacervation. Soft Matter 16(12):2885–2914
Slootbeek AD, Van Haren MHI, Smokers IBA, Spruijt E (2022) Growth, replication and division enable evolution of coacervate protocells. Chem Commun 58(80):11183–11200
Smith DE, Morowitz HJ (2016) The origin and nature of life on earth: the emergence of the fourth geosphere. Cambridge university press
Smith JM, Szathmary E (1997) The major transitions in evolution. Oxford University Press
Smith GP, Fraccia TP, Todisco M, Zanchetta G, Zhu C, Hayden E, Bellini T, Clark NA (2018) Backbone-free duplex-stacked monomer nucleic acids exhibiting Watson–Crick selectivity. Proc Natl Acad Sci 115(33)
Smokers IBA, Van Haren MHI, Lu T, Spruijt E (2022) Complex coacervation and compartmentalized conversion of prebiotically relevant metabolites. ChemSystemsChem 4(4):e202200004
Sojo V, Herschy B, Whicher A, Camprubi E, Lane N (2016) The origin of life in alkaline hydrothermal vents. Astrobiology 16(2):181–197. WOS:000371013300006
Strulson CA, Molden RC, Keating CD, Bevilacqua PC (2012) RNA catalysis through compartmentalization. Nat Chem 4(11):941–946. https://doi.org/10.1038/nchem.1466
Szathmáry E, Demeter L (1987) Group selection of early replicators and the origin of life. J Theor Biol 128(4):463–486
Takeuchi N, Hogeweg P (2009) Multilevel selection in models of prebiotic evolution II: a direct comparison of compartmentalization and spatial self-organization. PLoS Comput Biol 5(10): e1000542
Tang D, Zhu J, Wang H, Chen N, Wang H, Huang Y, Jiang L (2025) Universal membranization of synthetic coacervates and biomolecular condensates towards ultrastability and spontaneous emulsification. Nat Chem 17(6):911–923
Ter-Ovanessian LMP, Maurel MC, Lambert JF (2023) Ring-closure on the rocks in a prebiotic environment. Chembiochem 24(10):e202300143
Todisco M, Fraccia TP, Smith GP, Corno A, Bethge L, Klussmann S, Paraboschi EM, Asselta R, Colombo D, Zanchetta G, Clark NA, Bellini T (2018) Nonenzymatic polymerization into long linear RNA templated by liquid crystal self-assembly. ACS Nano 12(10):9750–9762
Uversky VN, Kuznetsova IM, Turoverov KK, Zaslavsky B (2015) Intrinsically disordered proteins as crucial constituents of cellular aqueous two phase systems and coacervates. FEBS Lett 589(1):15–22
Vieregg JR, Lueckheide M, Marciel AB, Leon L, Bologna AJ, Rivera JR, Tirrell MV (2018) Oligonucleotide–peptide complexes: phase control by hybridization. J Am Chem Soc 140(5): 1632–1638
Wächtershäuser G (1988) Before enzymes and templates: theory of surface metabolism. Microbiol Rev 52:452–484

Walter H, Brooks DE, Fisher D (1985) Partitioning in aqueous two-phase systems: theory, methods, uses, and application to biotechnology. Academic Press

Wang J, Abbas M, Wang J, Spruijt E (2023) Selective amide bond formation in redox-active coacervate protocells. Nat Commun 14(1):8492

Wu M, Higgs PG (2012) The origin of life is a spatially localized stochastic transition. Biol Direct 7(1):42

Zwicker D, Seyboldt R, Weber CA, Hyman AA, Jülicher F (2017) Growth and division of active droplets provides a model for protocells. Nat Phys 13(4):408–413

Chapter 26
The Challenge of Scale in Molecular Adaptation: Local Searches in Astronomical Genotype Networks

Susanna Manrubia, Luis F. Seoane, and José A. Cuesta

Abstract The exploration of vast genotype spaces poses fundamental challenges for evolving populations. As the number of genotypes encoding viable phenotypes grows exponentially with genome length, populations can only explore a tiny fraction of these immense spaces, a fact consistently supported by empirical and theoretical evidence. Paradoxically, local, mutation-driven searches near abundant sequences allow populations to generate phenotypic improvements and functional innovations despite this immense search space. In this contribution, we integrate insights from viral evolution with theoretical expectations derived from genotype–phenotype maps to reexamine how high-dimensional sequence spaces shape evolutionary dynamics. In resolving the paradox, abundant phenotypes play a crucial role because their combinatorial weight biases evolutionary trajectories. We discuss how this bias, together with limited accessibility of fitness peaks, modifies traditional metaphors—such as fitness landscapes—and challenges standard notions of evolutionary optimality. Our results underscore that adaptation is predominantly local yet remarkably efficient, providing a unifying perspective on the coexistence of robustness, innovation, and constrained exploration in molecular evolution.

S. Manrubia (✉)
Museo Nacional de Ciencias Naturales (CSIC), Madrid, Spain

Grupo Interdisciplinar de Sistemas Complejos (GISC), Madrid, Spain
e-mail: smanrubia@mncn.csic.es

L. F. Seoane
Institut de Biologia Evolutiva (UPF-CSIC), Barcelona, Spain

Grupo Interdisciplinar de Sistemas Complejos (GISC), Madrid, Spain
e-mail: lf.seoane@ibe.upf-csic.es

J. A. Cuesta
Departamento de Matemáticas, Universidad Carlos III de Madrid, Leganés, Spain

Grupo Interdisciplinar de Sistemas Complejos (GISC), Madrid, Spain
e-mail: cuesta@math.uc3m.es

G. Witzany (ed.), *Integrative Theory of Evolution - Aspects and Insights*,
https://doi.org/10.1007/978-3-032-24295-2_26

1 Introduction

Genetic diversity is a fundamental prerequisite for species survival and evolution (Agashe et al., 2023). Natural populations produce and maintain high levels of diversity through multiple mechanisms, ensuring efficient exploration of functional alternatives and rapid responses to changing environments. Our knowledge of intraspecies diversity is particularly extensive for viruses and microbial populations. High-throughput sequencing and improved sampling have repeatedly revealed substantial genetic diversity within single hosts (Lauring, 2020), within microbial species in complex environments, and across replicate experimental populations. Deep sequencing of SARS-CoV-2 (Jones, 2025) and other RNA viruses (Bordería et al., 2011) has identified measurable intra-host single-nucleotide variants (Seoane et al., 2026), persistent mixed infections in some patients, and lineage-specific differences in within-host diversity (Gu et al., 2023). These studies and many others indicate that viral populations commonly exist as diverse "clouds" of related genotypes consistent with a quasispecies-like structure (Domingo, 2006).

Despite the typically lower mutation rates of prokaryotes and smaller effective population sizes compared with viral populations, metagenomics and genome-resolved sequencing have revealed high standing diversity within bacterial species in natural microbiomes (Kim et al., 2024). Many species are not represented by a single dominant genotype but by coexisting strain lineages and within-strain polymorphisms (Viver et al., 2024). Long-read and single-cell approaches uncover fine-scale structures—haplotypes, mobile elements, and recombination tracts—that substantially increase inferred within-taxon diversity relative to earlier short-read-based snapshots. These data indicate that bacteria in situ maintain far more genetic variation than simple single-clone models predicted (Harris et al., 2021).

In vivo observations are complemented by in vitro evolution experiments, which show that, in large asexual populations, adaptation often involves multiple coexisting beneficial lineages (Lang et al., 2013), polygenic shifts, and frequent soft sweeps (Harris et al., 2021). Populations experience both rapid selective sweeps and extended periods of balanced polymorphism or competing mutations (Laguna-Castro et al., 2025), depending on population size, mutation supply, recombination, and ecological structure. These findings suggest that periods away from mutation–selection equilibrium may be the natural state of microbial (Cvijović et al., 2018) and viral (Braun et al., 2018) populations.

Several theoretical frameworks have been developed to explain these observations. Kimura's Neutral Theory (Kimura, 1968) proposed that most molecular substitutions are effectively neutral, with substitution rates equal to mutation rates—a powerful null model for sequence divergence. Ohta extended this to the nearly neutral regime (Ohta, 1973), emphasizing that the fate of weakly selected mutations depends on population size and drift. Maynard Smith's protein space metaphor (Maynard Smith, 1970) and subsequent work on neutral networks (Schuster et al., 1994) showed that many genotypes can have similar fitness and that connectivity via neutral mutations provides a mechanism to perform a sustained exploration of sequence space without fitness losses. In parallel, Eigen introduced the quasispecies framework (Eigen, 1971; Eigen et al., 1989), describing high-mutation-rate populations of replicators (Schuster and Sigmund, 1983; Schuster and Swetina, 1988), later applied to RNA viruses as clouds of related genotypes surrounding a master sequence (Domingo et al., 1978). In quasispecies populations, the effects of the underlying fitness landscape are attenuated by collective dynamics, allowing coexistence of mutants with diverse fitness values both in and out of mutation–selection equilibrium.

Efficient generation of genetic diversity is essential for long-term adaptability. Mechanisms that maintain extant diversity—such as high mutation rates or transposition bursts in response to unexpected selection pressures (Piacentini et al., 2014)—have likely been selected over evolutionary timescales (Ferrare and Good, 2024; Kumawat et al., 2025). In light of these observations, several key questions arise: To what extent are neutral spaces essential for evolvability? How constrained are populations to occupy fitness maxima? Are "valleys" in fitness landscapes true barriers to adaptation in high-dimensional, networked genotype spaces? And to what extent does finite population size limit the discovery of functional phenotypes?

2 Neutral Networks and Genotype-to-Phenotype Maps

2.1 Synthetic Genotype-to-Phenotype Maps

Maynard Smith's idea of protein networks has been highly influential in subsequent characterizations and descriptions of both the structure of sequence spaces (induced by a variety of genotype-to-phenotype (GP) maps (Schuster et al. 1994; Lipman and Wilbur 1991; Arias et al. 2014; Greenbury et al. 2014; Fortuna et al. 2017) and the supposed constraints operating along evolutionary dynamics. In all analyses of GP maps there is the underlying assumption that mutations causing significant variation in the expressed phenotype will be rare or absent in populations, since they are deleterious and therefore selected against. Two ideas should be kept in mind in this context. First, natural selection acts on phenotypes, so significant evolutionary changes will only occur when phenotypes change. Since the fitness value of a phenotype is context-dependent, it is generally not possible to measure the value of a phenotype without testing its performance in specific environments. More often than not, even in constant environments, non-neutral mutations whose adaptive value is difficult to quantify are fixed in populations (Villanueva et al., 2022), occasionally modifying phenotypes in unforeseeable ways (Escarmís et al., 2008; Grant et al., 2021). Second, the presence of deleterious mutations in a population results from a competition between two temporal scales: the rate of generation of such mutations and the strength of purifying selection. This tension between the two scales is particularly relevant in ensembles of mutationally linked replicators, such as viruses, where the fast mutational scale translates into a persistent coexistence of variants bearing mutations with dissimilar effects on fitness (Lauring, 2020) and even different phenotypes (Duarte et al., 1994; Vignuzzi et al., 2006).

The previous observations notwithstanding, simple GP models have been tremendously enlightening for developing a more accurate picture of how genotypes map onto phenotypes. There are some properties shared by all navigable GP maps (Garcia-Martin et al., 2018), such as a negative correlation between genotypic evolvability and genotypic robustness, a positive correlation between phenotypic evolvability and phenotypic robustness, and a linear growth of phenotypic robustness with the logarithm of the number of genotypes expressing the same phenotype (what is usually referred to as "phenotype size") (Manrubia et al., 2021). One of the most striking results of such models has been the quantification of phenotypic bias: Most of sequence space maps to a few highly abundant phenotypes, while an astronomical number of phenotypes are rare (few sequences map to them) and essentially invisible to evolution. These quantities are illustrated in Fig. 26.1 with the iconic example of the RNA sequence-to-secondary-structure GP map.

Phenotypic bias—the huge difference in phenotype sizes—hinted at by numerical analyses of GP maps, can be analytically derived in a general scenario where only one feature of the GP map matters: the fact that different sites in a sequence have different degrees of versatility (or neutrality) while yielding the same phenotype

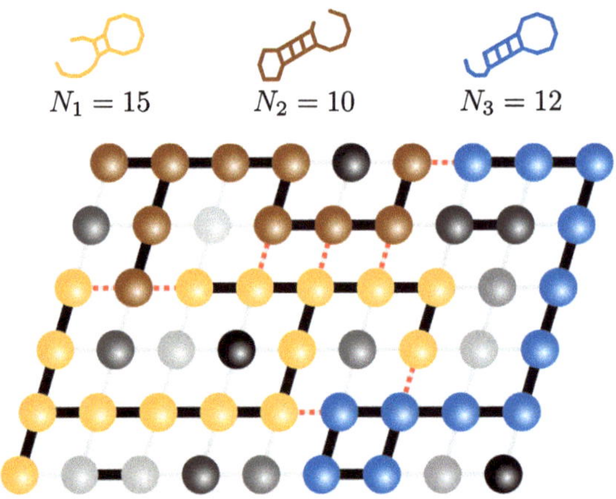

Fig. 26.1 Neutral networks in the RNA sequence-to-secondary structure genotype–phenotype map. This grid provides a simplified representation of genotype networks, which in reality are far more complex and not planar. Each node represents a unique RNA sequence, and a single point mutation corresponds to a displacement to a neighboring node. All nodes of the same color (connected by thick black edges) represent RNA sequences that fold into the same secondary structure—i.e., different genotypes mapping to the same phenotype—and together constitute that phenotype's *neutral network*. Three phenotypes (yellow, brown, and blue) occupy most of genotype space. The size of a neutral network ($N_1 = 15$ for the yellow folding, $N_2 = 10$ for the brown folding, and $N_3 = 12$ for the blue folding) is the number of sequences that produce the same phenotype. Many other, much smaller neutral networks (shown in various shades of gray) correspond to much rarer foldings, produced by only a few sequences (at most one or two in this example). Mutations within a neutral network allow divergence without losing a successful phenotype (hence without risking a fitness loss). Neutral networks usually traverse genotype space (e.g., the blue folding spans from top to bottom), and they get in touch with many other large phenotypes at one point or another (dashed red edges)

(Manrubia and Cuesta, 2017; Garcia-Martin et al., 2018; Martin and Ahnert, 2022; Catalán et al., 2023). The size of a phenotype can be approximated by the product of single-site versatility along the sequence, thus generating a log-normal distribution of phenotype sizes in almost any realistic GP map. Interestingly, various studies have shown that only large phenotypes (the rightmost tail of the log-normal distribution) are selected in natural systems (Jörg et al., 2008; Dingle et al., 2015). This highly relevant observation has two main consequences: The size of a phenotype must be incorporated as an important feature of its fitness, and the comparatively fewer large phenotypes are much more easily found through blind mutation than rare phenotypes—implying that highly frequent phenotypes dominate evolutionary dynamics (Cowperthwaite et al., 2008; Schaper and Louis, 2014; Catalán et al., 2020; Martin et al., 2024).

Neutral networks of large phenotypes have at least two additional evolutionary advantages: They are more robust on average, since the average number of neutral changes (the average degree of the neutral network $\langle k \rangle$) accepted by a sequence in a phenotype of size N is $\langle k \rangle \propto \log N$ (Aguirre et al., 2011; Catalán et al., 2023), and they also percolate the space of sequences, guaranteeing navigability and access to steadily growing novelty along evolution (Huynen et al., 1996; Huynen, 1996; Buchholz et al., 2017).

Still, what is the quantitative relevance of these potential advantages?

2.2 Phenotypic Robustness and Abundance

Let us begin by critically examining the effects of the increase in robustness with phenotype size, using the case of the RNA sequence-to-secondary structure map, where the relevant quantities have been analytically and numerically worked out. The goal of the following exercise is to estimate the fraction of single-point mutations that maintain the phenotype (thick black edges in Fig. 26.1) and the complementary fraction (i.e., those mutations causing a change in phenotype, light gray or dashed red edges in Fig. 26.1) for sequences of length L. This will quantitatively illustrate the difference between rare phenotypes (where most mutations are non-neutral, gray nodes in Fig. 26.1) and typically large phenotypes (colored nodes in Fig. 26.1), as found in natural systems.

On the basis of structural properties of RNA secondary structure, the expected number of neutral neighbors in a network of N genotypes mapping to the same secondary structure is, asymptotically in N (Catalán et al., 2023),

$$\langle k \rangle = c \log N + O((\log N)^{1/2}) , \tag{26.1}$$

with $c \simeq 1.88$ (Aguirre et al., 2011). Bounds to minimum and maximum values of the coefficient c can be analytically obtained, yielding $c^{min} = 1$ and $c^{max} = 3/\log 4 \simeq 2.16$, respectively (Aguirre et al., 2011). In order to estimate the average degree of frequent phenotypes, we need to estimate their average size N_{freq} as a function of sequence length. Analytical estimations show that the absolute number M_p of abundant phenotypes (those accumulating a fraction $p > 0.99$ of genotypes) grows as $M_p \simeq 1.33L^{-2}(1.68)^L$ (Catalán et al., 2023). A rough estimate of the typical size of those phenotypes is therefore given by

$$N_{\text{freq}} = \frac{4^L}{M_p} \simeq 0.75L^2(2.38)^L \ . \qquad (26.2)$$

Neither M_p nor N_{freq} is significantly sensitive to the specific value of p, as long as $p \simeq 1$. Substituting the latter expression in Eq. (26.1), we obtain the expected neutrality for typical frequent phenotypes

$$\langle k \rangle \simeq 0.87cL + O(L^{1/2}) \simeq 1.64L + O(L^{1/2}) \ . \qquad (26.3)$$

Now, since every sequence of length L has $3L$ possible neighbors one mutation away, the expected fraction of neutral neighbors becomes $\rho_a = \langle k \rangle/(3L) \simeq 1.64/3 = 0.55$, with a correction $O(L^{-1/2})$. Asymptotically, therefore, we expect that around 45% of mutations yield new RNA secondary structures.

The latter figure compares well with that inferred from numerical studies, where the typical size of large, abundant phenotypes has been calculated (Dingle et al., 2015). Using the distribution of phenotype sizes corresponding to natural non-coding RNAs (Kin et al., 2007), the numerically estimated fraction ρ_n of neutral mutations (for sequence lengths $40 < L < 126$) reaches about $\rho_n \simeq 0.6$, so around 40% of mutations cause changes in phenotype.[1]

Interestingly, both analytical ρ_a and numerical ρ_n estimations yield values independent of sequence length, revealing that robustness is not a quantity that can be tuned through insertions or deletions. In practice, the increase in neutrality of large phenotypes (with respect to small phenotypes, which are not observed in natural systems) implies that the mutation rate is slightly less effective in finding phenotypic novelty, requiring two or three attempts more than small phenotypes to escape the neutral network. It would be no surprise to learn that the mutation rate has been tuned along evolution to guarantee evolvability in large phenotypes—apparently the only ones occurring in natural populations.

[1] Data in Fig. 3 of Dingle et al. (2015) yield a numerical estimate of the size N for each length L; the average neutral degree of a phenotype of size N is then estimated as $\langle k \rangle_n = 1.88 \log N$, from which we get $\rho_n = \langle k \rangle_n/(3L)$; and ρ_n is a rough average over the values obtained for large L.

2.3 Percolation and Navigability in Neutral Networks

Multiple theoretical observations indicate that the neutral networks of large phenotypes contain sequences as distant as two random sequences could be. This property, together with the high dimensionality of sequence spaces (and the concomitant intertwining of phenotypes), conceptually guarantees that most (frequent) phenotypic alternatives are a few mutations away from at least some genomic region of the current phenotype. These properties result in network apposition—there are points of contact between any pair of phenotypes (Fontana, 2002) (dashed red edges in Fig. 26.1), a property empirically demonstrated (Schultes and Bartel, 2000)—and in the related property of shape space covering (novelty is a few mutations away) (Grüner et al., 1996a,b). What, then, is the role played by percolating phenotypes? Do populations need to explore neutral networks extensively in order to adapt successfully? This is not to suggest that neutral drift is irrelevant—on the contrary, it plays an important role (Bloom et al., 2007; Erdoğan et al., 2024). Rather, the question concerns the extent of neutral drift required for adaptation and innovation. We illustrate this point below with a quantitative example.

Consider one of the shortest functional genomes, that of *Avocado sunblotch viroid* (ASBVd), with 246 nucleotides (Symons, 1981). The total number of different sequences of length $L = 246$ is $4^L \simeq 10^{148}$. A smaller figure corresponds to the number of sequences compatible with the RNA secondary structure of the ASBVd folded genome, with about 180 paired and 66 unpaired nucleotides, which is about 5×10^{41} (Garcia-Martin et al., 2018; Catalán et al., 2019): Still, the mass of all possible sequences in the ASBVd neutral network is about 10^4 times the total biomass on Earth. How many different sequences could have been produced since this viroid appeared? An upper bound can be estimated if we assume (i) ASBVd appeared simultaneously with land plants, some 470 My ago; (ii) the number of infected plants at any time is 10^{12},[2] (iii) the viroid load of any infected plant is 10^{12},[3] and (iv) every replication of the genome includes on average one error (a new nucleotide). Assuming also that all generated sequences are different (though many of them might be identical or very similar and/or produce identical descent), we get an upper bound for the total number of explored sequences: 10^{36}. Though large, this number is still far from covering the complete neutral network of the ASBVd secondary structure, not to mention a significant fraction of the whole space

[2] Viroids infect a wide range of cultivated plants and ornamentals. The FAO estimates about 1.5×10^9 hectares of arable cropland; at a density of 10^5–10^6 plants per hectare, there are 10^{13}–10^{14} plants in cultivation at any time. Assuming that 1–10% are infected at any time yields the figure of 10^{12} plants infected used in the order-of-magnitude estimation.

[3] Although precise quantification of viroid copy numbers per plant is seldom reported, high-throughput sequencing and biochemical assays reveal that systemic viroid infections can produce extremely high intracellular titers. For large hosts such as fruit trees, the product of cell number and per-cell viroid load readily exceeds 10^{12} genomes, supporting the use of that scale in upper bound estimates.

of constant L. Analogous figures for other genes and genomes yield values that make a systematic exploration of the set of sequences compatible with a phenotype absolutely out of reach (Louis, 2016).

What do these numbers mean? In light of the high diversity and plasticity of life and of the ability of organisms (notably microorganisms) to adapt to harsh and novel environments, it is clear that exploring most of sequence space is not a requisite for a population to thrive. As early studies with RNA postulated and illustrated numerically (Huynen et al., 1996) and more recent analyses suggest (Dryden et al., 2008), the specific identity of nucleotides (or amino acids) is not a requirement to attain function. Local exploration of variants—rather than exhaustive exploration of the full sequence space or even of an entire neutral network—must suffice.

3 Fitness Landscapes, Phenotype Accessibility, and High-Dimensional Constraints

A fitness landscape is a mapping from the genomic, multidimensional space to a real value representing fitness (Wright, 1931). Traditionally, fitness landscapes have been viewed as topographic representations where genotypes lie on the horizontal plane and fitness on the z-axis; this view has illustrated evolutionary dynamics for almost a century (Pigliucci, 2012). These landscapes portray the adaptive process as a parsimonious movement of populations that "traverse valleys," "reach fitness peaks," and adapt by incorporating mutations that pull them "uphill," (see Fig. 26.2a). Inherent to these representations are concepts such as trapping in local maxima, peaks that can be attained, and, most of the time, fitness values with a sort of Platonic existence, independent of the state of the population. Although ample evidence shows that molecular evolution does not generally follow the iconic topographic fitness landscape, this metaphor remains deeply imprinted in our conceptual visualization of evolutionary dynamics.

3.1 Accessibility of Fitness Peaks in High Dimensions

Dynamics in multidimensional spaces differ fundamentally from two-dimensional projections. In high-dimensional genotype spaces, the accessibility of fitness peaks—i.e., the existence of monotonic mutational paths leading from arbitrary genotypes to the global fitness maximum—depends critically on the ruggedness and correlation structure of the landscape. Analytical and numerical studies of random fitness models have shown that in uncorrelated or weakly correlated landscapes, such as the House-of-Cards or NK models with fixed local interactions, the probability that any strictly increasing path reaches the global optimum decays exponentially with sequence length L (Franke et al., 2011; Krug, 2019; Schmiegelt and Krug, 2023). This loss of accessibility occurs because each beneficial step must satisfy a sequence of rare inequalities in fitness orderings, and in high dimensions these constraints multiply, making global fitness peaks effectively unreachable by purely uphill walks. Increasing the alphabet size K (allowing more allelic states per site) or introducing a global additive gradient (as in the Mount Fuji model) can substantially increase the number of accessible trajectories, yet these conditions are atypical of natural, epistatic landscapes (Zagorski et al., 2016; Franke et al., 2011).

While small empirical landscapes often exhibit multiple accessible routes (Szendro et al., 2013), the generic expectation for large genomes is that strictly monotonic adaptive paths are exceedingly rare. Reaching the global maximum typically requires transient declines in fitness or exploration of neutral or nearly neutral networks. Hence, the intuitive Wrightian picture of populations climbing continuous mountains is misleading in high-dimensional sequence spaces: Despite the astronomical number of possible mutational paths, most are blocked by sign epistasis, and purely uphill trajectories almost never connect distant genotypes (Krug, 2019).

3.2 *Invisible Peaks and Entropic Barriers*

A more accurate description of evolutionary dynamics on sequence spaces should combine rough fitness landscapes with the partition of the space induced by mapping genotypes to phenotypes. Although genotype-to-fitness landscapes are often the default representation, considering the intermediate mapping to phenotypes illustrates an additional potential difficulty in reaching high-fitness solutions. The unified picture is that of adaptive multiscapes (Catalán et al., 2017), where each genotype maps to one phenotype (in a simplified many-to-one GP map), and all genotypes belonging to the same phenotype take the same fitness value in a specific environment. In this scenario, the highly skewed distribution of phenotype sizes comes into play: Only the largest phenotypes are relevant to evolution; fitter phenotypes, if sufficiently smaller, are almost never found. In other words, even if the global fitness peak (understood as a degenerate "peak" corresponding to all genotypes mapping to that phenotype) is accessible, it is likely lost in a sea of astronomically more abundant alternatives.

3.3 *Phenotypic Accessibility and the Arrival of the Frequent*

Mean-field models of GP maps provide a remarkably compact description of the high-dimensional dynamics of molecular evolution. In such models, the complex local structure of neutral networks is averaged out, and the transition probability $\phi_{\xi \to \chi}$ from phenotype ξ to phenotype χ upon a point mutation can be computed as an average over all genotypes realizing ξ (Schaper and Louis, 2014; Greenbury et al., 2016; Manrubia and Cuesta, 2015; Huynen et al., 1996). This approach predicts that, to a first approximation,

$$\phi_{\xi \to \chi} \propto f_{\chi} , \tag{26.4}$$

where f_{χ} is the global frequency (or relative neutral set size) of phenotype χ, i.e., the fraction of sequence space that maps to it. Because the distribution of f_{χ} values typically spans many orders of magnitude and is approximately log-normally distributed in RNA and other GP maps, the corresponding transition probabilities vary comparably. As a result, most of the mutational flux from any given phenotype leads to a very small subset of abundant, highly connected phenotypes, while rare phenotypes—those with small size N—are practically inaccessible, even when they would confer large fitness advantages. This bias, termed the *arrival of the frequent* (Schaper and Louis, 2014), implies that the spectrum of available variation is determined primarily by the statistical geometry of the GP map rather than by selective gradients on fitness. Note that the mean-field estimation is remarkably optimistic regarding transition probabilities: If there is a small number of links between the two phenotypes, and considering the huge sizes of the latter (even small ones), populations may evolve for extremely long times away from the apposition regions where the two phenotypes are one mutation apart.

A simple numerical example illustrates the magnitude of this bias in the mean-field scenario. Consider sequences of length $L = 100$ with a four-letter alphabet (as in RNA), yielding $4^{100} \simeq 10^{60}$ possible genotypes. A typical large phenotype has size $N_{\text{freq}} \simeq 10^{42}$, the most abundant phenotype size is 10^{32}, and typical small phenotypes have size $N_{\text{rare}} \simeq 10^{22}$ (Dingle et al., 2015). For a rare phenotype with a neutral set size 10^{20} times smaller, both theory and simulation suggest that, to a first approximation, the probability of a mutation leading into this phenotype scales roughly with its relative size N, so $\phi_{\text{rare}}/\phi_{\text{freq}} \simeq 10^{-20}$.[4] Hence, even if a population explores 10^{12} *different* genotypes through mutation and drift, the expected number of encounters with that rare phenotype remains negligibly small. Conversely, neutral exploration within frequent phenotypes dominates: The population will perform an effectively *local* random walk within large, percolating neutral networks that connect vast regions of sequence space, even if large excursions from the initial sequence are uncommon.

Analytical results allow estimating the ratio between the sizes of frequent and rare phenotypes as the sequence length increases. For RNA, the distribution of the natural logarithm of phenotype sizes $\log N$ follows a normal distribution with mean $\mu_L = \mu_1 L + O(1)$ and standard deviation $\sigma_L = \sigma_1 L^{1/2} + O(L^{-1/2})$, where $\mu_1 = 0.2865$ and $\sigma_1 = 0.4434$ (Garcia-Martin et al., 2018). Then, the typical size of large phenotypes can be shown to be (Catalán et al., 2023)

[4] This ratio could be corrected by the fraction of neutral mutations for each phenotype size, but the correction is just a factor of 2, thus not modifying the order of magnitude.

$$\log N_{\text{freq}} \simeq \mu_L + \sigma_L^2 - z\sqrt{2}\sigma_L \,, \tag{26.5}$$

with $z \in (1.5, 2.7)$ for the fraction of genotypes included in the set of frequent phenotypes, $p \in (0.99, 0.9999)$. By symmetry, since the distribution of phenotype sizes peaks at μ_L, the typical size of rare phenotypes can be analogously estimated as

$$\log N_{\text{rare}} \simeq \mu_L - \sigma_L^2 + z\sqrt{2}\sigma_L \,. \tag{26.6}$$

Hence, the mean-field estimate of the ratio of rare to frequent phenotype sizes is

$$\frac{\phi_{\text{rare}}}{\phi_{\text{freq}}} \simeq \exp\left(-2\sigma_1^2 L + 2\sqrt{2}z\sigma_1\sqrt{L}\right) \,. \tag{26.7}$$

This estimation yields larger relative values for $L = 100$ than the rough numerical estimate obtained above, since it estimates smaller and larger phenotype sizes for frequent and rare phenotypes, respectively. However, the important result is that the ratio decreases exponentially with L, broadening the separation between the size of frequent and rare phenotypes and making rare phenotypes increasingly difficult to meet. For $L = 246$, as in the Avocado sunblotch viroid example, and taking $z = 2$, $\phi_{\text{rare}}/\phi_{\text{freq}} \simeq 4 \times 10^{-34}$.

In summary, small, isolated phenotypes remain essentially invisible. Canalization or selective bias could, in principle, guide trajectories toward fitter phenotypes, but only within the subset of variation that the GP map makes accessible. This subset is essentially covered by other frequent phenotypes.

Even if a population were artificially placed on one of these small, high-fitness phenotypes—analogous to fixing an adaptive RNA or protein structure in the laboratory—the phenotype would be rapidly lost under mutation pressure. Because most mutational trajectories leave the small neutral set, the population diffuses away in a few generations toward more abundant phenotypes with larger mutational robustness. This loss is fundamentally entropic: In the same way that systems in statistical physics evolve toward macrostates with more microstates, evolutionary dynamics naturally drift toward regions of sequence space with greater combinatorial weight. Thus, both the difficulty of locating rare phenotypes and the instability of remaining on them ensure that the "arrival of the frequent" dominates evolutionary exploration, limiting accessibility to the global fitness maximum (which, by probability, corresponds to the set of non-frequent phenotypes) except under extreme, nongeneric conditions—an outcome consistent with results from models of fitness landscapes in high dimensions, where ruggedness and dimensionality conspire to render global peaks exponentially rare and typically unreachable (Franke et al., 2011; Nowak and Krug, 2013).

4 Efficient Adaptation Within High-Abundance Phenotypes

The entropic trapping effect caused by the large size of abundant phenotypes was described long ago (Schuster and Swetina, 1988) and observed in multiple numerical studies (Wolff and Krug, 2009; Khatri et al., 2009; Wilke et al., 2001; Codoñer et al., 2006; Cowperthwaite and Meyers, 2007; Schaper and Louis, 2014; Catalán et al., 2020), where this effect manifests in various forms. Terms such as *phenotypic entropy*, *redundancy*, or *landscape flatness* have been used to conceptually describe the number of genotypes that realize a given phenotype and, consequently, its mutational robustness. A pioneering theoretical analysis (Schuster and Swetina, 1988) considered the competition between two quasispecies corresponding to distinct phenotypes and showed that the phenotype with the highest replication rate could nevertheless be outcompeted by a slightly less fit one if the latter occupied a larger and more connected neutral network. Too low robustness, they demonstrated, could be fatal at high mutation rates. Note that in this analysis, the phenotype size was not included as a feature of fitness, and only the replicative ability of the genotypes within the phenotype was considered. An analytical treatment (Wolff and Krug, 2009) later quantified the conditions for this crossover between fitness and robustness, showing that increasing mutation rates can drive the dominance of phenotypes with lower intrinsic fitness but broader mutational neighborhoods and that this transition is closely related to the onset of the error catastrophe.

Other studies have examined analogous effects in different biological and computational contexts. In models of spatial gene regulation, it was shown that finite populations often fail to converge to the globally optimal phenotype when multiple equivalent solutions exist, an outcome interpreted as entropic and arising from the multiplicity of equally viable configurations (Khatri et al., 2009). The same principle underlies the so-called survival of the flattest phenomenon (Wilke et al., 2001), where, at sufficiently high mutation rates, populations evolve toward flatter regions of the fitness landscape, prioritizing mutational robustness over maximal replication speed. This prediction was later confirmed experimentally with RNA viroids (Codoñer et al., 2006), where lineages displaying slower replication but higher neutrality outcompeted faster yet fragile variants, showing that fitness can be maximized by minimizing the mutational load rather than by maximizing growth rate.

Computational analyses of mutation–selection dynamics have further clarified the evolutionary implications of phenotype size. It has been shown that adaptation tends to proceed through abundant phenotypes—a process also termed the "ascent of the abundant" (Cowperthwaite and Meyers, 2007)—and that phenotypic innovation is biased toward more frequent phenotypes in genotype–phenotype maps, the effect already described as the "arrival of the frequent" (Schaper and Louis, 2014). More recent analytical work (Catalán et al., 2020) has demonstrated that this bias persists across a broad range of mutation–selection regimes, confirming that phenotypic accessibility, rather than differences in maximal fitness, largely determines the evolutionary trajectories available to finite populations.

In summary, only a vanishingly small fraction of highly abundant phenotypes is explored along evolution and thus presented to natural selection. The question of whether this restriction limits, and to what extent, the localization of phenotypes with sufficiently high fitness was addressed in a study where the relative weight of phenotype size versus replicative ability was analytically and numerically explored (Catalán et al., 2023). Returning to the case of RNA sequence-to-secondary structure GP maps, the fraction u of phenotypes visible to evolution diminishes exponentially with L as

$$u \simeq \frac{0.9}{L^{1/2}}(1.1)^{-L} . \tag{26.8}$$

However, since the total number of different open secondary structures for sequences of length L grows as $1.48L^{-3/2}(1.85)^{L}$, the absolute number of different, large phenotypes covered by a fraction $p \simeq 1$ of genotypes grows exponentially with L as $1.33L^{-2}(1.68)^{L}$.

Assuming, as in Catalán et al. (2023), that replicative ability values follow a normal distribution with average $\langle r \rangle$ and variance σ_r^2, even in the restricted set of large phenotypes visible to natural selection, the maximum value of r that can be attained is only about 10% lower than in the whole ensemble of phenotypes. In other words, limiting the evolutionary search to large phenotypes does not entail a large fitness loss, while it can yield a significant gain in robustness and navigability.

The exploration of phenotypic diversity is not only restricted to large phenotypes, but it is also local in sequence space. No better demonstration of the locality of evolution exists than gene phylogenies, where common ancestors occurred billions of years ago. Retaining memory of this distant past indicates that extant, successful functions were not far from the sequences where the search started. This observation relates to the idea of shape space covering discussed above, which appears to be a general feature of the broad sequence-to-function map: New functions may lie a few mutations away from current genotypes in high-dimensional spaces. Support for this expectation, originally derived from simple GP maps, has come from experimental results showing that functional genes can be rapidly generated from nongenic sequences (Carvunis et al., 2012; Vakirlis et al., 2020), that transposable elements can be domesticated to perform specific functions in their host (Sinzelle et al., 2009), or that promoters can emerge from random sequences (Yona et al., 2018), even in the absence of sequence diversity, simply through successive cycles of mutation, enrichment, and selection (Wachowius et al., 2021). Thus, high-dimensional sequence spaces allow the discovery and fixation of molecular function diversity observed in current organisms, speeding up adaptive rates so much as to allow organisms to thrive in steadily changing environments.

5 Hierarchical Genotype Networks in Empirical Populations

A conceptual metaphor that helps us understand how evolutionary dynamics unfold in sequence spaces needs to integrate the main elements discussed in previous sections: (i) Sequence spaces are so vast that only a minute fraction is explored by natural populations, (ii) neutral networks are astronomically large, and natural populations only visit a local, tiny fraction of the whole, (iii) spaces of high dimensionality are so deeply interwoven that most (large) phenotypes can be found a few mutations away, (iv) as sequence length increases, high-fitness maxima become increasingly difficult to attain, (v) the fraction of (large) phenotypes explored by genotypes of length L decreases exponentially with L, though at the same time, the absolute number of different phenotypes increases exponentially, and (vi) the latter restriction has a small impact on the replicative ability values that populations can achieve. Establishing a mindset within this scenario should also be helpful when interpreting empirical results and should prove especially relevant when heterogeneity is intrinsic to the population, such as in viruses and other organisms with high genotypic diversity. These populations should comply with the restrictions above while managing to explore sufficient evolutionary innovation or variants that allow their adaptation to changing conditions. A pictorial summary of how the elements above modify suitable representations of quasispecies evolution on sequence spaces is represented in Fig. 26.2.

Advances in sequencing technologies have boosted the detailed description of genetic diversity in viral populations in particular. This has allowed the characterization, to varying but deep detail, of the within-host structure of viral populations (Baaijens et al., 2017; Zanini et al., 2017; Delgado et al., 2021) and their phylodynamics (Attwood et al., 2022; Sun et al., 2025). Recently, deep sequencing of the Qβ bacteriophage has revealed a hierarchical structure with implications for viral evolutionary dynamics and, in particular, for the search process in sequence spaces (Seoane et al., 2026).

Qβ populations exhibit robust organizational patterns along evolution, even when new mutations with a supposed adaptive value become fixed. Figure 26.3 represents a snapshot of a genotype network of a deep-sequenced Qβ population. At any time point in a process of adaptation to new environments, genomes cluster around the most abundant sequence, which appears to continuously regenerate diversity, acting as a diversity source and populating its neighborhood at increasing mutational distance in a process dominated by simple diffusion: Sequence abundance decays exponentially with Hamming distance to the most abundant sequence. A remarkable observation is that, on average, no significant correlation between the abundance of variants and the type of mutation they carry (synonymous or non-synonymous) is found. In this out-of-equilibrium state, the search process is dominated by reaction (the most abundant sequence produces a larger number of offspring) and diffusion (through mutation), with a minor effect of the underlying fitness landscape. As new mutations fix in the population, the whole hierarchical structure is rapidly

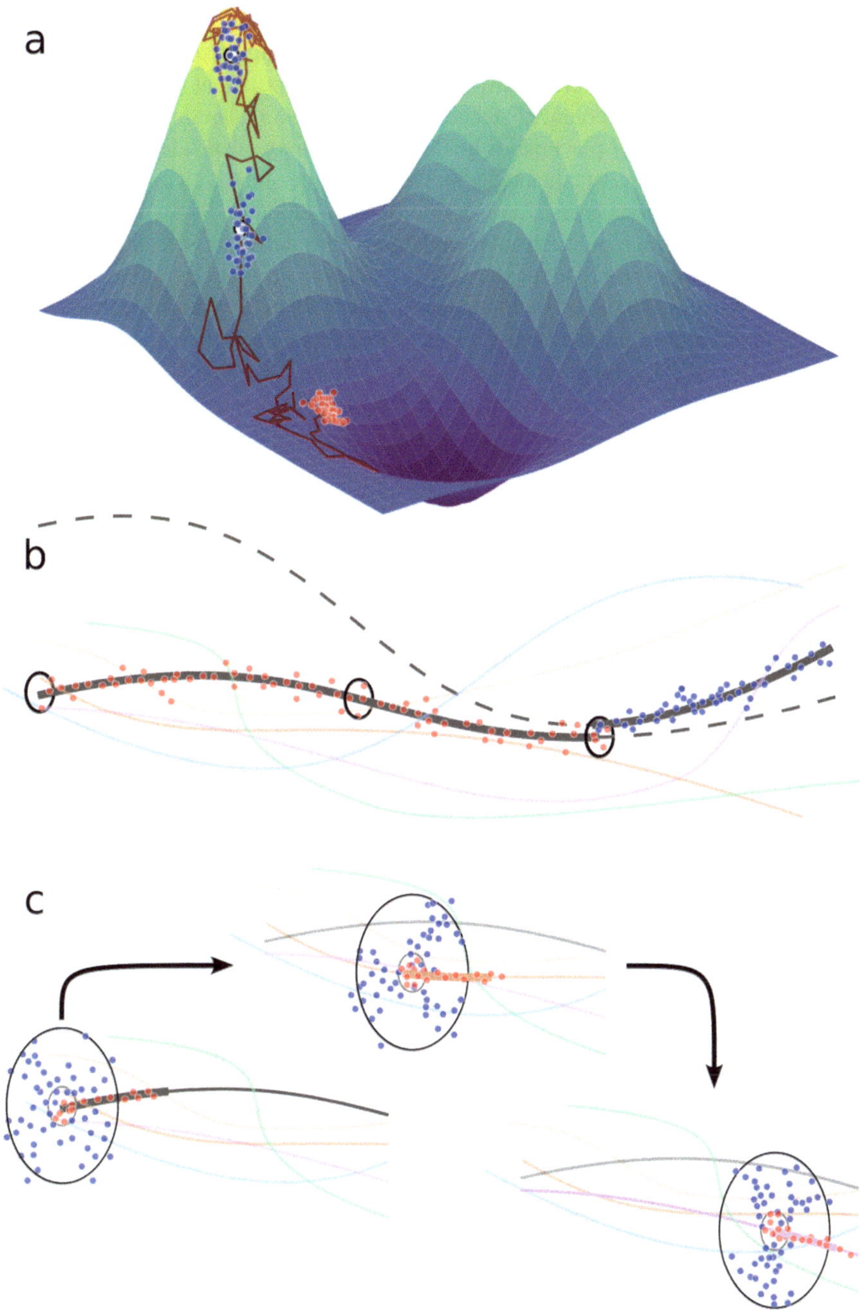

Fig. 26.2 Historical illustration of evolving viral quasispecies. Our understanding of how viral quasispecies explore sequence space has evolved over time as empirical evidence has accumulated. (**a**) In early fitness-landscape models, evolutionary dynamics were primarily described in terms of fitness increases. A simulated adaptive walk (brown curve with sampled red and blue points) illustrates this view. A viral population is assumed to start at low fitness, with individual variants independently sampling the local landscape. Fitter variants are selectively amplified, allowing the

reconstructed around the new center. Figure 26.3 illustrates the uncommon instance when the population transitions between two dominant haplotypes.

These observations illustrate how heterogeneous populations explore sequence space and, with it, new potential phenotypes: Little use of neutral paths seems to occur, with sequential displacements being apparently more dependent on mutations with at least some adaptive advantage. At the same time, almost all possible variants one- and two-mutations away, and a significant fraction at larger distances, are generated by a massive production of mutants before purifying selection can eliminate low-fitness variants. Therefore, fitness "valleys" no longer confine the population to high-fitness variants. In this scenario, the concept of a fitness peak becomes diffuse: Population dynamics only see a coarse representation of the landscape, where the fitness of individual sequences is no longer the unique determinant of their fate—whether maintained or discarded. This provides an alternative view of quasispecies as units of selection, since their structure and dynamics are in mutual feedback to guarantee persistence.

6 Conclusions

Evolution and adaptation occur primarily at a local level in sequence space. The astronomical size of sequence spaces, even of the subsets of genotypes mapping to the same phenotype, strongly suggests that this cannot be otherwise. Evidence for limited displacements in these vast spaces since functional solutions were first located is provided by gene phylogenies that can be traced hundreds of millions of years back in time: Long excursions through neutral drift are not essential for adaptation.

Fig. 26.2 (continued) population as a whole to climb toward higher fitness peaks. Stochastic effects may occasionally enable escape from local optima. **(b)** When most mutations are nonviable or incur large fitness losses, the simple fitness-ascent picture becomes problematic. This motivated a shift in emphasis from fitness progression to neutral navigation. Neutral networks (colored curves) span genotype space, connecting widely different sequences that produce the same phenotype through successive neutral mutations. A quasispecies can drift along such a network (thick black curve and red dots) without loss of fitness, while continually sampling nearby phenotypes. Occasionally, a previously distant neutral network (dashed black curve) comes into close proximity, offering access to higher fitness. The population then shifts to and explores this new network (blue dots). **(c)** Recent empirical evidence suggests that viral quasispecies do not primarily rely on neutral network connectivity to navigate fitness landscapes. Instead, they generate a large fraction of all possible variants within a relatively broad mutational neighborhood (outer black circles) around the most abundant sequence, which is itself highly populated (blue dots). Neutral variants (red dots) are explored within the same radius, rather than preferentially. The schematic separation between neutral and non-neutral variation is shown for clarity; in reality, neutral networks are interwoven with non-neutral neighborhoods. Shifts in the dominant (reference) sequence occur frequently (twice in this illustration), after which the statistical structure of the explored neighborhood is rapidly reestablished

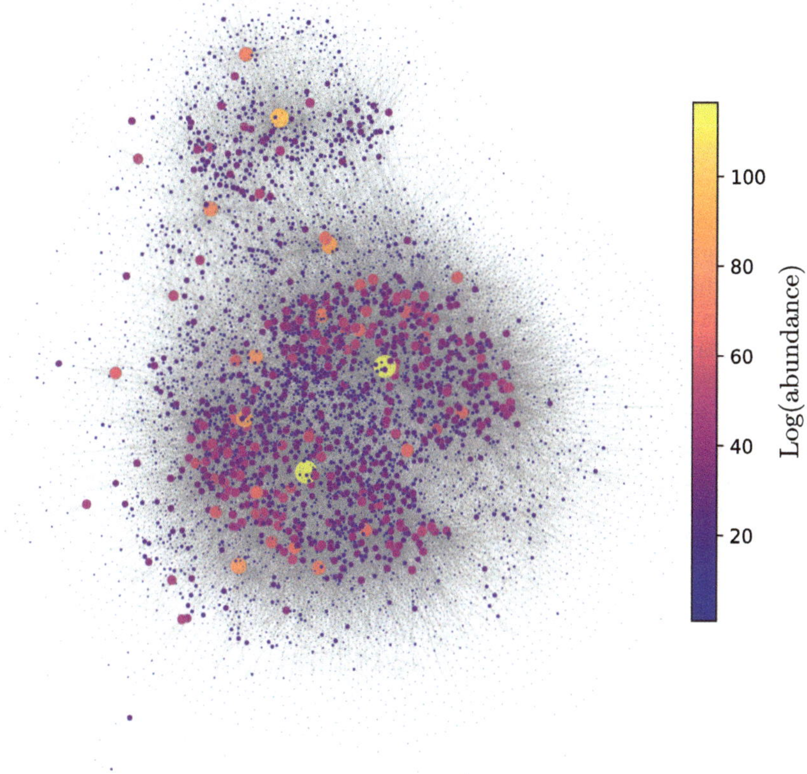

Fig. 26.3 An empirical genotype network. Each node in this network represents a unique sequence (haplotype) of a Qβ phage population; two haplotypes are connected if they differ in just one nucleotide. Data correspond to passage 60 of a population adapting to a temperature of 43 °C. A full description of the deep sequencing data and the curation protocol can be found in Seoane et al. (2026). The node size is proportional to the logarithm of the abundance of the haplotype (the number of sequences identical to that haplotype), and color conveys the same information. The spring layout (which positions nodes by treating them as masses connected by springs and minimizing tensions) suggests community structure around a series of salient nodes. These seem to act as *sources* of viral replicas which, through mutation, populate a neighborhood around each network hub. The self-similar structure of the network is reflected, among others, in a robust power-law degree distribution (Seoane et al., 2026)

Despite this local exploration, the high dimensionality of sequence spaces causes abundant phenotypes to be highly interwoven, with regions of contact that are frequent and occur in many separated regions of sequence space. Consequently, a limited exploration of genotype space does not imply a limited exploration of phenotype space. This conclusion extends to the exploration of fitness space, assuming that the distribution of fitness values has a well-defined average (Catalán et al., 2023)—an assumption supported by the variations in functionality observed in natural systems.

A particularly striking result is that Maynard Smith's notion of a connected protein space, which allows the sustained accumulation of mutations without substantial fitness loss, is not required for the evolution and adaptation of viral populations. Instead, it suffices that phenotype abundances be highly biased, ensuring that frequent phenotypes lie close to one another in sequence space. Viral populations generate and explore all available diversity at short mutational distances from fit, abundant sequences, largely independently of the detailed structure of the underlying fitness landscape. This finding is relevant for understanding the emergence of function in populations of replicators, both within cellular environments (as may have occurred during the emergence of viroids (Catalán et al. 2019)) and in pre-cellular evolutionary contexts (Manrubia, 2022).

We may even entertain the idea that percolation of sequence space by neutral networks and the notion that it facilitates navigation of genotype spaces are merely spandrels of the properties of large phenotypes that are actually relevant for evolution: They exist due to their abundance and persist due to entropic principles. Only a minute fraction of neutral networks is ever visited in evolutionary time, a fact that does not contradict the emergence of molecular function through repeated selection cycles on random sequences: Functional sequences are often just a few mutations away.

Acknowledgements The authors acknowledge the contributions of Jacobo Aguirre, Iker Atienza, Pablo Catalán, Juan Antonio García-Martín, Ester Lázaro, Samuel Martínez-Alcalá, Henry Secaira-Morocho, and Ariadna Villanueva to the development of the unified picture presented here. This work was funded by MICIU/AEI/10.13039/501100011033 and by ERDF/EU (PGE) "A way of making Europe," through grants PID2023-147963NB-C21 (SM), PID2023-153225NA-I00 (LFS), and PID2022-141802NB-I00 (BASIC) (JAC).

References

Agashe D, Sane M, Singhal S (2023) Revisiting the role of genetic variation in adaptation. Am Natural 202(4):486–502. https://doi.org/10.1086/726012

Aguirre J, Buldú JM, Stich M, Manrubia SC (2011) Topological structure of the space of phenotypes: the case of RNA neutral networks. PLoS ONE 6:e26324

Arias CF, Catalán P, Manrubia S, Cuesta JA (2014) toyLIFE: a computational framework to study the multi-level organisation of the genotype-phenotype map. Sci Rep 4:7549

Attwood SW, Hill SC, Aanensen DM, Connor TR, Pybus OG (2022) Phylogenetic and phylodynamic approaches to understanding and combating the early SARS-CoV-2 pandemic. Nat Rev Genet 23(9):547–562. https://doi.org/10.1038/s41576-022-00483-8

Baaijens JA, Aabidine AZE, Rivals E, Schönhuth A (2017) De novo assembly of viral quasispecies using overlap graphs. Genome Res 27(5):835–848

Bloom JD, Romero PA, Lu Z, Arnold FH (2007) Neutral genetic drift can alter promiscuous protein functions, potentially aiding functional evolution. Biol Dir 2:17

Bordería AV, Stapleford KA, Vignuzzi M (2011) RNA virus population diversity: implications for inter-species transmission. Current Opin Virol 1(6):643–648. https://doi.org/10.1016/j.coviro.2011.09.012. https://www.sciencedirect.com/science/article/pii/S1879625711001192. Innate immunity/Antivirals and resistance/Emerging viruses

Braun T, Bordería AV, Barbezange C, Vignuzzi M, Louzoun Y (2018) Long-term context-dependent genetic adaptation of the viral genetic cloud. Bioinformatics 35(11): 1907–1915https://doi.org/10.1093/bioinformatics/bty891

Buchholz PCF, Fademrecht S, Pleiss J (2017) Percolation in protein sequence space. PLOS ONE 12(12):1–11https://doi.org/10.1371/journal.pone.0189646

Carvunis AR, Rolland T, Wapinski I, Calderwood MA, Yildirim MA, Simonis N, Charloteaux B, Hidalgo CA, Barbette J, Santhanam B, Brar GA, Weissman JS, Regev A, Thierry-Mieg N, Cusick ME, Vidal M (2012) Proto-genes and de novo gene birth. Nature 487(7407):370–374

Catalán P, Arias CF, Cuesta JA, Manrubia S (2017) Adaptive multiscapes: an up-to-date metaphor to visualize molecular adaptation. Biol Direct 12:7

Catalán P, Elena SF, Cuesta JA, Manrubia S (2019) Parsimonious scenario for the emergence of viroid-like replicons de novo. Viruses 11:425

Catalán P, Manrubia S, Cuesta JA (2020) Populations of genetic circuits are unable to find the fittest solution in a multilevel genotype-phenotype map. J R Soc Interface 17:20190843

Catalán P, Garcia-Martin JA, Aguirre J, Cuesta JA, Manrubia S (2023) Entropic contribution to phenotype fitness. J Phys A Math Theoret 56(34):345601. https://doi.org/10.1088/1751-8121/ace8d6

Codoñer FM, Darós JA, Solé RV, Elena SF (2006) The fittest versus the flattest: experimental confirmation of the quasispecies effect with subviral pathogens. PLoS Pathog 2:e136

Cowperthwaite MC, Meyers LA (2007) How mutational networks shape evolution: lessons from RNA. Annu Rev Ecol Syst 38:203–230

Cowperthwaite MC, Economo EP, Harcombe WR, Miller EL, Meyers LA (2008) The ascent of the abundant: how mutational networks constrain evolution. PLoS Comput Biol 4:e1000110

Cvijović I, Nguyen Ba AN, Desai MM (2018) Experimental studies of evolutionary dynamics in microbes. Trends Genet 34(9):693–703. https://doi.org/10.1016/j.tig.2018.06.004

Delgado S, Perales C, García-Crespo C, Soria ME, Gallego I, de Ávila AI, Martínez-González B, Vázquez-Sirvent L, López-Galíndez C, Morán F, Domingo E (2021) A two-level, intramutant spectrum haplotype profile of hepatitis C virus revealed by self-organized maps. Microbiol Spectrum 9(3):e01459–21. https://doi.org/10.1128/Spectrum.01459-21. https://journals.asm.org/doi/abs/10.1128/spectrum.01459-21

Dingle K, Schaper S, Louis AA (2015) The structure of the genotype-phenotype map strongly constrains the evolution of non-coding RNA. Interface Focus 5:20150053

Domingo E (ed) (2006) Quasispecies: concept and implications for virology. Springer, Berlin

Domingo E, Sabo D, Taniguchi T, Weissmann C (1978) Nucleotide sequence heterogeneity of an RNA phage population. Cell 13(4):735–744. https://doi.org/10.1016/0092-8674(78)90223-4

Dryden DT, Thomson AR, White JH (2008) How much of protein sequence space has been explored by life on earth? J R Soc Interface 5(25):953–956. https://doi.org/10.1098/rsif.2008.0085. https://royalsocietypublishing.org/doi/abs/10.1098/rsif.2008.0085. https://royalsocietypublishing.org/doi/pdf/10.1098/rsif.2008.0085

Duarte EA, Novella IS, Ledesma S, Clarke DK, Moya A, Elena SF, Domingo E, Holland JJ (1994) Subclonal components of consensus fitness in an RNA virus clone. J Virol 68(7):4295–4301. https://doi.org/10.1128/jvi.68.7.4295-4301.1994. https://journals.asm.org/doi/abs/10.1128/jvi.68.7.4295-4301.1994. https://journals.asm.org/doi/pdf/10.1128/jvi.68.7.4295-4301.1994

Eigen M (1971) Selforganization of matter and the evolution of biological macromolecules. Naturwissenschaften 58:465–523

Eigen M, McCaskill J, Schuster P (1989) The molecular quasi-species. Adv Chem Phys 75:149–263

Erdoğan AN, Dasmeh P, Socha RD, Chen JZ, Life BE, Jun R, Kiritchkov L, Kehila D, Serohijos AWR, Tokuriki N (2024) Neutral drift upon threshold-like selection promotes variation in antibiotic resistance phenotype. Nature Communications 15(1):10813. https://doi.org/10.1038/s41467-024-55012-4

Escarmís C, Lázaro E, Arias A, Domingo E (2008) Repeated bottleneck transfers can lead to non-cytocidal forms of a cytopathic virus: implications for viral extinction. J Mol Biol 376(2): 367–379. https://doi.org/10.1016/j.jmb.2007.11.042. https://www.sciencedirect.com/science/article/pii/S0022283607015136

Ferrare JT, Good BH (2024) Evolution of evolvability in rapidly adapting populations. Nat Ecol Evol 8(11):2085–2096. https://doi.org/10.1038/s41559-024-02527-0

Fontana W (2002) Modelling 'evo-devo' with RNA. BioEssays 24:1164–1177

Fortuna MA, Zaman L, Ofria C, Wagner A (2017) The genotype-phenotype map of an evolving digital organism. PLoS Comput Biol 13(2):e1005414

Franke J, Klozer A, de Visser JAGM, Krug J (2011) Evolutionary accessibility of mutational pathways. PLoS Comput Biol 7(8):e1002134. https://doi.org/10.1371/journal.pcbi.1002134

Garcia-Martin JA, Catalán P, Cuesta JA, Manrubia S (2018) Statistical theory of phenotype abundance distributions: a test through exact enumeration of genotype spaces. Europhys Lett 123:28001

Grant NA, Magid AA, Franklin J, Dufour Y, Lenski RE (2021) Changes in cell size and shape during 50,000 generations of experimental evolution with escherichia coli. J Bacteriol 203(10): 10.1128/jb.00469–20. https://doi.org/10.1128/jb.00469-20. https://journals.asm.org/doi/abs/10.1128/jb.00469-20, https://journals.asm.org/doi/pdf/10.1128/jb.00469-20

Greenbury SF, Johnston IG, Louis AA, Ahnert SE (2014) A tractable genotype-phenotype map for the self-assembly of protein quaternary structure. J R Soc Interface 11:20140249

Greenbury SF, Schaper S, Ahnert SE, Louis AA (2016) Genetic correlations greatly increase mutational robustness and can both reduce and enhance evolvability. PLoS Comput Biol 12: e1004773

Grüner W, Giegerich R, Strothmann D, Reidys C, Weber J, Hofacker IL, Stadler PF, Schuster P (1996a) Analysis of RNA sequence structure maps by exhaustive enumeration. I. Neutral networks. Monatsh Chem 127:355–374

Grüner W, Giegerich R, Strothmann D, Reidys C, Weber J, Hofacker IL, Stadler PF, Schuster P (1996b) Analysis of RNA sequence structure maps by exhaustive enumeration II. Structures of neutral networks and shape space covering. Monatsh Chem 127:375–389

Gu H, Quadeer AA, Krishnan P, Ng DYM, Chang LDJ, Liu GYZ, Cheng SMS, Lam TTY, Peiris M, McKay MR, Poon LLM (2023) Within-host genetic diversity of SARS-CoV-2 lineages in unvaccinated and vaccinated individuals. Nat Commun 14(1):1793. https://doi.org/10.1038/s41467-023-37468-y

Harris KB, Flynn KM, Cooper VS (2021) Polygenic adaptation and clonal interference enable sustained diversity in experimental pseudomonas aeruginosa populations. Mol Biol Evol 38(12):5359–5375. https://doi.org/10.1093/molbev/msab248. https://academic.oup.com/mbe/article-pdf/38/12/5359/41714606/msab248.pdf

Huynen MA (1996) Exploring phenotype space through neutral evolution. J Mol Evol 43:165–169

Huynen MA, Stadler PF, Fontana W (1996) Smoothness within ruggedness: the role of neutrality in adaptation. Proc Natl Acad Sci USA 93:397–401

Jones LR (2025) Intra-host variability of SARS-CoV-2: patterns, causes and impact on COVID-19. Virology 603:110366, https://doi.org/10.1016/j.virol.2024.110366. https://www.sciencedirect.com/science/article/pii/S0042682224003908

Jörg T, Martin OC, Wagner A (2008) Neutral network sizes of biological RNA molecules can be computed and are not atypically small. BMC Bioinf 9:464

Khatri BS, McLeish TCB, Sear RP (2009) Statistical mechanics of convergent evolution in spatial patterning. Proc Natl Acad Sci USA 106(24):9564–9569

Kim N, Ma J, Kim W, Kim J, Belenky P, Lee I (2024) Genome-resolved metagenomics: a game changer for microbiome medicine. Exp Mol Med 56(7):1501–1512. https://doi.org/10.1038/s12276-024-01262-7

Kimura M (1968) Evolutionary rate at the molecular level. Nature 217:624–626

Kin T, Yamada K, Terai G, Okida H, Yoshinari Y, Ono Y, Kojima A, Kimura Y, Komori T, Asai K (2007) fRNAdb: a platform for mining/annotating functional RNA candidates from non-coding RNA sequences. Nucleic Acids Res 35(suppl 1):D145–D148

Krug J (2019) Accessibility percolation in random fitness landscapes. In: Statistical physics and evolutionary biology. Springer, Berlin. https://arxiv.org/abs/1903.11913. arXiv:1903.11913

Kumawat B, Lalejini A, Acosta MM, Zaman L (2025) Evolution takes multiple paths to evolvability when facing environmental change. Proc Natl Acad Sci 122(1):e2413930121. https://doi.org/10.1073/pnas.2413930121. https://www.pnas.org/doi/abs/10.1073/pnas.2413930121. https://www.pnas.org/doi/pdf/10.1073/pnas.2413930121

Laguna-Castro M, Somovilla P, López-Muñoz V, Pacios LF, Lázaro E (2025) Navigating the fitness landscape: host density, epistasis, and clonal interference drive divergent evolutionary pathways in phage Qβ. Int J Mol Sci 26(18). https://doi.org/10.3390/ijms26189020. https://www.mdpi.com/1422-0067/26/18/9020

Lang GI, Rice DP, Hickman MJ, Sodergren E, Weinstock GM, Botstein D, Desai MM (2013) Pervasive genetic hitchhiking and clonal interference in forty evolving yeast populations. Nature 500(7464):571–574. https://doi.org/10.1038/nature12344

Lauring AS (2020) Within-host viral diversity: a window into viral evolution. Ann Rev Virol 7:63–81. https://doi.org/10.1146/annurev-virology-010320-061642. https://www.annualreviews.org/content/journals/10.1146/annurev-virology-010320-061642

Lipman DJ, Wilbur WJ (1991) Modelling neutral and selective evolution of protein folding. Proc Roy Soc London B 245:7–11

Louis AA (2016) Contingency, convergence and hyper-astronomical numbers in biological evolution. Stud Hist Philos Sci C 58:107–116

Manrubia S (2022) The simple emergence of complex molecular function. Phil Trans R Soc A Math Phys Eng Sci 380(2227):20200422. https://doi.org/10.1098/rsta.2020.0422

Manrubia S, Cuesta JA (2015) Evolution on genotype networks accelerates the ticking rate of the molecular clock. J R Soc Interface 12:20141010

Manrubia S, Cuesta JA (2017) Distribution of genotype network sizes in sequence-to-structure genotype-phenotype maps. J R Soc Interface 14:20160976

Manrubia S, Cuesta JA, Aguirre J, Ahnert SE, Altenberg L, Cano AV, Catalán P, Diaz-Uriarte R, Elena SF, Garcia-Martin JA, Hogeweg P, Khatri BS, Krug J, Louis AA, Martin NS, Payne JL, Tarnowski MJ, WeißM (2021) From genotypes to organisms: state-of-the-art and perspectives of a cornerstone in evolutionary dynamics. Phys Life Rev 38:55–106. https://doi.org/10.1016/j.plrev.2021.03.004

Martin NS, Ahnert SE (2022) Fast free-energy-based neutral set size estimates for the rna genotype-phenotype map. J R Soc Interface 19(191):20220072. https://doi.org/10.1098/rsif.2022.0072

Martin NS, Camargo CQ, Louis AA (2024) Bias in the arrival of variation can dominate over natural selecti on in richard dawkins's biomorphs. PLOS Comput Biol 20(3):1–31. https://doi.org/10.1371/journal.pcbi.1011893

Maynard Smith J (1970) Natural selection and the concept of a protein space. Nature 225:563–564

Nowak S, Krug J (2013) Accessibility percolation on the hypercube and the phase transition of greedy adaptive walks. Europhys Lett 101:66004. https://doi.org/10.1209/0295-5075/101/66004

Ohta T (1973) Slightly deleterious mutant substitutions in evolution. Nature 246:96–98

Piacentini L, Fanti L, Specchia V, Bozzetti MP, Berloco M, Palumbo G, Pimpinelli S (2014) Transposons, environmental changes, and heritable induced phenotypic variability. Chromosoma 123(4):345–354. https://doi.org/10.1007/s00412-014-0464-y

Pigliucci M (2012) Landscapes, surfaces, and morphospaces: what are they good for? In: Svensson E, Calsbeek R (eds) The adaptive landscape in evolutionary biology, p 26. Oxford University Press, Oxford

Schaper S, Louis AA (2014) The arrival of the frequent: how bias in genotype-phenotype maps can steer populations to local optima. PLoS ONE 9:e86635

Schmiegelt B, Krug J (2023) Accessibility percolation on cartesian power graphs. J Math Biol 86: 46. https://doi.org/10.1007/s00285-023-01882-z

Schultes EA, Bartel DP (2000) One sequence, two ribozymes: implications for the emergence of new ribozyme folds. Science 289:448–452

Schuster P, Sigmund K (1983) Replicator dynamics. J Theor Biol 100:533–538

Schuster P, Swetina J (1988) Stationary mutant distributions and evolutionary optimization. Bull Math Biol 50:635–660

Schuster P, Fontana W, Stadler PF, Hofacker IL (1994) From sequences to shapes and back: a case study in RNA secondary structures. Proc Roy Soc London B 255:279–284

Seoane LF, Secaira-Morocho H, Somovilla P, Lázaro E, Manrubia S (2026) Hierarchical genotype networks and incipient speciation in $Q\beta$ phage quasispecies. Proc Natl Acad Sci 123(14) e2512150123. https://doi.org/10.1073/pnas.2512150123

Sinzelle L, Izsvák Z, Ivics Z (2009) Molecular domestication of transposable elements: from detrimental parasites to useful host genes. Cellular Mol Life Sci 66:1073–93. https://doi.org/10.1007/s00018-009-8376-3

Sun H, Liu H, Zhang J, Qu X, Pang Z, Xu F, Wu C, Jiang Y, Shi M, Liu Q, Liao M (2025) Genome-scale evolution and phylodynamics of swine influenza A viruses in China: a genomic epidemiology study. Lancet Microbe 6(6):101020. https://doi.org/10.1016/j.lanmic.2024.101020

Symons RH (1981) Avocado sunblotch viroid: primary sequence and proposed secondary structure. Nucleic Acids Res 9(23):6527–6537. https://doi.org/10.1093/nar/9.23.6527

Szendro IG, Schenk MF, Franke J, Krug J, de Visser JAGM (2013) Quantitative analyses of empirical fitness landscapes. J Stat Mech Theory Exp 2013(01):P01005. https://doi.org/10.1088/1742-5468/2013/01/P01005

Vakirlis N, Carvunis AR, McLysaght A (2020) Synteny-based analyses indicate that sequence divergence is not the main source of orphan genes. eLife 9:e53500. https://doi.org/10.7554/elife.53500

Vignuzzi M, Stone JK, Arnold JJ, Cameron CE, Andino R (2006) Quasispecies diversity determines pathogenesis through cooperative interactions in a viral population. Nature 439(7074): 344–348. https://doi.org/10.1038/nature04388

Villanueva A, Secaira-Morocho H, Seoane LF, Lázaro E, Manrubia S (2022) Genotype-to-protein map and collective adaptation in a viral population. Biophysica 2(4):381–399. https://doi.org/10.3390/biophysica2040034. https://www.mdpi.com/2673-4125/2/4/34

Viver T, Conrad RE, Rodriguez-R LM, Ramírez AS, Venter SN, Rocha-Cárdenas J, Llabrés M, Amann R, Konstantinidis KT, Rossello-Mora R (2024) Towards estimating the number of strains that make up a natural bacterial population. Nat Commun 15(1):544. https://doi.org/10.1038/s41467-023-44622-z

Wachowius F, Porebski BT, Johnson CM, Holliger P (2021) Emergence of function from single RNA sequences by darwinian evolution. bioRxiv. https://doi.org/10.1101/2021.03.03.433769

Wilke CO, Wang JL, Ofria C, Lenski RE, Adami C (2001) Evolution of digital organisms at high mutation rates leads to survival of the flattest. Nature 412:331–333

Wolff A, Krug J (2009) Robustness and epistasis in mutation-selection models. Phys Biol 6(3): 036007

Wright S (1931) Evolution in Mendelian populations. Genetics 16:97–159

Yona AH, Alm EJ, Gore J (2018) Random sequences rapidly evolve into de novo promoters. Nat Commun 9(1):1530

Zagorski M, Burda Z, Waclaw B (2016) Beyond the hypercube: evolutionary accessibility of fitness landscapes with realistic mutational networks. PLoS Comput Biol 12:e1005218

Zanini F, Puller V, Brodin J, Albert J, Neher RA (2017) In vivo mutation rates and the landscape of fitness costs of HIV-1. Virus Evol 3(1):vex003

Chapter 27
Nongenetic Mechanisms of Animal Evolution

Nelson R. Cabej

Abstract Nongenetic information is a form of biological information, alongside genetic information, and is in a close, complementary functional relationship with it. Everything we attribute to nongenetic information is ultimately implemented via genes, and, in a certain sense, genetic information is the raison d'être of nongenetic information. The discovery of DNA as the carrier of genetic information was one of the two greatest successes of natural sciences in the twentieth century. This success was so spectacular that it overshadowed several discoveries of heritable changes not involving genes or DNA nucleotide sequences, such as cortical inheritance (mid-60 s) and heritable epigenetic modifications in DNA (mid-70 s). Over time, other factors affecting inheritance across all living kingdoms, independent of DNA sequence, have been discovered, leading to the introduction of the overarching concept of nongenetic information two to three decades ago. As this chapter focuses on nongenetic information in animals, other organismal groups are considered out of scope.

The evolution of animals is a historical process, of which we observe only the current stage. Although a direct observation of past events is impossible, the past is not indelibly lost; previous stages left lasting imprints in the physical fossil record, in the embryonic and post-embryonic development. These are the objects of biological disciplines such as comparative anatomy, paleontology, comparative genetics, molecular biology, and evolutionary theory, which biologists use to reconstruct the evolutionary history of animals.

N. R. Cabej (✉)
Department of Biology, University of Tirana, Albania, Dumont, NJ, USA

G. Witzany (ed.), *Integrative Theory of Evolution - Aspects and Insights*,
https://doi.org/10.1007/978-3-032-24295-2_27

1 Darwin in the Age of Nongenetic Information

In *The Origin* (Darwin 1859), Darwin complained about the misrepresentation of his idea of natural selection: "it has been stated that I attribute the modification of species exclusively to natural selection, I may be permitted to remark that in the first edition of this work, and subsequently, I placed in a most conspicuous position—namely, at the close of the Introduction—the following words: I am convinced that natural selection has been the main but not the exclusive means of modification" (Darwin 1872, p. 421).

Unfortunately, another misrepresentation of Darwin continued even after he died, as the Modern Synthesis, unfairly omitted the principle of "definite," i.e., guided changes in evolution from his biological legacy. The accelerated increase in our time of experimental evidence on the role of nongenetic information in the induction of heritable, "definite" changes in Darwin's expression, in our time validates the accuracy of the Darwinian concept of the role of the definite changes in evolution that was ignored by Neo-Darwinism, while proclaiming the exclusive role of "indefinite" changes (identified with genic mutations).

It is often argued that Darwin said that evolution is based virtually on the accumulation of beneficial "indefinite" changes alone, while "definite" changes are rare or exceptional. For truth's sake, this might have been his initial opinion, but he corrected it in the revised edition of 1872, and in all later editions of *The Origin*, until his death in 1882. He admitted that earlier he had "underrated the frequency and value" of "definite" changes. He put it in plain English, straightforwardly, that he distinguished clearly between the two contrasting heritable variations:

1. "Indefinite" variation, random changes that are small and very rare, hence evolution is necessarily *gradual.*
2. "Definite" variation, or guided changes that commonly lead to the sudden emergence of new traits, which are illustrated herein with adequate experimental and observational evidence.

In the conclusions of the definitive 1872 edition, and consistently in the later publications of *The Origin*, Darwin proclaimed the prominent role of "definite," guided or directed changes, pointing out that "definite," directed changes were often considered to be spontaneous changes due to "our ignorance." With exemplary scientific candor, Darwin admitted having initially "underrated" the "frequency and value" of "definite," directed changes in earlier publications of his work. Recapitulating his theory, he wrote about "the facts and considerations which have thoroughly convinced me that species have been modified, during a long course of descent... by the direct action of external conditions, and by *variations which seem to us in our ignorance to arise spontaneously. It appears that I formerly underrated the frequency and value of these latter forms of variation, as leading to permanent modifications of structure independently of natural selection*" (italics mine! N.R.C.) (Darwin 1872. p. 421). He believed in the role of definite or directed evolutionary changes, and he was so convinced about it that even earlier, in 1868, he put forth his hypothesis of Pangenesis on the mechanism of inheritance of acquired traits (Darwin 1868) via gemmules, small particles released by all organs and parts of the body that accumulate in gametes and transmit parental traits to progeny.

Now we know that Darwin's "indefinite" changes are random changes in the genetic information caused by genic mutations, whereas "definite," directed changes are unmistakably distinct from genic mutations.

2 Biological Information

The concept of information, as one of the fundamental components of the real world, along with matter and energy, including biology, is a relatively new concept that was introduced in natural sciences in 1948 in the last century, after the publication of the works of Claude E. Shannon (1916–2001) and Norbert Wiener (1894–1964).

In biology, the concept of information was formally introduced in 1958 by F.H.C. Crick (1916–2004), when he theoretically argued the existence of the universal genetic triplet code and the unidirectional flow of information from DNA to RNA and proteins. This was known as the Central Dogma of biology, positing that a specific sequence of three nucleotides in DNA determines the amino acid that will be placed in the growing polypeptide chain (Crick 1958), a prediction that was experimentally verified during the 1960s of the last century.

Two distinct types of biological information are presently known: genetic information and nongenetic information, in the sense that the latter is used herein and elsewhere. Both types are unique to the living world.

According to Crick (1958), genetic information has a causal role (Griffiths 2017). This is exhibited in the special relationship between the information sender (DNA) and its receiver (protein-synthesizing machinery), in which the receiver deterministically responds to the received information by appropriately activating transcription and translation to produce proteins. All the intermediate elements between the sender and recipient are sender- or information-dependent. This type of information flow is unique to the living world. Thus, genes can be defined as action-oriented molecules that, under appropriate conditions, guide protein synthesis, although they are inactive by default situations because DNA is tightly wound into nucleosomes within chromatin (Phillips and Shaw 2008).

The discovery of heritable changes in 1975, without changing gene structure (nucleotide sequence) via placement of chemical groups (methyl, acetyl, etc.) in specific DNA nucleotide sequences and nucleosomal histones, led to the formation of the concept of epigenetic information that controls the function (switching on/off) of genes) (Holliday and Pugh 1975; Riggs 1975). After the discovery of the cortical inheritance in mid-60 s, this was a second empirical blow to the theoretical basis of Modern Synthesis that considered genes and DNA as the only carriers of heredity and biological information. Earlier in the 70 s, Niles Eldredge and Stephen Jay Gould (1941–2002) interpreted their fossil findings as proof of the real existence of the alternation of periods of short bursts of rapid evolutionary change and long periods of evolutionary stasis, challenging the conventional idea of gradualism in biological evolution. They published their study in 1972 as a chapter in the multi-authored book *Models in Paleobiology* (1972), which is known as the theory of punctuated equilibria (Eldredge and Gould 1972).

2.1 Nongenetic Information

The Origin of Nongenetic Information in Animals

In Darwin's time, it was impossible to talk about the material basis of heritable changes. From the vantage of the modern view, the Darwinian "indefinite" heritable changes are spontaneous and random genic mutations, whereas "definite," directed, changes in response to environmental changes are nongenetic changes determined by nongenetic information.

For any biologist, the real meaning of "definite" or directed is clear and obvious. A glimpse in English Dictionaries offers several synonyms for the word *definite*: a result that is *inevitable*, *instructed*, *certain*, *guided*, *channeled,* and *certain*. Nongenetic information produces a predictable effect, and, from this viewpoint, Darwin included a teleonomic element in his theory of evolution. The controversy about the possible theological implications of Darwin's concept on definite/directed changes may have been resolved with the introduction by Ernst Mayr (1904–2005) of the concept of teleonomy to describe the phenomenon of the apparent purposiveness of biological processes that are based on inherent sources of genetic information (Mayr 1961, 1974) and the evolutionary acquisition of the ability to generate developmental information. Teleonomy excludes instructions provided by external or supernatural agents as sources of Darwinian "definite" changes but sees apparent purposiveness in the evolution as a result of the historical process of acquisition of the ability to generate nongenetic information by information-generating structures like the cytoskeleton in single-celled organisms and the integrated system control, the centralized nervous system/brain, in animals.

The discovery of the structure and function of DNA as the material substrate of heredity in 1953 by the American biologist James Watson (1928–2025) and the English physicist Francis Harry Compton Crick (1916–2004) (Watson and Crick 1953) and the chemical nature of genic mutations in the early 60 s were seen as the ultimate confirmation of Neo-Darwinism. In the shade of these spectacular successes, other crucial discoveries were neglected like that of the cortical nongenetic inheritance from experiments by Janine Beisson (1931–2020) and Tracy Sonneborn (1905–1981) (Sonneborn and Beisson 1965) and David L. Nanney (1925–2016), who demonstrated that heritable changes in the mating types of the ciliate *Paramecium aurelia* occur independently of genes by nongenetic mechanisms (Nanney 1957). These heritable phenotypic changes challenged the basic tenet of Modern Synthesis that all heritable changes in the living world result from changes in genes or DNA nucleotide sequences.

Now, nongenetic inheritance and nongenetic information are objects of extensive study in all kingdoms of the living world, but most students of nongenetic information still speak of nongenetic information in quite an abstract form, without showing the origin and the material basis of nongenetic information (Cabej 2023).

3 Nongenetic Information Was Necessary to the Evolutionary Process

Conventionally, genic mutations are perceived as spontaneous (or induced), rare, but permanent, virtually irreversible changes in gene DNA sequences that cause random changes in the structure and/or function of relevant proteins. They are nondeterministic in the sense that where and when a mutation will occur in genes are unpredictable. For this reason, they are identified with Darwinian "indefinite" changes. Their frequency, fluctuating between different regions of the genome, is estimated to be ~7.97×10^{-9} mutations in mammals (approximately ~2.5×10^{-8} in humans) per nucleotide site per generation). Most of them are neutral or slightly harmful. Of the total number of mutations, the frequency of beneficial mutations (U_b) is "infinitesimally" low, ~0.0001–0.01 per genome per generation, so that evolution proceeds via the gradual accumulation of these tiny modifications" (Koonin 2009).

Accumulation of beneficial mutations (microevolutionary changes), under the action of natural selection over generations, may lead to the evolution of an adaptive phenotype. But, in cases when the environment reverses to its original state, or changes further, as it often occurs in nature, the evolved adaptation may no longer be beneficial and turn into a liability for the organism. The evolution of the living world from its unicellular beginning was complemented by another form of biological information: the nongenetic information (Cabej 2025). Nongenetic information offers emergency responses to a changed environment until genetic accommodation and genetic fixation may take place.

Nongenetic information enables us to understand mechanisms of rapid evolutionary adaptation in nature that cannot be explained by slow mechanisms of the accumulation of beneficial genic mutations, a conclusion supported by the ever-growing stock of experimental evidence on transgenerational inheritance.

The Darwinian "definite" changes, such as experimental transgenerational changes, emerge as rapid heritable changes involving no changes in genes or genetic information, but rather nongenetic changes in gene expression. They arise within a lifetime and are inherited by the progeny as adaptive responses to changed environmental conditions, implying that another form of information is invested in them for them to occur. The nongenetic information responsible for the heritable phenotypic changes in experimental transgenerational cases is parentally delivered in gametes, mainly in the form of miRNAs secreted by the brain in adaptive response to specific external or internal visual, olfactory, auditory, and electrical signals (in weakly electrical fish).

The transmission of both genetic and nongenetic information suggests the existence of a sender and a receiver of information that induces a change in the receiver, and both show how something can be done in the receiver. Genetic information is fixed in the nucleotide sequences of DNA and shows how a protein can be made. It is inherited by the progeny and is relatively stable, while the nongenetic information is not engraved on any material structure; it is generated as a transient entity in response to a specific stimulus and disappears when the stimulus is no longer sensed, if not embodied in the changed phenotype. Both the induction of "emergency" adaptive responses and the regulation of gene expression are the raison d'être of nongenetic information, and, at the risk of exaggeration, nongenetic information exists for the sake of genetic information (Cabej 2023). Despite the differences, all the effects of nongenetic information are carried out via gene activation/repression. Aphoristically said, genes can but don't know. In their close relationship, genetic and nongenetic information—the former is subordinate to the latter (Cabej 2025).

3.1 In Search of a Fitting Term

The term information in natural sciences, mathematics, physics, chemistry, linguistics, sociology, etc., is used to define different, although related, notions. This situation is sometimes confusing; hence, the need for a more specific term is demanding. Even within the field of biology, there are two distinct types of information: genetic information and nongenetic information that is clearly different in regard to origin, material carrier, and function.

The source of nongenetic information is different in single-celled organisms, plants, and animals, but I'm going to restrict my discussion to the source of nongenetic information in animals. Animals have uniquely evolved the nervous system.

From a linguistic and semantic view, the term nongenetic information is vague and unpromising: instead of defining what this type of information is, it shows what it is not. The word non-, a negating/denying Latin prefix used to give the opposite meaning to the following word in the compound, converts the term genetic into its antonym, something that excludes any genetic meaning. Non means *not* or *without*. In eumetazoans, where we have a relatively solid grasp of nongenetic inheritance and the direction of the flow of nongenetic information, the latter arises from the nervous system. While the "neural information" seems like a fitting term for the animal kingdom, it doesn't apply to other kingdoms of living organisms that lack a nervous system, like single-celled organisms and plants. This limitation led me to propose earlier the term *adaptive information* (Cabej 2025), which aligns better with the adaptive nature of the Darwinian heritable "definite" changes, as opposed to "indefinite" changes. The generally adaptive character of nongenetic changes contrasts with the nonadaptive and unpredictable nature of genic mutations, while being applicable to all kingdoms of living organisms.

For this reason, earlier I proposed to introduce the term adaptive information in agreement with the adaptive character of Darwinian heritable "definite" changes. In the realm of the animal kingdom, where we have a solid understanding of nongenetic inheritance and the flow of information, we see that this nongenetic information arises from the nervous system.

Nongenetic information in animals is delivered in four main forms:

(a) Parental mRNAs provided to gametes, egg and sperm cells, are necessary for early development. The information carried by these mRNAs is nongenetic information. In contrast to genes/DNA, mRNAs cannot replicate, are not part of the genome, are not passed on heritably to the progeny, and are transient nonheritable molecules because they are degraded during development. Nongenetic information in the zygote acts as the sole template for protein biosynthesis necessary for activating expression of zygotic genes during the initial stages of embryonic development, before the activation of the zygotic genome; it continues to play its developmental role beyond the formation of body axes and the phylotypic stage.
(b) Bioelectrical messages, generated impromptu for gene expression/repression in neurons and other somatic cells throughout the animal body (Cabej 2004, 2011, 2020, 2021).
(c) Neural miRNAs deployed to gametes that induce phenotypic heritable changes in animals (Berleth et al. 1988; Driever and Nüsslein-Volhard 1988).
(d) Via epigenetic marks (see Sects. 8.1 and 8.2).

4 The Integrated Control System (ICS) in Animals

Human experience shows that even the simplest man-made automatic devices, such as a thermostat, cannot function without a control system consisting of a temperature sensor and a setpoint to compare the actual temperature with the setpoint and, accordingly, send a signal to turn the switch on or off. Evidently, a living animal or a human being with trillions of cells of hundreds of different functions and structures, as incomparable marvels of complexity, are unimaginable to exist without a control system.

The extremely complex internal environment in animals is subject to continual change, but nonetheless, it remains constant within relatively narrow limits. Obviously, this implies the existence in animals of an integrated control system (ICS) that monitors and maintains thousands of variables of body fluids at a (relative) steady state for various periods of time, from days to many years. It is well known that the organ system that monitors the state of animal organisms, composed of millions to trillions of cells, is the nervous system (CNS) (Brown 1991; Cabej 2004).

The CNS, as the controller of ICS, discriminates between "noise," small changes in variables of the internal and external environments, which it neglects, whereas the changes that exceed species-specific thresholds (setpoints) it takes as stimuli to which it responds adaptively by generating nongenetic information to restore the normal values of variables. Based on the presence of physiological species-specific set points and thresholds, the CNS differentiates between the changes it must respond to and those it must not.

Stimuli exist only as perceived changes in external and internal environments, but false subjective senses may also act as stimuli, inducing stress responses even though they are not related to any material agents of the external and internal environments. As McEwen defined it for humans, stress is "a threat, real or implied, to the psychological or physiological integrity of an individual" (McEwen 1999). Stimuli are species-specific, so that what serves as a stimulus for an animal may be "noise" for another (Cabej 2012, 2019).

The fact that set points are species-specific and fluctuate around a fixed value indicates the existence of species-specific mechanisms that detect deviations of variables beyond a "restricted range" and, most importantly, activate mechanisms that restore the reference (set point) value.

More than 150 years ago, the French physiologist Claude Bernard (1813–1878) introduced the concept of the stability of the internal environment as a condition of life: "La fixité du milieu intérieur est la condition de la vie libre" (Bernard 1865), which came to be known as homeostasis (Cannon 1932).

Homeostasis—the maintenance of a relatively constant internal environment and the ability to restore it despite the continuous erosion of structure, loss of cells, molecules, and supramolecular structures—is an attribute of any living organism, including animals. Evidently, the control system of animals is incomparably more complex than that of single-celled organisms; it must maintain the homeostasis of every individual cell as well as the homeostasis of a multicellular organism composed of a myriad of cells of widely different types. Hence, it is an integrated control system (ICS) required to control, regulate, and coordinate in time and space the functions of millions to trillions of cells. Without this coordination, any living organism would soon descend into chaos and meet a rapid demise. The control of homeostasis in animals is a function of the nervous system (Brown 1991; Cabej 2004, 2012).

The idea of two separate systems, the nervous and endocrine systems, is misleading and is going to become obsolete. The central nervous system (CNS) functions as the controller of the system; being an inseparable part of ICS, the endocrine system, is a subordinate subsystem functioning as a mediator of instructions coming from the CNS; most endocrine hormones are secreted in response to neural signals coming from the central nervous system (CNS) in the form of hypothalamic and pituitary neurohormones via the hypothalamic–pituitary axis.

The pervasive presence of the controller of the ICS throughout the animal body, often at the cell level via the nerve endings (Toró et al. 1969; Chan 1992; Hara-Irie et al. 1996; Hegstrom et al. 1998; Maurin et al. 2004; Magloire et al. 2009; Pirzgalska et al. 2017; Yumoto et al. 2018; Wülfing et al. 2018; Noguchi et al. 2020; Yang et al. 2022), allows it to constantly monitor unavoidable changes taking place in the animal's internal environment at the level of body fluids and cell composition and direct the restoration of normal structure and function. These functions imply that it has information about the normal state and functions of the body to compare it with the actual state. On this basis, it detects changes occurring and generates *nongenetic information* to activate necessary biochemical pathways to restore the normal state (physiological variables in body fluids and structure of the body).

It is noteworthy that adult animals maintain their body mass for relatively long periods of time, even though they lose millions to billions of cells every day. Based on their experiments on planarian regeneration from fragments of hundredth parts of their bodies, biologists have shown that these simple animals are in possession of a "counting mechanism that has the ability to regulate both the absolute and relative numbers of different cell types in complex organs such as the brain during cell turnover, starvation, and regeneration" even when the body size changes markedly (Takeda et al. 2009).

The pervasive presence of nerve endings in the animal body enables neural activity to regulate expression and suppression of genes *not only in neurons* but also in nonneural somatic cells (Klarsfeld and Changeux 1985; Hegstrom et al. 1998; Tang and Goldman 2006; Cohen et al. 2007; Herbst et al. 2021), including the prostate (Magnon 2015; March et al. 2020), the pancreas (Owyang and Logsdon 2004; Thorens 2011), the adrenal gland (Tank et al. 1998; Ishida et al. 2005), the liver (Terazono et al. 2003), anterior pituitary cells (Fontaine et al. 2015) and sexual cells (Henrich and Denlinger 1982; Gapp et al. 2014; Posner et al. 2019; Deshe et al. 2023).

From the current state of knowledge, we can affirm that generation of new nongenetic information (including epigenetic modifications and transgenerational inducing effects of miRNAs, alternative splicing, and cultural evolution (Cabej 2023; 2025) in eumetazoans is a property of neural activity. The nervous system detects changes occurring in the animal organism by comparing the actual state with the norms or physiological set points. It processes the findings obtained to generate nongenetic information in the form of chemical signals to restore the lost order, lost cells, and molecular components of the body fluids (Cabej 2012, 2023).

4.1 Physiological Set Points Reside in the Nervous System

Set points for several variables of the human body are identified in brain areas of mammals, especially in the hypothalamus. One of the best-known set points in mammals, humans included, is that of the regulation of body temperature by neurons of the preoptic area (POA) of the hypothalamus that functions as the thermostat of the animal body (Hammel et al. 1960; Boulant 1998, 2006; Boulant and Hardy 1974), but temperature-sensitive neurons are identified in other brain areas and in the peripheral nervous system. The entirety of these temperature neurons establishes the set point for core body temperature, which is around 37 °C (98.6 °F) in humans.

Examples of experimentally determined set points in the brain abound. For instance, a subset of neurons in the hypothalamus acts as the main center that controls energy balance and thereby body weight/obesity (Affinati and Myers Jr 2021; Johansen et al. 2025). Among other known set points in the hypothalamus are the regulation of water and electrolyte balance in the body fluids. An increase in plasma osmolality is sensed by osmoreceptor neurons situated in the hypothalamus, which respond to this change by stimulating the posterior pituitary to secrete vasopressin. Vasopressin, in turn, stimulates reabsorption of water in the kidneys, thus decreasing water excretion, and also induces the feeling of thirst and drinking of water, thereby restoring normal osmolality (Knepper et al. 2015). The regulation of water and electrolyte balance is thus a function of three cellular levels, with the hypothalamic neural circuits at the top of the hierarchy.

4.2 Gene Regulatory Networks: GRNs

The indispensable role of genes in the development and evolution of animal phenotypes is indisputable. There is abundant evidence that the evolution of new phenotypic traits, as we see them in extant animals, is associated with the appearance of new or changed genes. However, to the best of my knowledge, we yet lack studies that clarify whether the changes preceded or followed the appearance of new traits, the new phenotypic traits. Solid evidence to resolve this problem is still missing, and, consequently, controversies arise when it comes to the role of genes in animal evolution.

Often, it is claimed that new phenotypic traits may arise via repurposing (coopting/recruiting) existing genes in new gene regulatory networks (GRNs), but proponents of the hypothesis have failed to demonstrate the mechanism. Only a neural mechanism has been demonstrated to switch genes in new GRNs. For example, the small crustacean *Daphnia magna* is a parthenogenetic planktonic organism, but in an adaptive response to the shortening of the photoperiod as winter approaches (and other environmental cues), it nongenetically activates a cholinergic system in the brain to induce juvenile hormone (JH) production, switching from parthenogenetic reproduction to sexual reproduction, and then switching back to parthenogenetic generation when conditions improve.

Similarly, in vertebrates, the huge eye GRN is recruited by neural signals from the forebrain and the optic vesicle/neural retina (Reza and Yasuda 2005; Weaver and Hogan 2001; Arresta et al. 2005; Cannata et al. 2008; Rétaux and Casane 2013). Another example of the indispensable nature of neural activity for recruiting the eye GRN during development is the loss within the last ~20,000 years of eyes in cavefish. This occurred even though they possess not only the complete set of crystallin genes and the key gene *Pax6*, but their "eye gene cascades are completely operational in cavefish embryos" (Jeffery 2005).

Neuronal activity, via Gaddb, recruits the epigenetic machinery (ten-eleven translocation (TET) + thymine DNA glycosylase (TDG)) that cooperates in the process of DNA demethylation (Fig. 27.1).

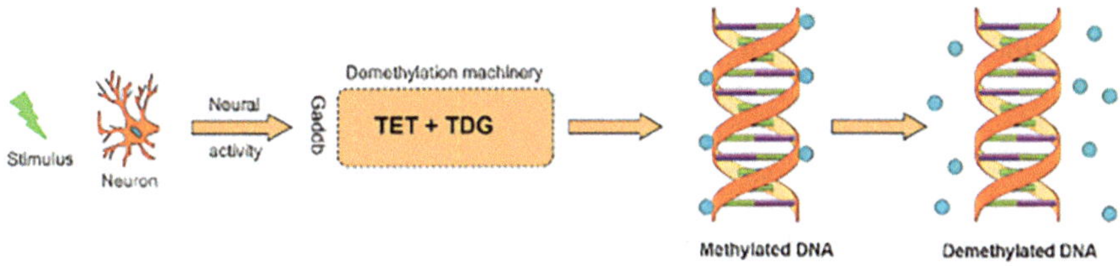

Fig. 27.1 Schematic representation of the flow of information for DNA demethylation in neurons. Neuronal activity in response to external/internal stimuli, via Gaddb, recruits the epigenetic machinery (TET + TDG) that cooperates in the process of demethylation. *Abbreviations*: Gaddb, growth arrest and DNA damage protein b; TDG, thymine DNA glycosylase; TET, ten-eleven translocation enzyme. (From Cabej (2023))

5 Genetic and Nongenetic Information in Animal Evolution

From the onset of the Modern Synthesis, it was assumed that the transition from single-celled organisms to complex animals involved the evolution of an increased number of genes. Consequently, it was predicted that the increased complexity of the structure and functions of animals in the course of evolution led to an increased number of genes. In the 1960s, it was thought that the information necessary to develop a human body could be comparable to one thousand volumes of the Encyclopedia Britannica. Ever since, estimations have kept going down; in the year 2000, it was thought that the number of protein-coding genes in humans was up to 120,000 (Liang et al. 2000), over 100,000 (Collins and Fink 1995) or 120,000 genes (11th International Genome Sequencing 1999).

Now we know that the number of protein-coding genes in the human genome is quite modest, ~ 20,000 and surprisingly smaller than the ~31,000 protein-coding genes of the 0.2 to 3.0 mm long microcrustacean *Daphnia pulex* (Colbourne et al. 2011) or equal to approximately 20,000 protein-coding genes and 24,812 noncoding genes in the genome of the 1 mm long worm, *Caenorhabditis elegans*, composed of just 959 cells in adult hermaphrodite worms, of which one third (302) of whose being neurons (Soares et al. 2017). All this empirical evidence obviously disproves the obsolete idea of a causal relationship between the number of genes and the growth of structural and functional complexity in the animal world. Complexity is related not to the number of genes but to the changed spatiotemporal patterns of gene expression. Among the major factors in the changes of gene expression patterns are nongenetic factors, such as epigenetic marks (DNA methylation, histone modifications) and noncoding RNAs, particularly miRNAs (Lee et al. 1993; Wightman et al. 1993).

The first generation of molecular geneticists believed that "the whole plan of growth. . .the order and the site of synthesis and their coordination are all written down in the nucleic acid message" (Jacob 1970). Now, half a century later, the statement remains unattested, and messages like this remain metaphors, hence, oversimplifications of the extremely complex process of animal development. As H.M. Harold (1995) put it, we don't know how one-dimensional nucleotide sequences can determine living forms in three dimensions.

5.1 The Paradox of the Genetic Toolkit and Gene Regulatory Networks

Three decades ago, Quiring et al. (1994) from the University of Basel, Switzerland, reported that the genes *ey* (eyeless) of *Drosophila*, *Small eye* of the mouse, and human *Aniridia* are encoded by homologous genes (Quiring et al. 1994) despite their long phylogenetic distance. Soon this pioneer study, and other similar studies after it, led evolutionary biologists to the idea of the existence of a common genetic toolkit, a finite common group of genes used for the development of widely different species and higher taxa, hinting at the idea that the evolution of animals depended not on evolving new or modified genes but rather than on a relatively small group of genes of the so-called common genetic toolkit, used in different expression patterns to build the basic body shape (from German Körpergrundgestalt) (Seidel 1960) at the phylotypic stage. The genetic toolkit is used to develop the body plans of widely different animal species and higher taxa that differ substantially in anatomy, behavior, and life histories (King and Wilson 1975).

In attempts to explain this surprising evidence, the American biologists Roy Britten (1919–2012) and Eric H. Davidson (1937–2015) proposed the gene-battery model positing that for each developmental stage a certain set of genes is transcriptionally activated (Britten and Davidson 1969) that came to be known as complex gene regulatory networks/GRNs (Davidson 1990), of coordinately interacting genes, mRNAs (Lee et al. 2002), proteins, especially transcription factors, and several other molecules (Ali and Alrashid 2025). GRNs have a hierarchical structure with clear beginning and terminal states that, consequently, have directionality: each state depends on the previous one (Davidson and Erwin 2006).

It took two decades to empirically identify GRNs in animals (Howard-Ashby et al. 2006; Steinke et al. 2007). This, however, led biologists to the question of the source of nongenetic information necessary for the activation of specific GRNs *at precise times and at precise sites* in the animal body.

5.2 Nongenetic Information Activates Gene Regulatory Networks (GRNs)

It is interesting to point out that the neural origin of signals for the activation of gene regulatory networks (GRNs) was discovered before the experimental identification of a GRN in the 1994 landmark paper by Greenberg and Ziff (1984). It was found that different neuronal patterns of neuronal activity induce distinct patterns of expression of GRN genes (Greenberg and Ziff 1984; Douglas et al. 1988; Worley et al. 1993). Neurally derived nongenetic information controls both the qualitative and quantitative aspects of GRN gene expression during the primary response genes (PRGs) and secondary response genes (SRGs) (Fowler et al. 2011; Tyssowski et al. 2018). The correlation between the duration of neural activity and patterns of gene expression is so strong that, from profiles of expression of GRN genes, biologists can infer the duration of neural activity.

Neural control of the eye GRN is demonstrated in the case of the loss of eyes in the blind cavefish *Astyanax mexicanus* (Mexican tetra). The development of the eyes in this fish and in other animals results from the activation of a complex GRN. It is demonstrated that the lens GRN is activated from signals from neural structures, the retina, and the optic vesicles.

The development of a complete *Drosophila* eye requires the activation of a gene regulatory network (GRN), comprising more than 2500 genes (Halder et al. 1995; Gehring 1996). In vertebrates, despite the large number of genes involved in eye development, within the lens developmental GRN, *Pax6* expression in the optic vesicle, the anterior part of the forebrain, plays the central role as a master gene.

The gene regulatory network for eye development in extant animals is recruited to the lens placode by neural signals released by the optic vesicle/neural retina (Reza and Yasuda 2004; Weaver and Hogan 2001; Arresta et al. 2005; Cannata et al. 2008; Rétaux and Casane 2013). Some biologists believe that the development of the lens in extant animals "recapitulates" its evolution in vertebrates. Summarily, paracrine secretions from the optic vesicle that direct the whole process from the development of the lens placode to the formation of a complete lens are elements of a large, complex GRN, which is responsible for eye development. *Pax6* expression in the optic vesicle, the anterior part of the forebrain, plays the central role in the lens GRN as a master gene.

The above examples and others described in other developmental papers unambiguously show that neural structures and activities deliver the nongenetic information that controls and regulates GRN expression patterns.

It is often observed that major transitions are associated with the appearance of new or changed genes, but we cannot exactly know whether changed or new genes evolved before or after the evolutionary transitions took place. It is believed that this question may be resolved with the help of comparative genomics and the molecular clock, but both have limitations, and even their combination cannot assure that the results will be dependable or trustworthy. In the absence of such studies, the long evolutionary times since transitions took place make it more likely that most new or changed genes associated with them occurred after the transitions. Because of this uncertainty and several other facts, in 2003, the great American evolutionary biologist, Mary Jane West-Eberhard, famously stated: "Genes are followers, not leaders, in evolution" (West-Eberhard 2003). It is more likely that major transitions are related mainly to nongenetic regulatory mechanisms, as is suggested, among other things, by the evidence on the association of major transitions with a remarkable increase in the number of both miRNAs and long noncoding RNAs (lncRNAs).

However, there is evidence that evolutionary transitions of lower taxa such as species, subspecies, or varieties experience phenotypic changes without changes in protein-coding genes or new genes but with changes in the nongenetic manipulation of gene expression (Jarvis and Nottebohm 1997; Strand 1999; Heideman et al. 2001; Anderson and Weir 2022; Morrison 2004; Cabej 2012, 2019). Sibling species commonly have the same protein-coding genes and produce the same proteins, but differ in behavior and ecological niche, suggesting that genes and their protein products are not involved in programming and executing animal behavior. Food depletion leads to the death of the amoeba, and while dying, it releases a glycoprotein that serves as a chemoattractant whose concentration grows with the growing numbers of dead amoebas, leading to the formation of a multicellular, migrating slug-like structure composed of up to 215 amoebas. Under favorable conditions, the "slug" forms a fruiting body, which releases spores that develop into new amoebas (Wang et al. 2021), thus reverting to unicellularity without any changes in genes.

6 Major Transitions

In 1868, the German biologist Ernst Haeckel (1834–1919) put forth the hypothesis that all animals descended from a hypothetical two-layered common ancestor, which he named *gastraea* and introduced the concepts of phylogeny for the species' evolutionary history and ontogeny for the embryonic development. He also formulated the biogenetic law, positing: "Die Keimesentwicklung ist eine gedrängte und verkürzte Wiederholung der Stammesentwicklung" (Ontogenesis is the short and fast recapitulation of phylogenesis) (Haeckel 1866, p. 300). Just two years later, in 1868, the American naturalist Henry James Clark (1826–1873) described the remarkable morphological similarities of single-celled choanoflagellates to the feeding choanocytes of the multicellular sponges (James-Clark 1868; Richter and Nitsche 2017). Now choanoflagellates are demonstrated to be the sister group of animals (Snell et al. 2001) and the closest unicellular organisms to animals (Lang et al. 2002).

The indispensable role of genes in the evolution of animal phenotypes is indisputable; they are providers of genetic information for the synthesis of proteins, the fundamental building blocks of the cell, and the most versatile of the known macromolecules. There is abundant evidence that major transitions and evolution of new phenotypic traits are followed by spontaneous and random gene mutations, but we lack studies that clarify whether the appearance of changes in these particular genes preceded or followed the evolution of new traits. Solid evidence to resolve this problem is still missing, and, consequently, controversies arise when it comes to the role of genes in major transitions and speciation processes.

The transition to animals implies cell differentiation and specialization. The undifferentiated cells of an animal have no incentive to differentiate; it is the multicellular organism that needs new cell types. It generates the relevant impulse and sends extracellular signals to stimulate cell differentiation; the order originates at higher levels of organization (Noble 2016). Additionally, the transition required the evolution of a special apparatus, an integrated control system (ICS), to control the continually changing contents of body fluids and of the extracellular fluid and to maintain homeostasis in general. Genetic factors are essentially involved in the process because there is no other way nongenetic information can affect the phenotype but by activating specific genes and GRNs.

It is consensually admitted that all evolutionary transitions are associated with the appearance of new or changed genes or nucleotide sequences. However, this, like any other correlation or association per se, cannot be taken as proof that these changes in DNA sequence are the cause of an evolutionary change. To verify this, one needs to prove that the new or mutated gene did not occur after the transition (to my best knowledge, such verifications are commonly not done); even if all the correct genes were in place before the transition, how can we determine which of the preexisting genes or GRNs is responsible for the transition? The most reliable way of determining this is by ascertaining the new/changed gene or GRN that appeared after the transition. Again, the only experimentally demonstrated mechanism of repurposing or recruiting GRNs for another use is a neural one (Reza and Yasuda 2005; Weaver and Hogan 2001; Arresta et al. 2005; Cannata et al. 2008; Rétaux and Casane 2013).

The transition to multicellularity was not a unique event in the history of life; it took place repeatedly (Rokas 2008). Depending on the species, different types of cells live for definite periods of time (from days to years), and animals succeed in maintaining, in a quasi-steady state, the number of cells and body mass for relatively long periods of time. It is impossible even to imagine that an animal organism, the most complex structure in the Universe, could survive without an integrated control system (ICS) to coordinate the functioning of the myriad cells it consists of.

Many biologists believe that nongenetic information was equally indispensable to genetic information in all major transitions since the early stages of the evolution of life and preceded the DNA-based genetic system. Two decades ago, E. Jablonka and M.J. Lamb stated: "transmissible non-DNA information not only existed but was a prerequisite for the stages that followed later and culminated in the evolution of a DNA-based genetic system.... information transmitted by nongenetic means has played a key role in the major transitions" (Jablonka and Lamb 2006). Indeed, there is no known genetic mechanism that would guide the precise cytokinesis during cell division in single-celled organisms, but empirical experimental evidence shows that even in prokaryotic microorganisms, this process is regulated by a nongenetic element, the cytoskeleton (Söderström et al. 2016; Du and Lutkenhaus 2017; Du et al. 2018).

6.1 *Transition from Single-Celled Organisms to Animals*

The transition from unicellular microbes to multicellular animals that is believed to have occurred several times around 600 million years ago and led to the evolution of the tremendous diversity of animal forms, from tiny worms like *Caenorhabditis elegans* composed of ~1000 cells (959 in hermaphrodite individuals and 1031 in males) to highly complex mammals, including humans, with billions to trillions of cells of hundreds of different forms and functions.

The unicellular organism, *Capsaspora owczarzaki*, discovered in 1980 (Owczarzak et al. 1980), but formally described and classified two decades later (Hertel et al. 2002), is a protist, a unicellular eukaryote. It, along with choanoflagellates, represents one of the closest relatives of animals. This microorganism with the same genes reversibly transitions from a single-celled stage of an aggregative multicellular organism, consisting of three different cell types over its life cycle (Sebé-Pedrós et al. 2016).

Surprisingly, this amoeba possesses several canonical signaling pathways, such as Wnt, Hedgehog, TGF-beta, Notch, and Hippo pathways, previously believed to be present only in animals (Phillips and Shaw 2008), that play key developmental roles. Like other close unicellular relatives of animals, *C. owczarzaki* has complete core animal Hippo pathway components (Sebé-Pedrós et al. 2012). In this amoeba, the Hippo pathway, via its nuclear transcriptional co-activator protein, determines cell shape, cytoskeletal dynamics, and the 3D structure of cell aggregates, similarly to the role it plays in controlling tissue size in animals.

Cytoplasmic tyrosine kinases, receptor tyrosine kinases, and nonreceptor kinases are among the most important enzymes involved in diverse functions in animal cells, including cell division, growth, metabolism, and cell communication as cell receptor kinases and nonreceptor kinases. They were indispensable for the evolutionary transition to multicellular life and cell adhesion, but kinases existed in unicellulars before the advent of multicellularity: the basic repertoire of metazoan cytoplasmic tyrosine kinases was established before the divergence of unicellular filastereans from the Metazoa and Choanoflagellata clades (Suga et al. 2012), about 600 to 800 million years ago.

Transcriptome data from another close relative of metazoans, the choanoflagellate *Salpingoeca rosetta*, indicate that a number of conserved cytokinesis genes, including septins, were involved in initiating the colony development of the unicellular organism. During colony maturation, genes unique to *S. rosetta* are activated (Fairclough et al. 2013).

Another example of a reversible transition from single-celled to colonial life is seen in the life history of the soil amoeba, *Dictyostelium discoideum* (Wang et al. 2021). Starvation because of food depletion causes the death of amoebas. The dying amoebas release a protein that, in turn, serves as an attractant for live amoebas to aggregate and form a migrating, slug-like structure. When food is replenished, the fruit body releases spores that develop into single-celled *Dictyostelia*.

Two essential nongenetic phenomena are observed in *Dictyostelia* phase transitions:

First, no changes in genes or new genes are involved in the reversible transitions from the unicellular to the multicellular stage of the amoeba, and

Second, the transition of *Dictyostelia* into the slug-like multicellular structure is correlated with the epigenetic opening of the chromatin structure to allow the binding sites of two genes to be accessed by their respective transcription factors (Wang et al. 2021).

Sphaeroforma arctica is another unicellular eukaryote and a close relative of animals that continues synchronous nuclear divisions, sharing a common cytoplasm until its divisions form a multicellular structure of 128 nuclei (Ondracka et al. 2018) before being cellularized. Cellularization of this organism again relies on a nongenetic mechanism involving both components of the cell skeleton, microtubules and the actin cytoskeleton, in a process of division of labor in which microtubules regulate the positioning of actin filaments, with the latter forming plasma membrane invaginations at the site of the actin cytoskeleton for cellularization of coenocytic nuclei. This specific division of labor is validated by experiments in which depolymerization of microtubules in the presence of actin filaments prevents cellularization, while depolymerization of microtubules alone does not prevent the cellularization of *S. arctica* (Dudin et al. 2019). It is true that an equally important role for the cellularization of the amoeban cenocyte is played by the Arp2/3 complex, which is another nongenetic compound formed of several proteins.

6.2 Transition from Asexual Reproduction to Dioecy

The onset of dioecy in some animal lineages preceded the Cambrian explosion (542 million years ago), and it evolved repeatedly, especially in the plant kingdom. Since sexual organs don't fossilize, we do not know exactly how many times sexuality in animals evolved.

That the transition from a parthenogenetic to a sexual mode of reproduction did not need the "invention" of new genes or changes in DNA nucleotide sequences, but the repurposing of existing genes is indicated by a well-known developmental and life history fact: under normal conditions, *Daphnia* reproduce asexually, but under conditions of increased density of the conspecific population, depleted food resources, and declining photoperiods, with the approach of winter, perceived respectively in the olfactory and photosensitive centers of the brain. The minute cladoceran responds adaptively to these environmental changes by activating signal transduction pathways and genes/GRNs to induce phenotypic changes necessary for the adaptive transition of its life history strategy. Thus, transitioning to the dioecious stage, the production of durable resting eggs (with ephippia, or encased chitin shells) allows it to survive cold and drought. The fact that *Daphnia* can reversibly transition to asexual and sexual reproduction without genetic changes strongly suggests that the evolutionary transition did not require genetic changes.

6.3 Transition to Bilaterians

With the evolution of cephalization and predation during the Cambrian explosion, primitive animals like sessile sponges and radially symmetrical cnidarians, as well as formless, slowly gliding placozoa, became vulnerable to predators, and a strong selective pressure arose for the evolution of structures like appendages, limbs, wings, etc., that would make animals capable of directional locomotion and offer more maneuverability. The transition was so successful that 99% of extant animal species are bilaterally symmetrical (Holló and Novák 2012).

The great transition to bilaterians also did not require the evolution of new or changed genes. On the contrary, studies show that a considerable fragment of the Urbialterian genes was lost in different animal lineages (Matsui et al. 2009). At least three decades ago, biologists understood that in macroevolution, "changes in the cis-regulatory systems of genes more often underlie the evolution of morphological diversity than do changes in gene number or protein function" (Carroll 2000; Davidson and Erwin 2006). The fact that no new genetic information was invested to produce phenotypic changes related to the evolution of bilateral body plans evidently indicates that nongenetic information of some type was used to evolve bilaterality and to accelerate the functional and structural complexity of bilaterians. But nongenetic information has no other way of inducing evolutionary changes than via genes, by adaptively changing patterns of expression of genes of the genetic toolkit and other preexisting genes. Thus, the transition to the bilateral body plan was performed nongenetically by changing patterns of expression of preexisting genes during the Cambrian explosion, which was made possible with the evolution of the centralized nervous system. Ample experimental and observational evidence shows that adaptive changes in expression patterns of genes are often a function of nongenetic information generated in the nervous system (Wayman et al. 2008; Eacker et al. 2013; Kiltschewskij and Cairns 2019).

In the search for the cause of the nongenetic evolution of transition to bilaterality, a correlation is observed between the beginning of the Cambrian explosion and the advent of Bilateria. The fact that no new or changed genes were involved in this transition evidently indicates that the cause is nongenetic. The transition also coincided with the centralization of the nervous system (CNS), while animals with nerve nets instead develop only radial symmetry (Cnidaria) or biradial symmetry (Ctenophores). It is noteworthy that the radial symmetry of cnidarians and the biradial symmetry of ctenophora reflect the respective radial and biradial symmetries of their neural nets. And the fact that both these phyla represent evolutionary "dead ends" as opposed to the brainy bilaterians seems to corroborate the idea that the nongenetic information necessary to produce the enormously complex and diversified bilaterian body plans may be generated by the CNS (Cabej 2004, 2011, 2012, 2019). The centralization of the nervous system in bilaterians significantly increased the computational capability of the brain, and by shortening the distances neuronal signals had to travel between neurons, saved energy (Gavilán et al. 2016). Indeed, a clear correlation is shown to exist in the animal world between the evolution of the nervous system and the degree of anatomical complexity of animal groups. Animals that have no neurons (placozoa and sponges) have no definite body plans. Cnidaria and Ctenophora, which evolved neurons and the neural net, but not ganglia or brains, remained "dead ends" of evolution, and only "brainy" Bilateria evolved the anatomically most complex and most diversified body plans that now dominate the animal kingdom. The correlation between the evolution of the brain and bilateral body plans may not have been a lucky coincidence.

The centralization of the nervous system amplified the computational capacity of bilaterians, making possible the transition from random drifting of cnidarians and crawling locomotion to placozoans to directed locomotion, thereby increasing their ability to explore and profitably use their environment, to assign positive or negative valence to environmental cues, and accordingly make informed decisions on whether to move in the direction of the cue or avoid it (Redondo et al. 2014).

7 Nongenetic Mechanisms of Major Developmental Transitions in Eusocial Insects

Animal eusociality implies the presence within a species of distinct castes with distinct morphologies, behaviors, and life histories. They display a reproductive division of labor, cooperative brood care, multigenerational cohabitation, and altruistic behavior (Lebedev et al. 2024). Eusociality is a widespread phenomenon in hexapods, including insects, especially in the large order of Hymenoptera, where it evolved 11 times (Wilson and Hölldobler 2005), which, from a Neo-Darwinian view, would have required an unlikely high frequency of an unknown beneficial "gene for eusociality" 11 times. Individuals of colonies of various eusocial insects of the order Hymenoptera, such as bees, ants, and wasps, share the same genetic constitution but develop highly diversified morphologies, behaviors and perform specialized functions for the colony.

Nongenetic factors are found to be indispensable in the development of animal eusociality. So, e.g., secretory neurons of the pars lateralis, by processing and interpreting specific internal and external cues, function as a command center for the synthesis and fine-tuning the production of juvenile hormone (JH). The command center projects to the corpora allata and via nerve endings releases neurohormones, allatostatins/allatotropins, respectively inhibiting and stimulating JH production. This is corroborated by the fact that the experimental use of JH analogs in damp-wood termites, *Zootermopsis nevadensis*, led to the production of a new intermediate caste, featuring both soldier and winged morphology, thus changing the species' caste structure (Miura et al. 2003).

Sex of individual hymenopterans depends on the number of chromosomes they inherit (Cowan and Stahlhut 2004). Fertilized eggs in hymenopterans develop into females, and from unfertilized eggs, males hatch, but other nongenetic factors, such as DNA methylation, histone modifications, and noncoding RNAs, are involved in the process (Sieber et al. 2021).

Pheidole megacephala has a winged queen caste, two wingless worker castes (major and minor workers), and one wingless soldier caste. Formation of castes in this ant is related to neurally induced secretion of JH. A first pulse in the JH level induces the formation of mesothoracic wing disks in both queens and workers, and a second pulse brings about the formation of wings in major workers but not in minor workers (Sameshima et al. 2004; Wheeler and Nijhout 2003).

In stingless bees of Melipona genus both workers and queens are born with the diploid set of chromosomes but the development into a queen or a worker is characterized by different patterns of gene expression, indicating that caste determination depends on the regulatory mechanism of gene expression. The caste determination in Melipona, as well as in other social bees, is function of endocrine developmental switches (Pinto et al. 2002) and these *JH and ecdysteroid switches are under neural control of neural mechanisms activated in response to the nature of the diet*. The feminizer gene (*Mi-fem*) mRNA in Melipona is sex-specifically spliced in which only the female splice variant encodes the full-length protein. It is observed that fem levels in pupae and young adults were significantly higher in queens than workers and males, but ectopic application of juvenile hormone (JH) at the time of divergence of worker and queen phenotypes turns 100% of females into queens from a normal rate of 10% (Kärcher et al. 2013), indicating the feminizing role of JH. Production and fine-tuning of JH are neurally controlled via two neuropeptides with opposing effects, allatostatins and allatotropins, respectively inhibiting and stimulating JH secretion by the corpora allata, situated behind the insect brain.

8 Conveyors of the Nongenetic Information

8.1 Neural Activity Determines Epigenetic Modifications in DNA

Animal cells are in possession of a DNA methylation machinery consisting of three types of DNA methyltransferases: DNMT3a and DNMT3b, which form de novo DNA methylation (Robert et al. 2003), and DNMT1, which adds a methyl group to the unmethylated DNA strand during DNA replication (Stricker and Götz 2018). The synthesis of DNMTs is regulated by neuronal activity (Levenson et al. 2004; Lubin et al. 2008; Sharma et al. 2008; Guo et al. 2011; Oliveira et al. 2012; Morris and Monteggia 2014).

DNA methylation is a major contributor to gene expression by typically repressing it at the transcriptional level (Siegfried and Cedar 1997). The deployment of epigenetic marks on DNA is observed in the case of the expression of immediate early genes (IEGs) in neurons during learning, when the neuronal activity precisely targets the promoters of those genes, causing double-strand breaks (DSBs) at IEG promoters, thus relieving them from insulation by CCCTC-binding factors (CTCF) and exposing them to their enhancers and transcription factors. This makes promoter–enhancer interaction possible, thereby enabling IEG transcription (Dileep and Tsai 2021; Madabhushi et al. 2015; Stott et al. 2021; Watson and Tsai 2017), preceded by neurally induced chromatin opening (Su et al. 2017).

Neural activity also regulates the process of DNA demethylation to enable the expression of repressed genes (Fig. 27.1) (Cabej 2023) from DNA, in a process that involves the neurally induced ten-eleven translocation enzyme (TET) (Tahiliani et al. 2009) and thymine DNA glycosylase (TDG), which excises 5-carboxylcytosine (5caC) from a cytosine-derived site (Cortellino et al. 2011).

It is demonstrated that the deployment of epigenetic marks in neuronal DNA requires neural activity, and in its absence, the deployment cannot take place (Champagne 2008; Meaney and Szyf 2005; Ma et al. 2009; Day et al. 2013; Jarome and Lubin 2014; Gupta-Agarwal et al. 2014; Pearce et al. 2017). The sense of the flow of information for gene suppression (DNA methylation) and expression (demethylation) in the figure illustrates that epigenetic marks are conveyors rather than authentic carriers of nongenetic information for DNA demethylation.

The evidence about the role of neuronal activity in controlling the brain epigenome is overwhelming. Unambiguously, it shows that the correlation between neuronal activity and the onset of DNA methylation and demethylation is causal rather than accidental; no wonder it is stated that "neuronal activity shapes the brain's epigenome" (Yap and Greenberg 2018).

8.2 Neural Activity Determines Epigenetic Modifications in Chromatin

An important aspect of epigenetic modifications is the deployment of various chemical groups (mainly methyl and acetyl groups) in nucleosomal histones. Nucleosomes consist of a nucleus of eight protein molecules of four types (H2A, H2B, H3, H4), wrapped around by a ~ 147 base pairs of double-stranded DNA. Nucleosomes are the elementary units of chromatin. In nucleosomes DNA is compacted and in default state it is repressed, inaccessible to transcription factors (TFs), but the nucleosome structure is dynamic; acetylation, the addition of an acetyl group (-$COCH_3$), to nucleosomal histones, exposes DNA to transcription factors to induce gene expression; the "opening" of chromatin to TFs in neurons is reversible, and deployment of methyl (CH_3), and other organic groups can reverse chromatin to the default repressive state.

Because of the coiling of DNA around nucleosomes and the tight binding of the positive charges of histones with the negative charges of DNA, making the latter inaccessible to (TFs), the default state of DNA in nucleosomes is repressed. In order for gene expression to take place, the demethylation of genes requires chromatin to "open" and become accessible to TFs, and gene expression (Geiman and Robertson 2002). Histone acetylation releases nucleosomes from DNA (Phillips and Shaw 2008).

Essentially involved in the opening of the chromatin structure are chromatin remodelers that use adenosine triphosphate (ATP) energy to change histone-DNA interactions and induce adaptive changes in chromatin. To this group of enzymes belong histone acetyltransferases (HATs) that, by adding an acetyl group to histones, relax wound DNA tightly and expose genes to TFs to induce gene expression. Histone deacetylases (HDACs) tighten the winding of DNA around the nucleosomal histone octamer. As chromatin remodelers, a few enzyme complexes [switch/sucrose nonfermentable (SWI/SNF), inositol-requiring 80 (INO80), imitation switch (ISWI) and chromodomain-helicase DNA-binding protein (CHD)] also serve to slide nucleosomes along the DNA to clear them from the promoter regions (Cosgrove et al. 2004).

Neural activity induces histone acetylation and gene expression (Tang et al. 2009; Méjat et al. 2005). Also, novel taste learning (Swank and Sweatt 2001) and contextual fear conditioning (Levenson et al. 2004) induce histone acetylation and chromatin remodeling in neurons, whereas denervation leads to histone deacetylation by the accumulation of histone deacetylases (HDAC) in chromatin (Cohen et al. 2009; Kawano et al. 2015). Neuronal activity determines adaptive changes not only in neuronal chromatin but in nonneural cells like glial cells of the CNS, nonneural Schwann cells, and the peripheral endothelial cells.

There is experimental evidence that neural activity or innervation is involved in the development of almost all tissues and organs: *muscles* (Hegstrom et al. 1998; Brancaccio and Palacios 2015; Sarkissian et al. 2016; Fernandes and Keshishian 1998), *blood vasculature* (Mukouyama et al. 2005; Li et al. 2013), *bones* (Zaidi 2005; Zaidi et al. 2018; Abeynayake et al. 2021), *eye development* (Graw 2010), *sex determination* (Willingham et al. 2000; Milnes Jr et al. 2002; Scholz et al. 2006), *lung development* (Pan et al. 2006; Cutz et al. 2007; Bower et al. 2014), *pituitary* (Takuma et al. 1998; Treier et al. 2001; Osmundsen et al. 2017), *pancreas* (Nekrep et al. 2008; Borden et al. 2013; Moullé et al. 2019), *thymus* (Kendall and al-Shawaf 1991; Bowden et al. 2005; Wülfing and Günther 2015), inner ear (Brigande et al. 2000; Bok et al. 2005).

8.3 Neural Activity Regulates miRNA Biogenesis and Expression

Less than 2 percent of the genes of the human genome are protein-coding genes. Of the total of 19,467 annotated protein-coding genes in humans, only 17,470 proteins are evident (Omenn et al. 2019; Kohshour et al. 2025). The rest of the genome is represented by regulatory segments such as promoters and enhancers, but also by DNA sequences coding for various forms of short noncoding RNAs (sncRNAs) of several types, such as circRNAs, miRNAs, dsRNAs, rRNAs, and long noncoding RNAs. The DNA sequences that do not code for proteins, which were previously considered to be functionless or "junk DNA," turned out to be responsible for a new, important regulatory level of gene expression. For our purpose herein, we are going to only succinctly discuss a single type of sncRNA, microRNAs (miRNAs), focusing on the nongenetic control of their biogenesis and their regulatory role in gene expression.

miRNAs are short, single-stranded RNAs consisting of 22–23 nucleotides (nt) encoded from extragenic DNA segments, or by a combination of mRNAs of other genes. They are transcribed in the cell nucleus in the form of pri-miRNA and processed there by the microprocessor complex, consisting of the RNAse II/III endonuclease, Drosha and double-stranded RNA-binding protein (DGCR8), into a shorter ~70 nt pre-miRNA that is exported to the cytoplasm, where the enzyme Dicer produces a mature miRNA. Thus, neuronal activity regulates the key enzymes of RNA biogenesis, RNA endonucleases II and III (Gonçalves et al. 2018) and Dicer (Sim et al. 2014), which remove the passenger strand from pre-miRNA to produce the mature ~22 nt long miRNA, which binds to members of the Ago (Argonaute) and the GW182 family of proteins, with the latter also regulated by neuronal activity (Nawalpuri et al. 2021). The Ago-miRNA complex represents the core of the miRNA-induced silencing complex (miRISC), which binds to target mRNAs to prevent their translation to protein or fine-tune protein biosynthesis by degrading their mRNAs. Different patterns of neuronal activity can interrupt or degrade the RNA-induced silencing complex (RISC).

miRNAs act as switches to regulate the development of the skeletal and cardiac muscles (Callis et al. 2007). Neuronal activity regulates miRNA activity (Portugal et al. 2025) not only in neurons and nonneural brain cells (microglia, astrocytes, oligodendrocytes) but also in peripheral cells such as immune cells and pericytes throughout the animal body. So, experimental evidence shows that neuronal activity is involved in all the stages and factors of miRNA biogenesis.

8.4 Neuronal Activity Determines Mirna Expression Patterns

miRNAs are single-stranded noncoding RNAs composed of ~22 nucleotides (nts). They are transcribed in the cell nucleus as two-stranded hairpin miRNA precursors (pri-miRs), which are processed by neural activity–driven (Gonçalves et al. 2018) microprocessor Drosha (consisting of the endonuclease RNAse 3 and the double-stranded RNA-binding protein DGCR8) into ~70 nt pre-miR hairpins that are transported to the cytoplasm. In the cytoplasm, Dicer, an RNAse enzyme, cleaves the hairpin loop to form a duplex miR that loses the antisense (passenger) strand (Macrae et al. 2006; Kiltschewskij and Cairns 2019). The remaining single-stranded ~22 nt miR binds to members of Argonaute (Ago) and the GW182 family of proteins to form the core of the miRNA-induced silencing complex (miRISC) (Fabian and Sonenberg 2012), which binds complementary segments of mRNAs, thereby preventing translation (rarely activating it) of mRNAs into proteins by breaking down mRNA into pieces or by pruning the mRNA poly(A) tail.

The key role of miRNAs in the evolution of animals is suggested by the strong correlation that exists between the increasing number of miRNAs and the functional and structural complexity of animals. So, e.g., the number of identified miRNAs in *C. elegans* is 253, in *Drosophila* ~ 469 (Griffiths-Jones et al. 2006), and in humans it is more than 2000 or 2300 (Todorov et al. 2024). The percentage of genes undergoing miRNA post-transcriptional regulation has also grown from 30% in *D. melanogaster* (Wessels et al. 2019) to 50% in humans (Agarwal et al. 2015).

miRNAs are negative gene regulators at the post-transcriptional level that repress the translation of mRNA molecules into proteins (Wightman et al. 1993; Lee et al. 1993) and facilitate mRNA degradation (Huntzinger and Izaurralde 2011; Kiltschewskij and Cairns 2019). Most miRNAs target several different mRNAs, thus blocking the production of respective proteins to fine-tune their levels in the interstitial fluid. The adaptive nature of miRNA spatiotemporal patterns and the site-specific expression of miRNAs indicate that information of some kind is invested for this to occur, and there is no evidence that the information for activating miRNA synthesis is in the miRNA itself or in any gene.

Vast evidence shows that miRNA expression patterns are regulated by neuronal activity (Wayman et al. 2008; Fiore et al. 2009; Krol et al. 2010; Eacker et al. 2011; Kiltschewskij and Cairns 2019; Sim et al. 2014). Neuronal activity also uses the degradation of mRNAs by another group of miRNAs as a complementary way to fine-tune protein production (Nudelman et al. 2010; van Spronsen et al. 2013; Rajasethupathy et al. 2009; Wibrand et al. 2010), based on the presence of simple nucleotide sequence elements (SSE) binding mRNAs. Increased neuronal activity by treatment with KCl and BDNF released by neurons regulates the transcription of 38 miRNAs of the miR-379-410 cluster during embryonic neurogenesis (Winter 2015; Fiore et al. 2009). When mice during the night are exposed to light, they increase the expression of four specific miRNAs that repress Rev-erba nuclear receptors in liver cells, leading to hepatic steatosis, an abnormally high accumulation of fat in the liver (Borck et al. 2018). miRNAs play a determining role in establishing traits of obligatory aestivation (summer dormancy) in the cabbage stem flea beetle (*Psylliodes chrysocephala*), a major pest of oilseed rape, by suppressing a specific subset of transcripts and proteins and conferring resilience to heat stress (Güney et al. 2025).

We know that this nongenetic information originates in the nervous system, but how it is converted sensory data into nongenetic information for regulating patterns of miRNA expression, thereby affecting animal physiology, is still a major enigma in biology.

Nongenetic Determination of Alternative Splicing

Alternative splicing (AS) showed us that genes are not what they were thought to be: an elementary, indivisible unit of inheritance containing genetic information for producing a protein. The concept of the gene evolved sequentially from the initial Mendelian concept "one gene—one phenotypic trait" to "one gene one enzyme (Beadle and Tatum 1941), to "one gene - one polypeptide chain" by the late 1950s.

Alternative splicing was discovered in 1977 by Phillip A. Sharp (Berget et al. 1977) and Richard J. Roberts (Chow et al. 1977) when, in studies on adenovirus, they observed that genes produced more than one protein by removing noncoding segments (introns) from pre-mRNAs and combining their coding segments (exons) in distinct ways. Thanks to the inclusion or exclusion of exons of pre-mRNA, the transcript can change a protein's function.

The discovery led to a major change in the concept of a gene. Multiexon genes in animals are potential carriers of genetic information for not just one, but a varying number of proteins. Transformation of this potentiality into actuality requires information of a type that genes lack and that is provided in the form of external stimuli to the cell. The selection of exons to form pre-mRNA is a complex process of "cutting-and-pasting" and stitching together selected exons after removing introns.

In *Drosophila*, the *Dscam* (Down syndrome cell adhesion molecule) gene is predicted to produce 38,016 alternatively spliced protein isoforms (Schmucker et al. 2000; Hattori et al. 2009). This number is almost twice that of the entire human genome. Furthermore, three neurexin genes, which are exclusively expressed in the brain, produce between 1000 and 3000 neurexin isoforms (Ullrich et al. 1995; Missler and Südhof 1998).

Neuronally induced AS regulates the selection of Neofibromin (NF) isoforms in rat neurons by controlling splicing at exons 26–29, but during different stages of development, it produces ~50 isoforms. Several neuromodulators, such as dopamine and glutamate, influence the AS machinery to produce different isoforms of the Ania-6a(CCNL1) protein. Neuronal activity stimulated by pilocarpine changes the AS pattern of the human transformer-2-beta gene in the brain.

An indication of the role of AS in the evolution of animals is the increase in the proportion of genes undergoing AS with the evolution of structural and functional complexity. So, e.g., while in *Caenorhabditis elegans* AS is estimated to affect ~10% (Hillier et al. 2009) of multiexon genes, the proportion of genes undergoing AS in *Drosophila melanogaster* surged to 20–37% (Gibilisco et al. 2016), and the rate of AS-experiencing genes in humans jumped to about 98% (Pan et al. 2008).

In neurons, precise site-specific AS is regulated by neuronal activity (Lee et al. 2007; Wang and Burge 2008; Schor et al. 2009; Oe and Yoneda 2010; Iijima et al. 2011; Irimia et al. 2014) in response to specific external stimuli.

AS is performed by spliceosomes, large ribonucleoprotein (RNP) complexes consisting of five ribonucleoprotein (RNP) subunits (Will and Lührmann 2011; Jurica and Moore 2003), and other proteins (Rino et al. 2007). Spliceosomes form at each intron-exon junction of the nuclear pre-mRNA (Matera et al. 2007; Matera and Wang 2014), where they catalyze the excision of introns (or specific exons) and the ligation of specific exons. RNA-binding proteins (RBPs), as active parts of the spliceosome, select sites where splicing events will take place and the exons that will be stitched together. One of the best-studied RBP families is the Rbfox family that comprises three splicing paralogs, Rbfox1, Rbfox2 and Rbfox3 (Raj and Blencowe 2015). It is responsible for the selection of neurofascin (NF) isoforms in rat neurons. By controlling splicing at exons 26–29 and K^+-induced depolarization triggers a shift in splicing from Nfasc140 to Nfasc186 in cerebellar GCs (Kriebel et al. 2012), as demonstrated by the evidence that Rbfox1/3 double-knockdown reduces the inclusion of exons 26–29 in Nfasc transcripts (Suzuki et al. 2017). Rbfox1 itself is regulated by neuronal activity (Lee et al. 2016).

9 Nongenetic Induction of the Predictable Transgenerational Inheritance

9.1 *Brain miRNAs Are Transported to Gametes: Nongenetic Inheritance*

Adequate evidence on transgenerational inheritance in animals shows that short noncoding RNAs (sncRNAs) are secreted by brain neurons and transported to the germ line to guide the development of acquired traits for a finite or definite number of generations. The contribution of epididymal sncRNAs is also under neural control. Earlier, it was observed that denervation of the sympathetic (inferior mesenteric ganglion and proximal hypogastric nerves) of the epididymis impaired the reproductive capacity of male rats heavily (Ricker et al. 1997; Ricker 1998).

Now, it is clear why. Epididymal linings release small vesicles containing, among other things, specific miRNAs that are transferred to sperm cells during their travel through the epididymal lumen fluid. It is believed that, in early development, these miRNAs play the leading role in the emergence of new or acquired traits.

The brain is the main producer of miRNAs in vertebrates and invertebrates. Of the total 2.300 miRNAs identified in humans, about 1600 (~ 70%) are expressed in the brain (Todorov et al. 2024; Kohshour et al. 2025; Smal et al. 2024). Hundreds of brain-derived miRNAs are found in gametes (sperm and oocytes), with many known functions in the early development of animals. But, in light of recent evidence on the presence and transport of neuronal precursor miRNAs (pre-miRNAs) and miRNAs (Corradi et al. 2020) in the cones at the tip of axons (precursor miRNAs), it is possible that they may be transported in other tissues and organs.

Before the recognition of the role of miRNAs in the evolution of the living world, many biologists believed that the accelerated increase in the morphological and functional complexity of vertebrates was related to a hypothetical increase in the number of genes due to gene duplications. Experimental evidence from the beginning of animal gene sequencing at the turn of the century has already rejected this hypothesis.

Biologists are now pointing to small noncoding RNAs, a group of regulatory transcripts, as possible factors that facilitated the accelerated functional and structural complexity of vertebrates. In 2008, it was reported that in lampreys, fish of the order Petromyzontiformes, the most primitive of extant vertebrates, around the Cambrian explosion ~500–535 million years ago, there was an exceptionally rapid increase of 41 new miRNA families occurred (Heimberg et al. 2008). Nothing of this kind occurred in closely related ascidians, amphioxus, or other nonchordates) An agent that has rejection of this hypothesis it is suggested that the correlation observed between the increase in the number of miRNAs and the morphophysiological complexity in vertebrates may indicate a possible role of miRNAs in the evolution of vertebrates. The authors also believe that "lying behind the origin of vertebrate complexity is the dramatic expansion of the noncoding RNA inventory, including miRNAs, rather than an increase in the protein-encoding inventory caused by gene duplication events" (Heimberg et al. 2008).

9.2 *Neural Control of Transgenerational Inheritance in Animals*

Evidence is continually accumulating on the role of the nervous system as an inducer of transgenerational adaptive changes in animals.

In Mammals

The new era of the study of the role of miRNAs in evolution began two decades ago, with experiments on transgenerational inheritance in mice by nongenetic information, involving no changes in genes, nucleotide sequences, or genetic information in general. This new field was pioneered by the French-Iranian biologist Minoo Rassoulzadegan and her group (Rassoulzadegan et al. 2006). They observed the accumulation of nonpolyadenylated Kit mRNA in mice heterozygous for the gene of the KIT tyrosinase kinase receptor ($Kit^{+/-}$), which could not be translated and was degraded, leading to the development of a mutant phenotype with white feet and tail tips. The same modified phenotype was obtained in ~50% of the progeny obtained by injection of one-celled embryos with miRNA from F1 $Kit^{+/+}$ homozygotes (because the mutant allele induced a heritable *change in the expression* of the normal) or F1 $Kit^{+/-}$ heterozygotes. Besides, injection of brain or sperm RNAs of heterozygous mice into the embryo produced mice with white feet and tail tips, suggesting a brain origin of these RNAs. More specifically, they demonstrated that injection into early mouse embryos of either one of two miRNAs, miR-221 and miR-222, produced the same heritable phenotypic effect, namely, progeny with white feet and tail tips (Rassoulzadegan et al. 2006).

Olfactory fear conditioning of 2-month-old sexually inexperienced and odor-naive mice with the odor acetophenone associated with a mild electrical shock delivered to the feet induced morphological and behavioral changes inherited in F1 and F2 generations. It was observed that dorsal and medial M71-specific glomeruli in the olfactory bulb of the F1 and F2 offspring, and the number of M71-specific mouse odorant receptors (OR) encoded by the Olfr151 gene, expressed by olfactory sensory neurons of the main olfactory epithelium of acetophenone-trained F0 males, were significantly increased in size. It was suggested that this increase in the size of the main olfactory epithelium and the olfactory bulb may be responsible for the inheritance of the F0-conditioned odor fear and enhanced olfactory sensitivity in the F1 and F2 to acetophenone. CpG (a DNA region where a cytosine (C) nucleotide is followed by a guanine (G) --(CpG)) hypomethylation in the Olfr151 gene in the progeny of the conditioned F0 males and F1 naive offspring was also observed (Dias and Ressler 2014).

In both cases, neurally induced phenotypic changes are transmitted in F1 and F2 mice in the absence of the original trigger without relevant “changes in DNA nucleotide sequences.” Both cases defy the Weismann barrier regarding the impossibility of somatic cells (brain neurons in these cases) influencing inherited traits.

In Invertebrates

In the same year, a group of researchers from the Netherlands demonstrated that injection of *Caenorhabditis elegans* worms with double-stranded miRNA (dsRNA) targeting the gene *ceh-13* led to the production of small and dumpy progeny with 30% penetrance, and the trait is inherited indefinitely in the absence of the initial trigger. The inherited phenotypic change was a result of RNA interference (RNAi), caused by the injection of dsRNA, which targets a matching sequence of the *C. elegans* gene *ceh-13*. The progeny of crosses between worms of the Ceh-13 phenotype and normal males displayed the same mutant phenotype with equal 30% penetrance, showing that the trait is dominant. No changes in genes are reported to be involved in this "acquired" trait that is inherited indefinitely (Vastenhouw et al. 2006).

- *Phase transition in locusts.* Two locust species, *Locusta migratoria* and *Schistocerca gregaria*, in response to environmental stimuli perceived in their brain and thoracic ganglia, change their behavior, size, morphology, and physiology rapidly. These locusts shift from a slow-moving, green-solitary stage to a gregarious stage. The shift is characterized by a tendency to join conspecific migrating locusts and to change their color from green to brown. Additional changes occur in brain size and body size.

  The proximate cause of the phase shift is the secretion of the neuromodulator serotonin by ~30 neurons of the thoracic ganglia. Two distinct neural pathways are involved in locust phase transition: a brain pathway and a peripheral neural pathway. The first is activated in response to visual and olfactory (increased levels of aggregation pheromones because of crowding) signals, while the peripheral, tactile pathway is activated by contact with conspecifics during crowding or, experimentally, by tickling the femur of solitary locusts. It is worthwhile noting that the serotonin in *Locusta migratoria* has an opposite effect, i.e., it induces the reverse transition from the gregarious to the solitary phase, indicating that these species give opposite valences to the same stimulus. The transition to the gregarious phase is inherited across many generations (Rogers and Ott 2015) in the absence of the inducing stimuli, and no gene changes are reported to be involved.
- *Neural induction of transgenerational inheritance of a behavioral trait.* Neuronal activity via double-stranded/miRNAs induces transgenerational inheritance of a behavioral trait in the tiny nematode *C. elegans* for three generations, involving no changes in genes. Neurons of the worm secrete the double-stranded RNA-binding protein, RDE-4, which induces the production of neuronal endogenous short interfering RNAs (endo-siRNAs) in neurons and in the germline. Endo-siRNA move to the germline, where they bind the nuclear Argonaute protein HRDE-1 (heritable RNAi deficient 1) and, in this form, cause RNAi-mediated silencing of the *saeg-2* gene, leading to transgenerational chemotaxis under neuronal activity or stress for three generations (Posner et al. 2019).
- The nematode *Caenorhabditis elegans* feeds on nonpathogenic microorganisms by avoiding many pathogenic microbes. This worm recognizes and avoids eating

pathogenic bacteria like *Pseudomonas aeruginosa* PA14 as a survival defense strategy. The pathogen is recognized by ciliated sensory neurons in the anterior part of the body, which leads to the development of avoidance behavior against the bacterium. The sensory neurons are involved in chemotaxis, mechanosensation, osmotaxis, and thermotaxis. It was found that when the worm was exposed to *P. aeruginosa* PA14, this avoidance behavior required the activation of the *C. elegans* mir-67. This microRNA targets the transmembrane cell adhesion receptor molecule SAX-7, as demonstrated by the evidence that knockdown of *sax-7* rescues pathogen avoidance behavior in mir-67(n4899) mutants (Ma et al. 2017).

10 Nongenetic Information in Sympatric Speciation

During the last three centuries, the notion of species has evolved from an essentialist concept of fixed and unchanged entities with Carl Linnaeus (1707–1778) to a changing and transient biological unit with Jean-Baptiste Lamarck (1744–1829) and Charles Darwin (1809–1882). Darwin saw species as "branches in the lines of descent with modification" (Darwin 1859) that continuously diverge from their ancestors. The most widespread definition of species is that of Ernst Mayr: "a group of actually or potentially interbreeding natural populations that are reproductively isolated from other such groups" (Mayr 1942). Given that at the time of the introduction of the definition, no other reproductive barriers except geographical barriers such as mountains, rivers, seas, etc., were known, biologists thought that the main, if not the only, form of reproductive isolation and formation of new species was allopatric speciation that splits and isolates the original population into two sides of the barrier. Accordingly, in the absence of gene flow, each population, over time, accumulates genotypic and phenotypic changes that may lead to prezygotic or postzygotic isolation when individuals of the incipient species from both sides of the barrier may come into contact again.

The problem with Mayr's strictly allopatric definition of reproductive isolation via physical barriers (new rivers or mountains) is that they are so rare and gradual that they could hardly account for the mind-boggling diversity of species on earth; it also fails to keep with the ample evidence on the presence of sympatric speciation. Studies of the last two to three decades have shown that the concept of allopatry as the primary source of speciation is outdated.

According to the gene-centric view, the formation of new species is a result of the action of natural selection on gradual, random genic mutations accumulated over long periods of time, which lead to reproductive isolation caused by geographic barriers (mountains, rivers, seas). From this viewpoint, allopatric speciation has been considered the main, if not the sole, mechanism of speciation in animals, and sympatric speciation (speciation with gene flow), the formation of new species in the absence of geographic barriers, has been considered virtually impossible. But not only is there observational evidence of the formation of hundreds of fish species within incredibly short periods of time in lakes, but it is argued that speciation can proceed under conditions of gene flow in sympatry (Via 2001; Schluter et al. 2025; Yusuf et al. 2025). Allopatry can no longer be considered the "null hypothesis" of speciation.

Darwin addressed sexual selection in evolution and speciation and distinguished between intrasexual competition (male–male combat) and intersexual selection (mate choice). In the animal world, both male and female choosiness occurs, but female choosiness is by far more prevalent and even influences the evolution of male traits, especially male ornaments. In Darwin's time, it was impossible even to speculate about the biological underpinnings of mate choice. Now we know that mate choice and mate preferences have a neurosensory basis. Sensory (visual, olfactory, auditory, tactile, and low-electrical) cues are perceived in the brain and analyzed in specific neural circuits.

Mate preferences are determined in specific neural circuits, and they can heritably change even within the lifetime of an animal, thus excluding the involvement of any genetic change. Neurally determined changes in mate choice can create reproductive isolation between populations in sympatry.

Cichlid fish species of Lake Victoria belong to the subfamily Haplochrominae, consisting of several genera and > 500 endemic species. How can we account for the evolution of 500 endemic and reproductively isolated cichlid fish species within an evolutionarily very short period of 15,000 years in Lake Victoria? Tens of thousands of years are obviously inadequate to allow for the substantial accumulation of exceptional beneficial mutations to produce 500 new species in a lake; there is no option but to accept a nongenetic regulatory mechanism that modifies patterns of gene expression. Obviously, this is impossible to occur by the evolution of changed or new genes, nor by other genetic means (existing genetic variation or gene drift), especially in light of recent studies confirming that the genomes of these species, even when they belong to different genera, "are very similar within same lake species". Cichlid species that are tested in experiments can breed successfully with other cichlid species and produce viable progeny, but hybridization in the wild is exceptional because of neurocognitive reproductive barriers to be dealt with in the following section.

The concept of allopatric speciation was predominant in biology from the onset of the Modern Synthesis, and Mayr's criticism of the sympatric model made sympatric speciation virtually impossible until three decades ago, when empirical evidence on the reality of sympatric speciation forced even the ardent supporters of allopatric speciation to admit the important role of sympatry in the speciation of animals. Two decades ago, Michael Doebeli pointed out: "It has become clear that the traditional geographic classification of speciation modes is no longer appropriate to capture the essential complexity of many speciation processes. By emphasizing adaptive processes rather than restricting attention to biogeographical patterns of diversification, theoretical and experimental speciation research has taken off again to new shores" (Doebeli 2005). Moreover, recent studies show that even "speciating and recently speciated allopatric taxa seem to overwhelmingly evolve under similar rather than divergent macro-selective pressures (Anderson and Weir 2022). Now we have evidence that the speciation process could have occurred even under conditions of gene flow between populations in sympatry (Perdeck 1958; Seehausen and van Alphen 1998; Streelman et al. 2003; Deily and Schul 2004; Saarikettu et al. 2005; Arnegard et al. 2005; Billeter et al. 2009; Elmer et al. 2010; Arnegard et al. 2010; Shahandeh et al. 2020; Seeholzer et al. 2018; Di 2025).

10.1 Visual-Cognitive Mechanisms of Reproductive Isolation

Visual perception of signaling traits is a major means of discriminating between conspecific and heterospecific individuals and for mate choice decisions. Visual signals received by the retina are processed in the lateral geniculate nucleus and the visual cortex for interpretation to produce visual perception of colors, sizes, levels of activity, etc. Females of cichlid species of Lake Victoria prefer larger, more active and colored individuals, and these traits are believed to have played the most important role in the exceptional speciation (Seehausen and van Alphen 1998; Danley and Kocher 2001).

In Nicaragua's oldest crater lake, the Lake Apoyo, a lake formed in a volcanic crater that for the last time erupted about 1800 years ago but is populated by the *Amphilophus citrinellus* only recently, about one century ago (Elmer et al. 2010). Nevertheless, now it has evolved into six clearly distinct species with distinct body colors and patterns: *A. zaliosus, A. astorquii, A. chancho, A. flaveolus, A. supercilius*, and *A. globosus*.

No changed genes are reported to have been involved in the evolution of the visual-cognitive reproductive isolation of cichlid fishes. What occurred instead were regulatory changes in miRNA and transcription factor–binding sites modifying the GRN and contributing to the accelerated evolution of adaptive traits in cichlid fish. Let us remember here that not only is miRNA expression a nongenetic function of neural activity, but behavior-related GRNs also are regulated by neural networks (Sinha et al. 2020).

10.2 Olfactory-Cognitive Mechanisms of Reproductive Isolation

In insects, olfactory cues are important factors for the recognition of conspecific individuals and mate choice. Olfactory signals, especially pheromones released by mating partners, bind receptors of olfactory receptor neurons (ORNs) and, via second-order projection neurons of antennal lobe glomeruli, reach the lateral protocerebrum, where they are processed and interpreted for mating decisions. Two fly sibling species, and morphologically quite similar, *Drosophila melanogaster* and *D. simulans* (Shahandeh et al. 2020), evolved in sympatry in the

Seychelles archipelago, Indian Ocean, about 3 million years ago (Billeter et al. 2009). Despite having the same behavioral neural circuit, males of these two species assign opposing courtship meanings to the female-specific cuticular hydrocarbon, pheromone 77,11-HD and respond to it with opposite behaviors (Seeholzer et al. 2018): *D. melanogaster* males respond to it with male courtship and reproduction, whereas females of *D. simulans* repel males of the sibling species. Thus, the brain, in sympatry, established a reproductive barrier between the two sibling species by assigning different courtship meanings, by "reweighting" excitatory and inhibitory inputs in the brain, to the same pheromone (Seeholzer et al. 2018; Shahandeh et al. 2020) without evolving any new or changed genes.

Olfactory determined reproductive isolation and the evolution of two morphotypes occurred in the marine fish (threespine stickleback) *Gasterosteus aculeatus* in sympatry in two lakes in British Columbia, Canada. The females of the benthic morphotype recognize and prefer the conspecific males based on both olfactory (odor) and visual (body shape) cues (Mobley et al. 2016). This suffices to establish and maintain reproductive isolation in sympatry, although females of the limnetic morphotype cannot discriminate between conspecific and heterospecific males.

10.3 Auditory-Cognitive Mechanisms of Reproductive Isolation

Hearing in most animals is a function of a brain center. Arthropods receive auditory information via neurons of the sensillae of the chordotonal organs within body segments, commonly near joints, detecting movement, vibration, and gravity and via specialized sensory neurons convert it into electrical spike trains that are transmitted and integrated in specific auditory neuropiles located, depending on the species, in specific regions of the brain to generate corresponding sensation (Di 2025).

In 1958, the Dutch biologist Albert Christiaan Perdeck (1923–2009) described the first case of the reproductive isolation of two sibling grasshopper species living in sympatry. They have the same courtship behaviors and can hybridize, but nevertheless, they are reproductively isolated. He explained this not by any genetic or morphological factor, but by a nongenetic ethological barrier, with their different songs as "the only effective isolating mechanism" (Perdeck 1958).

Another neural mechanism responsible for reproductive isolation in sympatry evolved between two sibling cricket species, *Teleogryllus commodus* and *T. oceanicus*, in Australia, although they can produce viable offspring. After diverging from their common ancestor, *T. commodus* evolved two chirp filters, one for the chirp and the other for the trill of the male song, while *T. oceanicus* evolved only one for the chirp part of the male song (Hennig and Weber 1997). Females of *T. commodus* recognize conspecific male calling based on the duration of pulses, whereas those of *T. oceanicus* recognize it based on the rate of pulses (Hennig 2003). The reproductive isolation of the two sibling species is based on distinct modes of processing of auditory signals in their brains.

Two *Drosophila* species, *D. lummei* and *D. montana* females, use male courtship song both for recognizing conspecific individuals and mate choice (Saarikettu et al. 2005). *Drosophila repleta*, a group of ~100 species, uses the A-type courtship song to recognize conspecifics from heterospecific flies and to attract female flies, whereas the B-type song is used to sexually stimulate females (Ewing and Miyan 1986).

Two insect sibling species, *Neoconocephalus robustus* and *N. bivocatus,* are morphologically so indistinguishable that they were considered as a single species (Deily and Schul 2004). Males of these species produce advertisement calls with fast pulse rates of 200/s (which was the ancestral pulse rate) and 175/s, respectively. The brain of females of *N. bivocatus* evolved the ability to halve the pulse rate of conspecific males, reducing two consecutive pulses to an 87 pulse-pair rate (Deily and Schul 2004), enabling them to discriminate between conspecific and heterospecific males, thus ensuring the reproductive isolation.

10.4 Electro-cognitive Mechanisms of Sympatric Speciation

Radiation within the fish Paramormyrops (formerly Brienomyrus) genus has given rise to more than 20 species over the last 0.5–2 million years in the drainages of West-Central Africa (Sullivan et al. 2002; Lavoué et al. 2008). The weak electric mormyrid fish *Paramormyrops kingsleyae* in rivers of Gabon, Central West Africa, emits species-specific electric organ discharges (EODs). These fish use EOD waveforms for recognition of conspecific fish, electrolocation and communication, mate discrimination and choice, thus establishing a reproductive barrier between themselves and other similar species that use the same habitat but cannot be recognized otherwise. EOD variation in *P. kingsleyae* could be a cue for assortative mating. Its populations show remarkable variation in EOD waveforms. Populations of this fish are still evolving explosively (Picq et al. 2020).

The waveforms of mormyrid EODs are species–specific and sex-specific and are used to distinguish conspecific from heterospecific fish (Arnegard et al. 2005). EODs released by the sender are received by electroreceptors (Knollenorgans) distributed over fish skin and analyzed in the receiver's electrosensory lateral line lobe (ELL) neurons in the midbrain to identify small differences in spike timing (Baker et al. 2013; Lyons-Warren et al. 2013). Neuron-derived electrical signals evolve much faster than morphology, size and environment and are early drivers of speciation processes and the species diversification in the genus *Paramormyrops* through sexual selection (Arnegard et al. 2010).

In the Gabon rivers, 20 *Paramormyrops weak* electric fish morphs, species in the making, are an example of evolution in action. They are morphologically indistinguishable to the human eye, and each produces a distinct EOD waveform while displaying remarkable genetic similarities, suggesting that the reproductive isolation of these indistinguishable morphs derives from the different waveforms each morph generates.

References

11th International Genome Sequencing and Analysis Conference (1999) Abstracts. Microb Comp Genomics 4:75–135. https://doi.org/10.1089/omi.1.1999.4.75

Abeynayake N, Arthur A, Gronthos S (2021) Crosstalk between skeletal and neural tissues is critical for skeletal health. Bone 142:115645. https://doi.org/10.1016/j.bone.2020.115645

Affinati AH, Myers MG Jr (2021) Neuroendocrine control of body energy homeostasis. In: Endotext MDText.com, South Dartmouth. Available from: https://www.ncbi.nlm.nih.gov/books/NBK570658/

Agarwal V, Bell GW, Nam JW, Bartel DP (2015) Predicting effective microRNA target sites in mammalian mRNAs. elife 4:e05005. https://doi.org/10.7554/eLife.05005

Ali SIM, Alrashid SZ (2025) A review of methods for gene regulatory networks reconstruction and analysis. Artif Intell Rev 58:256. https://doi.org/10.1007/s10462-025-11257-z

Anderson SAS, Weir JT (2022) The role of divergent ecological adaptation during allopatric speciation in vertebrates. Science 378:1214–1218. https://doi.org/10.1126/science.abm9124

Arnegard ME, Bogdanowicz SM, Hopkins CD (2005) Multiple cases of striking genetic similarity between alternate electric fish signal morphs in sympatry. Evolution 59:324–343. https://doi.org/10.1111/j.0014-3820.2005.tb00993.x

Arnegard ME, McIntyre PB, Harmon LJ, Zelditch ML, Crampton WGR, Davis JK et al (2010) Sexual signal evolution outpaces ecological divergence during electric fish species radiation. Am Nat 176:335–356. https://doi.org/10.1086/655221

Arresta E, Bernardini S, Gargioli C, Filoni S, Cannata SM (2005) Lens-forming competence in the epidermis of Xenopus laevis during development. J Exp Zool A Comp Exp Biol 303:1–12. https://doi.org/10.1002/jez.a.138

Baker CA, Kohashi T, Lyons-Warren AM, Ma X, Carlson BA (2013) Multiplexed temporal coding of electric communication signals in mormyrid fishes. J Exp Biol 216:2365–2379. https://doi.org/10.1242/jeb.082289

Beadle GW, Tatum EL (1941) Genetic control of biochemical reactions in neurospora. Proc Natl Acad Sci USA 27:499–506. https://doi.org/10.1073/pnas.27.11.499

Berget SM, Moore C, Sharp PA (1977) Spliced segments at the 5′ terminus of adenovirus 2 late mRNA. Proc Natl Acad Sci USA 74:3171–3175. https://doi.org/10.1073/pnas.74.8.3171

Berleth T, Burri M, Thoma G, Bopp D, Richstein S, Frigerio G et al (1988) The role of localization of bicoid RNA in organizing the anterior pattern of the Drosophila embryo. EMBO J 7:1749–1756. https://doi.org/10.1002/j.1460-2075.1988.tb03004.x

Bernard C (1865) Introduction à l'étude de la médecine expérimentale. J.-B. Baillière et fils, Paris

Billeter JC, Atallah J, Krupp J, Millar JG, Levine JD (2009) Specialized cells tag sexual and species identity in Drosophila melanogaster. Nature 461:987–991. https://doi.org/10.1038/nature08463

Bok J, Bronner-Fraser M, Wu DK (2005) Role of the hindbrain in dorsoventral but not anteroposterior axial specification of the inner ear. Development 132:2115–2124. https://doi.org/10.1242/dev.01796

Borck PC, Batista TM, Vettorazzi JF, Soares GM, Lubaczeuski C, Guan D et al (2018) Nighttime light exposure enhances Rev-erbα-targeting microRNAs and contributes to hepatic steatosis. Metabolism 85:250–258. https://doi.org/10.1016/j.metabol.2018.05.002

Borden P, Houtz J, Leach SD, Kuruvilla R (2013) Sympathetic innervation during development is necessary for pancreatic islet architecture and functional maturation. Cell Rep 4:287–301. https://doi.org/10.1016/j.celrep.2013.06.019

Boulant JA (1998) Hypothalamic neurons. Mechanisms of sensitivity to temperature. Ann N Y Acad Sci 856:108–115. https://doi.org/10.1111/j.1749-6632.1998.tb08319.x

Boulant JA (2006) Counterpoint: heat-induced membrane depolarization of hypothalamic neurons: an unlikely mechanism of central thermosensitivity. Am J Physiol Regul Integr Comp Physiol 290:R1481–R1484. https://doi.org/10.1152/ajpregu.00655.2005

Boulant JA, Hardy JD (1974) The effect of spinal and skin temperatures on the firing rate and thermosensitivity of preoptic neurones. J Physiol 240:639–660. https://doi.org/10.1113/jphysiol.1974.sp010627
Bowden TJ, Cook P, Rombout JH (2005) Development and function of the thymus in teleosts. Fish Shellfish Immunol 19:413–427. https://doi.org/10.1016/j.fsi.2005.02.003
Bower DV, Lee H-K, Lansford R, Zinn K, Warburton D, Fraser SE et al (2014) Airway branching has conserved needs for local parasympathetic innervation but not neurotransmission. BMC Biol 12:92. https://doi.org/10.1186/s12915-014-0092-2
Brancaccio A, Palacios D (2015) Chromatin signaling in muscle stem cells: interpreting the regenerative microenvironment. Front Aging Neurosci 7:36. https://doi.org/10.3389/fnagi.2015.00036
Brigande JV, Kiernan AE, Gao X, Iten LE, Fekete DM (2000) Molecular genetics of pattern formation in the inner ear: do compartment boundaries play a role? Proc Natl Acad Sci USA 97:11700–11706. https://doi.org/10.1073/pnas.97.22.11700
Britten RJ, Davidson EH (1969) Gene regulation for higher cells: a theory. Science 165:349–357. https://doi.org/10.1126/science.165.3891.349
Brown AG (1991) The nervous system and homeostasis — interactions with the internal and external environments. In: Nerve cells and nervous systems. Springer, London. https://doi.org/10.1007/978-1-4471-3345-2_16
Cabej NR (2004) Neural control of development. Albanet, Dumont
Cabej NR (2011) Neural control of gene recruitment in metazoans. Dev Dyn 240:1–8. https://doi.org/10.1002/dvdy.22505
Cabej NR (2012) Epigenetic principles of evolution. Elsevier, London
Cabej NR (2019) Epigenetic principles of evolution. Academic Press, London
Cabej NR (2020) A neural mechanism of nuclear receptor expression and regionalization. Dev Dyn 249:1172–1181. https://doi.org/10.1002/dvdy.195
Cabej NR (2021) A mechanism of inheritance of acquired traits in animals. Dev Biol 475:106–117. https://doi.org/10.1016/j.ydbio.2021.03.004
Cabej NR (2023) On the origin and nature of nongenetic information in eumetazoans. Ann N Y Acad Sci 1525:104–127. https://doi.org/10.1111/nyas.15001
Cabej NR (2025) Nongenetic information and evolution. Academic Press, London
Callis TE, Chen JF, Wang DZ (2007) MicroRNAs in skeletal and cardiac muscle development. DNA Cell Biol 26:219–225. https://doi.org/10.1089/dna.2006.0556
Cannata SM, Bernardini S, Filoni S, Gargioli C (2008) The optic vesicle promotes cornea to lens transdifferentication in larval Xenopus laevis. J Anat 212:621–626. https://doi.org/10.1111/j.1469-7580.2008.00886.x
Cannon WB (1932) The wisdom of the body. WW Norton, New York
Carroll SB (2000) Endless forms: the evolution of gene regulation and morphological diversity. Cell 101:577–580. https://doi.org/10.1016/s0092-8674(00)80868-5
Champagne FA (2008) Epigenetic mechanisms and the transgenerational effects of maternal care. Front Neuroendocrinol 29:386–397. https://doi.org/10.1016/j.yfrne.2008.03.003
Chan AS (1992) Association of nerve fibres with myoid cells in the chick thymus. J Anat 181:509–512
Chow LT, Gelinas RE, Broker TR, Roberts RJ (1977) An amazing sequence arrangement at the 5′ ends of adenovirus 2 messenger RNA. Cell 12:1–8. https://doi.org/10.1016/0092-8674(77)90180-5
Cohen S, Janicki-Deverts D, Miller GE (2007) Psychological stress and disease. JAMA 298:1685–1687. https://doi.org/10.1001/jama.298.14.1685
Cohen TJ, Barrientos T, Hartman ZC, Garvey SM, Cox GA, Yao TP (2009) The deacetylase HDAC4 controls myocyte enhancing factor-2-dependent structural gene expression in response to neural activity. FASEB J 23:99–106. https://doi.org/10.1096/fj.08-115931

Colbourne JK, Pfrender ME, Gilbert D, Thomas WK, Tucker A, Oakley TH et al (2011) The ecoresponsive genome of *Daphnia pulex*. Science 331:555–561. https://doi.org/10.1126/science.1197761

Collins FS, Fink L (1995) The human genome project. Alcohol Health Res World 19:190–195

Corradi E, Dalla Costa I, Gavoci A, Iyer A, Roccuzzo M, Otto TA et al (2020) Axonal precursor miRNAs hitchhike on endosomes and locally regulate the development of neural circuits. EMBO J 39:e102513. https://doi.org/10.15252/embj.2019102513

Cortellino S, Xu J, Sannai M, Moore R, Caretti E, Cigliano A et al (2011) Thymine DNA glycosylase is essential for active DNA demethylation by linked deamination-base excision repair. Cell 146:67–79. https://doi.org/10.1016/j.cell.2011.06.020

Cosgrove MS, Boeke JD, Wolberger C (2004) Regulated nucleosome mobility and the histone code. Nat Struct Mol Biol 11:1037–1043. https://doi.org/10.1038/nsmb851

Cowan DP, Stahlhut JK (2004) Functionally reproductive diploid and haploid males in an inbreeding hymenopteran with complementary sex determination. Proc Natl Acad Sci USA 101: 10374–10379. https://doi.org/10.1073/pnas.0402481101

Crick FHC (1958) On protein synthesis. Symp Soc Exp Biol 12:138–163

Cutz E, Yeger H, Pan J (2007) Pulmonary neuroendocrine cell system in pediatric lung disease-recent advances. Pediatr Dev Pathol 10:419–435. https://doi.org/10.2350/07-04-0267.1

Danley PD, Kocher TD (2001) Speciation in rapidly diverging systems: lessons from Lake Malawi. Mol Ecol 10:1075–1086. https://doi.org/10.1046/j.1365-294X.2001.01283.x

Darwin C (1859) The origin of species by means of natural selection or the preservation of favored races in the struggle for life. John Murray, London

Darwin C (1868) The variation of animals and plants under domestication. Orange Judd & Co, New York

Darwin C (1872) The origin of species by means of natural selection or the preservation of favored races in the struggle for life, 6th edn. John Murray, London

Davidson EH (1990) How embryos work: a comparative view of diverse modes of cell fate specification. Development 108:365–389. https://doi.org/10.1242/dev.108.3.365

Davidson EH, Erwin DH (2006) Gene regulatory networks and the evolution of animal body plans. Science 311:796–800. https://doi.org/10.1126/science.1113832

Day JJ, Childs D, Guzman-Karlsson MC, Kibe M, Moulden J, Song E et al (2013) DNA methylation regulates associative reward learning. Nat Neurosci 16:1445–1452. https://doi.org/10.1038/nn.3504

Deily JA, Schul J (2004) Recognition of calls with exceptionally fast pulse rates: female phonotaxis in the genus Neoconocephalus (Orthoptera: Tettigoniidae). J Exp Biol 207:3523–3529. https://doi.org/10.1242/jeb.01193

Deshe N, Eliezer Y, Hoch L, Itskovits E, Bokman E, Ben-Ezra S et al (2023) Inheritance of associative memories and acquired cellular changes in *C. elegans*. Nat Commun 14:4232. https://doi.org/10.1038/s41467-023-39804-8

Di C (2025) Insect auditory system: from nervous system mechanisms to practical applications. Theor Nat Sci 101:245–250

Dias BG, Ressler KJ (2014) Parental olfactory experience influences behavior and neural structure in subsequent generations. Nat Neurosci 17:89–96. https://doi.org/10.1038/nn.3594

Dileep V, Tsai LH (2021) Three-dimensional chromatin organization in brain function and dysfunction. Curr Opin Neurobiol 69:214–221. https://doi.org/10.1016/j.conb.2021.04.006

Doebeli M (2005) Adaptive speciation when assortative mating is based on female preference for male marker traits. J Evol Biol 18:1587–1600. https://doi.org/10.1111/j.1420-9101.2005.00897.x

Douglas RM, Dragunow M, Robertson HA (1988) High-frequency discharge of dentate granule cells, but not long-term potentiation, induces c-fos protein. Brain Res 464:259–262. https://doi.org/10.1016/0169-328x(88)90033-2

Driever W, Nüsslein-Volhard C (1988) Gradient of bicoid protein in *drosophila*. Cell 54:83–93. https://doi.org/10.1016/0092-8674(88)90182-1

Du S, Lutkenhaus J (2017) Assembly and activation of the *Escherichia coli* divisome. Mol Microbiol 105:177–187. https://doi.org/10.1111/mmi.13696

Du S, Pichoff S, Kruse K, Lutkenhaus J (2018) FtsZ filaments have the opposite kinetic polarity of microtubules. Proc Natl Acad Sci USA 115:10768–10773. https://doi.org/10.1073/pnas.1811919115

Dudin O, Ondracka A, Grau-Bové X, Haraldsen AA, Toyoda A, Suga H et al (2019) A unicellular relative of animals generates a layer of polarized cells by actomyosin-dependent cellularization. elife 8:e49801. https://doi.org/10.7554/eLife.49801

Eacker SM, Keuss MJ, Berezikov E, Dawson VL, Dawson TM (2011) Neuronal activity regulates hippocampal miRNA expression. PLoS One 6:e25068. https://doi.org/10.1371/journal.pone.0025068

Eldredge N, Gould SJ (1972) Punctuated equilibria: an alternative to phyletic gradualism. In: Schopf TJM (ed) Models in Paleobiology. Freeman Cooper, San Francisco, pp 82–115

Elmer KR, Lehtonen TK, Kautt AF, Harrod C, Meyer A (2010) Rapid sympatric ecological differentiation of crater lake cichlid fishes within historic times. BMC Biol 8:60. https://doi.org/10.1186/1741-7007-8-60

Ewing AW, Miyan JA (1986) Sexual selection, sexual isolation and the evolution of song in the *Drosophila repleta* group of species. Anim Behav 34:421–429. https://doi.org/10.1016/S0003-3472(86)80110-1

Fabian MR, Sonenberg N (2012) The mechanics of miRNA-mediated gene silencing: a look under the hood of miRISC. Nat Struct Mol Biol 19:586–593. https://doi.org/10.1038/nsmb.2296

Fairclough SR, Chen Z, Kramer E, Zeng Q, Young S, Robertson HM et al (2013) Premetazoan genome evolution and the regulation of cell differentiation in the choanoflagellate *Salpingoeca rosetta*. Genome Biol 14:R15. https://doi.org/10.1186/gb-2013-14-2-r15

Fernandes JJ, Keshishian H (1998) Nerve-muscle interactions during flight muscle development in *Drosophila*. Development 125:1769–1779. https://doi.org/10.1242/dev.125.9.1769

Fiore R, Khudayberdiev S, Christensen M, Siegel G, Flavell SW, Kim TK et al (2009) Mef2-mediated transcription of the miR379-410 cluster regulates activity-dependent dendritogenesis by fine-tuning Pumilio2 protein levels. EMBO J 28:697–710. https://doi.org/10.1038/emboj.2009.10

Fontaine R, Affaticati P, Bureau C, Colin I, Demarque M, Dufour S et al (2015) Dopaminergic neurons controlling anterior pituitary functions: anatomy and ontogenesis in zebrafish. Endocrinology 156:2934–2948. https://doi.org/10.1210/en.2015-1091

Fowler T, Sen R, Roy AL (2011) Regulation of primary response genes. Mol Cell 44:348–360. https://doi.org/10.1016/j.molcel.2011.09.014

Gapp K, Jawaid A, Sarkies P, Bohacek J, Pelczar P, Prados J et al (2014) Implication of sperm RNAs in transgenerational inheritance of the effects of early trauma in mice. Nat Neurosci 17:667–669. https://doi.org/10.1038/nn.3695

Gavilán B, Perea-Atienza E, Martínez P (2016) *Xenacoelomorpha*: a case of independent nervous system centralization? Philos Trans R Soc Lond Ser B Biol Sci 371:20150039. https://doi.org/10.1098/rstb.2015.0039

Gehring WJ (1996) The master control gene for morphogenesis and evolution of the eye. Genes Cells 1:11–15. https://doi.org/10.1046/j.1365-2443.1996.11011.x

Geiman TM, Robertson KD (2002) Chromatin remodeling, histone modifications, and DNA methylation-how does it all fit together? J Cell Biochem 87:117–125. https://doi.org/10.1002/jcb.10286

Gibilisco L, Zhou Q, Mahajan S, Bachtrog D (2016) Alternative splicing within and between *drosophila* species, sexes, tissues, and developmental stages. PLoS Genet 12:e1006464. https://doi.org/10.1371/journal.pgen.1006464

Gonçalves ICG, Brecht J, Thelen MP, Rehorst WA, Peters M, Lee HJ et al (2018) Neuronal activity regulates DROSHA via autophagy in spinal muscular atrophy. Sci Rep 8:7907. https://doi.org/10.1038/s41598-018-26347-y

Graw J (2010) Eye development. Curr Top Dev Biol 90:343–386. https://doi.org/10.1016/S0070-2153(10)90010-0

Greenberg ME, Ziff EB (1984) Stimulation of 3T3 cells induces transcription of the c-fos proto-oncogene. Nature 311:433–438. https://doi.org/10.1038/311433a0

Griffiths PE (2017) Genetic, epigenetic and exogenetic information in development andevolution. 7(5):20160152. https://doi.org/10.1098/rsfs.2016.0152

Griffiths-Jones S, Grocock RJ, van Dongen S, Bateman A, Enright AJ (2006) miRBase: microRNA sequences, targets and gene nomenclature. Nucleic Acids Res 34(Database issue):D140–D144. https://doi.org/10.1093/nar/gkj112

Güney G, Schmitt K, Zicola J, Toprak U, Rostás M, Scholten S et al (2025) The microRNA pathway regulates obligatory aestivation in the cabbage stem flea beetle *Psylliodes chrysocephala*. Commun Biol 8:1288. https://doi.org/10.1038/s42003-025-08721-5

Guo J, Ma D, Mo H, Ball MP, Jang M-H, Bonaguidi MA et al (2011) Neuronal activity modifies the DNA methylation landscape in the adult brain. Nat Neurosci 14:1345–1351. https://doi.org/10.1038/nn.2900

Gupta-Agarwal S, Jarome TJ, Fernandez J, Lubin FD (2014) NMDA receptor- and ERK-dependent histone methylation changes in the lateral amygdala bidirectionally regulate fear memory formation. Learn Mem 21:351–362. https://doi.org/10.1101/lm.035105.114

Haeckel E (1866) Generelle Morphologie II-Allgemeine Anatomie der Organismen. II: Allgemeine Entwickelungsgeschichte der Organismen. Verlag von Georg Reimer, Berlin

Halder G, Callaerts P, Gehring WJ (1995) Induction of ectopic eyes by targeted expression of the eyeless gene in *drosophila*. Science 267:1788–1792. https://doi.org/10.1126/science.7892602

Hammel HT, Hardy JD, Fusco MM (1960) Thermoregulatory responses to hypothalamic cooling in unanesthetized dogs. Am J Phys 198:481–486. https://doi.org/10.1152/ajplegacy.1960.198.3.481

Hara-Irie F, Amizuka N, Ozawa H (1996) Immunohistochemical and ultrastructural localization of CGRP-positive nerve fibers at the epiphyseal trabecules facing the growth plate of rat femurs. Bone 18:29–39. https://doi.org/10.1016/8756-3282(95)00425-4

Hattori D, Chen Y, Matthews BJ, Salwinski L, Sabatti C, Grueber WB et al (2009) Robust discrimination between self and non-self neurites requires thousands of Dscam1 isoforms. Nature 461:644–648. https://doi.org/10.1038/nature08431

Hegstrom CD, Riddiford LM, Truman JW (1998) Steroid and neuronal regulation of ecdysone receptor expression during metamorphosis of muscle in the moth, *Manduca sexta*. J Neurosci 18:1786–1794. https://doi.org/10.1523/JNEUROSCI.18-05-01786.1998

Heideman PD, Bierl CK, Sylvester CJ (2001) Photoresponsive Fischer 344 rats are reproductively inhibited by melatonin and differ in 2-[125I]iodomelatonin binding from nonphotoresponsive Sprague-Dawley rats. J Neuroendocrinol 13:223–232. https://doi.org/10.1046/j.1365-2826.2001.00617.x

Heimberg AM, Sempere LF, Moy VN, Donoghue PC, Peterson KJ (2008) MicroRNAs and the advent of vertebrate morphological complexity. Proc Natl Acad Sci USA 105:2946–2950. https://doi.org/10.1073/pnas.0712259105

Hennig RM (2003) Acoustic feature extraction by cross-correlation in crickets. J Comp Physiol A Neuroethol Sens Neural Behav Physiol 189:589–598. https://doi.org/10.1007/s00359-003-0438-7

Hennig R, Weber T (1997) Filtering of temporal parameters of the calling song by cricket females of two closely related species: a behavioral analysis. J Comp Physiol A 180:621–630. https://doi.org/10.1007/s003590050074

Henrich VC, Denlinger DL (1982) A maternal effect that eliminates pupal diapause in progeny of the flesh fly, *Sarcophaga bullata*. J Insect Physiol 28:881–884. https://doi.org/10.1016/0022-1910(82)90102-0

Herbst WA, Deng W, Wohlschlegel JA, Achiro JM, Martin KC (2021) Neuronal activity regulates the nuclear proteome to promote activity-dependent transcription. J Cell Biol 220:e202103087. https://doi.org/10.1083/jcb.202103087

Hertel LA, Bayne CJ, Loker ES (2002) The symbiont *Capsaspora owczarzaki*, nov. gen. nov. sp., isolated from three strains of the pulmonate snail *Biomphalaria glabrata* is related to members of the Mesomycetozoea. Int J Parasitol 32:1183–1191. https://doi.org/10.1016/s0020-7519(02)00066-8

Hillier LW, Reinke V, Green P, Hirst M, Marra MA, Waterston RH (2009) Massively parallel sequencing of the polyadenylated transcriptome of *C. Elegans*. Genome Res 19:657–666. https://doi.org/10.1101/gr.088112.108

Holliday R, Pugh JE (1975) DNA modification mechanisms and gene activity during development: developmental clocks may depend on the enzymic modification of specific bases. Science 187: 226–232. https://doi.org/10.1126/science.187.4173.226

Holló G, Novák M (2012) The manoeuvrability hypothesis to explain the maintenance of bilateral symmetry in animal evolution. Biol Direct 7:22

Howard-Ashby M, Materna SC, Brown CT, Chen L, Cameron RA, Davidson EH (2006) Gene families encoding transcription factors expressed in early development *of Strongylocentrotus purpuratus*. Dev Biol 300:90–107. https://doi.org/10.1016/j.ydbio.2006.08.033

Huntzinger E, Izaurralde E (2011) Gene silencing by microRNAs: contributions of translational repression and mRNA decay. Nat Rev Genet 12:99–110. https://doi.org/10.1038/nrg2936

Iijima T, Wu K, Witte H, Hanno-Iijima Y, Glatter T, Richard S (2011) SAM68 regulates neuronal activity-dependent alternative splicing of neurexin-1. Cell 147:1601–1614. https://doi.org/10.1016/j.cell.2011.11.028

Irimia M, Weatheritt RJ, Ellis JD, Parikshak NN, Gonatopoulos-Pournatzis T, Babor M et al (2014) A highly conserved program of neuronal microexons is misregulated in autistic brains. Cell 159: 1511–1523. https://doi.org/10.1016/j.cell.2014.11.035

Ishida A, Mutoh T, Ueyama T, Bando H, Masubuchi S, Nakahara D et al (2005) Light activates the adrenal gland: timing of gene expression and glucocorticoid release. Cell Metab 2:297–307. https://doi.org/10.1016/j.cmet.2005.09.009

Jablonka E, Lamb MJ (2006) The evolution of information in the major transitions. J Theor Biol 239:236–246. https://doi.org/10.1016/j.jtbi.2005.08.038

Jacob F (1970) The logic of life. Pantheon Books, New York

James-Clark H (1868) On the Spongiae Ciliatae as Infusoria Flagellata; or, observations on the structure, animality and relationship of *Leucosolenia botryoides* Bowerbank. Ann Mag Nat Hist 1:133–142

Jarome TJ, Lubin FD (2014) Epigenetic mechanisms of memory formation and reconsolidation. Neurobiol Learn Mem 115:116–127. https://doi.org/10.1016/j.nlm.2014.08.002

Jarvis E, Nottebohm F (1997) Motor-driven gene expression. Annu Rev Genet 42:235–251

Jeffery WR (2005) Adaptive evolution of eye degeneration in the Mexican blind cavefish. J Hered 96:185–196. https://doi.org/10.1093/jhered/esi028

Johansen VBI, Petersen J, Lund J, Mathiesen CV, Fenselau H, Clemmensen C (2025) Brain control of energy homeostasis: implications for anti-obesity pharmacotherapy. Cell 188:4178–4212. https://doi.org/10.1016/j.cell.2025.06.010

Jurica MS, Moore MJ (2003) Pre-mRNA splicing: awash in a sea of proteins. Mol Cell 12:5–14. https://doi.org/10.1016/s1097-2765(03)00270-3

Kärcher MH, Menezes C, Alves DA, Beveridge OS, Imperatriz-Fonseca VL, Ratnieks FL (2013) Factors influencing survival duration and choice of virgin queens in the stingless *bee Melipona quadrifasciata*. Naturwissenschaften 100:571–580. https://doi.org/10.1007/s00114-013-1053-2

Kawano F, Suzuki H, Furuya A, Sato M (2015) Engineered pairs of distinct photoswitches for optogenetic control of cellular proteins. Nat Commun 6:6256. https://doi.org/10.1038/ncomms7256

Kendall MD, al-Shawaf AA (1991) Innervation of the rat thymus gland. Brain Behav Immun 5:9–28. https://doi.org/10.1016/0889-1591(91)90004-t

Kiltschewskij D, Cairns MJ (2019) Temporospatial guidance of activity-dependent gene expression by microRNA: mechanisms and functional implications for neural plasticity. Nucleic Acids Res 47:533–545. https://doi.org/10.1093/nar/gky1235

King MC, Wilson AC (1975) Evolution at two levels in humans and chimpanzees. Science 188: 107–116. https://doi.org/10.1126/science.1090005

Klarsfeld A, Changeux JP (1985) Activity regulates the levels of acetylcholine receptor alpha-subunit mRNA in cultured chicken myotubes. Proc Natl Acad Sci USA 82:4558–4562. https://doi.org/10.1073/pnas.82.13.4558

Knepper MA, Kwon TH, Nielsen S (2015) Molecular physiology of water balance. N Engl J Med 372:1349–1358. https://doi.org/10.1056/NEJMra1404726

Kohshour M, Heilbronner U, Mueller T, Rossner M, Papiol S, Schulze TG (2025) The microRNA profile of brain-derived extracellular vesicles: a promising step forward in developing pharmacodynamic biomarkers for psychiatric disorders. Eur Neuropsychopharmacol 90:62–68. https://doi.org/10.1016/j.euroneuro.2024.10.002

Koonin EV (2009) Darwinian evolution in the light of genomics. Nucleic Acids Res 37:1011–1034. https://doi.org/10.1093/nar/gkp089

Kriebel M, Wuchter J, Trinks S, Volkmer H (2012) Neurofascin: a switch between neuronal plasticity and stability. Int J Biochem Cell Biol 44:694–697. https://doi.org/10.1016/j.biocel.2012.01.012

Krol J, Busskamp V, Markiewicz I, Stadler MB, Ribi S, Richter J et al (2010) Characterizing light-regulated retinal microRNAs reveals rapid turnover as a common property of neuronal microRNAs. Cell 141:618–631. https://doi.org/10.1016/j.cell.2010.03.039

Lang BF, O'Kelly C, Nerad T, Gray MW, Burger G (2002) The closest unicellular relatives of animals. Curr Biol 12:1773–1778. https://doi.org/10.1016/s0960-9822(02)01187-9

Lavoué S, Sullivan JP, Arnegard ME, Hopkins CD (2008) Differentiation of morphology, genetics and electric signals in a region of sympatry between sister species of African electric fish (Mormyridae). J Evol Biol 21:1030–1045. https://doi.org/10.1111/j.1420-9101.2008.01549.x

Lebedev E, Smutin D, Timkin P, Kotelnikov D, Taldaev A, Panushev N et al (2024) The eusocial non-code: unveiling the impact of noncoding RNAs on hymenoptera eusocial evolution. Noncoding RNA Res 11:48–59. https://doi.org/10.1016/j.ncrna.2024.10.007

Lee RC, Feinbaum RL, Ambros V (1993) The *C. elegans* heterochronic gene lin-4 encodes small RNAs with antisense complementarity to lin-14. Cell 75:843–854. https://doi.org/10.1016/0092-8674(93)90529-y

Lee TI, Rinaldi NJ, Robert F, Odom DT, Bar-Joseph Z, Gerber GK et al (2002) Transcriptional regulatory networks in *Saccharomyces cerevisiae*. Science 298:799–804. https://doi.org/10.1126/science.1075090

Lee JA, Xing Y, Nguyen D, Xie J, Lee CJ, Black DL (2007) Depolarization and CaM kinase IV modulate NMDA receptor splicing through two essential RNA elements. PLoS Biol 5:e40. https://doi.org/10.1371/journal.pbio.0050040

Lee JA, Damianov A, Lin CH, Fontes M, Parikshak NN, Anderson ES et al (2016) Cytoplasmic Rbfox1 regulates the expression of synaptic and autism-related genes. Neuron 89:113–128. https://doi.org/10.1016/j.neuron.2015.11.025

Levenson JM, O'Riordan KJ, Brown KD, Trinh MA, Molfese DL, Sweatt JD (2004) Regulation of histone acetylation during memory formation in the hippocampus. J Biol Chem 279:40545–40559. https://doi.org/10.1074/jbc.M402229200

Li W, Kohara H, Uchida Y, James JM, Soneji K, Cronshaw DG et al (2013) Peripheral nerve-derived CXCL12 and VEGF-A regulate the patterning of arterial vessel branching in developing limb skin. Dev Cell 24:359–371. https://doi.org/10.1016/j.devcel.2013.01.009

Liang F, Holt I, Pertea G, Karamycheva S, Salzberg SL, Quackenbush J (2000) Gene index analysis of the human genome estimates approximately 120,000 genes. Nat Genet 25:239–240. https://doi.org/10.1038/76126

Lubin FD, Roth TL, Sweatt JD (2008) Epigenetic regulation of BDNF gene transcription in the consolidation of fear memory. J Neurosci 28:10576–10586. https://doi.org/10.1523/JNEUROSCI.1786-08.2008

Lyons-Warren AM, Kohashi T, Mennerick S, Carlson BA (2013) Detection of submillisecond spike timing differences based on delay-line anticoincidence detection. J Neurophysiol 110: 2295–2311. https://doi.org/10.1152/jn.00461.2013

Ma DK, Guo JU, Ming GL, Song H (2009) DNA excision repair proteins and Gadd45 as molecular players for active DNA demethylation. Cell Cycle 8:1526–1531. https://doi.org/10.4161/cc.8.10.8500

Ma YC, Zhang L, Dai LL, Khan RU, Zou CG (2017) mir-67 regulates *P. aeruginosa* avoidance behavior in *C. elegans*. Biochem Biophys Res Commun 494:120–125. https://doi.org/10.1016/j.bbrc.2017.10.069

Macrae IJ, Zhou K, Li F, Repic A, Brooks AN, Cande WZ et al (2006) Structural basis for double-stranded RNA processing by Dicer. Science 311:195–198. https://doi.org/10.1126/science.1121638

Madabhushi R, Gao F, Pfenning AR, Pan L, Yamakawa S, Seo J et al (2015) Activity-induced DNA breaks govern the expression of neuronal early-response genes. Cell 161:1592–1605. https://doi.org/10.1016/j.cell.2015.05.032

Magloire H, Couble ML, Thivichon-Prince B, Maurin JC, Bleicher F (2009) Odontoblast: a mechano-sensory cell. J Exp Zool B Mol Dev Evol 312B:416–424. https://doi.org/10.1002/jez.b.21264

Magnon C (2015) Role of the autonomic nervous system in tumorigenesis and metastasis. Mol Cell Oncol 2:e975643. https://doi.org/10.4161/23723556.2014.975643

March B, Faulkner S, Jobling P, Steigler A, Blatt A, Denham J et al (2020) Tumour innervation and neurosignalling in prostate cancer. Nat Rev Urol 17:119–130. https://doi.org/10.1038/s41585-019-0274-3

Matera AG, Wang Z (2014) A day in the life of the spliceosome. Nat Rev Mol Cell Biol 15:108–121. https://doi.org/10.1038/nrm3742

Matera AG, Terns RM, Terns MP (2007) Non-coding RNAs: lessons from the small nuclear and small nucleolar RNAs. Nat Rev Mol Cell Biol 8:209–220. https://doi.org/10.1038/nrm2124

Matsui T, Yamamoto T, Wyder S, Zdobnov EM, Kadowaki T (2009) Expression profiles of urbilaterian genes uniquely shared between honey bees and vertebrates. BMC Genomics 10: 17. https://doi.org/10.1186/1471-2164-10-17

Maurin JC, Couble ML, Didier-Bazes M, Brisson C, Magloire H, Bleicher F (2004) Expression and localization of reelin in human odontoblasts. Matrix Biol 23:277–285. https://doi.org/10.1016/j.matbio.2004.06.005

Mayr E (1942) Systematics and the origin of species. Columbia University Press, New York

Mayr E (1961) Cause and effect in biology. Science 134:1501–1506. https://doi.org/10.1126/science.134.3489.1501

Mayr E (1974) Teleological and teleonomic, a new analysis. In: Cohen RS, Wartofsky MW (eds) A portrait of twenty-five years. Boston Studies in the Philosophy of Science. Springer, Dordrecht, pp 133–159

McEwen BS (1999) Stress and hippocampal plasticity. Annu Rev Neurosci 22:105–122. https://doi.org/10.1146/annurev.neuro.22.1.105

Meaney MJ, Szyf M (2005) Maternal care as a model for experience-dependent chromatin plasticity? Trends Neurosci 28:456–463. https://doi.org/10.1016/j.tins.2005.07.006

Méjat A, Ramond F, Bassel-Duby R, Khochbin S, Olson EN, Schaeffer L (2005) Histone deacetylase 9 couples neuronal activity to muscle chromatin acetylation and gene expression. Nat Neurosci 8:313–321. https://doi.org/10.1038/nn1408

Milnes MR Jr, Roberts RN, Guillette LJ Jr (2002) Effects of incubation temperature and estrogen exposure on aromatase activity in the brain and gonads of embryonic alligators. Environ Health Perspect 110:393–396. https://doi.org/10.1289/ehp.02110s3393

Missler M, Südhof TC (1998) Neurexins: three genes and 1001 products. Trends Genet 14:20–26. https://doi.org/10.1016/S0168-9525(97)01324-3

Mobley RB, Tillotson ML, Boughman JW (2016) Olfactory perception of mates in ecologically divergent stickleback: population parallels and differences. Evol Ecol Res 17:551–564

Morris MJ, Monteggia LM (2014) Role of DNA methylation and the DNA methyltransferases in learning and memory. Dialogues Clin Neurosci 16:359–371. https://doi.org/10.31887/DCNS.2014.16.3/mmorris
Morrison SF (2004) Central pathways controlling brown adipose tissue thermogenesis. News Physiol Sci 19:67–74. https://doi.org/10.1152/nips.01502.2003
Moullé VS, Tremblay C, Castell AL, Vivot K, Ethier M, Fergusson G et al (2019) The autonomic nervous system regulates pancreatic β-cell proliferation in adult male rats. Am J Physiol Endocrinol Metab 317:E234–E243. https://doi.org/10.1152/ajpendo.00385.2018
Mukouyama YS, Gerber HP, Ferrara N, Gu C, Anderson DJ (2005) Peripheral nerve-derived VEGF promotes arterial differentiation via neuropilin 1-mediated positive feedback. Development 132: 941–952. https://doi.org/10.1242/dev.01675
Nanney D (1957) Mating type inheritance at conjugation in variety 4 of *Paramecium aurelia*. J Protozool 4:89–95. https://doi.org/10.1111/j.1550-7408.1957.tb02492.x
Nawalpuri B, Sharma A, Chattarji S, Muddashetty RS (2021) Distinct temporal expression of the GW182 paralog TNRC6A in neurons regulates dendritic arborization. J Cell Sci 134:jcs258465. https://doi.org/10.1242/jcs.258465
Nekrep N, Wang J, Miyatsuka T, German MS (2008) Signals from the neural crest regulate beta-cell mass in the pancreas. Development 135:2151–2160. https://doi.org/10.1242/dev.015859
Noble D (2016) Dance to the tune of life. Biological Relativity. Cambridge University Press, Cambridge
Noguchi M, Furukawa KT, Morimoto M (2020) Pulmonary neuroendocrine cells: physiology, tissue homeostasis and disease. Dis Model Mech 13:dmm046920. https://doi.org/10.1242/dmm.046920
Nudelman AS, DiRocco DP, Lambert TJ, Garelick MG, Le J, Nathanson NM et al (2010) Neuronal activity rapidly induces transcription of the CREB-regulated microRNA-132, in vivo. Hippocampus 20:492–498. https://doi.org/10.1002/hipo.20646
Oe S, Yoneda Y (2010) Cytoplasmic polyadenylation element-like sequences are involved in dendritic targeting of BDNF mRNA in hippocampal neurons. FEBS Lett 584:3424–3430. https://doi.org/10.1016/j.febslet.2010.06.040
Oliveira AM, Hemstedt TJ, Bading H (2012) Rescue of aging-associated decline in Dnmt3a2 expression restores cognitive abilities. Nat Neurosci 15:1111–1113. https://doi.org/10.1038/nn.3151
Omenn GS, Lane L, Overall CM, Corrales FJ, Schwenk JM, Paik YK et al (2019) Progress on identifying and characterizing the human proteome: 2019 metrics from the HUPO human proteome project. J Proteome Res 18:4098–4107. https://doi.org/10.1021/acs.jproteome.9b00434
Ondracka A, Dudin O, Ruiz-Trillo I (2018) Decoupling of nuclear division cycles and cell size during the coenocytic growth of the Ichthyosporean *Sphaeroforma arctica*. Curr Biol 28:1964–1969.e2. https://doi.org/10.1016/j.cub.2018.04.074
Osmundsen AM, Keisler JL, Taketo MM, Davis SW (2017) Canonical WNT signaling regulates the pituitary organizer and pituitary gland formation. Endocrinology 158:3339–3353. https://doi.org/10.1210/en.2017-00581
Owczarzak A, Stibbs HH, Bayne CJ (1980) The destruction of *Schistosoma mansoni* mother sporocysts in vitro by amoebae isolated from *Biomphalaria glabrata*: an ultrastructural study. J Invertebr Pathol 35:26–33. https://doi.org/10.1016/0022-2011(80)90079-8
Owyang C, Logsdon CD (2004) New insights into neurohormonal regulation of pancreatic secretion. Gastroenterology 127:957–969. https://doi.org/10.1053/j.gastro.2004.05.002
Pan J, Copland I, Post M, Yeger H, Cutz E (2006) Mechanical stretch-induced serotonin release from pulmonary neuroendocrine cells: implications for lung development. Am J Physiol Lung Cell Mol Physiol 290:L185–L193. https://doi.org/10.1152/ajplung.00237.2005
Pan Q, Shai O, Lee LJ, Frey BJ, Blencowe BJ (2008) Deep surveying of alternative splicing complexity in the human transcriptome by high-throughput sequencing. Nat Genet 40:1413–1415. https://doi.org/10.1038/ng.259

Pearce K, Cai D, Roberts AC, Glanzman DL (2017) Role of protein synthesis and DNA methylation in the consolidation and maintenance of long-term memory in Aplysia. elife 6:e18299. https://doi.org/10.7554/eLife.18299

Perdeck AC (1958) The isolating value of specific Song patterns in two sibling species of grasshoppers *(Chorthippus Brunneus* Thunb and *C. Biguttulus* L). Ardea 46:1–37

Phillips T, Shaw K (2008) Chromatin remodeling in eukaryotes. Nature Education 1:209

Picq S, Sperling J, Cheng CJ, Carlson BA, Gallant JR (2020) Genetic drift does not sufficiently explain patterns of electric signal variation among populations of the mormyrid electric fish *Paramormyrops kingsleyae*. Evolution 74:911–935. https://doi.org/10.1111/evo.13956

Pinto LZ, Hartfelder K, Gentile Bitondi MM, Simões ZLP (2002) Ecdysteroid titers in pupae of highly social bees relate to distinct modes of caste development. J Insect Physiol 48:783–790. https://doi.org/10.1016/S0022-1910(02)00103-8

Pirzgalska RM, Seixas E, Seidman JS, Link VM, Sánchez NM, Mahú I et al (2017) Sympathetic neuron-associated macrophages contribute to obesity by importing and metabolizing norepinephrine. Nat Med 23:1309–1318. https://doi.org/10.1038/nm.4422

Portugal R, Rodrigues B, Leitão RA, Silva M, Pinheiro PS, Carvalho AL (2025) Shaping the synapse through neuronal activity-regulated miRNAs. Trends Neurosci 48:679–690. https://doi.org/10.1016/j.tins.2025.06.001

Posner R, Toker IA, Antonova O, Star E, Anava S, Azmon E et al (2019) Neuronal small RNAs control behavior Transgenerationally. Cell 177:1814–1826.e15. https://doi.org/10.1016/j.cell.2019.04.029

Quiring R, Walldorf U, Kloter U, Gehring WJ (1994) Homology of the eyeless gene of *drosophila* to the small eye gene in mice and *Aniridia* in humans. Science 265:785–789. https://doi.org/10.1126/science.7914031

Raj B, Blencowe BJ (2015) Alternative splicing in the mammalian nervous system: recent insights into mechanisms and functional roles. Neuron 87:14–27. https://doi.org/10.1016/j.neuron.2015.05.004

Rajasethupathy P, Fiumara F, Sheridan R, Betel D, Puthanveettil SV, Russo JJ (2009) Characterization of small RNAs in *Aplysia* reveals a role for miR-124 in constraining synaptic plasticity through CREB. Neuron 63:803–817. https://doi.org/10.1016/j.neuron.2009.05.029

Rassoulzadegan M, Grandjean V, Gounon P, Vincent S, Gillot I, Cuzin F (2006) RNA-mediated non-mendelian inheritance of an epigenetic change in the mouse. Nature 441:469–474. https://doi.org/10.1038/nature04674

Redondo RL, Kim J, Arons AL, Ramirez S, Liu X, Tonegawa S (2014) Bidirectional switch of the valence associated with a hippocampal contextual memory engram. Nature 513:426–430. https://doi.org/10.1038/nature13725

Rétaux S, Casane D (2013) Evolution of eye development in the darkness of caves: adaptation, drift, or both? EvoDevo 4:26. https://doi.org/10.1186/2041-9139-4-26

Reza HM, Yasuda K (2004) The involvement of neural retina pax6 in lens fiber differentiation. Dev Neurosci 26:318–327. https://doi.org/10.1159/000082273

Reza HM, Yasuda K (2005) The involvement of neural retina pax6 in lens fiber differentiation. Dev Neurosci 26:318–327. https://doi.org/10.1159/000082273

Richter DJ, Nitsche F (2017) Choanoflagellatea. In: Handbook of the Protists, pp 1479–1496

Ricker DD (1998) The autonomic innervation of epididymis: its effects on epididymal function and fertility. J Androl 19:1–4. https://doi.org/10.1002/j.1939-4640.1998.tb02463.x

Ricker DD, Crone JK, Chamness SL, Klinefelter GR, Chang TS (1997) Partial sympathetic denervation of the rat epididymis permits fertilization but inhibits embryo development. J Androl 18:131–138. https://doi.org/10.1002/j.1939-4640.1997.tb01893.x

Riggs AD (1975) X inactivation, differentiation, and DNA methylation. Cytogenet Genome Res 14:9–25. https://doi.org/10.1159/000130315

Rino J, Carvalho T, Braga J, Desterro JM, Lührmann R, Carmo-Fonseca M (2007) A stochastic view of spliceosome assembly and recycling in the nucleus. PLoS Comput Biol 3:e201. https://doi.org/10.1371/journal.pcbi.0030201

Robert MF, Morin S, Beaulieu N, Gauthier F, Chute IC, Barsalou A et al (2003) DNMT1 is required to maintain CpG methylation and aberrant gene silencing in human cancer cells. Nat Genet 33: 61–65. https://doi.org/10.1038/ng1068

Rogers SM, Ott SR (2015) Differential activation of serotonergic neurons during short- and long-term gregarization of desert locusts. Proc Biol Sci 282:20142062. https://doi.org/10.1098/rspb.2014.2062

Rokas A (2008) The origins of multicellularity and the early history of the genetic toolkit for animal development. Annu Rev Genet 42:235–251. https://doi.org/10.1146/annurev.genet.42.110807.091513

Saarikettu M, Liimatainen JO, Hoikkala A (2005) Intraspecific variation in mating behaviour does not cause sexual isolation between *Drosophila virilis* strains. Anim Behav 70:417–426. https://doi.org/10.1016/j.anbehav.2004.12.009

Sameshima SY, Miura T, Matsumoto T (2004) Wing disc development during caste differentiation in the ant *Pheidole megacephala* (Hymenoptera: Formicidae). Evol Dev 6:336–341. https://doi.org/10.1111/j.1525-142X.2004.04041.x

Sarkissian T, Arya R, Gyonjyan S, Taylor B, White K (2016) Cell death regulates muscle fiber number. Dev Biol 415:87–97. https://doi.org/10.1016/j.ydbio.2016.04.018

Schluter D, Veen T, Thompson KA, Owens GL, Rennison DJ (2025) Coexistence in sympatry with gene flow before speciation has completed. Mol Ecol 34:e70079. https://doi.org/10.1111/mec.70079

Schmucker D, Clemens JC, Shu H, Worby CA, Xiao J, Muda M et al (2000) *Drosophila* Dscam is an axon guidance receptor exhibiting extraordinary molecular diversity. Cell 101:671–684. https://doi.org/10.1016/s0092-8674(00)80878-8

Scholz B, Kultima K, Mattsson A, Axelsson J, Brunström B, Halldin K et al (2006) Sex-dependent gene expression in early brain development of chicken embryos. BMC Neurosci 7:12. https://doi.org/10.1186/1471-2202-7-12

Schor IE, Rascovan N, Pelisch F, Alló M, Kornblihtt AR (2009) Neuronal cell depolarization induces intragenic chromatin modifications affecting NCAM alternative splicing. Proc Natl Acad Sci USA 106:4325–4330. https://doi.org/10.1073/pnas.0810666106

Sebé-Pedrós A, Zheng Y, Ruiz-Trillo I, Pan D (2012) Premetazoan origin of the hippo signaling pathway. Cell Rep 1:13–20. https://doi.org/10.1016/j.celrep.2011.11.004

Sebé-Pedrós A, Ballaré C, Parra-Acero H, Chiva C, Tena JJ, Sabidó E et al (2016) The dynamic regulatory genome of *Capsaspora* and the origin of animal multicellularity. Cell 165:1224–1237. https://doi.org/10.1016/j.cell.2016.03.034

Seehausen O, van Alphen J (1998) The effect of male coloration on female mate choice in closely related Lake Victoria cichlids (*Haplochromis nyererei* complex). Behav Ecol Sociobiol 42:1–8. https://doi.org/10.1007/s002650050405

Seeholzer LF, Seppo M, Stern DL, Ruta V (2018) Evolution of a central neural circuit underlies *Drosophila* mate preferences. Nature 559:564–569. https://doi.org/10.1038/s41586-018-0322-9

Seidel F (1960) Körpergrundgestalt und Keimstruktur. Eine Erörterung über die Grundlagen der vergleichenden und experimentellen Embryologie und deren Gültigkeit by phylogenetischen Überlegungen. Zool Anz 164:245–305

Shahandeh MP, Pischedda A, Rodriguez JM, Turner TL (2020) The genetics of male pheromone preference difference between *Drosophila melanogaster* and *Drosophila simulans*. G3 10:401–415. https://doi.org/10.1534/g3.119.400877

Sharma RP, Tun N, Grayson DR (2008) Depolarization induces downregulation of DNMT1 and DNMT3a in primary cortical cultures. Epigenetics 3:74–80. https://doi.org/10.4161/epi.3.2.6103

Sieber KR, Dorman T, Newell N, Yan H (2021) (Epi)genetic mechanisms underlying the evolutionary success of eusocial insects. Insects 12:498. https://doi.org/10.3390/insects12060498

Siegfried Z, Cedar H (1997) DNA methylation: a molecular lock. Curr Biol 7:R305–R307. https://doi.org/10.1016/s0960-9822(06)00144-8

Sim SE, Bakes J, Kaang BK (2014) Neuronal activity-dependent regulation of MicroRNAs. Mol Cells 37:511–517. https://doi.org/10.14348/molcells.2014.0132
Sinha S, Jones BM, Traniello IM, Bukhari SA, Halfon MS, Hofmann HA et al (2020) Behavior-related gene regulatory networks: a new level of organization in the brain. Proc Natl Acad Sci USA 117:23270–23279. https://doi.org/10.1073/pnas.2009947117
Smal M, Memoli D, Alexandrova E, Di Rosa D, D'Agostino Y, Russo F (2024) Small non-coding RNA transcriptomic profiling in adult and fetal human brain. Sci Data 11:767. https://doi.org/10.1038/s41597-024-03604-6
Snell EA, Furlong RF, Holland PW (2001) Hsp70 sequences indicate that choanoflagellates are closely related to animals. Curr Biol 11:967–970. https://doi.org/10.1016/s0960-9822(01)00275-5
Soares FA, Fagundez DA, Avila DS (2017) Neurodegeneration induced by metals in *Caenorhabditis elegans*. Adv Neurobiol 18:355–383. https://doi.org/10.1007/978-3-319-60189-2_18
Söderström B, Mirzadeh K, Toddo S, von Heijne G, Skoglund U, Daley DO (2016) Coordinated disassembly of the divisome complex in *Escherichia coli*. Mol Microbiol 101:425–438. https://doi.org/10.1111/mmi.13400
Sonneborn TM, Beisson J (1965) Cytoplasmic inheritance of the organization of the cortex in *Paramecium aurelia*. Proc Natl Acad Sci USA 53:275–282. https://doi.org/10.1073/pnas.53.2.275
Steinke F, Seeger M, Tsuda K (2007) Experimental design for efficient identification of gene regulatory networks using sparse Bayesian models. BMC Syst Biol 1:51. https://doi.org/10.1186/1752-0509-1-51
Stott RT, Kritsky O, Tsai LH (2021) Profiling DNA break sites and transcriptional changes in response to contextual fear learning. PLoS One 16:e0249691. https://doi.org/10.1371/journal.pone.0249691
Strand FL (1999) Neuropeptides. MIT Press, Cambridge
Streelman JT, Webb JF, Albertson RC, Kocher TD (2003) The cusp of evolution and development: a model of cichlid tooth shape diversity. Evol Dev 5:600–608. https://doi.org/10.1046/j.1525-142X.2003.03065.x
Stricker SH, Götz M (2018) DNA-methylation: master or slave of neural fate decisions? Front Neurosci 12:5. https://doi.org/10.3389/fnins.2018.00005
Su Y, Shin J, Zhong C, Wang S, Roychowdhury P, Lim J et al (2017) Neuronal activity modifies the chromatin accessibility landscape in the adult brain. Nat Neurosci 20:476–483. https://doi.org/10.1038/nn.4494
Suga H, Dacre M, de Mendoza A, Shalchian-Tabrizi K, Manning G, Ruiz-Trillo I (2012) Genomic survey of premetazoans shows deep conservation of cytoplasmic tyrosine kinases and multiple radiations of receptor tyrosine kinases. Sci Signal 5:ra35. https://doi.org/10.1126/scisignal.2002733
Sullivan JP, Lavoué S, Hopkins CD (2002) Discovery and phylogenetic analysis of a riverine species flock of African electric fishes (Mormyridae, Teleostei). Evolution 56:597–616. https://doi.org/10.1111/j.0014-3820.2002.tb01370.x
Suzuki S, Ayukawa N, Okada C, Tanaka M, Takekoshi S, Iijima Y et al (2017) Spatio-temporal and dynamic regulation of neurofascin alternative splicing in mouse cerebellar neurons. Sci Rep 7:11405. https://doi.org/10.1038/s41598-017-11319-5
Swank MW, Sweatt JD (2001) Increased histone acetyltransferase and lysine acetyltransferase activity and biphasic activation of the ERK/RSK cascade in insular cortex during novel taste learning. J Neurosci 21:3383–3391. https://doi.org/10.1523/JNEUROSCI.21-10-03383.2001
Tahiliani M, Koh KP, Shen Y, Pastor WA, Bandukwala H, Brudno Y et al (2009) Conversion of 5-methylcytosine to 5-hydroxymethylcytosine in mammalian DNA by MLL partner TET1. Science 324:930–935. https://doi.org/10.1126/science.1170116
Takeda H, Nishimura K, Agata K (2009) Planarians maintain a constant ratio of different cell types during changes in body size by using the stem cell system. Zool Sci 26:805–813. https://doi.org/10.2108/zsj.26.805

Takuma N, Sheng HZ, Furuta Y, Ward JM, Sharma K, Hogan BL et al (1998) Formation of Rathke's pouch requires dual induction from the diencephalon. Development 125:4835–4840. https://doi.org/10.1242/dev.125.23.4835

Tang H, Goldman D (2006) Activity-dependent gene regulation in skeletal muscle is mediated by a histone deacetylase (HDAC)-Dach2-myogenin signal transduction cascade. Proc Natl Acad Sci USA 103:16977–16982. https://doi.org/10.1073/pnas.0601565103

Tang H, Macpherson P, Marvin M, Meadows E, Klein WH, Yang XJ et al (2009) A histone deacetylase 4/myogenin positive feedback loop coordinates denervation-dependent gene induction and suppression. Mol Biol Cell 20:1120–1131. https://doi.org/10.1091/mbc.e08-07-0759

Tank AW, Osterhout CA, Sterling CR (1998) Regulation of tyrosine hydroxylase activity by muscarinic agonists in rat adrenal medulla. J Pharmacol Exp Ther 286:848–854

Terazono H, Mutoh T, Yamaguchi S, Kobayashi M, Akiyama M, Udo R et al (2003) Adrenergic regulation of clock gene expression in mouse liver. Proc Natl Acad Sci USA 100:6795–6800. https://doi.org/10.1073/pnas.0936797100

Thorens B (2011) Brain glucose sensing and neural regulation of insulin and glucagon secretion. Diabetes Obes Metab 13:82–88. https://doi.org/10.1111/j.1463-1326.2011.01453.x

Todorov H, Weißbach S, Schlichtholz L, Mueller H, Hartwich D, Gerber S et al (2024) Stage-specific expression patterns and co-targeting relationships among miRNAs in the developing mouse cerebral cortex. Commun Biol 7:1366. https://doi.org/10.1038/s42003-024-07092-7

Toró I, Oláh I, Röhlich P, Virágh S (1969) Electron microscopic observations on myoid cells of the frog's thymus. Anat Rec 165:329–341. https://doi.org/10.1002/ar.1091650302

Treier M, O'Connell S, Gleiberman A, Price J, Szeto DP, Burgess R et al (2001) Hedgehog signaling is required for pituitary gland development. Development 128:377–386. https://doi.org/10.1242/dev.128.3.377

Tyssowski KM, DeStefino NR, Cho JH, Dunn CJ, Poston RG, Carty CE et al (2018) Different neuronal activity patterns induce different gene expression programs. Neuron 98:530–546.e11. https://doi.org/10.1016/j.neuron.2018.04.001

Ullrich B, Ushkaryov YA, Südhof TC (1995) Cartography of neurexins: more than 1000 isoforms generated by alternative splicing and expressed in distinct subsets of neurons. Neuron 14:497–507. https://doi.org/10.1016/0896-6273(95)90306-2

van Spronsen M, van Battum EY, Kuijpers M, Vangoor VR, Rietman ML, Pothof J (2013) Developmental and activity-dependent miRNA expression profiling in primary hippocampal neuron cultures. PLoS One 8:e74907. https://doi.org/10.1371/journal.pone.0074907

Vastenhouw NL, Brunschwig K, Okihara KL, Müller F, Tijsterman M, Plasterk RH (2006) Gene expression: long-term gene silencing by RNAi. Nature 442:882. https://doi.org/10.1038/442882a

Via S (2001) Sympatric speciation in animals: the ugly duckling grows up. Trends Ecol Evol 16: 381–390. https://doi.org/10.1016/s0169-5347(01)02188-7

Wang Z, Burge CB (2008) Splicing regulation: from a parts list of regulatory elements to an integrated splicing code. RNA 14:802–813. https://doi.org/10.1261/rna.876308

Wang SY, Pollina EA, Wang IH, Pino LK, Bushnell HL, Takashima K et al (2021) Role of epigenetics in unicellular to multicellular transition in *Dictyostelium*. Genome Biol 22:134. https://doi.org/10.1186/s13059-021-02360-9

Watson J, Crick F (1953) Molecular structure of nucleic acids: a structure for deoxyribose nucleic acid. Nature 171:737–738. https://doi.org/10.1038/171737a0

Watson LA, Tsai LH (2017) In the loop: how chromatin topology links genome structure to function in mechanisms underlying learning and memory. Curr Opin Neurobiol 43:48–55. https://doi.org/10.1016/j.conb.2016.12.002

Wayman GA, Davare M, Ando H, Fortin D, Varlamova O, Cheng HY et al (2008) An activity-regulated microRNA controls dendritic plasticity by down-regulating p250GAP. Proc Natl Acad Sci USA 105:9093–9098. https://doi.org/10.1073/pnas.0803072105

Weaver M, Hogan B (2001) Powerful ideas driven by simple tools: lessons from experimental embryology. Nat Cell Biol 3:E165–E167. https://doi.org/10.1038/35083125

Wessels HH, Lebedeva S, Hirsekorn A, Wurmus R, Akalin A, Mukherjee N et al (2019) Global identification of functional microRNA-mRNA interactions in *Drosophila*. Nat Commun 10: 1626. doi:https://doi.org/10.1038/s41467-019-09586-z
West-Eberhard MJ (2003) Developmental plasticity and evolution. Oxford University Press, New York
Wheeler DE, Nijhout HF (2003) A perspective for understanding the modes of juvenile hormone action as a lipid signaling system. BioEssays 25:994–1001. https://doi.org/10.1002/bies.10337
Wibrand K, Panja D, Tiron A, Ofte ML, Skaftnesmo KO, Lee CS et al (2010) Differential regulation of mature and precursor microRNA expression by NMDA and metabotropic glutamate receptor activation during LTP in the adult dentate gyrus in vivo. Eur J Neurosci 31:636–645. https://doi.org/10.1111/j.1460-9568.2010.07112.x
Wightman B, Ha I, Ruvkun G (1993) Posttranscriptional regulation of the heterochronic gene lin-14 by lin-4 mediates temporal pattern formation in *C. elegans*. Cell 75:855–862. https://doi.org/10.1016/0092-8674(93)90530-4
Will CL, Lührmann R (2011) Spliceosome structure and function. Cold Spring Harb Perspect Biol 3:a003707. https://doi.org/10.1101/cshperspect.a003707
Willingham E, Baldwin R, Skipper JK, Crews D (2000) Aromatase activity during embryogenesis in the brain and adrenal-kidney-gonad of the red-eared slider turtle, a species with temperature-dependent sex determination. Gen Comp Endocrinol 119:202–207. https://doi.org/10.1006/gcen.2000.7505
Wilson EO, Hölldobler B (2005) Eusociality: origin and consequences. Proc Natl Acad Sci USA 102:13367–13371. https://doi.org/10.1073/pnas.0505858102
Winter J (2015) MicroRNAs of the miR379-410 cluster: new players in embryonic neurogenesis and regulators of neuronal function. Neurogenesis 2:e1004970. https://doi.org/10.1080/23262133.2015.1004970
Worley PF, Bhat RV, Baraban JM, Erickson CA, McNaughton BL, Barnes CA et al (1993) Thresholds for synaptic activation of transcription factors in hippocampus: correlation with long-term enhancement. J Neurosci 13:4776–4788. https://doi.org/10.1523/JNEUROSCI.13-11-04776.1993
Wülfing C, Günther HS (2015) Dendritic cells and macrophages neurally hard-wired in the lymph node. Sci Rep 5:16866. https://doi.org/10.1038/srep16866
Wülfing C, Schuran FA, Urban J, Oehlmann J, Günther HS (2018) Neural architecture in lymphoid organs: hard-wired antigen presenting cells and neurite networks in antigen entrance areas. Immun Inflamm Dis 6:354–370. https://doi.org/10.1002/iid3.223
Yang YHC, Briant LJB, Raab CA, Mullapudi ST, Maischein HM, Kawakami K (2022) Innervation modulates the functional connectivity between pancreatic endocrine cells. elife 11:e64526. https://doi.org/10.7554/eLife.64526
Yap EL, Greenberg ME (2018) Activity-regulated transcription: bridging the gap between neural activity and behavior. Neuron 100:330–348. https://doi.org/10.1016/j.neuron.2018.10.013
Yumoto H, Hirao K, Hosokawa Y, Kuramoto H, Takegawa D, Nakanishi T et al (2018) The roles of odontoblasts in dental pulp innate immunity. Jpn Dent Sci Rev 54:105–117. https://doi.org/10.1016/j.jdsr.2018.03.001
Yusuf LH, Laetsch DR, Lohse K, Ritchie MG (2025) Genomic analyses in *Drosophila* do not support the classic allopatric model of speciation. bioRxiv. https://doi.org/10.1101/2024.05.20.595063
Zaidi M (2005) Neural surveillance of skeletal homeostasis. Cell Metab 1:219–221. https://doi.org/10.1016/j.cmet.2005.03.006
Zaidi M, Yuen T, Sun L, Rosen CJ (2018) Regulation of skeletal homeostasis. Endocr Rev 39:701–718. https://doi.org/10.1210/er.2018-00050

Chapter 28
Biological Senomic Thermodynamics in the Evolution of Life on Earth

F. Baluška, W. B. Miller Jr., A. S. Reber, and P. Slijepčević

Life is Represented by Living Cells and All Their Constructs.

Abstract Autopoietic living systems have arisen from simple prokaryotic cells and are scaling up to complex eukaryotic cells that assemble multicellular organisms, with trillions of coordinating cellular participants. All cells act as Kantian wholes by achieving catalytic and constraint closures to physical laws and performing thermodynamic work to construct and maintain themselves. According to Stuart Kauffman, cells and multicellular organisms act beyond previously identified physical laws. Cells have closed loops of causation based on biological codes, among which the redox code is the most important. They assemble and maintain biological organization through biological thermodynamics, as proposed by Ervin Bauer almost a hundred years ago. This biological law of thermodynamics, or Kauffman's Fourth Law of Thermodynamics, allows cells and cellular organisms to maintain their living state and to evolve uninterruptedly across billions of years. The electrically charged

---

The author "Reber AS" was deceased on the 2nd September 2025.

F. Baluška (✉)
Institute of Cellular and Molecular Botany, University of Bonn, Bonn, Germany
e-mail: baluska@uni-bonn.de

W. B. Miller Jr.
Bioverse Foundation, Paradise Valley, AZ, USA
e-mail: wbmiller1@cox.net

A. S. Reber
Department of Psychology, University of British Columbia, Vancouver, BC, Canada

P. Slijepčević
Department of Life Sciences, College of Health, Medicine and Life Sciences, University of Brunel, Uxbridge, UK
e-mail: Predrag.Slijepcevic@brunel.ac.uk

G. Witzany (ed.), *Integrative Theory of Evolution - Aspects and Insights*,
https://doi.org/10.1007/978-3-032-24295-2_28

plasma membrane represents the border between the cellular inside and outside, generating the cellular senome. This senome includes a constituent cell-wide bioelectric field acting as a bioplasma-like medium for cellular sentience. Cellular senomes confer normativity to biochemistry and cell metabolism by integrating ionomes, proteomes, glycomes, lipidomes, redoxomes, and electromes into the coherent and agential units that permit living cells to act as the Kantian wholes in accordance with a prospective Fourth Law of Thermodynamics. Here, we propose that the bioplasma-based cellular senomes act as an anti-entropic force, enabling creative biological evolution driven by sentient cells.

1 The Sentíent Cell and the Emergence of Biological Consciousness

> All species, extant and extinct, from the simplest unicellular prokaryotes to humans, have an existential consciousness. (Reber et al. 2023)

> If you're not talking about a sentient cell, you are not discussing biology.

Life on Earth started with the first cells assembling from prebiotic systems some 3–4 billion years ago, with ancient archaea and bacteria representing the prokaryotic realm of life. These ancient archaea and bacteria still live around and within us, but they are invisible due to their minuscule sizes. Nevertheless, they constitute the major portion of the Earth's biomass, occupying all niches available for life (Whitman et al. 1998; Bar-On et al. 2018). They exist not only as single cells but also as highly integrated biofilms, which represent one of the most successful life forms (Flemming and Wuertz 2018; van Wolferen et al. 2018). Bacteria are well known for their prokaryotic intelligence (Ben-Jacob et al. 2004, 2005; Hellingwerf 2005; Shapiro 2007; Lyon 2015; McMillen and Levin 2024), and bioelectric signals first emerged in bacteria, serving as bacterial communication in their biofilms (Prindle et al. 2015). It can be proposed that both sentience and cognition (Reber et al. 2023) first evolved, similar to most basic metabolic processes (Lane 2022) in archaea and bacteria. Much later, all these processes were adapted by more complex life forms. We have recently characterized and elaborated the major cellular components underlying cell sentience (Baluška and Reber 2019, 2021a, b; Baluška et al. 2021; Reber et al. 2023), including the plasma membrane enclosing bioplasma, organelles, vesicles, cytoskeleton, biomembranes, critical protein complexes, and diverse charged radical molecules. Importantly, all the basic cellular structures and metabolic processes that first evolved in archaea and bacteria were maintained and further complexified in all subsequent life forms.

2 The Senome: The Bioelectromagnetic Apparatus of Cellular Sentience

The concept of the cellular senome (Baluška and Miller Jr. 2018) has been introduced as an electron-based, bioelectromagnetic bioplasma that orchestrates cellular life (Baluška and Reber 2021a, b; Baluška et al. 2021; Reber et al. 2023; Baluška 2025; Miller Jr. et al. 2020, 2026). The cell's interior is rich in charged particles, molecules, and mobile free electrons, behaving as a dynamic and excitable plasma-like entity (Reber et al. 2023; Miller et al. 2026). As in the case of physical plasma, the bioplasma is an exotic state of matter that is extremely dynamic due to the charged nature of its components. This bioplasma generates senomic fields through various sources, including excitable membranes, organellar electron transport chains, and the collective activity of diverse charged particles, molecules, and biophotons. Senomic fields are central to cellular sensing, decision-making, and agency, operating at rapid timescales (milliseconds to seconds). The senome's unique biophysical properties (Baluška et al. 2021; Reber et al. 2023; Miller et al. 2026) permit its self-assembly and integration into supracellular episenomes. Ultimately, the senome acts as a fundamental, evolutionarily conserved apparatus that underpins cellular sentience (Reber et al. 2023) and drives biological evolution (Miller Jr. et al. 2020, 2026). The forgotten concept of bioplasma was originally proposed by Lionel S. Beale, who stated that cellular life acts as plasma-like matter (Beale 1872).

The biological ionic plasma of living cells is based on high concentrations of charged ions within and around cells (Eisenberg 1996, 2003, 2013a, b). The cellular inside is electrolytic in nature, organized by electrostatic stabilizations of charged intracellular surfaces. The living nature of excited organic molecules is based on free electrons accomplishing electronic charge transfers, giving a bioelectric nature to living cells (Eisenberg 1996, 2003, 2013a, b; Wong and Pollack 2010; De Loof 2016; Funk and Scholkmann 2023). The free flow of electrons within and across macromolecules is critical for cellular energy handling in ancient processes of photosynthesis and cellular respiration, invented by bacteria billions of years ago. Free electron flows are also important for biogeology (Falkowski and Godfrey 2008; Jelen et al. 2016).

The senome is based on inherent bioelectromagnetic fields emanating from all charged particles, molecules, and macromolecules. These senomic cellular fields are based on jostling and vibrating charges that assemble the cell-wide senomic fields through Abraham-Lorentz forces and other exotic physics underlying the cellular bioplasma. Senomic fields emanate from every cell, allowing development, morphogenesis, and wound healing of organisms (Funk and Monsees 2006). All senomic fields have an inherent property to coherently synchronize (Jaeken 2006; Sun et al. 2022; Madl and Renati 2023) and form higher-order senomic fields at the level of tissues, organs, and organisms (Levin and Martyniuk 2017; Levin et al. 2017; Funk and Scholkmann 2023), and the entire biosphere (Falkowski 2015; Falkowski and Godfrey 2008; Baluška 2025).

3 Cells as First-Order Organismal Kantian Wholes

Living Cells and Organisms Are Kantian Wholes. (Kauffman 2022)

In order to survive, cells must have agency and sentience (Reber et al. 2023, 2024), which allow them to act as a first-order organismal Kantian wholes (Kauffman and Roli 2021; Kauffman 2022; Wilks 2023). Although typically not viewed in this framework, living cells can be considered the primary living substrate whose organic form and competencies construct diverse supracellular organisms, such as prokaryotic biofilms, eukaryotic cells, and all multicellular organisms, including fungi, plants, and animals. In fact, during the first two billion years of biological evolution, all organisms living on the planet were unicellular prokaryotes. These simple organisms evolved all the basic features of life, including organismal agency, sentience, cognition, and most metabolic processes (Lane 2022; Reber et al. 2023).

Some two billion years ago, prokaryotic organisms developed symbiotic relationships, generating the first eukaryotic cells (organisms), which continue to enjoy independent life in the present moment as diverse protists. Although these are traditionally considered unicellular organisms, they are, in fact, obscured multicellular consortia that preserve the genomes of their prokaryotic predecessors. In order to act as unitary organisms with agency, their original prokaryotic senomes assembled into the coherent integrated senome of the eukaryotic cell. The plasma membrane of the hypothetical primary host cells generated the primary senome, which encompasses and controls the senomes of their symbiotic organelles. For this form of unitary agency, the eukaryotic composite senome must be cognitively closed, and this is accomplished by the plasma membrane delimiting the cellular *subjective* inside from the environmental *objective* outside. Intracellular organelles assemble only partially autonomous organellar senomes. Cells represent the primary living systems, and they generate different kinds of supracellular and multicellular living systems, from simple colonies, such as prokaryotic bacterial films, as well as eukaryotic animal-like choanoflagellates, and plant-like volvox algae (Ben-Jacob et al. 2004, 2005; Ueki et al. 2010; Prindle et al. 2015; Baluška et al. 2023a, b; Larson 2023; Ros-Rocher and Brunet 2023). Cells invented sentience mainly through their discriminating, limiting plasma membranes, which, as is discussed below, have unique properties relevant for the internal ordering of macromolecules that allow their survival and reproduction.

4 Membranes Define Inside Space: The Holographic Perspective of Vesicles and Cells

Given any closed surface, we can represent all that happens inside it by degrees of freedom on this surface itself. (Hooft 2000)

The limiting plasma membrane defines the subjective inside from the external environment. In some sense, there is a surprising analogy between the information of biological living cells and physical black holes in the Universe. For outside observers, the information inside the cellular interior of the black hole is not accessible. In other words, cellular subjectivity (sentience) is reminiscent of the black hole information paradox: although the resident information is not accessible, that information is not lost. Moreover, both living cells (Meijer and Geesink 2018) and black holes seem to follow the holographic principle, in which the boundary border encodes their interior spaces. As stated by Gerard `t Hooft: "Given any closed surface, we can represent all that happens inside it by degrees of freedom on this surface itself" (Hooft 2000; Fields et al. 2022; Stoll 2022). The first vesicles, believed to have emerged some four billion years ago, had limiting lipid-bilayer membranes without any protein-based channels or transporters, with their central zone being hydrophobic and devoid of any water molecules. This hydrophobic layer effectively sealed the cellular inside from the environmental outside (Tanford 1978), making the inside (subjective) space inaccessible for any outside observer. It is from the crucial membranes of cells that cells and vesicles construct their holographic representation of their outer environment (Baluška and Miller Jr. 2018). This holographic view of vesicles and cells, both enclosed with biological membranes, invokes some informational similarities between vesicles/cells and black holes.

Interestingly, hydrophobic forces are not only behind holographic features of vesicles and cells but are also instrumental in shaping the 3D structures of proteins dictated by the sequence of amino acids and the biophysical features of intracellular space (Tanford 1997; Dyson et al. 2006; Baldwin 2007; Deamer 2011). As the plasma membrane of cells tightly controls all the physicochemical features of the inside space through their senomic fields, these membrane-based fields also dictate conformations of all proteins and other macromolecules. We will discuss this important issue in detail below.

5 The Lipid Bilayer–Based Senome as the Source of Biological Thermodynamics, Allowing Cells to Act Creatively Against the Second Law of Thermodynamics

It Is an Astonishing Fact That This Incredibly Thin Membrane, Just Two Molecules Thick, Is Essential for Life to Exist. (Deamer 2011, page 136)

Living Cells Achieve Constraint Closure and Construct Themselves. (Kauffman and Roli 2023)

The crucial role in the surprising properties of life, which do not fit into the usual physical theories, is based on their biological membranes, which are dependent on their lipid bilayers. Interestingly, although the plasma membrane plays a central role in the life of cells, it was the last structure discovered microscopically and finally acknowledged as an essential feature of cell theory (Lombard 2014; Campos Muñiz and Fernández Perrino 2023). Contentious debates, which lasted around 175 years, were resolved by the direct visualization of the plasma membrane with electron microscopy, especially using a freeze-etching method, which finally excluded any possible artifacts due to chemical fixation (Moor and Mühlethaler 1963; Mühlethaler 1971; Branton 2016). Perhaps not surprisingly, then, the essential functions of the plasma membrane that control intracellular physicochemical phenomena are still underappreciated.

Besides providing constraint closure and protection, the plasma membrane is essential for sensory perception, which is first translated into bioelectric signals before being transduced into biochemical cascades. The important aspect here is the electric features of the plasma membrane, which transduce all sensory events into a bioelectromagnetic language of senomic fields. The significance of the cell surface, just as with the plasma membrane, was neglected due to a focus on DNA, the genome, and the nucleus. Pertinent to a gradual shift, the epigenetic philosophy of Thomas Henry Huxley (Lyons 1999; Richmond 2000, 2001) served as a critical conceptual source for the emergence of the existence of an epigenome (Waddington 1942; Noble 2015; Byrnes 2020). Later, this same critical line of thought led to the concept of the senome (Baluška and Miller Jr. 2018), which is intimately associated with the cell surface, especially relating to its excitable, electrogenic plasma membrane.

Beyond the protection afforded by the basic features of a biologically limiting, external membrane, the plasma membrane generates gradients and fields, which allow cells to not only control cellular energy fluxes but also to act as an ordering apparatus, generating cellular order and locally decreasing entropy. This basic life feature can also be considered in a thermodynamic framework, as adhering to Kauffman's 4th law of thermodynamics. This thermodynamic law was discovered by Ervin Bauer more than a hundred years ago (Bauer 1920, 1930; Igamberdiev 2023, 2024; Elek and Müller 2024; Müller and Elek 2024), and biological thermodynamics was elaborated further by Stuart Kauffman and his coauthors (Kauffman and Roli 2021, 2023; Kauffman 2022). In living systems, the laws of physics and chemistry are subordinated to biological thermodynamics (Bauer 1930; Elek and Müller 2024; Mizrai 2024). This is because the living systems have closed loops of causation through biological codes, which act to assemble and maintain biological order (Igamberdiev 2024). Upstream of all these codes is the senome, which composes and orchestrates all these codes to generate and maintain cellular sentience, allowing cellular cognition, survival, and evolution. The senome is the most ancient cellular system, predating the emergence of the cellular epigenome and genome. As we discuss below, the proto-senome emerged during the precellular vesicular stage of biological evolution, when dynamic, unstable vesicles and coacervates evolved into the first competent cells. In those cells, vesicles play a central role in cellular homeostasis and cell–cell communication.

6 The Senome as the Cellular Sensorium: A Bioelectromagnetic Apparatus for Cell Sentience

What Coordinates All This? What Unifies the Whole? Once You Think About Electrical Fields, It Is Hard to Imagine Anything Else. (Nick Lane 2022, page 281)

A unique feature of the senome is that it reacts very quickly to environmental changes since all sensory events at the plasma membrane elicit bioelectrical responses within the plasma membrane in a millisecond-to-second time delay and transforms these bioelectrical messages into a chemical one, often using diverse second messengers that are compatible with the epigenome apparatus, which then manipulates the genome within minutes and hours after the initial senomic stimulation. The senome allows the maintenance of cellular non-equilibrium dynamics. The biological law proposed by Ervin Bauer implies that external energy is not used by biological systems to perform direct work but rather to support the perpetuation of non-equilibrium processes. For the senome, the most important physical parameters are gradients and fields that pulsate from the limiting plasma membrane deeper into the cellular interior. These bioelectromagnetic fields act as the cellular composers and conductors, orchestrating all the codes and metabolic processes, based on charged molecules, proteins, and diverse macromolecules (Lane 2022).

7 Tissues, Organs, and Organisms as Higher-Order Kantian Wholes

In Multicellular Organisms, Individual Cellular Information Fields Aggregate into a Collective Information Architectural Matrix, Termed a N-space Episenome. (Miller Jr. et al. 2020)

Senomic fields not only permeate the entire cellular interior but also radiate beyond cellular boundaries, which allows adjacent cells, if sufficiently close together, to assemble supracellular N-space episenome fields. The evolutionary predecessor of such an integration of several overlapping senomic fields is the eukaryotic cellular form, which assembled approximately two billion years ago after the emergence of the last universal common ancestor (LUCA), producing the last eukaryotic ancestor (LECA) via merger and symbiotic fusion of several prokaryotic cells (Mikhailovsky and Gordon 2018; Baluška and Lyons 2021; Baluška et al. 2023b).

Thereafter, it took another 1.5 billion years before a true multicellular organism slowly emerged beyond biofilms. In other words, it took nearly 2 billion years to merge the prokaryotic senomes to generate a fully effective eukaryotic composite senome. It can be expected that the first cell and LUCA emerged from communities of dynamic and short-lived vesicles. It can be expected that these vesicles had already started to assemble a kind of proto-senomes, which were instrumental in the evolution of the fully assembled senomes of the first cells. David Deamer proposed that the lipid bilayer membranous outer border of these ancient and short-lived vesicles had a hydrophobic middle part, which was not sufficiently permeable for the free transport of molecules in and out of these proto-vesicles (Deamer 2011). However, the integrity of their limiting biomembranes was not perfect, permitting the transient transport of some molecules, thereby allowing their survival and further evolution. At some later stage, these micropores, which opened and closed on a nanosecond timescale, were transformed into transport complexes under the control of vesicular senomes, which merged and generated the first viable proto-cells, representing the most ancient units of the nascent Earth biosphere.

8 The Senome and Biological Thermodynamics

> The Laws of Physics and Chemistry Remain Valid in the Living System. However, They Are Subordinated to the Laws of the Higher Entity of the Biological System and Manifest Themselves in Particular Form. (Bauer 1930, p. 101; Elek and Müller 2024)

In presently living cells, senomes are highly evolved and guide, together with epigenomes and genomes, self-organization, morphogenesis, development, and the evolution of cells and all their constructs based on their cellular senomically centered sentience. To survive and evolve, living cells rely on senomic electromagnetic fields, which orchestrate all cellular activities and processes in such a manner that living cells effectively resist the second law of thermodynamics. As Ervin Bauer stated in his book, in living systems (such as cells and all cellular constructs from archaea and bacteria up to the biosphere), physical and chemical laws are still valid but are subordinated to the biological law of thermodynamics (Bauer 1930; Elek and Müller 2024), which acts as a kind of anti-entropic force. This biological thermodynamics strives toward non-equilibrium, which is the opposite direction with respect to the physical second law of thermodynamics, which, in cellular terms, strives toward equilibrium. Moreover, biological thermodynamics supports the permanent and uninterrupted continuity of living systems, which implies the historical nature of living systems. Cellular senomes allow all cells and their cellular constructs to act as agents supported by semiomes and their scaffolding (Hoffmeyer 2014, 2015). Semiosis, based on cellular senomes, is critical for the maintenance of evolutionary knowledge and meaningfulness (Kull 2021). Importantly, semiomes underlie the continuity and permanence of life, based on cells and their cellular senome-based constructs. These embody the evolutionary (historical) memory of cells directly connected to the ur-cells via uninterrupted chains of cell divisions. All living systems constructed by living cells, from the unicellular and multicellular organisms up to the whole Earth biosphere, are part of this continuously unbroken chain.

9 Vesicles as Basic Units of Kantian Wholes Are Behind Inter-kingdom Communication and Integration, Allowing Cells to Assemble the Earth's Biosphere

The Evolving Biosphere Is Outside the Newtonian Paradigm. (Kauffman and Roli 2023)

Life on Earth probably started with some form of primordial proto-vesicles, which likely evolved from dynamic coacervates. How these proto-vesicles evolved their competencies to further generate proto-cells and, finally, the first fully competent cells still remains mysterious. However, vesicles remain vital aspects of both prokaryotic and eukaryotic cells.

Eukaryotic cells generate different kinds of vesicles. Previously, only two major classes of vesicles had been identified: so-called endosomal vesicles, originating from the plasma membrane and traveling into the cellular interior, and secretory vesicles, generated from the Golgi apparatus vesicles, which typically fuse with the plasma membrane. More recently, this rather simple picture has become much more complex. All kinds of membranous organelles are now known to generate a diversity of vesicles, which serve for intracellular communication (Schlacht et al. 2014; Gu et al. 2020; More et al. 2024). Moreover, it has also been discovered that eukaryotic cells release vesicles into their extracellular space. These extracellular vesicles serve not only for cell–cell communication among diverse multicellular organisms, but also for the inter-kingdom communication (Woith et al. 2019; Makarova and May 2021; Jahnke and Staufer 2024). As vesicles were the first living systems, predating proto-cells, they represent the most ancient proto-lifeforms, acting as primordial Kantian wholes, and are well-suited to act as cell–cell and inter-kingdom messengers, integrating all organisms into the wholeness and integrity of Earth's complex biosphere.

10 Coda: Arthur Reber (1940–2025) and Life-Sentience Continuity

Arthur Reber introduced the revolutionary concept of the Cellular Basis of Consciousness in his book The First Minds (Reber 2019). The first author of this chapter met him at "The Other Minds Problem: Animal Sentience and Cognition" summer school organized by Stevan Harnad at the Université du Québec à Montréal on 3 July 2018. We discussed common interests and started our collaboration almost immediately (Baluška and Reber 2019). Since this first paper, we have teamed up with William Miller, Jr. and Predrag Slijepčević in a fruitful collaboration, resulting in 25 papers and chapters published in seven years, followed by two books. In these seven years, that collaboration turned into friendship. Arthur died on the 2nd of September 2025, and we deeply miss his unique personality, unusual insights, and keen sense of humor.

* https://sites.grenadine.uqam.ca/sites/isc/en/summer18/schedule

Author Contributions FB conceptualization and first draft. WM, ASR, PS –revisions, writing, editing and conceptualization.

Conflicts of Interest The authors declare that they have no known competing financial interests or personal relationships that could have appeared to influence the work reported in this paper. The authors have no conflicts of interest to declare.

Grants None.

References

Baldwin RL (2007) Energetics of protein folding. J Mol Biol 371:283–301
Baluška F (2025) Cognitive cells: from cellular senomic spheres to Earth's biosphere. Biol Theory:1–11. https://doi.org/10.1007/s13752-025-00489-5
Baluška F, Lyons S (2021) Archaeal origins of eukaryotic cell and nucleus. Biosystems 203: 10437510437
Baluška F, Miller WB Jr (2018) Senomic view of the cell: Senome versus genome. Commun Integr Biol 11:1–9
Baluška F, Reber A (2019) Sentience and consciousness in single cells: how the first minds emerged in unicellular species. BioEssays 41:e1800229
Baluška F, Reber A (2021a) Senomic and ephaptic principles of cellular consciousness: the biomolecular basis for plant and animal sentience. J Consc Stud 28:31–49
Baluška F, Reber AS (2021b) CBC-Clock Theory of Life - integration of cellular circadian clocks and cellular sentience is essential for cognitive basis of life. BioEssays 43:e2100121
Baluška F, Miller WB, Reber AS (2021) Biomolecular basis of cellular consciousness via subcellular nano-brains. Int J Mol Sci 22:2545
Baluška F, Miller WB, Reber A (2023a) Chapter 3: Cellular basis of cognition and evolution: from protists and fungi up to animals, plants, and root-fungal networks. In: Corning PA, Kauffman SA, Noble D, Shapiro JA, Vane-Wright RI, Pross A (eds) Evolution on purpose: teleonomy in living systems. MIT Press, pp 33–58
Baluška F, Miller WB Jr, Reber A (2023b) Cellular and evolutionary perspectives on organismal cognition: from unicellular to multicellular organisms. Biol J Linnean Soc 139:503–513
Bar-On YM, Phillips R, Milo R (2018) The biomass distribution on earth. Proc Natl Acad Sci USA 115:6506–6511
Bauer E (1920) Die Definition des Lebewesens auf Grund seiner thermodynamischen Eigenschaften und die daraus folgenden biologischen Grundprinzipien. Naturwissenschaften 8:338–340
Bauer E (1930) Theoretical biology. Republished by Akademiai Kiado, Budapest in 1982, Moscow and Leningrad. VIEM
Beale SL (1872) Bioplasm. Churchill JA/Lindsay & Blakiston, London/Philadelphia
Ben-Jacob E, Becker I, Shapira Y, Levine H (2004) Bacterial linguistic communication and social intelligence. Trends Microbiol 12:366–372
Ben-Jacob E, Shapira Y, Tauber AI (2005) Seeking the foundations of cognition in bacteria: from Schrödinger's negative entropy to latent information. Phys A 359:495–524
Branton D (2016) Fracture faces of frozen membranes: 50th anniversary. Mol Biol Cell 27:421–423
Byrnes WM (2020) E. E. Just's broad, yet hidden, influence on modern cell and developmental biology. Mol Repr Dev 87:387–391
Campos Muñiz C, Fernández Perrino FL (2023) Evolution of the concepts of architecture and supramolecular dynamics of the pasma membrane. Membranes (Basel) 13:547

De Loof A (2016) The cell's self-generated "electrome": the biophysical essence of the immaterial dimension of life? Commun Integr Biol 9:e1197446
Deamer D (2011) First life. Discovering the connections between stars, cells and how life began. University of California Press
Dyson HJ, Wright PE, Scheraga HA (2006) The role of hydrophobic interactions in initiation and propagation of protein folding. Proc Natl Acad Sci USA 103:1357–1361
Eisenberg B (1996) Computing the field in proteins and channels. J Membr Biol 150:1–25
Eisenberg B (2003) Proteins, channels and crowded ions. Biophys Chem 100:507–517
Eisenberg B (2013a) Interacting ions in biophysics: real is not ideal. Biophys J 104:1849–1866
Eisenberg B (2013b) Ionic interactions are everywhere. Physiology 28:28–38
Elek G, Müller M (2024) Ervin Bauer's concept of biological thermodynamics and its different evaluations. Biosystems 235:105090
Falkowski PG (2015) Life's engines. How microbes made earth habitavle. Princeton University Press, Princeton and Oxford
Falkowski PG, Godfrey LV (2008) Electrons, life and the evolution of earth's oxygen cycle. Philos Trans R Soc Lond Ser B Biol Sci 363:2705–2716
Fields J, Glazebrook JF, Marciano A (2022) The physical meaning of the holographic principle. Quanta 11:72–96
Flemming H-C, Wuertz S (2018) Bacteria and archaea on earth and their abundance in biofilms. Nat Rev Microbiol 17:247–260
Funk RW, Monsees TK (2006) Effects of electromagnetic fields on cells: physiological and therapeutical approaches and molecular mechanisms of interaction. Cells Tissues Organs 182: 59–78
Funk RWH, Scholkmann F (2023) The significance of bioelectricity on all levels of organization of an organism. From the subcellular level to cells. Progr Biophys Mol Biol 177:185–201
Gu X, Brennan A, Lidnsey K (2020) Vesicle transport in plants: a revised phylogeny of SNARE proteins. Evol Bioinforma 16:1176934320956575
Hellingwerf KJ (2005) Bacterial observations: a rudimentary form of intelligence? Trends Microb 13:152–158
Hoffmeyer J (2014) The semiome: from genetic to semiotic scaffolding. Semiotica 198:11–31
Hoffmeyer J (2015) Introduction: semiotic scaffolding. Biosemiotics 8:153–158
Hooft GT (2000) The holographic principle. https://arxiv.org/abs/hep-th/0003004
Igamberdiev AU (2023) Toward the relational formulation of biological thermodynamics. Entropy (Basel) 26:43
Igamberdiev AU (2024) Biological thermodynamics: Ervin Bauer and the unification of life sciences and physics. Biosystems 235:105089
Jaeken J (2006) Linking physiological mechanisms of coherent cellular behaviour with more general physical approaches towards the coherence of life. IUBMB Life 58:642–646
Jahnke K, Staufer O (2024) Membranes on the move: the functional role of the extracellular vesicle membrane for contact-dependent cellular signalling. J Extracell Vesicles 13:e12436
Jelen B, Giovannelli D, Falkowski PG (2016) The role of microbial electron transfer in the coevolution of the biosphere and geosphere. Ann Rev Microbiol 70:45–62
Kauffman S (2022) Is there a fourth law for non-ergodic systems that do work to construct their expanding phase space? Entropy 24:1383
Kauffman S, Roli A (2021) The world is not a theorem. Entropy 23:1467
Kauffman S, Roli A (2023) A third transition in science? Interface Focus 13:20220063
Kull K (2021) Natural selection and self-organization do not make meaning, while the agent's choice does. Biosemiotics 14:49–53
Lane N (2022) Transformer - the deep chemistry of life and death. Profile Books Ltd, London
Larson BT (2023) Perspectives on principles of cellular behavior from the biophysics of protists. Integr Comp Biol 63:1405–1421
Levin M, Martyniuk CJ (2017) The bioelectric code: an ancient computational medium for dynamic control of growth and form. Biosystems 164:76–93

Levin M, Pezzulo G, Finkelstein JM (2017) Endogenous bioelectric signaling networks: exploiting voltage gradients for control of growth and form. Annu Rev Biomed Eng 19:353–387
Lombard J (2014) Once upon a time the cell membranes: 175 years of cell boundary research. Biol Direct 9:32
Lyon P (2015) The cognitive cell: bacterial behavior reconsidered. Front Microbiol 6:264
Lyons SL (1999) Thomas Henry Huxley: the evolution of a scientist. Prometheus Books, Amhurst
Madl P, Renati P (2023) Quantum electrodynamics coherence and hormesis: foundations of quantum biology. Int J Mol Sci 24:1400
Makarova M, May RC (2021) Fungal extracellular vesicles in interkingdom communication. In: Rodrigues M, Janbon G (eds) Fungal extracellular vesicles, Curr Topics Microbiol Immunol, vol 432, pp 81–88
McMillen P, Levin M (2024) Collective intelligence: a unifying concept for integrating biology across scales and substrates. Commun Biol 7:378
Meijer DKF, Geesink HJH (2018) Revisiting the quantum brain hypothesis: toward quantum (neuro) biology? Open J Biophys 8:117–154
Mikhailovsky G, Gordon R (2018) Symbiosis: why was the transition from microbial prokaryotes to eukaryotic organisms a cosmic gigayear event? In: Gordon R, Sharov AA (eds) Habitability of the universe before earth. Astrobiology exploring life on earth and beyond. Academic Press, Cambridge, MA, pp 355–405
Miller WB Jr, Torday JS, Baluška F (2020) The N-space Episenome unifies cellular information space-time within cognition-based evolution. Progr Biophys Mol Biol 150:112–139
Miller WB, Baluška F, Slijepčević P, Reber AS (2026) Cognition-based biology: cellular sentience, communication, and Symbiosis. Springer-Nature
Mizrai E (2024) Homeostasis and information processing: the key frames for the thermodynamics of biological systems. Biosystems 236:05115
Moor H, Mühlethaler K (1963) Fine structure in frozen-etched yeast cells. J Cell Biol 17:609–628
More K, Klinger CM, Barlow LD, Dacks JB (2024) Evolution and natural history of membrane trafficking in eukaryotes. Curr Biol 30:R553–R564
Mühlethaler K (1971) Studies on freeze-etching of cell membranes. Int Rev Cytol 31:1–19
Müller M, Elek G (2024) The history of Ervin Bauer's publications on the theory of life. Biosystems 235:105212
Noble D (2015) Conrad Waddington and the origin of epigenetics. J Exp Biol 218:816–818
Prindle A, Liu J, Asally M, Ly S, Garcia-Ojalvo J, Süel GM (2015) Ion channels enable electrical communication in bacterial communities. Nature 527:59–63
Reber AS (2019) The first minds: caterpillars, karyotes, and consciousness. Oxford University Press, New York
Reber AS, Baluška F, Miller WB (2023) The sentient cell: the cellular foundations of consciousness. Oxford University Press
Reber AS, Baluška F, Miller WB, Slijepčević P (2024) The sensual cell: feeling and affect in unicellular species. Biosystems 238:105197
Richmond ML (2000) T.H. Huxley's criticism of German cell theory: an epigenetic and physiological interpretation of cell structure. J Hist Biol 33:247–289
Richmond ML (2001) Thomas Henry Huxley's developmental view of the cell. Nat Rev Mol Cell Biol 3:61–65
Ros-Rocher N, Brunet T (2023) What is it like to be a choanoflagellate? Sensation, processing and behavior in the closest unicellular relatives of animals. Anim Cogn 26:1767–1782
Schlacht A, Herman EK, Klute MJ, Field MC, Dacks JB (2014) Missing pieces of an ancient puzzle: evolution of the eukaryotic membrane-trafficking system. Cold Spring Harb Persp Biol 6:a016048
Shapiro JA (2007) Bacteria are small but not stupid: cognition, natural genetic engineering and socio-bacteriology. Stud Hist Philos Sci Part C 38:807–819

Stoll EA (2022) Modeling electron interference at the neuronal membrane yields a holographic projection of representative information content. BioRxiv Preprint. https://doi.org/10.1101/2022.12.03.518989

Sun G, Li J, Zhou W, Hoyle RG, Zhao Y (2022) Electromagnetic interactions in regulations of cell behaviors and morphogenesis. Front Cell Dev Biol 10:1014030

Tanford C (1978) The hydrophobic effect and the organization of living matter. Science 200:1012–1018

Tanford C (1997) How protein chemists learned about the hydrophobic factor. Prot Sci 6:1358–1366

Ueki N, Matsunaga S, Inouye I, Hallmann A (2010) How 5000 independent rowers coordinate their strokes in order to row into the sunlight: Phototaxis in the multicellular green alga Volvox. BMC Biol 8:103

van Wolferen M, Orell A, Albers S-V (2018) Archaeal biofilm formation. Nat Rev Microbiol 16: 699–713

Waddington CH (1942) The epigenotype. Int J Epidemiol 41:10–13

Whitman WB, Coleman DC, Wiebe WJ (1998) Prokaryotes: the unseen majority. Proc Natl Acad Sci USA 95:6578–6583

Wilks AF (2023) The 'whole' truth about biological individuality in Kant's account of living nature. Kantian Rev 28:429–446

Woith E, Fuhrmann G, Melzig MF (2019) Extracellular vesicles - connecting kingdoms. Int J Mol Sci 20:5695

Wong GCL, Pollack L (2010) Electrostatics of strongly charged biological polymers: ion-mediated interactions and self-organization in nucleic acids and proteins. Annu Reb Phys Chem 61:171–189

MIX
Papier aus verantwortungsvollen Quellen
Paper from responsible sources
FSC® C105338

If you have any concerns about our products,
you can contact us on
ProductSafety@springernature.com

In case Publisher is established outside the EU,
the EU authorized representative is:
Springer Nature Customer Service Center GmbH
Europaplatz 3, 69115 Heidelberg, Germany

Printed by Libri Plureos GmbH
in Hamburg, Germany